# Oxidative Stress in Applied Basic Research and Clinical Practice

**Editor-in-Chief**
Donald Armstrong

For other titles published in this series, go to
http://www.springer.com/series/8145

## Note from the Editor-in-Chief

All books in this series illustrate point-of-care testing and critically evaluate the potential of antioxidant supplementation in various medical disorders associated with oxidative stress. Future volumes will be updated as warranted by emerging new technology, or from studies reporting clinical trials.

Donald Armstrong

Editor-in-Chief

Heinrich Sauer · Ajay M. Shah ·
Francisco R.M. Laurindo
Editors

# Studies on Cardiovascular Disorders

Humana Press

*Editors*
Heinrich Sauer
Universität Gießen
Physiologisches Institut
Aulweg 129
35392 Gießen
Germany
heinrich.sauer@physiologie.med.unigiessen.de

Ajay M. Shah
King's College London
James Black Centre
Coldharbour Lane 125
SE5 9NU London
King's Denmark Hill Campus
United Kingdom
ajay.shah@kcl.ac.uk

Francisco R.M. Laurindo
Universidade of São Paulo
Fac. Medicina
Instituto do Coração (INCOR)
Lab. Biologia Molecular
Av. Enéas de Carvalho Aguiar 44
05403-000 São Paulo, São Paulo
Subsolo
Brazil
expfrancisco@incor.usp.br

ISBN 978-1-60761-599-6 e-ISBN 978-1-60761-600-9
DOI 10.1007/978-1-60761-600-9
Springer New York Dordrecht Heidelberg London

Library of Congress Control Number: 2010934121



Printed on acid-free paper

Humana Press is part of Springer Science+Business Media (www.springer.com)

# Preface

The role of reactive oxygen species (ROS) in the cardiovascular system is Janus-faced. Whereas low concentrations of ROS are involved in variety of physiological signalling events, oxidative stress resulting from deregulated overproduction of ROS and/or impaired antioxidant defences contributes to cardiovascular disease. The actions of ROS in the cardiovascular system are a fascinating topic, not only for the basic science researcher but also for the clinician who is interested in seeking new therapies for his patients suffering from cardiovascular disease. The current book provides a comprehensive overview of the molecular mechanisms and pathophysiological settings in which chronic and detrimental oxidative stress arises within the heart and vasculature. The book also considers currently discussed strategies in avoiding chronic redox stress resulting from exposure to risk factors or various cardiovascular interventions.

The series starts with an overview by *Denise de Castro Fernandes, Diego Bonatto and Francisco Laurindo* of redox signaling models that could underlie the development of redox-associated cardiovascular disorders. The interactions of proteins within signalling cascades with ROS and the regulation of such interactions by the anti-oxidative capacity of the cell are discussed. *Rebecca Charles, Joseph Burgoyne* and *Philip Eaton* report on redox-mediated modifications of proteins under physiological and pathophysiological conditions and the variety of post-translational oxidative modifications that explain redox sensing and signal transduction by proteins at the molecular level.

ROS are generated during embryogenesis and may be involved in the proper development of the cardiovascular system. This is underscored by the increasing evidence that ROS regulate the cardiomyogenesis and vascular differentiation processes of stem cells, which mimic essential events occurring during normal embryogenesis of the cardiovascular system. *Heinrich Sauer* and *Maria Wartenberg* outline the signalling pathways in cardiovascular development during embryogenesis and their meaning in differentiation processes of resident cardiac stem cells and embryonic stem cells derived from the inner cell mass of blastocysts.

Sensory nerves act via perivascular neuronal networks to release potent vasoactive neuropeptides that work in combination with the autonomic nervous system to regulate both physiological vascular tone and pathophysiological disease processes. Sensory nerve endings can be in contact with vascular smooth muscle cells and also in intimate contact with endothelial cells. In the article by *Rabea Graepel*, *Jennifer Bodkin* and *Susan Brain*, current knowledge of the sensory nervous system in terms of its influence on the cardiovascular system and the established and putative links between the sensory nervous system and ROS generation relevant to the cardiovascular system are outlined.

A major source of ROS is the mitochondrial respiratory chain where ROS are generated in the electron transport chain complexes I and III. Mitochondria-derived ROS are known to participate in cardiac reperfusion injury but paradoxically – as outlined in the article of *Ariel Cardoso, Bruno Queliconi* and *Alicia Kowaltowski* – also contribute to cardioprotection in myocardial pre- and postconditioning. Mitochondrial ROS generation is closely coupled to coenzyme $Q_9/Q_{10}$, which acts as an electron carrier between the nicotinamide adenine dinucleotide (NADH) and succinate dehydrogenases and the cytochrome system. The article by *Samarjit Das*, *Somak Das* and *Dipak Das* presents the intriguing hypothesis that increased ROS generation in mitochondria with abundance of CoQ could represent a novel mechanism of cardioprotection through the potentiation of redox signaling, thereby preventing oxidative damage and dysfunction of mitochondria under excess ROS-generating conditions. Furthermore, ROS derived from mitochondria are involved in homocysteine (HCY)-related cardiovascular diseases. As pointed out in the study of *Karni Moshal* and coworkers, HCY causes activation and the mitochondrial translocation of calpain-1 (calcium-dependent cysteine protease) thereby increasing intramitochondrial oxidative stress and leading to the induction of MMP-9. In their study, the authors summarize current knowledge on hydrogen sulphide in myocardial protection as well as the role that HCY-induced oxidative stress in the mitochondria plays during the regulation of myocyte contractility.

Nicotinamide adenine dinucleotide phosphate (NADPH) oxidases are another important source of ROS in the cardiovascular system that have been shown to be involved in many human diseases, such as metabolic syndrome, hypertension, diabetes, left ventricular hypertrophy, heart failure, renal disease, atherosclerosis, and cerebrovascular disease. *Tomasz Guzik* reviews the important vascular roles of these complex enzymes in human circulation. *Guillermo Zalba* and *Javier Diez* summarize the experimental evidence supporting a pathophysiological role for polymorphisms in the $p22^{phox}$ gene (the CYBA gene), some of which are able to influence NADPH oxidase gene expression and activity in the context of cardiovascular diseases. The theme of genetic variation is also the subject of the article by *Christian Delles* and *Anna Dominiczak*, who report on strategies to unravel the genetics of redox-related cardiovascular diseases and describe the interactions of redox-regulated genes and the environment. *Timo Kahles*, *Sabine Heumüller* and *Ralf Brandes* focus their article on the role of NADPH oxidase in blood-brain barrier dysfunction, which occurs during ischemic stroke as well as during ischemia/reperfusion.

The likelihood of adverse cardiovascular events has been associated with risk factors related to a "typical western lifestyle" such as physical inactivity, obesity and smoking, which all appear to be associated with oxidative stress. The link between smoking and increased oxidative stress is reviewed by *David Bernhard*.

Elevated levels of ROS have also been linked with increasing age and vascular aging (reviewed by *Anna Csiszar* and *Zoltan Ungvari*), heart failure, diabetes mellitus (reviewed by *Divya Gupta, Kathy Griendling* and *Robert Taylor*), coronary artery disease, hypertension (reviewed by *Rhian Touyz, Andreia Chignalia*, and *Mona Sedeek*), as well as with relatively rare cardiac diseases such as peripartum cardiomyopathy, which has been associated with increased oxidative stress during pregnancy (reviewed by *Denise Hilfiker-Kleiner, Arash Haghikia* and *Andres Hilfiker*). However, oxidative stress not only arises in the sequence of cardiovascular diseases but also in response to cardiovascular interventions such as coronary angiography (reviewed by *Raymond Farah*) or during cardiac transplantation (reviewed by *Galen Pieper* and *Ashwani Khanna*). Interestingly, conditions of chronically elevated ROS within the heart are associated with atrial fibrillation, which among other problems may cause stroke and peripheral embolization (reviewed by *Ali Sovari* and *Samuel Dudley*).

Acute myocardial infarction due to atherosclerotic coronary artery disease often results in remodeling responses of the myocardium that may culminate in congestive heart failure. *Yao Sun* describes the current knowledge on oxidative stress arising during cardiac infarction and its role in influencing the severity of cellular apoptosis, the inflammation process and development of hypertrophy. *Min Zhang, Alex Sirker* and *Ajay Shah* report on the process of cardiac remodelling with an emphasis on cardiomyocyte hypertrophy, apoptosis, interstitial fibrosis, contractile dysfunction and chamber dilatation through specific modulation of redox-sensitive signalling pathways that alter gene and protein expression and function. A deepened insight into cardiovascular fibrosis is provided by the article by *Subramaniam Pennathur, Louise Hecker* and *Victor Thannickal*, who describe the role of NADPH oxidases in the initiation of fibrotic processes and outline therapeutic strategies to inhibit oxidative stress in cardiovascular fibrosis.

Cardiovascular disease is not uniformly distributed between the sexes. Risk factors specific to women include parity, oophorectomy, pre-eclampsia and menopause. In the article by *Manuela Gago-Dominguez, Xuejuan Jiang, and Jose Esteban Castelao,* the oxidation hypothesis of reproductive factor-cardiovascular disease association is developed, which is based on the observation that pregnant, oophorectomized, and postmenopausal women exhibit higher levels of lipid peroxidation than nonpregnant, nonoophorectomized and premenopausal women, respectively. The authors propose that the increased levels of lipid peroxidation during these states are responsible, at least in part, for the increased risk of cardiovascular disease in women.

The well-established connection between cardiovascular disease and oxidative stress has led to the investigation of various antioxidative strategies for patient treatment. The most natural way to cope with cardiovascular disease is perhaps by prevention. *Alfonso Giovane*, and *Claudio Napoli* report on the French paradox of cardiovascular disease and consider the potential beneficial effects of the Mediterranean diet, which could be related to antioxidants contained in red wine or vegetable, fruit and olive oil. During recent years, novel synthetic antioxidants such as hybrid compounds designed to improve the efficacy of natural antioxidants have been developed. *Gloria López* and *Homero Rubbo* describe novel hybrid antioxidants (tocopherol analogs-nitric oxide donors) that share nitric oxide-releasing properties and LDL incorporation capacity, demonstrating the importance of this site-specific release of nitric oxide in the cascade of events involved in the inhibition of LDL oxidation. This may offer novel approaches for the prevention of atherosclerosis and related disorders that involve reactive oxygen and nitrogen species, although this remains to be demonstrated in clinical trials. Alternative approaches could utilize the antioxidative capacity of the cell, e.g. thioredoxin (TRX), which catalyzes the conversion of disulfide oxidized proteins to their thiol-reduced forms, and has been shown to exert protective effects when intravenously administered in laboratory animals (reviewed by *Bradford Berk*). A further substance produced naturally in the body is the pineal gland hormone melatonin, which besides regulating circardian rhythms is a strong antioxidant and – as elaborated on by *Amanda Lochner* – ameliorates tissue damage in ischaemia/reperfusion in a number of organs. A wealth of recent studies demonstrate that the physiological stimulus of endurance exercise is overwhelmingly cardioprotective. In their article, *Karyn Hamilton* and *John Quindry* focus their discussion on the role of endogenous antioxidants in mediating protection and secondarily on the protective mechanisms peripheral to redox control. The overall benefits observed with the lipid-lowering HMG CoA reductase inhibitors (statins) appear to be greater than might be expected from changes in lipid levels alone. *Oliver Adam* and *Ulrich Laufs* review the current knowledge on the action of statins regarding endothelial NO synthase (eNOS), endothelin, free oxygen radicals, MHC-II, the protein kinase Akt and metalloproteinases.

The present series of articles on oxidative stress in clinical practice summarizes the current knowledge in a rapidly evolving field. Its intention is both to provide a mechanistic overview of the ways in which oxidative stress impacts cardiovascular disease and to consider potential therapeutic options to target such pathways. Although large clinical trials of "simple" antioxidant approaches, such as vitamin C and E, have not demonstrated significant benefit for cardiovascular end points, the data discussed in this book should make quite clear that such an approach is too simplistic. Understanding the complexity of the cellular redox system may in the future allow the development of better-targeted interventions to facilitate the path of patients from disease back to health.

# Contents

# Contributors

**Oliver Adam** Klinik für Innere Medizin III, Kardiologie, Angiologie und Internistische Intensivmedizin, Universitätsklinikum des Saarlandes, D-66421 Homburg/Saar, Germany, o.adam@freenet.de

**Bradford C. Berk** University of Rochester Medical Center, Rochester, NY 14642, USA, bradford_berk@urmc.rochester.edu

**David Bernhard** Cardiac Surgery – Research Laboratories, Department of Surgery, Medical University of Vienna/AKH, Ebene 8, G09/07 Währinger Gürtel 18-20, A-1090 Vienna, Austria, david.bernhard@meduniwien.ac.at

**Jennifer Victoria Bodkin** Cardiovascular Division, King's College London BHF Centre of Excellence, London SE1 9NH, UK, jennifer.bodkin@kcl.ac.uk

**Diego Bonatto** Instituto de Biotecnologia, Universidade de Caxias do Sul (UCS), Caxias do Sul, RS, Brazil, diegobonatto@gmail.com

**Susan Diana Brain** Cardiovascular Division, King's College London BHF Centre of Excellence, London SE1 9NH, UK, sue.brain@kcl.ac.uk

**Ralf P. Brandes** Institut für Kardiovaskuläre Physiologie, Fachbereich Medizin der Goethe-Universität, 60596 Frankfurt am Main, Germany, brandes@zphys1.uni-frankfurt.de

**Joseph R. Burgoyne** Cardiovascular Division, King's College London BHF Centre of Excellence, The Rayne Institute, St Thomas' Hospital, London SE1 7EH, UK, joseph.burgoyne@kcl.ac.uk

**Ariel R. Cardoso** Departamento de Bioquímica, Instituto de Química, Universidade de São Paulo, São Paulo, SP, Brazil

**Jose Esteban Castelao** Complexo Hospitalario Universitario de Vigo, CHUVI Genetic Oncology Unit, CHUVI, Meixoeiro s/n, Vigo, Spain; USC/Norris Comprehensive Cancer Center, Keck School of Medicine of the University of Southern California, Los Angeles, CA 90089-9175, USA, castelao@usc.edu

**Rebecca L. Charles** Cardiovascular Division, King's College London BHF Centre of Excellence, The Rayne Institute, St Thomas' Hospital, London SE1 7EH, UK, rebecca.charles@kcl.ac.uk

**Andreia Chignalia** Ottawa Hospital Research Institute, Kidney Research Centre, University of Ottawa, Ottawa, ON K1H 8M5, Canada

**Anna Csiszar** Reynolds Oklahoma Center on Aging, Department of Geriatric Medicine, University of Oklahoma Health Sciences Center, 975 NE 10th Street, BRC-1313, Oklahoma City, OK 73104, anna-csiszar@ouhsc.edu

**Dipak K. Das** School of Medicine, Cardiovascular Research Center, University of Connecticut, Farmington, CT, USA, ddas@neuron.uchc.edu

**Samarjit Das** School of Medicine, Cardiovascular Research Center, University of Connecticut, Farmington, CT, USA

**Somak Das** School of Medicine, Cardiovascular Research Center, University of Connecticut, Farmington, CT, USA

**Denise de Castro Fernandes** Vascular Biology Laboratory, School of Medicine, Heart Institute (InCor), University of São Paulo, CEP 05403-000 São Paulo, SP, Brazil, denisef@usp.br

**Christian Delles** BHF Glasgow Cardiovascular Research Centre, University of Glasgow, Glasgow G12 8TA, Scotland, UK, c.delles@clinmed.gla.ac.uk

**Javier Díez** Center for Applied Medical Research, 31008 Pamplona, Spain, jaimar@unav.es

**Anna F. Dominiczak** BHF Glasgow Cardiovascular Research Centre, University of Glasgow, Glasgow G12 8TA, Scotland, UK, ad7e@clinmed.gla.ac.uk

**Samuel C. Dudley** Section of Cardiology, University of Illinois at Chicago, Chicago, IL 60612, USA; Jesse Brown VA Medical Center, Chicago, IL, USA, scdudley@uic.edu

**Philip Eaton** Cardiovascular Division, King's College London BHF Centre of Excellence, The Rayne Institute, St Thomas' Hospital, London SE1 7EH, UK, philip.eaton@kcl.ac.uk

**Raymond Farah** Department of Internal Medicine B, Ziv Medical Center, Safed, Israel, raymond.f@ziv.health.gov.il

**Manuela Gago-Dominguez** Department of Preventive Medicine, USC/Norris Comprehensive Cancer Center, Keck School of Medicine of the University of Southern California, Los Angeles, CA 90089-9175, USA, mgago@usc.edu

**Alfonso Giovane** Department of Biochemistry and Biophysics, 1st School of Medicine, Second University of Naples, Naples, Italy, alfonso.giovane@unina2

**Rabea Graepel** Cardiovascular Division, King's College London BHF Centre of Excellence, London SE1 9NH, UK, rabea.graepel@kcl.ac.uk

**Kathy K. Griendling** Departments of Medicine, The Atlanta VA Medical Center, Emory University School of Medicine, Atlanta, GA, USA, kgriend@emory.edu

**Divya Gupta** Departments of Medicine, The Atlanta VA Medical Center, Emory University School of Medicine, Atlanta, GA, USA, dgupta2@emory.edu

**Tomasz J. Guzik** Translational Medicine Laboratory, Department of Internal and Agricultural Medicine and Department of Pharmacology Jagiellonian, University School of Medicine, Cracow 31-121, Poland, tguzik@cm-uj.krakow.pl

**Arash Haghikia** Department of Cardiology and Angiology, Department of Cardiac, Thoracic, Transplantation, and Vascular Surgery, Hannover Medical School, 30625 Hannover, Germany, haghikia.arash@mh-hannover.de

**Karyn L. Hamilton** Human Performance Clinical Research Laboratory, Applied Human Sciences, Colorado State University, Fort Collins, CO 80523-1582, USA, karynh@cahs.colostate.edu

**Louise Hecker** Division of Pulmonary and Critical Care Medicine, Department of Internal Medicine, University of Michigan Medical Center, Ann Arbor, MI 48109, USA, lhecker@uab.edu

**Sabine Heumüller** Institut für Kardiovaskuläre Physiologie, Fachbereich Medizin der Goethe-Universität, 60596 Frankfurt am Main, Germany, heumueller@zphys1.uni-frankfurt.de

**Andres Hilfiker** Department of Cardiology and Angiology, Department of Cardiac, Thoracic, Transplantation, and Vascular Surgery, Hannover Medical School, 30625 Hannover, Germany, hilfiker.andres@mh-hannover.de

**Denise Hilfiker-Kleiner** Department of Cardiology and Angiology, Department of Cardiac, Thoracic, Transplantation, and Vascular Surgery, Hannover Medical School, 30625 Hannover, Germany, hilfiker.denise@mh-hannover.de

**Xuejuan Jiang** Department of Preventive Medicine, USC/Norris Comprehensive Cancer Center, Keck School of Medicine of the University of Southern California, Los Angeles, CA 90089-9175, USA, xuejuanj@usc.edu

**Timo Kahles** Institut für Kardiovaskuläre Physiologie, Fachbereich Medizin der Goethe-Universität, 60596 Frankfurt am Main, Germany, kahles@zphys1.uni-frankfurt.de

**Ashwani K. Khanna** Division of Cardiology, Department of Medicine, University of Maryland, Baltimore, MD, USA, akhanna@medicine.umaryland.edu

**Alicia J. Kowaltowski** Departamento de Bioquímica, Instituto de Química, Universidade de São Paulo, São Paulo, SP, Brazil, alicia@iq.usp.br

**Munish Kumar** Department of Physiology and Biophysics, School of Medicine University of Louisville, Louisville, KY 40202, USA

**Saumi Kundu** Department of Physiology and Biophysics, School of Medicine University of Louisville, Louisville, KY 40202, USA

**Ulrich Laufs** Klinik für Innere Medizin III, Kardiologie, Angiologie und Internistische Intensivmedizin, Universitätsklinikum des Saarlandes, D-66421 Homburg/Saar, Germany, ulrich@laufs.com

**Francisco R.M. Laurindo** Vascular Biology Laboratory, School of Medicine, Heart Institute (InCor), University of São Paulo, CEP 05403-000 São Paulo, SP, Brazil, francisco.laurindo@incor.usp.br

**Amanda Lochner** Division of Medical Physiology, Department of Biomedical Sciences, Faculty of Health Sciences, University of Stellenbosch, Tygerberg 7505, Republic of South Africa, alo@sun.ac.za

**Gloria V. López** Laboratorio de Química Orgánica, Facultad de Ciencias, Universidad de la República, 11400 Montevideo, Uruguay; Departamento de Bioquímica, Facultad de Medicina, Universidad de la República, 11800 Montevideo, Uruguay, vlopez@fq.edu.uy

**Paras Kumar Mishra** Department of Physiology and Biophysics, School of Medicine, University of Louisville, Louisville, KY 40202, USA

**Karni S. Moshal** Department of Physiology and Biophysics, School of Medicine, University of Louisville, Louisville, KY 40202, USA

**Claudio Napoli** Department of General Pathology, 1st School of Medicine, Second University of Naples, Naples, Italy, claudio.napoli@unina2.it

**Subramaniam Pennathur** Division of Nephrology, Department of Internal Medicine, University of Michigan Medical Center, Ann Arbor, MI 48109, USA, spennath@umich.edu

**Galen M. Pieper** Division of Transplant Surgery, Department of Surgery, Medical College of Wisconsin, Cardiovascular Research Center and the Free Radical Research Center, Milwaukee, WI, USA, gmpieper@mcw.edu

**Bruno B. Queliconi** Departamento de Bioquímica, Instituto de Química, Universidade de São Paulo, São Paulo, SP, Brazil

**John C. Quindry** Cardioprotection Laboratory, Department of Kinesiology, Auburn University, Auburn, AL 36849, USA, jcq0001@auburn.edu

**Homero Rubbo** Departamento de Bioquímica, Facultad de Medicina, Universidad de la República, 11800 Montevideo, Uruguay, hrubbo@fmed.edu.uy

**Heinrich Sauer** Department of Physiology, Justus Liebig University Giessen, Giessen 35392, Germany, heinrich.sauer@physiologie.med.uni-giessen.de

**Mona Sedeek** Ottawa Hospital Research Institute, Kidney Research Centre, University of Ottawa, Ottawa, ON K1H 8M5, Canada

**Ajay M. Shah** Cardiovascular Division, King's College London British Heart Foundation Centre of Research Excellence, London SE5 9NU, UK, ajay.shah@kcl.ac.uk

**Alex Sirker** Cardiovascular Division, King's College London British Heart Foundation Centre of Excellence, London SE5 9NU, UK, alex.sirker@kcl.ac.uk

**Ali A. Sovari** Section of Cardiology, University of Illinois at Chicago, Jesse Brown VA Medical Center, Chicago, IL 60612, USA, asovari@med.net.ucla.edu

**Yao Sun** Division of Cardiovascular Diseases, Department of Medicine, University of Tennessee, Health Science Center, Memphis, TN 38163, USA, yasun@uthsc.edu

**W. Robert Taylor** Departments of Medicine and Biomedical Engineering, The Atlanta VA Medical Center, Emory University School of Medicine, Atlanta, GA, USA, wtaylor@emory.edu

**Victor J. Thannickal** Division of Pulmonary, Allergy and Critical Care Medicine, University of Alabama at Birmingham, Birmingham, Alabama, USA; Department of Internal Medicine, University of Michigan Medical Center, Ann Arbor, MI 48109, USA, vjthan@uab.edu

**Rhian M. Touyz** Ottawa Hospital Research Institute, Kidney Research Centre, University of Ottawa, Ottawa, ON K1H 8M5, Canada, rtouyz@uottawa.ca

**Neetu Tyagi** Department of Physiology and Biophysics, School of Medicine, University of Louisville, Louisville, KY 40202, USA, neetu.tyagi@louisville.edu

**Suresh C. Tyagi** Department of Physiology and Biophysics, School of Medicine, University of Louisville, Louisville, KY 40202, USA, suresh.tyagi@louisville.edu

**Zoltan Ungvari** Reynolds Oklahoma Center on Aging, Department of Geriatric Medicine, University of Oklahoma Health Sciences Center, 975 NE 10th Street, BRC-1313, Oklahoma City, OK 73104, zoltan-ungvari@ouhsc.edu

**Maria Wartenberg** Cardiology Division, Department of Internal Medicine I, Friedrich Schiller University Jena, Jena 07743, Germany, maria.wartenberg@med.uni-jena.de

**Cameron World** Aab Cardiovascular Research Institute and Department of Medicine, University of Rochester, Rochester, NY, USA

**Guillermo Zalba** Center for Applied Medical Research, 31008 Pamplona, Spain, gzalba@unav.es

**Min Zhang** Cardiovascular Division, King's College London British Heart Foundation Centre of Excellence, London SE5 9NU, UK, min.zhang@kcl.ac.uk

# Chapter 1
# The Evolving Concept of Oxidative Stress

**Denise de Castro Fernandes, Diego Bonatto, and Francisco R.M. Laurindo**

**Abstract** The metaphoric concept of oxidative stress has been fundamental to knowledge systematization in the field of redox processes in biomedicine. Oxidative stress has evolved over recent years to account for a disruption of redox signaling and equilibrium, rather than just a plain imbalance between prooxidants and antioxidants causing molecular damage. Redox signaling has been documented as a potent and ubiquitous mode of regulation of several important physiological events, and its dysregulation accounts for disease pathophysiology. However, there are as yet several unclear aspects regarding the mechanisms whereby redox-related intermediates modulate signaling targets at the required level of specificity and robustness. Thus, the redox signaling concept itself is also an evolving entity. The model of ROS-mediated differential regulation of thiol targets solely on the basis of distinct chemical reactivities of thiol groups has not been able to fully account for the variety and sophistication of redox-dependent responses. Thus, current models of redox signaling have to take into account additional hierarchical levels of regulation in the cell biology realm. The notion of compartmentalization is an important example in this direction, and here we have tied it to the systems biology–based idea of modularity. In this context, oxidative stress may be viewed as a disruption of redox modular architecture and the consequent emergence of supramodular secondary signaling. Further contextualizing these mechanisms is essential in order to allow meaningful progress in strategies aiming at improving detection of disrupted redox signaling or redox-related therapeutic interventions. These considerations indicate that, while having lost some its metaphorical strength with respect to mechanistical insights, the dynamically reformulated concept of oxidative stress remains powerful as an operational tool to communicate and contextualize science in the field.

---

F.R.M. Laurindo (✉)
Vascular Biology Laboratory, School of Medicine, Heart Institute (InCor), University of São Paulo, CEP 05403-000 São Paulo, SP, Brazil
e-mail: francisco.laurindo@incor.usp.br

H. Sauer et al. (eds.), *Studies on Cardiovascular Disorders*,
Oxidative Stress in Applied Basic Research and Clinical Practice,
DOI 10.1007/978-1-60761-600-9_1, 

**Keywords** oxidative stress · redox signaling · redox systems biology · modelling · thiol proteins · free radical molecular damage · reactive oxygen species

The importance of redox processes in biology and medicine lies in at least two facts. First, redox processes are ancestral and ubiquitous, playing a relevant role in the homeostasis of virtually every prokaryotic and eukaryotic cell tested so far. Second, redox processes are powerful biological effectors; i.e., within ranges that can reasonably be achieved in physiological or pathophysiological scenarios, redox processes can robustly affect essentially all aspects of cellular function, metabolism, and structure. Consequently, the interest of investigators in this field is not only intense but unusually long-lived [1–3]. A very large body of studies has focused on the role of redox-dependent mechanisms in a broad variety of disease conditions of different natures. Over the past several years, significant attention has been directed to the integrative role of redox processes in cell signaling, gradually switching the focus of redox processes from toxicology to physiology.

The metaphoric concept of oxidative stress has been fundamental to knowledge systematization in the field. However, a key attribute of metaphors, and all tools for conceptual synthesis in general, is that they have to evolve and adapt to new knowledge in the area until they lose efficacy and are best left aside. Might that be the case with the concept of oxidative stress? This chapter makes use of this discussion to briefly summarize current knowledge on aspects involved in the chemical and biological basis of homeostatic and disruptive redox-centered signaling processes.

## 1.1 A Brief Historical Note and Some Definitions

The investigative field of redox processes in biology and medicine started to mature in the late 1960s with the discovery that the enzyme erythrocuprein had the specific function of promoting the dismutation of the superoxide free radical, the mono-electronic product of oxygen reduction [4]. This provided then unclear evidence for a biological role exerted by free radicals. It was soon recognized that a host of other related intermediates, generically termed *reactive oxygen species* (ROS), were likely to be generated in vivo and that some ROS were able to induce powerful cellular effects due to damage to lipids, proteins, and carbohydrates. An important outcome of such investigations was the notion of a transition-metal catalyzed Fenton chemistry generating a hydroxyl radical, a strong oxidant. The concept of *oxidative stress as the imbalance between prooxidants and antioxidants* was then established [5]. Some beneficial effects of free radicals in host defenses were also recognized in professional phagocytes [6]. In the late 1980s, particularly in the cardiovascular and immunological areas, another free radical, nitric oxide, was identified as a major autocrine and paracrine mediator, able to induce vascular relaxation, immunological regulation, and also many other effects. This was followed by the notion that superoxide radical interact with nitric oxide and not only regulate NO bioavailability, but also lead to secondary by-products such as peroxynitrite and related intermediates [7, 8], which can yield strongly reactive radicals such as nitrogen dioxide and carbonate [9]. In parallel, the identification of nitric oxide and of intracellular growth factor–dependent hydrogen peroxide production [10–13] prompted the concept of *redox-mediated signaling,* which evolved to comprise *cellular signal transduction networks in which the integrative element is a series of interconnected electron transfer reactions involving free radicals or nonradical oxidant species* (modified from [14]). Thus, chemically simple intermediates such as ROS are able to exert specific intracellular second messenger effects that regulate major cellular functions.

Recently, progress has occurred in the elucidation of the chemical biology of reactive intermediates, in understanding the regulation and structure of signaling ROS generators such as NADPH oxidases, in the elucidation of multiple mitochondrial functions, and in the integration of oxidative stress with other forms of stress, such as nutrient deprivation or endoplasmic reticulum stress. Important advances in high-throughput methods have also been extended to the redox arena, prompting not only an increased investigative capacity but also the impending development of redox systems biology.

In parallel with these developments, the concept of oxidative stress has carried from the outset an intrinsic connection with investigating the effects of antioxidant interventions. Such interventions have been explored largely as a tool to understand pathophysiology and of course to exert therapeutic effects against an array of clinical problems. At the same time, the complexities raised in understanding the multiple pathways involved in redox signaling indicate that even the definition of what is expected to be an antioxidant strategy must be considerably expanded from the *strictly chemical definition that an antioxidant is a compound that halts the oxidation of a substrate at concentrations significantly lower than that of the substrate* [15]. Finally, the redox area is a prototypical situation in which the general principle that scientific developments closely follow advances in investigative techniques holds true. Improvements in EPR methods, fluorescent indicators, mass spectrometry biomarkers, and proteomic techniques have been significant steps forward, but this is still an area of limited conceptual advances and practical applications.

## 1.2 Molecular Damage by Free Radicals and Oxidant Species

The early classical idea of oxidative stress lies in the foundations of toxicology in which free radical research was born [5] and is strongly linked to the notion that *molecular damage promoted by oxidizing free radical species would be a major factor underlying the pathophysiology of many disease conditions*. In this paradigm, the genesis of free radicals was at first viewed as a somewhat exogenous or accidental process [1, 3], even when it was enzyme-mediated, e.g., by xanthine oxidase following an ischemic insult [16]. Free radicals and oxidant species can indeed promote damage to essentially every cell constituent. This model of oxidative stress, therefore, relies heavily on the chemical reactivity of such species. The properties that determine the reactivity of free radicals or oxidant species have been reviewed in excellent texts [9, 17, 18] and are defined by two main factors: kinetics and thermodynamics (Box 1.1).

## Box 1.1 Thermodynamics and Kinetics

Reactive species reactivity can be estimated based on thermodynamic and kinetic parameters, which in biological systems depend mainly in reactant/product concentrations, since factors that affect reaction rates such as temperature or pressure tend to be constant.

Thermodynamics deals with the possibility that a specific reaction occurs, i.e., a given reaction is spontaneous when free energy ($\Delta G^{\circ\prime}$) between products and reactants is negative ($\Delta G^{\circ\prime} < 0$). For reactive species, which transfer electrons, the common thermodynamic parameter employed is the redox potential ($E^{\circ\prime}$), that measures the tendency of a chemical species to accept (reduction) or donate (oxidation) electrons. Free energy for redox reactions can be converted to electrochemical potential ($\Delta G^{\circ\prime} = -nF\Delta E^{\circ\prime}$), which can be transformed into the Nernst redox potential, which takes into account the estimated initial concentration of the redox pair and their products ($E^{\circ\prime} = \Delta E^{\circ\prime} - RT/nF\ (\ln K_{eq})$), in which n is the number of electrons and F the Faraday constant. The relative positions of redox pairs in redox potential tables allow prediction of the direction of electron flow from one redox couple to another; for example, considering the reduction potential for tocopheryl radical/tocopherol (+ 480 mV) and ascorbyl radical/ascorbate (+ 282 mV), the electron will flow from tocopheryl radical to ascorbate, which in turn will form an ascorbate radical, and not the opposite way [16].

The other parameter, the rate constant (k, $M^{-1} \times s^{-1}$), implies how fast two species will react. Several rate constants were measured mainly by pulse radiolysis or stopped flow experiments for radical species, and their values are easily found, e.g., on the site of the Chemical Kinetics Rate Constants from Notre Dame Radiation Lab. (http://hamill.rad.nd.edu/compilations/solnkin.html). Although all oxidants are called *reactive* species, their reactivities are very distinct: glutathione, the main low molecular thiol compound in cytosol, reacts with the hydroxyl radical at a rate near to the diffusion limit ($1.4 \times 10^{10}\ M^{-1}.s^{-1}$), while with the superoxide radical the rate is less than $10\ M^{-1} \times s^{-1}$. On the other hand, if the superoxide anion is in its protonated form ($HO_2^{\bullet}$), which takes place at a lower pH (such as that in phagolysosomes), the reaction rate rises to $1.4 \times 10^{10}\ M^{-1}.s^{-1}$. Rate constants are also very useful for comparing which biological targets will preferentially react with many reactive species, when the concentrations of targets can be estimated. For example, based on estimated cytosolic concentrations for peroxiredoxin (20 μM) and glutathione (2 mM), and considering their rate constants for hydrogen peroxide (Prx, $10^7$ and GSH, 0.89 $M^{-1} \times s^{-1}$), less than ~1% of such oxidant will react with glutathione, even if the latter is 100-fold more concentrated than peroxiredoxin [33]. This, among other considerations, provides a basis for the necessity of compartmentalization and modularity considering redox signaling specificity.

Over the years, however, this model has been increasingly challenged in several aspects. First, the majority of studies examining free radical damage to biomolecules utilized exogenous oxidants in high concentrations, which are unlikely to exist in vivo under normal physiological conditions and perhaps under at least some pathological conditions as well [19]. More recently, it has been suggested that the majority of oxidants generated under a prooxidant challenge are two-electron non-radical oxidants such as hydrogen peroxide, the aldehydes, and peroxynitrite, among others, many of them indeed unlikely to promote extensive molecular damage at their usual concentrations [20]. Moreover, kinetical and other constraints can be raised regarding the occurrence of Fenton reaction in vivo [21], questioning the major mechanism of generation of the hydroxyl radical—the main oxidant species under this model. While several of these assumptions still have to be demonstrated in vivo, they pose additional obstacles to this oxidative stress model.

## 1.3 The Redox Signaling Concept

Much evidence over the recent several years has increasingly indicated that ROS are normally produced at low levels under basal conditions by essentially every cell, and in addition can undergo increased generation in the course of a number of physiological events. These evidences led to the concept of redox signaling, which itself evolved from a vague theoretical proposal to the demonstration that low-level intracellular oxidant generation was not only able but also necessary to mediate cellular signal transduction [2, 22–24]. Some examples include: tumor necrosis factor-α [25], platelet-derived growth factor [24], epidermal growth factor [26], angiotensin II [27], interleukin (IL) IL-1β [28], and insulin [29], which are all reported to transiently increase intracellular levels of species such as hydrogen peroxide. Downstream effects include enhanced stress resistance, cell proliferation, cytokine release, cell adhesion, growth arrest, and apoptosis. Thus, the classical model of redox signaling proposes the generation of ROS as an intracellular second messenger of such mediators (Model I from Fig. 1.1). It has now been shown that redox signaling can occur in the absence of overall changes in the redox status of the major intracellular reductants glutathione and thioredoxin [30], and thus reflects localized compartmental cell events. The main conceptual revolution underlying the redox signaling notion was that vicious signaling circuits in several

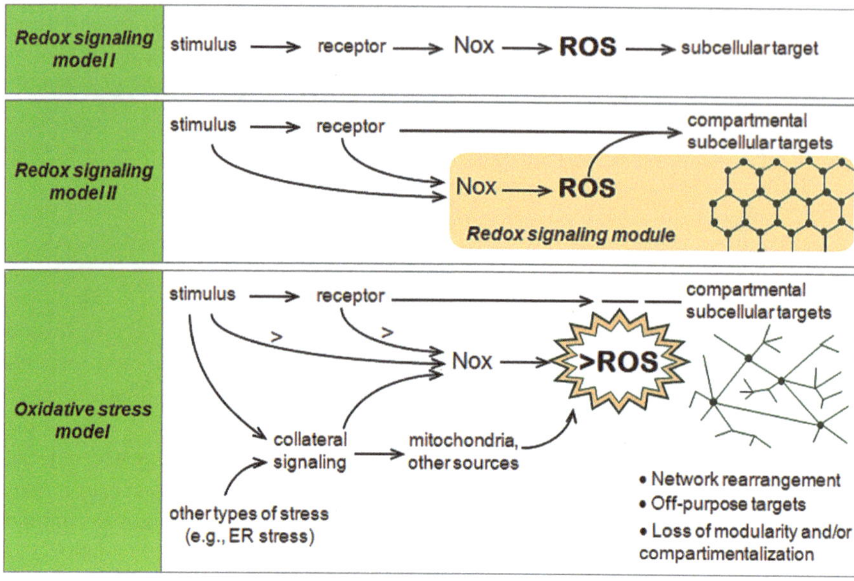


**Fig. 1.1** Models of redox signaling and oxidative stress

disease conditions were modeled to occur *not only as a result of direct free radical-mediated chemical damage to biomolecules, but—perhaps mostly—from disordered activation and/or expression of subcellular signaling targets due to excessive, uncompensated, or decompartmentalized reactive oxygen species (ROS) generation*. In this context, a redefinition of oxidative stress has recently been proposed as "an imbalance between oxidants and antioxidants in favor of the oxidants, leading to a disruption of redox signaling and control and/or molecular damage" [31]. A curious detail is that such conceptual evolution, while widely disseminated, has not yet been paralleled by updates in graphic design of oxidative stress representations, with nonnegligible use in recent reviews of symbols associated with disequilibrium or molecular damage (e.g., shifted balances, explosion diagrams, etc.). This might denote that the concept shift has not yet been completely assimilated through the investigative field.

A primary requisite of redox signaling is not only that the radical or oxidant generation should occur at controlled conditions regarding amount, time, and space; but also that the oxidant is not too reactive, which would preclude its diffusion, undermining efficient signal communication. This creates considerable difficulties in ascribing a signaling role to powerfully reactive oxidants such as the hydroxyl radical. In fact, most evidence is consistent with signaling roles for less reactive species such as superoxide, hydrogen peroxide, and nitric oxide [18, 32, 33].

Another simultaneous requisite and corollary of redox signaling is the fact that cellular ROS generation is mainly a nonaccidental regulated process, controlled via enzyme-dependent sources. In fact, it has become clear that enzyme-dependent ROS sources account for ROS production in most (patho)physiological situations, even under exposure to exogenous oxidants [23, 34]. Thus, such enzymatic sources of ROS are intrinsic components of redox-signaling cascades, and their regulation is likely to be just as important as the regulation of the targets themselves in order to allow the transduction and flow of cellular signals [2, 23]. Among the main sources of ROS, mitochondria are likely the most important quantitative source, but their role in the fine-tuning of redox signaling is less evident [35]. On the other hand, isoforms of the phagocyte NADPH oxidase multisubunit complex appear to be the most prominent and studied source of basal as well as agonist-induced signaling ROS in a number of different cell types.

A further attribute of redox signaling is the presence, in target proteins, of redox-sensitive structural domains, which essentially sum up to redox-active metals and particularly thiol groups. Although about 40% of biologically important enzymes depend on catalytic metal centers [36], the importance of metals for signaling is yet unclear and will not be discussed further here (for a review, see [37]. On the other hand, a review of basic mechanisms underlying chemical reactivity of thiol groups is important for the comprehension of possible integrative redox pathways.

## 1.4 Reactivity of Thiols: A Chemical Route for Redox-Dependent Messages

Oxidant signaling can involve, as intermediates, free radicals (e.g., superoxide, nitric oxide) which promote one-electron oxidations, or two-electron oxidants (e.g., hydrogen peroxide, peroxynitrite, aldehydes). The quantitative importance of two-electron oxidants may be modeled as being significantly more important than that of free radicals, a fact that has pathophysiological and therapeutic implications [20]. In oxidant-mediated signaling pathways, thiol proteins have been considered the major mechanism by which intracellular changes in redox state integrate biochemical processes [18, 33]. In this context, thiols have been proposed to account for specificity with respect to signaling, given their wide array of post-translational modifications and particularly of distinct forms of oxidation. In addition, thiols generally allow reversibility—an essential assumption of any form of signaling—on the basis of abundant intracellular reductase systems [18, 38].

### *1.4.1 Thiol Oxidation Pathways*

A thiol is any organic compound that contains the functional group composed of a sulfur and a hydrogen atom (–SH); among the amino acids, cysteine is the only one that shows –SH in its side chain. Methionine is another sulfur-containing amino acid, but the sulfur atom is covalently linked to a carbon atom, which characterizes its side chain as a thioether. The primary function of cysteine in biomolecules involves the maintenance of correct protein folding in tertiary structures, by forming structural disulfide bonds. Besides giving rise to disulfides, reduced cysteines also appear in many proteins, and when they are found in an active site of enzymes and participate in a catalytic cycle, these proteins are named thiolproteins. Their redox modifications, such as oxidation of critical cysteine(s), alter protein activity and/or post-translational modifications, as a widely disseminated paradigm of redox-sensitive signaling proteins.

One specific cysteine among others in a protein is defined as "critical" when it appears in a deprotonated state or thiolate form (P-S$^-$). This occurs in cysteines with a low acid dissociation constant (pKa), which itself depends on the cysteine residue molecular environment, with neighboring positive amino acids facilitating ionization within the three-dimensional conformation or quaternary structure. Thiols and thiolate anions react with almost all physiological oxidants, but low pKa is a key property for enhancing the reactivity of any cysteine. As seen in Table 1.1, the cysteine of glutathione shows a pKa ~8, while the cysteine from an active site of human peroxiredoxin has a pKa ~5–6 [41], which is enough to increase the rate constant with hydrogen peroxide $10^6$-fold. For some proteins involved in redox signaling, the cysteine was clearly shown to be a thiolate, as in thioredoxin [43] and protein tyrosine phosphatase (PTP) [12]; in other cases, evidences that the critical cysteines are in thiolate form are also strong (bacterial transcriptional factor OxyR [44], eukaryotic transcriptional factors AP-1 [45] and NF-κβ [46], and caspases [47]). However, low pKa of a protein thiol alone is not enough to confer selectivity, as can be clearly seen comparing values of pKa and rate constant of reaction with $H_2O_2$ for PTP1B and peroxiredoxins in Table 1.1.

**Table 1.1** pKa of the critical cysteine of physiologically relevant protein or nonprotein thiols and respective rate constants of their reactions with hydrogen peroxide at physiological pH and 37°C

| Thiol compound/protein | pKa | Rate constant (k; $M^{-1}$ $s^{-1}$) | References |
|---|---|---|---|
| Glutathione (GSH) | 8.8 | 0.89 | [38] |
| Cystein | 8.3 | 2.9 | [38] |
| Thioredoxin | 6.5 | 1.05 | [39] |
| PTP1B ($Cys^{215}$) | 5.4 | 20 | [12] |
| Peroxiredoxins (resolving Cys) | 5–6 | $1–4 \times 10^7$ | [40, 41] |

Reactivity of thiols is quite complex (Fig. 1.2). First of all, thiols can be oxidized by 2-electron oxidants to sulfenic acid (as hydrogen peroxide, peroxynitrite, hypochlorous acid, haloamides, etc.):

$$RS^- + H_2O_2 \rightarrow RSO^- + H_2O + H^+ \qquad \text{(reaction 1)}$$

or by 1-electron oxidants to thiyl radical (superoxide anion, carbonate radical, hydroxyl radical, etc.):

$$RSH + {}^{\bullet}OH \rightarrow RS^{\bullet} + H_2O \qquad \text{(reaction 2)}$$

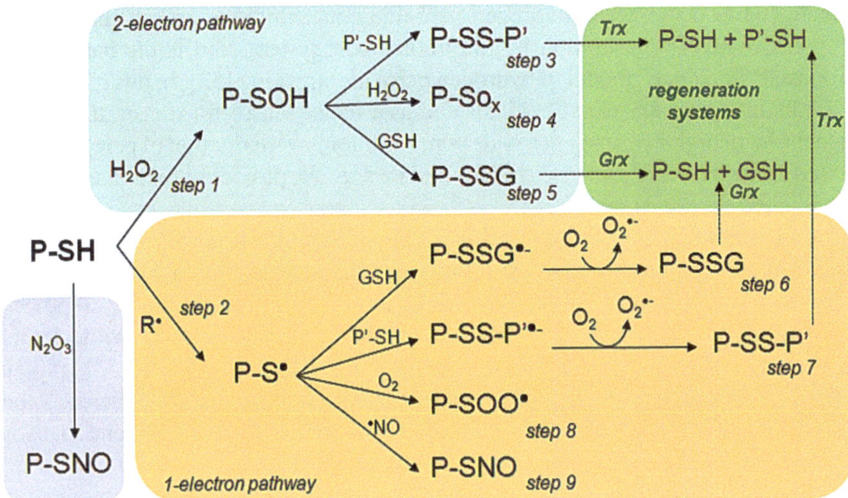


**Fig. 1.2** Pathways for protein thiol oxidation. Thiols can be oxidized by 2- (upper panel, blue) or 1-electron oxidants (down panel, orange), to reactive intermediates, respectively, sulfenic acid or thiyl radical. Secondary reactions for 2-electron oxidation pathways form mixed disulfides with GSH (P-SS-G); vicinal thiols favor intramolecular disulfide bonds formation; or higher oxidation products, such as sulfinic acid, sulfonic acid, sulfinamide, among others. Glutathionylation and disulfide bonds can be reversed to reduced thiols by regeneration systems (green panel), glutaredoxin (Grx) and thioredoxin (Trx) systems. Thiyl radical formed by 1-electron oxidation pathways form, whether in presence of oxygen (aerobic conditions) or disulfide anion radical ($P\text{-}SS\text{-}P^{\bullet-}$ or $P\text{-}SS\text{-}G^{\bullet-}$), which are strong reducing agents, and promote superoxide ($O_2^{\bullet-}$) formation by reducing molecular oxygen and thus amplifying oxidative reactions. Thiyl and sulfinyl (P-SOO•) radicals can propagate radical chain reactions. Finally, nitrosylated thiols can be formed by radical recombination of thiyl radical and nitric oxide (1-electron oxidation pathway) or by direct reaction with the nitrosating species dinitrogen trioxide ($N_2O_3$). Please see text and cited references for more detailed discussion

Once formed, sulfenic acid (RSOH, step 1, Fig. 1.2) can either be overoxidized, form mixed disulfides with GSH, or form inter- or intramolecular disulfide bonds. Sulfenic acid is very unstable, so it is considered preferentially a reaction intermediate; however, in some proteins it was possible to isolate sulfenic acid due to proper stabilizing microenvironment conditions [48]. One important example is the formation of sulfenic acid during the catalytic cycle of peroxiredoxins, ubiquitous and abundant multicompartmental proteins present from bacteria to eukaryotes, which decompose hydrogen peroxide at high rate constants (Table 1.1). Peroxiredoxins contain two cysteines in their active sites, one being a thiolate residue. Based on site-specific mutagenesis experiments, it was shown that after hydrogen peroxide oxidation, the thiolate forms a stable sulfenate, which in turn forms an intramolecular disulfide bond with the second cysteine (step 2, Fig. 1.2) [49–51]. The same example can be extended to overoxidation of sulfenic acid to RSOx (representing sulfinic acid, sulfonic acid, sulfinamide, or sulfonamide; step 3, Fig. 1.2), since peroxiredoxins can be inactivated by overoxidation to sulfinic acid. This mechanism is believed to occur when hydrogen peroxide concentrations exceed the capacity of peroxiredoxin regeneration by the thioredoxin system, and is the basis of the proposed "floodgate" model of hydrogen peroxide signaling [52]. In this model, peroxiredoxins could act as antioxidants and as a redox sensor for transmitting redox signals. With high hydrogen peroxide concentration, overoxidation of peroxiredoxin would allow localized increases in this oxidant species, providing focus and possible specificity of redox signals.

A most significant reaction of sulfenic acid is its interaction with thiols, forming disulfides, with particular relevance for glutathione, present in high concentrations in cells, and the consequent formation of a mixed disulfide, a process called glutathionylation (step 4, Fig. 1.2). Glutathionylation was shown to inhibit some enzymes (phospho-fructokinase [53], GAPDH [54], PTP1B [55], protein kinase Cα [56], NFκβ [57], mitochondrial complex I [50], etc.); whereas other enzymes are activated (matrix metalloproteinase [58], hRas [59], sarco/endoplasmic reticulum calcium ATPase (SERCA) [60], mitochondrial complex II [61], etc.). Although the identification of pathways underlying the protein glutathionylation mechanism are still under investigation (1- vs. 2-electron oxidation, as shown in Fig. 1.2), the primary mechanism of deglutathionylation has been well characterized and attributed to the glutathione/glutaredoxin system. Reversibility allows glutathionylated proteins to act as redox signaling proteins [62].

Thiyl radicals may be formed by hydrogen abstraction from oxidizing radical species, such as hydroxyl radical ($^{\bullet}OH$), nitrogen dioxide ($^{\bullet}NO_2$), carbonate radical ($CO_3^{\bullet-}$), tyrosyl radical ($Tyr^{\bullet}$); by transition metal-catalyzed thiol oxidation; or by the action of peroxidases (horseradish peroxidase, myeloperoxidase, etc.) [63]. Thiyl radicals undergo distinct sets of reactions, the most favored with the thiolate anion [64, 65]. Although at the end there is disulfide formation (steps 6–8, Fig. 1.2), the strong reductant disulfide intermediates can produce superoxide anions [66] and increase the oxidant response. Radical chain reactions are likely to be inhibited by ascorbic acid or phenolic antioxidants such as vitamin E and flavonoids.

Nitric oxide can also interact with thiols and alter protein function by forming nitrosothiols that also may contribute to redox signaling. The generation of nitrosothiols can occur via several mechanisms that are dictated by the cellular environment; however, these mechanisms have not been clearly shown to occur in vivo [32]. The direct reaction between nitric oxide and the thiolate group is too slow to operate physiologically [67]. Two species are able to nitrosylate thiols, nitrogen dioxide ($^{\bullet}NO_2$) and dinitrogen trioxide ($N_2O_3$). The first oxidizes thiols to the thiyl radical (step 2, Fig. 1.2), which recombines with nitric oxide (radical-radical recombination) very fast $k = (2–3) \times 10^9\ M^{-1}\ s^{-1}$ (step 9, Fig. 1.2) [68]. Dinitrogen trioxide formation depends on nitric oxide and oxygen stationary concentrations (due to reactions 3–4), and it is favored in lipid membranes, where both $^{\bullet}NO$ and $O_2$ can accumulate [69]. Thus, there are two possible mechanisms for protein nitrosylation, one preferred to occur in hydrophobic environments such as membranes (mediated by $N_2O_3$) and the other possibly favored in cytoplasm, involving $^{\bullet}NO_2$ radical formation. It is important to note that there is increasing evidence that the thiyl radical is part of the catalytic cycle of many enzymes [70].

$$2\,{}^{\bullet}NO + {}^{\bullet}O_2 \rightarrow 2\,{}^{\bullet}NO_2 \quad \text{(reaction 3)}$$

$$^{\bullet}NO_2 + {}^{\bullet}NO \rightarrow N_2O_3 \quad \text{(reaction 4)}$$

## *1.4.2 Mechanism for Thiol-Mediated Signal Transduction*

Signal transduction occurs as the oxidized thiol-containing protein transmits a signal to the cellular environment, while the transmission should be easily turned on/off. When compared to phosphorylation-mediated signaling, thiol oxidation presents unique features, the main one being the lack of enzymatic catalysis of formation and degradation of its products, with the exception of glutathionylated protein by the glutathione/glutaredoxin system and protein disulfide by the thioredoxin/thioredoxin reductase system [71, 72]. Higher thiol oxidizing states usually are irreversible, with the exception of peroxiredoxins, where slow enzymatic reduction, mediated by sestrins and sulfiredoxins, has been demonstrated [73, 74].

The first evidence of thiol oxidation reversibility in signaling proteins was described with protein tyrosine phosphatase 1B (PTP1B), which was reversibly inactivated by endogenous hydrogen peroxide in cells stimulated with epidermal growth factor (EGF) [75], and a further oxidizing site was identified as the cysteine 215 [76]. Nowadays, reversible inactivation of phosphatases (PTPs and the lipid phosphatase PTEN) by hydrogen peroxide is a prime example of the activation of phosphorylation pathways, although the mechanisms involving reduction of phosphatases remains under investigation [18]. Similar to phosphatases, some protein kinase C isoforms show cysteines in the regulatory site that are susceptible to hydrogen peroxide oxidation, thereby altering their regulation [77].

Thiols can also transmit signals by disulfide bond formation, which alters the protein tertiary structure and influences its functional properties and possible interactions with other proteins. This is a case of bacterial transcription factor OxyR, whose disulfide bond formation between vicinal thiols changes its conformation, leading OxyR to bind to DNA and activate antioxidant genes, including glutaredoxin 1. Interestingly, glutaredoxin 1 deactivates OxyR by reducing its disulfide bond, providing an autoregulatory mechanism [78]. In glucose-starved mammalian cells, transcriptional factor ATF6 was shown to translocate from the endoplasmic reticulum to the Golgi apparatus only in its reduced form, where it is cleaved to release its cytoplasmic domain and able to activate unfolded protein response genes after nuclear translocation [79]. Another interesting example is the formation of an intermolecular disulfide in a regulatory region of cGMP-dependent protein kinase (PKG) in mammalian myocytes exposed to hydrogen peroxide, which increases its affinity for substrates, and constitutes an alternative mechanism for cGMP-independent vasorelaxation in response to hydrogen peroxide [80].

Oxidations that interfere in protein-protein interactions with signaling consequences have been emerging in the literature. The Nrf2/Keap1 system plays an important role during oxidative stress and xenobiotic detoxification metabolism by upregulating phase 2 enzymes. Both proteins are associated in the cytoplasm, which maintains the rapid Nrf2 turnover by facilitating its ubiquitination and degradation [81]; after Keap1 oxidation, protein-protein association is disrupted and Nrf2 translocates to the nucleus, regulating transcription of phase 2 genes including thioredoxin and thioredoxin reductase genes [82, 83]. In fact, as mentioned above, the thioredoxin/thioredoxin reductase system modulates several redox-sensitive transcription factors, such as NF-κβ [84], the tumor repressor p53 [85], the hypoxia-induced factor 1α (HIF-1α) [86], and the AP-1 protein complex [87].

With the development of new techniques, such as proteomics or subcellular localization studies, there is accumulating evidence that several proteins are glutathionylated and nitrosylated in (patho)physiological conditions, mainly in the cardiovascular system. An intriguing observation is that many proteins were demonstrated to be nitrosylated and glutathionylated in the same cysteine in different experimental conditions, suggesting that there is a close relationship between both of these post-translational modifications. Some examples are: (i) PTP1B, whose glutathionylation/nitrosylation decreases its activity, thereby enhancing phosphorylation events [55, 88]; (ii) p21Ras, in which both thiol modifications increase its activity and phosphorylation of downstream targets [59, 60, 89, 90]; and (iii) caspase-3, which is protected from cleavage by thiolation/nitrosylation, in the context of apoptosis [91–93]. One interesting exception is the ryanodine receptor channel (RyR1), which is essential for striated muscle contraction and contributes to diverse neuronal functions, including synaptic plasticity, by controlling calcium release from intracellular stores. Different cysteines of the ryanodine receptor are selectively nitrosylated [94] or glutathionylated [95, 96], each leading to specific functional consequences. It is important also to note that NO-derived oxidants and GSNO are able to promote glutathionylation, as for example described for sarco(endo)plasmic reticulum calcium ATPase (SERCA), a key protein regulating the intracellular storage of calcium. Similarly to RyR1, glutathionylation is mediated by peroxynitrite-promoted SERCA activation in the context of arterial relaxation [60]. The causal relationship between both thiol modifications remains unclear, with some arguing that nitrosothiol itself is another activated form of protein cysteines. Indeed, S-nitrosylated proteins show high lability, while S-glutathionylation is more stable, especially in the presence of thiols like glutathione [38]. The relative selectivity of each protein to a particular modification will distinguish which of the cellular proteins will be more easily or more stably modified by one or the other modification [38]. Finally, evidences for protein S-nitrosylation usually need further confirmation, as the evidences of nitrosylation are often based on one method, the biotin switch, which is an assay with some interferences/artifacts, thus requiring additional complementary methods [97, 98].

Overall, such a wide profile of thiol modifications provide many differential and to some extent ROS-specific routes for transducing redox-modulated signals to particular targets. Whether such features are sufficient to account for redox signaling specificity will be discussed in the next sections.

## 1.5 The Evolving Characteristic of Redox Signaling Models: Critical Analysis

Despite the overwhelming evidence supporting the role of ROS as intracellular signaling intermediates, discussed in great detail throughout this book, there are still many uncertainties as to *how low intracellular levels of ROS may account for specific target modulation in the rather sophisticated profile of physiological effects.* This is certainly one of the most recurrent and debated topics in the field [2, 18]. In addition, how robustness is achieved with regard to redox-dependent signal networks is still far from established. The model of intracellular ROS generation or changes in redox status acting as a second messenger at a cell-level scale is possibly still prevalent in the minds of many investigators. Several pieces of evidence as well as theoretical considerations, however, indicate that this concept is unlikely. Objections for this model rely on the poor signal specificity that can be achieved through non-targeted low-level increases in ROS levels [3, 18, 33]. In addition, a less considered but important critique is that the design of this model is not robust with respect to unaccountable cell conditions. Even though some systems do display an apparent behavior in the way predicted by Model I from Fig. 1.1, this may be influenced by factors such as poor sensitivity and specificity of ROS indicators and the temporal dissociation between ROS production and late cellular effects dictated by parallel signaling. In addition, studies using exogenous oxidants such as hydrogen peroxide tend to promote mass activation of signaling targets in an incoherent temporal or topological fashion. Also, many studies are performed in cultured cells, bringing about potential limitations [3]. In particular, agonist concentrations necessary to trigger detectable oxidative stress in this condition may differ from physiological ones, with the remarkable example of angiotensin II in vascular smooth muscle cells, for which the usual concentrations of 100 nM (e.g. [99]), are 2–3 orders of magnitude above physiological levels [100]. In fact, the poor success of antioxidant therapy, potentially due to many causes, does indicate at least that unidimensional models of redox signaling are unlikely in the pathophysiological scenario.

A major attribute of Model I (Fig. 1.1) is that it has to assume that most of the specificity of redox signaling would result from the pattern of chemical reactivity, e.g., of thiol groups [18]. As discussed in the previous section, thiol groups do display a varied profile of reactions, which might thus confer a possible menu of specific effects. However, the evidence that these thiol modifications do provide specificity to cell signaling in vivo appears still insufficient, since consistent models for how each ROS interact with thiols and potentially other redox-sensitive groups are yet imprecise. Before we refer to such possible models of ROS signaling, it is important to note that the term "reactive oxygen species" (ROS) is meant to designate *a general array of chemical species arising from oxygen reduction and their related precursors and/or reactive reaction products.* However, ROS are a very heterogeneous group of intermediates which differ widely with respect to reactivity, cellular location, partition, solubility, and diffusibility [33]. This makes the physiological consequences of each specific ROS substantially distinct and emphasizes the importance of accurately understanding the precise intermediates being analyzed.

Although the superoxide radical is clearly an important primary ROS generated by mitochondria, as well as Nox1 and Nox2 (although likely not Nox4) [101, 102] and other enzymatic sources, whether and how superoxide promotes direct signaling is uncertain. Superoxide is not very reactive at neutral pH, although a decrease in pH, e.g., at the level found in some cell compartments such as secretory vesicles and lysosomes, promotes its deprotonation to the hydroperoxyl radical ($^{\bullet}OOH$), given the pKa ~4.8 of the reaction $^{\bullet}O_2H \leftrightarrow H^+ + O_2^{\bullet -}$. The hydroperoxyl radical not only is more oxidizing but also becomes uncharged and thus can also more freely permeate membranes. Superoxide is able to oxidize Fe-S proteins, such as aconitase, yielding hydrogen peroxide as a byproduct [103]. Superoxide can also specifically displace iron from ferritin [104] and reduce quinones or oxidize diphenols to semiquinones [105], although the significance of these reactions is unclear regarding signaling. Recent suggestions that superoxide dismutases (SODs) may act as superoxide sensors may provide novel paradigms of superoxide signaling (Box 1.2). An intriguing aspect of superoxide signaling is that is difficult to reconcile such a role with the exceedingly low rate constants of the direct reaction between superoxide and thiols [39]. Superoxide may also signal indirectly via removal of nitric oxide and generation of peroxynitrite and related oxidants. Peroxynitrite, in turn, is not usually considered as a signaling species, given the highly oxidizing characteristic of its derived products such as carbonate radicals and nitrogen dioxide [9], which make reversibility unlikely. However, such products may S-nitrosylate/glutathionylate, or oxidize protein thiols, all such reactions being potentially reversible, as discussed above. On the other hand, there is no consistent evidence for reversibility of tyrosine nitration.

## Box 1.2 Superoxide Dismutases as Possible Superoxide Sensors?

The major enzymatic scavengers of $O_2^{\bullet -}$ are superoxide dismutases (SODs), which promote $O_2^{\bullet -}$ dismutation to $H_2O_2$ and $O_2$. SODs are widely distributed among aerobic prokaryotic and eukaryotic organisms [189], and inactivation of *sod* genes perturbs cell viability [190, 191]. In mammalian cells, overexpression of Sod1p induces genomic instability in cells deficient in genes involved in nonhomologous end joining (NHEJ) recombinational repair [192]. In addition, spinal motor neurons from Sod1-null mice show reduced expression and activity of redox factor-1 [193], a key enzyme in the DNA base excision repair pathway [194]. These data might suggest that SODs contribute to control and/or induce DNA repair pathways under physiological conditions where protection against oxidative damage is required. We recently applied systems biology tools in order to investigate the interplay between SODs and DNA repair mechanisms in yeast [195]. The large amount of biological data available from high-throughput experiments can be used to identify thousands of pairwise protein-protein interactions in different biological models and to predict different cellular behaviors under specific physiological conditions [196]. The observed pattern of Sod1p interactions [195] allow the proposal of a model in which a pulse of $O_2^{\bullet-}$ under nonphysiological conditions in *S. cerevisiae* induces Sod1p to activate the oxidative "DNA damage detection circuitry" composed mainly by yeast cell-cycle checkpoint kinases. In fact, activation of oxidative damage-responsive cell-cycle checkpoint kinase Mec1p requires a functional Sod1p [194]. Although hydrogen peroxide increase is a possible mediator of Sod1p effects, this is not straightforward, because SOD activity does not uniformly promote increases in hydrogen peroxide output [198]. This will happen only if superoxide is redirectioned from prior reactions that do not yield hydrogen peroxide [3]. Indeed, repeated overexpression of Sod1 in yeast provided conflicting data regarding an increase in steady-state hydrogen peroxide concentrations [199].

Our model, furthermore, supports the idea that both Sod1p and Sod2p could act as sensors of intracellular $O_2^{\bullet-}$, interacting with and inducing different DNA repair pathways, cell-cycle checkpoints, chromatin remodeling, and synthesis of dNTPs in a quasi-hierarchical mode of action. It is noteworthy that cancer and aging are associated with diminished SOD activity [200], while transfection of malignant tumor cells with MnSOD can reverse the malignant phenotype, suggesting that MnSOD functions as a tumor suppressor [201, 202]. The mechanism for this effect is still unknown, but the model of SOD-sensing molecules suggests that functional SODs could restore the activity of DNA repair and cell cycle checkpoints, reducing tumor invasiveness. Also, many DSB repair-associated genes are specifically down-regulated by hypoxia [203], known to reduce SOD activity [204]. In addition, it was recently shown in glial cells that SOD1 at endosomal surfaces physically interacts with the small GTPase Rac1 in a redox-inhibitable fashion as a regulatory mechanism to reversibly sustain the active NADPH oxidase complex [139]. Accordingly, SOD1-deficient cells fail to activate NF-κB in response to IL-1β stimulus [205]. Whether each of these effects depends exclusively on the dismutase activity of SOD or involves other mechanisms such as SOD thiol oxidase activity [206] is unclear. Together, these considerations are consistent with the proposal of SODs functioning not only as scavengers but also as superoxide sensors, with implications for models of superoxide signaling.

Hydrogen peroxide is usually regarded as the prototypical signaling ROS, given its permeability and diffusibility (see Box 1.3), as well as its generally moderate reactivity with regard to biological targets at its usual concentrations. A number of studies have shown the potential for hydrogen peroxide-mediated modulation of thiol redox state and, in particular, the sensitivity of specific proteins with critical low pKa thiols to oxidation by exogenous hydrogen peroxide [26, 75, 106].

## Box 1.3 Membrane Permeability of Signaling Species

Among less reactive species (see Box 1.1), hydrogen peroxide is considered freely diffusible across membranes [207] because it is an uncharged small molecule. Highly reactive species, such as hydroxyl radical, which reacts with biomolecules at near-diffusion rates, does not outlive enough time for crossing lipid bilayers. On the other hand, superoxide anion is not able to easily permeate membranes due to its negative charge, although some anion channels were described to facilitate superoxide crossing in endosomes [205] and endothelial plasma membrane [113]. Another important characteristic of superoxide anion is that it can be protonated at low pH (pKa = 4.8), and thus the uncharged form (hydroperoxyl radical) becomes membrane-permeable. Recently it was reported that hydrogen peroxide permeability is dependent on membrane lipid composition, especially during cellular development [208], and that some aquaporins facilitate hydrogen peroxide diffusion across membranes [209]. This could indicate that even the permeability to hydrogen peroxide can potentially be regulated.

This provides a potential basis for signaling specifity of this species, a concept that is at the basis of most current paradigms of redox signaling [33, 35]. However, several considerations indicate that low cysteine pKa is insufficient to confer specificity for a given target protein, particularly when considering the usual concentrations of hydrogen peroxide assumed within the paradigms of cell signaling [33, 35]. The main concern in this regard is the quite low range of rate constants for the direct reaction with thiol compounds from target proteins or regulatory buffers (Table 1.1). Thus, a postulated oxidation of protein thiols by signaling concentrations of hydrogen peroxide is unlikely to provide significant or efficient signal transduction [33]. Although hydrogen peroxide can give rise to more powerful oxidants, such as hydroxyl radical via the Fenton mechanism, the in vivo occurrence and significance of the Fenton mechanism is not clear [21], particularly in a context of cell signaling. Given these considerations, enhancing mechanisms have been postulated to account for the increased efficiency of hydrogen peroxide signaling. In the next section, we discuss two such mechanisms: compartmentalization (the ability to promote local transient increases in ROS concentrations), and the possible existence of adaptors that couple ROS production to target protein redox modifications. Novel and improved paradigms have been proposed to allow the modeling of these as yet unclear mechanisms, each one emphasizing particular aspects such a comprehensive role of thiols and/or local redox buffers [2, 18, 33, 107].

These considerations, taken together, suggest that issues beyond chemical reactivity alone must be considered in order to allow understanding redox signaling specificity. In the next sections, we discuss further steps at the cell biology level which can help compose a picture of how redox processes affect (patho)physiology and what can be expected from related therapeutic interventions.

## 1.6 Compartmentalization: One of Nature's Solutions for Redox Signaling Specificity and Robustness

Mitochondria are a prototype of a highly efficient compartmentalization of redox processes [108], allowing to a substantial extent the intraorganelle confinement of ROS generated during electron transfer. Emerging evidence indicates that compartmentalization is also an important way to localize ROS signaling, while providing a circumscribed ROS generation that can potentially prevent their overflow into the cytosol or to other collateral extracompartmental targets. The definition of compartment in this context is not yet precise, but *in terms of signal transduction, compartments may be assumed to be any platform for optimized signal communication to a target or a group of targets.* Spatial compartmentalization is well known to occur in other nonredox signaling networks, not only with respect to organelles and their derived structures such as vesicles, endosomes, etc., but also in structures including scaffold proteins; lipid-rich domains such as caveolae and lipid rafts; protein complexes; and nanoclusters [109, 110]. Redox-dependent signal transduction associated with Nox1 has been shown to occur in the absence of overall changes in glutathione or thioredoxin buffers [111]. The association between signaling and compartmental ROS generation and NADPH oxidase is consistent with the highly focal mode of activation of this enzyme complex, while signaling from mitochondrial ROS will tend to display a less focal pattern [35]. The most well studied examples of Nox-associated compartmentalization include endosomes, caveolae and lamellipodia. In response to cytokine stimulation, endosomes are dynamically formed and recruit active Nox2 in macrophage-like cells [112] or Nox1 in vascular smooth muscle cells [113]. Together with Nox2 at the endosomal membrane, scaffold and signaling proteins are assembled at the endosomal surface in a Rac and ROS-dependent fashion to provide output signals culminating in NF-κB activation [114]. Superoxide can exit endosomes via anion channels (see Box 1.3). Nox1 endosomal signaling requires the ClC-3 ion channel [113]. Caveolae are well known to compartmentalize endothelial NOS [115] and likely also Nox1 [116], in line with the role of this Nox isoform in angiotensin-II signaling, given that AT1 receptors are also dynamically recruited to caveolae [117]. Both Nox1 or Nox2 are known to localize at lamellipodia [110] to provide localized ROS bursts that positively regulate cell migration. The subunit p47phox is known to bind moesin and WAVE1, 2 proteins that are enriched at leading edge lamellipodia [118, 119]. Rac1 targetting to these subcompartments is essential for ROS localization. One mechanism of Rac1 targeting is possibly its interaction with the actin-binding scaffold protein IQGAP [120]. Nox1 enrichment at the lamellipodia promotes integrin switch in a Rac-dependent fashion, while promoting disruption of stress fibers and focal adhesions via Rho, thus allowing directional cell motility [121, 122].

Very little is known regarding concentrations of ROS achieved within compartments such as endosomes, due to technical limitations, as well as to the fact that transient instantaneous ROS flow within those compartments is more likely than steady-state ROS accumulation [112]. A recent estimate regarding superoxide in endosomes yielded figures of $\sim$8 μM for steady-state levels (variable according to endosome size and pH) and ~100 $\mu M \cdot s^{-1}$ for flow [112], values about 25 times lower than similar estimates for the phagosome [123], but larger than the nanomolar level estimates previously calculated on a whole cell basis [124]. Thus, further advances in understanding intracompartmental ROS fluxes may bring the somewhat ironic conclusion that some redox reactions judged as unlikely on the basis of artificially high oxidant concentrations may turn out to be physiologically relevant. In fact, while indirect evidences and theoretical constraints argue against proposed direct oxidation of some signaling proteins (e.g., protein tyrosine phosphatase 1B) by low-level hydrogen peroxide, proteomic analysis does reveal that such proteins are indeed oxidized [3].

One characteristic of compartmentalization is that signal transduction is not only dependent on protein targeting to such compartments, but also on their quite dynamical rates of formation, intracellular traffic, and destruction, as shown, e.g., with caveolae and its many associated proteins [125, 126]. Additional examples include the fact that blockade of lipid raft formation precludes Nox1 activation [127] and that NADPH oxidase subunit p47phox associates to cortactin, which regulates the persistence of lamellipodia [128]. Moreover, endosomes are well-known to present dynamic cycles of migration to and from the plasma membrane [113, 129, 130]. This dynamic behavior provides a basis for the interesting and emerging perspective of temporal signal compartmentalization. Some examples in this regard are the well-known cycles of membrane integrin traffic [131], as well as the oscillatory stochastic pattern of NF-κB activation [132] and Ras activation [109] reported in single-cell studies. Thus, at least in part, cell signals may be transduced in digital, rather than analogical models [133]. Whether redox signals also behave in this way is yet unknown, but it is noteworthy that NADPH oxidase has been shown to undergo a localized, phasic, and periodic mode of activation; and ROS release in phagocytes [134] and ROS generation in mitochondria display temporal random bursts [135]. Interestingly, rapid recycling of signaling proteins might potentially substitute for lack of reversibility of some of their redox modifications.

## 1.7 Redox Modularity: A Systems Biology–Based Version of Compartmentalization

The notion of signal compartmentalization is closely related to and merges with the concept of modularity, which permeates the design of most biological systems [136]. A *module may be defined as a set of nodes that have strong interactions and a common function, having defined input and output nodes that control the interaction with the rest of the network, while some internal nodes do not communicate with external elements* [136]. Biological modules are reminiscent of those used in engineering systems, e.g., subroutines in software or exchangeable circuits [136]. In this context, a compartment is a likely part of any signaling module and/or may contain one or more signaling modules. The term module may be viewed in some aspects as a systems biology–based analog of *signalosome*, proposed previously [35].

Although discussing a proposed redox signaling module is still just a theoretical exercise, there is no reason to suspect that redox signaling would preclude such a modular structure, as supported by identifications of redox-active endosomes or *redoxosomes* [137]. The input of the module is a growth factor or cytokine receptor–derived signal, for example, while the integrative element within the module is the reactivity of a free radical or a two-electron oxidant species. A module should contain an enzymatic ROS generator (in this example, usually a highly localized enzyme such as NADPH oxidase), which, as we saw above, is usually associated with a compartment. Thiol redox buffers, such as glutathione, thioredoxin, or cysteine are very likely to act as local modulators and particularly as adaptors controlling the flow of reducing equivalents [30]. This possibility is supported by evidences that these buffers achieve independent regulation of their equilibrium redox potential, thus conferring to each of them the capacity to specifically modulate a redox ambient, as opposed to just an overall plain buffering of excess oxidants [30]. A similar adaptor role may be exerted by thioredoxin family proteins such as protein disulfide isomerase(s), which were described by us to associate with NADPH oxidase subunits and to assist in its redox-mediated signaling in response to angiotensin II in vascular smooth muscle cells [138]. Moreover, several inducible or constitutive antioxidant enzymes are an intrinsic part of the modular arrangement and are expected to lie at the modular periphery (which in Fig. 1.3 is the compartment surface) and in normal conditions act to prevent significant ROS flows outside the area of interest. In fact, SOD1 is recruited to the endosome surface during cytokine signaling [139]. An essential part of a redox module would be a redox sensor capable of interacting with other proteins, as well as of providing feedback information to the modular structure about ROS flows. Although this is still unclear,

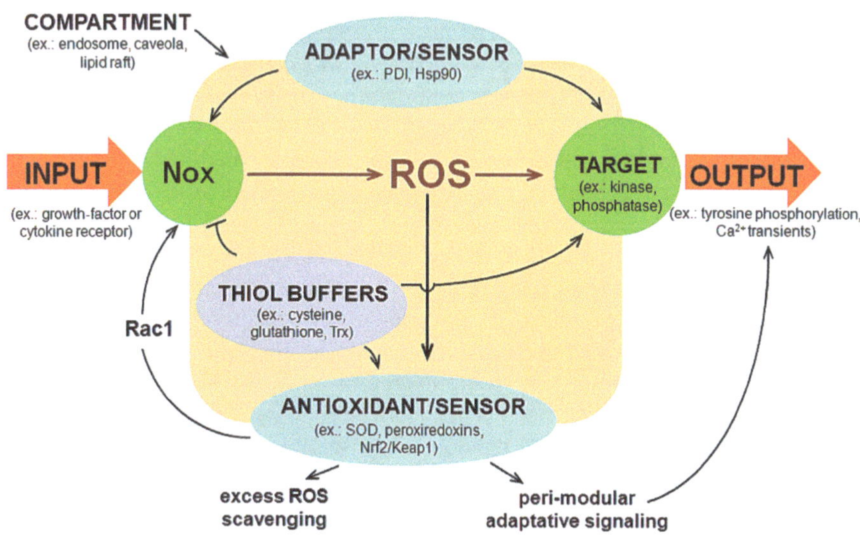


**Fig. 1.3** Proposed redox signaling module

some proposed ROS sensors include thiol proteins or thiol buffers [30, 33], particularly peroxiredoxins [52]. Moreover, SOD1 and SOD2 have been shown to display a behavior consistent with a superoxide sensor, both in yeast models and in neuronal cells (Box 1.2). The protein target to be modulated is also part of the module and is predicted to be recruited to the compartmental structure in a dynamic fashion, as suggested, e.g., by findings such as transient recruitment of TRAF6 to endosomes during cytokine signaling [129]. In the simple design of Fig. 1.3, the target represents the output of the module, leading to downstream signaling events such as changes in tyrosine phosphorylation or calcium fluxes, which modulate the consequences of the input signal, as in Model II from Fig. 1.1. The modular structure modeled here would allow a transient flow of ROS to exert targeted and localized effects under a more robust design.

## 1.8 Oxidative Stress as Collateral Supra-Modular Signaling: A Proposal

The conclusion that Model I from Fig. 1.1 is unlikely is supported by much experimental evidences from the literature [3, 33, 110]. An improved way to model redox signaling is shown as Model II in Fig. 1.1. The concepts of compartmentalization and modularity drive our thoughts towards the opposite direction of ROS as true second messengers such as, e.g., cyclic nucleotides. Rather, the structure that regulates signal transduction is the whole redox module, which acts in a way to enhance or inhibit the stimulus to targets embedded in a specific compartment as a component of the redox module. As we discussed above, the architecture of this redox signaling model is centered in the input and output of the module, which coincide, respectively, with the stimulus and target activation. Thus, the main characteristic of successful redox signaling is to fit within a given purposeful transduction, as predicted from the input signal. That is, *a purposeful transduction is here regarded as one that preserves coherence between module input signals and output responses.* The second characteristic is that all adaptations to ROS flows should occur in an intra- or perimodular way (otherwise the module concept would lose significance). Oxidative stress, in this context, can be *defined to represent a disruption of the modular architecture of signaling, with loss of purposeful transduction and emergence of collateral supra-modular secondary signaling*, which may represent an adaptation or response to excess ROS flows or a convergence with other types of stress (Fig. 1.4). Such supra-modular adaptation may frequently lead in its advanced stages to suppression of cell propagation—apoptosis, senescence, autophagy—or necrosis, likely reflecting at least some degree of oxidative modification ("damage") of biomolecules. Therefore, this paradigm helps to put together in a simplified fashion ideas expressed in many previous proposals [33, 35, 107]. One possible thought derived from compartmental/modularity models is that signaling cascades, especially in the case of redox signaling, behave as flattened two-dimensional oversimplifications of a multidimensional process connecting distinct hierarchical levels of regulation [136].

**Fig. 1.4** Convergence between ER stress and oxidative stress. ER stress in the course of several diseases triggers the signaling cascade known as unfolded protein response (UPR), which comprises three main arms derived from ER sensors PERK, IRE1, and ATF6. ROS production is downstream to the UPR triggers, but also contributes to feed forward the UPR itself. ROS production during the UPR can be due to ER sources (such as the oxidase Ero1), mitochondria, and the NADPH oxidase isoform Nox4

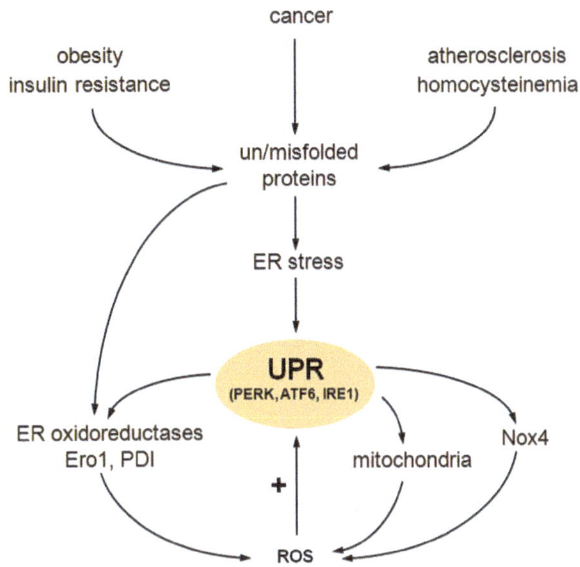


## 1.9 Intermediate States of Redox Signaling vs. Oxidative Stress

Despite many previous attempts to contextualize distinct states of redox signaling and to discriminate them from oxidative stress, including our proposal of oxidative stress as collateral supramodular signaling, some common situations are consistently difficult to be adequately modeled.

One of these situations is signaling associated with mitochondrial ROS production. Clearly, ROS production from mitochondria is much less compartmentalized and more abundant than that, e.g., of NADPH oxidases, which are primarily involved in localized signaling [22, 140]. Thus, it is doubtful whether mitochondria truly exert *strictu sensu* redox signaling [35], since mitochondrial ROS production will result in mass activation of signaling programs rather than discrete targets. Consequently, in several instances mitochondrial ROS production will lead to oxidative stress, with nonspecific secondary signaling, as in Fig. 1.4. On the other hand, some recent evidence suggests that mitochondria may provide a rather fine control of hypoxia signaling [141] and perhaps also of signaling resulting from physiological metabolic routes via the AMPK sensor mechanism [142]. Therefore, at least in such situations, mitochondria may represent a case in which the redox modular arrangement is particularly large and involves several targets, while purposeful signaling is maintained. We have come to term this phenomenon "redox macrosignaling," as opposed to the usual "redox microsignaling"[1] modeled in other instances.

[1] The concepts of "macro" and "micro" signaling were co-developed together with Rafael Radi and Homero Rubbo, from Universidade de la Republica, Montevideo, Uruguay.

Another situation can be termed *latent or adapted oxidative stress*, in which ROS production is noticeably increased; but increased adaptive signaling, e.g., Nrf2/Keap or transcription factors such as ATF4 [143, 144] and XBP1 [145] succeed in increasing several antioxidant enzymes to the point that redox signaling becomes preserved, at least at baseline. This situation may be quite common during many chronic sublethal forms of oxidant challenge. Indeed, failure to effectively upregulate antioxidant defenses (glutathione peroxidases, catalase, SOD2, uncoupling protein-2, and transcriptional pathways DJ-1 and FOXO) has been shown to account for increased oxidative stress in aged cholesterol-fed LDL receptor-deficient mice [146]. This situation suggests that to some extent it is possible to maintain redox modularity at increased rates of ROS generation, in accordance with the expected ideas of compartmentalization/modularity. The notion of latent oxidative stress has some bearing in the concept of *hormesis, in which adaptive processes arising from a given stressor are able to (over)compensate such stress response or to prevent it upon a repeated challenge* [147].

## 1.10 Reduction-Dependent Signaling and Reductive Stress

Despite the widespread use of the term oxidative stress, and sometimes oxidant signaling, cellular redox-dependent (patho)physiological signaling events can be mediated by oxidizing as well reducing reactions. The most illustrative example in this regard is redox-dependent activation of transcription factors such as NF-κB, in which the initial step of IKK-α phosphorylation and degradation is dependent on oxidizing species, but the subsequent nuclear transport and DNA binding require a reductive step mediated by thioredoxin family enzymes [114]. Thus, the effects of antioxidant compounds on NF-κB activation can be quite variable depending on cell type and conditions. In addition, oxidant generation can often trigger antioxidant pathways (e.g., Nfr-2/Keap), which culminate in increased synthesis of glutathione and a reductive shift of cell redox status.

A more extreme reductive challenge promotes a situation known as reductive stress, characterized by accumulation of reduced metabolites such as cysteine, glutathione, and NAD(P)H. Reductive stress has been well characterized in yeast, in which exposure to reductants such as DTT promotes oxidative protein folding stress [148] or markedly affects the viability of thioredoxin mutants bearing increased glutathione levels [149]. Both examples reflect a major mechanism of reductive stress toxicity, i.e., perturbed redox protein folding due to endoplasmic reticulum underoxidation [148]. In addition, disruption of transcription factor signaling, mitochondrial dysfunction, and proteasome dysregulation [150] are additional effects of reductive stress. Moreover, increased NADPH levels may feed reducing equivalents to oxidant-generating NADPH oxidases, thus providing a link between reductive and oxidative stresses. Emerging, though less organized, information is available on the role of reductive stress in upper eukaryotic cells, particularly in disease models. Recent work suggests that in a model of protein misfolding–associated cardiomyopathy, increased activity of G6PD (glucose 6-phosphate dehydrogenase, which accounts for and regulates NADPH production) is responsible for reductive stress–mediated myocardial dysfunction [151, 152]. Reductive stress may also contribute to the pathogenesis of diabetes mellitus [153].

## 1.11 Integration of Oxidative Stress at the Cellular Level: Convergence with Other Types of Stress

The discussions so far in this chapter clearly indicate that cellular platforms and circuits converge with redox signaling in a significant interactive two-directional way. In this context, it is increasingly evident that oxidative stress can occur as a component of cellular response to other types of stress, which is in line with the concepts of compartmentalization and modularity. Redox-related proteins are important upstream and downstream components of the conserved core cellular stress response [143, 154, 155]. Indeed, stresses such as heat shock [158] and osmotic shock are associated with oxidative stress, at least in part from mitochondrial origin. Overexpression of stress protein(s) such as p53 promotes oxidative stress, possibly as its main mechanism of apoptosis [156].

Particularly, a number of studies has provided evidence that ROS generation is an intrinsic part of the unfolded protein response (UPR), a complex signaling cascade that is triggered by endoplasmic reticulum stress, a situation in which there is a mismatch between the ER protein synthesis load and the capacity of this organelle to process them at any level, including folding, post-translational modifications, and traffic to the secretory system [144, 157, 158]. As with any form of stress, the UPR encompasses both proadaptive and prosurvival pathways. The main arms of the UPR are dependent on ER transmembrane kinases/transcription factors that trigger nuclear transcription of genes coding for chaperones, metabolic changes, and, in the later phases, apoptosis. Particularly, antioxidant responses are also activated during the UPR as an adaptive change, including the PERK/Nrf2/Keap pathway [159] and genes coded by transcription factors ATF4 [160] and XBP1 [145]. ROS generation causes not only downstream UPR triggering; but ROS also provide a feed-forward mechanism sustaining both proadaptive and proapoptotic UPR responses [144, 161] (Fig. 1.4). In fact, many oxidants can trigger the UPR, although not uniformly [144]. ROS generation during the UPR unravels a quantitatively neglected, but potentially important source. This pathway is related to the ER oxidoreductase Ero1, which transfers oxidizing equivalents to protein disulfide isomerase, enabling it for redox protein folding characterized by introduction of disulfides. At the same time, reduced Ero1 transfers electrons via FAD to molecular oxygen, generating hydrogen peroxide [162]. Overactivation or malfunction of this otherwise normal physiological mechanism may account for substantial rates of ROS output, particularly in secretory cells, with estimates at the level of 25% of cellular ROS [162]. Mitochondria and NADPH oxidase isoform Nox4 also account for ROS generation during the UPR, the latter a particularly important source in vascular smooth muscle cells [144].

One essential aspect of stress integration is the adaptation of the protein interaction network driven by stressful stimuli, including oxidative stress itself. Segregated physical compartments or the components of a signaling module can be ultimately decomposed into highly elaborated protein-protein interaction networks (*interactome*), the profile of which displays, under normal conditions, an architecture that tends to simplify signal communication by using a few highly populated hubs. During stress, the network is rearranged, with the emergence of several hubs that are poorly populated and only a few central hubs, making the system as a whole less agile but more resistant [163]. Molecular chaperones are important both to provide hub connections and for interactome rearrangement [164]. Some hubs that remain prominent during stress include proteins related to proteasome, nuclear transport and actin regulation [164]. In terms of redox signaling, this aspect, schematized in Fig. 1.1, is likely a network representation of modular signaling disruption. An additional characteristic of stress signaling in general is the increased level of stochasticity in gene expression. This is known to occur in aged tissues [165], during mitochondrial dysfunction [166], as well as under an oxidative challenge [165], and seems to represent a strategy for cell survival [167, 168].

## 1.12 Assessment of Disrupted Signaling Due to Oxidative Stress: Problems and Perspectives

One of the main challenges to understand redox signaling models in vivo is the accurate assessment of what one would define as *disrupted signaling*. So far, detection of oxidative stress is based on assessment of ROS production rates and footprints of redox-induced modifications in a number of targets or redox buffers. More recently, proteomic techniques have brought about the possibility of understanding at a high-throughput level the organization of normal and redox-modified sets of proteins [3, 169]. Here, we will provide only a brief summary of strategies for redox status assessment and how methodological issues possibly influence conceptual assumptions in the field.

The crucial difference between normal and diseased states is the intracellular steady-state concentrations of ROS, which are in micromolar levels, e.g., in phagosome during a neutrophil oxidative burst, but only at low nanomolar levels for signal transduction in the majority of cells. This statement neglects possibly high compartmental ROS concentrations, in the same line as that of most detection methods. Therefore, measurement of these species is technically difficult because of their relatively short half-lives added to their scavenging by small antioxidant molecules (e.g., ascorbate) and antioxidant proteins (e.g., catalase and peroxiredoxins).

Basically, there are three strategies for oxidant species measurement: (i) oxidant trapping and quantification of its levels, (ii) identification of cellular damage done by the oxidant, and (iii) measurement of redox state, such as glutathione ratio (GSH/GSSG) in tissue extracts and total antioxidant capacity in body fluids.

Regardless of the approach, the chosen method should be sensitive, specific, and reliable for detected changes in reactive species. In the case of detection probes, desirable characteristics are: (i) adequate intracellular access to truly reflect intracellular conditions, (ii) lack of overlap with reactive species reactivity, leading to unequivocal identification and quantification of the specific intermediate, and (iii) sensitivity to effectively outcompete intra- or extracellular antioxidants/scavengers [170].

### *1.12.1 Approaches for Reactive Species Detection and Oxidative Stress Measurement*

The unequivocal identification of any free radical is only obtained with the electron paramagnetic resonance (EPR) technique, because it detects the presence of unpaired electrons. However, direct EPR detects only "unreactive" radicals, such as the ascorbate radical in plasma [171] or nitrosylhemoglobin in blood [172], since reactive ones do not accumulate enough to be directly measured. This is overcome with spin trap molecules, which react with radicals and form a stable radical that accumulates and can then be detected by EPR, even in in vivo experiments [173]. One interesting derivation from spin traps was the recent development of polyclonal antibodies that bind to protein adducts of the nitrone spin trap 5,5-dimethyl-1-pyrroline N-oxide (DMPO), making possible the analysis of free radical production by immunoassays such as the western blot or ELISA [174].

With the exception of EPR, other methodologies for reactive species detection are based on indirect mechanisms. Table 1.1 shows some methods for identification/quantification of reactive species, with comments about advantages and disadvantages for each one. The most widely used probe for oxidative stress in cells and tissues is 2′,7′-dichlorodihydrofluorescein diacetate (DCFH-DA), which is wrongly considered specific for $H_2O_2$ measurement (Table 1.2). The DCF assay is probably a useful indicator of a cell shifting to a more oxidizing state, without specific identification of cellular oxidants or mechanisms involved in their generation [3]. Dihydrorhodamine (DHR) also can be oxidized by several oxidants (see Table 1.2), although some consider DHR a qualitative probe for peroxynitrite, provided exhaustive controls are performed [175]. The third probe in Table 1.2 is dihydroethidium (DHE), which was for a long time employed as a superoxide marker, especially by tissue and cellular confocal fluorescence microscopy. In recent years, however, Zhao and co-workers [176] showed that this probe is oxidized to several products, of which two are more easily identifiable, have very similar fluorescent emission spectra, but are formed by different oxidants: 2-hydroxyethidium, formed mainly by superoxide, but also ONOO– in the presence of $CO_2$; and ethidium, formed in the presence of hemeproteins plus $H_2O_2$. Accordingly, while analysis of total fluorescence derived from DHE oxidation is an oxidative stress marker (Table 1.2), the analysis of fluorescent DHE-oxidation products by HPLC can be considered a semiquantitative method for superoxide production in vivo (Table 1.3).

**Table 1.2** Common methods to measure reactive species in biological samples

| Method | How it works | Pros | Cons |
|---|---|---|---|
| *Hydrogen peroxide* | | | |
| Horseradish peroxidase plus substrates | $H_2O_2$ is oxidized by HRP, which in turn forms compound I that oxidizes substrates as Amplex Red (N-acetyl-3,7-dihydroxyphenoxazine) and scopoletin to fluorescent products | Simple to perform | Ascorbate and thiols can cause artefactual inhibition, as they are substrates for HRP<br>Superoxide decreases HRP activity |
| *Peroxides* | | | |
| FOX (ferrous oxidation xylenol orange) assay | $H_2O_2$ or other peroxides oxidize $Fe^{2+}$ to $Fe^{3+}$, detected by xylenol orange (colorimetric assay) | (µM range) simple to perform<br>Distinguishes protein and lipid peroxides by perchloric acid addition | Amplification of signal if peroxide is oxidized to peroxyl radical by $Fe_2^+$ |
| Gas chromatography or HPLC/mass spectrometry | Peroxides are extracted from samples, usually reduced to alcohols, separated by gas chromatography or HPLC, and identified by mass spectrometry | Low µM range<br>All peroxides can be identified, and also isoprostanes, aldehydes, cholesterol, etc. | Lipid hydroxides can be absorbed from diet |
| *Superoxide radical* | | | |
| Citochrome $c^{3+}$ reduction | $O_2^{\bullet-}$ reduces citochrome $c^{3+}$ to citochrome $c^{2+}$, measured colorimetrically<br>Nitroblue tetrazolium (NBT) | µM range | Others substances can reduce $cit^{3+}$, such as ascorbate and thiols |
| Aconitase assay | $O_2^{\bullet-}$ displaces iron from $[4Fe\text{-}4S]^{2+}$ cluster of aconitase, causing loss of enzyme activity | Very sensitive (pM) | Other reactive species can affect aconitase activity (e.g., $ONOO^-$, $N_2O_3$) |
| Dihydroethidine (DHE) oxidation coupled to HPLC | $O_2^{\bullet-}$ oxidizes DHE specifically to 2-hydroethidium (2-EOH), a compound that fluoresces at the same region of other DHE-derived oxidant products, such as ethidium. By HPLC, 2-EOH can be separated from other products and quantified (fluorescent or electrochemical detection) | Very sensitive (pM) | 2-EOH can be formed by $ONOO^-$ in the presence of bicarbonate |

**Table 1.2** (continued)

| Method | How it works | Pros | Cons |
|---|---|---|---|
| *Nitric oxide* | | | |
| Light emission | •NO reacts with ozone and produces light, via excited-state nitrogen dioxide | Very sensitive (nM range) | nitrosocompounds interference (such as NO synthase inhibitor L-NAME) |
| Hemoglobin trapping | •NO reacts with oxyhemoglobin, eventually converting it to methaemoglobin, measured by absorbance | Simple to perform (μM range) | hemoglobin can be oxidized also by $ONOO^-$ or $NO_2^-$ (more slowly) |
| 4,5-diaminofluorescein diacetate (DAF) | •NO oxidation products ($N_2O_3$ or $NO^+$) react with hydrolyzed DAF, in which acetyl group was removed by intracellular esterases | Easily accumulates to mM intracellular levels; gives insights about •NO production/compartimentalization in cells and tissue slices (nM range) | Also reacts with ONOO– and peroxidase/$H_2O_2$ systems. Dependent on esterases activity |
| Griess reaction | $NO_2^-$ formed from •NO oxidation reacts with sulfanilamide in acidic solution of N-(1-naphthyl)ethylenediamine to give a purple azo compound (measured colorimetrically). In body fluids, $NO_2^-$ is rapidly oxidized to $NO_3^-$, which can be reduced to $NO_2^-$ by nitrate reductase | Simple to perform (μM range) | $NO_2^-$ can come from diet |

**Table 1.3** Some probes widely used to assess oxidative stress in biological samples

| Probe | How it works | Comments |
|---|---|---|
| 2′,7′-dichlorofluorescein diacetate (DCFH-DA) | Reactive species react with hydrolyzed DCFH, in which acetyl group was removed by intracellular esterases. The product DCF is fluorescent | It is not a specific probe for $H_2O_2$; peroxidases can oxidize DCFH (in the presence or absence of $H_2O_2$), as well as transition metal complexes, $ONOO^-$, and thiyl radicals [180, 181]. Is oxidized by cytochrome *c* during apoptosis [182] |
| Dihydrorhodamine 123 (DHR) | DHR is oxidized to fluorescent rhodamine by $ONOO^-$, HOCl, heme-peroxidases in the presence of $H_2O_2$ | Selective application of SOD, catalase, various NOS inhibitors, and $ONOO^-$ scavengers are required to provide more precise identification of the substances responsible for DHR oxidation |
| Dihydroethidium (Hydroethidine) (DHE) | DHE is oxidized to fluorescent products (ethidium and 2-hydroxyethidium) and nonfluorescent products (such as dimmers [183]) | It is not a specific probe for superoxide detection; $O_2^{\bullet -}$, hemeproteins/$H_2O_2$, $ONOO^-$ in the presence of $CO_2$, cytochrome $c^{3+}$ are able to oxidize DHE to fluorescent compounds ethidium and 2-hydroxyethidium |

Another approach is the identification of specific biomolecular damage, which permits an inference about which reactive species has been formed. Such *biomarkers* can be used to investigate the effects of antioxidants or others agents in oxidative damage; but currently no available biomarker meets either the key criterion of *predicting the later development of disease*, or all necessary technical criteria (e.g., to show low variation between assays/subjects, not to be confounded by diet, to be stable during storage, and to be easily measured in samples like urine, saliva, and blood), criteria suggested by Halliwell and Gutteridge [1]. For example, 3-nitrotyrosine identification in tissues is considered a nitrosative stress marker, associated mainly with high levels of $^{\bullet}NO_2$, $NO_2^-$ plus peroxidases or $ONOO^-$, but not a biomarker of any of these specific species. There are several examples of protein damage that can be found in a number of pathologies, besides other components of biological systems, such as DNA, lipids, and carbohydrates [1]. It is important to note that is currently unclear which biomolecule damage (and its intensity) is critical to a particular pathologic event, as opposed to being a modification without cellular consequences [177].

## *1.12.2 Approaches for Redox State Measurement*

Changes in the intracellular thiol-disulfide (GSH/GSSG) balance within the cell can be used as an indicator of the redox status of the cell or body fluids, such as plasma. The advantage is that GSH/GSSG ratio and the reduction potential (mV) are two ways to indicate oxidative stress, because they reflect the availability of GSH to protect against oxidative reactions and the generation of GSSG from oxidative reactions [2]. For example, while more reduced states of intracellular cysteine (reduced Cys/oxidized Cys) were measured in proliferating cells [178], more oxidized states were associated with increased monocyte adhesion in endothelial cells [179] and increased sensitivity to oxidant-induced apoptosis [180]. In addition, progressive declines in GSH/GSSG ratios were documented in individuals aged above 45 years [178]. Another common measurement in body fluids and erythrocytes is total antioxidant capacity (TAC), whose data should be interpreted with care, since they reflect contributions from urate, ascorbate, and sometimes thiol groups from albumin, depending on the method used, and can be influenced by diet [181].

### *1.12.3 How to Choose a Particular Method for Detection of Reactive Species or Oxidative Stress*

Because of methodological limitations (for example, Table 1.2), it is essential to understand thoroughly how each method works and to adapt it for each specific experimental situation, almost always with exhaustive controls. Moreover, it is important to ask *what really is being measured with that technique*, so as to adequately interpret results. It is also recommended to perform at least two different methods for measuring the same reactive species to get more reliable results. Erroneous interpretations can also be minimized by taking into account eventual interferences regarding sample type; for example, fluorescence derived from dihydroethidium oxidation can be used for measuring NADPH-triggered oxidase activity in isolated cellular membrane fraction, while the same probe measures only total oxidant production when samples are cellular or tissue homogenates [182]. Furthermore, the use of antioxidants to probe the role of specific ROS should be considered with care. While SOD and to some extent catalase can be assumed to be reasonably specific probes for superoxide and hydrogen peroxide, respectively (in a context of adequate controls), compounds such as ascorbic acid and thiol-based antioxidants such as N-acetylcysteine or dithiocarbamates are too nonspecific to serve as probes for ROS effects. The only thing they can show is that a particular process is redox- or thiol-dependent, but they allow no conclusion regarding direct ROS effects.

Overall, these considerations should not be taken to conclude only that "such particular probes cannot be used for ROS measurements." Rather, better knowledge of their effects and limitations is important to prevent such methodological imperfections from shaping paradigms that carry inadequate assumptions, even though such imperfect methods may provide conclusions that are operational and even have physiological correlations.

## 1.13 Redefining Antioxidants and Antioxidant Therapy in a Redox Signaling Scenario

The scenario of redox signaling, compartmentalization, and modularity poses the need for expanding and reformulating the concept of antioxidants and antioxidant interventions. Under such novel paradigms, antioxidants may be viewed as any compound or enzymatic pathway that contributes to maintain redox signaling modularity and/or prevent or attenuate secondary supramodular signaling. In this context, therapeutic antioxidant interventions should have aims that are much broader than just providing a general balance in favor of a less prooxidant tendency of the cell. Such aims include: (i) preserving modularity in redox signaling; (ii) restoring or maintaining coherence between input and output module signals; (iii) scavenging, metabolizing, or redirecting reactive species—particularly 2-electron oxidants—that are formed in excess within signaling compartments and/or that escape from compartmental restriction; and (iv) correcting or compensating for secondary supramodular signaling. Clearly, the current portfolio of antioxidant therapy falls significantly short of these goals and the current models of redox pathophysiology are still insufficient to provide advances in these directions. Rather, current models of antioxidants are still heavily based on their properties of scavenging 1-electron free radical oxidants [20] and essentially neglect antioxidant compartmentalization [30, 35]. Possibly, paradigms of antioxidant therapy may have to involve a host of converging interventions, some of them even of primary nonredox nature. Emerging advances in this direction are being provided by, among others, interventions such as caloric restriction mimetics [187] and natural compounds able to trigger hormetic responses [188]. Importantly, the use of vitamins such as alpha-tocopherol and ascorbate can abrogate potentially beneficial effects of exercise, possibly via the inhibition of mito-hormetic mechanisms [147].

## 1.14 Concluding Remarks

The concept of oxidative stress has evolved over recent years to account for a disruption of redox signaling and equilibrium, rather than a plain imbalance between prooxidants and antioxidants causing molecular damage. Redox signaling has emerged as an extremely important and potentially powerful mode of regulation of several physiological events, with its dysregulation accounting for disease pathophysiology. However, the considerations put forth in the present chapter indicate that the redox signaling concept itself is also an evolving entity. The model of ROS-mediated differential regulation of thiol targets solely on the basis of distinct chemical reactivities of thiol groups has not been able to fully account for the variety and sophistication of redox-dependent responses. Thus, current models of redox signaling have to take into account additional hierarchical levels of regulation at the cell biology level. The notion of compartmentalization is an important example in this direction and here we have tied it to the idea of modularity. Thus, oxidative stress may be viewed as a disruption of such redox modular architecture and the consequent emergence of supramodular secondary signaling. These considerations indicate that, while having lost some of its metaphorical strength with respect to mechanistical insights, the dynamically reformulated concept of oxidative stress remains powerful as an operational tool to communicate and contextualize science in the field.

## References

1. Halliwell B, Gutteridge JMC (2007) Free Radicals in Biology and Medicine. Biosciences, Oxford
2. Jones DP (2006) Redefining oxidative stress. Antioxid Redox Signal 8(9–10):1865–1879
3. Winterbourn CC (2008) Reconciling the chemistry and biology of reactive oxygen species. Nat Chem Biol 4(5):278–286
4. McCord JM, Fridovich I (1969) Superoxide dismutase. An enzymic function for erythrocuprein (hemocuprein). J Biol Chem 244(22):6049–6055
5. Sies H (1985) Oxidative Stress. London, Academic Press
6. Behe P, Segal AW (2007) The function of the NADPH oxidase of phagocytes, and its relationship to other NOXs. Biochem Soc Trans 35(Pt 5):1100–1103
7. Beckman JS, Beckman TW, Chen J, Marshall PA, Freeman BA (1990) Apparent hydroxyl radical production by peroxynitrite: implications for endothelial injury from nitric oxide and superoxide. Proc Natl Acad Sci USA 87(4):1620–1624
8. Radi R, Beckman JS, Bush KM, Freeman BA (1991) Peroxynitrite oxidation of sulfhydryls. The cytotoxic potential of superoxide and nitric oxide. J Biol Chem 266(7):4244–4250
9. Augusto O, Bonini MG, Amanso AM, Linares E, Santos CC, De Menezes SL (2002) Nitrogen dioxide and carbonate radical anion: two emerging radicals in biology. Free Radic Biol Med 32(9):841–859
10. Furchgott RF, Zawadzki JV (1980) The obligatory role of endothelial cells in the relaxation of arterial smooth muscle by acetylcholine. Nature 288(5789):373–376
11. Palmer RM, Ferrige AG, Moncada S (1987) Nitric oxide release accounts for the biological activity of endothelium-derived relaxing factor. Nature 327(6122):524–526
12. Denu JM, Tanner KG (1998) Specific and reversible inactivation of protein tyrosine phosphatases by hydrogen peroxide: evidence for a sulfenic acid intermediate and implications for redox regulation. Biochemistry 37:5633–5642
13. Reth M (2002) Hydrogen peroxide as second messenger in lymphocyte activation. Nat Immunol. 3(12):1129–34
14. Jones DP, Go YM, Anderson CL, Ziegler TR, Kinkade JM Jr, Kirlin WG (2004) Cysteine/cystine couple is a newly recognized node in the circuitry for biologic redox signaling and control. FASEB J 18(11):1246–1248
15. Azzi A, Davies KJ, Kelly F (2004) Free radical biology – terminology and critical thinking. FEBS Lett 558(1–3):3–6
16. Berry CE, Hare JM (2004) Xanthine oxidoreductase and cardiovascular disease: molecular mechanisms and pathophysiological implications. J Physiol 555(Pt 3):589–606
17. Buettner GR (1993) The pecking order of free radicals and antioxidants: lipid peroxidation, alpha-tocopherol, and ascorbate. Arch Biochem Biophys 300(2):535–543
18. Forman HJ, Fukuto JM, Torres M (2004) Redox signaling: thiol chemistry defines which reactive oxygen and nitrogen species can act as second messengers. Am J Physiol Cell Physiol 287(2):C246–C256
19. Chiu DT, Monteiro HP, Stern A (1996) The intracellular reducing environment modulates cytoregulation and cytotoxicity by reactive oxygen species. Biochem Soc Trans 24(3): 884–887
20. Jones DP (2008) Radical-free biology of oxidative stress. Am J Physiol Cell Physiol 295(4):C849–C868

21. Winterbourn CC (1995) Toxicity of iron and hydrogen peroxide: the Fenton reaction. Toxicol Lett 82–83:969–974
22. Clempus RE, Griendling KK (2006) Reactive oxygen species signaling in vascular smooth muscle cells. Cardiovasc Res 71(2):216–225
23. Finkel T (2003) Oxidant signals and oxidative stress. Curr Opin Cell Biol 15(2):247–254
24. Sundaresan M, Yu ZX, Ferrans VJ, Irani K, Finkel T (1995) Requirement for generation of $H_2O_2$ for platelet-derived growth factor signal transduction. Science 270(5234):296–299
25. Meier B, Radeke HH, Selle S, Younes M, Sies H, Resch K, Habermehl GG (1989) Human fibroblasts release reactive oxygen species in response to interleukin-1 or tumour necrosis factor-alpha. Biochem J 263(2):539–545
26. Bae YS, Kang SW, Seo MS, Baines IC, Tekle E, Chock PB, Rhee SG (1997) Epidermal growth factor (EGF)-induced generation of hydrogen peroxide. Role in EGF receptor-mediated tyrosine phosphorylation. J Biol Chem 272(1):217–221
27. Zafari AM, Ushio-Fukai M, Akers M, Yin Q, Shah A, Harrison DG, Taylor WR, Griendling KK (1998) Role of NADH/NADPH oxidase-derived $H_2O_2$ in angiotensin II-induced vascular hypertrophy. Hypertension 32(3):488–495
28. Robinson KA, Stewart CA, Pye QN, Nguyen X, Kenney L, Salzman S, Floyd RA, Hensley K (1999) Redox-sensitive protein phosphatase activity regulates the phosphorylation state of p38 protein kinase in primary astrocyte culture. J Neurosci Res 55(6):724–732
29. Mahadev K, Wu X, Zilbering A, Zhu L, Lawrence JT, Goldstein BJ (2001) Hydrogen peroxide generated during cellular insulin stimulation is integral to activation of the distal insulin signaling cascade in 3T3-L1 adipocytes. J Biol Chem 276(52):48662–48669
30. Kemp M, Go YM, Jones DP (2008) Nonequilibrium thermodynamics of thiol/disulfide redox systems: a perspective on redox systems biology. Free Radic Biol Med 44(6):921–937
31. Jones EP, Sies H (2007) Encyclopedia of Stress, Hardbound. Academic Press, New York
32. Bindoli A, Fukuto JM, Forman HJ (2008) Thiol chemistry in peroxidase catalysis and redox signaling. Antioxid Redox Signal 10(9):1549–1564
33. Winterbourn CC, Hampton MB (2008) Thiol chemistry and specificity in redox signaling. Free Radic Biol Med 45(5):549–561
34. Colston JT, de la Rosa SD, Strader JR, Anderson MA, Freeman GL (2005) $H_2O_2$ activates Nox4 through PLA2-dependent arachidonic acid production in adult cardiac fibroblasts. FEBS Lett 579(11):2533–2540
35. Terada LS (2006) Specificity in reactive oxidant signaling: think globally, act locally. J Cell Biol 174(5):615–623
36. Cowan JA (1993) Inorganic Biochemistry – An Introduction. VCH Publishers, Berlin
37. Krezel A, Hao Q, Maret W (2007) The zinc/thiolate redox biochemistry of metallothionein and the control of zinc ion fluctuations in cell signaling. Arch Biochem Biophys 463(2): 188–200
38. Martínez-Ruiz A, Lamas S (2007) Signalling by NO-induced protein S-nitrosylation and S-glutathionylation: convergences and divergences. Cardiovasc Res 75(2):220–228
39. Winterbourn CC, Metodiewa D (1999) Reactivity of biologically relevant thiol compounds with superoxide and hydrogen peroxide. Free Radic Biol Med 27:322–328
40. Goldman R, Stoyanovsky DA, Day BW, Kagan VE (1995) Reduction of phenoxyl radicals by thioredoxin results in selective oxidation of its SH-groups to disulfides. An antioxidant fuction of thioredoxin. Biochemistry 34:4765–4772
41. Peskin AV, Low FM, Paton LN, Maghzal GJ, Hampton MB, Winterbourn CC (2007) The high reactivity of peroxiredoxin 2 with $H_2O_2$ is not reflected in its reaction with other oxidants and thiol reagents. J Biol Chem 282:11885–11892
42. Trujillo M, Clippe A, Manta B, Ferrer-Sueta G, Smeets A, Declercq JP, Knoops B, Radi R (2007) Pre-steady state kinetic characterization of human peroxiredoxin 5: taking advantage of Trp84 fluorescence increase upon oxidation. Arch Biochem Biophys 467:95–106
43. Holmgren A (1995) Thioredoxin structure and mechanism: conformational changes on oxidation of the active-site sulfhydryls to a disulfide. Structure 3:239–243

44. Aslund F, Zheng M, Beckwith J, Storz G (1999) Regulation of the OxyR transcription factor by hydrogen peroxide and the cellular thiol-disulfide status. Proc Natl Acad Sci USA 96:6161–6165
45. Klatt P, Molina EP, García De Lacoba M, Padilla CA, Martínez-Galesteo E, Bárcena JA, Lamas S (1999) Redox regulation of c-Jun DNA binding by reversible S-glutathiolation. FASEB J 13:1481–1490
46. Nishi T, Shimizu N, Hiramoto M, Sato I, Yamaguchi Y, Hasegawa M, Aizawa S, Tanaka H, Kataoka K, Watanabe H, Handa H (2002) Spatial redox regulation of a critical cysteine residue of NF-κβ in vivo. J Biol Chem 277:44548–44556
47. Brady KD, Giegel DA, Grinnell C, Lunney E, Talanian RV, Wong W, Walker N (1999) A catalytic mechanism for caspase-1 and for bimodal inhibition of caspase-1 by activated aspartic ketones. Bioorg Med Chem 7:621–631
48. Claiborne A, Yeh JI, Mallett TC, Luba J, Crane EJ 3rd, Charrier V, Parsonage D (1999) Protein-sulfenic acids: diverse roles for an unlikely player in enzyme catalysis and redox regulation. Biochemistry 38(47):15407–15416
49. Ellis HR, Poole LB (1997) Roles for the two cysteine residues of AhpC in catalysis of peroxide reduction by alkyl hydroperoxide reductase from *Salmonella typhimurium*. Biochemistry 36:13349–13356
50. Poole LB, Karplus PA, Claiborne A (2004) Protein sulfenic acids in redox signaling. Annu Rev Pharmacol Toxicol 44:325–347
51. Rhee SG, Kang SW, Chang TS, Jeong W, Kim K (2001) Peroxiredoxin, a novel family of peroxidases. IUBMB Life 52:35–41
52. Wood ZA, Poole LB, Karplus PA (2003) Peroxiredoxin evolution and the regulation of hydrogen peroxide signaling. Science 300(5619):650–653
53. Mieyal JJ, Starke DW, Gravina SA, Dothey C, Chung JS (1991) Thioltransferase in human red blood cells: purification and properties. Biochemistry 30:6088–6097
54. Lind C, Gerdes R, Schuppe-Koistinen I, Cotgreave IA (1998) Studies on the mechanism of oxidative modification of human glyceraldehyde-3-phosphate dehydrogenase by glutathione: catalysis by glutaredoxin. Biochem Biophys Res Commun 247:481–486
55. Barrett WC, DeGnore JP, Konig S, Fales HM, Keng YF, Zhang ZY, Yim MB, Chock PB (1999) Regulation of PTP1B via glutathionylation of the active site cysteine 215. Biochemistry 38:6699–6705
56. Ward NE, Stewart JR, Ioannides CG, O'Brian CA (2000) Oxidant-induced S-glutathiolation inactivates protein kinase C-alpha (PKC-alpha): a potential mechanism of PKC isozyme regulation. Biochemistry 39:10319–10329
57. Pineda-Molina E, Klatt P, Vazquez J, Marina A, García De Lacoba M, Perez-Sala D, Lamas S (2001) Glutathionylation of the p50 subunit of NF-kappaB: a mechanism for redox-induced inhibition of DNA binding. Biochemistry 40:14134–14142
58. Okamoto T, Akaike T, Sawa T, Miyamoto Y, van der Vliet A, Maeda H (2001) Activation of matrix metalloproteinases by peroxynitrite-induced protein S-glutathiolation via disulfide S-oxide formation. J Biol Chem 276:29596–29602
59. Adachi T, Pimentel DR, Heibeck T, Hou X, Lee YJ, Jiang B, Ido Y, Cohen RA (2004) S-glutathiolation of Ras mediates redox-sensitive signaling by angiotensin II in vascular smooth muscle cells. J Biol Chem 279:29857–29862
60. Adachi T, Weisbrod RM, Pimentel DR, Ying J, Sharov VS, Schoneich C, Cohen RA (2004) S-Glutathiolation by peroxynitrite activates SERCA during arterial relaxation by nitric oxide. Nat Med 10:1200–1207
61. Chen YR, Chen CL, Pfeiffer DR, Zweier JL (2007) Mitochondrial complex II in the post-ischemic heart: oxidative injury and the role of protein S-glutathionylation. J Biol Chem 282:32640–32654
62. Mieyal JJ, Gallogly MM, Qanungo S, Sabens EA, Shelton MD (2008) Molecular mechanisms and clinical implications of reversible protein S-glutathionylation. Antioxid Redox Signal 10(11):1941–1988

63. D'Aquino M, Bullion C, Chopra M, Devi D, Devi S, Dunster C, James G, Komuro E, Kundu S, Niki E, Raza F, Robertson F, Sharma J, Willson R (1994) Sulfhydryl free radical formation enzymatically, by sonolysis, by radiolysis, and thermally: vitamin A, curcumin, muconic acid, and related conjugated olefins as references. Methods Enzymol 233:34–46
64. Ross D, Norbeck K, Moldéus P (1985) The generation and subsequent fate of glutathionyl radicals in biological systems. J Biol Chem 260(28):15028–15032
65. Wardman P, von Sonntag C (1995) Kinetic factors that control the fate of thiyl radicals in cells. Methods Enzymol 251:31–45
66. Wardman P (1998) Evaluation of the "radical sink" hypothesis from a chemical-kinetic viewpoint. J Radioanal Nucl Chem 232:23–27
67. Folkes LK, Wardman P (2004) Kinetics of the reaction between nitric oxide and glutathione: implications for thiol depletion in cells. Free Radic Biol Med 37(4):549–556
68. Madej E, Folkes LK, Wardman P, Czapski G, Goldstein S (2008) Thiyl radicals react with nitric oxide to form S-nitrosothiols with rate constants near the diffusion-controlled limit. Free Radic Biol Med 44(12):2013–2018
69. Liu X, Miller MJ, Joshi MS, Thomas DD, Lancaster JR Jr (1998) Accelerated reaction of nitric oxide with $O_2$ within the hydrophobic interior of biological membranes. Proc Natl Acad Sci USA 95:2175–2179
70. Stubbe J, van der Donk WA (1998) Protein radicals in enzyme catalysis. Chem Rev 98: 705–762
71. Berndt C, Lillig CH, Holmgren A (2007) Thiol-based mechanisms of the thioredoxin and glutaredoxin systems: implications for diseases in the cardiovascular system. Am J Physiol Heart Circ Physiol 292(3):H1227–H1236
72. Gallogly MM, Mieyal JJ (2007) Mechanisms of reversible protein glutathionylation in redox signaling and oxidative stress. Curr Opin Pharmacol 7(4):381–391
73. Budanov AV, Sablina AA, Feinstein E, Koonin EV, Chumakov PM (2004) Regeneration of peroxiredoxins by p53-regulated sestrins, homologs of bacterial AhpD. Science 304(5670):596–600
74. Woo HA, Chae HZ, Hwang SC, Yang KS, Kang SW, Kim K, Rhee SG (2003) Reversing the inactivation of peroxiredoxins caused by cysteine sulfinic acid formation. Science 300(5619):653–656
75. Lee SR, Kwon KS, Kim SR, Rhee SG (1998) Reversible inactivation of protein-tyrosine phosphatase 1B in A431 cells stimulated with epidermal growth factor. J Biol Chem 273:15366–15372
76. Kim JR, Yoon HW, Kwon KS, Lee SR, Rhee SG (2000) Identification of proteins containing cysteine residues that are sensitive to oxidation by hydrogen peroxide at neutral pH. Anal Biochem 283:214–221
77. Gopalakrishna R, Jaken S (2000) Protein kinase C signaling and oxidative stress. Free Radic Biol Med 28:1349–1361
78. Zheng M, Aslund F, Storz G (1998) Activation of the OxyR transcription factor by reversible disulfide bond formation. Science 279:1718–1721
79. Nadanaka S, Yoshida H, Mori K (2006) Reduction of disulfide bridges in the lumenal domain of ATF6 in response to glucose starvation. Cell Struct Funct 31(2):127–134
80. Burgoyne JR, Madhani M, Cuello F, Charles RL, Brennan JP, Schroder E, Browning DD, Eaton P (2007) Cysteine redox sensor in PKGIa enables oxidant induced activation. Science 317:1393–1397
81. Zhang DD, Hannink M (2003) Distinct cysteine residues in Keap1 are required for Keap1-dependent ubiquitination of Nrf2 and for stabilization of Nrf2 by chemopreventive agents and oxidative stress. Mol Cell Biol 23:8137–8151
82. Kim YC, Yamaguchi Y, Kondo N, Masutani H, Yodoi J (2003) Thioredoxin-dependent redox regulation of the antioxidant responsive element (ARE) in electrophile response. Oncogene 22:1860–1865
83. Wakabayashi N, Dinkova-Kostova AT, Holtzclaw WD, Kang MI, Kobayashi A, Yamamoto M, Kensler TW, Talaly P (2004) Protection against electrophiles and oxidant stress by

induction of phase 2 response: fate of cysteines of the Keap1 sensor modified by inducers. Proc Natl Acad Sci USA 101:2040–2045
84. Kabe Y, Ando K, Hirao S, Yoshida M, Handa H (2005) Redox regulation of NF-kB activation: distinct redox regulation between the cytoplasm and the nucleus. Antioxid Redox Signal 7:395–403
85. Ueno M, Masutani H, Arai RJ, Yamauchi A, Hirota K, Sakai T, Inamoto T, Yamaoka Y, Yodoj J, Nikaido T (1999) Thioredoxin-dependent redox regulation of p53-mediated p21 activation. J Biol Chem 274:35809–35815
86. Welsh SJ, Bellamy WT, Briehl MM, Powis G (2002) The redox protein thioredoxin-1 (Trx-1) increases hypoxia-inducible factor 1alpha protein expression: Trx-1 overexpression results in increased vascular endothelial growth factor production and enhanced tumor angiogenesis. Cancer Res 62:5089–5095
87. Hirota K, Matsui M, Iwata S, Nishiyama A, Mori K, Yodoi J (1997) AP-1 transcriptional activity is regulated by a direct association between thioredoxin and Ref-1. Proc Natl Acad Sci USA 94:3633–3638
88. Barrett DM, Black SM, Todor H, Schmidt-Ullrich RK, Dawson KS, Mikkelsen RB (2005) Inhibition of protein-tyrosine phosphatases by mild oxidative stresses is dependent on S-nitrosylation. J Biol Chem 280(15):14453–14461
89. Clavreul N, Adachi T, Pimental DR, Ido Y, Schöneich C, Cohen RA (2006) S-glutathiolation by peroxynitrite of p21ras at cysteine-118 mediates its direct activation and downstream signaling in endothelial cells. FASEB J 20(3):518–520
90. Lander HM, Hajjar DP, Hempstead BL, Mirza UA, Chait BT, Campbell S, Quilliam LA (1997) A molecular redox switch on p21(ras). Structural basis for the nitric oxide-p21(ras) interaction. J Biol Chem 272(7):4323–4326
91. Mannick JB, Hausladen A, Liu L, Hess DT, Zeng M, Miao QX, Kane LS, Gow AJ, Stamler JS (1999) Fas-induced caspase denitrosylation. Science 284(5414):651–654
92. Mitchell DA, Marletta MA (2005) Thioredoxin catalyzes the S-nitrosation of the caspase-3 active site cysteine. Nat Chem Biol 1(3):154–158
93. Pan S, Berk BC (2007) Glutathiolation regulates tumor necrosis factor-alpha-induced caspase-3 cleavage and apoptosis: key role for glutaredoxin in the death pathway. Circ Res 100(2):213–219
94. Sun J, Xin C, Eu JP, Stamler JS, Meissner G (2001) Cysteine-3635 is responsible for skeletal muscle ryanodine receptor modulation by NO. Proc Natl Acad Sci USA 98:11158–11162
95. Aracena P, Sanchez G, Donoso P, Hamilton SL, Hidalgo C (2003) S-glutathionylation decreases $Mg^{2+}$ inhibition and S-nitrosylation enhances $Ca^{2+}$ activation of RyR1 channels. J Biol Chem 278:42927–42935
96. Aracena-Parks P, Goonasekera SA, Gilman C, Dirksen RT, Hidalgo C, Hamilton SL (2006) Identification of cysteines involved in S-nitrosylation, S-glutathionylation, and oxidation to disulfides in RyR1. J Biol Chem 281:40354–40368
97. Lancaster JR Jr, Gaston B (2004) NO and nitrosothiols: spatial confinement and free diffusion. Am J Physiol Lung Cell Mol Physiol 287:L
98. Martínez-Ruiz A, Lamas S (2004) S-nitrosylation: a potential new paradigm in signal transduction. Cardiovasc Res 62:43–52
99. Fernandes DC, Manoel AH, Wosniak J Jr, Laurindo FR (2009) Protein disulfide isomerase overexpression in vascular smooth muscle cells induces spontaneous preemptive NADPH oxidase activation and Nox1 mRNA expression: effects of nitrosothiol exposure. Arch Biochem Biophys 484(2):197–204
100. Gonzalez-Villalobos RA, Satou R, Seth DM, Semprun-Prieto LC, Katsurada A, Kobori H, Navar LG (2008) Angiotensin-converting enzyme-derived angiotensin II formation during angiotensin II-induced hypertension. Hypertension 53:351–355
101. Helmcke I, Heumüller S, Tikkanen R, Schröder K, Brandes RP (2009) Identification of structural elements in Nox1 and Nox4 controlling localization and activity. Antioxid Redox Signal 11(6):1279–1287

102. Serrander L, Cartier L, Bedard K, Banfi B, Lardy B, Plastre O, Sienkiewicz A, Fórró L, Schlegel W, Krause KH (2007) NOX4 activity is determined by mRNA levels and reveals a unique pattern of ROS generation. Biochem J 406(1):105–114
103. Gardner PR, Fridovich I (1991) Superoxide sensitivity of the Escherichia coli aconitase. J Biol Chem 266(29):19328–19333
104. Biemond P, Swaak AJ, Beindorff CM, Koster JF (1986) Superoxide-dependent and -independent mechanisms of iron mobilization from ferritin by xanthine oxidase. Implications for oxygen-free-radical-induced tissue destruction during ischaemia and inflammation. Biochem J 239(1):169–173
105. Cadenas E, Mira D, Brunmark A, Lind C, Segura-Aguilar J, Ernster L (1988) Effect of superoxide dismutase on the autoxidation of various hydroquinones—a possible role of superoxide dismutase as a superoxide:semiquinone oxidoreductase. Free Radic Biol Med 5(2):71–79
106. Kamata H, Honda S, Maeda S, Chang L, Hirata H, Karin M (2005) Reactive oxygen species promote TNFalpha-induced death and sustained JNK activation by inhibiting MAP kinase phosphatases. Cell 120(5):649–661
107. Janssen-Heininger YM, Mossman BT, Heintz NH, Forman HJ, Kalyanaraman B, Finkel T, Stamler JS, Rhee SG, van der Vliet A (2008) Redox-based regulation of signal transduction: principles, pitfalls, and promises. Free Radic Biol Med 45(1):1–17
108. Kowaltowski AJ, de Souza-Pinto NC, Castilho RF, Vercesi AE (2009) Mitochondria and reactive oxygen species. Free Radic Biol Med. doi:10.1016/j.freeradbiomed.2009.05.004
109. Plowman SJ, Muncke C, Parton RG, Hancock JF (2005) H-ras, K-ras, and inner plasma membrane raft proteins operate in nanoclusters with differential dependence on the actin cytoskeleton. Proc Natl Acad Sci USA 102(43):15500–15505
110. Ushio-Fukai M (2006) Localizing NADPH oxidase-derived ROS. Sci STKE (349):re8
111. Go YM, Gipp JJ, Mulcahy RT, Jones DP (2004) $H_2O_2$-dependent activation of GCLC-ARE4 reporter occurs by mitogen-activated protein kinase pathways without oxidation of cellular glutathione or thioredoxin-1. J Biol Chem 279(7):5837–5845
112. Li Q, Spencer NY, Oakley FD, Buettner GR, Engelhardt J (2009) Endosomal Nox2 Facilitates Redox-Dependent Induction of NFκB by TNFalpha. Antioxid Redox Signal 11(6):1249–1263
113. Miller FJ Jr, Filali M, Huss GJ, Stanic B, Chamseddine A, Barna TJ, Lamb FS (2007) Cytokine activation of nuclear factor kappa B in vascular smooth muscle cells requires signaling endosomes containing Nox1 and ClC-3. Circ Res 101(7):663–671
114. Li Q, Engelhardt JF (2006) Interleukin-1beta induction of NFkappaB is partially regulated by $H_2O_2$-mediated activation of NFkappaB-inducing kinase. J Biol Chem 281(3):1495–1505
115. García-Cardeña G, Oh P, Liu J, Schnitzer JE, Sessa WC (1996) Targeting of nitric oxide synthase to endothelial cell caveolae via palmitoylation: implications for nitric oxide signaling. Proc Natl Acad Sci USA 93(13):6448–6453
116. Hilenski LL, Clempus RE, Quinn MT, Lambeth JD, Griendling KK (2004) Distinct subcellular localizations of Nox1 and Nox4 in vascular smooth muscle cells. Arterioscler Thromb Vasc Biol 24(4):677–683
117. Ishizaka N, Griendling KK, Lassègue B, Alexander RW (1998) Angiotensin II type 1 receptor: relationship with caveolae and caveolin after initial agonist stimulation. Hypertension 32(3):459–466
118. Wientjes FB, Reeves EP, Soskic V, Furthmayr H, Segal AW (2001) The NADPH oxidase components p47(phox) and p40(phox) bind to moesin through their PX domain. Biochem Biophys Res Commun 289(2):382–388
119. Wu RF, Gu Y, Xu YC, Nwariaku FE, Terada LS (2003) Vascular endothelial growth factor causes translocation of p47phox to membrane ruffles through WAVE1. J Biol Chem 278(38):36830–36840

120. Kuroda S, Fukata M, Kobayashi K, Nakafuku M, Nomura N, Iwamatsu A, Kaibuchi K (1996) Identification of IQGAP as a putative target for the small GTPases, Cdc42 and Rac1. J Biol Chem 271(38):23363–23367
121. Sadok A, Pierres A, Dahan L, Prévôt C, Lehmann M, Kovacic H (2009) NADPH Oxidase 1 controls the persistence of directed cell migration by a Rho-dependent switch of {Alpha}2/{Alpha}3 Integrins. Mol Cell Biol 29(14):3915–3928
122. Shinohara M, Shang WH, Kubodera M, Harada S, Mitsushita J, Kato M, Miyazaki H, Sumimoto H, Kamata T (2007) Nox1 redox signaling mediates oncogenic Ras-induced disruption of stress fibers and focal adhesions by down-regulating Rho. J Biol Chem 282(24):17640–17648
123. Winterbourn CC, Hampton MB, Livesey JH, Kettle AJ (2006) Modeling the reactions of superoxide and myeloperoxidase in the neutrophil phagosome: implications for microbial killing. J Biol Chem 281(52):39860–39869
124. Souza HP, Liu X, Samouilov A, Kuppusamy P, Laurindo FR, Zweier JL (2002) Quantitation of superoxide generation and substrate utilization by vascular NAD(P)H oxidase. Am J Physiol Heart Circ Physiol 282(2):H466–H474
125. Anderson RG, Jacobson K (2002) A role for lipid shells in targeting proteins to caveolae, rafts, and other lipid domains. Science 296(5574):1821–1825
126. Gratton JP, Bernatchez P, Sessa WC (2004) Caveolae and caveolins in the cardiovascular system. Circ Res 94(11):1408–1417
127. Zuo L, Ushio-Fukai M, Hilenski LL, Alexander RW (2004) Microtubules regulate angiotensin II type 1 receptor and Rac1 localization in caveolae/lipid rafts: role in redox signaling. Arterioscler Thromb Vasc Biol 24(7):1223–1228
128. Touyz RM, Yao G, Quinn MT, Pagano PJ, Schiffrin EL (2005) p47phox associates with the cytoskeleton through cortactin in human vascular smooth muscle cells: role in NAD(P)H oxidase regulation by angiotensin II. Arterioscler Thromb Vasc Biol 25(3): 512–518
129. Li Q, Harraz MM, Zhou W, Zhang LN, Ding W, Zhang Y, Eggleston T, Yeaman C, Banfi B, Engelhardt JF (2006) Nox2 and Rac1 regulate $H_2O_2$-dependent recruitment of TRAF6 to endosomal interleukin-1 receptor complexes. Mol Cell Biol 26(1):140–154
130. Oakley FD, Abbott D, Li Q, Engelhardt J (2009) Signaling Components of Redox Active Endosomes: The Redoxosomes. Antioxid Redox Signal 11(6):1313–1333
131. Streuli CH, Akhtar N (2009) Signal co-operation between integrins and other receptor systems. Biochem J 418(3):491–506
132. Lipniacki T, Kimmel M (2007) Deterministic and stochastic models of NFkappaB pathway. Cardiovasc Toxicol 7(4):215–234
133. Harding A, Hancock JF (2008) Ras nanoclusters: combining digital and analog signaling. Cell Cycle 7(2):127–134
134. Kindzelskii AL, Petty HR (2002) Apparent role of traveling metabolic waves in oxidant release by living neutrophils. Proc Natl Acad Sci USA 99(14):9207–9212
135. Wang W, Fang H, Groom L, Cheng A, Zhang W, Liu J, Wang X, Li K, Han P, Zheng M, Yin J, Wang W, Mattson MP, Kao JP, Lakatta EG, Sheu SS, Ouyang K, Chen J, Dirksen RT, Cheng H (2008) Superoxide flashes in single mitochondria. Cell 134(2):279–290
136. Alon U (2006) An introduction to systems biology: design principles of biological circuits. Chapman and Hall/CRC, Boca Raton, FL
137. Oakley FD, Abbott D, Li Q, Engelhardt J (2009) Signaling components of redox active endosomes: the redoxosomes. Antioxid Redox Signal 11(6):1313–1333
138. Janiszewski M, Lopes LR, Carmo AO, Pedro MA, Brandes RP, Santos CX, Laurindo FR (2005) Regulation of NAD(P)H oxidase by associated protein disulfide isomerase in vascular smooth muscle cells. J Biol Chem 280(49):40813–40819
139. Harraz MM, Marden JJ, Zhou W, Zhang Y, Williams A, Sharov VS, Nelson K, Luo M, Paulson H, Schöneich C, Engelhardt JF (2008) SOD1 mutations disrupt redox-sensitive Rac regulation of NADPH oxidase in a familial ALS model. J Clin Invest 118(2):659–670

140. Nauseef WM (2008) Biological roles for the NOX family NADPH oxidases. J Biol Chem 283(25):16961–16965
141. Archer SL, Gomberg-Maitland M, Maitland ML, Rich S, Garcia JG, Weir EK (2008) Mitochondrial metabolism, redox signaling, and fusion: a mitochondria-ROS-HIF-1alpha-Kv1.5 $O_2$-sensing pathway at the intersection of pulmonary hypertension and cancer. Am J Physiol Heart Circ Physiol 294(2):H570–H578
142. Hardie DG, Hawley SA, Scott JW (2006) AMP-activated protein kinase – development of the energy sensor concept. J Physiol 574(Pt 1):7–15
143. Rutkowski DT, Kaufman RJ (2007) That which does not kill me makes me stronger: adapting to chronic ER stress. Trends Biochem Sci 32(10):469–476
144. Santos CX, Tanaka LY, Wosniak JJ, Laurindo FR (2009) Mechanisms and implications of reactive oxygen species generation during the unfolded protein response: roles of endoplasmic reticulum oxidoreductases, mitochondrial electron transport and NADPH oxidase. Antioxid Redox Signal. doi:10.1089/ARS.2009.2625
145. Liu Y, Adachi M, Zhao S, Hareyama M, Koong AC, Luo D, Rando TA, Imai K, Shinomura Y (2009) Preventing oxidative stress: a new role for XBP1. Cell Death Differ 16(6): 847–857
146. Collins AR, Lyon CJ, Xia X, Liu JZ, Tangirala RK, Yin F, Boyadjian R, Bikineyeva A, Praticò D, Harrison DG, Hsueh WA (2009) Age-accelerated atherosclerosis correlates with failure to upregulate antioxidant genes. Circ Res 104(6):e42–e54
147. Ristow M, Zarse K, Oberbach A, Klöting N, Birringer M, Kiehntopf M, Stumvoll M, Kahn CR, Blüher M (2009) Antioxidants prevent health-promoting effects of physical exercise in humans. Proc Natl Acad Sci USA 106(21):8665–8670
148. Merksamer PI, Trusina A, Papa FR (2008) Real-time redox measurements during endoplasmic reticulum stress reveal interlinked protein folding functions. Cell 135(5):933–947
149. Trotter EW, Grant CM (2002) Thioredoxins are required for protection against a reductive stress in the yeast Saccharomyces cerevisiae. Mol Microbiol 46(3):869–878
150. Kang SW, Hegde RS (2008) Lighting up the stressed ER. Cell 135(5):787–789
151. Dimmeler S, Zeiher AM (2007) A "reductionist" view of cardiomyopathy. Cell 130(3): 401–402
152. Rajasekaran NS, Connell P, Christians ES, Yan LJ, Taylor RP, Orosz A, Zhang XQ, Stevenson TJ, Peshock RM, Leopold JA, Barry WH, Loscalzo J, Odelberg SJ, Benjamin IJ (2007) Human alpha B-crystallin mutation causes oxido-reductive stress and protein aggregation cardiomyopathy in mice. Cell 130(3):427–439
153. Tilton RG (2002) Diabetic vascular dysfunction: links to glucose-induced reductive stress and VEGF. Microsc Res Tech 57(5):390–407
154. Kültz D (2005) Molecular and evolutionary basis of the cellular stress response. Annu Rev Physiol 67:225–257
155. Ron D, Walter P (2007) Signal integration in the endoplasmic reticulum unfolded protein response. Nat Rev Mol Cell Biol 8(7):519–529
156. Preston GA, Lyon TT, Yin Y, Lang JE, Solomon G, Annab L, Srinivasan DG, Alcorta DA, Barrett JC (1996) Induction of apoptosis by c-Fos protein. Mol Cell Biol 16(1):211–218
157. Malhotra JD, Kaufman RJ (2007) The endoplasmic reticulum and the unfolded protein response. Semin Cell Dev Biol 18(6):716–731
158. Ron D, Walter P (2007) Signal integration in the endoplasmic reticulum unfolded protein response. Nat Rev Mol Cell Biol 8(7):519–529
159. Cullinan SB, Diehl JA (2006) Coordination of ER and oxidative stress signaling: the PERK/Nrf2 signaling pathway. Int J Biochem Cell Biol 38(3):317–332
160. Lewerenz J, Maher P (2009) Basal levels of eIF2alpha phosphorylation determine cellular antioxidant status by regulating ATF4 and xCT expression. J Biol Chem 284(2):1106–1115
161. Malhotra JD, Miao H, Zhang K, Wolfson A, Pennathur S, Pipe SW, Kaufman RJ (2008) Antioxidants reduce endoplasmic reticulum stress and improve protein secretion. Proc Natl Acad Sci USA 105(47):18525–18530

162. Tu BP, Weissman JS (2004) Oxidative protein folding in eukaryotes: mechanisms and consequences. J Cell Biol 164(3):341–346
163. Szalay MS, Kovács IA, Korcsmáros T, Böde C, Csermely P (2007) Stress-induced rearrangements of cellular networks: consequences for protection and drug design. FEBS Lett 581(19):3675–3680
164. Palotai R, Szalay MS, Csermely P (2008) Chaperones as integrators of cellular networks: changes of cellular integrity in stress and diseases. IUBMB Life 60(1):10–18
165. Bahar R, Hartmann CH, Rodriguez KA, Denny AD, Busuttil RA, Dollé ME, Calder RB, Chisholm GB, Pollock BH, Klein CA, Vijg J (2006) Increased cell-to-cell variation in gene expression in ageing mouse heart. Nature 441(7096):1011–1014
166. Passos JF, Saretzki G, Ahmed S, Nelson G, Richter T, Peters H, Wappler I, Birket MJ, Harold G, Schaeuble K, Birch-Machin MA, Kirkwood TB, von Zglinicki T (2007) Mitochondrial dysfunction accounts for the stochastic heterogeneity in telomere-dependent senescence. PLoS Biol 5(5):e110
167. Raj A, van Oudenaarden A (2008) Nature, nurture, or chance: stochastic gene expression and its consequences. Cell 135(2):216–226
168. Shahrezaei V, Swain PS (2008) The stochastic nature of biochemical networks. Curr Opin Biotechnol 19(4):369–374
169. Leichert LI, Gehrke F, Gudiseva HV, Blackwell T, Ilbert M, Walker AK, Strahler JR, Andrews PC, Jakob U (2008) Quantifying changes in the thiol redox proteome upon oxidative stress in vivo. Proc Natl Acad Sci USA 105(24):8197–8202
170. Tarpey MM, Wink DA, Grisham MB (2004) Methods for detection of reactive metabolites of oxygen and nitrogen: in vitro and in vivo considerations. Am J Physiol Regul Integr Comp Physiol 286(3):R431–R444
171. Vásquez-Vivar J, Santos AM, Junqueira VB, Augusto O (1996) Peroxynitrite-mediated formation of free radicals in human plasma: EPR detection of ascorbyl, albumin-thiyl and uric acid-derived free radicals. Biochem J 314(Pt 3):869–876
172. Linares E, Nakao LS, Augusto O, Kadiiska MB (2003) EPR studies of in vivo radical production by lipopolysaccharide: potential role of iron mobilized from iron-nitrosyl complexes. Free Radic Biol Med 34(6):766–773
173. He G, Samouilov A, Kuppusamy P, Zweier JL (2002) In vivo imaging of free radicals: applications from mouse to man. Mol Cell Biochem 234–235(1–2):359–367
174. Mason RP (2004) Using anti-5,5-dimethyl-1-pyrroline N-oxide (anti-DMPO) to detect protein radicals in time and space with immuno-spin trapping. Free Radic Biol Med 36(10):1214–1223
175. Tarpey MM, Fridovich I (2001) Methods of detection of vascular reactive species: nitric oxide, superoxide, hydrogen peroxide, and peroxynitrite. Circ Res 89(3):224–236
176. Zhao H, Kalivendi S, Zhang H, Joseph J, Nithipatikom K, Vásquez-Vivar J, Kalyanaraman B (2003) Superoxide reacts with hydroethidine but forms a fluorescent product that is distinctly different from ethidium: potential implications in intracellular fluorescence detection of superoxide. Free Radic Biol Med 34(11):1359–1368
177. Hawkins CL, Morgan PE, Davies MJ (2009) Quantification of protein modification by oxidants. Free Radic Biol Med 46(8):965–988
178. Jones DP, Mody VC Jr, Carlson JL, Lynn MJ, Sternberg P Jr (2002) Redox analysis of human plasma allows separation of pro-oxidant events of aging from decline in antioxidant defenses. Free Radic Biol Med 33(9):1290–1300
179. Go YM, Jones DP (2005) Intracellular proatherogenic events and cell adhesion modulated by extracellular thiol/disulfide redox state. Circulation 111(22):2973–2980
180. Jiang S, Moriarty-Craige SE, Orr M, Cai J, Sternberg P Jr, Jones DP (2005) Oxidant-induced apoptosis in human retinal pigment epithelial cells: dependence on extracellular redox state. Invest Ophthalmol Vis Sci 46(3):1054–1061
181. Halliwell B, Whiteman M (2004) Measuring reactive species and oxidative damage in vivo and in cell culture: how should you do it and what do the results mean? Br J Pharmacol 142(2):231–255

182. Fernandes DC, Wosniak J Jr, Pescatore LA, Bertoline MA, Liberman M, Laurindo FR, Santos CX (2007) Analysis of DHE-derived oxidation products by HPLC in the assessment of superoxide production and NADPH oxidase activity in vascular systems. Am J Physiol Cell Physiol 292(1):C413–C422
183. Glebska J, Koppenol WH (2003) Peroxynitrite-mediated oxidation of dichlorodihydrofluorescein and dihydrorhodamine. Free Radic Biol Med 35(6):676–682
184. Wrona M, Patel KB, Wardman P (2008) The roles of thiol-derived radicals in the use of 2',7'-dichlorodihydrofluorescein as a probe for oxidative stress. Free Radic Biol Med 44(1):56–62
185. Burkitt MJ, Wardman P (2001) Cytochrome C is a potent catalyst of dichlorofluorescin oxidation: implications for the role of reactive oxygen species in apoptosis. Biochem Biophys Res Commun 282(1):329–333
186. Zielonka J, Srinivasan S, Hardy M, Ouari O, Lopez M, Vasquez-Vivar J, Avadhani NG, Kalyanaraman B (2008) Cytochrome c-mediated oxidation of hydroethidine and mito-hydroethidine in mitochondria: identification of homo- and heterodimers. Free Radic Biol Med 44(5):835–846
187. Oliveira GA, Tahara EB, Gombert AK, Barros MH, Kowaltowski AJ (2008) Increased aerobic metabolism is essential for the beneficial effects of caloric restriction on yeast life span. J Bioenerg Biomembr 40(4):381–388
188. Howitz KT, Sinclair DA (2008) Xenohormesis: sensing the chemical cues of other species. Cell 133(3):387–391
189. McCord JM, Keele BB Jr, Fridovich I (1971) An enzyme-based theory of obligate anaerobiosis: the physiological function of superoxide dismutase. Proc Natl Acad Sci USA 68(5):1024–1027
190. Carlioz A, Touati D (1986) Isolation of superoxide dismutase mutants in Escherichia coli: is superoxide dismutase necessary for aerobic life? EMBO J 5:623–630
191. Li Y, Huang TT, Carlson EJ, Melov S, Ursell PC, Olson JL, Noble LJ,, Yoshimura MP, Berger C, Chan PH, Wallace DC, Epstein CJ (1995) Dilated cardiomyopathy and neonatal lethality in mutant mice lacking manganese superoxide dismutase. Nat Genet 11:376–381
192. Karanjawala ZE, Hsieh CL, Lieber MR (2003) Overexpression of Cu/Zn superoxide dismutase is lethal for mice lacking double-strand break repair. DNA Repair 2:285–294
193. Nagano I, Murakami T, Manabe Y, Abe K (2002) Early decrease of survival factors and DNA repair enzyme in spinal motor neurons of presymptomatic transgenic mice that express a mutant SOD1 gene. Life Sci 72:541–548
194. Walton M, Lawlor P, Sirimanne E, Williams C, Gluckman P, Dragunow M (1997) Loss of Ref-1 protein expression precedes DNA fragmentation in apoptotic neurons. Brain Res Mol Brain Res 44:167–170
195. Bonatto D (2007) A systems biology analysis of protein-protein interactions between yeast superoxide dismutases and DNA repair pathways. Free Radic Biol Med 43:557–567
196. Ideker T, Thorsson V, Ranish JA, Christmas R, Buhler J, Eng JK, Bumgarner R, Goodlett DR, Aebersold R, Hood L (2001) Integrated genomic and proteomic analyses of a systematically perturbed metabolic network. Science 292:929–934
197. Carter CD, Kitchen LE, Au WC, Babic CM, Basrai MA (2005) Loss of SOD1 and LYS7 sensitizes Saccharomyces cerevisiae to hydroxyurea and DNA damage agents and downregulates MEC1 pathway effectors. Mol Cell Biol 25:10273–10285
198. Liochev SI, Fridovich I (2007) The effects of superoxide dismutase on $H_2O_2$ formation. Free Radic Biol Med 42(10):1465–1469
199. Teixeira HD, Schumacher RI, Meneghini R (1998) Lower intracellular hydrogen peroxide levels in cells overexpressing CuZn-superoxide dismutase. Proc Natl Acad Sci USA 95(14):7872–7875
200. St. Clair DK, Holland JC (1991) Complementary DNA encoding human colon cancer manganese superoxide dismutase and the expression of its gene in human cells. Cancer Res 51:939–943

201. Safford SE, Oberley TD, Urano M, St. Clair DK (1994) Suppression of fibrosarcoma metastasis by elevated expression of manganese superoxide dismutase. Cancer Res 54: 4261–4265
202. Zhong W, Oberley LW, Oberley TD, Yan T, Domann FE, St. Clair DK (1996) Inhibition of cell growth and sensitization to oxidative damage by overexpression of manganese superoxide dismutase in rat glioma cells. Cell Growth Differ 7:1175–1186
203. Menga AX, Jalalia F, Cuddihya A, Chana N, Bindrab RS, Glazerb PM, Bristow RG (2005) Hypoxia down-regulates DNA double strand break repair gene expression in prostate cancer cells. Radiother Oncol 76:168–176
204. Lievre V, Becuwe P, Bianchi A, Koziel V, Franck P, Schroeder H, Nabet P, Dauca M, Daval JL (2000) Free radical production and changes in superoxide dismutases associated with hypoxia/reoxygenation-induced apoptosis of embryonic rat forebrain neurons in culture. Free Radic Biol Med 29:1291–1301
205. Mumbengegwi DR, Li Q, Li C, Bear CE, Engelhardt JF (2008) Evidence for a superoxide permeability pathway in endosomal membranes. Mol Cell Biol 28(11):3700–3712
206. Winterbourn CC, Peskin AV, Parsons-Mair HN (2002) Thiol oxidase activity of copper, zinc superoxide dismutase. J Biol Chem 277(3):1906–1911
207. Chance B, Sies H, Boveris A (1979) Hydroperoxide metabolism in mammalian organs. Physiol Rev 59(3):527–605
208. Sousa-Lopes A, Antunes F, Cyrne L, Marinho HS (2004) Decreased cellular permeability to H2O2 protects Saccharomyces cerevisiae cells in stationary phase against oxidative stress. FEBS Lett 578(1–2):152–156
209. Bienert GP, Møller AL, Kristiansen KA, Schulz A, Møller IM, Schjoerring JK, Jahn TP (2007) Specific aquaporins facilitate the diffusion of hydrogen peroxide across membranes. J Biol Chem 282(2):1183–1192

# Chapter 2
# Mechanisms of Redox Signaling in Cardiovascular Disease

**Rebecca L. Charles, Joseph R. Burgoyne, and Philip Eaton**

**Abstract** Arrays of chemical oxidants are produced in healthy cells, where they function as important signaling molecules that are crucial in homeostatic regulation and cellular adaptation. The molecular basis of "redox signaling" is a series of oxido-reductive chemical reactions in which oxidants or reductants post-translationally alter the structure of proteins. These modifications equate to signal sensing events, in which an alteration in protein redox status may couple to a change in its function. This coupling of sensing to function is a true transduction event, allowing conversion of the cellular redox state into altered enzymatic activities. Here we review redox signaling in the cardiovascular system, considering the variety of post-translational oxidative modifications that explain redox sensing and signal transduction by proteins at the molecular level.

**Keywords** Cardiovascular disease · Redox signaling · Oxidant stress · Cysteine · Thiol · Post-translational oxidative modification

## 2.1 Overview of Cardiovascular Disease

Diseases of the cardiovascular system are common, broadly encompassing pathologies involving dysfunction of blood vessels and the heart. The consequences of aberrant blood vessel and cardiac function are complex and multiple, potentially affecting most tissues and organs in the body. This is expected, as the supply of blood is crucial to healthy cellular function; so when this becomes compromised, system-wide problems may be anticipated.

---

P. Eaton (✉)
Cardiovascular Division, King's College London BHF Centre of Excellence,
The Rayne Institute, St Thomas' Hospital, London SE1 7EH, UK
e-mail: philip.eaton@kcl.ac.uk

H. Sauer et al. (eds.), *Studies on Cardiovascular Disorders*,
Oxidative Stress in Applied Basic Research and Clinical Practice,
DOI 10.1007/978-1-60761-600-9_2, © Springer Science+Business Media, LLC 2010

A number of well-established risk factors predict the likelihood of an adverse cardiovascular event. Many of these are nonmodifiable such as age, sex, ethnicity, and genetics. Other risk factors such as elevated serum lipids (cholesterol, triglycerides), high blood pressure, physical inactivity, obesity, and smoking are modifiable, either by lifestyle changes or with pharmaceuticals.

In the absence of an established genetic predisposition to cardiovascular disease, the "typical" Western lifestyle is itself associated with increased risk. This is because this lifestyle is linked with increases in the modifiable risk factors outlined above. A typical scenario involves a diet that is excessively calorific in which the amount of fat intake is too high. These factors alone, but especially when coupled with lack of exercise or with smoking, commonly compromise blood vessel function [1]. This dysfunction involves blood vessels not dilating appropriately in response to typical biological cues to do so. Thus the vessels become more constricted, resulting in elevated blood pressure. At the same time, arterial blood vessels also tend to develop atherosclerosis [2], a process whereby elevated levels of serum fats become modified and deposited in the vessel walls, initiating a complex inflammatory process which ultimately damages the vessel. The accumulation of these atheromatous plaques narrows the lumen of the arteries to impede blood flow, resulting in an inadequate blood supply to meet the metabolic demands of many tissues (ischemia). When sustained ischemia occurs in the coronary blood vessels, the myocardium can die (or infarct), which is known as a heart attack. If an infarction is not fatal, a typical scenario is subsequent progression to heart failure, a condition in which the heart cannot pump adequate blood to meet the body's demand. Heart failure can also manifest independently of infarction, most notably in cases of sustained elevations in blood pressure (hypertension) [3].

## 2.2 Oxidative Stress—A Recurrent Hallmark of Cardiovascular Pathologies

Clearly, a modern Western lifestyle increases the risk of the events outlined above that lead to heart failure, as well as other cardiovascular diseases (angina, heart attack, stroke, peripheral vascular and renal dysfunction) associated with loss of a regulated blood supply. In this chapter we consider the role of oxidants in the pathogenesis of these cardiovascular diseases. When each of the individual components of the complex, multifactorial processes that lead to cardiovascular disease is dissected, it is clear that alterations in cellular redox (especially oxidative stress) is a common theme at every level. For example, recent reviews consider and highlight the importance of redox alterations in smoking [4], hyperlipidemia and atherosclerosis [5], hypertension [6], ischemia [7], cell death during and after infarction [8], and hypertrophy and heart failure [9]. Intriguingly, oxidant stress is also associated with postischemic reperfusion injury despite the resupply of blood being ultimately essential for tissue survival [10]. Similarly, cardioprotective interventions such as ischemic preconditioning and postconditioning [11], as well as many drug interventions that limit damage during ischemia and reperfusion, require redox-dependent signaling events [12].

## 2.3 Nondeleterious Roles for Oxidants

Whilst it is clear that cellular redox is altered at many points during scenarios that culminate in cardiovascular disease, it is tempting to generalize that all adverse events are explained by oxidation. The traditional view of most diseases, including those of the cardiovascular system, is that aberrant generation of oxidant molecules is a major mechanism of injury [13]. The idea is that oxidants oxidise biomolecules (equated to damage) within cells to render them dysfunctional, providing a mechanism of damage. Whilst there is a wealth of evidence supporting oxidant-mediated damage, it is increasingly appreciated that oxidants can play important regulatory roles. The failure of antioxidant therapy trials, which have generally shown no benefit, or indeed in many cases have been harmful [14, 15], may be because they interrupt the fundamental need for oxidant production and sensing to maintain homeostasis. Broad spectrum antioxidant treatment may block important fundamental regulatory pathways, as well as attenuate adaptation to cellular stress. One answer to this problem could be to selectively remove damaging oxidants, whilst leaving the homeostatic species. However, this may not be possible if the damaging species is the same as the regulatory (albeit present at higher abundance). Despite this, if the damaging species were formed at specific cellular locations, such as the mitochondria, one possibility might be to use targeted antioxidants designed to accumulate only there [16], perhaps leaving the regulatory oxidants in other locations to carry out their homeostatic functions.

Overall, the case for oxidants in mediating disease has likely been overstated, with certain antioxidant regimes potentially causing "reductive stress," an often overlooked potential perpetrator of dysfunction. Lack of oxygen availability, as occurs during hypoxia and ischemia, is commonly assumed to induce oxidative stress. Whilst there is evidence for this [10], there is also evidence for reductive stress under these conditions (NADH accumulation) [17, 18]. Elevations in the abundant cellular reducing equivalent glutathione (GSH) appear crucial to development of cardiomyopathy during overexpression of mutant chaperone proteins [19]. Increases in cellular reducing equivalents can potentially enhance free radical accumulation, consistent with them being electron donors [17]. Similarly, antioxidant therapies ultimately supply cells with a source of electrons which could feed into detrimental free radical–generating pathways. Antioxidants may also prevent adaptive pathways that are triggered in response to oxidant stress, such as ischaemic preconditioning [20]. In some scenarios, such as ischaemic preconditioning, oxidants may signal a concomitant or impending change to the cell, triggering an appropriate adaptive response. This response requires the oxidants to be sensed and transduced into a functional adaptive response. Thus, covering these warning signs with an "antioxidant blanket" may prevent adaptation, offering an explanation for how these reducing agents can be harmful.

## 2.4 Cellular Oxidants

A change in the balance between oxidants and antioxidants towards a pro-oxidising (beyond certain limits) environment culminates in oxidative stress. However, not only is it difficult to precisely define these limits, but perhaps the use of the term "stress" is not always appropriate or helpful. This is because the word may infer a negative consequence, which is not always the case. "Oxidative stress" may reflect the historical misconception that oxidants are simply harmful, playing no positive or regulatory roles. Oxidative stress may occur as a result of increased formation of reactive oxygen species (ROS) or reduction in antioxidant species [21]. When the reverse occurs and reducing equivalents accumulate, reductive stress occurs. Accumulation of reducing equivalents may feed into pro-oxidant pathways, highlighting a scenario of concomitant oxido-reductive stress. Consequently, the redox state of tissue is heavily dependent on the parameter measured, especially as not all redox endpoints are in equilibrium or co-locate within cells. Thus, whilst ischemia increases NADH:$NAD^+$ (i.e., reduction), enhanced free radical production (and oxidation) can also simultaneously occur [22].

Cellular redox is a term that reflects the net state as the cell generates reducing and oxidising equivalents. This redox state is governed by the amount, rate of production and consumption of these agents, and the equilibria between the various redox couples. Many oxidants and antioxidants are not in equilibrium, and so not reacting efficiently with one another. A number of biologically important oxidant and antioxidant molecules, and their reaction chemistry, are described below. ROS form when electrons add to oxygen, producing various reduced states. A single electron addition to $O_2$ forms the superoxide anion radical ($O_2^{\bullet-}$). Donation of a second electron, as occurs during superoxide dismutation, forms hydrogen peroxide ($H_2O_2$), the properties of which allow it to function as an efficient second messenger signaling molecule. If a third electron is donated to $O_2$, the highly reactive hydroxyl radical (OH•) is formed, which occurs when superoxide reacts via Fenton chemistry with iron ($Fe^{2+}$) or by peroxynitrite ($OONO^-$) decomposition. $OONO^-$ is generated when $O_2^{\bullet-}$ reacts with nitric oxide (NO), and mediates both oxidant and nitrating reactions. Catecholamines can generate oxidants by auto-oxidation or via the enzymatic action of monoamine oxidase, which produces $H_2O_2$. Although oxidants can form spontaneously, there are specific oxidase enzymes whose function is to generate these various oxidant species, such as myeloperoxidase, which converts $H_2O_2$ to hypochlorous acid (HOCl), xanthine oxidase, NAD(P)H oxidases, cytochrome P450, and uncoupled NO synthases (NOS) [23]. The mitochondria also generate ROS, which appear to be of regulatory importance [24].

Whilst NO exerts many of its effects through the NO-cGMP-PKG pathway, it also functions via covalently adducting to protein thiols (S-nitrosylation) [25]. NO can alter protein activity by adducting to noncysteine residues or by weak binding to aromatic side chains of proteins [26]. However, NO may potentially react with other small redox-active species to produce related molecules that exert alternate functional effects. For example, the one electron reduced form of NO is nitroxyl (HNO), which more readily reacts with thiols than NO. It is notable that nitroxyl has cardioprotective properties offering a similar degree of protection to ischaemic preconditioning [27]. The bioactivity of lipids can be modulated by nitration, which can alter their interactions with proteins and so control function [28]. Although previously thought to be a biologically inert oxidation product of NO, nitrite ($NO_2^-$) has recently been shown to be bioactive, providing protection against injury during ischemia and reperfusion [29]. Indeed, nitrite has now been shown to be enzymatically reduced to NO [30], which likely mediates PKG signaling and protection [31], as well as other functional consequences such as vasorelaxation [32].

## 2.5 Protein Oxidation Involved in Redox Signaling

Integral to the transduction of a molecular oxidant signal into a cellular response is the post-translational oxidative modification of redox-active proteins. Oxidative modification of a great many amino acids can occur, but perhaps the best studied are methionine, tyrosine, tryptophan, histidine, lysine, and, most notably, cysteine, which is considered in detail below.

Some modes of protein oxidation (especially those that can occur at a cysteinyl thiol) have the essential elements of a post-translational regulatory system, including sensitivity, specificity, and reversibility (see Fig. 2.1). The stoichiometry of protein oxidation can be directly and proportionately coupled to the cellular concentration of oxidants, the biosynthesis of which may also be carefully regulated. This can involve phosphoregulation of oxidase activity [33], although oxidant generation may also be controlled by post-translational oxidative modifications themselves. When a protein becomes oxidised, its function may be altered, perhaps most simply serving as an on or off switch. In more complex scenarios, protein redox alterations may serve as a rheostat, to modulate protein-protein interactions (see Fig. 2.1). Cysteinyl thiols, especially those that ionise to the thiolate state (i.e., those with a low pKa) are especially disposed to oxidative addition reactions, and thereafter, depending on the precise modification, their reversal back to the basal state. A variety of oxidative alterations of thiols can occur, depending on the species and concentration of oxidants that they encounter. The most studied of these modifications include sulfoxidation, S-thiolation, S-nitrosylation, and inter- and intra-disulfide formation. A number of electrophilic lipids can also form adducts with thiols, although such species can also target lysine and histidine residues [34]. S-thiolation involves a disulfide bond forming between a protein and a low molecular weight thiol, resulting in a mixed disulfide. S-glutathiolation is the most common form of S-thiolation, due to glutathione's abundance. Other small thiols such as cysteine, homocysteine, and lipoic acid can also form mixed disulfides with proteins. This diversity of

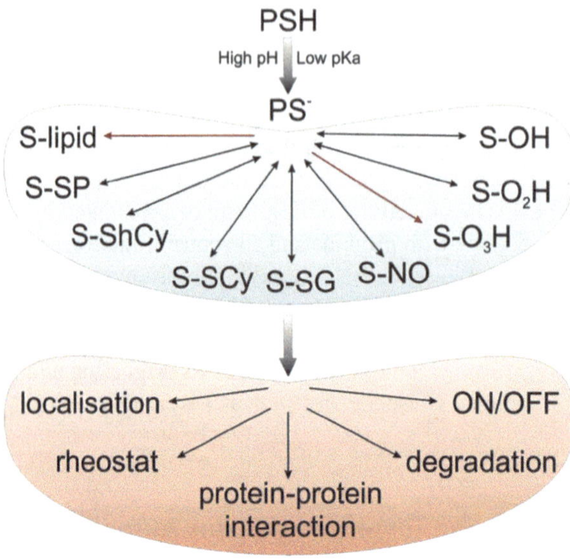


**Fig. 2.1** Oxidant molecules can function as signals and be transduced into a regulatory response. Protein thiols that ionise to the thiolate anion state can undergo a range of post-translational oxidative modifications, some of which are shown here. The thiolate state is present at neutral pH in cysteine residues with a low pKa, which is promoted by proximity to amino acids with basic side changes such as arginine or lysine. Increases in cellular pH will also increase the thiolate state, and enhance the likelihood of thiol oxidation. These redox state–controlled structural alterations couple to changes in enzymatic activities. Thus, depending on the protein modified and the precise oxidative modifications that occur, protein activities and interactions can be altered in a number of ways to enable homeostatic regulation and cellular adaptation to stress

oxidative modification provides a potential mechanism for allowing a graded or differential functional effect, depending on the precise structural change, as established in H-Ras [35]. S-oxidation by glutathione can inactivate a protein, especially when the thiol is catalytically essential, as occurs with protein phosphatase 1B [36, 37], cAMP-dependent protein kinase [38], and tyrosine hydroxylase [39]. In contrast, S-glutathiolation activates HIV-1 protease [40], the microsomal glutathione S-transferase [21], and SERCA calcium pump function [41]. S-thiolation may also serve as a protective mechanism, because disulfide formation prevents overoxidation (to sulfinic and then sulfonic acid), and the possibility of eventual recovery back to the basal reduced state.

When thiols are directly oxidised by oxidants such as molecular oxygen or peroxides, a principal product formed is a sulfenic acid (PSOH). Whilst sulfenates are reversible back to the reduced state, their instability renders them susceptible to hyperoxidation to form sulfinic ($PSO_2H$) and then sulfonic ($PSO_3H$) acids. Sulfenic acids may also undergo recycling, normally forming a transient disulfide intermediate with reducing enzymes fuelled by reducing equivalents derived from glutathione, thioredoxin, NADH, or NADPH. Cysteine sulfinic or sulfonic acids were historically believed to be irreversible modifications, but recent work showed the 2-Cys enzymes can be enzymatically reduced back to thiol form by sulphiredoxin [42]. Likewise, the sulfinated form of peroxiredoxin 1 can also be retro-reduced by p53-regulated sestrin 2 [43]. Protein tyrosine phosphatase 1B (PTP1B) is also regulated by peroxide-induced sulfenation. Furthermore, this phosphatase can form a stable sulfenyl-amide, a newly identified molecular bond in which a sulfenic acid reacts with the amide nitrogen of the protein backbone [44]. Sulfenyl-amides are readily reversed by reducing equivalents such as glutathione, and as with disulfide this bond may serve to prevent cysteine overoxidation. Also, there is some evidence showing reversible sulfination of PTP1B [45].

S-nitrosylation or S-nitrosation is the covalent adduction of NO to a thiol, and is considered by many as a major mechanism by which NO modulates protein activity [46], independent of the classical NO-cGMP-PKG pathway. Potentially, S-nitrosylation may occur because of the direct reaction of free NO with a protein thiol. However, classical NO donors do not generally promote S-nitrosylation efficiently [47], perhaps because the free NO rapidly binds heme centres. NO may first adduct to other abundant thiols in the system (such as free cysteine or glutathione), before undergoing a transnitrosylation exchange reaction culminating in target protein S-nitrosylation. Indeed, S-nitrosylated cysteine or glutathione are efficient S-nitrosylating agents. Like S-glutathiolation, some cysteines are preferentially nitrosylated over others, giving rise to target specificity. The determinants of specificity include the thiol's pKa, the reaction chemistry of the thiol with a specific NO donor, as well as access of the NO donor to the target cysteine [25]. An "S-nitrosylation motif" has been identified, in which the cysteinyl thiol is flanked by acidic (Asp, Glu) and basic (Arg, His, Lys) residues and is located in a hydrophobic pocket [48]. This is consistent with proteins like the ryanodine receptor, which despite having many cysteines, is S-nitrosylated primarily by endogenous NO only at Cys3635. Ryanodine receptor S-nitrosylation status is functionally important, as reduction in this modification under basal conditions increases sarcoplasmic reticulum calcium leak and arrhythmias in cardiomyocytes [49]. Both the S-nitrosylation and the NO-cGMP-PKG pathways can integrate; for example, to modulate cardiac contractility [50]. Although NO-dependent cGMP production activates PKG, recent work has added further complexity, showing that cGMP can also be nitrated [51]. This newly identified signaling molecule can S-guanylate proteins, such as Keap-1, to alter their activity. Another recent observation is that sGC (which NO normally binds to produce cGMP), can be inactivated by S-nitrosylation [52], which also occurs during glycerol trinitrate tolerance [53].

Disulfide bonds can also form between two proteins (interprotein) or between two thiols within the same protein (intraprotein) during oxidative stress. Again these structural alterations can couple to alter protein function. Intermolecular protein disulfide formation during cardiac oxidative stress may lead to the modification of many proteins [54], potentially changing the function of molecular chaperone proteins, growth factors, and signal transduction proteins [55]. Intraprotein disulfide bond formation between vicinal thiols also occurs in many proteins [56]. Affinity capture of proteins with vicinal thiols (on phenylarsine oxide columns) has demonstrated that this mode of redox regulation is widespread [56].

Dityrosine is probably the most prevalent modification of tyrosine residues, although its nitration and chlorination may also occur [57]. Tyrosine nitration is a product of reaction with peroxynitrite, and may synergise with phosphoregulation of this residue. Nitration may irreversibly lock an enzyme into a fixed state [58], although some studies show a cellular denitrase activity that reduces nitrotyrosine back to basal [59]. However, this is controversial, and definitive identification of a denitrase enzyme would greatly enhance the case for tyrosine nitration being an important regulatory mechanism. Similarly, the recent demonstration of a denitrosylating activity of cytosolic and mitochondrial thioredoxins has added significant weight to S-nitrosylation being a fundamentally important regulatory process [60].

Methionine residues can also be reversibly oxidised, forming a sulfoxide [61], although the sulfone, which is not readily reducible, can also be formed during severe or chronic oxidative stress. Methionine sulfoxidation is reversible, either by chemical reduction or by methionine sulfoxide reductases. Because methionine redox state may alter protein conformation, it may also serve as a post-translational regulator [62]. Methionine may serve as a sacrificial antioxidant, protecting other residues from oxidation [63].

Carbonylation is an irreversible protein modification, generally leading to the protein's degradative removal by the cell. Carbonyls can be introduced at several amino acid side chains, including proline, arginine, lysine, and threonine, via multiple mechanisms. For example, carbonyls can be generated by oxidative cleavage of proteins, often during metal catalysed reactions. Protein carbonylation also forms via reactions with reactive oxidised lipids, such as hydroxy-trans-2-nonenal, a lipid peroxidation product. Hydroxynonenal forms adducts with histidine and lysine, but preferentially with cysteinyl thiols. Lysine reacts to form a Schiff base product, whereas thiols undergo Michael additions with the αβ-unsaturated double bonds of electrophilic lipids [64]. Adduction of reactive lipids may alter protein function, in some cases serving a regulatory role. For example, 15-deoxy-$\Delta^{12,14}$-prostaglandin $J_2$ (15d-$PGJ_2$) adducts to Keap-1, which up-regulates transcription and subsequently antioxidant gene expression [65–67].

Hydrogen sulfide ($H_2S$) is a dithiol compound produced by cells, which promotes important biological responses such as vasorelaxation and cardioprotection against ischemia. Indeed, recent studies in which cystathionine gamma-lyase (which makes $H_2S$) was knocked out resulted in murine hypertension [68]. $H_2S$ is anticipated to interact with protein thiols. For example, it may reduce their disulfide, sulphenated, or S-nitrosylated states. Alternatively, given its low pKa, it may directly (or via reaction with other cellular components such as $O_2$ or peroxide) form regulatory protein disulfide adducts.

## 2.6 Techniques for Monitoring Thiol Redox State

A multitude of methods are available for monitoring protein oxidation state, many of which are based on determining the reduced thiol status of cysteine residues. Antibody tools allow specific oxidative modifications of proteins to be assessed, such as glutathiolation, homocysteinylation, carbonylation (via DNPH derivatisation), HNE, malondialdehyde, lipid peroxide (and other reactive lipid adducts), and also nitration, nitrosylation, sulfination, and sulfonation [34, 69–73]. Bifunctional compounds equipped with one of many different thiol reactive groups, together with a reporter moiety, enable detection and quantification of the thiol oxidation state. For example, molecules with maleimide, iodoacetamide, iodoacetate, disulfides, or mercurial functionalities will often react efficiently with reduced thiols. By coupling these functionalities to reporter labels such as biotin, fluorophores, radionucleotides, peroxidase enzymes, or molecular weight tags, these "functionalised" molecules allow the oxidation state of a protein or tissue to be monitored. The principle of these "difference methods" is attenuated labelling following oxidative loss of a thiol. There are additional methods that can be used to detect specific thiol oxidation states. Protein sulfenic acids can be indexed spectrophotometrically using NBD-Cl or dimedone. Arsenite-selective reduction with subsequent thiol biotinylation, as well as generation of functionalised or radiolabeled dimedone molecules allows protein sulphenate quantitation [72, 74–77]. A biotin switch method has been developed to detect S-nitrosylated proteins [78], and is based on ascorbate-reduction of S-nitrosylated proteins with their subsequent biotinylation. N-labelled cysteine or glutathione can serve as a redox probe, for detection and purification of proteins into which they form mixed disulfides during oxidative stress [79]. N-labelled GSSG and cystine have also been developed as tools for detecting proteins that can undergo S-glutathiolation or S-cysteinylation [69, 79]. Diagonal electrophoresis is a method that allows detection and identification of constitutive interprotein disulfide bonds, and those that form nascently during oxidative stress [54, 80].

## 2.7 Proteins in the Cardiovascular System That Are Thiol Redox Modulated

Models of oxidative stress are routinely utilised, concomitantly altering signaling fluxes, especially phosphorylation cascades. Such changes in cell signaling following oxidant treatment are broadly termed "redox signaling." At this time, this seems an inadequate synopsis, because the reported signaling change may be quite distal from the redox sensor and transducer, and may be one of a great many concomitant signaling events. For example, whilst ischemia and reperfusion cause oxidative stress, this scenario also triggers a plethora of concomitant biochemical changes that alter signaling flux. Thus, in oxidant-focused ischemia and reperfusion studies any alterations detected can inappropriately be assigned as redox-dependent. Clearly, a multitude of other nonredox events occur during ischemia and reperfusion, such as ionic imbalance, cell swelling, energy depletion, and alterations in pH. The cell comes well equipped not only for redox sensing, but also for monitoring changes in these parameters so as to enable homeostatic control. Some studies provide supporting evidence that a signaling event is redox modulated by demonstrating that the same pathway is independently triggered by oxidants alone or blunted by antioxidant interventions. However, as signaling cascades are complex, with intricate webs involving multiple inputs, sensors, end-effectors, second messenger and phosphorylation cascades, commonly with nodes of cross-talk, this information may still have a limited value. Of greater value perhaps is to definitively determine the primary signal sensor and transducer, because this provides a strong platform on which to base subsequent studies.

Our laboratory and others have used proteomic methodologies to identify proteins that are susceptible to oxidant modification, and so may participate in redox signaling. A summary of potentially redox-active proteins is given in Table 2.1, which highlights their broad variety, including metabolic enzymes, ion channels, molecular chaperones, structural proteins, and signaling molecules. When low abundance proteins (such as signaling molecules) are identified using these proteomic screens, this may strongly indicate them being truly oxidant sensitive, likely reflecting stoichiometric post-translational modifications. This is in contrast to the routine presence of high abundance proteins, which are often false positive in proteomic screens. We have studied interprotein disulfide bond formation in cardiac myocytes during oxidative stress [54], which identified the RI regulatory subunit of protein kinase A (PKA) as a disulfide forming protein. Subsequently, we found that this oxidation was associated with kinase activation [94]. We also found that protein kinase G (PKG) Iα forms disulfide dimers, activating it to induce coronary vasodilation [93]. Human protein kinase C (PKC) isozymes contain 16–28 cysteine residues, and they have been implicated in the redox regulation of some its isoforms [95]. Redox control of PKC activity is complex, with evidence for both oxidant-induced inactivation as well as activation [96–101]. PKC function is modulated in an isoform specific manner by S-glutathiolation and S-cysteinylation. Whilst PKC-α can be oxidatively inactivated by S-glutathiolation [99, 101], the same group also showed that the δ isoform is activated by S-cysteinylation, with concomitant ε inactivation in the same cell type [98]. The sarcoplasmic reticulum calcium pump is also redox sensitive, being susceptible to S-glutathiolation [54, 41]. Other calcium-handling proteins are redox sensitive, such as the $Na^+$-$Ca^{2+}$ exchanger which is activated by intramolecular disulfide formation [102]. Similarly, the calcium release channel is also activated by oxidation, although some studies show irreversible inactivation by oxidation [103–106]. Oxidation of $Ca^{2+}$ handling proteins during ischemia and reperfusion injury may causatively facilitate the loss of ionic homeostasis which occurs at this time, thus contributing to injury.

## 2.8 Conclusions

Although oxidative stress contributes to the pathogenesis of a number of cardiovascular diseases, emerging data support a role in homeostatic regulatory or adaptive pathways. Indeed, oxidative stress can initiate pathways that actually limit injury, as well as having integral roles in the normal functioning of cells and tissue. Therefore,

**Table 2.1** Proteins that can be redox modified and the post-translational-oxidative modification(s) that they can form

| Protein | Reactive thiol | Thiolated | Nitrosylation | Disulfide | Sulfenated Sulfinated Sulfonated | Methionine oxidation | Carbonylation |
|---|---|---|---|---|---|---|---|
| 14-3-3- | | | [48, 81] | | | | |
| Aconitase | | [82] | [83] | | [72, 77] | | |
| Actin | | [82, 84] | [81, 82, 85] | [54, 84] | [72, 77] | | [84] |
| Acyl-CoA dehydrogenase | | | [83] | [54] | [77] | | [86] |
| ANT | | | | [54] | [77] | | |
| ATP synthase | | | [85] | [54] | [72, 77] | | [86] |
| Calmodulin | | | | | | [87] | |
| Complex 1 | [88] | | | [54] | | | |
| Creatine kinase | [88] | [69] | [83] | [54] | [72] | | [86, 89] |
| Cytochrome c oxidase | | | | | [77] | | [86] |
| Desmin | | | | [54] | [72, 77] | | |
| Glyceraldehyde 3-phosphate dehydrogenase | [88] | [69, 82] | [81] | [54] | | | |
| G-protein Ras | | [82] | | | | | |
| Haemoglobin | | | [90] | | | | |
| Heat shock proteins | | [82] | [83, 85] | | | | |
| Lactate dehydrogenase | | | | | [77] | | |
| Malate dehydrogenase | [88] | | [83] | [54] | [72, 77] | | [86] |
| Myoglobin | | | [91] | [54] | [77] | | |
| Myosin heavy chain | | [69] | [48] | [54] | [72, 77] | | |
| Myosin light chain | | | [83] | [54] | | | |
| NDPKB | | | | [54] | | | |
| Peroxiredoxins | | | [85] | [54] | | | |
| Phosphatidyl-cholinesterol acyltransferase | [88] | [82] | | | | | |
| Phosphofructose kinase | [88] | | | | | | |
| Phosphorylase B Kinase | [88] | | | | | | |

**Table 2.1** (continued)

| Protein | Reactive thiol | Thiolated | Nitrosylation | Disulfide | Sulfenated Sulfinated Sulfonated | Methionine oxidation | Carbonylation |
|---|---|---|---|---|---|---|---|
| PKA | [92] | | | [54] | | | |
| PKCs | | [82] | | [54] | | | |
| PKG | | | | [93] | | | |
| Plasma retinol binding protein | [88] | | | | | | |
| Protein tyrosine phosphatase 1B | | | | | [44, 45] | | |
| Prx | | | [81, 85] | | [70] | | |
| Ryanodin Recpetor | | | [49] | | | | |
| SERCA | [88] | [41] | | [54] | | | |
| Soluble Guanylate Cyclase | | | [52, 53] | | | | |
| Succinate dehydrogenase | [88] | | | [54] | | | [86] |
| Superoxide Dismutase | | | | [54] | | | |
| Triosephosphate isomerase | | [82] | | [54] | | | |
| Tropomyosin | | [84] | [81, 85] | [54, 84] | [72, 77] | | |
| Troponin | | | | [54] | [72, 77] | | |
| Tubulin | | | | [54] | | | |

oxidants may simultaneously stimulate redox-active homeostatic pathways, but at the same time cause damage by biomolecule oxidation. Consequently, oxidant species may be expected to promote simultaneously a mixture of both protective and injurious components, the net effect of which will depend on species, concentration, duration, and site of production of the oxidant, as well as the underlying health or disease state of the tissue. This complexity may help explain why oxidant stress is reported to mediate damage in some studies, but to be crucial to protection in others. Clearly, a better understanding of the role of oxidative stress and molecular redox signaling may increase the likelihood of new effective therapies against cardiovascular diseases.

## References

1. Graham I, Atar D, Borch-Johnsen K, Boysen G et al (2007) European guidelines on cardiovascular disease prevention in clinical practice: full text. Fourth Joint Task Force of the European Society of Cardiology and other societies on cardiovascular disease prevention in clinical practice (constituted by representatives of nine societies and by invited experts). Eur J Cardiovasc Prev Rehabil 14(Suppl 2):S1–S113
2. Stephens JW, Khanolkar MP, Bain SC (2008) The biological relevance and measurement of plasma markers of oxidative stress in diabetes and cardiovascular disease. Atherosclerosis 202(2):321–329
3. Kostis JB, Davis BR, Cutler J, Grimm RH Jr et al (1997) Prevention of heart failure by antihypertensive drug treatment in older persons with isolated systolic hypertension. SHEP Cooperative Research Group. J Am Med Assoc 278:212–216
4. Barnoya J, Glantz SA (2005) Cardiovascular effects of secondhand smoke: nearly as large as smoking. Circulation 111:2684–2698
5. Le NA (2006) Hyperlipidaemia and cardiovascular disease: oxidative damage and atherosclerosis. Curr Opin Lipidol 17:92–94
6. Schulz E, Jansen T, Wenzel P, Daiber A, Munzel T (2008) Nitric oxide, tetrahydrobiopterin, oxidative stress, and endothelial dysfunction in hypertension. Antioxid Redox Signal 10:1115–1126
7. Vanden Hoek TL, Li C, Shao Z, Schumacker PT, Becker LB (1997) Significant levels of oxidants are generated by isolated cardiomyocytes during ischemia prior to reperfusion. J Mol Cell Cardiol 29:2571–2583
8. Hearse DJ (1990) Ischemia, reperfusion, and the determinants of tissue injury. Cardiovasc Drugs Ther 4(Suppl 4):767–776
9. Seddon M, Looi YH, Shah AM (2007) Oxidative stress and redox signalling in cardiac hypertrophy and heart failure. Heart 93:903–907
10. Hearse DJ (1998) Myocardial protection during ischemia and reperfusion. Mol Cell Biochem 186:177–184
11. Hausenloy DJ, Yellon DM (2007) Preconditioning and postconditioning: united at reperfusion. Pharmacol Ther 116:173–191
12. Venardos KM, Perkins A, Headrick J, Kaye DM (2007) Myocardial ischemia-reperfusion injury, antioxidant enzyme systems, and selenium: a review. Curr Med Chem 14: 1539–1549
13. Griendling KK, FitzGerald GA (2003) Oxidative stress and cardiovascular injury: part I: basic mechanisms and in vivo monitoring of ROS. Circulation 108:1912–1916
14. Heart Protection Study Collaborative Group (2002) MRC/BHF Heart Protection Study of antioxidant vitamin supplementation in 20,536 high-risk individuals: a randomised placebo-controlled trial. Lancet 360:23–33

15. Yusuf S, Dagenais G, Pogue J, Bosch J, Sleight P, Vitamin E (2000) supplementation and cardiovascular events in high-risk patients. The Heart Outcomes Prevention Evaluation Study Investigators. N Engl J Med 342:154–160
16. Smith RA, Porteous CM, Coulter CV, Murphy MP (1999) Selective targeting of an antioxidant to mitochondria. Eur J Biochem 263:709–716
17. Ido Y, Kilo C, Williamson JR (1997) Cytosolic NADH/NAD+, free radicals, and vascular dysfunction in early diabetes mellitus. Diabetologia 40(Suppl 2):S115–S117
18. Kannurpatti SS, Joshi NB (1999) Energy metabolism and NAD-NADH redox state in brain slices in response to glutamate exposure and ischemia. Metab Brain Dis 14:33–43
19. Rajasekaran NS, Connell P, Christians ES, Yan LJ et al (2007) Human alpha B-crystallin mutation causes oxido-reductive stress and protein aggregation cardiomyopathy in mice. Cell 130:427–439
20. Tanaka M, Fujiwara H, Yamasaki K, Sasayama S (1994) Superoxide dismutase and N-2-mercaptopropionyl glycine attenuate infarct size limitation effect of ischaemic preconditioning in the rabbit. Cardiovasc Res 28:980–986
21. Sies H, Dafre AL, Ji Y, Akerboom TP (1998) Protein S-thiolation and redox regulation of membrane-bound glutathione transferase. Chem Biol Interact 111–112:177–185
22. Robin E, Guzy RD, Loor G, Iwase H et al (2007) Oxidant stress during simulated ischemia primes cardiomyocytes for cell death during reperfusion. J Biol Chem 282:19133–19143
23. Sauer H, Wartenberg M, Hescheler J (2001) Reactive oxygen species as intracellular messengers during cell growth and differentiation. Cell Physiol Biochem 11:173–186
24. Murphy E, Steenbergen C (2007) Preconditioning: the mitochondrial connection. Annu Rev Physiol 69:51–67
25. Hess DT, Matsumoto A, Kim SO, Marshall HE, Stamler JS, Protein S- (2005) Nitrosylation: purview and parameters. Nat Rev Mol Cell Biol 6:150–166
26. Akhter S, Vignini A, Wen Z, English A et al (2002) Evidence for S-nitrosothiol-dependent changes in fibrinogen that do not involve transnitrosation or thiolation. Proc Natl Acad Sci U S A 99:9172–9177
27. Pagliaro P, Mancardi D, Rastaldo R, Penna C et al (2003) Nitroxyl affords thiol-sensitive myocardial protective effects akin to early preconditioning. Free Radic Biol Med 34: 33–43
28. O'Donnell VB, Eiserich JP, Bloodsworth A, Chumley PH et al (1999) Nitration of unsaturated fatty acids by nitric oxide-derived reactive species. Methods Enzymol 301:454–470
29. Gonzalez FM, Shiva S, Vincent PS, Ringwood LA et al (2008) Nitrite anion provides potent cytoprotective and antiapoptotic effects as adjunctive therapy to reperfusion for acute myocardial infarction. Circulation 117:2986–2994
30. Hendgen-Cotta UB, Merx MW, Shiva S, Schmitz J et al (2008) Nitrite reductase activity of myoglobin regulates respiration and cellular viability in myocardial ischemia-reperfusion injury. Proc Natl Acad Sci U S A 105:10256–10261
31. Burley DS, Ferdinandy P, Baxter GF, Cyclic GMP (2007) Protein kinase-G in myocardial ischaemia-reperfusion: opportunities and obstacles for survival signaling. Br J Pharmacol 152:855–869
32. Webb AJ, Patel N, Loukogeorgakis S, Okorie M et al (2008) Acute blood pressure lowering, vasoprotective, and antiplatelet properties of dietary nitrate via bioconversion to nitrite. Hypertension 51:784–790
33. Li JM, Shah AM (2003) Mechanism of endothelial cell NADPH oxidase activation by angiotensin II. Role of the p47phox subunit. J Biol Chem 278:12094–12100
34. Hashimoto M, Sibata T, Wasada H, Toyokuni S, Uchida K (2003) Structural basis of protein-bound endogenous aldehydes. Chemical and immunochemical characterizations of configurational isomers of a 4-hydroxy-2-nonenal-histidine adduct. J Biol Chem 278: 5044–5051
35. Mallis RJ, Buss JE, Thomas JA (2001) Oxidative modification of H-ras: S-thiolation and S-nitrosylation of reactive cysteines. Biochem J 355:145–153

36. Barrett WC, DeGnore JP, Keng YF, Zhang ZY et al (1999) Roles of superoxide radical anion in signal transduction mediated by reversible regulation of protein-tyrosine phosphatase 1B. J Biol Chem 274:34543–34546
37. Barrett WC, DeGnore JP, Konig S, Fales HM et al (1999) Regulation of PTP1B via glutathionylation of the active site cysteine 215. Biochemistry 38:6699–6705
38. Humphries KM, Juliano C, Taylor SS (2002) Regulation of cAMP-dependent protein kinase activity by glutathionylation. J Biol Chem 277:43505–43511
39. Sadidi M, Geddes TJ, Kuhn DM (2005) S-thiolation of tyrosine hydroxylase by reactive nitrogen species in the presence of cysteine or glutathione. Antioxid Redox Signal 7: 863–869
40. Davis DA, Dorsey K, Wingfield PT, Stahl SJ et al (1996) Regulation of HIV-1 protease activity through cysteine modification. Biochemistry 35:2482–2488
41. Adachi T, Weisbrod RM, Pimentel DR, Ying J et al (2004) S-Glutathiolation by peroxynitrite activates SERCA during arterial relaxation by nitric oxide. Nat Med 10:1200–1207
42. Biteau B, Labarre J, Toledano MB (2003) ATP-dependent reduction of cysteine-sulphinic acid by S. cerevisiae sulphiredoxin. Nature 425:980–984
43. Budanov AV, Sablina AA, Feinstein E, Koonin EV, Chumakov PM (2004) Regeneration of peroxiredoxins by p53-regulated sestrins, homologs of bacterial AhpD. Science 304: 596–600
44. Salmeen A, Andersen JN, Myers MP, Meng TC et al (2003) Redox regulation of protein tyrosine phosphatase 1B involves a sulphenyl-amide intermediate. Nature 423:769–773
45. Groen A, Lemeer S, van der Wijk T, Overvoorde J et al (2005) Differential oxidation of protein-tyrosine phosphatases. J Biol Chem 280:10298–10304
46. Stamler JS, Lamas S, Fang FC (2001) Nitrosylation. the prototypic redox-based signaling mechanism. Cell 106:675–683
47. Hogg N, Broniowska KA, Novalija J, Kettenhofen NJ, Novalija E (2007) Role of S-nitrosothiol transport in the cardioprotective effects of S-nitrosocysteine in rat hearts. Free Radic Biol Med 43:1086–1094
48. Greco TM, Hodara R, Parastatidis I, Heijnen HF et al (2006) Identification of S-nitrosylation motifs by site-specific mapping of the S-nitrosocysteine proteome in human vascular smooth muscle cells. Proc Natl Acad Sci USA 103:7420–7425
49. Gonzalez DR, Beigi F, Treuer AV, Hare JM (2007) Deficient ryanodine receptor S-nitrosylation increases sarcoplasmic reticulum calcium leak and arrhythmogenesis in cardiomyocytes. Proc Natl Acad Sci USA 104:20612–20617
50. Gonzalez DR, Fernandez IC, Ordenes PP, Treuer AV et al (2008) Differential role of S-nitrosylation and the NO-cGMP-PKG pathway in cardiac contractility. Nitric Oxide 18: 157–167
51. Sawa T, Zaki MH, Okamoto T, Akuta T et al (2007) Protein S-guanylation by the biological signal 8-nitroguanosine 3′,5′-cyclic monophosphate. Nat Chem Biol 3:727–735
52. Sayed N, Baskaran P, Ma X, van den Akker F, Beuve A (2007) Desensitization of soluble guanylyl cyclase, the NO receptor, by S-nitrosylation. Proc Natl Acad Sci USA 104: 12312–12317
53. Sayed N, Kim DD, Fioramonti X, Iwahashi T et al (2008) Nitroglycerin-induced S-nitrosylation and desensitization of soluble guanylyl cyclase contribute to nitrate tolerance. Circ Res 103:606–614
54. Brennan JP, Wait R, Begum S, Bell JR et al (2004) Detection and mapping of widespread intermolecular protein disulfide formation during cardiac oxidative stress using proteomics with diagonal electrophoresis. J Biol Chem 279:41352–41360
55. Linke K, Jakob U (2003) Not every disulfide lasts forever: disulfide bond formation as a redox switch. Antioxid Redox Signal 5:425–434
56. Gitler C, Zarmi B, Kalef E (1997) General method to identify and enrich vicinal thiol proteins present in intact cells in the oxidized, disulfide state. Anal Biochem 252: 48–55

57. Shacter E (2000) Quantification and significance of protein oxidation in biological samples. Drug Metab Rev 32:307–326
58. Berlett BS, Stadtman ER (1997) Protein oxidation in aging, disease, and oxidative stress. J Biol Chem 272:20313–20316
59. Irie Y, Saeki M, Kamisaki Y, Martin E, Murad F (2003) Histone H1.2 is a substrate for denitrase, an activity that reduces nitrotyrosine immunoreactivity in proteins. Proc Natl Acad Sci USA 100:5634–5639
60. Benhar M, Forrester MT, Hess DT, Stamler JS (2008) Regulated protein denitrosylation by cytosolic and mitochondrial thioredoxins. Science 320:1050–1054
61. Vogt W (1995) Oxidation of methionyl residues in proteins: tools, targets, and reversal. Free Radic Biol Med 18:93–105
62. Hoshi T, Heinemann S (2001) Regulation of cell function by methionine oxidation and reduction. J Physiol 531:1–11
63. Levine RL, Mosoni L, Berlett BS, Stadtman ER (1996) Methionine residues as endogenous antioxidants in proteins. Proc Natl Acad Sci USA 93:15036–15040
64. Schaur RJ (2003) Basic aspects of the biochemical reactivity of 4-hydroxynonenal. Mol Aspects Med 24:149–159
65. Ceaser EK, Moellering DR, Shiva S, Ramachandran A et al (2004) Mechanisms of signal transduction mediated by oxidized lipids: the role of the electrophile-responsive proteome. Biochem Soc Trans 32:151–155
66. Levonen AL, Landar A, Ramachandran A, Ceaser EK et al (2004) Cellular mechanisms of redox cell signalling: role of cysteine modification in controlling antioxidant defences in response to electrophilic lipid oxidation products. Biochem J 378:373–382
67. Sanchez-Gomez FJ, Cernuda-Morollon E, Stamatakis K, Perez-Sala D (2004) Protein thiol modification by 15-deoxy-Delta12,14-prostaglandin J2 addition in mesangial cells: role in the inhibition of pro-inflammatory genes. Mol Pharmacol 66:1349–1358
68. Yang G, Wu L, Jiang B, Yang W et al (2008) $H_2S$ as a physiologic vasorelaxant: hypertension in mice with deletion of cystathionine gamma-lyase. Science 322:587–590
69. Brennan JP, Miller JI, Fuller W, Wait R et al (2006) The utility of N,N-biotinyl glutathione disulfide in the study of protein S-glutathiolation. Mol Cell Proteomics 5:215–225
70. Schroder E, Brennan JP, Eaton P (2008) Cardiac peroxiredoxins undergo complex modifications during cardiac oxidant stress. Am J Physiol Heart Circ Physiol 295: H425–H433
71. Soderling AS, Hultman L, Delbro D, Hojrup P, Caidahl K (2007) Reduction of the nitro group during sample preparation may cause underestimation of the nitration level in 3-nitrotyrosine immunoblotting. J Chromatogr B Analyt Technol Biomed Life Sci 851:277–286
72. Charles RL, Schroder E, May G, Free P et al (2007) Protein sulfenation as a redox sensor: proteomics studies using a novel biotinylated dimedone analogue. Mol Cell Proteomics 6:1473–1484
73. Eaton P, Hearse DJ, Shattock MJ (2001) Lipid hydroperoxide modification of proteins during myocardial ischaemia. Cardiovasc Res 51:294–303
74. Ellis HR, Poole LB (1997) Novel application of 7-chloro-4-nitrobenzo-2-oxa-1,3-diazole to identify cysteine sulfenic acid in the AhpC component of alkyl hydroperoxide reductase. Biochemistry 36:15013–15018
75. Poole LB, Karplus PA, Claiborne A (2004) Protein sulfenic acids in redox signaling. Annu Rev Pharmacol Toxicol 44:325–347
76. Poole LB, Zeng BB, Knaggs SA, Yakubu M, King SB (2005) Synthesis of chemical probes to map sulfenic acid modifications on proteins. Bioconjug Chem 16:1624–1628
77. Saurin AT, Neubert H, Brennan JP, Eaton P (2004) Widespread sulfenic acid formation in tissues in response to hydrogen peroxide. Proc Natl Acad Sci USA 101:17982–17987
78. Jaffrey SR, Snyder SH (2001) The biotin switch method for the detection of S-nitrosylated proteins. Sci STKE 2001:L1

79. Sullivan DM, Wehr NB, Fergusson MM, Levine RL, Finkel T (2000) Identification of oxidant-sensitive proteins: TNF-alpha induces protein glutathiolation. Biochemistry 39: 11121–11128
80. Sommer A, Traut RR (1974) Diagonal polyacrylamide-dodecyl sulfate gel electrophoresis for the identification of ribosomal proteins crosslinked with methyl-4-mercaptobutyrimidate. Proc Natl Acad Sci USA 71:3946–3950
81. Martinez-Ruiz A, Lamas S (2004) Detection and proteomic identification of S-nitrosylated proteins in endothelial cells. Arch Biochem Biophys 423:192–199
82. Eaton P, Byers HL, Leeds N, Ward MA, Shattock MJ (2002) Detection, quantitation, purification, and identification of cardiac proteins S-thiolated during ischemia and reperfusion. J Biol Chem 277:9806–9811
83. Sun J, Morgan M, Shen RF, Steenbergen C, Murphy E (2007) Preconditioning results in S-nitrosylation of proteins involved in regulation of mitochondrial energetics and calcium transport. Circ Res 101:1155–1163
84. Canton M, Neverova I, Menabo R, Van Eyk J, Di Lisa F (2004) Evidence of myofibrillar protein oxidation induced by postischemic reperfusion in isolated rat hearts. Am J Physiol Heart Circ Physiol 286:H870–H877
85. Reinartz M, Ding Z, Flogel U, Godecke A, Schrader J (2008) Nitrosative stress leads to protein glutathiolation, increased s-nitrosation, and up-regulation of peroxiredoxins in the heart. J Biol Chem 283:17440–17449
86. Han B, Stevens JF, Maier CS (2007) Design, synthesis, and application of a hydrazide-functionalized isotope-coded affinity tag for the quantification of oxylipid-protein conjugates. Anal Chem 79:3342–3354
87. Galeva NA, Esch SW, Williams TD, Markille LM, Squier TC (2005) Rapid method for quantifying the extent of methionine oxidation in intact calmodulin. J Am Soc Mass Spectrom 16:1470–1480
88. Sethuraman M, McComb ME, Huang H, Huang S et al (2004) Isotope-coded affinity tag (ICAT) approach to redox proteomics: identification and quantitation of oxidant-sensitive cysteine thiols in complex protein mixtures. J Proteome Res 3:1228–1233
89. Musatov A, Carroll CA, Liu YC, Henderson GI et al (2002) Identification of bovine heart cytochrome c oxidase subunits modified by the lipid peroxidation product 4-hydroxy-2-nonenal. Biochemistry 41:8212–8220
90. Wolzt M, MacAllister RJ, Davis D, Feelisch M et al (1999) Biochemical characterization of S-nitrosohemoglobin. Mechanisms underlying synthesis, no release, and biological activity. J Biol Chem 274:28983–28990
91. Rayner BS, Wu BJ, Raftery M, Stocker R, Witting PK, Human S- (2005) Nitroso oxymyoglobin is a store of vasoactive nitric oxide. J Biol Chem 280:9985–9993
92. Humphries KM, Pennypacker JK, Taylor SS (2007) Redox regulation of cAMP-dependent protein kinase signaling: kinase versus phosphatase inactivation. J Biol Chem 282: 22072–22079
93. Burgoyne JR, Madhani M, Cuello F, Charles RL et al (2007) Cysteine redox sensor in PKGIa enables oxidant-induced activation. Science 317:1393–1397
94. Brennan JP, Bardswell SC, Burgoyne JR, Fuller W et al (2006) Oxidant-induced activation of type I protein kinase A is mediated by RI subunit interprotein disulfide bond formation. J Biol Chem 281:21827–21836
95. Gopalakrishna R, Jaken S (2000) Protein kinase C signaling and oxidative stress. Free Radic Biol Med 28:1349–1361
96. Chu F, Chen LH, O'Brian CA (2004) Cellular protein kinase C isozyme regulation by exogenously delivered physiological disulfides – implications of oxidative protein kinase C regulation to cancer prevention. Carcinogenesis 25:585–596
97. Chu F, Ward NE, O'Brian CA (2001) Potent inactivation of representative members of each PKC isozyme subfamily and PKD via S-thiolation by the tumor-promotion/progression antagonist glutathione but not by its precursor cysteine. Carcinogenesis 22:1221–1229

98. Chu F, Ward NE, O'Brian CA (2003) PKC isozyme S-cysteinylation by cystine stimulates the pro-apoptotic isozyme PKC delta and inactivates the oncogenic isozyme PKC epsilon. Carcinogenesis 24:317–325
99. Ward NE, Chu F, O'Brian CA (2002) Regulation of protein kinase C isozyme activity by S-glutathiolation. Methods Enzymol 353:89–100
100. Ward NE, Pierce DS, Chung SE, Gravitt KR, O'Brian CA (1998) Irreversible inactivation of protein kinase C by glutathione. J Biol Chem 273:12558–12566
101. Ward NE, Stewart JR, Ioannides CG, O'Brian CA, Oxidant-induced S- (2000) Glutathiolation inactivates protein kinase C-alpha (PKC-alpha): a potential mechanism of PKC isozyme regulation. Biochemistry 39:10319–10329
102. Santacruz-Toloza L, Ottolia M, Nicoll DA, Philipson KD (2000) Functional analysis of a disulfide bond in the cardiac Na(+)-Ca(2+) exchanger. J Biol Chem 275:182–188
103. Aghdasi B, Reid MB, Hamilton SL (1997) Nitric oxide protects the skeletal muscle Ca2+ release channel from oxidation induced activation. J Biol Chem 272:25462–25467
104. Boraso A, Williams AJ (1994) Modification of the gating of the cardiac sarcoplasmic reticulum Ca(2+)-release channel by $H_2O_2$ and dithiothreitol. Am J Physiol 267:H1010–H1016
105. Sun J, Xin C, Eu JP, Stamler JS, Meissner G (2001) Cysteine-3635 is responsible for skeletal muscle ryanodine receptor modulation by NO. Proc Natl Acad Sci U S A 98:11158–11162
106. Xie H, Zhu PH (2006) Biphasic modulation of ryanodine receptors by sulfhydryl oxidation in rat ventricular myocytes. Biophys J 91:2882–2891

# Chapter 3
# Reactive Oxygen and Nitrogen Species in Cardiovascular Differentiation of Stem Cells

**Heinrich Sauer and Maria Wartenberg**

**Abstract** Reactive oxygen species (ROS) and nitric oxide (NO) are involved in a variety of signalling events that regulate physiological and pathophysiological processes in the cardiovascular system. NO also undergoes reactions with oxygen, superoxide ions, and reducing agents to create products that themselves show distinctive reactivity toward particular targets, sometimes with the manifestation of toxic effects, such as nitrosative stress. During early embryogenesis, NADPH oxidases and nitric oxide synthases are already expressed in the growing embryo, suggesting that gradients of ROS and NO may exist in the developing organs and be involved in proper functioning of differentiation programs. During pathophysiological insults of the cardiovascular system, e.g., during hypertension, atherosclerosis, and cardiac infarction, high levels of ROS and NO are generated, thus creating an inflammatory microenvironment which on the one hand contributes to cell damage, apoptosis, and remodeling; but which on the other hand may activate repair processes that involve recruitment and differentiation of stem cells of the cardiovascular cell lineage. In this chapter the current knowledge about activation, recruitment, and differentiation of various cardiovascular stem cell populations by ROS and NO within inflamed tissues and the involved signal transduction cascades is reviewed. Furthermore, the specific microenvironmental requirements for proper stem cell engraftment and maintenance are outlined.

**Keywords** Mesenchymal stem cells · Embryonic stem cells · Endothelial progenitor cells · Reactive oxygen species · Reactive nitrogen species · Redox-regulated signaling pathways

---

H. Sauer (✉)
Department of Physiology, Justus Liebig University Giessen, Giessen 35392, Germany
e-mail: heinrich.sauer@physiologie.med.uni-giessen.de

Grant sponsor: Excellence Cluster "Cardiopulmonary System" (ECCPS) of the German Research Foundation

H. Sauer et al. (eds.), *Studies on Cardiovascular Disorders*,
Oxidative Stress in Applied Basic Research and Clinical Practice,
DOI 10.1007/978-1-60761-600-9_3, 

## 3.1 Introduction

According to the free radical theory of development established by Allen and Balin more than 20 years ago [1], metabolic gradients exist in embryos and may influence developmental processes. Most decisive amongst these gradients are those of oxygen, which could influence the expression and activity of ROS and NO generating enzymes like NADPH oxidases or NO synthases. Both ROS and NO can interact to form reactive peroxynitrite. An excessive formation of peroxynitrite represents an important mechanism contributing to cell death and dysfunction in multiple cardiovascular pathologies, such as myocardial infarction, heart failure, and atherosclerosis [2]. However, increasing evidence suggests that peroxynitrite in concert with ROS and NO regulates the activity of enzymes and the expression of a variety of genes involved in cardiovascular differentiation [2]. Gradients of ROS in the organism are balanced by the antioxidative defense which differs in the respective organs, thus separating distinct areas of organ-restricted redox microenvironments. Stem cells are crucial regulators of organ formation. In the blastocyst, embryonic (ES) stem cells are constituents of the inner cell mass, which during later development variegates into the different cell types of the organs, where single cells may persist in their undifferentiated state, thus forming tissue-specific stem cells of so far not well-defined (patho)physiological function. It is currently not known whether mesenchymal stem cells are descendants of embryonic stem cells and whether comparable signaling pathways are involved in the initiation of differentiation programs in distinct subpopulations of stem cells, but some clues point to this direction. A number of studies of ours and others have outlined ROS and NO as crucial signaling molecules involved in cardiovascular differentiation of embryonic stem cells. Recently it has been demonstrated that ROS/NO may be likewise involved in the activation of differentiation programs in mesenchymal stem cells. Signaling pathways that involve ROS and NO to regulate enzyme functions and initiate differentiation programs are legion. Deciphering these pathways and delineating the tissue microenvironment arising during tissue injury and inflammation will support our understanding of the cellular regenerative processes occurring during wound and tissue healing, and will enable us to specifically design biotechnical protocols to generate differentiated tissue-specific stem cells that may be used for patient treatment in cell transplantation approaches.

## 3.2 Oxygen and ROS Generation During Embryogenesis

The prenatal period is divided into the embryonic and the fetal stages. In the embryonic stage organogenesis takes place, i.e., tissues and organs are developed; whereas in the fetal stage the organs grow and mature and take over their adult functions. It is well established that the embryo during early pregnancy lives in an environment of low oxygen tension within the uterus [3, 4]. This hypoxic microenvironment appears to be crucial during the period of organogenesis, where the embryo is most sensitive to environmental oxidative stress, the latter being discussed as the teratogenic principle of a variety of known environmental teratogens [5]. During later stages, when the utero-placental circulation is established, the embryo is more capable of coping with oxidative stress because of a stronger antioxidative stress response and at least partially because of the metabolic switch from glycolysis to oxidative phosphorylation which occurs at times when the embryonic heart starts to contract, thus requesting more energy for heart performance. Despite the sensitivity of the early embryo towards oxidative stress, few studies have demonstrated that ROS at very low concentrations are actively generated during the blastocyste state in rabbits and in postimplantation mouse embryos harvested on day 8 of pregnancy, when ROS generation is localized to the trophoblast cell layers [6]. Moreover, placental NADPH oxidase-mediated ROS generation occurs in women during early pregnancy and may contribute to elevated ROS levels in embryos [7]. These data suggest that very low but physiologically relevant concentrations of ROS may be involved in very early developmental processes during organogenesis and differentiation of stem cells of the inner cell mass. The meaning of ROS during later stages of organ maturation and morphogenesis is not well defined but may at least be involved in neuronal, cardiac, and vascular growth, in which ROS have been shown in several studies to be involved in growth factor and cytokine-mediated signaling pathways such as the vascular endothelial growth factor/flk-1 (VEGF/flk-1) [8], platelet-derived growth factor BB (PDGF-BB) [9], cardiotrophin-1 (CT-1) [10], and nerve growth factor (NGF)-mediated signaling pathways [11] associated with vasculogenesis, angiogenesis, and the development of the central and peripheral nerve system, in which ROS may be involved in the regulation of axon guidance through semaphorin 3A [12]. Furthermore, high levels of ROS have been implicated in site-specific cell death in interdigital regions of the developing limb [13], where peroxidase activity and glutathione peroxidase-4 gene (Gpx4) expression were restricted to the nonapoptotic tissue (e.g., digits) of the developing autopod, thus suggesting that differential tissue growth may be regulated by redox gradients which are determined by distinct expression patterns of antioxidant molecules.

## 3.3 Oxidative Stress During Myocardial Infarction—A Potential Stimulus for Stem Cell Activation

During cardiovascular repair processes embryonic genes are activated, suggesting that comparable signaling pathways are involved in embryonic development of the cardiovascular system and in cardiac repair during adult life. During hypertension and hypertrophic cardiac growth [14, 15], but also in acute myocardial infarction [16–18], ROS are generated in the ischemic myocardium, especially after reperfusion. ROS in high concentrations directly injure the cell membrane and cause cell death. However, ROS in low concentrations also stimulate signal transduction to elaborate inflammatory cytokines, e.g., tumour necrosis factor-α (TNF-α) and interleukin (IL)-1β and -6, in the ischemic region and surrounding myocardium as a host reaction. These inflammatory cytokines regulate cell survival and cell death in the chain reaction with ROS [19]. Other cytokines like transforming growth factor-β (TGF-β) are upregulated upon inflammation [20], and recent evidence suggests that TGF-β signaling may be crucial for repression of inflammatory gene synthesis in healing infarcts mediating resolution of the inflammatory infiltrate. Furthermore, TGF-β may play an important role in modulating fibroblast phenotype and gene expression, promoting extracellular matrix deposition in the infarct by upregulating collagen and fibronectin synthesis, and by decreasing matrix degradation through induction of protease inhibitors [19]. TGF-β is also a key mediator in the pathogenesis of hypertrophic and dilative ventricular remodeling by stimulating cardiomyocyte growth and by inducing interstitial fibrosis [21]. Furthermore, TGF-β has been demonstrated to enhance cardiomyogenesis of mouse embryonic stem cells, thus suggesting that stem cell differentiation requires a paracrine pathway within the heart [22].

Cardiac repair following myocardial injury is restricted because of the limited proliferative potential of adult cardiomyocytes. The ability of mammalian cardiomyocytes to proliferate is lost shortly after birth, as cardiomyocytes withdraw from the cell cycle and differentiate. However, recent research using integration of carbon-14, generated by nuclear bomb tests during the Cold War, into DNA to establish the age of cardiomyocytes in humans revealed that cardiomyocytes indeed renew, with a gradual decrease from 1% turning over annually at the age of 25 to 0.45% at the age of 75. Fewer than 50% of cardiomyocytes are exchanged during a normal life span [23]. In contrast, Hsieh et al. did not find significant cardiac repopulation to occur during normal aging in mice; however, they found cardiomyocyte repopulation, albeit modest, by endogenous progenitors following injury, e.g., during cardiac infarction [24], thus suggesting that cardiac repair and renewal processes may occur through stem cell–mediated cell replacement.

## 3.4 Stem Cells Within the Heart and Potential Redox-Regulated Signaling Pathway Involved in Stem Cell Proliferation and Specification

The cellular basis for the exchange of cardiomyocytes during human life is not yet known but could be comparable to mice because of the mobilization of bone marrow-derived stem cells (BMSC) and/or the activation of resident stem cells in the heart. Several studies on patients have shown that myocardial infarction results in the mobilization of various populations of BMSCs which may be involved in cardiac repair processes [25–28]. Besides BMSCs and circulating multipotent progenitor cells [29], several populations of resident cardiac stem cells have been described during recent years. In the early embryo, progenitor cells in the pharyngeal mesoderm contribute to the rapid growth of the heart tube during looping morphogenesis. These progenitor cells constitute the second heart field and were first identified in 2001 [30]. Side population (SP) cells residing within the adult heart and comprising about 1% of all cells were identified in 2002 by Hierlihy et al., who used the

Hoechst 33342 dye exclusion procedure which was previously used to isolate stem cell populations expressing ATP-binding cassette (ABC) membrane transporters, e.g., P-glycoprotein, which confers multidrug resistance in cancer disease [31]. Upon coculture of SP cells from $GFP^+$ mice with adult cardiac cells from wild type mice, this cell population gained positive α-actinin immunoreactivity, suggesting that a cardiac phenotype was attained [32]. A subpopulation of SP cells comprising approximately 10% of the total SP cells expressing the stem cell marker Sca-1 was identified by Pfister et al. in 2005. This cell population was negative for the endothelial cell marker CD31, expressed Nkx2.5 and GATA-4, but not α-actinin or α-MHC. The cells could be differentiated into a more mature cardiac phenotype upon coculture with ventricular cardiomyocytes [33]. Upon cardiac infarction, the CD31 negative cell population in the heart was depleted within both the infarct and noninfarct areas. SP pools were subsequently reconstituted to baseline levels within seven days after myocardial infarction, both through proliferation of resident SP cells, as well as through homing of BMSCs to specific areas of myocardial injury and immunophenotypic conversion of BMCs to adopt an SP phenotype [34]. Besides the SP cell population, $Sca\text{-}1^+$ c-Kit– cells have been reported to be present in the mouse heart [35], and so-called cardiospheres were isolated by mild enzymatic digestion of mouse and human heart tissues [36]. A further resident stem cell population within the heart are $Isl1^+$ cells, which express the islet-1 (Isl1) LIM homeodomain transcription factor [37]. $Isl1^+$ cells give rise to cardiomyocyte, endothelial, and smooth muscle lineages in vitro and may be involved in embryonic development of the coronary artery tree and in coronary artery growth. Previously it was shown that $Isl1^+$ cells with the transcriptional signature of $Isl1^+/Nkx2.5^+/flk1^+$ define a multipotent cardiovascular progenitor which is capable of differentiating not only into cardiac cells, but also into smooth muscle and endothelial cells, which may participate in coronary artery formation [38]. During embryonic development Isl1 is expressed by progenitor cells of the second heart field, which gives rise to the formation of the outflow tract, the atria, and the right ventricle, and which is required for proliferation, survival, and migration of these progenitors into the forming heart [39]. Isl1 also marks cardiac progenitors found within postnatal hearts of rodents and humans [37]. Recently it has been shown that β-catenin directly regulates Isl1 expression in cardiovascular progenitors and is required for multiple aspects of cardiogenesis [40]. β-catenin is also required upstream of a number of genes required for pharyngeal arch, outflow tract, and/or atrial septal morphogenesis, including *Tbx2*, *Tbx3*, *Wnt11*, *Shh*, and *Pitx2* [40].

The signaling pathways that regulate differentiation of BMSCs and resident cardiac stem cells and/or stimulate proliferation of cardiac progenitor cells are just emerging. Potentially, inflammation and elevation of ROS levels following cardiac infarction are involved in the initiation of signaling pathways that activate quiescent resident cardiac stem cells and BMSCs (see Fig. 3.1). A beneficial effect of proinflammatory signals during bone marrow stem cell therapy has been recently outlined [41]. In the latter study, transplanted BMSCs increased heart tissue inflammation, and elevated TNF-α, TGF-β, and fibroblast growth factor-2 (FGF-2) levels, which resulted in improved heart function and capillary density in the border zone of the myocardial infarct [41]. Many answers on signaling pathways involved in stem cell

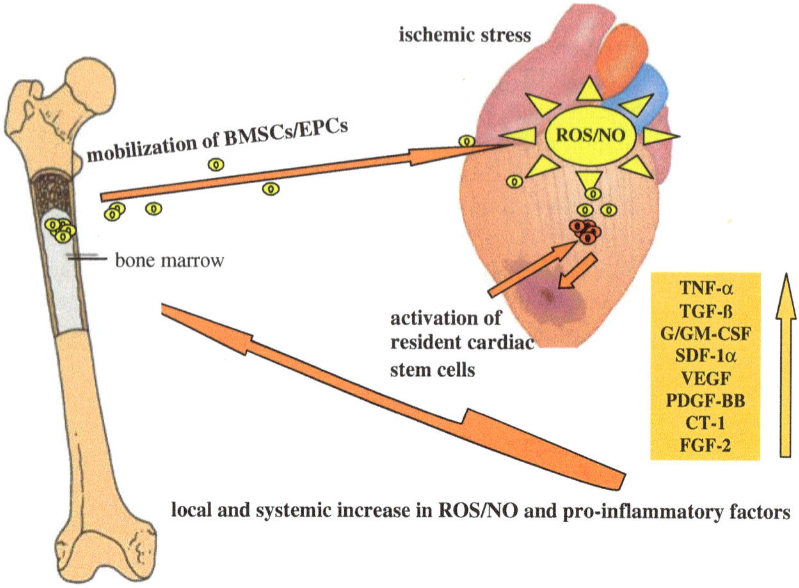


**Fig. 3.1** Under ischemic conditions, e.g., during cardiac infarction, a plethora of inflammatory cytokines as well as growth factors are up-regulated not only within the site of tissue injury but also within the systemic circulation. Key signaling molecules in the upregulation cytokines/growth factors are NO/ROS, which are likewise involved in the activation of different stem cell subtypes. Tissue injury is "sensed" by BMSCs, EPCs, and resident cardiac stem cells, which under conditions of moderate and transient oxidative and nitrosative stress migrate into the injured tissue and initiate cardiovascular repair processes

activation can be given from lessons in cardiac embryology where several signaling pathways that are involved in the development of the first heart field and the second heart field have been recently deciphered [30]. One of the main features of the second heart field is the control of cardiac progenitor cell proliferation. The latter has recently been shown to be regulated by β-catenin, the intracellular mediator of the canonical Wnt pathway, which is likewise known to be involved in the regulation of several stem cell populations [42]. Wnt signaling displays positive as well as negative effects on early mesoderm commitment and cardiac specification, depending on the developmental stage of the embryo [30]. In embryonic stem cells, Wnt signals are required for early mesoderm differentiation [40], whereas during later stages of cardiomyogenesis, Wnt signaling restricts cardiac differentiation to the lateral splanchnic mesoderm [43, 44]. Recently it was shown that the Wnt/β-catenin pathway is essential for cardiac myogenesis to occur in embryonic stem cells, acting at a gastrulation-like stage, mediating mesoderm formation and patterning. Among genes associated temporally with this step was Sox17, encoding an endodermal HMG-box transcription factor [45]. β-catenin interacts with TCF/LEF-1 transcription factors to activate the expression of Wnt target genes. In the absence of Wnt signaling, β-catenin function is blocked by a destruction complex consisting of Axin, APC, and the kinases GSK3ß and CK1α, which targets β-catenin for destruction by the proteasome. Binding of Wnt to its receptors Frizzled and LRP leads to inhibition of the destruction complex and allows β-catenin signaling. The cytoplasmic protein Dishevelled (Dvl) is involved in this process by binding to the redox-sensitive protein nucleoredoxin (NRX), which belongs to the thioredoxin protein family known to be involved in the regulation of a variety of ROS mediated signaling pathways [46]. ROS are presumably involved in a variety of signaling pathways that are crucial for heart development. Recently it was shown that ROS can modulate signaling by the Wnt/β-catenin pathway [47]. Oxidative stress inhibits the interaction between NRX and Dvl, thus stabilizing β-catenin and leading to an increase in the expression of endogenous Wnt target genes. Further studies have demonstrated that ROS can also inhibit Wnt/β-catenin signaling [48], which suggests that a specific time frame and concentration of ROS may be necessary for redox-mediated modulation of the Wnt/β-catenin signaling pathway. Another important pathway known to be crucial for cardiac mesoderm specification and differentiation is the bone morphogenic protein (BMP) pathway. *BMP-4* overexpression promotes a cardiac cell lineage in the cranial mesoderm [49]. BMP-4 is known to be regulated by Wnt/β-catenin and FGF signaling and is involved in outflow tract septation which includes smooth muscle and endocardial cushion development [50]. Furthermore BMP-2, another member of the BMP family is essential for cardiac cushion epithelial-mesenchymal transition and myocardial patterning [51]. Proinflammatory cytokine TNF-α and $H_2O_2$ significantly increased endothelial expression of BMP-2 but not BMP-4, and induced a proinflammatory endothelial phenotype [52]. In further studies, the same group demonstrated that BMP-4 exerts prooxidant, prohypertensive, and proinflammatory effects, but only in the systemic circulation; whereas pulmonary arteries are protected from these adverse effects of BMP-4 [53]. BMP-4 by itself may increase ROS generation, which has been shown in endothelial cells where oscillatory shear stress elevates BMP-4 and induces monocyte adhesion by stimulating ROS production from a Nox-1-based NADPH oxidase [54]. In malformed embryos from diabetic rats which exert elevated levels of systemic ROS, sonic hedgehog homolog (Shh) expression was decreased, and BMP-4 was increased, thus pointing to a redox sensitive regulation of the Shh/BMP-4 pathway. Recently it has been shown that Shh, which is secreted by stem cells in the amphibian intestine, induces BMP-4 in subepithelial fibroblasts, suggesting that both Shh and BMP-4 are involved in the development of the cell-renewable epithelium [55].

## 3.5 Impact of Redox-Regulated Pro-angiogenic Signals During Cardiac Infarction

During cardiac insults, growth factors and cytokines which are involved in the proliferation and differentiation of resident cardiac stem cells towards cardiac cells are upregulated. In addition, the healing of infarction is also grossly dependent on proper revascularization, which may itself depend on redox-mediated expression/release of pro-angiogenic growth factors like FGF-2 [56], VEGF [57], and

PDGF [58], which have been demonstrated to occur after cardiac infarction. Pro-angiogenic factors are also released by monocytes and neutrophils [59] which are migrating to the area of infarction, where they induce the formation of granulation tissue, containing myofibroblasts and neovessels [60]. Increasing angiogenic growth factors in the infarcted hearts has therefore been recently used for cardioprotection and/or to improve cardiac healing [61–66]. Conversely, inhibition of pro-angiogenic signaling, e.g., PDGF-signaling in infarcted hearts of mice, resulted in impaired maturation of the infarct vasculature, enhanced capillary density, and formation of dilated uncoated vessels. Defective vascular maturation in antibody-treated mice was associated with increased and prolonged extravasation of red blood cells and monocyte/macrophages [58]. VEGF is critical for stem cell–mediated cardioprotection, which was shown in experiments where VEGF was downregulated in mesenchymal stem cells by siRNA approaches. When these cells were infused in the coronary circulation, the increase in postischemic myocardial recovery after ischemia reperfusion injury was significantly impaired [67]. Furthermore, bone marrow mesenchymal stem cells by themselves release VEGF as a potentially beneficial paracrine response, which is enhanced by TGF-$\alpha$ and TNF-$\alpha$ [68]. The angiogenic factors VEGF, PDGF-BB, and FGF-2 are all upregulated by exogenous ROS [69, 70] and exert cardioprotective effects under conditions of ischemia-reperfusion injury [64, 71]. Furthermore, VEGF upregulation has also been observed under tissue stress conditions associated with ROS generation, e.g., physical exercise [8] and cardiac infarction where not only the *VEGF* gene but also the VEGF receptors flt-1 and flk-1 were upregulated [72]. Exogenous FGF-2 increased endogenous FGF-2 promoter activity and protein levels in ovine pulmonary arterial smooth muscle cells (PASMC). These increases in FGF-2 expression were mediated by elevations in superoxide levels via NADPH oxidase activation. In addition, FGF-2–mediated increases in FGF-2 expression and PASMC proliferation were attenuated by inhibition of phosphatidylinositol 3-kinase, Akt, and NADPH oxidase [73]. Comparably exogenous ROS increased VEGF and VEGFR expression [74, 75] and stimulated endothelial cell proliferation and migration [76] as well as cytoskeletal reorganization [77] and tubular morphogenesis [78], which all utilize ROS within their signal transduction pathways. The addition of PDGF-BB, FGF-2, and VEGF to nonphagocytic cells has been shown to rapidly increase ROS generation [79], which may likewise occur in stem cells, thus stimulating cardiovascular differentiation. Taken together these data suggest that the inflammatory tissue state following cardiac infarction induces the expression of cardioprotective and pro-angiogenic factors which not only are upregulated through ROS, but are themselves utilizing ROS for proper functioning of their signaling pathways.

## 3.6 Redox-Regulated Pathways Involved in Mobilization of Stem Cells from the Bone Marrow

Stem cells and progenitor cells are mobilized from the bone marrow in response to inflammation, tissue injury, and cytokines [80]. A cytokine playing a prominent role in stem cell mobilization, endothelial cell differentiation, and vascular repair is stromal cell-derived factor-1α (SDF-1α), a CXC chemokine known to play a critical role in the trafficking of hematopoietic, lymphopoietic cells as well as stem cell progenitors, and in maintaining hematopoietic stem cell niches in bone marrow [81]. The high SDF-1α in the bone marrow creates a concentration gradient, which retains hematopoietic stem cells within the stem cell niche. Disruption of this SDF-1α gradient results in mobilization of stem cells into the circulation. This degradation occurs after upregulation of G-CSF levels during systemic stress or injury. Under these conditions elastase is secreted from neutrophils, which cleaves membrane-bound SDF-1/CXCR4 complexes on the surface of bone marrow stem cells in the marrow [82, 83]. SDF-1 is released by stromal cells and binds to its CXCR4 receptor on stem and progenitor cells. The signaling cascade following interaction between SDF-1 and CXCR4 may involve the generation of ROS. This has been recently evidenced in studies on B-lymphocytes in which ROS were involved in CXCR4-induced Akt activation [84]. If high concentration gradients of circulating SDF-1 exist, CXCR4-positive cells are leaving the bone marrow to be directed to sites of tissue injury. During tissue damage, ischemia, and inflammation, plasma and tissue levels of SDF-1α are upregulated [85]. Consequently SDF-1α expression is significantly upregulated in experimental rat and mouse models of infarction [86], and in the plasma and cardiac tissue of patients with myocardial infarction [87]. Furthermore, SDF-1α expression has been shown to increase under hypoxic conditions [88], and thus may serve to attract stem cells to sites of tissue injury and ischemia. Recently it has been shown that expression of SDF-1α on circulating platelets is increased in patients with acute coronary syndrome and correlates with the number of $CD34^+$ progenitor cells [89]. Expression of SDF-1α appears to be correlated to the expression of eNOS in the heart since $eNOS^{-/-}$ mice displayed reduced SDF-1α levels in isolated cardiomyocytes. eNOS in the host myocardium promoted mesenchymal stem cell migration to the ischemic myocardium and improved cardiac function through cGMP-dependent increases in SDF-1α expression [90]. The local inflammatory response implying adhesion molecule expression and eNOS-dependent signaling was required for SDF-1α-induced adhesion of $c\text{-}kit^+$ cells to the vascular endothelium [91]. Furthermore, oxidative stress from lactate metabolism by circulating stem/progenitor cells accelerated further stem cell recruitment and differentiation through thioredoxin-1 (Trx1)–mediated elevations in hypoxia-inducible factor–1 (HIF-1) levels and the subsequent synthesis of HIF-1–dependent growth factors, including VEGF and SDF-1α [92]. Taken together, these data suggest a model in which, in response to tissue injury and inflammation, stem cells within the bone marrow are expanded and primed through G-CSF, which then results in mobilization of stem cells via degradation of SDF-1α in the marrow and recruitment of the stem cells to sites of elevated SDF-1α levels within the injured, inflamed, or ischemic tissues. Mobilization is then terminated when the increased SDF-1α gradient in the marrow is re-established, and retains newly formed or nonmobilized stem cells as a reserve for future emergency signals [93]. Interestingly, G-CSF stimulation induced ROS generation in bone marrow neutrophils correlating with activation of Lyn, PI3-kinase, and Akt; whereas the antioxidant N-acetyl cysteine diminished G-CSF–induced ROS production and cell proliferation [94]. Further research on the priming function of G-CSF on ROS generation by neutrophils revealed that mitogen-activated protein kinase (MAPK) pathways are involved in the phosphorylation of Ser345 of p47phox, a cytosolic component of NADPH oxidase in human neutrophils [95]. Previously it was shown that several hematopoietic growth factors including G-CSF signal through the formation of ROS [96], which has been associated with a stimulation of cell proliferation of hematopoetic stem cells upon treatment with G-CSF [97]. Furthermore, the blood oxidative status was found to be significantly increased in healthy hematopoetic stem cell donors receiving G-CSF, which indicated that during stem cell mobilization a transient inflammatory status is generated [98], which may facilitate further stem cell mobilization. ROS-mediated stem cell mobilization and recruitment may be used in therapeutic angiogenesis approaches. In this respect, hyperbaric oxygen has been shown to stimulate recruitment and differentiation of circulating stem/progenitor cells in subcutaneous Matrigel which was inhibited by antagonists of NADPH oxidase and free radical scavengers [99].

Mostly, ROS elicited by growth factor and cytokine signaling act only within a narrow time window. Recently the interesting concept of the redox window of coronary collateral growth was formulated. This concept suggests that the redox window constitutes a range in the redox state of cells, which not only is permissive for the actions of growth factors but amplifies their actions as well. Initial changes in cellular redox arise from different events, e.g., from the oxidative burst during reperfusion following ischemia, to recruitment of various types of inflammatory cells capable of producing ROS. Any event that upsets the normal redox equilibrium is capable of amplifying growth. However, extremes of the redox window, oxidative and reductive stresses, are associated with diminished growth factor signaling and reduced activation of redox-dependent kinases [100]. Previously the same group had demonstrated that ROS are involved in human coronary artery endothelial cell (HCAEC) tube formation, coronary collateral growth in vivo, and signaling (p38 MAP kinase), by which ROS may stimulate vascular growth [100].

## 3.7 NO and ROS in EPC Mobilization and Function

EPCs derived from the bone marrow and released to the blood stream have been identified as an important source of vascular cells that may be potentially involved in cardiac repair and neovascularization of ischemic tissue [101]. EPCs are a subset of BMSCs that can readily differentiate into mature endothelial cells under appropriate micro-environmental stimulations. Asahara et al. [102] promoted a novel paradigm, referred to as postnatal vasculogenesis, when they reported that progenitor cells for the endothelial lineage could be found in the circulation of human subjects and rodents, and that the cells displayed the ability to localize to areas of vascular ischemia in vivo. After an ischemic injury such as myocardial infarction or unstable angina, or angioplastic balloon endothelial denudation, more EPCs are detected in the circulating blood [103, 104]. However, the efficiency of participation of EPCs in vascular repair and neoangiogenesis in the heart is still a matter of debate [105].

The stem cell niche which represents the local microenvironment of fibroblasts, osteoblasts, and endothelial cells within the bone marrow plays a critical cue for the mobilization of EPCs [106]. Mobilization of EPCs occurs upon stimulation by cytokines, which alter the interaction between stem cells and bone marrow stromal cells, thus allowing the stem cells to disengage the bone marrow and to pass through the sinusoidal endothelium to enter the blood stream [80, 107]. The Wnt signaling antagonist Dickkopf (Dkk)-1 is involved in the mobilization of vasculogenic progenitor cells. Using TOP-GAL transgenic mice to determine activation of β-catenin, it was demonstrated that Dkk-1 regulates endosteal cells in the bone marrow stem cell niche and subsequently mobilizes vasculogenic and hematopoietic progenitor cells without concomitant mobilization of inflammatory neutrophils. The mobilization of vasculogenic progenitors requires the presence of functionally active osteoclasts, as demonstrated in PTPepsilon-deficient mice with defective osteoclast function. Dkk-1 induced the osteoclast differentiation factor RANKL, which subsequently stimulated the release of the major bone-resorbing protease cathepsin K [108]. Mobilization of EPCs is induced by physiological and pathophysiological events via a variety of growth factors, hormones, and cytokines, including VEGF [109], SDF-1α [110], PDGF-CC [111], brain-derived neurotrophic factor (BNDF) [112], placental growth factor (PIGF) [113], as well as the hormones estrogen [114] and erythropoietin [115]. Recently it has been shown that pretreatment of mice with VEGF did not disrupt the CXCR4/SDF-1alpha chemokine axis, but stimulated entry of hematopoietic stem cells into the cell cycle via VEGFR1, reducing their migratory capacity in vitro and suppressing their mobilization in vivo. In contrast, VEGF pretreatment enhanced EPC mobilization via VEGFR2 in response to CXCR4 antagonism. Stromal progenitor cell (SPC) mobilization was detected when the CXCR4 antagonist was administered to mice pretreated with VEGF, but not G-CSF. The authors suggested that differential mobilization of progenitor cell subsets is dependent upon the cytokine milieu that regulates cell retention and proliferation [109].

The mobilization of EPCs appears to be closely linked to NO availability. The EPC mobilization cascade starts with peripheral hypoxia-induced tissue release of VEGF-A and the subsequent activation of bone marrow stromal NOS, resulting in increased bone marrow NO levels [103]. In this process, eNOS is essential in the bone marrow microenvironment, and increases in bone marrow NO levels result in the mobilization of EPCs from bone marrow niches to circulation, ultimately allowing for their participation in tissue-level vasculogenesis and wound healing [116]. At the tissue level, EPC recruitment depends on ischemia-induced upregulation of SDF-1α [88]. Defective mobilization of EPCs in response to different stimuli such as estrogen [114] has been obtained in $eNOS^{-/-}$ knockout mice supporting a role of NO in stem cell mobilization. In these mice VEGF, statins, exercise, and estrogen failed to mobilize EPCs. Furthermore, NO is involved in the mobilization of EPCs by SDF-1α, which acts via an enhancement of protein kinase B (Akt) and eNOS activity [110]. One of the key features for EPC mobilization is tissue inflammation, which occurs in the sequence of cardiac infarction and vascular trauma. Inflammatory cytokines released during trauma, sepsis [117], bone fracture healing [118], and cardiac infarction [119] stimulate EPC mobilization (see Fig. 3.1). A further stimulus for EPC mobilization is physical exercise of mice [120], healthy humans [121], and patients with cardiovascular disease [122], which has been utilized to support cardiac rehabilitation [123]. Under these conditions, increased NO levels in the peripheral blood were observed [120, 123] in addition to the well-documented feature that both resting and contracting skeletal muscles produce ROS [124]. A possible role of NO for stem cell mobilization was evidenced in studies where pressure-induced, cardiac overload–induced upregulation of EPCs was abolished in eNOS double-knockout mice [125]. The notion that ROS/NO are the mediators of the beneficial effects for human health during physical exercise was recently substantiated by the observation that antioxidants prevent the health promoting effects of exercise in humans, which suggests that a transient generation of ROS is necessary, e.g., for ameliorating insulin resistance in diabetic patients [126]. Recently it has been pointed out that a transient restricted inflammatory response possibly associated with low levels of ROS generation may constitute a stimulus for EPC mobilization, whereas persistent or excessive inflammatory stimuli may have deleterious effects, resulting in decreased EPCs in the circulation [127, 128]. Low levels of ROS have been implicated in bone marrow and progenitor cell function in a hindlimb ischemia model. In this study, it was shown that hindlimb ischemia in mice significantly increased Nox-2 expression and ROS generation in bone marrow-mononuclear cells, which was associated with an increase in circulating EPC-like cells. Mice lacking Nox-2 showed reduction of ischemia-induced flow recovery, and Nox-2 deficient c-kit$^{+}$/Lin$^{-}$ bone marrow stem/progenitor cells displayed impaired chemotaxis and invasion in response to SDF-1α [129]. In the early postinfarction period a reduced EPC mobilization was observed, which was correlated to increased oxidative stress within the bone marrow and impaired MMP-9 activity [130]. Recently it was shown that enhanced mechanical stretch in renovascular hypertension induces EPC mobilization in a p47phox-dependent manner, involving bone marrow SDF-1α and MMP-9, thus suggesting a role of NADPH oxidase in EPC mobilization [131]. ROS may also be involved in hemin-induced neovascularization at the sites of hematoma formation, since it has been shown that hemin promotes proliferation and differentiation of EPCs via activation of AKT and ERK and elevation of intracellular ROS levels [132]. However, under conditions of high and chronic stress and inflammation, e.g., under conditions of hypertension, hypercholesterolemia, diabetes, and cigarette smoking, EPC numbers and function are severely impaired [133]. These disease states are associated with excessive and longlasting oxidative stress, which may either exhaust EPC mobilization from the bone marrow, or may accelerate EPC aging and EPC function. In vitro oxidant treatment decreased the clonogenic capacity of EPCs, increased apoptosis, and diminished tube-forming ability in vitro and in vivo in response to oxidative stress, which was directly linked to activation of a redox-dependent stress-induced kinase pathway [134]. Standard post–myocardial infarction drugs, such as angiotensin converting enzyme (ACE) and 3-hydroxy-3-methylglutaryl-coenzyme A (HMG-CoA) reductase inhibitors (statins), increase levels of EPCs [135, 136], presumably by decreasing intracellular ROS, elevating NO levels, and enhancing MMP9 activity. Conversely, asymmetric dimethylarginine (ADMA), an endogenous inhibitor of NO-synthases, decreased EPC mobilization. It was demonstrated that the plasma concentration of ADMA was related to the severity of coronary artery disease and correlated inversely with the number of circulating $CD34^+/CD133^+$ progenitor cells and endothelial colony forming units [137]. Altered stem cell differentiation towards inflammatory cells such as macrophages was observed during hyperglycemia, but could be reversed by treatment with statins, which are known to exert antioxidant properties [138, 139].

## 3.8 ROS and NO Generation in Bone Marrow–Derived Stem Cells

Besides the role of ROS and NO generated during states of tissue inflammation, ischemia and injury stem cells per se are generating ROS as well as NO, which may be involved in proliferation and differentiation processes. ROS and NO generation in stem cells could occur in response to transient changes in systemic redox balance and could initiate a feedforward cycle of ROS/NO generation and elaboration of a balanced antioxidative response system that may be the basis of stem cell proliferation, migration, and differentiation. An increasing number of studies has reported on the crucial role of ROS/NO for mesenchymal stem cell differentiation. It was shown that neuronal differentiation of mesenchymal stem cells involved upregulation of NADPH oxidase and increased ROS generation [140]. Furthermore, physical shockwave treatment was shown to increase osteogenic activity of human umbilical cord blood (HUCB) mesenchymal progenitor cells through superoxide-mediated TGF-β1 induction [141]. ROS generation through the activity of the Nox-2 and Nox-4 isoform of NADPH oxidase has been demonstrated in human $CD34^+$ cells, which may contribute to the activation of intracellular signaling pathways leading to mitochondriogenesis, cell survival, and differentiation in hematopoietic stem cells [142]. In the latter study, the authors suggest that the coordinated activity of the Nox isoforms in hematopoietic stem cells functions as an environmental oxygen sensor and generates low levels of ROS, which likely serve as second messengers. The pro-oxidant setting, entering into play when hematopoietic stem and progenitor cells leave the hypoxic bone marrow niche, would enable them to be more responsive to proliferative/differentiative stimuli. Moreover, it is suggested that enhanced ROS elicit mitochondrial "differentiation" in a precommitment phase needed to match the bioenergetic request in the oncoming proliferation/differentiation process [143]. Mesenchymal stem cells from the bone marrow have been shown to express iNOS [144] as well as eNOS [145]. Recently it was shown that hematopoietic stem cell development is dependent on blood flow and is closely associated to NO generation, since intrauterine NOS inhibition or embryonic eNOS deficiency resulted in a reduction of hematopoietic clusters and transplantable murine hematopoietic stem cells [146]. Generation of NO by eNOS has been reported in mouse EPCs and was utilized to identify the EPC population [147]. Administration of Angiotensin II (Ang II) significantly promoted NO release, inhibited EPC apoptosis, and enhanced EPC adhesion potential [148]. In a recent study it was shown that two NO agents (SNAP and DEA/NO), able to activate both cGMP-dependent and -independent pathways, were increasing the cardiomyogenic potential of bone marrow–derived mesenchymal stem cells and adipose tissue–derived stem cells (ADSCs) [149]. In contrast to the knowledge about the NO requirement for EPC mobilization and function, evidence about the expression of ROS generating enzymes in EPCs is scarce. In vitro, Nox-2-deficient c-kit$^+$Lin$^-$ bone marrow stem/progenitor cells were shown to display impaired chemotaxis and invasion, as well as polarization of actins in response to SDF-1α, which is associated with blunted SDF-1α-mediated phosphorylation of Akt [129]. Previously it was shown that Ang II accelerates EPC senescence by oxidative stress through peroxynitrite, suggesting an interplay between ROS and NOS. The authors demonstrated that Ang II increased the expression of gp91phox mRNA and protein in a dose-dependent manner, which was attenuated by the Ang II type 1 (AT1) receptor antagonist valsartan [150].

## 3.9 ROS and NO in Cardiovascular Differentiation of Embryonic Stem Cells

Most evidence about the role of NO and ROS in cardiovascular differentiation has been obtained in mouse embryonic stem cells. It was shown that undifferentiated self-renewing stem cells are devoid of endogenous ROS generation and expression of NADPH oxidase. Undifferentiated embryonic stem cells were demonstrated to be equipped with highly efficient mechanisms to defend themselves against various stresses and to prevent or repair DNA damage. One of these mechanisms is high activity of a verapamil-sensitive multidrug efflux pump. During the differentiation process, antioxidative genes are downregulated, which should result in increased ROS generation [151]. Consequently, during the differentiation process the gp91-phox homologues Nox-1, Nox-2, and Nox-4 are upregulated in a distinct time frame, starting with Nox-1 and followed by Nox-4 [152]; whereas Nox-2 is closely correlated to the differentiation of phagocytic cells from embryonic stem cells, which occurs subsequent to cardiovascular differentiation [153]. During the early stages of embryonic stem cell differentiation, i.e., between day 4 and day 10 of cell culture, ROS generation is elevated and downregulated during later stages. The stages of active ROS generation are just those where cardiovascular differentiation occurs, i.e., between day 4 and day 9 of cell culture. Any approaches to increase intracellular ROS, e.g., by the addition of nanomolar concentrations of $H_2O_2$ to differentiating embryoid bodies [152, 154, 155], treatment with direct current electrical fields [154, 156], application of mechanical strain [157], treatment with cardiotrophin-1 (CT-1)

[10], PDGF-BB [9], or peroxisome proliferator-activated receptor α (PPARα) [158] resulted in prominent stimulation of cardiovascular differentiation of embryonic stem cells. Interestingly, elevation of intracellular ROS by exogenous stimulators resulted in upregulation of Nox-1 and Nox-4, thus initiating a feed-forward stimulation of prolonged ROS generation [152, 157]. Consequently, siRNA inactivation of Nox-4 resulted in complete inhibition of embryonic stem cell–derived cardiomyogenesis [159]. Stimulation of ROS generation by different means resulted in activation of the MAPK pathways ERK1,2, p38 and JNK. Furthermore, stimulation of embryoid bodies by ROS resulted in activation of the cardiogenic transcription factors BMP-10, MEF2C, GATA-4, DTEF-1, and Nkx-2.5 [152]. Interestingly, vasculogenesis required activation of ERK1,2 and JNK, whereas p38 activation was dispensable. Cardiomyogenesis, however, required the activation of all three pathways, since pharmacological inhibition of either pathway abolished cardiac cell differentiation [157] (see Fig. 3.2). When cardiomyogenesis was stimulated with CT-1, activation of NF-κB and the JAK/STAT signaling pathway in a redox-sensitive manner was additionally observed [10]. CT-1 has been previously shown to exert cardioprotective effects, which may be related to the activation of anti-apoptotic

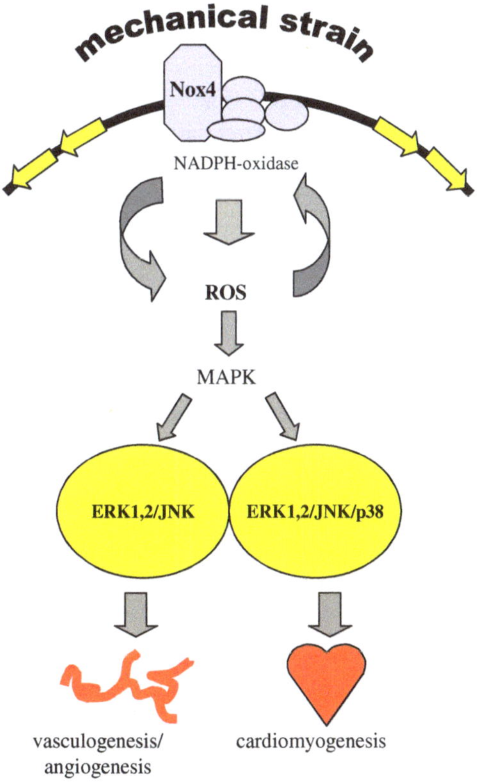


**Fig. 3.2** Diagram of the involvement of ROS in signalling cascades resulting in cardiovascular commitment of ES cells. ROS are generated through the activity of a presumably membrane-bound NADPH oxidase that is upregulated, e.g., following mechanical strain application. ROS initiate phosphorylation of the MAPKs ERK1,2, JNK, and p38. Vasculogenesis/angiogenesis requires the activation of ERK1,2 and JNK, whereas the activity of ERK1,2, JNK, and p38 is necessary for cardiomyogenesis

signaling pathways [160]. CT-1 is expressed in the post-myocardial infarct heart, and may play an important role in infarct scar formation and ongoing remodeling of the scar [161]. Furthermore, CT-1 is a cytokine that induces hypertrophy and has been shown to be increased in hypertensive patients [162]. An additional role of CT-1 may involve the activation and differentiation of resident cardiac stem cells. In this respect, it has been recently shown that CT-1 signaling through glycoprotein 130 (gp130) regulates the endothelial differentiation of cardiac stem cells [163]. Recently CT-1 in combination with 5-azacytidine, which is an inhibitor of DNA methylation, was shown to induce cardiac gene expression in mesenchymal stem cells [164]. In human embryonic stem cells, telomere maintenance, oxidative stress generation, and genes involved in antioxidant defense and DNA repair were investigated during spontaneous differentiation of two human embryonic stem cell lines. Telomerase activity was quickly downregulated during differentiation, probably because of deacetylation of histones H3 and H4 at the hTERT promoter, and deacetylation of histone H3 at the hTR promoter. Telomere length decreased accordingly. Mitochondrial superoxide production and cellular levels of ROS increased as a result of stimulated mitochondrial biogenesis. The expression of major antioxidant genes was downregulated despite this increased oxidative stress. DNA damage levels increased during differentiation, whereas the expression of genes involved in different types of DNA repair decreased [165]. Besides the evident role of ROS for cardiovascular differentiation, a prominent involvement of NO in cardiomyogenesis of embryonic stem cells has been evidenced. In murine undifferentiated embryonic stem cells, NOS-1, NOS-3, and sGCβ(1) were detected, while NOS-2, sGCα(1), and PKG were very low or undetectable. When embryonic stem cells were subjected to differentiation, NOS-1 abruptly decreased within one day, NOS-2 mRNA became detectable after several days, and NOS-3 increased after 7–10 days [166]. Components of NO signaling were likewise expressed in human embryonic stem cells [166]. Nkx2.5 and myosin light chain (MLC2) mRNA expression was increased on exposure of mouse and human embryonic stem cells to NO donors, and a decrease in mRNA expression of both cardiac-specific genes was observed with nonspecific NOS inhibitor [167]. In several studies it was reported that NO is acting as a signaling molecule during cardiomyogenesis of embryonic stem cells [167–169]. Studies on NO generating agents revealed that sGC activators alone exhibited an increase in mRNA expression of cardiac genes (MLC2 and Nkx2.5). Robust inductions of mRNA and protein expression of marker genes were observed when NO donors and sGC activators were combined. Measurement of NO metabolites demonstrated an increase in the nitrite levels in the conditioned media and cell lysates on exposure of cells to the different concentrations of NO donors. cGMP analysis in undifferentiated stem cells revealed a lack of stimulation with NO donors. Differentiated cells however, acquired the ability to be stimulated by NO donors [167]. Generation of NO is apparently also the mediator of cardiomyogenesis of mouse embryonic stem cells achieved with the hormone oxytocin [125] and arginine vasopressin [170]. Hence these data suggest that an interplay of ROS and NO is required to direct undifferentiated embryonic stem cells into the cardiovascular cell lineage.

## 3.10 Summary and Conclusions

Tissue injury results in upregulation of ROS and NO, which not only mediate the expression of pro-inflammatory growth factors and cytokines, but are also involved in the regulation of a variety of signaling pathways regulating cell differentiation, proliferation, and apoptotic cell death. The bone marrow as well as various tissues including the heart contain stem cells that may be mobilized and activated following tissue injury and tissue inflammation. Although the mediators of inflammation as well as the pathophysiological changes in the cardiac microenvironment during cardiovascular disease are well known, their significance for the differentiation of the various stem cell species resident in the organs is not well established. Reviewing the literature, it becomes apparent that virtually all effectors of tissue injury and inflammation exert stimulatory effects on the cardiovascular differentiation of stem cells. Almost all signaling pathways involve ROS/NO, which activate differentiation signals by currently unknown means. Most studies so far have focused their investigations on either NO or ROS. However, the well known interaction of NO and ROS to form peroxynitrite has been frequently neglected. This is so much the more disadvantageous since it has been recently pointed out that as part of the normal physiological process, superoxide anion and NO function separately and interactively as second messengers [171].

NO and ROS release may occur at a distinct site of anatomical localization within cells and organs, which may influence stem cell differentiation patterns. Therefore, future research has to unravel the time-concentration– and time-location–dependent changes in ROS/NO occurring during cardiovascular disease in order to estimate the critical concentrations, time durations, and sites of action of oxidative and nitrosative stress that is determining the balance between cell injury and tissue destruction versus stem cell activation and tissue repair.

## References

1. Allen RG, Balin AK (1989) Oxidative influence on development and differentiation: an overview of a free radical theory of development. Free Radic Biol Med 6:631–661
2. Liaudet L, Vassalli G, Pacher P (2009) Role of peroxynitrite in the redox regulation of cell signal transduction pathways. Front Biosci 14:4809–4814
3. Schneider H (2009) Tolerance of human placental tissue to severe hypoxia and its relevance for dual ex vivo perfusion. Placenta 30(Suppl A):S71–S76
4. Webster WS, Abela D (2007) The effect of hypoxia in development. Birth Defects Res C Embryo Today 81:215–228
5. Kovacic P, Somanathan R (2006) Mechanism of teratogenesis: electron transfer, reactive oxygen species, and antioxidants. Birth Defects Res C Embryo Today 78:308–325
6. Gagioti S, Colepicolo P, Bevilacqua E (1995) Post-implantation mouse embryos have the capability to generate and release reactive oxygen species. Reprod Fertil Dev 7: 1111–1116
7. Raijmakers MT, Burton GJ, Jauniaux E et al (2006) Placental NAD(P)H oxidase mediated superoxide generation in early pregnancy. Placenta 27:158–163

8. Roy S, Khanna S, Sen CK (2008) Redox regulation of the VEGF signaling path and tissue vascularization: hydrogen peroxide, the common link between physical exercise and cutaneous wound healing. Free Radic Biol Med 44:180–192
9. Lange S, Heger J, Euler G et al (2009) Platelet-derived growth factor BB stimulates vasculogenesis of embryonic stem cell-derived endothelial cells by calcium-mediated generation of reactive oxygen species. Cardiovasc Res 81:159–168
10. Sauer H, Neukirchen W, Rahimi G et al (2004) Involvement of reactive oxygen species in cardiotrophin-1-induced proliferation of cardiomyocytes differentiated from murine embryonic stem cells. Exp Cell Res 294:313–324
11. Suzukawa K, Miura K, Mitsushita J et al (2000) Nerve growth factor-induced neuronal differentiation requires generation of Rac1-regulated reactive oxygen species. J Biol Chem 275:13175–13178
12. Schwamborn JC, Fiore R, Bagnard D et al (2004) Semaphorin 3A stimulates neurite extension and regulates gene expression in PC12 cells. J Biol Chem 279:30923–30926
13. Schnabel D, Salas-Vidal E, Narvaez V et al (2006) Expression and regulation of antioxidant enzymes in the developing limb support a function of ROS in interdigital cell death. Dev Biol 291:291–299
14. Akki A, Zhang M, Murdoch C et al (2009) NADPH oxidase signaling and cardiac myocyte function. J Mol Cell Cardiol 47:15–22
15. Anilkumar N, Sirker A, Shah AM (2009) Redox sensitive signaling pathways in cardiac remodeling, hypertrophy and failure. Front Biosci 14:3168–3187
16. Hori M, Nishida K (2009) Oxidative stress and left ventricular remodelling after myocardial infarction. Cardiovasc Res 81:457–464
17. Di Lisa F, Canton M, Menabo R et al (2007) Mitochondria and cardioprotection. Heart Fail Rev 12:249–260
18. Webster KA, Graham RM, Thompson JW et al (2006) Redox stress and the contributions of BH3-only proteins to infarction. Antioxid Redox Signal 8:1667–1676
19. Frangogiannis NG (2008) The immune system and cardiac repair. Pharmacol Res 58:88–111
20. Czarkowska P, Przybylski J, Marciniak A et al (2004) Proteolytic enzymes activities in patients after myocardial infarction correlate with serum concentration of TGF-beta. Inflammation 28:279–284
21. Ellmers LJ, Scott NJ, Medicherla S et al (2008) Transforming growth factor-beta blockade down-regulates the renin-angiotensin system and modifies cardiac remodeling after myocardial infarction. Endocrinology 149:5828–5834
22. Behfar A, Zingman LV, Hodgson DM et al (2002) Stem cell differentiation requires a paracrine pathway in the heart. FASEB J 16:1558–1566
23. Bergmann O, Bhardwaj RD, Bernard S et al (2009) Evidence for cardiomyocyte renewal in humans. Science 324:98–102
24. Hsieh PC, Segers VF, Davis ME et al (2007) Evidence from a genetic fate-mapping study that stem cells refresh adult mammalian cardiomyocytes after injury. Nat Med 13:970–974
25. Leone AM, Rutella S, Bonanno G et al (2005) Mobilization of bone marrow-derived stem cells after myocardial infarction and left ventricular function. Eur Heart J 26:1196–1204
26. Leone AM, Rutella S, Bonanno G et al (2006) Endogenous G-CSF and CD34+ cell mobilization after acute myocardial infarction. Int J Cardiol 111:202–208
27. Wojakowski W, Tendera M, Kucia M et al (2009) Mobilization of bone marrow-derived Oct-4+ SSEA-4+ very small embryonic-like stem cells in patients with acute myocardial infarction. J Am Coll Cardiol 53:1–9
28. Wojakowski W, Tendera M, Zebzda A et al (2006) Mobilization of CD34(+), CD117(+), CXCR4(+), c-met(+) stem cells is correlated with left ventricular ejection fraction and plasma NT-proBNP levels in patients with acute myocardial infarction. Eur Heart J 27:283–289
29. Cesselli D, Beltrami AP, Rigo S et al (2009) Multipotent progenitor cells are present in human peripheral blood. Circ Res 104:1225–1234

30. Rochais F, Mesbah K, Kelly RG (2009) Signaling pathways controlling second heart field development. Circ Res 104:933–942
31. Hierlihy AM, Seale P, Lobe CG et al (2002) The post-natal heart contains a myocardial stem cell population. FEBS Lett 530:239–243
32. Martin CM, Meeson AP, Robertson SM et al (2004) Persistent expression of the ATP-binding cassette transporter, Abcg2, identifies cardiac SP cells in the developing and adult heart. Dev Biol 265:262–275
33. Pfister O, Mouquet F, Jain M et al (2005) CD31– but not CD31+ cardiac side population cells exhibit functional cardiomyogenic differentiation. Circ Res 97:52–61
34. Mouquet F, Pfister O, Jain M et al (2005) Restoration of cardiac progenitor cells after myocardial infarction by self-proliferation and selective homing of bone marrow-derived stem cells. Circ Res 97:1090–1092
35. Tallini YN, Greene KS, Craven M et al (2009) c-kit expression identifies cardiovascular precursors in the neonatal heart. Proc Natl Acad Sci USA 106:1808–1813
36. Messina E, De Angelis L, Frati G et al (2004) Isolation and expansion of adult cardiac stem cells from human and murine heart. Circ Res 95:911–921
37. Laugwitz KL, Moretti A, Lam J et al (2005) Postnatal isl1+ cardioblasts enter fully differentiated cardiomyocyte lineages. Nature 433:647–653
38. Moretti A, Caron L, Nakano A et al (2006) Multipotent embryonic isl1+ progenitor cells lead to cardiac, smooth muscle, and endothelial cell diversification. Cell 127: 1151–1165
39. Cai CL, Liang X, Shi Y et al (2003) Isl1 identifies a cardiac progenitor population that proliferates prior to differentiation and contributes a majority of cells to the heart. Dev Cell 5:877–889
40. Lin L, Cui L, Zhou W et al (2007) Beta-catenin directly regulates Islet1 expression in cardiovascular progenitors and is required for multiple aspects of cardiogenesis. Proc Natl Acad Sci USA 104:9313–9318
41. Sun J, Li SH, Liu SM et al (2009) Improvement in cardiac function after bone marrow cell thearpy is associated with an increase in myocardial inflammation. Am J Physiol Heart Circ Physiol 296:H43–H50
42. Reya T, Clevers H (2005) Wnt signalling in stem cells and cancer. Nature 434:843–850
43. Tzahor E, Lassar AB (2001) Wnt signals from the neural tube block ectopic cardiogenesis. Genes Dev 15:255–260
44. Nakamura T, Sano M, Songyang Z et al (2003) A Wnt- and beta-catenin-dependent pathway for mammalian cardiac myogenesis. Proc Natl Acad Sci USA 100:5834–5839
45. Liu Y, Asakura M, Inoue H et al (2007) Sox17 is essential for the specification of cardiac mesoderm in embryonic stem cells. Proc Natl Acad Sci USA 104:3859–3864
46. Korswagen HC (2006) Regulation of the Wnt/beta-catenin pathway by redox signaling. Dev Cell 10:687–688
47. Funato Y, Michiue T, Asashima M et al (2006) The thioredoxin-related redox-regulating protein nucleoredoxin inhibits Wnt-beta-catenin signalling through dishevelled. Nat Cell Biol 8:501–508
48. Shin SY, Kim CG, Jho EH et al (2004) Hydrogen peroxide negatively modulates Wnt signaling through downregulation of beta-catenin. Cancer Lett 212:225–231
49. Tirosh-Finkel L, Elhanany H, Rinon A et al (2006) Mesoderm progenitor cells of common origin contribute to the head musculature and the cardiac outflow tract. Development 133:1943–1953
50. Liu W, Selever J, Wang D et al (2004) Bmp4 signaling is required for outflow-tract septation and branchial-arch artery remodeling. Proc Natl Acad Sci USA 101:4489–4494
51. Ma L, Lu MF, Schwartz RJ et al (2005) Bmp2 is essential for cardiac cushion epithelial-mesenchymal transition and myocardial patterning. Development 132:5601–5611
52. Csiszar A, Ahmad M, Smith KE et al (2006) Bone morphogenetic protein-2 induces proinflammatory endothelial phenotype. Am J Pathol 168:629–638

53. Csiszar A, Labinskyy N, Jo H et al (2008) Differential proinflammatory and prooxidant effects of bone morphogenetic protein-4 in coronary and pulmonary arterial endothelial cells. Am J Physiol Heart Circ Physiol 295:H569–H577
54. Sorescu GP, Song H, Tressel SL et al (2004) Bone morphogenic protein 4 produced in endothelial cells by oscillatory shear stress induces monocyte adhesion by stimulating reactive oxygen species production from a Nox-1-based NADPH oxidase. Circ Res 95:773–779
55. Ishizuya-Oka A, Hasebe T (2008) Sonic hedgehog and bone morphogenetic protein-4 signaling pathway involved in epithelial cell renewal along the radial axis of the intestine. Digestion 77(Suppl 1):42–47
56. Detillieux KA, Sheikh F, Kardami E et al (2003) Biological activities of fibroblast growth factor-2 in the adult myocardium. Cardiovasc Res 57:8–19
57. Wojakowski W, Maslankiewicz K, Ochala A et al (2004) The pro- and anti-inflammatory markers in patients with acute myocardial infarction and chronic stable angina. Int J Mol Med 14:317–322
58. Zymek P, Bujak M, Chatila K et al (2006) The role of platelet-derived growth factor signaling in healing myocardial infarcts. J Am Coll Cardiol 48:2315–2323
59. Lambert JM, Lopez EF, Lindsey ML (2008) Macrophage roles following myocardial infarction. Int J Cardiol 130:147–158
60. Nahrendorf M, Swirski FK, Aikawa E et al (2007) The healing myocardium sequentially mobilizes two monocyte subsets with divergent and complementary functions. J Exp Med 204:3037–3047
61. Lahteenvuo JE, Lahteenvuo MT, Kivela A et al (2009) Vascular endothelial growth factor-B induces myocardium-specific angiogenesis and arteriogenesis via vascular endothelial growth factor receptor-1- and neuropilin receptor-1-dependent mechanisms. Circulation 119:845–856
62. Zhang J, Ding L, Zhao Y et al (2009) Collagen-targeting vascular endothelial growth factor improves cardiac performance after myocardial infarction. Circulation 119:1776–1784
63. Harada K, Grossman W, Friedman M et al (1994) Basic fibroblast growth factor improves myocardial function in chronically ischemic porcine hearts. J Clin Invest 94:623–630
64. Hsieh PC, MacGillivray C, Gannon J et al (2006) Local controlled intramyocardial delivery of platelet-derived growth factor improves postinfarction ventricular function without pulmonary toxicity. Circulation 114:637–644
65. Hsieh PC, Davis ME, Gannon J et al (2006) Controlled delivery of PDGF-BB for myocardial protection using injectable self-assembling peptide nanofibers. J Clin Invest 116:237–248
66. House SL, Bolte C, Zhou M et al (2003) Cardiac-specific overexpression of fibroblast growth factor-2 protects against myocardial dysfunction and infarction in a murine model of low-flow ischemia. Circulation 108:3140–3148
67. Markel TA, Wang Y, Herrmann JL et al (2008) VEGF is critical for stem cell-mediated cardioprotection and a crucial paracrine factor for defining the age threshold in adult and neonatal stem cell function. Am J Physiol Heart Circ Physiol 295:H2308–H2314
68. Wang Y, Crisostomo PR, Wang M et al (2008) TGF-alpha increases human mesenchymal stem cell-secreted VEGF by MEK- and PI3-K- but not JNK- or ERK-dependent mechanisms. Am J Physiol Regul Integr Comp Physiol 295:R1115–R1123
69. Sen CK, Khanna S, Babior BM et al (2002) Oxidant-induced vascular endothelial growth factor expression in human keratinocytes and cutaneous wound healing. J Biol Chem 277:33284–33290
70. Eyries M, Collins T, Khachigian LM (2004) Modulation of growth factor gene expression in vascular cells by oxidative stress. Endothelium 11:133–139
71. Iwai-Kanai E, Hasegawa K, Fujita M et al (2002) Basic fibroblast growth factor protects cardiac myocytes from iNOS-mediated apoptosis. J Cell Physiol 190:54–62
72. Li J, Brown LF, Hibberd MG et al (1996) VEGF, flk-1, and flt-1 expression in a rat myocardial infarction model of angiogenesis. Am J Physiol 270:H1803–H1811

73. Black SM, DeVol JM, Wedgwood S (2008) Regulation of fibroblast growth factor-2 expression in pulmonary arterial smooth muscle cells involves increased reactive oxygen species generation. Am J Physiol Cell Physiol 294:C345–C354
74. Gonzalez-Pacheco FR, Deudero JJ, Castellanos MC et al (2006) Mechanisms of endothelial response to oxidative aggression: protective role of autologous VEGF and induction of VEGFR2 by $H_2O_2$. Am J Physiol Heart Circ Physiol 291:H1395–H1401
75. Chua CC, Hamdy RC, Chua BH (1998) Upregulation of vascular endothelial growth factor by $H_2O_2$ in rat heart endothelial cells. Free Radic Biol Med 25:891–897
76. Luczak K, Balcerczyk A, Soszynski M et al (2004) Low concentration of oxidant and nitric oxide donors stimulate proliferation of human endothelial cells in vitro. Cell Biol Int 28: 483–486
77. Vepa S, Scribner WM, Parinandi NL et al (1999) Hydrogen peroxide stimulates tyrosine phosphorylation of focal adhesion kinase in vascular endothelial cells. Am J Physiol 277:L150–L158
78. Shono T, Ono M, Izumi H et al (1996) Involvement of the transcription factor NF-kappaB in tubular morphogenesis of human microvascular endothelial cells by oxidative stress. Mol Cell Biol 16:4231–4239
79. Thannickal VJ, Day RM, Klinz SG et al (2000) Ras-dependent and -independent regulation of reactive oxygen species by mitogenic growth factors and TGF-beta1. FASEB J 14: 1741–1748
80. Aicher A, Zeiher AM, Dimmeler S (2005) Mobilizing endothelial progenitor cells. Hypertension 45:321–325
81. Kucia M, Jankowski K, Reca R et al (2004) CXCR4-SDF-1 signalling, locomotion, chemotaxis and adhesion. J Mol Histol 35:233–245
82. Heissig B, Hattori K, Dias S et al (2002) Recruitment of stem and progenitor cells from the bone marrow niche requires MMP-9 mediated release of kit-ligand. Cell 109:625–637
83. Jin F, Zhai Q, Qiu L et al (2008) Degradation of BM SDF-1 by MMP-9: the role in G-CSF-induced hematopoietic stem/progenitor cell mobilization. Bone Marrow Transplant 42: 581–588
84. Lee RL, Westendorf J, Gold MR (2007) Differential role of reactive oxygen species in the activation of mitogen-activated protein kinases and Akt by key receptors on B-lymphocytes: CD40, the B cell antigen receptor, and CXCR4. J Cell Commun Signal 1:33–43
85. Schober A (2008) Chemokines in vascular dysfunction and remodeling. Arterioscler Thromb Vasc Biol 28:1950–1959
86. Pillarisetti K, Gupta SK (2001) Cloning and relative expression analysis of rat stromal cell derived factor-1 (SDF-1)1: SDF-1 alpha mRNA is selectively induced in rat model of myocardial infarction. Inflammation 25:293–300
87. Yamani MH, Ratliff NB, Cook DJ et al (2005) Peritransplant ischemic injury is associated with up-regulation of stromal cell-derived factor-1. J Am Coll Cardiol 46:1029–1035
88. Ceradini DJ, Kulkarni AR, Callaghan MJ et al (2004) Progenitor cell trafficking is regulated by hypoxic gradients through HIF-1 induction of SDF-1. Nat Med 10:858–864
89. Stellos K, Bigalke B, Langer H et al (2009) Expression of stromal-cell-derived factor-1 on circulating platelets is increased in patients with acute coronary syndrome and correlates with the number of CD34+ progenitor cells. Eur Heart J 30:584–593
90. Li N, Lu X, Zhao X et al (2009) Endothelial nitric oxide synthase promotes bone marrow stromal cell migration to the ischemic myocardium via upregulation of stromal cell-derived factor-1alpha. Stem Cells 27:961–970
91. Kaminski A, Ma N, Donndorf P et al (2008) Endothelial NOS is required for SDF-1alpha/CXCR4-mediated peripheral endothelial adhesion of c-kit+ bone marrow stem cells. Lab Invest 88:58–69
92. Milovanova TN, Bhopale VM, Sorokina EM et al (2008) Lactate stimulates vasculogenic stem cells via the thioredoxin system and engages an autocrine activation loop involving hypoxia-inducible factor 1. Mol Cell Biol 28:6248–6261

93. Mays RW, van't Hof W, Ting AE et al (2007) Development of adult pluripotent stem cell therapies for ischemic injury and disease. Expert Opin Biol Ther 7:173–184
94. Zhu QS, Xia L, Mills GB et al (2006) G-CSF induced reactive oxygen species involves Lyn-PI3-kinase-Akt and contributes to myeloid cell growth. Blood 107:1847–1856
95. Dang PM, Stensballe A, Boussetta T et al (2006) A specific p47phox-serine phosphorylated by convergent MAPKs mediates neutrophil NADPH oxidase priming at inflammatory sites. J Clin Invest 116:2033–2043
96. Sattler M, Winkler T, Verma S et al (1999) Hematopoietic growth factors signal through the formation of reactive oxygen species. Blood 93:2928–2935
97. Pyatt DW, Stillman WS, Irons RD (1996) Reactive oxygen species mediate stem cell factor synergy with granulocyte/macrophage colony-stimulating factor in a subpopulation of primitive murine hematopoietic progenitor cells. Mol Pharmacol 49:1097–1103
98. Cella G, Marchetti M, Vignoli A et al (2006) Blood oxidative status and selectins plasma levels in healthy donors receiving granulocyte-colony stimulating factor. Leukemia 20:1430–1434
99. Milovanova TN, Bhopale VM, Sorokina EM et al (2009) Hyperbaric oxygen stimulates vasculogenic stem cell growth and differentiation in vivo. J Appl Physiol 106:711–728
100. Yun J, Rocic P, Pung YF et al (2009) Redox-Dependent Mechanisms in Coronary Collateral Growth: The Redox Window Hypothesis. Antioxid Redox Signal, in press
101. Yoder MC, Ingram DA (2009) Endothelial progenitor cell: ongoing controversy for defining these cells and their role in neoangiogenesis in the murine system. Curr Opin Hematol 16:269–273
102. Asahara T, Murohara T, Sullivan A et al (1997) Isolation of putative progenitor endothelial cells for angiogenesis. Science 275:964–967
103. Takahashi T, Kalka C, Masuda H et al (1999) Ischemia- and cytokine-induced mobilization of bone marrow-derived endothelial progenitor cells for neovascularization. Nat Med 5: 434–438
104. Huang L, Hou D, Thompson MA et al (2007) Acute myocardial infarction in swine rapidly and selectively releases highly proliferative endothelial colony forming cells (ECFCs) into circulation. Cell Transplant 16:887–897
105. Prater DN, Case J, Ingram DA et al (2007) Working hypothesis to redefine endothelial progenitor cells. Leukemia 21:1141–1149
106. Frisch BJ, Porter RL, Calvi LM (2008) Hematopoietic niche and bone meet. Curr Opin Support Palliat Care 2:211–217
107. Besler C, Doerries C, Giannotti G et al (2008) Pharmacological approaches to improve endothelial repair mechanisms. Expert Rev Cardiovasc Ther 6:1071–1082
108. Aicher A, Kollet O, Heeschen C et al (2008) The Wnt antagonist Dickkopf-1 mobilizes vasculogenic progenitor cells via activation of the bone marrow endosteal stem cell niche. Circ Res 103:796–803
109. Pitchford SC, Furze RC, Jones CP et al (2009) Differential mobilization of subsets of progenitor cells from the bone marrow. Cell Stem Cell 4:62–72
110. Hiasa K, Ishibashi M, Ohtani K et al (2004) Gene transfer of stromal cell-derived factor-1alpha enhances ischemic vasculogenesis and angiogenesis via vascular endothelial growth factor/endothelial nitric oxide synthase-related pathway: next-generation chemokine therapy for therapeutic neovascularization. Circulation 109:2454–2461
111. Li X, Tjwa M, Moons L et al (2005) Revascularization of ischemic tissues by PDGF-CC via effects on endothelial cells and their progenitors. J Clin Invest 115:118–127
112. Kermani P, Rafii D, Jin DK et al (2005) Neurotrophins promote revascularization by local recruitment of TrkB+ endothelial cells and systemic mobilization of hematopoietic progenitors. J Clin Invest 115:653–663
113. Li B, Sharpe EE, Maupin AB et al (2006) VEGF and PlGF promote adult vasculogenesis by enhancing EPC recruitment and vessel formation at the site of tumor neovascularization. FASEB J 20:1495–1497

114. Iwakura A, Luedemann C, Shastry S et al (2003) Estrogen-mediated, endothelial nitric oxide synthase-dependent mobilization of bone marrow-derived endothelial progenitor cells contributes to reendothelialization after arterial injury. Circulation 108:3115–3121
115. Heeschen C, Aicher A, Lehmann R et al (2003) Erythropoietin is a potent physiologic stimulus for endothelial progenitor cell mobilization. Blood 102:1340–1346
116. Aicher A, Heeschen C, Mildner-Rihm C et al (2003) Essential role of endothelial nitric oxide synthase for mobilization of stem and progenitor cells. Nat Med 9:1370–1376
117. Becchi C, Pillozzi S, Fabbri LP et al (2008) The increase of endothelial progenitor cells in the peripheral blood: a new parameter for detecting onset and severity of sepsis. Int J Immunopathol Pharmacol 21:697–705
118. Lee DY, Cho TJ, Kim JA et al (2008) Mobilization of endothelial progenitor cells in fracture healing and distraction osteogenesis. Bone 42:932–941
119. Massa M, Rosti V, Ferrario M et al (2005) Increased circulating hematopoietic and endothelial progenitor cells in the early phase of acute myocardial infarction. Blood 105: 199–206
120. Laufs U, Werner N, Link A et al (2004) Physical training increases endothelial progenitor cells, inhibits neointima formation, and enhances angiogenesis. Circulation 109: 220–226
121. Laufs U, Urhausen A, Werner N et al (2005) Running exercise of different duration and intensity: effect on endothelial progenitor cells in healthy subjects. Eur J Cardiovasc Prev Rehabil 12:407–414
122. Steiner S, Niessner A, Ziegler S et al (2005) Endurance training increases the number of endothelial progenitor cells in patients with cardiovascular risk and coronary artery disease. Atherosclerosis 181:305–310
123. Paul JD, Powell TM, Thompson M et al (2007) Endothelial progenitor cell mobilization and increased intravascular nitric oxide in patients undergoing cardiac rehabilitation. J Cardiopulm Rehabil Prev 27:65–73
124. Powers SK, Jackson MJ (2008) Exercise-induced oxidative stress: cellular mechanisms and impact on muscle force production. Physiol Rev 88:1243–1276
125. Muller P, Kazakov A, Semenov A et al (2008) Pressure-induced cardiac overload induces upregulation of endothelial and myocardial progenitor cells. Cardiovasc Res 77:151–159
126. Ristow M, Zarse K, Oberbach A et al (2009) Antioxidants prevent health-promoting effects of physical exercise in humans. Proc Natl Acad Sci USA 106:8665–8670
127. Tousoulis D, Andreou I, Antoniades C et al (2008) Role of inflammation and oxidative stress in endothelial progenitor cell function and mobilization: therapeutic implications for cardiovascular diseases. Atherosclerosis 201:236–247
128. Andreou I, Tousoulis D, Tentolouris C et al (2006) Potential role of endothelial progenitor cells in the pathophysiology of heart failure: clinical implications and perspectives. Atherosclerosis 189:247–254
129. Urao N, Inomata H, Razvi M et al (2008) Role of Nox-2-based NADPH oxidase in bone marrow and progenitor cell function involved in neovascularization induced by hindlimb ischemia. Circ Res 103:212–220
130. Thum T, Fraccarollo D, Galuppo P et al (2006) Bone marrow molecular alterations after myocardial infarction: impact on endothelial progenitor cells. Cardiovasc Res 70:50–60
131. Salguero G, Akin E, Templin C et al (2008) Renovascular hypertension by two-kidney one-clip enhances endothelial progenitor cell mobilization in a p47phox-dependent manner. J Hypertens 26:257–268
132. Wang JY, Lee YT, Chang PF et al (2009) Hemin promotes proliferation and differentiation of endothelial progenitor cells via activation of AKT and ERK. J Cell Physiol 219:617–625
133. Yao EH, Yu Y, Fukuda N (2006) Oxidative stress on progenitor and stem cells in cardiovascular diseases. Curr Pharm Biotechnol 7:101–108
134. Ingram DA, Krier TR, Mead LE et al (2007) Clonogenic endothelial progenitor cells are sensitive to oxidative stress. Stem Cells 25:297–304

135. Yao EH, Fukuda N, Matsumoto T et al (2008) Effects of the antioxidative beta-blocker celiprolol on endothelial progenitor cells in hypertensive rats. Am J Hypertens 21: 1062–1068
136. Yao EH, Fukuda N, Matsumoto T et al (2007) Losartan improves the impaired function of endothelial progenitor cells in hypertension via an antioxidant effect. Hypertens Res 30:1119–1128
137. Thum T, Tsikas D, Stein S et al (2005) Suppression of endothelial progenitor cells in human coronary artery disease by the endogenous nitric oxide synthase inhibitor asymmetric dimethylarginine. J Am Coll Cardiol 46:1693–1701
138. Stoll LL, McCormick ML, Denning GM et al (2005) Antioxidant effects of statins. Timely Top Med Cardiovasc Dis 9:E1
139. Williams HC, Griendling KK (2007) NADPH oxidase inhibitors: new antihypertensive agents? J Cardiovasc Pharmacol 50:9–16
140. Wang N, Xie K, Huo S et al (2007) Suppressing phosphatidylcholine-specific phospholipase C and elevating ROS level, NADPH oxidase activity and Rb level induced neuronal differentiation in mesenchymal stem cells. J Cell Biochem 100:1548–1557
141. Wang FS, Yang KD, Wang CJ et al (2004) Shockwave stimulates oxygen radical-mediated osteogenesis of the mesenchymal cells from human umbilical cord blood. J Bone Miner Res 19:973–982
142. Piccoli C, D'Aprile A, Ripoli M et al (2007) Bone-marrow derived hematopoietic stem/progenitor cells express multiple isoforms of NADPH oxidase and produce constitutively reactive oxygen species. Biochem Biophys Res Commun 353:965–972
143. Piccoli C, D'Aprile A, Scrima R et al (2007) Role of reactive oxygen species as signal molecules in the pre-commitment phase of adult stem cells. Ital J Biochem 56:295–301
144. Sato K, Ozaki K, Oh I et al (2007) Nitric oxide plays a critical role in suppression of T-cell proliferation by mesenchymal stem cells. Blood 109:228–234
145. Klinz FJ, Schmidt A, Schinkothe T et al (2005) Phospho-eNOS Ser-114 in human mesenchymal stem cells: constitutive phosphorylation, nuclear localization and upregulation during mitosis. Eur J Cell Biol 84:809–818
146. North TE, Goessling W, Peeters M et al (2009) Hematopoietic stem cell development is dependent on blood flow. Cell 137:736–748
147. Loomans CJ, Wan H, de Crom R et al (2006) Angiogenic murine endothelial progenitor cells are derived from a myeloid bone marrow fraction and can be identified by endothelial NO synthase expression. Arterioscler Thromb Vasc Biol 26:1760–1767
148. Yin T, Ma X, Zhao L et al (2008) Angiotensin II promotes NO production, inhibits apoptosis and enhances adhesion potential of bone marrow-derived endothelial progenitor cells. Cell Res 18:792–799
149. Rebelatto CK, Aguiar AM, Senegaglia AC et al (2009) Expression of cardiac function genes in adult stem cells is increased by treatment with nitric oxide agents. Biochem Biophys Res Commun 378:456–461
150. Imanishi T, Hano T, Nishio I (2005) Angiotensin II accelerates endothelial progenitor cell senescence through induction of oxidative stress. J Hypertens 23:97–104
151. Saretzki G, Armstrong L, Leake A et al (2004) Stress defense in murine embryonic stem cells is superior to that of various differentiated murine cells. Stem Cells 22:962–971
152. Buggisch M, Ateghang B, Ruhe C et al (2007) Stimulation of ES-cell-derived cardiomyogenesis and neonatal cardiac cell proliferation by reactive oxygen species and NADPH oxidase. J Cell Sci 120:885–894
153. Hannig M, Figulla HR, Sauer H et al (2010) Control of leukocyte differentiation from embryonic stem cells upon vasculogenesis and confrontation with tumour tissue. J Cell Mol Med 14:303–312
154. Sauer H, Rahimi G, Hescheler J et al (1999) Effects of electrical fields on cardiomyocyte differentiation of embryonic stem cells. J Cell Biochem 75:710–723

155. Sauer H, Rahimi G, Hescheler J et al (2000) Role of reactive oxygen species and phosphatidylinositol 3-kinase in cardiomyocyte differentiation of embryonic stem cells. FEBS Lett 476:218–223
156. Sauer H, Bekhite MM, Hescheler J et al (2005) Redox control of angiogenic factors and CD31-positive vessel-like structures in mouse embryonic stem cells after direct current electrical field stimulation. Exp Cell Res 304:380–390
157. Schmelter M, Ateghang B, Helmig S et al (2006) Embryonic stem cells utilize reactive oxygen species as transducers of mechanical strain-induced cardiovascular differentiation. FASEB J 20:1182–1184
158. Sharifpanah F, Wartenberg M, Hannig M et al (2008) Peroxisome proliferator-activated receptor alpha agonists enhance cardiomyogenesis of mouse ES cells by utilization of a reactive oxygen species-dependent mechanism. Stem Cells 26:64–71
159. Li J, Stouffs M, Serrander L et al (2006) The NADPH oxidase Nox-4 drives cardiac differentiation: role in regulating cardiac transcription factors and MAP kinase activation. Mol Biol Cell 17:3978–3988
160. Calabro P, Limongelli G, Riegler L et al (2009) Novel insights into the role of cardiotrophin-1 in cardiovascular diseases. J Mol Cell Cardiol 46:142–148
161. Freed DH, Moon MC, Borowiec AM et al (2003) Cardiotrophin-1: expression in experimental myocardial infarction and potential role in post-MI wound healing. Mol Cell Biochem 254:247–256
162. Lopez B, Gonzalez A, Querejeta R et al (2009) Association of plasma cardiotrophin-1 with stage C heart failure in hypertensive patients: potential diagnostic implications. J Hypertens 27:418–424
163. Mohri T, Fujio Y, Obana M et al (2009) Signals through glycoprotein 130 regulate the endothelial differentiation of cardiac stem cells. Arterioscler Thromb Vasc Biol 29:754–760
164. Xinyun C, Zhi Z, Bin Z et al (2009) Effects of cardiotrophin-1 on differentiation and maturation of rat bone marrow mesenchymal stem cells induced with 5-azacytidine in vitro. Int J Cardiol, in press
165. Saretzki G, Walter T, Atkinson S et al (2008) Downregulation of multiple stress defense mechanisms during differentiation of human embryonic stem cell. Stem Cells 26:455–464
166. Mujoo K, Krumenacker JS, Wada Y et al (2006) Differential expression of nitric oxide signaling components in undifferentiated and differentiated human embryonic stem cells. Stem Cells Dev 15:779–787
167. Mujoo K, Sharin VG, Bryan NS et al (2008) Role of nitric oxide signaling components in differentiation of embryonic stem cells into myocardial cells. Proc Natl Acad Sci USA 105:18924–18929
168. Kanno S, Kim PK, Sallam K et al (2004) Nitric oxide facilitates cardiomyogenesis in mouse embryonic stem cells. Proc Natl Acad Sci USA 101:12277–12281
169. Bloch W, Fleischmann BK, Lorke DE et al (1999) Nitric oxide synthase expression and role during cardiomyogenesis. Cardiovasc Res 43:675–684
170. Gassanov N, Jankowski M, Danalache B et al (2007) Arginine vasopressin-mediated cardiac differentiation: insights into the role of its receptors and nitric oxide signaling. J Biol Chem 282:11255–11265
171. Linnane AW, Kios M, Vitetta L (2007) The essential requirement for superoxide radical and nitric oxide formation for normal physiological function and healthy aging. Mitochondrion 7:1–5

# Chapter 4
# Reactive Oxygen Species (ROS) and the Sensory Neurovascular Component

**Rabea Graepel, Jennifer Victoria Bodkin, and Susan Diana Brain**

**Abstract** A dense perivascular network of C- and Aδ-sensory nerve fibers innervate the vascular system and are ideally situated to influence vascular events. The nerves release potent vasodilator neuropeptides including substances P, CGRP and a range of other agents, depending on their location and the nature of nerve activation. A number of interactions between neuropeptides and ROS have been described and are discussed here. We particularly emphasize the roles of ROS as signaling molecules that have the potential to influence cardiovascular events in an important manner. We also provide evidence of recent findings involving the transient receptor potential (TRP) channels that activate sensory nerves. It is now realized that the sensory nerve-derived TRPA1 channel is directly activated by hydrogen peroxide and a range of lipid peroxidation products. The influence of this on the cardiovascular system is only now beginning to emerge, but a range of exciting, recent findings are summarized in this review.

**Keywords** Sensory nerves · CGRP · Substance P · Neuropeptides · Inflammation · Oxidant stress · Channels

## 4.1 Introduction

The sensory nervous system is well described in the literature [1–3]. It is primarily known for its role in pain processing, in transporting nociceptive information to the central nervous system. The nerves link peripheral tissues with the central nervous system, resulting in an extensive neuronal network which amplifies and regulates nociceptive and sensory information. However, the peripheral sensory nervous system also has another important role, which is not adequately described by

S.D. Brain (✉)
Cardiovascular Division, King's College London BHF Centre of Excellence, London SE1 9NH, UK
e-mail: sue.brain@kcl.ac.uk

H. Sauer et al. (eds.), *Studies on Cardiovascular Disorders*, Oxidative Stress in Applied Basic Research and Clinical Practice, DOI 10.1007/978-1-60761-600-9_4, © Springer Science+Business Media, LLC 2010

the word "sensory". Sensory nerves act via perivascular neuronal networks to release potent vasoactive neuropeptides that work in combination with the autonomic nervous system to regulate both physiological vascular tone and pathophysiological disease processes. Sensory nerve endings can be in contact with vascular smooth muscle cells and also in intimate contact with endothelial cells [4, 5]. Indeed, it has been suggested that microvascular endothelial cells may produce nerve growth factor, which influences sensory nerve growth and activity in certain circumstances [6]. They are therefore ideally placed to influence both the physiological and pathophysiological control of the heart and blood vessels. Such close association with vascular tissue also implies that these nerves are perfectly placed to be influenced by ROS derived from both inflammatory and signaling mechanisms. Taking this into consideration, it is perhaps surprising that a relationship between these two powerful mediators and signaling systems is only now starting to be unraveled. This chapter reviews current knowledge of the sensory nervous system in terms of its influence on the cardiovascular system and then describes the established and putative links between the sensory nervous system and ROS generation, relevant to the cardiovascular system.

## 4.2 The Sensory Neurogenic Component and Vascular Innervation

The sensory nervous system comprises two types of nerves: the slowly conducting unmyelinated C-fibers and the faster conducting thinly myelinated Aδ-fibers. The nerves were first realized to exist by Goltz and Stricker in the 1880s and then confirmed by Bayliss in 1901 from studies involving stimulation of dorsal roots that triggered increased blood flow in the skin [7]. Thomas Lewis (1927) investigated the response to intradermal injections of histamine [8]. The response, which has become commonly known as the triple response to injury, consists of a wheal at the site of injection, due to histamine $H_1$ receptor-induced plasma extravasation, local reddening, and a flare. The flare is mediated by increased blood flow and can spread for up to several centimeters around the site of injection. It is enabled via special terminal arborisations of the sensory nerves. For some time it was considered that these anatomical structures were only found in the skin. However, it is now realized that this system of nerve terminals, which can signal simultaneously to adjacent tissues, operates in almost all tissues of the body, albeit with substantially smaller innervating fields than those that occur in the skin.

### *4.2.1 Nerve Activating Mechanisms and Cardiovascular Consequences of Neuropeptide Action*

The realization that sensory nerves could be stimulated to mediate vascular effects led to a wide search for activating factors for these nerves. Jancso showed that topical administration of the chili pepper extract capsaicin causes pain and reddening of the skin, and that this response was lost with repeated application [9, 10]. The response was associated with sensory nerve activity because it was lost on denervation and upon treatment with a local anesthetic. This work led to the concept of the capsaicin receptor. Importantly, the use of capsaicin depletion techniques has become a common and widely used protocol, by which experiments can be carried out in the absence of the sensory neurogenic component in laboratory species [2].

Molecular biological techniques were used to clone the capsaicin receptor. It is now identified as the transient receptor potential vanilloid 1 (TRPV1) receptor and TRPV1 knockout (KO) mice have been developed and are used in a range of experimental studies [11, 12]. This led to the realization that the TRPV1 receptor is involved in mediating thermal hyperalgesia (a key finding for the many ongoing drug development programs) as well as local vasodilatation and edema formation. Surprisingly, however, TRPV1 appears to also possess protective mechanisms in murine models of myocardial injury [13], atopic dermatitis [14], and sepsis [15]. It is currently debated whether these protective effects are mainly due to immune or vascular protective mechanisms. Many other receptors are also present on sensory nerve endings, which can activate and in some cases modulate sensory nerve responses (see Fig. 4.1). There is also a growing realization that other TRP receptors (e.g., TRPA1) have important roles in activating sensory nerve systems [16]. The influence of these receptors on the cardiovascular system is at present largely unknown, but current findings will be discussed in 4.71, 4.72, and 4.73.

## 4.3 ROS and Localization Within Sensory Nerves

Rat pheochyromocytoma (PC-12) cells resemble rat sensory neurons and are used as a model for in vitro studies. They respond to increased glucose levels by enhancing ROS generation and decreasing nerve growth factor (NGF) activity, which is essential for capsaicin receptor activity. The increase in ROS can in turn lead to apoptosis in vitro [17]. These findings are supported by studies in dorsal root ganglion cells, where a short (2 h) exposure to hyperglycemia was found to promote ROS production and lipid peroxidation [18, 19]. More recently it has been suggested that ROS potentiate the sensation of pain by accelerating the translocation of PKCε in dorsal root ganglion neurons, thus promoting TRPV1 activity and pain sensitivity [20]. It was realized from the study of dorsal root ganglion cells in vitro that NOX1/NADPH oxidase was involved in this response [20].

There is little direct evidence at this time of functional interactions between ROS and the neurogenic vascular system. Oltman and colleagues have recently studied Zucker rats that develop vascular and neuronal impairment independently of hyperglycemia [21]. They observed a loss of microvascular relaxation which preceded a loss of thermal hyperalgesia. Superoxide and nitrotyrosine levels could be measured in vascular tissue as the obesity developed. Interestingly, both ROS levels and thermal sensitivity returned towards normal values when rats were treated with an ACE inhibitor or a statin.

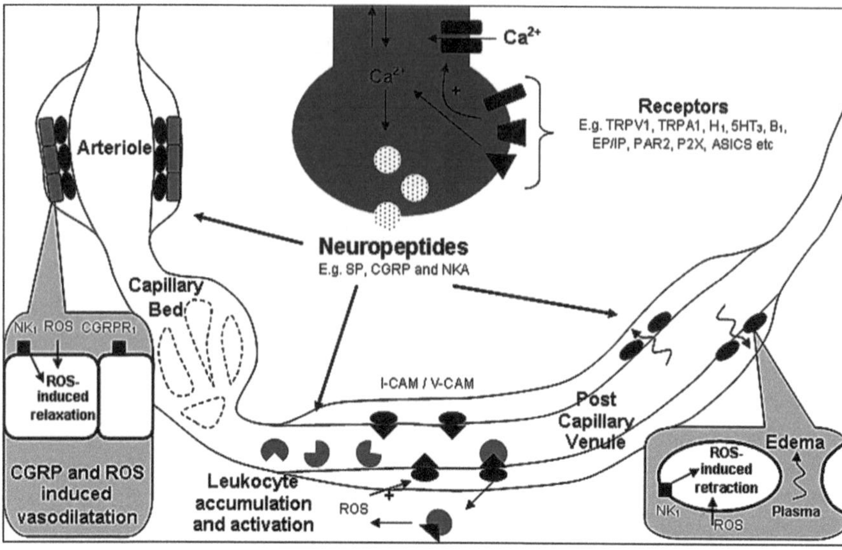


**Fig. 4.1** The major actions of sensory nerve-derived neuropeptides on the microvasculature and interactions with ROS. Sensory nerves can be activated by a range of endogenous and exogenous mediators. Receptors for some of these are shown in the diagram, including histamine ($H_1$), 5HT ($5HT_3$), bradykinin ($B_1$), prostaglandins (EP/IP), tryptases ($PAR_2$), and ATP (P2X). Activation of both membrane and intracellularly expressed receptors leads to production of secondary messengers, depolarization of the nerve, influx of calcium and in turn, release of vesicularly stored neuropeptides. A schematic of a microvascular bed is shown. Vasodilatation in the arteriole is mediated by CGRP and substance P. Intracellularly produced ROS can also contribute downstream of the $NK_1$ receptor in smooth muscle cells, along with membrane permeable extracellular ROS. In the postcapillary venule, leukocyte accumulation is enhanced by both substance P–derived and extracellular ROS, inducing adhesion molecule expression on endothelial cells. Production of ROS following activation of these inflammatory cells potentiates the system. Edema is mediated by substance P actions on post capillary venule endothelial cell $NK_1$ receptors; similarly, extracellular and intracellularly derived ROS can participate

## 4.4 Vascular Effects of ROS

Superoxide is a common precursor for most ROS and can be formed from molecular oxygen by multiple enzymes that are present in all cells of the vasculature. NADPH oxidases are thought to be the primary source of ROS in blood vessels, but lipoxygenases, cyclooxygenases, the mitochondrial respiratory chain, and uncoupled eNOS can also mediate the one-electron reduction of molecular oxygen to superoxide [22, 23]. It is generally thought that high concentrations of ROS are involved in various cardiovascular pathologies; whereas low level, continuous production of ROS has a physiological role in controlling vascular functions. ROS can influence various intracellular signaling pathways and thus have widespread effects on the vasculature, such as modulating vascular tone, cell growth, apoptosis, and inflammation [24, 25]. This review will focus on the effects of $H_2O_2$ and superoxide on vascular tone and edema formation, which are key features of neurogenic inflammation. Superoxide is highly reactive and has a short half-life because it rapidly reacts with nitric oxide (NO) to form peroxynitrite ($ONOO^-$), or it forms hydrogen peroxide ($H_2O_2$) either spontaneously or through a superoxide dismutase (SOD)–mediated reaction. $H_2O_2$ is considered to be a fairly stable ROS and unlike superoxide it is cell-permeable and thus represents a more important signaling molecule. $H_2O_2$ metabolites include the highly reactive hydroxyl radical (•OH), formed via a metal ion-catalyzed Fenton reaction, and hypochlorous acid (HOCL), formed by myeloperoxidase, an enzyme present in the phagosomes of neutrophils. $H_2O_2$ is broken down to water and oxygen by catalase, glutathione peroxidase, and peroxiredoxins.

There is good evidence in the literature to show that $H_2O_2$ acts as a vasodilator in small and large arteries of the systemic and cerebral vasculature [24]. Exogenous $H_2O_2$ induces relaxation of rabbit aortas by stimulating endothelial NO production [26]. Indeed, it was demonstrated that $H_2O_2$ upregulates eNOS gene expression in bovine endothelial cells [27, 28]. Endogenous $H_2O_2$ is produced in response to multiple stimuli such as bradykinin, ischemia/reperfusion, and ACh, mediating vasodilatation in a variety of vascular beds [29–33]. The mechanism of action of $H_2O_2$ is not yet clear, but both endothelium-dependent and -independent pathways have been described. It was suggested that $H_2O_2$ acts as an EDHF in the canine coronary circulation as well as human and mouse mesenteric arteries [29, 34]. On the other hand, production of NO as well as activation of various potassium channels were implicated in mediating the vasodilator effects of $H_2O_2$ in a variety of systemic and cerebral arteries [26, 32, 35]. In addition to the effects of $H_2O_2$ on vascular tone, larger concentrations of $H_2O_2$ have been shown to disrupt the barrier function of endothelial cells in vitro and thus lead to edema formation in vivo [24, 36, 37]. The pathophysiological relevance of this is demonstrated, for example, in a mouse model of carrageenan-induced hindpaw inflammation where $H_2O_2$ was shown to be involved in mediating both edema and hyperalgesia [37]. In vitro studies using endothelial cell lines aimed to define the complex signaling mechanisms that are involved in $H_2O_2$-induced barrier dysfunction and actin cytoskeleton reorganization of the endothelial cells. This work is well summarized by Cai (2005) [24].

It is generally agreed that systemically produced superoxide acts as a vasoconstrictor because it rapidly reacts with and thus inactivates the vasodilator NO to form $ONOO^-$ [38]. The rate of this reaction is three times faster than that between superoxide and its endogenous scavenger SOD [38]. This pathway may be of importance in pathologies where vasoconstriction or decreased vasodilatation contributes to the disease progression, for example, in atherosclerosis, hypertension, and diabetes [39–41]. However, it has also been shown that superoxide mediates constriction of rat renal arteries in response to AngII independently of its inactivating effects on NO, suggesting that superoxide can directly influence vascular tone [42]. In the cerebral vasculature, superoxide has been shown to induce both vasoconstriction and vasodilatation. For example, in the cat pial microcirculation, in vivo generation of superoxide and $H_2O_2$ triggered reversible vasodilatation that was attributed to both ROS [43]. However, in excised canine basilar arteries it was shown that endothelium-dependent contractions induced by a calcium ionophore (A23187) were mediated by superoxide [44]. Indeed, in the rabbit basilar artery, superoxide generation in response to exogenous NADH induced both relaxations and contractions in vivo and in vitro. These responses were seen to be dependent on the dose of NADH, with low concentrations (0.1–10 μM) causing relaxations whilst higher concentrations (>10 μM) triggered contractions [45]. In addition, superoxide, like $H_2O_2$, has the ability to induce edema formation. In a model of carrageenan-induced hindpaw inflammation, pretreatment with the potent SOD mimetic, M40403, inhibited edema formation, implicating superoxide as a mediator of edema formation [46].

## 4.5 Neuropeptides and Interactions with Vascular-Derived ROS

To date, sensory nerves have been described to be releasing many neuropeptides. There is now good evidence that the major vasoactive neuropeptides that are released are substance P and calcitonin gene–related peptide [47, 48]. Both neuropeptides have potent vasodilator effects, and substance P was also shown to be a potent mediator of increased microvascular permeability in many species. CGRP also possesses vascular protective properties [5, 49, 50]. These peptides have been widely studied in terms of their cardiovascular activities, which will be outlined before discussing their interactions with ROS.

## 4.6 CGRP

CGRP is a 37 amino acid peptide that is heavily conserved among the species. CGRP was discovered as a consequence of gene splicing of the calcitonin gene. Whilst CGRP is found in patients with medullary thyroid carcinoma and in the thyroids of aging rats, it is most commonly localized to sensory nerves [51]. The major vascular form is α-CGRP, although a β-CGRP also exists, with >90% structural homology with α-CGRP [52]. CGRP is now considered to be the primary member of the CGRP family of peptides which also includes adrenomedullin and intermedin. They are 3–10 times less potent as vasodilators than CGRP, but show some protective effects in the cardiovascular system. The CGRP receptor is composed of a G-protein component, named a calcitonin-like receptor (CLR), and also a receptor activity modifying protein (RAMP). The primary vasoactive CGRP receptor is composed of CLR and RAMP1 [5].

CGRP acts via an endothelium-dependent NO system to stimulate relaxation and can also act directly on vascular smooth muscle cells to mediate relaxation via cAMP-dependent mechanisms and possibly EDHF [5]. CGRP induces hypotension when administered intravenously in humans, but does not contribute to the regulation of basal blood pressure. This has been confirmed following the development and clinical testing of two nonpeptide selective CGRP receptor antagonists [53, 54]. The very potent vasodilator effect of CGRP was initially discovered when it was injected into skin and it was soon realized that this potency was twinned with a long duration of action [55]. It is now realized that the most important effects of CGRP occur at the local level, in the tissue where CGRP is released. Indeed, some have suggested that CGRP is released when vascular stress is exerted within the tissue [5]. There is evidence of protective effects that are a direct consequence of vasodilator activity, in addition to others that appear to be due to direct protective effects.

### *4.6.1 CGRP and Protection Against Oxidative Stress as a Consequence of Vasodilator Networks*

CGRP plays an important protective role in the gastric mucosa, acting to increase blood flow and protect against ulcer formation [47, 56]. It has been long established that there is a functional interaction between sensory nerves, their ability to increase blood flow, and reactive oxygen species. A good example of this is given in a stress-induced model of gastric mucosal lesions [57]. Lower levels of lipid peroxidation products were measured when the sensory nervous system was intact and contributing in a physiologically important manner to increasing blood flow through the release of neuropeptides. However, capsaicin pretreatment to ablate the sensory nerves led to increased levels of lipid peroxidation products such as malondialdehyde and 4-hydroxynonenal (4-HNE) in tissue, in association with a decrease in SOD activity [57]. The complex role of ROS is also emphasized by a more recent paper from Gazzieri and colleagues, who show that ethanol-induced gastric ulcers in rodents are mediated via substance P-dependent ROS formation, following activation of the $NK_1$ receptor. Here, the ROS production was thought to be mediated via epithelial gastric cells [58], rather than a vascular source.

Infusion of CGRP into patients with stable angina pectoris delayed the onset of myocardial ischemia, a disease strongly associated with oxidative stress, and enabled an increased workload on the heart during exercise [59, 60]. These effects were presumed to be directly due to CGRP-induced vasodilatation, although vasodilator-independent mechanisms may also be involved (see below). Studies in the rat show that endogenous CGRP is released during ischemic preconditioning in hearts in vitro, and a cardioprotective role has been suggested because the CGRP antagonist $CGRP_{(8-37)}$ can block this protection [61]. Another protective role of CGRP has also been revealed through the study of intestinal preconditioning in rats [62].

### *4.6.2 CGRP and Protection via Vasodilator-Independent Mechanisms Against ROS-Mediated Vascular Injury*

It has been known for some time that CGRP can protect against macrophage activation and oxidant release [63]. In addition, CGRP (10 nM) is also able to protect against oxidative stress-induced vascular smooth muscle cell apoptosis [64]. In a later study this inhibition was shown to be a CGRP receptor-dependent mechanism that triggers Bcl-2 mRNA increases and an inhibition of caspase 3 activation [65].

### *4.6.3 Substance P*

Substance P is an extremely potent vasodilator acting mainly by NO-dependent mechanisms in large blood vessels, although vasodilator responses in microvascular beds are less clear and may be species- and tissue-dependent. Substance P is the primary member of the tachykinin family of peptides. Other members include neurokinin A, which can be formed from the same gene as substance P in sensory nerves; and neurokinin B, which is mainly found in the central nervous system. In addition, hemokinin-1 is a nonneuronally derived member of the tachykinin family [66], that appears to relax blood vessels in a similar manner to substance P. Substance P, in a similar manner to CGRP, does not appear important in the regulation of blood pressure since antagonists of the neurokinin 1 ($NK_1$), the major vasoactive substance P receptor, have no effect on basal blood pressure.

Substance P is a NO-dependent vasodilator in large blood vessels; however, it has also been shown to act via hyperpolarizing mechanisms, with little evidence of a role for vasodilator prostaglandins. Indeed, it has been demonstrated in tissues such as human mesenteric arteries, that substance P acts via both endothelium-derived NO and EDHF [67]. $H_2O_2$ has been proposed as an EDHF, as discussed above. The ability of substance P to release $H_2O_2$ has been investigated in pig coronary arteries, where catalase had no effect. This suggests that if substance P does mediate vascular relaxation via release of a hyperpolarizing factor, it is distinct from $H_2O_2$ [68]. It is now realized that the contribution of EDHF to substance P-induced vascular relaxation is most probably mediated by small- and intermediate-conductance $Ca^{2+}$-activated $K^+$ channels [69–71].

### *4.6.4 Influence of Vascular-Derived ROS on Substance P–Induced Vasodilatation*

Substance P is a potent vasodilator in large blood vessels acting via the tachykinin $NK_1$ receptor. It is best known for its activity as a NO-dependent vasodilator as discussed above. This renders its vasodilator potential susceptible to reactive oxygen species regulation, because of the ability of superoxide to rapidly react with NO preventing relaxation; see 4.4.

The ability of substance P to stimulate NO-dependent cardiac vascular relaxation has been shown in terms of its impact on left ventricular contractile function [72, 73]. This effect was lost in a model of hypertrophy, but restored in the presence of a SOD mimetic [74]. This finding provides experimental evidence for the concept that ROS interactions can substantially impact on substance P–induced NO-dependent vasodilatation.

Khodr and colleagues have studied the role of oxidants in a wound healing model, where the vasodilator response to added substance P was determined [75]. The results suggested a differential involvement of ROS, depending on age. However, SOD and catalase potentiated the vasodilator response, suggesting that there is a functional interaction between substance P and ROS, possibly as a consequence of NO generation.

### *4.6.5 Influence of Substance P on Inflammatory ROS Production*

There are tachykinin receptors on inflammatory cells, and substance P (100 nM–10 μM) has been suggested to prime human neutrophils to release ROS and NO [76–78]. However, the peptide did not appear potent in its own right, as a high dose (30 μM) was required to induce superoxide release in human neutrophils [79]. The mechanisms for this are thought to be related to activation of $IP_3$ and PKC [78, 79]. Interestingly, an aldehyde product of lipid peroxidation, 4-HNE, also has the ability to potentiate substance P–induced release of superoxide from neutrophils [80]. In addition, substance P can act via both $NK_1$ and $NK_2$ receptors to mediate ROS generation in monocytes and macrophages, although it should be noted that there are many agents that are more potent [81].

## 4.7 TRP Receptor and Localization on Sensory Nerves

The sensory nerves that innervate the skin, soft tissues, and blood vessels of the entire body are either polymodal, that is, they are activated by chemical, mechanical, and thermal stimuli; or they are unimodal, and are activated exclusively by one of these modalities. To be able to respond to and integrate such a variety of stimuli, sensory nerve endings express a large number of receptors (see Fig. 4.1). Amongst them, two members of the superfamily of TRP receptors will be discussed in more detail, because they were shown to be key sensory nerve activating systems, acting as molecular integrators of various noxious stimuli, including reactive oxygen species (see Fig. 4.2).

The TRPV1 receptor is a nonselective cation channel that is activated by a range of agonists, including vanilloids such as capsaicin, the pungent component of chili peppers, and its ultrapotent analogue resiniferatoxin (RTX). TRPV1 channels are activated directly by other noxious stimuli such as heat (>43°C), extracellular protons, ($pH < 6$) and ethanol; and indirectly by inflammatory mediators such as bradykinin, prostaglandins, and serotonin, which activate their respective receptors and produce downstream signaling molecules that sensitize TRPV1 [82, 83]. Therefore, TRPV1 receptors act as molecular integrators of noxious stimuli, and their critical role in nociception is highlighted in TRPV1 KO mice that are deficient

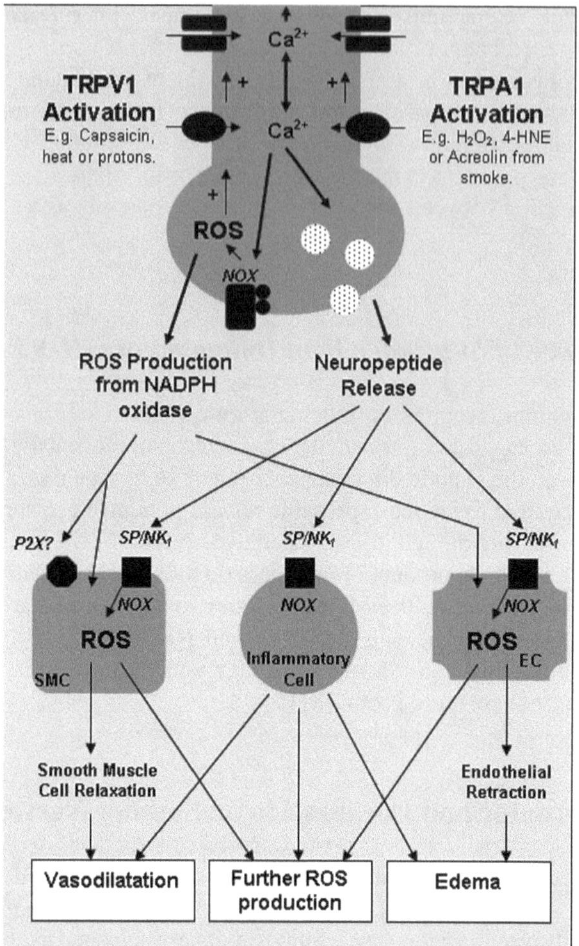


**Fig. 4.2** The major roles of ROS in neurogenic inflammation. TRPV1 and TRPA1 are both activated by a wide range of stimuli, leading to depolarization of the sensory nerve terminal, influx of calcium and release of stored neuropeptides. Increased intracellular calcium levels can also activate neuronal NADPH oxidase (Nox), producing intracellular ROS, which are known to sensitize and upregulate TRPV1 and directly activate TRPA1. Some ROS dismutate and become membrane permeable, leaking into the extracellular space. Here, neuropeptides and ROS act on various cell types, including smooth muscle cells (SMC), inflammatory cells (including monocytic and haematopoietic derived cells), and endothelial cells, creating the key features of neurogenic inflammation. Neuropeptides, particularly substance P acting on $NK_1$ receptors, lead to production of intracellular ROS via activation of NADPH oxidase (Nox). Membrane permeable extracellular ROS also activate some cells directly, and possibly via P2X receptors. Intracellular ROS produce smooth muscle cell relaxation and endothelial cell retraction, inducing vasodilatation and edema, two of the classic hallmarks of neurogenic inflammation. Further ROS pass in to the extracellular space via leakage from vascular cells and activation of inflammatory cells, potentiating the system

in their responses to capsaicin, protons, and heat, and no longer develop thermal hyperalgesia in several models of inflammation [11, 12].

The transient receptor potential ankyrin repeat 1 (TRPA1) receptor is the only mammalian member of the TRPA subfamily of receptors that has been identified so far [84]. The TRPA1 receptor is also a nonselective cation channel and has been found to be expressed in 50% of all TRPV1 positive sensory nerves [85]. Little or no TRPA1 expression was found in nonsensory neurons or a variety of organs and tissues [86]. The TRPA1 receptor has only fairly recently become the focus of research efforts; however, it is already known that it can be activated by oxidant stimuli, and that activation of TRPA1 triggers the development of neurogenic inflammation and pain [87, 88]. It has been shown that natural pungent compounds from mustard plants and cinnamon oil, as well as ginger and garlic extracts, directly activate the TRPA1 receptor [89–92]. In addition, several endogenous TRPA1 receptor agonists have been identified that activate the TRPA1 receptor directly or indirectly via the activation of downstream signaling mechanisms such as PKA and PLC [89, 93, 94]. It has also been suggested that TRPA1 is activated by noxious cold and mechanical stimuli; however, this is strongly debated. Unfortunately, these issues have not been resolved, despite the development of better TRPA1 receptor antagonists and the use of TRPA1 receptor KO mice that were cloned in 2006 by two separate groups [91, 95].

## *4.7.1 TRPV1 Receptors and Links with ROS*

Links between the TRPV1 receptor and ROS at the neurovascular junction have been demonstrated at several levels (see Fig. 4.2). It is becoming increasingly recognized that ROS are important signaling molecules, and it has transpired that there is a close interaction between ROS and the TRPV1 receptor in several pathophysiological conditions. It is well known that TRPV1 receptor expression is upregulated in response to peripheral inflammation, and that phosphorylation of the receptor by PKA or PKC sensitizes it to other stimuli. For example, intraplantar injections of NGF induce thermal hyperalgesia and moderate edema, and increases the expression of TRPV1 receptors in the sensory nerve ending without any changes in mRNA levels [96]. Puntambeckar and colleagues demonstrated this to be dependent on Rac1/NADPH oxidase activation and ROS production, which in turn activates the p38 MAPK pathway to trigger upregulation of TRPV1 [97]. Similarly, it was shown that streptozotocin, an agent commonly used to induce diabetes in rodents, also increases expression of TRPV1 in DRG neurons via a ROS- and p38 MAPK–dependent pathway [98]. Kitahara and colleagues observed that kanamycin, which was given to mice as a ROS-generating challenge, induced a marked increase in TRPV1 mRNA, and protein levels in inner ear ganglia that was attenuated by pretreatment with an antioxidant, demonstrating a role for ROS signaling [99]. Even though these effects are not related to the cardiovascular system, they still demonstrate the clear link between ROS and TRPV1 receptors. A critical role for ROS generated by NOX1 has previously been discussed in the development of thermal and mechanical hyperalgesia. NOX1-derived ROS oxidises PKCε to promote phosphorylation of TRPV1 and thus enhance the receptor activity and sensitivity [20]. This study demonstrates that ROS can regulate TRPV1 receptor expression and activity at the neurovascular junction in conditions that are of relevance to inflammation.

Intriguingly, it has been shown that ROS can be produced following the activation of both neuronal and non-neuronal TRPV1 receptors where they might elicit protective as well as deleterious effects. For example, Starr and colleagues recently demonstrated a novel role for ROS in a model of capsaicin-induced neurogenic inflammation in the mouse ear [100]. Capsaicin activates TRPV1 receptors to trigger the release of substance P and CGRP from perivascular sensory nerves, which in turn leads to vasodilatation and edema formation. The authors confirmed the classic neurogenic nature of the response; however, they also demonstrated that ROS are produced following neuropeptide release and are essential for mediating vasodilatation, but not edema formation. Finally, the authors were also able to identify the Nox2-containing NADPH oxidase isoform as the source of ROS in this response. This novel mechanism of TRPV1 receptor-mediated vasodilatation might be relevant to the protective role of TRPV1 receptors in myocardial infarction, where TRPV1-mediated vasodilatation is thought to preserve tissue function, and the release of substance P and CGRP was implicated in this protection [13, 101]. Similarly, the plant extract cannabidiol was shown to induce apoptosis of human breast cancer cell lines via activation of TRPV1 and possibly the $CB_2$ receptor, and production of ROS [102]. The relevance of this to the cardiovascular system is not known at this stage.

Neurogenic inflammation is a component of many diseases, including asthma, migraine, inflammatory bowel disease, and rheumatoid arthritis [16]. In a mouse model of Complete Freund's Adjuvant–induced knee joint inflammation, the TRPV1 receptor was shown to be essential for induction of thermal hyperalgesia and edema formation, demonstrating its central role in a disease with a strong neurogenic inflammatory component [103]. It was also shown that TRPV1 receptor-dependent ROS production triggers apoptosis of synoviocytes, which might be a novel therapeutic target for the treatment of rheumatoid arthritis because uncontrolled differentiation of synoviocytes contributes to the disease progress [104]. The newly identified roles for ROS, both in regulating TRPV1 receptor expression and activity, and in acting as signaling molecules downstream of TRPV1 receptor activation, raise the possibility of novel therapeutic approaches for diseases with a neurogenic inflammatory component. However, it is important to keep in mind that ROS have widespread effects that can be protective as well as detrimental.

### 4.7.2 $H_2O_2$ as a TRPA1 Receptor Agonist

The potent vasoactive effects of $H_2O_2$ were discussed in detail in 4.4; however, in this section we will focus on its potential action as a TRPA1 receptor agonist. The

TRPA1 receptor is activated directly by a range of structurally diverse chemicals such as mustard oil and cinnamaldehyde. The central characteristic that unites the majority of these direct agonists is not their structure, but their chemical reactivity. Two independent groups have shown that electrophilic agonists activate TRPA1 through covalent modifications of nucleophilic cysteine side chains in the intracellular N-terminus of the receptor [92, 105]. Depending on the agonist, there are different types of chemical reactions that occur between the agonist and the receptor, such as Michael addition or conjugation reactions, to form a receptor adduct and activate TRPA1 [92]. Thus it seems that the TRPA1 receptor can act as a molecular sensor for reactive, electrophilic agents (see Fig. 4.2).

$H_2O_2$ is a reactive chemical that was previously shown to be able to oxidise cysteine residues in proteins [106]. In addition, it is known to be cell permeable and can reach the intracellular N-terminus, which means that it fulfils some essential requirements of a direct TRPA1 receptor agonist. In fact, $H_2O_2$ can directly activate TRPA1 in vitro and in vivo, and thus represents a potent and endogenously produced TRPA1 receptor agonist [87]. In vitro $H_2O_2$ activates TRPA1 receptors expressed in CHO cells or HEK293 cells, as well as in isolated TRPA1 expressing sensory nerves [87, 107, 108]. Andersson and colleagues demonstrated that $H_2O_2$ promotes the formation of disulphide bonds to activate TRPA1 [87]. UVA light exposure activates TRPA1 receptors expressed in a HEK293 cell line, as well as in TRPA1 expressing cultured DRG neurons [109]. UVA light exposure is known to result in oxidative stress, and indeed the effects of UVA light exposure were mimicked by $H_2O_2$ [109]. In vivo intraplantar injections of $H_2O_2$ leads to acute nocifensive behaviors, mechanical and thermal hyperalgesia, as well as edema formation [37, 87]. TRPA1 KO mice displayed significantly reduced acute nocifensive behaviors compared to their WT counterparts, demonstrating that the TRPA1 receptor is a major target for $H_2O_2$ [87]. Indeed, a recent study by Keeble et al. examined a possible role for the TRPV1 receptor in mediating the effects of $H_2O_2$, and demonstrated that the TRPV1 receptor only plays a role in the maintenance of $H_2O_2$-induced thermal hyperalgesia until 24 h postinjection, which is probably not due to a direct activation of TRPV1 by $H_2O_2$ [37].

The upper airway is densely innervated by vagal and trigeminal sensory C-fibers that monitor inhaled air for potential threats and insults. They can elicit a protective reflex that consists of respiratory depression, nasal obstruction, sneezing, and coughing, and that is associated with a neurogenic inflammatory component that is mediated by the release of neuropeptides from the nerve endings. The respiratory tract can be exposed to oxidative chemicals that are present, for example, in cigarette smoke and exhaust fumes; as well as oxidative stress associated with various chronic inflammatory diseases of the airways, such as asthma [107, 110, 111]. Recently, Bessac and colleagues showed that $H_2O_2$ and hypochlorite, which are found in cigarette smoke and exhaust fumes, directly trigger respiratory depression via activation of TRPA1 receptors in the trigeminal neurons of the nasal passage in the mouse [107].

These data demonstrate that the TRPA1 receptor can act as a molecular sensor for $H_2O_2$ in vivo and can directly trigger the development of neurogenic inflammation via vascular mechanisms as well as other functional responses (see Fig. 4.2).

The emerging understanding of this novel role of the TRPA1 receptor could mean that TRPA1, like TRPV1, may have a role in modulating vascular responses in pathophysiological conditions.

### *4.7.3 Products of Oxidative Stress as TRPA1 Receptor Agonists*

Inflammation is closely associated with an increased production of ROS, which can overcome endogenous antioxidant defenses and lead to oxidative damage to proteins, lipids, and DNA [46]. As already discussed above, $H_2O_2$ can directly activate TRPA1; however, products of oxidative stress are often highly reactive and might contain electrophilic moieties, thus representing potential TRPA1 receptor agonists [87]. The lipid peroxidation products 4-HNE, 4-oxo-2-nonenal (4-ONE), and 4-hydroxyhexenal (4-HHE), and the prostaglandin metabolite 15-deoxy-$\Delta^{12,14}$-prostaglandin $J_2$ (15d-$PGJ_2$), which is formed nonenzymatically during oxidative stress, were all shown to activate TRPA1 receptors in vitro [87, 88, 108, 110]. 4-ONE, 4-HNE, and 4-HHE activate TRPA1 receptors expressed in CHO cells and HEK cells and in dissociated sensory neurons, most likely via the formation of a Michael adduct with the receptor [87, 110]. The potency of the agonists varied, with 4-ONE being the most potent, followed by 15d-$PGJ_2$, 4-HNE, and 4-HHE [87, 112].

4-HNE was the first lipid peroxidation product that was shown to activate TRPA1 receptors in vitro and trigger neurogenic inflammation and pain in vivo [88]. Trevisani and colleagues demonstrated that 4-HNE is an endogenously produced agonist of TRPA1 that induces the release of substance P and CGRP from peripheral and central sensory nerve endings in vitro (see Fig. 4.2). When injected into the hindpaws of rodents, 4-HNE triggers the development of acute nocifensive behaviors and mechanical hyperalgesia as well as edema formation. The pain-related behaviors were inhibited by pretreatment with TRPA1 receptor antagonists and were absent in TRPA1 receptor KO mice. Similarly, intraplantar injections of 15d-$PGJ_2$ induced acute nocifensive behaviors in WT but not TRPA1 KO mice [87].

4-ONE activates mouse bronchopulmonary C-fibers and triggers the contraction of isolated guinea pig bronchi in vitro [110]. This response was mediated by the release of tachykinins from nerve endings following the activation of TRPA1 [110]. Incidentally, 4-HNE could not elicit any functional responses here, demonstrating again that 4-ONE is a more potent agonist of TRPA1. We observed that 4-ONE induces long lasting mechanical hyperalgesia and edema formation at a dose of 10 nmol/50 μl, whereas Trevisani and colleagues gave 150 nmol/50 μl 4-HNE to induce mechanical hyperalgesia and edema [113]. Interestingly, it was shown that 4-ONE could also activate TRPV1 receptors at higher concentrations, albeit without eliciting a functional response [110].

These data demonstrate that the TRPA1 receptor serves a unique function in the detection of the oxidative state of a tissue, since it is directly activated by ROS and their downstream products. Moreover, activation of this receptor by ROS leads to the activation of sensory nerves and subsequent development of neurogenic inflammation and pain. This demonstrates that the TRPA1 receptor is an important molecular sensor for oxidising chemicals that integrates information and is essential for eliciting physiological and pathophysiological responses to ROS in vivo.

## 4.8 Conclusions and Therapeutic Implications

In conclusion, detailed studies over the years have revealed the intimate links of sensory nerve fibers with the vasculature and the dense nature of the perivascular nerve network in some vascular beds. We now know that the major sensory neuropeptides, despite their potent vasodilator activity, do not contribute to the baseline control of blood pressure in any major way. Instead, they are more likely to be involved in the regional regulation of blood flow and vascular inflammation in the dysfunctional or stressed vascular bed in both damaging and protective roles, depending on the site and tissue involved. This is in keeping with the knowledge that during oxidative stress ROS or lipid peroxidation products are upregulated in the cardiovascular system in disease (e.g., as observed in a rodent aortic constriction model as soon as three days after initiation [114]). Thus the nerves and the oxidants are ideally placed to interact with each other to influence the onset of cardiovascular disease. We provide evidence of interactions between TRPV1 receptors, the major neuropeptides, and ROS. Most recently a novel TRPV1 receptor agonist, N-oleoyldopamine, has been shown to protect the isolated heart against cardiac ischemic-reperfusion injury, demonstrating the protective potential of TRPV1 agonists in the heart [115]. On the other hand, the potent role of ROS and lipid peroxidation mechanisms in activating the TRPA1 channel in sensory nerves is only now being realized. Whilst we are beginning to understand the potential of this mechanism in influencing pain and neurogenic inflammation, the importance to the cardiovascular field has yet to be realized. This is a new and exciting area of cardiovascular research, in which we believe that there is potential for influential interactions between sensory nerves and ROS.

**Acknowledgments** This work was supported by the British Heart Foundation.

## References

1. Holzer P (1991) Capsaicin: cellular targets, mechanisms of action, and selectivity for thin sensory neurons. Pharmacol Rev 43:143–201
2. Maggi CA, Meli A (1988) The sensory-efferent function of capsaicin-sensitive sensory neurons. Gen Pharmacol 19:1–43
3. Szolcsanyi J (2004) Forty years in capsaicin research for sensory pharmacology and physiology. Neuropeptides 38:377–384
4. Brain SD, Cox HM (2006) Neuropeptides and their receptors: innovative science providing novel therapeutic targets. Br J Pharmacol 147(Suppl 1):S202–S211
5. Brain SD, Grant AD (2004) Vascular actions of calcitonin gene-related peptide and adrenomedullin. Physiol Rev 84:903–934

6. Gibran NS, Tamura R, Tsou R et al (2003) Human dermal microvascular endothelial cells produce nerve growth factor: implications for wound repair. Shock 19:127–130
7. Bayliss WM (1901) On the origin from the spinal cord of the vaso-dilator fibres of the hind-limb, and on the nature of these fibres. J Physiol 26:173–209
8. Lewis T (1927) The blood vessels of the human skin and their responses. Shaw and Sons, London
9. Jancso N, Jancso-Gabor A, Szolcsanyi J (1967) Direct evidence for neurogenic inflammation and its prevention by denervation and by pretreatment with capsaicin. Br J Pharmacol Chemother 31:138–151
10. Jancso G, Kiraly E, Jancso-Gabor A (1977) Pharmacologically induced selective degeneration of chemosensitive primary sensory neurones. Nature 270:741–743
11. Caterina MJ, Leffler A, Malmberg AB et al (2000) Impaired nociception and pain sensation in mice lacking the capsaicin receptor. Science 288:306–313
12. Davis JB, Gray J, Gunthorpe MJ et al (2000) Vanilloid receptor-1 is essential for inflammatory thermal hyperalgesia. Nature 405:183–187
13. Zhong B, Wang DH (2007) TRPV1 gene knockout impairs preconditioning protection against myocardial injury in isolated perfused hearts in mice. Am J Physiol Heart Circ Physiol 293:H1791–H1798
14. Banvolgyi A, Palinkas L, Berki T et al (2005) Evidence for a novel protective role of the vanilloid TRPV1 receptor in a cutaneous contact allergic dermatitis model. J Neuroimmunol 169:86–96
15. Clark N, Keeble J, Fernandes ES et al (2007) The transient receptor potential vanilloid 1 (TRPV1) receptor protects against the onset of sepsis after endotoxin. FASEB J 21: 3747–3755
16. Szallasi A, Cortright DN, Blum CA et al (2007) The vanilloid receptor TRPV1: 10 years from channel cloning to antagonist proof-of-concept. Nat Rev Drug Discov 6:357–372
17. Lelkes E, Unsworth BR, Lelkes PI (2001) Reactive oxygen species, apoptosis and altered NGF-induced signaling in PC12 pheochromocytoma cells cultured in elevated glucose: an in vitro cellular model for diabetic neuropathy. Neurotox Res 3:189–203
18. Russell JW, Golovoy D, Vincent AM et al (2002) High glucose-induced oxidative stress and mitochondrial dysfunction in neurons. FASEB J 16:1738–1748
19. Vincent AM, Stevens MJ, Backus C et al (2005) Cell culture modeling to test therapies against hyperglycemia-mediated oxidative stress and injury. Antioxid Redox Signal 7:1494–1506
20. Ibi M, Matsuno K, Shiba D et al (2008) Reactive oxygen species derived from NOX1/NADPH oxidase enhance inflammatory pain. J Neurosci 28:9486–9494
21. Oltman CL, Davidson EP, Coppey LJ et al (2008) Attenuation of vascular/neural dysfunction in Zucker rats treated with enalapril or rosuvastatin. Obesity (Silver Spring) 16:82–89
22. Touyz RM (2005) Reactive oxygen species as mediators of calcium signaling by angiotensin II: implications in vascular physiology and pathophysiology. Antioxid Redox Signal 7:1302–1314
23. Miller AA, Drummond GR, Sobey CG (2006) Reactive oxygen species in the cerebral circulation: are they all bad? Antioxid Redox Signal 8:1113–1120
24. Cai H (2005) Hydrogen peroxide regulation of endothelial function: origins, mechanisms, and consequences. Cardiovasc Res 68:26–36
25. Lee MY, Griendling KK (2008) Redox signaling, vascular function, and hypertension. Antioxid Redox Signal 10:1045–1059
26. Zembowicz A, Hatchett RJ, Jakubowski AM et al (1993) Involvement of nitric oxide in the endothelium-dependent relaxation induced by hydrogen peroxide in the rabbit aorta. Br J Pharmacol 110:151–158
27. Drummond GR, Cai H, Davis ME et al (2000) Transcriptional and posttranscriptional regulation of endothelial nitric oxide synthase expression by hydrogen peroxide. Circ Res 86:347–354

28. Cai H, Davis ME, Drummond GR et al (2001) Induction of endothelial NO synthase by hydrogen peroxide via a Ca(2+)/calmodulin-dependent protein kinase II/janus kinase 2-dependent pathway. Arterioscler Thromb Vasc Biol 21:1571–1576
29. Yada T, Shimokawa H, Hiramatsu O et al (2003) Hydrogen peroxide, an endogenous endothelium-derived hyperpolarizing factor, plays an important role in coronary autoregulation in vivo. Circulation 107:1040–1045
30. Yada T, Shimokawa H, Hiramatsu O et al (2006) Cardioprotective role of endogenous hydrogen peroxide during ischemia-reperfusion injury in canine coronary microcirculation in vivo. Am J Physiol Heart Circ Physiol 291:H1138–H1146
31. Yada T, Shimokawa H, Hiramatsu O et al (2007) Important role of endogenous hydrogen peroxide in pacing-induced metabolic coronary vasodilation in dogs in vivo. J Am Coll Cardiol 50:1272–1278
32. Xu Y, Liu B, Zweier JL et al (2008) Formation of hydrogen peroxide and reduction of peroxynitrite via dismutation of superoxide at reperfusion enhances myocardial blood flow and oxygen consumption in postischemic mouse heart. J Pharmacol Exp Ther 327:402–410
33. Capettini L, Cortes SF, Gomes MA et al (2008) Neuronal nitric oxide synthase-derived hydrogen peroxide is a major endothelium-dependent relaxing factor. Am J Physiol Heart Circ Physiol 295:H2503–H2511
34. Matoba T, Shimokawa H (2003) Hydrogen peroxide is an endothelium-derived hyperpolarizing factor in animals and humans. J Pharmacol Sci 92:1–6
35. Sobey CG, Heistad DD, Faraci FM (1997) Mechanisms of bradykinin-induced cerebral vasodilatation in rats. Evidence that reactive oxygen species activate K+ channels. Stroke 28:2290–2294; discussion 2295
36. Kevil CG, Ohno N, Gute DC et al (1998) Role of cadherin internalization in hydrogen peroxide-mediated endothelial permeability. Free Radic Biol Med 24:1015–1022
37. Keeble JE, Bodkin JV, Liang L et al (2008) Hydrogen peroxide is a novel mediator of inflammatory hyperalgesia, acting via Transient Receptor Potential Vanilloid 1-dependent and independent pathways. Pain. doi: 10.1016/j.pain.2008.10.025
38. Cai H, Harrison DG (2000) Endothelial dysfunction in cardiovascular diseases: the role of oxidant stress. Circ Res 87:840–844
39. Busija DW, Miller AW, Katakam P et al (2006) Adverse effects of reactive oxygen species on vascular reactivity in insulin resistance. Antioxid Redox Signal 8:1131–1140
40. Kerr S, Brosnan MJ, McIntyre M et al (1999) Superoxide anion production is increased in a model of genetic hypertension: role of the endothelium. Hypertension 33:1353–1358
41. Warnholtz A, Nickenig G, Schulz E et al (1999) Increased NADH-oxidase-mediated superoxide production in the early stages of atherosclerosis: evidence for involvement of the renin-angiotensin system. Circulation 99:2027–2033
42. Just A, Olson AJ, Whitten CL et al (2007) Superoxide mediates acute renal vasoconstriction produced by angiotensin II and catecholamines by a mechanism independent of nitric oxide. Am J Physiol Heart Circ Physiol 292:H83–H92
43. Wei EP, Christman CW, Kontos HA et al (1985) Effects of oxygen radicals on cerebral arterioles. Am J Physiol 248:H157–H162
44. Cosentino F, Sill JC, Katusic ZS (1994) Role of superoxide anions in the mediation of endothelium-dependent contractions. Hypertension 23:229–235
45. Didion SP, Faraci FM (2002) Effects of NADH and NADPH on superoxide levels and cerebral vascular tone. Am J Physiol Heart Circ Physiol 282:H688–H695
46. Wang ZQ, Porreca F, Cuzzocrea S et al (2004) A newly identified role for superoxide in inflammatory pain. J Pharmacol Exp Ther 309:869–878
47. Holzer P (1998) Implications of tachykinins and calcitonin gene-related peptide in inflammatory bowel disease. Digestion 59:269–283
48. Foreman JC, Jordan CC, Oehme P et al (1983) Structure-activity relationships for some substance P-related peptides that cause wheal and flare reactions in human skin. J Physiol 335:449–465

49. Lembeck F, Holzer P, Substance P (1979) as neurogenic mediator of antidromic vasodilation and neurogenic plasma extravasation. Naunyn Schmiedebergs Arch Pharmacol 310:175–183
50. Cao T, Gerard NP, Brain SD (1999) Use of NK(1) knockout mice to analyze substance P-induced edema formation. Am J Physiol 277:R476–R481
51. Amara SG, Jonas V, Rosenfeld MG et al (1982) Alternative RNA processing in calcitonin gene expression generates mRNAs encoding different polypeptide products. Nature 298:240–244
52. Rosenfeld MG, Mermod JJ, Amara SG et al (1983) Production of a novel neuropeptide encoded by the calcitonin gene via tissue-specific RNA processing. Nature 304:129–135
53. Salvatore CA, Hershey JC, Corcoran HA et al (2008) Pharmacological characterization of MK-0974 [N-[(3R,6S)-6-(2,3-difluorophenyl)-2-oxo-1-(2,2,2-trifluoroethyl)azepan-3-yl]-4-(2-oxo-2,3-dihydro-1H-imidazo[4,5-b]pyridin-1-yl)piperidine-1-carbox amide], a potent and orally active calcitonin gene-related peptide receptor antagonist for the treatment of migraine. J Pharmacol Exp Ther 324:416–421
54. Rudolf K, Eberlein W, Engel W et al (2005) Development of human calcitonin gene-related peptide (CGRP) receptor antagonists. 1. Potent and selective small molecule CGRP antagonists. 1-[N2-[3,5-dibromo-N-[[4-(3,4-dihydro-2(1H)-oxoquinazolin-3-yl)-1-piperidinyl]carbonyl]-D-tyrosyl]-l-lysyl]-4-(4-pyridinyl)piperazine: the first CGRP antagonist for clinical trials in acute migraine. J Med Chem 48:5921–5931
55. Brain SD, Williams TJ, Tippins JR et al (1985) Calcitonin gene-related peptide is a potent vasodilator. Nature 313:54–56
56. Holzer P (2007) Role of visceral afferent neurons in mucosal inflammation and defense. Curr Opin Pharmacol 7:563–569
57. Kwiecien S, Brzozowski T, Konturek PC et al (2002) The role of reactive oxygen species in action of nitric oxide-donors on stress-induced gastric mucosal lesions. J Physiol Pharmacol 53:761–773
58. Gazzieri D, Trevisani M, Springer J et al (2007) Substance P released by TRPV1-expressing neurons produces reactive oxygen species that mediate ethanol-induced gastric injury. Free Radic Biol Med 43:581–589
59. Mair J, Lechleitner P, Langle T et al (1990) Plasma CGRP in acute myocardial infarction. Lancet 335:168
60. Uren NG, Seydoux C, Davies GJ (1993) Effect of intravenous calcitonin gene related peptide on ischaemia threshold and coronary stenosis severity in humans. Cardiovasc Res 27:1477–1481
61. Lu R, Li YJ, Deng HW (1999) Evidence for calcitonin gene-related peptide-mediated ischemic preconditioning in the rat heart. Regul Pept 82:53–57
62. Tang ZL, Dai W, Li YJ et al (1999) Involvement of capsaicin-sensitive sensory nerves in early and delayed cardioprotection induced by a brief ischaemia of the small intestine. Naunyn Schmiedebergs Arch Pharmacol 359:243–247
63. Nong YH, Titus RG, Ribeiro JM et al (1989) Peptides encoded by the calcitonin gene inhibit macrophage function. J Immunol 143:45–49
64. Schaeffer C, Thomassin L, Rochette L et al (2003) Apoptosis induced in vascular smooth muscle cells by oxidative stress is partly prevented by pretreatment with CGRP. Ann N Y Acad Sci 1010:733–737
65. Sueur S, Pesant M, Rochette L et al (2005) Antiapoptotic effect of calcitonin gene-related peptide on oxidative stress-induced injury in H9c2 cardiomyocytes via the RAMP1/CRLR complex. J Mol Cell Cardiol 39:955–963
66. Zhang Y, Lu L, Furlonger C et al (2000) Hemokinin is a hematopoietic-specific tachykinin that regulates B lymphopoiesis. Nat Immunol 1:392–397
67. Tottrup A, Kraglund K (2004) Endothelium-dependent responses in small human mesenteric arteries. Physiol Res 53:255–263
68. Beny JL, von der Weid PY (1991) Hydrogen peroxide: an endogenous smooth muscle cell hyperpolarizing factor. Biochem Biophys Res Commun 176:378–384

69. Edwards G, Feletou M, Gardener MJ et al (2001) Further investigations into the endothelium-dependent hyperpolarizing effects of bradykinin and substance P in porcine coronary artery. Br J Pharmacol 133:1145–1153
70. Burnham MP, Bychkov R, Feletou M et al (2002) Characterization of an apamin-sensitive small-conductance Ca(2+)-activated K(+) channel in porcine coronary artery endothelium: relevance to EDHF. Br J Pharmacol 135:1133–1143
71. Bychkov R, Burnham MP, Richards GR et al (2002) Characterization of a charybdotoxin-sensitive intermediate conductance Ca2+-activated K+ channel in porcine coronary endothelium: relevance to EDHF. Br J Pharmacol 137:1346–1354
72. Grocott-Mason R, Anning P, Evans H et al (1994) Modulation of left ventricular relaxation in isolated ejecting heart by endogenous nitric oxide. Am J Physiol 267:H1804–H1813
73. Yatani A, Yokoyama M, Akita H et al (1990) Endothelium-dependent vasodilating effect of substance P during flow-reducing coronary stenosis in the dog. J Am Coll Cardiol 15: 1374–1384
74. MacCarthy PA, Grieve DJ, Li JM et al (2001) Impaired endothelial regulation of ventricular relaxation in cardiac hypertrophy: role of reactive oxygen species and NADPH oxidase. Circulation 104:2967–2974
75. Khodr B, Khalil Z (2001) Modulation of inflammation by reactive oxygen species: implications for aging and tissue repair. Free Radic Biol Med 30:1–8
76. Brunelleschi S, Tarli S, Giotti A et al (1991) Priming effects of mammalian tachykinins on human neutrophils. Life Sci 48:PL1–PL5
77. Serra MC, Calzetti F, Ceska M et al (1994) Effect of substance P on superoxide anion and IL-8 production by human PMNL. Immunology 82:63–69
78. Sterner-Kock A, Braun RK, van der Vliet A et al (1999) Substance P primes the formation of hydrogen peroxide and nitric oxide in human neutrophils. J Leukoc Biol 65:834–840
79. Tanabe T, Otani H, Bao L et al (1996) Intracellular signaling pathway of substance P-induced superoxide production in human neutrophils. Eur J Pharmacol 299:187–195
80. Dianzani C, Parrini M, Ferrara C et al (1996) Effect of 4-hydroxynonenal on superoxide anion production from primed human neutrophils. Cell Biochem Funct 14:193–200
81. Hartung HP, Toyka KV (1983) Activation of macrophages by substance P: induction of oxidative burst and thromboxane release. Eur J Pharmacol 89:301–305
82. Caterina MJ, Schumacher MA, Tominaga M et al (1997) The capsaicin receptor: a heat-activated ion channel in the pain pathway. Nature 389:816–824
83. Trevisani M, Gazzieri D, Benvenuti F et al (2004) Ethanol causes inflammation in the airways by a neurogenic and TRPV1-dependent mechanism. J Pharmacol Exp Ther 309:1167–1173
84. Nilius B, Owsianik G, Voets T et al (2007) Transient receptor potential cation channels in disease. Physiol Rev 87:165–217
85. Geppetti P, Nassini R, Materazzi S et al (2008) The concept of neurogenic inflammation. BJU Int 101(Suppl 3):2–6
86. Maher M, Ao H, Banke T et al (2008) Activation of TRPA1 by farnesyl thiosalicylic acid. Mol Pharmacol 73:1225–1234
87. Andersson DA, Gentry C, Moss S et al (2008) Transient receptor potential A1 is a sensory receptor for multiple products of oxidative stress. J Neurosci 28:2485–2494
88. Trevisani M, Siemens J, Materazzi S et al (2007) 4-Hydroxynonenal, an endogenous aldehyde, causes pain and neurogenic inflammation through activation of the irritant receptor TRPA1. Proc Natl Acad Sci U S A 104:13519–13524
89. Bandell M, Story GM, Hwang SW et al (2004) Noxious cold ion channel TRPA1 is activated by pungent compounds and bradykinin. Neuron 41:849–857
90. Bautista DM, Movahed P, Hinman A et al (2005) Pungent products from garlic activate the sensory ion channel TRPA1. Proc Natl Acad Sci U S A 102:12248–12252
91. Bautista DM, Jordt SE, Nikai T et al (2006) TRPA1 mediates the inflammatory actions of environmental irritants and proalgesic agents. Cell 124:1269–1282

92. Macpherson LJ, Dubin AE, Evans MJ et al (2007) Noxious compounds activate TRPA1 ion channels through covalent modification of cysteines. Nature 445:541–545
93. Dai Y, Wang S, Tominaga M et al (2007) Sensitization of TRPA1 by PAR2 contributes to the sensation of inflammatory pain. J Clin Invest 117:1979–1987
94. Wang S, Dai Y, Fukuoka T et al (2008) Phospholipase C and protein kinase A mediate bradykinin sensitization of TRPA1: a molecular mechanism of inflammatory pain. Brain 131:1241–1251
95. Kwan KY, Allchorne AJ, Vollrath MA et al (2006) TRPA1 contributes to cold, mechanical, and chemical nociception but is not essential for hair-cell transduction. Neuron 50: 277–289
96. Schuligoi R (1998) Effect of colchicine on nerve growth factor-induced leukocyte accumulation and thermal hyperalgesia in the rat. Naunyn Schmiedebergs Arch Pharmacol 358:264–269
97. Puntambekar P, Mukherjea D, Jajoo S et al (2005) Essential role of Rac1/NADPH oxidase in nerve growth factor induction of TRPV1 expression. J Neurochem 95: 1689–1703
98. Pabbidi RM, Cao DS, Parihar A et al (2008) Direct role of streptozotocin in inducing thermal hyperalgesia by enhanced expression of transient receptor potential vanilloid 1 in sensory neurons. Mol Pharmacol 73:995–1004
99. Kitahara T, Li HS, Balaban CD (2005) Changes in transient receptor potential cation channel superfamily V (TRPV) mRNA expression in the mouse inner ear ganglia after kanamycin challenge. Hear Res 201:132–144
100. Starr A, Graepel R, Keeble J et al (2008) A reactive oxygen species-mediated component in neurogenic vasodilatation. Cardiovasc Res 78:139–147
101. Wang L, Wang DH (2005) TRPV1 gene knockout impairs postischemic recovery in isolated perfused heart in mice. Circulation 112:3617–3623
102. Ligresti A, Moriello AS, Starowicz K et al (2006) Antitumor activity of plant cannabinoids with emphasis on the effect of cannabidiol on human breast carcinoma. J Pharmacol Exp Ther 318:1375–1387
103. Keeble J, Russell F, Curtis B et al (2005) Involvement of transient receptor potential vanilloid 1 in the vascular and hyperalgesic components of joint inflammation. Arthritis Rheum 52:3248–3256
104. Hu F, Sun WW, Zhao XT et al (2008) TRPV1 mediates cell death in rat synovial fibroblasts through calcium entry-dependent ROS production and mitochondrial depolarization. Biochem Biophys Res Commun 369:989–993
105. Hinman A, Chuang HH, Bautista DM et al (2006) TRP channel activation by reversible covalent modification. Proc Natl Acad Sci USA 103:19564–19568
106. Poole LB, Karplus PA, Claiborne A (2004) Protein sulfenic acids in redox signaling. Annu Rev Pharmacol Toxicol 44:325–347
107. Bessac BF, Sivula M, von Hehn CA et al (2008) TRPA1 is a major oxidant sensor in murine airway sensory neurons. J Clin Invest 118:1899–1910
108. Sawada Y, Hosokawa H, Matsumura K et al (2008) Activation of transient receptor potential ankyrin 1 by hydrogen peroxide. Eur J Neurosci 27:1131–1142
109. Hill K, Schaefer M (2008) Ultraviolet light and photosensitising agents activate TRPA1 via generation of oxidative stress. Cell Calcium:
110. Taylor-Clark TE, McAlexander MA, Nassenstein C et al (2008) Relative contributions of TRPA1 and TRPV1 channels in the activation of vagal bronchopulmonary C-fibres by the endogenous autacoid 4-oxononenal. J Physiol 586:3447–3459
111. Simon SA, Liedtke W (2008) How irritating: the role of TRPA1 in sensing cigarette smoke and aerogenic oxidants in the airways. J Clin Invest 118:2383–2386
112. Taylor-Clark TE, Undem BJ, Macglashan DW Jr et al (2008) Prostaglandin-induced activation of nociceptive neurons via direct interaction with transient receptor potential A1 (TRPA1). Mol Pharmacol 73:274–281

113. Graepel R, Andersson DA, Bevan S et al. (2008) 4-ONE, a product of lipid peroxidation, induces mechanical hyperalgesia and oedema formation. Joint European Neuropeptide Club and European Opioid Conference, Ferrara.
114. Kai H, Mori T, Tokuda K et al (2006) Pressure overload-induced transient oxidative stress mediates perivascular inflammation and cardiac fibrosis through angiotensin II. Hypertens Res 29:711–718
115. Zhong B, Wang DHN- (2008) Oleoyldopamine, a novel endogenous capsaicin-like lipid, protects the heart against ischemia-reperfusion injury via activation of TRPV1. Am J Physiol Heart Circ Physiol 295:H728–H735

# Chapter 5
# Mitochondrial Reactive Oxygen Species in Myocardial Pre- and Postconditioning

**Ariel R. Cardoso, Bruno B. Queliconi, and Alicia J. Kowaltowski**

**Abstract** Myocardial ischemia followed by reperfusion is a well established condition of medical importance in which reactive oxygen species (ROS) are determinant for the pathological outcome. Indeed, oxidative damage during reperfusion is causative of many of the complications found after ischemia. ROS leading to postischemic myocardial damage come from many sources, including mitochondria, NADPH oxidase, xanthine oxidase, and infiltrated phagocytes [1]. ROS also can act as signaling molecules in the cardiovascular system, including protecting the heart against myocardial ischemic damage, secondarily to ischemic pre- and postconditioning. In this case, there is ample evidence that the source of signaling ROS is mitochondrial [2–7]. This chapter will briefly review aspects of mitochondrial ROS signaling relevant to myocardial ischemic protection by pre- and postconditioning.

**Keywords** Electron transport chain · Oxidative phosphorylation · Uncoupling proteins · Mitochondrial $K_{ATP}$ channels · Mitochondrial membrane potential · Mitochondrial free radical production

## 5.1 Mitochondrial ROS Generation in the Heart

Mitochondrial ROS generation differs from that in other cellular compartments because it occurs mainly as a byproduct of energy metabolism, and not by enzymes specifically controlled by signaling pathways to produce these species. As a result, it occurs at high rates relative to other cardiovascular sources of ROS, such as NADPH oxidases [1, 8–10].

---

A.J. Kowaltowski (✉)
Departamento de Bioquímica, Instituto de Química, Universidade de São Paulo, São Paulo, SP, Brazil
e-mail: alicia@iq.usp.br

Ariel R. Cardoso and Bruno B. Queliconi have contributed equally.

H. Sauer et al. (eds.), *Studies on Cardiovascular Disorders*,
Oxidative Stress in Applied Basic Research and Clinical Practice,
DOI 10.1007/978-1-60761-600-9_5, © Springer Science+Business Media, LLC 2010

In heart mitochondria, the best documented sources of mitochondrial ROS are electron transport chain complexes I and III. One-electron redox transfers occur continuously within these complexes, and, in the event of $O_2$ access to the electron transferring components, a one-electron reduction of $O_2$ may occur, producing superoxide radical anions ($O_2^{-\bullet}$). The source of electrons promoting $O_2^{-\bullet}$ formation varies according to the substrate provided. In the heart, which relies heavily on fatty acids and ketone bodies as energy sources, electrons are provided mainly from citric acid cycle–generated NADH and flavoenzymes such as acyl CoA and succinate dehydrogenase. Succinate is a particularly important ROS source in heart mitochondria, leading to $O_2^{-\bullet}$ formation because of reverse electron transfer from succinate to complex I, where electron leakage occurs [10, 11]. Because of differences in redox potentials between mitochondrial respiratory complexes, reverse electron transfer is only thermodynamically possible if the mitochondrial inner membrane potential is high. Thus, conditions such as enhanced oxidative phosphorylation, which decrease the mitochondrial inner membrane potential, are efficient methods by which to prevent the generation of mitochondrial ROS in the heart. Conversely, the inhibition of the mitochondrial respiratory chain can lead to electron accumulation at points in which $O_2^{-\bullet}$ is formed, thus increasing mitochondrial ROS release. As a result, in a generalized manner, the faster mitochondrial respiratory rates are, the lower the ROS production by this organelle tends to be [12, 14].

Unfortunately, few studies quantify ROS production in heart mitochondria, because of methodological difficulties. At least in vitro, heart mitochondrial $O_2^{-\bullet}$ production can account for almost 2% of oxygen consumed when respiratory rates are low and succinate is used as a substrate. However, under physiologically relevant conditions such as when oxidative phosphorylation occurs, this production falls under 0.1% when succinate is present, and even lower in its absence [15]. From this simple example, it is clear that the quantities of ROS produced and, consequently, the results of this release, are strongly determined by metabolic conditions, and can vary intensely with changes in energy metabolism.

## 5.2 Regulation of Mitochondrial ROS Generation by Mild Uncoupling Pathways

Although a large amount of focus is placed on antioxidants in cardioprotection, it has increasingly become clear that the regulation of the generation of mitochondrial ROS, rather than their removal after they are already formed, is a crucial process for maintaining cellular redox balance. A highly effective method in which to decrease mitochondrial ROS formation in the heart is to increase $O_2$ consumption rates by uncoupling respiration from oxidative phosphorylation [16]. Uncoupling decreases the reduction of complexes I and III, decreases $O_2$ concentrations in the mitochondrial microenvironment, and prohibits reverse electron transfer, because of the low inner membrane potentials, strongly preventing mitochondrial ROS generation in the heart [12, 14]. If the uncoupling promoted is mild, ROS release can be substantially prevented without seriously hampering oxidative phosphorylation. Indeed, in recent years transport pathways in the inner mitochondrial membrane which lead to mild uncoupling have been identified as important regulators of ROS generation. We will discuss three of these transporters here (summarized in Fig. 5.1): uncoupling proteins, the adenine nucleotide translocator, and ATP-sensitive potassium channels ($mitoK_{ATP}$).

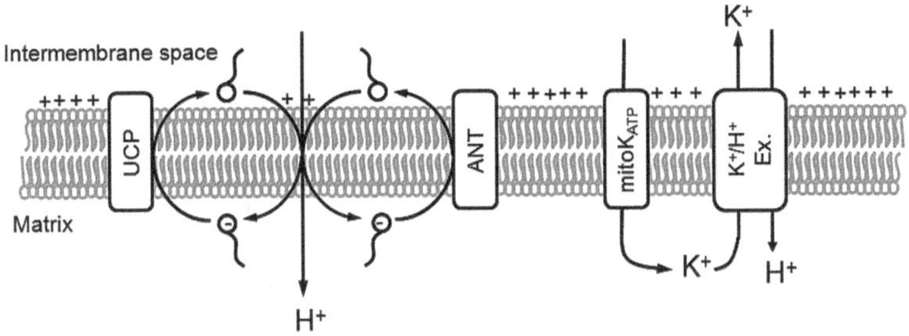


**Fig. 5.1** Mild uncoupling pathways in mitochondria. Uncoupling proteins (UCP) and the adenine nucleotide translocator (ANT) transport fatty acid anions from the mitochondrial matrix into the intermembrane space, where the proton gradient stimulates fatty acid protonation. The protonated fatty acid flip-flops through the lipid bilayer, releasing $H^+$ in the matrix. Mitochondrial ATP-sensitive $K^+$ channels ($mitoK_{ATP}$) transport $K^+$ into the matrix, which is exchanged for $H^+$ by the $K^+/H^+$ exchanger

The first mitochondrial inner membrane transporter described to promote uncoupling was uncoupling protein 1 (UCP1), which is found in brown adipose tissue; it dissipates the membrane potential significantly and can affect ATP synthesis [17–20]. UCP1 has been assigned an important role as a thermogenic protein and as a mechanism to control energy metabolism [21, 22]. Heart mitochondria do not express UCP1, but may present low quantities of UCP2 and UCP3, which are much less active than UCP1 and promote mild uncoupling [23]. Some conditions, such as exercise training and diet, can alter the expression of heart UCP2, which is regulated by peroxisome proliferator–activated receptors (PPARs), suggesting an important metabolic role for these proteins [24, 25]. Unfortunately, it is difficult to determine the activity of UCPs in vivo, and the majority of papers published demonstrate only differences in mRNA or protein levels, but not UCP activity. As a result, little is known to date about the functional consequences of changes in UCP expression. Although their metabolic effects remain to be directly demonstrated, these proteins have been strongly related to the control of the redox state, because of the prevention of ROS release promoted by uncoupling [12, 14, 26].

The most accepted hypothesis regarding the function of UCPs as uncouplers is that they are anion carriers, using fatty acids as physiological substrates. UCPs transport fatty acid anions from the mitochondrial matrix into the intermembrane space, where they are readily protonated. The protonated fatty acid then diffuses across the lipid bilayer and is dissociated into the fatty acid anion plus a proton in the matrix, generating fatty acid cycling that promotes proton leakage across the inner mitochondrial membrane [27] (see Fig. 5.1).

Another protein that can uncouple respiration is the adenine nucleotide translocator (ANT). The mechanism by which this protein decreases the inner membrane potential is similar to that involving UCPs, because it can also translocate fatty acids and promote proton leakage [28–30]. The treatment of isolated mitochondria with carboxyatractyloside, an ANT inhibitor, increases the generation of reactive species in state 4 mitochondria (without oxidative phosphorylation) and decreases oxygen consumption in the presence of free fatty acids, suggesting that uncoupling through ANT could be an important step to downregulate ROS production [31]. Furthermore, experiments with Ant1 null mice show that manganese SOD and glutathione peroxidase are augmented in the heart and muscle to compensate for the increases of $H_2O_2$ generation in the mitochondria [32].

A third regulated mild uncoupling pathway in mitochondria involves potassium cycling. Mitochondrial inner membrane ATP–sensitive potassium channels (mito$K_{ATP}$) allow for $K^+$ uptake into the matrix because of the electrochemical gradient, while the $K^+/H^+$ exchanger promotes electroneutral $K^+$ extrusion at the expense of the proton gradient (see Fig. 5.1). The activity of mito$K_{ATP}$ thus determines uncoupling, which is mild in most tissues because of limited $K^+$ transport [33–36]. Indeed, we have found that mito$K_{ATP}$ is an important regulator of ROS generation in infarction and ischemic preconditioning, as will be discussed below.

## 5.3 Mitochondrial Permeability Transition: A Cell Death–Inducing Consequence of Mitochondrial Oxidative Stress

Under physiological conditions, mitochondrially-generated ROS are in balance with antioxidant systems. However, when ROS generation increases or ROS removal is impaired, these species can lead to substantial alterations of mitochondrial biomolecules. The mitochondrial inner membrane is a specifically vulnerable target to oxidative damage, both because of its role in the generation of respiratory chain–derived ROS and because of the importance in maintaining its impermeability in order to sustain oxidative phosphorylation. Interestingly, the inner mitochondrial membrane is unusual in the sense that it contains more protein than lipids in its composition [9, 37], and thus protein oxidative alterations of the inner mitochondrial membrane are an expected result of excessive mitochondrially-generated ROS.

Mitochondrial permeability transition (MPT) is a consequence of inner mitochondrial membrane protein oxidation and excessive $Ca^{2+}$ uptake by this organelle which leads to a nonselective permeabilization of the inner membrane and loss of phosphorylating ability [38–41]. The permeabilization promoted by MPT involves alterations in specific membrane proteins, as indicated by the ability to regulate this process. Cyclophylins, for example, are of known importance because of the ability of ligands such as cyclosporin A to inhibit MPT. However, the specific composition for the MPT pore is not conserved and specific. Instead, this process seems to be the result of different subsets of oxidized, misfolded, and aggregated membrane proteins, leading to changes in membrane permeability [39, 42]. The differing compositions of MPT pores explain changes in conductance and pore sizes in distinct experimental settings and over time [39, 43, 44].

The result of a loss in oxidative phosphorylation capacity in a large subset of mitochondria within a cell is the failure to maintain ATP levels and necrotic cell death. Indeed, myocardial reperfusion, a condition in which mitochondrial $Ca^{2+}$ uptake and excessive ROS formation occur, has been extensively demonstrated to be accompanied by MPT [2, 4, 38, 40]. MPT is a causal event in myocardial postischemic damage, as indicated by the extensive cardioprotective effects of MPT inhibitors [45, 46].

MPT can also lead to apoptotic cell death, since it promotes the release of mitochondrial pro-apoptotic proteins. Although MPT is not the canonic pathway through which these proteins are released, it is widely believed to participate in "accidental apoptosis," or apoptosis resulting from less extensive damaging stimuli, such as that which is observed in border infarct areas [38, 47]. Clearly, for apoptosis to occur as a result of MPT in a subset of mitochondria, sufficient organelles within that cell must be preserved functionally in order to maintain the high energy phosphate levels necessary to organize apoptotic cell death [38, 40].

## 5.4 Preconditioning and Mitochondrial Redox Signaling

Myocardial preconditioning was first described in 1986, when Murray and coauthors noted that small ischemic periods preceding experimental index myocardial infarction significantly improved the outcome of the tissue [48]. Later, seminal work by Schumacker's group [6] determined that preconditioning depended on moderate increases in ROS generated by mitochondria during the brief ischemic episodes. These signaling increments in ROS levels protected against oxidative stress observed during reperfusion after the index ischemia [49]. Interestingly, ischemic preconditioning is also dependent on the activation of PKCε [50–57], which is regulated by ROS [58]. This indicates that ischemic preconditioning involves a signaling sequence that includes enhancement of mitochondrial ROS release and PKCε activation [59], followed by a prevention of mitochondrial ROS release at reperfusion [3, 49] (see Fig. 5.2).

In parallel, many groups were studying pharmacological mechanisms to protect the ischemic heart, and identified $K^+$ channel openers as highly efficient cardioprotective drugs [60–63]. Initially, the effect was attributed to plasma membrane $K^+$ channel activation, but the work of Garlid and Grover demonstrated that the main targets for these drugs were mito$K_{ATP}$ channels [60]. The finding that mitochondrial ROS were involved in the signaling pathway of preconditioning [3, 6], associated with the recognition of the cardioprotective properties of mito$K_{ATP}$ activation,

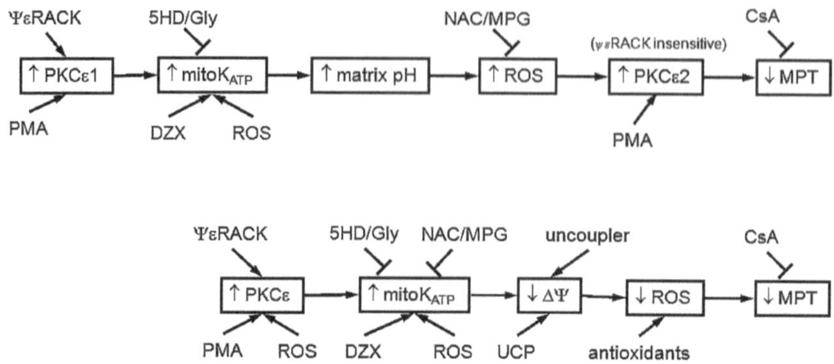


**Fig. 5.2** Two proposals for the sequence of events relating PKCε, mito$K_{ATP}$, ROS, and ischemic cardioprotection: The upper sequence of events indicates the signaling pathway proposed by Garlid and coworkers, in which ROS release increases upon activation of mitochondrial ATP-sensitive $K^+$ channels (mito$K_{ATP}$) channels, and two distinct mitochondrial PKCε pools are present. The lower sequence of events depicts the pathway proposed by our group, in which mito$K_{ATP}$ activation prevents mitochondrial ROS release. PMA, phorbol 12-myristate 13-acetate (a PKCε activator); DZX, diazoxide (a mito$K_{ATP}$ agonist); 5-HD, 5-hydroxydecanoate; Gly, glyburide/glibenclamide; ROS, reactive oxygen species; NAC, N-acetylcysteine; MPG, 2-mercapto-propionyl-glycine; UCP, uncoupling protein; MPT, mitochondrial permeability transition; CsA, cyclosporin A (an MPT inhibitor)

brought a strong focus on this organelle within studies of myocardial preservation. Indeed, the activation of mito$K_{ATP}$, in a process that involves upstream activation of PKCε, is widely recognized as a seminal event in ischemic preconditioning today.

However, the relationship between mitochondrial ROS release and mito$K_{ATP}$ activation during ischemic preconditioning remains controversial. Some groups support the concept that mitochondrial ROS production occurs downstream of mito$K_{ATP}$ activation in preconditioning [64, 65]; while others, including ourselves, have demonstrated that ROS increments in preconditioning occur upstream of mito$K_{ATP}$ activation, and that the activation of these channels involves redox signaling [3, 52, 66–68]. The proposed sequence of events in either case is outlined in Fig. 5.2.

The idea that mito$K_{ATP}$ activation could lead to increased ROS release by mitochondria was constructed upon the finding that cardioprotection by the mito$K_{ATP}$ agonist diazoxide was reversed by the concomitant presence of antioxidants such as N-acetylcysteine and 2-mercaptopropionyl glycine [65, 69, 70]. Unfortunately, these antioxidants, the only ones that to our knowledge were capable of reversing the beneficial effects of diazoxide, are thiol reagents that can interfere directly with the activity of the mito$K_{ATP}$ channel, inhibiting its activation [3, 67, 68, 71]. The idea that mito$K_{ATP}$ promoted mitochondrial ROS release was further supported by measurements of mitochondrial ROS using a new alleged mitochondrial ROS probe, MitoTracker, which was more fluorescent upon the addition of diazoxide [65, 72]. Regrettably, MitoTracker probes turned out to be an unreliable tool [73, 74], and present no response to additions of respiratory inhibitors or uncouplers, classic regulators of mitochondrial ROS release [9, 10], under the same experimental conditions as those in which the effects of mito$K_{ATP}$ agonists were studied [72]. Two groups also reported an increase in the fluorescence of dichlorofluorescein (DCF), a classic, albeit controversial, ROS probe upon treatment with diazoxide in cells [69] and isolated mitochondria [64]. Unfortunately, the results in cells could not be reproduced in many distinct groups [3, 75, 76], and may be attributable to an artifactual increase in DCF fluorescence promoted by diazoxide [3, 75]. In isolated mitochondria, we have been unable to see an increase in DCF fluorescence upon mito$K_{ATP}$ activation and, in fact, see a small decrease [77]. Indeed, the results in which increases in DCF fluorescence were measured are questionable because the probe, which is quite sensitive to changes in pH [78] was loaded into the mitochondrial matrix, which may suffer significant pH changes when mito$K_{ATP}$ is activated [79].

Perhaps the most significant problem with the hypothesis that mito$K_{ATP}$ increases ROS release is the complicated sequence of events necessary to explain the experimental results related to the cardioprotective effects of preconditioning within this standpoint (summarized in Fig. 5.2, upper sequence): The signaling pathway proposed includes an early increase in ROS and nitric oxide levels within the cardiomyocyte, resulting in increased cGMP levels and PKG activation and translocation to mitochondria [50]. Mitochondrial PKCε is then activated, and induces the opening of mito$K_{ATP}$ channels. As a result of the alleged increase in ROS resulting from channel activation, a second, functionally distinct pool of PKCε (dubbed PKCε2 by the authors) is activated, and this activation promotes the inhibition of MPT in a manner determined by changes in phosphorylation [80]. In addition to the complexity created by this pathway, which requires the existence of two functionally distinct yet structurally indistinguishable PKCε pools in mitochondria, as well as two distinct ROS-mediated signaling events, several points remain inconsistent: First, no explanation is offered as to why effects downstream of the alleged increase in ROS release promoted by the mito$K_{ATP}$ opening, including MPT inhibition, are observed in mitochondria treated with rotenone, which functionally dissociates complex I from coenzyme Q, impeding ROS formation through the mechanism the group has described mito$K_{ATP}$ to act through [80]. Second, PKCε2 would have to be insensitive or inaccessible to the activator ψεRACK, since the inhibitory effects of this peptide on MPT are fully reversed by mito$K_{ATP}$ antagonists [80]. The hypothesis is also inconsistent with careful studies demonstrating that MPT inhibition in the preconditioned heart is not related to changes in mitochondrial phosphorylation levels, but instead to an improvement in redox state [81]. Furthermore, it is widely accepted that MPT is inhibited by thiol reduction, and activated by oxidants [7, 9, 39, 40], the exact opposite of the proposal described above. Finally, independent studies have measured increases in ROS release during preconditioning, and demonstrate that they are not inhibited by mito$K_{ATP}$ antagonists, which are, nonetheless, efficient inhibitors of the beneficial effects of preconditioning [3, 6]. This last result clearly demonstrates that ROS release occurs upstream of mito$K_{ATP}$ activation in ischemic preconditioning.

Indeed, many groups have demonstrated that, in addition to being activated by phosphorylation, mito$K_{ATP}$ channels are also triggered by different kinds of ROS [67, 68, 71, 82] and also by nitric oxide [82, 83]. This finding is in line with our idea that mitochondrially-generated ROS occur upstream of mito$K_{ATP}$

activation during preconditioning [3]. There are many reasons why preconditioning can increase ROS release by mitochondria, including changes in oxygen tensions, respiratory inhibition, signaling by NO· (which can promote respiratory inhibition, among other effects), depletion of mitochondrial redox sources such as NADPH, and increased intracellular $Ca^{2+}$ levels [7, 34, 84]. A result of enhanced ROS formation in the mitochondrial microenvironment during preconditioning is the activation of mild mitochondrial uncoupling pathways, including mito$K_{ATP}$, as discussed above. Mito$K_{ATP}$ opening has many consequences which decrease the probability of MPT occurrence during reperfusion: We have consistently found, using different techniques, substrates, and tissues, that mito$K_{ATP}$ activation prevents mitochondrial ROS formation, a result compatible with its mild uncoupling effects [34, 67, 71, 77]. Indeed, other groups have associated the opening of mitochondrial $K^+$ channels with decreases in ROS formation in vivo [49, 85]. Furthermore, mito$K_{ATP}$ opening during ischemia prevents the loss of intracellular ATP, an MPT inhibitor [41, 66]. Mito$K_{ATP}$ opening may also prevent mitochondrial $Ca^{2+}$ uptake during ischemia [66, 86], a necessary stimulus for MPT.

In addition to activating mito$K_{ATP}$, increases in ROS levels and PKCε activation have other important mitochondrial effects which are involved in cardioprotection. A new study suggests PKCε activates mitochondrial aldehyde dehydrogenase, which is important in removing toxic aldehydes which accumulate during ischemia [87]. Although the activation of this enzyme is not surprising in a study that induced protection through treatment with ethanol, the authors were able to show that small molecule activators of this enzyme were sufficient to induce cardioprotection, a highly interesting finding, which opens the possibility of a novel mitochondrial cardioprotective target. Other studies have shown that other mild uncoupling pathways distinct from mito$K_{ATP}$ in mitochondria are activated by ischemic preconditioning, including the adenine nucleotide translocator activity of transporting fatty acids and, possibly, uncoupling proteins [88, 89].

Indeed, many studies demonstrate that promoting mitochondrial uncoupling is in itself cardioprotective. Both treatment with uncouplers and expression of uncoupling proteins have been found to be protective to the ischemic heart and brain [90, 91]. Altogether, these studies indicate that decreasing the efficiency of mitochondrial energy metabolism and, hence, the generation of ROS, is a highly interesting target for cardioprotective interventions.

## 5.5 Postconditioning and Mitochondrial Redox Signaling

While ischemic preconditioning attracted a lot of attention because a comprehension of the mechanisms underlying this process could uncover interesting cardioprotective targets, a new form of cardioprotection, postconditioning, may present immediate clinical applicability. Postconditioning consists in promoting 2–3 discontinued reperfusion periods after the index ischemic event, and provides significant protection against reperfusion injury. It was first described in 1996 [92], but gained significant attention only in the last few years.

Since this is a relatively new finding, the mechanisms involved are still poorly understood, but there is evidence that many pathways involved in preconditioning also participate in postconditioning. Postconditioning is inhibited by mito$K_{ATP}$ and PKC antagonists, suggesting it involves activation of these proteins [93, 94]. The process also prevents mitochondrial oxidative stress associated with reperfusion [5, 95, 96]. Furthermore, postconditioning has been suggested to prevent MPT [97–99], although some experimental approaches used in these studies are questionable, and direct in situ measurements have not yet been conducted [66]. One study suggests postconditioning is also dependent on increments in ROS levels based on the effects of antioxidants [100]. Unfortunately, this study used only thiol antioxidants, and the effects can therefore be ascribed to mito$K_{ATP}$ inhibition (and possibly PKCε, which is also regulated by thiol redox state [101], inhibition). Indeed, the effects of thiol antioxidants in postconditioning further confirm that these compounds have cellular effects unrelated to ROS, since reperfusion is widely associated with largely enhanced ROS release rates, and postconditioning prevents oxidative myocardial damage [95, 96].

Altogether, it is inviting to speculate that postconditioning may reduce oxidative stress at reperfusion because of its intermittent nature and, perhaps, by allowing for the activation of mitochondrial uncoupling pathways. As a result, consequences of mitochondrial oxidative stress such as MPT would decrease in the tissue.

## 5.6 Concluding Remarks

A large collection of data shows that oxidative damage during reperfusion is related to changes in mitochondrial ROS release. It is thus not surprising that mitochondrially-generated ROS also are being uncovered as signaling molecules within cardioprotective settings such as ischemic pre- and postconditioning. Altogether, these data demonstrate that the regulation of mitochondrial redox metabolism is an important target for therapeutic strategies in cardioprotection. Importantly, these data demonstrate that, because ROS can be both protective and damaging for molecules, there is no simple one-for-all solution, and antioxidant therapies must be cautiously evaluated.

## References

1. Dröge W (2002) Free radicals in the physiological control of cell function. Physiol Rev 82:47–95
2. Di Lisa F, Bernardi P (2006) Mitochondria and ischemia-reperfusion injury of the heart: fixing a hole. Cardiovasc Res 70:191–199
3. Facundo HTF, Carreira RS, de Paula JG, Santos CCX, Ferranti R, Laurindo FRM, Kowaltowski AJ (2006) Ischemic preconditioning requires increases in reactive oxygen release independent of mitochondrial $K^+$ channel activity. Free Radic Biol Med 40:469–479

4. Halestrap AP, Clarke SJ, Javadov SA (2004) Mitochondrial permeability transition pore opening during myocardial reperfusion-a target for cardioprotection. Cardiovasc Res 61:372–385
5. Sun H, Wang N, Kerendi F, Halkos M, Kin H, Guyton RA, Vinten-Johansen J, Zhao Z (2005) Hypoxic postconditioning reduces cardiomyocyte loss by inhibiting ROS generation and intracellular $Ca^{2+}$ overload. Am J Physiol Heart Circ Physiol 288:H1900–H1908
6. Vanden Hoek TL, Becker LB, Shao Z, Li C, Schumacker PT (1998) Reactive oxygen species released from mitochondria during brief hypoxia induce preconditioning in cardiomyocytes. J Biol Chem 273:18092–18098
7. da Silva MM, Sartori A, Belisle E, Kowaltowski AJ (2003) Ischemic preconditioning inhibits mitochondrial respiration, increases $H_2O_2$ release, and enhances $K^+$ transport. Am J Physiol Heart Circ Physiol 285:H154–H162
8. Brookes PS, Levonen A, Shiva S, Sarti P, Darley-Usmar VM (2002) Mitochondria: regulators of signal transduction by reactive oxygen and nitrogen species. Free Radic Biol Med 33:755–764
9. Kowaltowski AJ, Vercesi AE (1999) Mitochondrial damage induced by conditions of oxidative stress. Free Radic Biol Med 26:463–471
10. Turrens JF (2003) Mitochondrial formation of reactive oxygen species. J Physiol (Lond) 552:335–344
11. Liu Y, Fiskum G, Schubert D (2002) Generation of reactive oxygen species by the mitochondrial electron transport chain. J Neurochem 80:780–787
12. Brookes PS (2005) Mitochondrial $H^+$ leak and ROS generation: an odd couple. Free Radic Biol Med 38:12–23
13. Korshunov SS, Skulachev VP, Starkov AA (1997) High protonic potential actuates a mechanism of production of reactive oxygen species in mitochondria. FEBS Lett 416:15–18
14. Skulachev VP (1998) Uncoupling: new approaches to an old problem of bioenergetics. Biochim Biophys Acta 1363:100–124
15. Tahara EB, Navarete FD, Kowaltowski AJ (2009) Tissue-, substrate-, and site-specific characteristics of mitochondrial reactive oxygen species generation. Free Radic Biol Med 46:1283–1297
16. Caldeira da Silva CC, Cerqueira FM, Barbosa LF, Medeiros MHG, Kowaltowski AJ (2008) Mild mitochondrial uncoupling in mice affects energy metabolism, redox balance and longevity. Aging Cell 7:552–560
17. Huang SG, Klingenberg M (1996) Chloride channel properties of the uncoupling protein from brown adipose tissue mitochondria: a patch-clamp study. Biochemistry 35:16806–16814
18. Jacobsson A, Stadler U, Glotzer MA, Kozak LP (1985) Mitochondrial uncoupling protein from mouse brown fat. Molecular cloning, genetic mapping, and mRNA expression. J Biol Chem 260:16250–16254
19. Klingenberg M (1988) Nucleotide binding to uncoupling protein. Mechanism of control by protonation. Biochemistry 27:781–791
20. Saito S, Saito CT, Shingai R (2008) Adaptive evolution of the uncoupling protein 1 gene contributed to the acquisition of novel nonshivering thermogenesis in ancestral eutherian mammals. Gene 408:37–44
21. Cannon B, Nedergaard J (2004) Brown adipose tissue: function and physiological significance. Physiol Rev 84:277–359
22. Ricquier D, Bouillaud F (2000) The uncoupling protein homologues: UCP1, UCP2, UCP3, StUCP and AtUCP. Biochem J 345(Pt 2):161–179
23. Murray AJ, Anderson RE, Watson GC, Radda GK, Clarke K (2004) Uncoupling proteins in human heart. Lancet 364:1786–1788
24. Bo H, Jiang N, Ma G, Qu J, Zhang G, Cao D, Wen L, Liu S, Ji LL, Zhang Y (2008) Regulation of mitochondrial uncoupling respiration during exercise in rat heart: role of reactive oxygen species (ROS) and uncoupling protein 2. Free Radic Biol Med 44:1373–1381

25. Villarroya F, Iglesias R, Giralt M (2007) PPARs in the Control of Uncoupling Proteins Gene Expression. PPAR Res 2007:74364
26. Nègre-Salvayre A, Hirtz C, Carrera G, Cazenave R, Troly M, Salvayre R, Pénicaud L, Casteilla L (1997) A role for uncoupling protein-2 as a regulator of mitochondrial hydrogen peroxide generation. FASEB J 11:809–815
27. Garlid KD, Jabůrek M, Jezek P, Varecha M (2000) How do uncoupling proteins uncouple? Biochim Biophys Acta 1459:383–389
28. Arvier M, Lagoutte L, Johnson G, Dumas J, Sion B, Grizard G, Malthièry Y, Simard G, Ritz P (2007) Adenine nucleotide translocator promotes oxidative phosphorylation and mild uncoupling in mitochondria after dexamethasone treatment. Am J Physiol Endocrinol Metab 293:E1320–E1324
29. Schönfeld P (1990) Does the function of adenine nucleotide translocase in fatty acid uncoupling depend on the type of mitochondria? FEBS Lett 264:246–248
30. Tikhonova IM, Andreyev AYu, Antonenko YuN, Kaulen AD, Komrakov AYu, Skulachev VP (1994) Ion permeability induced in artificial membranes by the ATP/ADP antiporter. FEBS Lett 337:231–234
31. Shabalina IG, Kramarova TV, Nedergaard J, Cannon B (2006) Carboxyatractyloside effects on brown-fat mitochondria imply that the adenine nucleotide translocator isoforms ANT1 and ANT2 may be responsible for basal and fatty-acid-induced uncoupling respectively. Biochem J 399:405–414
32. Esposito LA, Melov S, Panov A, Cottrell BA, Wallace DC (1999) Mitochondrial disease in mouse results in increased oxidative stress. Proc Natl Acad Sci USA 96:4820–4825
33. Garlid KD, Paucek P (2003) Mitochondrial potassium transport: the $K^+$ cycle. Biochim Biophys Acta 1606:23–41
34. Facundo HTF, Fornazari M, Kowaltowski AJ (2006) Tissue protection mediated by mitochondrial $K^+$ channels. Biochim Biophys Acta 1762:202–212
35. Kowaltowski AJ, Seetharaman S, Paucek P, Garlid KD (2001) Bioenergetic consequences of opening the ATP-sensitive $K^+$ channel of heart mitochondria. Am J Physiol Heart Circ Physiol 280:H649–H657
36. Bajgar R, Seetharaman S, Kowaltowski AJ, Garlid KD, Paucek P (2001) Identification and properties of a novel intracellular (mitochondrial) ATP-sensitive potassium channel in brain. J Biol Chem 276:33369–33374
37. Daum G (1985) Lipids of mitochondria. Biochim Biophys Acta 822:1–42
38. Crompton M (1999) The mitochondrial permeability transition pore and its role in cell death. Biochem J 341:233–249
39. Kowaltowski AJ, Castilho RF, Vercesi AE (2001) Mitochondrial permeability transition and oxidative stress. FEBS Lett 495:12–15
40. Lemasters JJ, Nieminen AL, Qian T, Trost LC, Elmore SP, Nishimura Y, Crowe RA, Cascio WE, Bradham CA, Brenner DA, Herman B (1998) The mitochondrial permeability transition in cell death: a common mechanism in necrosis, apoptosis and autophagy. Biochim Biophys Acta 1366:177–196
41. Zoratti M, Szabò I (1995) The mitochondrial permeability transition. Biochim Biophys Acta 1241:139–176
42. Rodriguez-Enriquez S, He L, Lemasters JJ (2004) Role of mitochondrial permeability transition pores in mitochondrial autophagy. Int J Biochem Cell Biol 36:2463–2472
43. Castilho RF, Kowaltowski AJ, Vercesi AE (1996) The irreversibility of inner mitochondrial membrane permeabilization by $Ca^{2+}$ plus prooxidants is determined by the extent of membrane protein thiol cross-linking. J Bioenerg Biomembr 28:523–529
44. Ichas F, Mazat JP (1998) From calcium signaling to cell death: two conformations for the mitochondrial permeability transition pore. Switching from low- to high-conductance state. Biochim Biophys Acta 1366:33–50
45. Di Lisa F, Menabò R, Canton M, Barile M, Bernardi P (2001) Opening of the mitochondrial permeability transition pore causes depletion of mitochondrial and cytosolic $NAD^+$ and is

a causative event in the death of myocytes in postischemic reperfusion of the heart. J Biol Chem 276:2571–2575

46. Griffiths EJ, Halestrap AP (1995) Mitochondrial non-specific pores remain closed during cardiac ischaemia, but open upon reperfusion. Biochem J 307:93–98
47. Gottlieb RA, Burleson KO, Kloner RA, Babior BM, Engler RL (1994) Reperfusion injury induces apoptosis in rabbit cardiomyocytes. J Clin Invest 94:1621–1628
48. Murry CE, Jennings RB, Reimer KA (1986) Preconditioning with ischemia: a delay of lethal cell injury in ischemic myocardium. Circulation 74:1124–1136
49. Vanden Hoek T, Becker LB, Shao ZH, Li CQ, Schumacker PT (2000) Preconditioning in cardiomyocytes protects by attenuating oxidant stress at reperfusion. Circ Res 86:541–548
50. Costa ADT, Garlid KD, West IC, Lincoln TM, Downey JM, Cohen MV, Critz SD (2005) Protein kinase G transmits the cardioprotective signal from cytosol to mitochondria. Circ Res 97:329–336
51. Dorn GW 2nd, Souroujon MC, Liron T, Chen CH, Gray MO, Zhou HZ, Csukai M, Wu G, Lorenz JN, Mochly-Rosen D (1999) Sustained in vivo cardiac protection by a rationally designed peptide that causes epsilon protein kinase C translocation. Proc Natl Acad Sci USA 96:12798–12803
52. Lebuffe G, Schumacker PT, Shao Z, Anderson T, Iwase H, Vanden Hoek TL (2003) ROS and NO trigger early preconditioning: relationship to mitochondrial $K_{ATP}$ channel. Am J Physiol Heart Circ Physiol 284:H299–H308
53. Mitchell MB, Meng X, Ao L, Brown JM, Harken AH, Banerjee A (1995) Preconditioning of isolated rat heart is mediated by protein kinase C. Circ Res 76:73–81
54. Ping P, Zhang J, Qiu Y, Tang XL, Manchikalapudi S, Cao X, Bolli R (1997) Ischemic preconditioning induces selective translocation of protein kinase C isoforms epsilon and eta in the heart of conscious rabbits without subcellular redistribution of total protein kinase C activity. Circ Res 81:404–414
55. Ping P, Takano H, Zhang J, Tang XL, Qiu Y, Li RC, Banerjee S, Dawn B, Balafonova Z, Bolli R (1999) Isoform-selective activation of protein kinase C by nitric oxide in the heart of conscious rabbits: a signaling mechanism for both nitric oxide-induced and ischemia-induced preconditioning. Circ Res 84:587–604
56. Sato T, O'Rourke B, Marbán E (1998) Modulation of mitochondrial ATP-dependent K+ channels by protein kinase C. Circ Res 83:110–114
57. Saurin AT, Pennington DJ, Raat NJH, Latchman DS, Owen MJ, Marber MS (2002) Targeted disruption of the protein kinase C epsilon gene abolishes the infarct size reduction that follows ischaemic preconditioning of isolated buffer-perfused mouse hearts. Cardiovasc Res 55:672–680
58. Klann E, Roberson ED, Knapp LT, Sweatt JD (1998) A role for superoxide in protein kinase C activation and induction of long-term potentiation. J Biol Chem 273:4516–4522
59. Tritto I, D'Andrea D, Eramo N, Scognamiglio A, De Simone C, Violante A, Esposito A, Chiariello M, Ambrosio G (1997) Oxygen radicals can induce preconditioning in rabbit hearts. Circ Res 80:743–748
60. Garlid KD, Paucek P, Yarov-Yarovoy V, Murray HN, Darbenzio RB, D'Alonzo AJ, Lodge NJ, Smith MA, Grover GJ (1997) Cardioprotective effect of diazoxide and its interaction with mitochondrial ATP-sensitive $K^+$ channels. Possible mechanism of cardioprotection. Circ Res 81:1072–1082
61. Grover GJ (1994) Protective effects of ATP-sensitive potassium-channel openers in experimental myocardial ischemia. J Cardiovasc Pharmacol 24:S18–S27
62. Richer C, Pratz J, Mulder P, Mondot S, Giudicelli JF, Cavero I (1990) Cardiovascular and biological effects of $K^+$ channel openers, a class of drugs with vasorelaxant and cardioprotective properties. Life Sci 47:1693–1705
63. Yao Z, Gross GJ (1994) Effects of the $K_{ATP}$ channel opener bimakalim on coronary blood flow, monophasic action potential duration, and infarct size in dogs. Circulation 89: 1769–1775

64. Andrukhiv A, Costa AD, West IC, Garlid KD (2006) Opening mito$K_{ATP}$ increases superoxide generation from complex I of the electron transport chain. Am J Physiol Heart Circ Physiol 291:H2067–H2074
65. Carroll R, Gant VA, Yellon DM (2001) Mitochondrial $K_{ATP}$ channel opening protects a human atrial-derived cell line by a mechanism involving free radical generation. Cardiovasc Res 51:691–700
66. Facundo HTF, Kowaltowski AJ (2005) Letter regarding article by Argaud et al, "postconditioning inhibits mitochondrial permeability transition". Circulation 111:e442; author reply e442
67. Facundo HTF, de Paula JG, Kowaltowski AJ (2007) Mitochondrial ATP-sensitive $K^+$ channels are redox-sensitive pathways that control reactive oxygen species production. Free Radic Biol Med 42:1039–1048
68. Zhang DX, Chen YF, Campbell WB, Zou AP, Gross GJ, Li PL (2001) Characteristics and superoxide-induced activation of reconstituted myocardial mitochondrial ATP-sensitive potassium channels. Circ Res 89:1177–1183
69. Forbes RA, Steenbergen C, Murphy E (2001) Diazoxide-induced cardioprotection requires signaling through a redox-sensitive mechanism. Circ Res 88:802–809
70. Pain T, Yang XM, Critz SD, Yue Y, Nakano A, Liu GS, Heusch G, Cohen MV, Downey JM (2000) Opening of mitochondrial $K_{ATP}$ channels triggers the preconditioned state by generating free radicals. Circ Res 87:460–466
71. Fornazari M, de Paula JG, Castilho RF, Kowaltowski AJ (2008) Redox properties of the adenoside triphosphate-sensitive $K^+$ channel in brain mitochondria. J Neurosci Res 86:1548–1556
72. Krenz M, Oldenburg O, Wimpee H, Cohen MV, Garlid KD, Critz SD, Downey JM, Benoit JN (2002) Opening of ATP-sensitive potassium channels causes generation of free radicals in vascular smooth muscle cells. Basic Res Cardiol 97:365–373
73. Scorrano L, Petronilli V, Colonna R, Di Lisa F, Bernardi P (1999) Interactions of chloromethyltetramethylrosamine (Mitotracker Orange) with isolated mitochondria and intact cells. Ann N Y Acad Sci 893:391–395
74. Scorrano L, Petronilli V, Colonna R, Di Lisa F, Bernardi P (1999) Chloromethyltetramethylrosamine (Mitotracker Orange) induces the mitochondrial permeability transition and inhibits respiratory complex I. Implications for the mechanism of cytochrome c release. J Biol Chem 274:24657–24663
75. Dröse S, Brandt U, Hanley PJ (2006) $K^+$-independent actions of diazoxide question the role of inner membrane $K_{ATP}$ channels in mitochondrial cytoprotective signaling. J Biol Chem 281:23733–23739
76. Oldenburg O, Cohen MV, Yellon DM, Downey JM (2002) Mitochondrial $K_{ATP}$ channels: role in cardioprotection. Cardiovasc Res 55:429–437
77. Ferranti R, da Silva MM, Kowaltowski AJ (2003) Mitochondrial ATP-sensitive $K^+$ channel opening decreases reactive oxygen species generation. FEBS Lett 536: 51–55
78. Wrona M, Wardman P (2006) Properties of the radical intermediate obtained on oxidation of 2′,7′-dichlorodihydrofluorescein, a probe for oxidative stress. Free Radic Biol Med 41: 657–667
79. Costa ADT, Quinlan CL, Andrukhiv A, West IC, Jabůrek M, Garlid KD (2006) The direct physiological effects of mito$K_{ATP}$ opening on heart mitochondria. Am J Physiol Heart Circ Physiol 290:H406–H415
80. Costa ADT, Jakob R, Costa CL, Andrukhiv K, West IC, Garlid KD (2006) The mechanism by which the mitochondrial ATP-sensitive $K^+$ channel opening and $H_2O_2$ inhibit the mitochondrial permeability transition. J Biol Chem 281:20801–20808
81. Clarke SJ, Khaliulin I, Das M, Parker JE, Heesom KJ, Halestrap AP (2008) Inhibition of mitochondrial permeability transition pore opening by ischemic preconditioning is probably mediated by reduction of oxidative stress rather than mitochondrial protein phosphorylation. Circ Res 102:1082–1090

82. Costa ADT, Garlid KD (2008) Intramitochondrial signaling: interactions among mito$K_{ATP}$, PKCepsilon, ROS, and MPT. Am J Physiol Heart Circ Physiol 295:H874–H882
83. Ljubkovic M, Shi Y, Cheng Q, Bosnjak Z, Jiang MT (2007) Cardiac mitochondrial ATP-sensitive potassium channel is activated by nitric oxide in vitro. FEBS Lett 581:4255–4259
84. Burwell LS, Nadtochiy SM, Tompkins AJ, Young S, Brookes PS (2006) Direct evidence for S-nitrosation of mitochondrial complex I. Biochem J 394:627–634
85. Heinen A, Aldakkak M, Stowe DF, Rhodes SS, Riess ML, Varadarajan SG, Camara AKS (2007) Reverse electron flow-induced ROS production is attenuated by activation of mitochondrial $Ca^{2+}$-sensitive $K^+$ channels. Am J Physiol Heart Circ Physiol 293:H1400–H1407
86. Wang Y, Ashraf M (1999) Role of protein kinase C in mitochondrial $K_{ATP}$ channel-mediated protection against $Ca^{2+}$ overload injury in rat myocardium. Circ Res 84:1156–1165
87. Chen C, Budas GR, Churchill EN, Disatnik M, Hurley TD, Mochly-Rosen D (2008) Activation of aldehyde dehydrogenase-2 reduces ischemic damage to the heart. Science 321:1493–1495
88. Carreira RS, Miyamoto S, Di Mascio P, Gonçalves LM, Monteiro P, Providência LA, Kowaltowski AJ (2007) Ischemic preconditioning enhances fatty acid-dependent mitochondrial uncoupling. J Bioenerg Biomembr 39:313–320
89. Nadtochiy SM, Tompkins AJ, Brookes PS (2006) Different mechanisms of mitochondrial proton leak in ischaemia/reperfusion injury and preconditioning: implications for pathology and cardioprotection. Biochem J 395:611–618
90. Korde AS, Pettigrew LC, Craddock SD, Maragos WF (2005) The mitochondrial uncoupler 2,4-dinitrophenol attenuates tissue damage and improves mitochondrial homeostasis following transient focal cerebral ischemia. J Neurochem 94:1676–1684
91. Mattiasson G, Shamloo M, Gido G, Mathi K, Tomasevic G, Yi S, Warden CH, Castilho RF, Melcher T, Gonzalez-Zulueta M, Nikolich K, Wieloch T (2003) Uncoupling protein-2 prevents neuronal death and diminishes brain dysfunction after stroke and brain trauma. Nat Med 9:1062–1068
92. Na HS, Kim YI, Yoon YW, Han HC, Nahm SH, Hong SK (1996) Ventricular premature beat-driven intermittent restoration of coronary blood flow reduces the incidence of reperfusion-induced ventricular fibrillation in a cat model of regional ischemia. Am Heart J 132:78–83
93. Philipp S, Yang X, Cui L, Davis AM, Downey JM, Cohen MV (2006) Postconditioning protects rabbit hearts through a protein kinase C-adenosine A2b receptor cascade. Cardiovasc Res 70:308–314
94. Yang X, Proctor JB, Cui L, Krieg T, Downey JM, Cohen MV (2004) Multiple, brief coronary occlusions during early reperfusion protect rabbit hearts by targeting cell signaling pathways. J Am Coll Cardiol 44:1103–1110
95. Kin H, Zhao Z, Sun H, Wang N, Corvera JS, Halkos ME, Kerendi F, Guyton RA, Vinten-Johansen J (2004) Postconditioning attenuates myocardial ischemia-reperfusion injury by inhibiting events in the early minutes of reperfusion. Cardiovasc Res 62:74–85
96. Serviddio G, Di Venosa N, Federici A, D'Agostino D, Rollo T, Prigigallo F, Altomare E, Fiore T, Vendemiale G (2005) Brief hypoxia before normoxic reperfusion (postconditioning) protects the heart against ischemia-reperfusion injury by preventing mitochondria peroxyde production and glutathione depletion. FASEB J 19:354–361
97. Argaud L, Gateau-Roesch O, Raisky O, Loufouat J, Robert D, Ovize M (2005) Postconditioning inhibits mitochondrial permeability transition. Circulation 111:194–197
98. Bopassa JC, Vandroux D, Ovize M, Ferrera R (2006) Controlled reperfusion after hypothermic heart preservation inhibits mitochondrial permeability transition-pore opening and enhances functional recovery. Am J Physiol Heart Circ Physiol 291: H2265–H2271
99. Cohen MV, Yang X, Downey JM (2008) Acidosis, oxygen, and interference with mitochondrial permeability transition pore formation in the early minutes of reperfusion are critical to postconditioning's success. Basic Res Cardiol 103:464–471
100. Penna C, Rastaldo R, Mancardi D, Raimondo S, Cappello S, Gattullo D, Losano G, Pagliaro P (2006) Post-conditioning induced cardioprotection requires signaling through a

redox-sensitive mechanism, mitochondrial ATP-sensitive $K^+$ channel and protein kinase C activation. Basic Res Cardiol 101:180–189

101. Chu F, Ward NE, O'Brian CA (2003) PKC isozyme S-cysteinylation by cystine stimulates the pro-apoptotic isozyme PKC delta and inactivates the oncogenic isozyme PKC epsilon. Carcinogenesis 24:317–325

# Chapter 6
# Coenzyme $Q_9/Q_{10}$ and the Healthy Heart

**Samarjit Das, Somak Das, and Dipak K. Das**

**Abstract** The mitochondrial respiratory chain consists of several coenzymes (CoQ), including $CoQ_1$, $CoQ_2$, $CoQ_4$, $CoQ_6$, $CoQ_7$, $CoQ_8$, $CoQ_9$, and $CoQ_{10}$. $Q_{10}$ is the most prevalent form in humans and most of the mammals, and $Q_9$ is the primary form found in rats, mice, and guinea pigs. On the other hand, $Q_6$, $Q_7$, and $Q_8$ are found in yeast and bacteria. Most of the literature concerning the importance of $CoQ_{10}$ in attenuating various health problems has been reviewed; it demonstrates the importance of $Q_{10}$ nutritional supplementation to combat against various diseases. The safety profile of 100–200 mg of regular $Q_{10}$ supplementation is quite promising, as no adverse effects have been reported from the clinical trials using daily supplements of up to 200 mg $Q_{10}$ for 6–12 months and 100 mg daily for up to 6 years. In cardiovascular diseases, including cardiomyopathy, the significantly low levels of $Q_{10}$ in myocardial tissues proved the importance of nutritional supplementation of $CoQ_{10}$ against various heart diseases. It is shown that, unlike $CoQ_{10}$, the other coenzymes have not been extensively studied. So the purpose of this review is to highlight whether the other CoQs, especially $CoQ_9$, are equally as cardioprotective as $CoQ_{10}$ and provide similar health benefits.

**Keywords** Coenzyme $Q_9$ · Coenzyme $Q_{10}$ · Heart · Ischemia · Nutritional supplement · Oxidative stress

## 6.1 Introduction

A lipid soluble benzoquinone, coenzyme Q (CoQ), is an essential component for electron transport in oxidative phosphorylation of the mitochondria. Also called ubiquinone, its principal function is to act as an electron carrier between the NADH and succinate dehydrogenases and the cytochrome system [1]. During mitochondrial

D.K. Das (✉)
School of Medicine, Cardiovascular Research Center, University of Connecticut, Farmington, CT, USA
e-mail: ddas@neuron.uchc.edu

H. Sauer et al. (eds.), *Studies on Cardiovascular Disorders*,
Oxidative Stress in Applied Basic Research and Clinical Practice,
DOI 10.1007/978-1-60761-600-9_6, 

electron transport, ubiquinone also occurs as semiquinone and ubiquinol, the fully reduced form of ubiquinone. Semiquinone has a role in the generation of superoxide anions during mitochondrial respiration [2], whereas ubiquinol functions as an intracellular antioxidant, presumably by preventing both the initiation and propagation of lipid peroxidation [3].

CoQ represents "substrate-like molecules" linking successive enzymes as in a metabolic pathway, and in this respect it may represent the controlling devices of the overall rate of electron transfer [4]. Thus, under a normal aerobic environment, abundance of CoQ in mitochondria is an important determinant for ATP synthesis. Synthesised ATP can be utilized for maintaining intracellular ionic homeostasis by activating ATP-requiring ion pumps, thereby alleviating myocardial injury induced by various noxious stimuli [4].

In addition to an antioxidative role, CoQ may also act as a pro-oxidant. Redox cycling of CoQ in the mitochondrial electron transfer chain has been shown to be involved in $O_2^-$ generation [5]. ROS released into cytosol from mitochondria can trigger intracellular signal transduction pathways that may mediate cytoprotection [6, 7] and gene expression through activation of redox-sensitive transcriptional factors such as nuclear transcription factor *k*B and activating protein-1 [8]. It is, therefore, anticipated that increased ROS generation in mitochondria with an abundance of CoQ could represent a novel mechanism of cardioprotection through the potentiation of redox signaling. Thus the objective of the present study is to test the hypothesis that CoQ increases ROS generation, but prevents oxidative damage and dysfunction of mitochondria under excess ROS-generating conditions.

Oxidative stress caused by free radicals plays a crucial role in the pathophysiology associated with atherosclerosis, neoplasia, and neurodegenerative diseases. Therefore, extensive attention is being focused on the naturally occurring antioxidative phytochemicals. $CoQ_{10}$ appears to be involved in the coordinated regulation between oxidative stress and the antioxidant capacity of heart tissue. When the heart is subjected to oxidative stress in various pathogenic conditions [9], the amount of $CoQ_{10}$ is decreased, which triggers a signal for increased $CoQ_{10}$ synthesis. It has been reported that in patients with cardiac disease such as chronic heart failure, the myocardium becomes deficient in $CoQ_{10}$ and $CoQ_{10}$ reductase [1]. $CoQ_{10}$ level is also reduced in other cardiovascular diseases such as cardiomyopathy [10]. $CoQ_{10}$ can protect human low-density lipoprotein (LDL) from lipid peroxidation, suggesting its role in atherosclerosis [11]. Several reports exist in the literature indicating cardioprotective effects of $CoQ_{10}$ against ischemia-reperfusion injury [4, 10, 12–15]. However, none of these studies has attempted to evaluate the mechanism(s) of $CoQ_{10}$-mediated cardioprotection, and none demonstrated whether postischemic improvement of myocardial function was caused by the improvement of an endogenous defense system.

## 6.2 A Quick Look Back

The history of coenzymes is not very long. About six decades ago, in 1955, Festenstein et al. first identified a new substance with a role in electron transport in the cells, and they named this substance *ubiquinone* [16]. The name ubiquinone means "ubiquitous quinone," which relates to its presence in all the cells.

But the name coenzyme Q, and its real role was established by another group from the University of Wisconsin, almost two years after the discovery of ubiquinone. Crane et al. isolated a new compound, which was capable of undergoing reversible oxidation and reduction, from lipid extracts of beef heart mitochondria [17]. For convenience, they referred to it as $Q_{275}$. Later on, they chose the name "coenzyme Q," from the chemical structure of the compound. During that same year (1960) Professor Morton, from Britain, also discovered $CoQ_{10}$ in the livers of vitamin A–deficient rats [18]. During the following year researchers at Merck, Inc., determined its chemical structure and became the first to produce it [19].

It was neither the British nor the Americans that first found a practical use for the CoQ compounds. Professor Yamamura from Japan first used a related compound ($CoQ_7$) in the treatment of congestive heart failure [19]. Other practical uses then followed. $CoQ_6$ was used as an effective antioxidant in the mid 1960s [19]. In 1972 in Italy, a deficiency of $CoQ_{10}$ was linked to heart disease [20]. The Japanese, however, were the first to perfect the technology necessary to produce $CoQ_{10}$ in sizeable enough quantities to make large clinical trials a reality [19].

After Peter Mitchell won the Nobel Prize in 1978 for defining the biological energy transfer that occurs at the cellular level (for which $CoQ_{10}$ is essential), there was a considerable increase in the number of clinical studies performed in relation to $CoQ_{10}$'s usefulness [21–24]. This was due in part to the large amounts of pharmaceutical grade $CoQ_{10}$ that were now available from Japan and the ability to measure $CoQ_{10}$ in blood and body tissues. $CoQ_{10}$ has since become known for its importance as a powerful antioxidant and free radical scavenger and as a treatment in many chronic illnesses, especially heart disease. Lars Ernster of Sweden enlarged upon $CoQ_{10}$'s importance as an antioxidant and free radical scavenger [25].

All coenzymes differ by the number of isoprenyl units on one quinone group (Fig. 6.1); the most abundant and important form of coenzyme, $CoQ_{10}$, contains one quinone group and 10 isoprenyl units. Chemically, $Q_{10}$ is designated 2,3-dimethoxy-5-methyl-6-decaprenyl-1,4-benzoquinone.

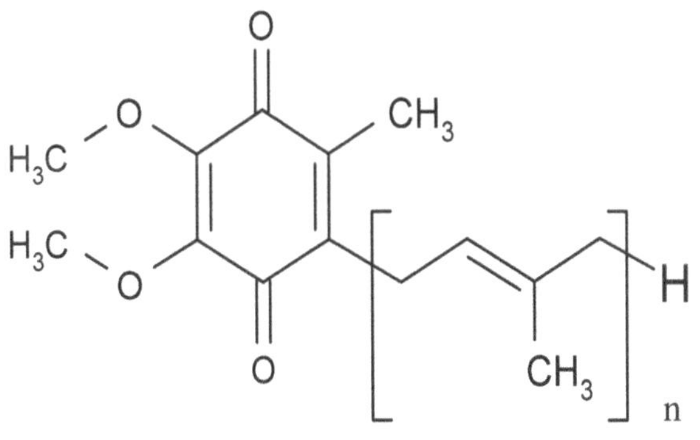


**Fig. 6.1** Structure of coenzyme Q. "n" denotes number of isoprenyl units present. $CoQ_{10}$ contains 10 isoprene units ($n = 10$)

## 6.3 Natural Occurrence and Distribution

Animals, plants, and microorganisms consist of coenzymes. $Q_{10}$ is the most prevalent form in humans and most of the mammals, and $Q_9$ is the primary form found in rats, mice, and guinea pigs. On the other hand, $Q_6$, $Q_7$, and $Q_8$ are found in yeast and bacteria [26].

Varying amounts of $CoQ_{10}$ are present in different human tissues. In the heart, the concentration is higher than among all the other organs. Normally, the concentration of $Q_{10}$ in the human heart is 110 μg/g of tissue [27]. $CoQ_{10}$ has been found in considerable amounts in the liver and kidney; the lowest concentration of $CoQ_{10}$ is in the lung tissue [27]. The major portion of $CoQ_{10}$ found in different human tissues is in reduced form, except for the human brain and lung tissue [27]. In human plasma, the range of $CoQ_{10}$ is from 0.75 to 1.00 μg/ml, of which 70% is in reduced form [27]. The total $CoQ_{10}$ content in the human body is approximately 1.0–1.5 g, most of it in the muscle cells [28, 29]. The level of $CoQ_{10}$ declines in humans with age [27]. Bowry et al. showed $CoQ_{10}$ is normally bound to LDL [30].

Tissue gets its $CoQ_{10}$ by endogenous synthesis as well as from food intake and oral supplements. Although the manufacturers of oral supplements recommend an intake of 10–30 mg/d for $CoQ_{10}$ [31] and around 1 mg/d for $CoQ_9$ [32], the recommended daily intake has not yet been determined by the FDA. Karlsson et al. came up with a survey of the amount of both $CoQ_{10}$ and $CoQ_9$ in regular food intake, as shown in Table 6.1 [28]. But Weber et al. showed on an experimental basis that the coenzyme content in cooked foods is almost 15–30% less than in similar raw products [32].

**Table 6.1** The content of $CoQ_9$ and $CoQ_{10}$ in regular food intake

| Food group | Food item | *n* | Cooking | $Q_{10}$ (μg/g food)[a] | $Q_9$ (μg/g food)[a] |
|---|---|---|---|---|---|
| Meat and poultry | Pork heart | 9 | Fried | 203 (151–282) | 3.9 (1.7–6.1) |
| | Beef | 1 | Fried | 31 | 2.6 |
| | Chicken | 1 | Fried | 17 | 0.8 |
| | Pork chop | 3 | Fried | 14 (9.0–17.8) | 1.0 |
| | Ham | 3 | Boiled | 7.7 (5.4–9.4) | 0.3 |
| Fish | Herring | 1 | Marinated | 27 | n.d. |
| | Rainbow trout | 1 | Streamed | 11 | n.d. |
| | Salmon | 1 | Smoked | 4.3 | n.d. |
| Cereals | Bread (rye) | 1 | None | <0.2 | 4.7 |
| | Bread (wheat) | 1 | None | <0.1 | 1.1 |
| | Rice | 1 | Boiled | 0.2 | n.d. |
| Vegetables | Broccoli | 1 | Boiled | 6.6 (5.9–7.7) | 0.6 (0.6–0.7) |
| | Cauliflower | 3 | Boiled | 4.9 | n.d. |
| | Potato | 1 | Boiled | 0.5 | n.d. |
| | Tomato | 1 | None | 0.2 | n.d. |
| | Carrot | 1 | None | <0.2 | n.d. |

Table 6.1 (continued)

| Food group | Food item | *n* | Cooking | $Q_{10}$ (μg/g food)[a] | $Q_9$ (μg/g food)[a] |
|---|---|---|---|---|---|
| | Carrot | 1 | Boiled | <0.2 | n.d. |
| | Cucumber | 1 | None | <0.2 | 0.1 |
| Fruit | Orange | 1 | None | 2.2 | 0.4 |
| | Apple | 1 | None | 1.1 | n.d. |
| | Kiwi | 1 | None | 0.5 | 0.3 |
| Dairy products | Yoghurt | 1 | None | 1.2 | 0.3 |
| | Hard cheese | 1 | None | <0.2 | n.d. |
| | Cream cheese | 1 | None | <0.3 | n.d. |
| Egg | Hen's egg | 2 | None | 1.5 (1.0–2.1) | 0.4 (0.3–0.5) |
| | Hen's egg | 2 | Boiled | 2.3 (1.7–2.9) | 0.3 (0.2–0.4) |

Adapted from Karlsson et al. [28]

Deficiency symptoms of coenzymes in the general world population have not yet been reported. Thus it can be assumed that a regular diet and endogenous biosynthesis will supply sufficient coenzymes to any healthy individual. But extensive studies showed that the oral supplementation of $CoQ_{10}$ alone among all the other coenzymes has immense beneficial effects. Recent studies have also shown that $CoQ_9$ is also very protective, which indicates that $CoQ_9$ by itself, or after being converted into $CoQ_{10}$, reduces various ischemia/reperfusion–induced injury, especially myocardial ischemia/reperfusion injury [12].

## 6.4 The Biochemical Background

The biosynthesis of $CoQ_{10}$ from the amino acid tyrosine is a process involving three major steps—in the Golgi apparatus, synthesis of the ring structure from essential amino acids tyrosine or phenylalanine; formation of the isoprenoid side chain from acetyl-CoA reduced via the mevanolate pathway; and finally condensation of these structures by means of the enzyme polyprenyl-transferase [33]. During this complex biochemical process at least eight vitamins and several trace elements are essential [33]. In Fig. 6.2, these steps are highlighted along with the possible role of hydroxymethylglutaryl (HMG)-coenzyme A reductase involvement also indicated.

## 6.5 Physiological Effects

Coenzymes are cofactors upon which larger and complex enzymes absolutely depend for their function. The cell's energy-producing engine is called the mitochondria. $CoQ_{10}$ is the coenzyme for at least three mitochondrial enzymes (complexes I, II, and III) as well as enzymes in other parts of the cell [34]. Mitochondrial

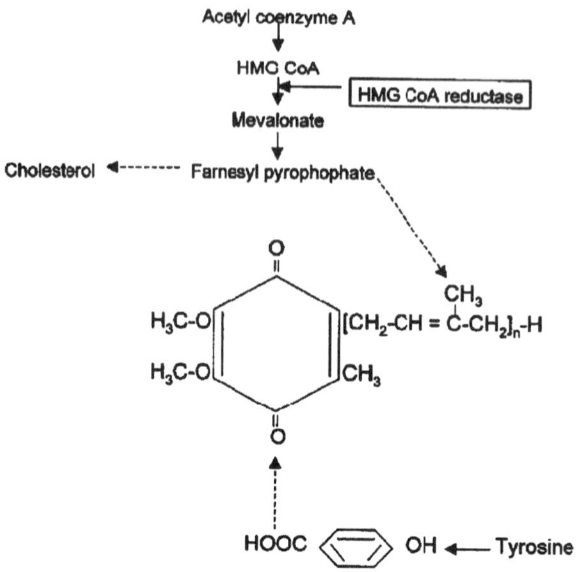


**Fig. 6.2** Biosynthetic pathways of formation of coenzymes. Role of HMG CoA reductase is also highlighted, leading to cholesterol. The involvement of CoQ is also shown in cholesterol formation

enzymes of the oxidative phosphorylation pathway are essential for the production of the high-energy phosphate, adenosine triphosphate (ATP), upon which all cellular functions depend [34]. Cells do not use the nutrients in the diet for their immediate supply of energy. Instead, they prepare an energy-rich compound called adenosine triphosphate, or ATP. ATP is the fuel used for all the energy-requiring processes within the cell. In turn, the energy in food is extracted to build more ATP for use by the body.

Coenzyme Q exists both in an oxidized form, ubiquinone, and a reduced form, ubiquinol. $CoQ_{10}$ has been shown to have an important role as a lipid soluble antioxidant [4, 13, 14]. Several other studies also describe $CoQ_{10}$ as having an in vitro as well as an in vivo role in the attenuation of LDL oxidation [35, 36]. A sparing effect on vitamin E as well as a direct antioxidative effect has also been reported by Stocker et al. [37].

## 6.6 Pharmacokinetics

The pharmacokinetics of the coenzymes is very difficult to study, because of the fact that coenzymes are synthesised in the body and obtained from nutritional sources. Tomono et al. [38] successfully overcome this complication with deuterium-labelled $Q_{10}$, of which 16 volunteers took about 100 mg. Blood values reached a maximum after approximately 6 h, indicating a relatively slow absorption, which relates to the high molecular weight and the lipid solubility of $Q_{10}$. In most of the subjects a secondary peak was found in the blood 24 h after intake, which is explained by absorption in the liver and subsequent redistribution via VLDL. The plasma half-life was about 33 h. The administered $Q_{10}$ dose did not affect the plasma levels of endogeneous synthesised $Q_{10}$ for up to 72 h after intake. The formulation of $Q_{10}$ was not described. After intakes of 3 × 20 mg $Q_{10}$ per day for 4 days, in six different formulations, an increase in the plasma level of up to 2.8 times the baseline level was observed. $Q_{10}$ dissolved in oil and formulated as capsules appears to have the best bioavailability [38]. Mohr et al. [39] found increased levels of Q10 in lipoproteins after oral supplementation and also an increased resistance of LDL to incipient lipid peroxidation.

Increased myocardial tissue content of $Q_{10}$ via exogenously administered $Q_{10}$ has been shown both in animal experiments [40] and in man [41]. After a supplementation period of 5 months with $Q_{10}$ 100 mg/d in patients with cardiomyopathy, the percentage increases of $Q_{10}$ in the myocardium ranged from 19% to 86% (endomyocardial biopsies). Case reports have suggested interaction between $Q_{10}$ and warfarin. In pigs, a thrombocyte anti-aggregating effect has been found after chronic dosing of 200 ± 40 mg/day.

Some studies suggest relatively low $Q_{10}$ concentration in the myocardium after administration of anthracyclines; other studies suggest that $Q_{10}$ administration may reduce the cardiotoxicity of anthracyclines [28]. Most of the widely used HMG-coenzyme A-reductase inhibitors, which inhibit cholesterol synthesis, result in reduced synthesis and thereby reduced $Q_{10}$ levels in plasma [42]. This dose-related decrease is true for, among others, lovastatin and pravastatin [43]. The significance of this is not known; however, the impaired biosynthesis and consequent $Q_{10}$ down-regulation deserves further evaluation in long-term trials considering the oxidative theory for the development of atherosclerosis.

In placebo-controlled long-term studies no serious or frequent side effects to intake of doses of 200 mg $Q_{10}$/d for up to 12 months and 100 mg $Q_{10}$/d for up to 6 y have been found. However, none of the controlled studies were specifically designed to examine possible side effects.

## 6.7 Cardioprotective Effects

Since cardiovascular disease contributes in a major way to morbidity and mortality, it is becoming a strain on the economy of many countries worldwide. Coenzyme $Q_{10}$ (CoQ) has long been known as a cardioprotective agent that has been utilized in treatment for ischemic heart disease, heart failure, and conditions such as adriamycin toxicity [44]. One of the important mechanisms by which CoQ exerts cardioprotection has been attributed to its role as a mobile electron carrier in the mitochondrial electron transport process of respiration and coupled phosphorylation [45].

The ability of CoQ to afford myocardial protection is also attributed to its antioxidant property. It is apparent that reactive oxygen species (ROS) are a common mediator of cytotoxic stress. Biochemical mechanisms underlying the toxicity of ROS include their ability to peroxidize membrane phospholipids with unsaturated free fatty acid and their interaction with certain sulfhydryl proteins. The net result of ROS-induced damage appears to be altered membrane function and structure. Eventually, the altered handling of the ionic gradient results in intracellular $Ca^{2+}$ overload, the major pathogenesis of myocardial reperfusion injury [46]. Mitochondria are especially vulnerable to increased $Ca^{2+}$. Intracellular $Ca^{2+}$ overload facilitates energy-linked uptake of $Ca^{2+}$ by mitochondria that dissipates mitochondrial membrane potential and inhibits ATP synthesis. Elevation of intracellular $Ca^{2+}$ is a prerequisite for increased ROS generation by mitochondria [47].

Furthermore, accumulation of $Ca^{2+}$ in the mitochondrial matrix above the critical level is known to result in the activation of cell death cascade by provoking permeability transition (PT) and releasing cytochrome *c* from the intermembrane space [48, 49]. Cytochrome *c* in the presence of Apaf (apoptotic protease activating factors) and dATP in the cytosol activates caspases leading to the degradation phase of apoptosis [50, 51].

### *6.7.1 Heart Failure*

There are numerous risk factors involved in the pathophysiology of heart failure, most notably oxidative stress, mitochondrial dysfunction, and energy starvation. From our discussion so far it appears quite promising that coenzymes, especially $CoQ_{10}$, will have a significant influence against all the major risk factors of heart failure. Mortensen et al. revealed the fact that heart failure patients have lower amounts of $CoQ_{10}$ compared with normal human beings at the same age; they also showed that $CoQ_{10}$ content is significantly lower even in endomyocardial tissue biopsies [52].

Clinicians are considering $CoQ_{10}$ as a therapeutic agent for all the populations above 40 years of age. The incidence of heart failure is found to be more common at that age: about 30–100 cases per million per year in the western world [31].

Some clinical trials have suggested using nutritional $Q_{10}$ supplements in patients with severe heart failure (NYHA classes III and IV). Between 1985 and 1995, numerous published studies have supported the treatment of heart failure with $CoQ_{10}$, in nine relevant randomised double-blind placebo-controlled studies [31]. In eight of these studies signs of improvement in clinical parameters, haemodynamic parameters, and/or exercise capacity were registered when conventional treatment was accompanied by $Q_{10}$ supplementation. The studies primarily included patients with advanced heart failure. The existing randomised, controlled studies are not sufficiently large to indicate any changes in mortality in the $Q_{10}$-treated group and studies with the primary endpoint of mortality are required. However, $Q_{10}$ is devoid of significant side effects; and for patients with heart failure, an improvement of 1–2 NYHA classes may constitute a significantly improved quality of life.

### 6.7.2 Atherosclerosis

Atherosclerosis is the process by which the deposition of cholesterol plaques on the walls of blood vessels makes those vessels narrow and ultimately blocked by the fatty deposits. Atherosclerosis finally leads to ischemia in the heart muscle and can cause muscle damage. The complete blockage of the arteries leads to myocardial infarction (MI). According to the World Health Organization, the major cause of death in the world as a whole by the year 2020 will be acute coronary occlusion [53].

In a randomised, double-blind, placebo-controlled trial of $CoQ_{10}$ in patients with myocardial infarction, Singh et al. [54] found a significant reduction of angina pectoris, arrhythmias, and ventricular dysfunction. In another study, Hiasa et al. used 1.5 mg/kg/day for a week with 18 patients and found improved work tolerance, a tendency to less ST-depression, with pulse and blood pressure remaining unaltered [55]. A very similar trial, Kamikawa et al. [56], used 150 mg/day $CoQ_{10}$ for 4 weeks and found that the 12 patients on which the trial was performed had fewer angina pectoris attacks and decreased nitroglycerin consumption. In another study of 15 patients with stable angina pectoris, when 600 mg/day $CoQ_{10}$ was administered for 4 days, these patients also had fewer angina pectoris attacks and less consumption of nitroglycerin [57].

Although all these human studies were extended on the basis of the animal studies, it was found that $CoQ_{10}$ could protect low-density lipoprotein (LDL) from lipid peroxidation, suggesting its role in artherosclerosis, the underlying mechanism (s) yet to be discover.

### 6.7.3 Hypertension

Several noncontrolled trials were conducted to check the role of $CoQ_{10}$ on high blood pressure.All but one trial successfully showed that blood pressure decreased markedly with $CoQ_{10}$ [58, 59]. In these studies both systolic and diastolic blood pressure were reduced. Further studies are necessary to confirm this observation.

### 6.7.4 Cardiac and Vascular Surgery

Many experimental approaches have illustrated the role of $CoQ_{10}$ with different clinical, haemodynamical, and biochemical parameters in cardiac surgery or valvular heart disease [60, 61]. Chello et al. [62], in a double-blind study, found that there is less enzyme leakage between creatine kinase and lactate dehydrogenase, and a lower level of split products of the oxidative burst, in the $CoQ_{10}$ treated patients. This clearly indicates that $CoQ_{10}$ pretreatment, at least 1 week prior, may have some favorable effects.

### 6.7.5 Pharmacological Preconditioning Effects

A state of the art technique, preconditioning, was initially reported by Murry et al. in 1986, in which a brief period of ischemia followed by a short interval of reperfusion for a few cycles provided cardioprotection against ischemic heart injury [63]. There are numerous attempts to mimic the protective effects of preconditioning by exogenous agents, and thereby to produce pharmacological preconditioning. There are several studies supporting the involvement of mitochondria in the preconditioning effect [64]. Being an endogenously synthesised provitamin present in the mitochondrial electron transport chain, $CoQ_{10}$ has been found to be cardioprotective and used as adjunct therapy for ischemic heart disease [4, 12–14]. One of the most important mechanisms by which CoQ exerts cardioprotection is aerobic ATP production as a mobile electron carrier in the mitochondrial electron transfer chain. The ability of CoQ to afford myocardial protection is also attributed to its antioxidant property. However, CoQ may also act as a pro-oxidant through the generation of ROS. Although excess oxidative stress is known to induce death signaling via cytochrome *c* release from mitochondria, it is now apparent that a brief exposure to oxidative stress stimulates redox signaling for acquisition of tolerance to oxidative stress. Yamamura et al. [4] showed that CoQ plays dual roles in the mitochondrial generation of intracellular signaling. CoQ acts as a pro-oxidant that participates in redox signaling. CoQ also acts as an antioxidant that inhibits permeability transition and cytochrome *c* release, and increases ATP synthesis, thereby attenuating death signaling toward apoptosis and necrosis. In another study, in swine, nutritional supplementation of $CoQ_{10}$ reduced myocardial ischemia-reperfusion injury by reducing the amount of malonaldehyde in the coronary effluent and providing a higher content of the endogenous antioxidants ascorbate and thiol [13]. Significant induction of the expression of ubiquitin mRNA was also found in the hearts of the $CoQ_{10}$-fed group [13]. The results of this study demonstrate that nutritional supplementation of $CoQ_{10}$ renders the heart resistant to ischemia-reperfusion injury, probably by reducing the oxidative stress.

An in vivo study reported that $CoQ_{10}$ reduces myocardial ischemia-reperfusion injury induced by oxidative stress through the suppression of the formation of ROS [10]. Ferrara et al. [65] showed that long-term $CoQ_{10}$ supplementation renders protection against oxidative stress induced by ischemia-reperfusion. On the other hand, Hano et al. [66] showed that a beneficial effect is not observed when $CoQ_{10}$ is added at the onset of reperfusion. Our study showed that long-term administration might enable the heart to increase $CoQ_{10}$, which may be beneficial in protecting it from ischemia-reperfusion injury. In another similar study [14], in pigs weighing 20–25 kg, 2 mg/kg/day $CoQ_{10}$ treatment for 3 weeks significantly improved the postischemic ventricular functions. Another unique study [12] demonstrated for the first time that nutritional supplementation of $CoQ_9$ could reduce myocardial ischemia-reperfusion injury to the same extent as $CoQ_{10}$, as evidenced by the comparable degree of the postischemic ventricular recovery and the reduction of myocardial infarct size and cardiomyocyte apoptosis [12]. However, whether the cardioprotection was achieved from $CoQ_9$ or after its bioconversion into $CoQ_{10}$ was not established from this study. Nevertheless, the finding that $CoQ_9$ and $CoQ_{10}$ can provide the same degree of cardioprotection appears to be important, because only very little exogenous $CoQ_{10}$ is taken up by the heart, while a significant amount of $CoQ_9/CoQ_{10}$ was detected in the heart after 4 weeks of $CoQ_9$ feeding. It is tempting to speculate that the heart may be able to better utilize $CoQ_9$ than $CoQ_{10}$.

## 6.8 Conclusion

CoQ may play dual roles in myocardial protection against oxidative stress by directly acting as an antioxidant that inhibits death signaling, i.e., PT, cytochrome *c* release, and ATP depletion; and by potentiating ROS release from mitochondria that may be participating in generation of the redox signaling for cellular acquisition of tolerance to oxidant stress. $CoQ_{10}$ has been extensively studied as an adjunctive therapy for ischemic heart disease, and its cardioprotective ability is well established. It is very important to study whether other CoQs, especially $CoQ_9$, are equally as cardioprotective as $CoQ_{10}$. If so, CoQ will become a regular therapeutic option in cardiovascular health.

**Acknowledgments** This study was supported by NIH grants HL 34360, HL 22559, HL 33889, and HL 56803.

## References

1. Sunamori M, Tanaka H, Maruyama T, Sultan I, Sakamoto T, Suzuki A (1991) Clinical experience of coenzyme $Q_{10}$ to enhance intraoperative myocardial protection in coronary artery revascularization. Cardiovasc Drugs Ther 5:297–300
2. Kagan VE, Arroyo A, Tyurin VA, Tyurina YY, Villalba JM, Navas P (1998) Plasma membrane NADH-coenzyme $Q_{10}$ reductase generates semiquinone radicals and recycles vitamin E homologue in a superoxide-dependent reaction. FEBS Lett 428:43–46
3. Ernster L (1993) Ubiquinol as a biological antioxidant. In: Yagi K (ed) Active Oxygens, Lipid Peroxides and Antioxidants. Japan Science Society, Tokyo, pp 1–38
4. Yamamura T, Otani H, Nakao Y, Hattori R, Osako M, Imamura H, Das DK (2001) Dual involvement of coenzyme Q10 in redox signaling and inhibition of death signaling in the rat heart mitochondria. Antioxid Redox Signal 3:103–112
5. Cadenas E, Boveris A, Ragan CI, Stoppani AO (1977) Production of superoxide radicals and hydrogen peroxide by NADH-ubiquinone reductase and ubiquinol-cytochrome *c* reductase from beef-heart mitochondria. Arch Biochem Biophys 180:248–257
6. Duranteau J, Chandel NS, Kulisz A, Shao Z, Schumacker PT (1998) Intracellular signaling by reactive oxygen species during hypoxia in cardiomyocytes. J Biol Chem 273:11619–11623
7. Vanden Hoek TL, Becker LB, Shao Z, Li C, Schumacker PT (1998) Reactive oxygen species released from mitochondria during brief hypoxia induce preconditioning in cardiomyocytes. J Biol Chem 273:18092–18098
8. Li X, Song L, Jope RS (1996) Cholinergic stimulation of AP-1 and NF*k*B transcription factor is differentially sensitive to oxidative stress in SH-SY5Y neuroblastoma: relationship to phosphoinositide hydrolysis. J Neurosci 16:5914–5922
9. Das DK, Maulik N (1995) Protection against free radical injury in the heart and cardiac performance. In: Sen CK, Packer L, Hänninen O (eds) Exercise and Oxygen Toxicity. Elsevier Science, Amsterdam, pp 359–388

10. Mortensen SA (1993) Perspectives on therapy of cardiovascular diseases with coenzyme Q10 (ubiquinone). Clin Investig 71(Suppl 8):S116–S123
11. Singal PK, Kirchenbaum LA (1990) A relative deficit in antioxidant reserve may contribute in cardiac failure. Can J Cardiol 6:47–49
12. Lekli I, Das S, Das S, Mukherjee S, Bak I, Juhasz B, Bagchi D, Trimurtulu G, Krishnaraju AV, Sengupta K, Tosaki A, Das DK (2008): Coenzyme $Q_9$ provides cardioprotection after converting into coenzyme $Q_{10}$. J Agric Food Chem, 56:5331–5337.
13. Maulik N, Yoshida T, Engelman RM, Bagchi D, Otani H, Das DK (2000) Dietary coenzyme Q(10) supplement renders swine hearts resistant to ischemia-reperfusion injury. Am J Physiol Heart Circ Physiol 278:H1084–H1090
14. Yoshida T, Maulik G, Bagchi D, Dash SK, Engelman RM, Das DK (1996) Increased myocardial tolerance to ischemia-reperfusion injury by feeding pigs with coenzyme Q10. Ann N Y Acad Sci 793:414–418
15. Stocker R, Bowry VW, Frei B (1991) Ubiquinol-10 protects human low-density lipoprotein more efficiently against lipid peroxidation than does alpha-tocopherol. Proc. Natl. Acad. Sci. USA 88:1646–1650
16. Festenstein GN, Heaton RW, Lowe JS, Morton RA (1955) A constituent of the unsaponifiable portion of animal tissue lipid (1max 272 mm). Biochem J 59:558–566
17. Crane FL, Hatefi Y, Lester RL, Widmer C (1957) Isolation of quinine from beef heart and beef mitochondria. Biochem Biophys Acta 25:220–221
18. Morton RA, Wilson GM, Lowe JS, Leat WMF (1957) Ubiquinone. Chemical Industry, London, p 1649
19. Greg Post: The History and Usefulness of Coenzyme Q10. http://EzineArticles.com/?expert=Greg Post
20. Littarru GP, Ho L, Folkers K (1972) Deficiency of Coenzyme Q10 in human heart disease. Part I and II. Int J Vitam Nutr Res 42(2):291; 42(3):413
21. Mitchell P (1976) Possible molecular mechanisms of the protonmotive function of cytochrome systems. J Theoret Biol 62:327–367
22. Mitchell P (1991) The vital protonmotive role of coenzyme Q. In: Folkers K, Littarru GP, Yamagami T (eds) Biomedical and Clinical Aspects of Coenzyme Q, vol 6. Elsevier, Amsterdam, pp 3–10
23. Mitchell P (1988) Respiratory chain systems in theory and practice. In: Kim CH et al (eds) Advances in Membrane Biochemistry and Bioenergetics. Plenum Press, New York, NY, pp 25–52
24. Mitchell P (1979) Kelin's respiratory chain concept and its chemiosmotic consequences. Science 206:1148–1159
25. Ernster L (1977) Facts and ideas about the function of coenzyme Q10 in the Mitochondria. In: Folkers K, Yamamura Y (eds) Biomedical and Clinical Aspects of Coenzyme Q. Elsevier, Amsterdam, pp 15–18
26. Ramasarma T (1985) Natural occurrence and distribution of coenzyme Q. In: Lenaz G (ed) Coenzyme Q. Biochemistry, Biogenetics and Clinical Applications of Uniquinone 1985. John Wiley and Sons, New York, NY, pp 67–81
27. Ernster L, Forsmark-AndreÂe P (1993) Uniquinol: an endogenous antioxidant in aerobic organisms. Clin Investig 71:60–65
28. Karlsson J, Folkers K, AstroÈm H, Jansson E, Pernow B, Holmgren A (1986) Effect of adriamycin on heart and skeletal muscle coenzyme Q (CoQ10) in man. In: Folkers K, Yamamura Y (eds) Biomedical and Clinical Aspects of Coenzyme Q, vol 5. Elsevier, Amsterdam, pp 241–246
29. Gale PH, Koniuszy FR, Page AC, Folkers K (1961) Coenzyme Q. XXIV. On the significance of coenzyme Q10 in human tissues. Arch Biochem Biophys 93:211–213
30. Bowry VW, Stanley KK, Stocker R (1992) High density lipoprotein is the major carrier of lipid hydroperoxides in human blood plasma from fasting donors. Proc Natl Acad Sci USA 89:10316–10320

31. Overvad K, Diamant B, Holm L, Holmer G, Mortensen SA, Stender S (1999) Coenzyme $Q_{10}$ in health and disease. Euro J Clin Nutr 53:764–770
32. Weber C, Bysted A, Hùlmer G (1997) Coenzyme Q10 in normal Danish food ingredients. Int J Vitam Nutr Res 67:123–129
33. Olson RE, Rudney H (1983) Biosynthesis of ubiquinone. Vitam Horm 40:1–43
34. Peter H Langsjoen: Introduction to Coenzyme Q10. http://faculty.washington.edu/ely/coenzq10.html
35. Alleva R, Tomasetti M, Battino M, Curatola G, Littarru GP, Folkers K (1995) The roles of coenzyme Q10 and vitamin E on the peroxidation of human low density lipoprotein subfractions. Proc Natl Acad Sci USA 92:9388–9391
36. Kaikkonen J, Nyyssönen K, Porkkala-Sarataho E, Poulsen HE, Metsä-Ketelä T, Hayn M, Salonen R, Salonen JT (1997) Effect of oral coenzyme Q10 supplementation on the oxidation resistance of human VLDL. LDL fraction: absorption and antioxidative properties of oil and granule-based preparations. Free Radic Biol Med 22:1195–1202
37. Stocker R, Thomas SR, Neuzil J (1996) Cosupplementation with coenzyme Q prevents the prooxidant effect of alpha-tocopherol and increases the resistance of LDL to transistion metal-dependent oxidation initiation. Arterioscler Thromb Vasc Biol 16:687–696
38. Tomono Y, Hasegawa J, Seki T, Moteggi K, Morishita N (1986) Pharmacokinetic study of deuterium-labelled coenzyme Q10 in man. Int J Clin Pharmac Ther Toxicol 24:536–541
39. Mohr D, Bowry VW, Stocker R (1992) Dietary supplementation with coenzyme Q10 results in increased levels of uniquinol-10 within circulating lipoproteins and increased resistance of human low-density lipoprotein to the initiation of lipid peroxidation. Biochem Biophys Acta 1126(3):247–254
40. Crestanello JA, Kamelgard J, Lingle DM, Mortensen SA, Rohde M, Whitman GJR (1996) Elucidation of a tripartite mechanism underlying the improvement in cardiac tolerance to ischemia by coenzyme Q10 pretreatment. J Thorac Cardiovasc Surg 111(2): 443–450
41. Folkers K, Vadhanavikit S, Mortensen SA (1985) Biochemical rationale and myocardial tissue data on the effective therapy of cardiomyopathy with coenzyme Q10. Proc. Nat. Acad. Sci. Amsterdam. Elsevier 82:901–904
42. Bargossi AM, Battino M, Gaddi A, Fiorella PL, Grossi G, Barozzi G et al (1994) Exogenous coenzyme Q10 preserves plasma ubiquinone levels in patients with 3-hydroxy-3-methylglutaryl coenzyme A reductase inhibitors. Int J Clin Lab Res 24:171–176
43. Mortensen SA, Leth A, Agner E, Rohde M (1997) Dose-related decrease of serum coenzyme Q10 during treatment with HMG-CoA reductase inhibitors. Mol Aspects Med 18(Suppl):S137–S144
44. Greenberg S, Frishman WH (1990) Co-enzyme Q10: a new drug for cardiovascular disease. J Clin Pharmacol 30:596–608
45. Mitchell P (1975) Protonmotive redox mechanism of the cytochrome b – c1 complex in the respiratory chain: protonmotive ubiquinone cycle. FEBS Lett 56:1–6
46. Otani H (1993) Role of calcium in the pathogenesis of myocardial reperfusion injury. In: Das DK (ed) Pathophysiology of Reperfusion Injury. CRC Press Inc., Boca Raton, FL, pp 181–219
47. Goldman R, Mosshonov S, Chen X, Berchansky A, Furstenberger G, Zor U (1998) Crosstalk between elevation of [Ca21]i, reactive oxygen species generation and phospholipase A2 stimulation in a human keratinocyte cell line. In: Sinzinger H et al (ed) Recent Advances in Prostaglandin, Thromboxane, and Leukotriene Research. Plenum Press, New York, NY, pp 41–45
48. McConkey DJ, Orrenius S (1997) The role of calcium in the regulation of apoptosis. Biochem Biophys Res Commun 239:357–366
49. Richter C (1993) Pro-oxidants and mitochondrial Ca21: their relationship to apoptosis and oncogenesis. FEBS Lett 325:104–107
50. Barinaga M (1998) Death by dozens of cuts. Science 280:32–34

51. Zou H, Henzel WJ, Liu X, Lutschg A, Wang X (1997) Apaf-1, a human protein homologous to *C. elegans* CED-4, participates in cytochrome *c*-dependent activation of caspase-3. Cell 90:405–413
52. Mortensen SA (1989) Endomyocardial biopsy. Technical aspects and indications. Dan Med Bull 36:507–532
53. Murray CJ, Lopez AD (1997) Alternate projections of mortality and disability by cause 1990–2020: global burden of disease study. Lancet 349:1498–1504
54. Singh RB, Wander GS, Rastogi A, Shukla PK, Mittal A, Sharma JP, Mehrotra SK, Kapoor R, Chopra RK (1998) Randomized, double-blind placebo-controlled trial of coenzyme Q10 inpatients with acute myocardial infarction. Cardiovasc Drugs Ther 12:347–353
55. Hiasa Y, Ishida T, Maeda T, Iwanc K, Aihara T, Mori H (1984) Effects of coenzyme Q10 in patients with stable angina pectoris. In: Folkers K, Yamamura Y (eds) Biomedical and Clinical Aspects of Coenzyme Q, vol 4. Elsevier, Amsterdam, pp 291–301
56. Kamikawa T, Kobayashi A, Yamashita T, Hayashi H, Yamazaki N (1985) Effects of coenzyme Q10 on exercise tolerance in chronic stable angina pectoris. Am J Cardio 56:247–251
57. Schardt F, Welzel D, SchiessW& Toda K (1986) Effect of coenzyme Q10 on ischemia-induced ST-segment depression: a double-blind, placebo-controlled crossover study. In: Folkers K, Yamamura Y (eds) Biomedical and Clinical Aspects of Coenzyme Q, vol 5. Elsevier, Amsterdam, pp 358–394
58. Yamagami T, Takagi M, Akagami H, Kubo H, Toyama S, Okamoto T (1986) Effect of coenzyme Q10 on essential hypertension, a double blind controlled study. In: Folkers K, Yamamura Y (eds) Biochemical and Clinical Aspects of Coenzyme Q, vol 5. Elsevier, Amsterdam, pp 337–343
59. Digiesi V, Cantini F, Brodbeck B (1990) Effect of coenzyme Q10 on essential arterial hypertension. Curr Ther Res 47:841–845
60. Chello M, Mastroroberto P, Romano R, Bevacqua E, Pantaleo D, Ascione R (1994) Protection by coenzyme Q10 from myocardial reperfusion injury during coronary artery bypass grafting. Ann Thorac Surg 58:1427–1432
61. Judy WV, Stogsdill WWI, Folkers K (1993) Myocardial preservation by therapy with coenzyme Q10 during heart surgery. Clin. Invest. Amsterdam. Elsevier 71:155–161
62. Chello M, Mastroroberto P, Romano R, Castaldo P, Bevacqua E, Marchese AR (1996) Protection by coenzyme Q10 of tissue reperfusion injury during abdominal aortic cross-clamping. J Cardiovasc Surg 37:229–235
63. Murry CE, Jennings RB, Reimer KA (1986) Preconditioning with ischemia: a delay of lethal cell injury in ischemic myocardium. Circulation 74:1124–1136
64. Das S, Wong R, Rajapakse N, Murphy E, Steenbergen C (2008): Glycogen synthase kinase 3 inhibition slows mitochondrial adenine nucleotide transport and regulates voltage-dependent anion channel phosphorylation. Circ Res. 103, 983–991.
65. Ferrara N, Abete P, Ambrosio G, Landino P, Caccese P, Cirillo P, Oradei A, Littarru GP, Chiariello M, Rengo F (1995) Protective role of chronic ubiquinone administration on acute cardiac oxidative stress. J Pharmacol Exp Ther 274:858–865
66. Hano O, Thompson-Gorman S, Zweier J, Lakatta E (1994) Coenzyme $Q_{10}$ enhances cardiac functional and metabolic recovery and reduces $Ca^{2+}$ overload during postischemic reperfusion. Am J Physiol Heart Circ Physiol 266:H2174–H2181

# Chapter 7
# Oxidative and Proteolytic Stress in Homocysteine-Associated Cardiovascular Diseases

**Karni S. Moshal, Munish Kumar, Neetu Tyagi, Paras Kumar Mishra, Saumi Kundu, and Suresh C. Tyagi**

**Abstract** Homocysteine (HCY) is a risk factor for atherosclerosis-related cardiovascular diseases. HCY increases oxidative stress, activates MMP, and alters calcium homeostasis, thereby leading to vascular dysfunction. The major source of oxidative stress in HCY-induced vascular remodeling is the mitochondria. HCY causes activation and the mitochondrial translocation of calpain-1 (calcium-dependent cysteine protease), thereby increasing intramitochondrial oxidative stress leading to the induction of MMP-9. Ample studies have focused on the role of HCY in vascular dysfunction. HCY increases fibrosis and causes cardiac contraction dysfunction; however, the role of HCY on the cardiomyocytes is not understood. HCY increases the expression of mitochondrial MMP-9, and induces mitochondrial permeability transition, leading to a decline in cardiomyocyte contractile function by agonizing the NMDA receptor. In the present review, the role of hydrogen sulfide in HCY-induced myocardium protection is summarized. Furthermore, the role of HCY-induced oxidative stress in the mitochondria in the regulation of myocyte contractility is discussed.

**Keywords** Matrix metalloproteinase · Endothelial-myocyte disconnection · Calcium · Homocysteine · Mitochondria · Hydrogen sulfide · Contraction uncoupling

## 7.1 Introduction

Remodeling by its very nature implies synthesis and degradation of connective extracellular matrix (ECM) proteins. A coordination between ECM fibrosis, muscular hypertrophy, and angiogenesis is a hallmark of compensatory response to

---

S.C. Tyagi (✉)
Department of Physiology and Biophysics, School of Medicine, University of Louisville, Louisville, KY 40202, USA
e-mail: suresh.tyagi@louisville.edu

**Grants**: Supported in part by NIH Grants HL-71010, HL-74185, and HL-88012

H. Sauer et al. (eds.), *Studies on Cardiovascular Disorders*,
Oxidative Stress in Applied Basic Research and Clinical Practice,
DOI 10.1007/978-1-60761-600-9_7, 

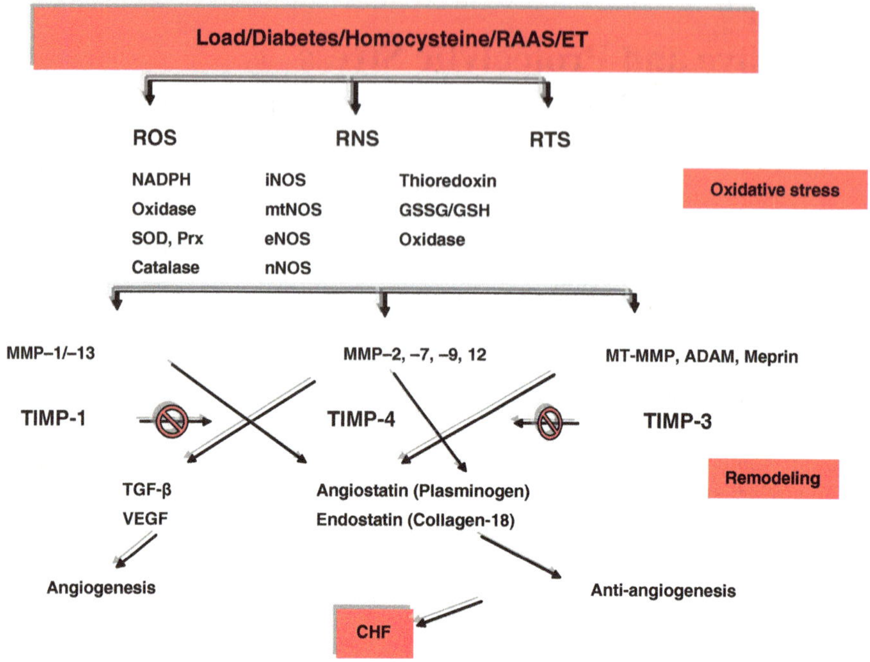


**Fig. 7.1** Mechanism of remodeling: Latent myocardial MMP activated by ROS, RNS, and RTS induced by load, Hcy, endothelin (ET), and RAAS. The differential role of MMP and TIMP in compensatory (hypertrophy and angiogenesis) to decompensatory (hypertrophy and no-angiogenesis) congestive heart failure (CHF)

heart failure; however, a disruption in coordination between hypertrophy and angiogenesis contributed to the transition to cardiac fibrosis and failure (Fig. 7.1). In the absence of blood supply to the myocardium at risk, the lack of angiogenesis by an increase in angiostatic factors leads to fibrosis and hypertrophy. Although it is a paradox that activation of matrix metalloproteinase (MMP) was needed to disrupt the matrix during angiogenesis and to create new blood vessels, or open collaterals, the peptides generated by MMP activation were angiostatic. The increase in MMP (Fig. 7.1) was consistent with the notion that MMPs cause discoordinated matrix degradation leading to decompensatory hypertrophy and heart failure by generating angiostatic factors, explaining how MMP, tissue inhibitor of metalloproteinase (TIMP) and hypertrophy coexist. In the basement membrane of endothelial cells and myocytes there is activation of latent MMP by nitric oxide (NO) and peroxynitrite during increases in oxidative stress and hyperhomocysteinemia (Fig. 7.2).

Cardiovascular disease (CVD) is a major cause of mortality and morbidity in the United States. It is estimated that hundreds of thousands of people die of cardiovascular disease, which includes coronary heart disease, congenital heart disease, and

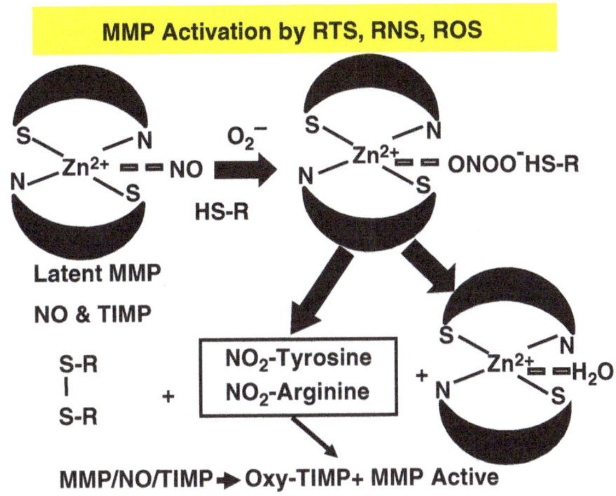


**Fig. 7.2** Mechanism of latent MMP activation in the basement membrane of endothelial cells by NO and peroxinitrite-thiol intermediate during oxidative stress and hyperhomocysteinemia

elevated blood pressure. It is very well appreciated that the major risk factors leading to the CVD-related deaths include hyperlipidemia (high cholesterol) and hypertension. However, in some cases the CVD-associated mortality cannot be explained based on major risk factors. Herein lies the concept of *hyperhomocysteinemia.* Hyperhomocysteinemia (HHCY) refers to elevated levels of homocysteine (HCY) in the blood plasma. HCY is a nonprotein amino acid containing a free sulfhydryl group. It is derived from the metabolic conversion of the essential amino acid methionine [1]. Compounds with a free sulfhydryl group are called thiols. Another biologically relevant low-molecular-weight thiol is cysteine, which in its oxidized state, cystine, is the most abundant thiol in plasma. Human plasma contains several reduced and oxidized HCY species. Oxidized HCY species comprise 98–99% of the total human plasma homocysteine, 80–90% of which is bound to proteins [2]. In humans, the normal concentration of HCY in the blood plasma is 10 μmol/L; moderate HHCY is reflected by 10–30 μmol/L HCY, and severe HHCY is reflected by >200 μmol/L of HCY in blood plasma. Based on clinical and basic research findings, it is suggested that elevated levels of HCY are an independent graded risk factor for atherosclerotic vascular disease [3, 4], which includes peripheral arterial disease, coronary artery disease, and cerebrovascular disease. The mechanism is not well elucidated, but one hypothesis is that HHCY leads to increased oxidant stress in the vasculature [5], which, besides other effects, leads to reduced bioavailability of nitric oxide. Besides oxidative stress, the roles of endoplasmic reticulum–associated stress and alterations in signal transduction pathways and activation of inflammatory factors are also suggested. Other pathologies manifesting the role of elevated homocysteine are in neurological disorders and renal dysfunction and nephrosclerosis. We have summarized our research findings on how the elevated levels of HCY induce the expression and activation of MMP-9, leading to vascular endothelial remodeling in Fig. 7.3.

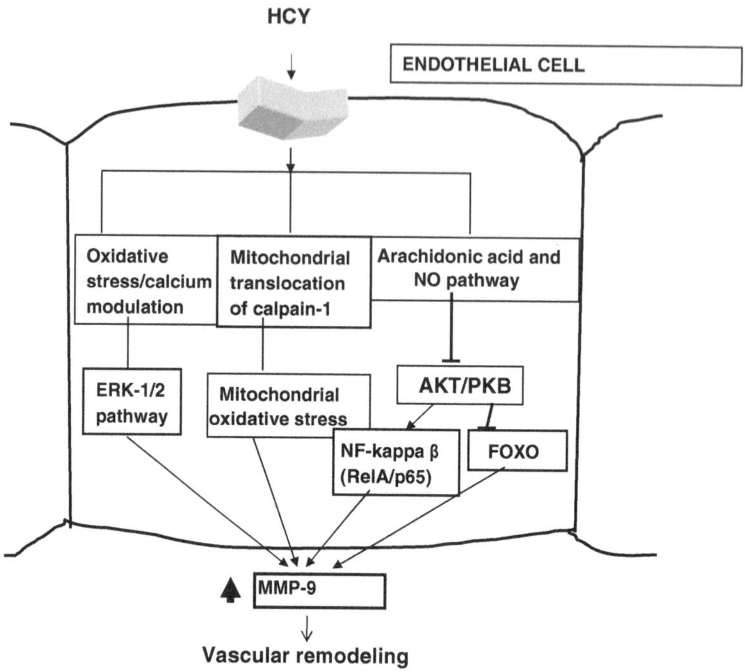


**Fig. 7.3** HCY signaling cascade leading to vascular dysfunction. HCY activates ERK-1/2 signal leading to MMP-9 induction. HCY causes activation and translocation of calpain-1 in the mitochondria, increases the mitochondrial oxidative stress leading to the activation of MMP-9. HCY modulates arachidonic acid pathway, causes the nuclear accumulation of Nf-kappa b, leading to the increase in MMP-9 expression and activity. The increase in MMP-9 activation in the setting of HHCY leads to vascular remodeling

## 7.2 HCY Mechanism of Oxidative and Nitrosative Stress

Experiments using animal models with genetic or diet-induced hyperhomocysteinemia or endothelial cells cultured under conditions that lead to elevated homocysteine levels showed an increased accumulation of reactive oxygen species (ROS), especially superoxide anion. ROS consist of superoxide anion, hydroxyl radical, peroxynitrite, hydrogen peroxide, or other peroxides, and hypochlorous acid, and their organic analogues.

ROS at moderate concentrations act as signaling molecules and thereby play important roles in the regulation of various cellular functions. ROS, however, can be toxic to cells and tissues through the promotion of lipid peroxidation of membranes (loss of membrane function and increased permeability) and generation of lipid autoperoxidation reactions, and through oxidant damage to low-density lipoproteins. DNA damage leads to mutation and death, and crosslinking of sulfhydryl-rich proteins (leading to stiff aged proteins, specifically, collagen of the extracellular matrix) [6]. Therefore, a supraphysiologic concentration of ROS may result in the loss of control of cell signaling.

Homocysteine induces increased formation of peroxynitrite, which initiates lipid peroxidation. Peroxynitrite, though, has a short half-life but reacts with cellular tyrosine residues to form nitrosylated products. The precise mechanism by which homocysteine may induce increased peroxynitrite formation remains to be elucidated. Two potential mechanisms include endothelial nitric oxide synthase as a source of superoxide (and peroxynitrite) and increased superoxide production from other enzymatic sources, in which some of the excess superoxide reacts with nitric oxide to form peroxynitrite [7]. Endothelial cells treated with homocysteine show increased formation of peroxynitrite, detected by the formation of nitrotyrosine and nitrotyrosine-modified proteins.

The accumulation of homocysteine leads to increased cellular oxidant stress also in mitochondria. Homocysteine has been shown to promote mitochondrial damage and alter mitochondrial gene expression and function, suggesting the participation of oxidant stress [8]. Biogenesis of mitochondria has been shown to be affected by homocysteine [9]. Incubation with homocysteine increased the intracellular ROS content and resulted in a significant increase in the mitochondrial mass of endothelial cells.

## 7.3 HHCY, Oxidative Stress, and Myocyte Dysfunction

Cardiac failure is linked to contractile and electrical uncoupling and is associated with reduced levels of connexin43 (Cxn43). Although the function ascribed to Cxn43 is limited to providing electrical conduction and contractile coupling between the cardiomyocytes, its role outside of traditional gap junctions is not well understood. HHCY is a risk factor for cardiac arrhythmias and causes cardiac contractile dysfunction. Cardiac arrhythmias and neurological disorders lead to sudden death, and the blockade of N-methyl-D-aspartate receptor-1 (NMDA-R1) mitigates sudden death.

The mechanisms of cardiac arrhythmias in HHCY are largely unknown. The NMDA receptor is expressed in cardiomyocytes and its activation induces mitochondrial dysfunction, oxidative stress, and calcium mishandling, leading to myocyte dysfunction.

Matrix metalloproteinases (MMPS) are the Zn-containing endopeptidases involved in extracellular matrix (ECM) remodeling leading to arrhythmogenesis. The colocalization and the activation of MMPs in sarcomere and the myofibrils of the myocyte leads to functional contractile and conduction uncoupling. Matrix metalloproteinases (MMPs) are present in the mitochondria of fibroblasts and cardiomyocytes. However, the physiological significance of the mitochondrial MMP remains obscure. Cxn43 is present in the myocyte mitochondria and maintains the mitochondrial permeability transition pore (MPTP) in a closed state leading to cardioprotection. Cxn43 is the novel target of MMP. HCY agonizes N-methyl-D-aspartate receptor-1 (NMDA-R1) and alters $Ca^{2+}$ homeostasis. The putative role of myocyte dysfunction in the setting of HHCY has been summarized. Recently we have shown that HCY decreases myocyte contractility, in part by inducing MMP-9 in the mitochondria, thereby causing mitochondrial permeability and leading to contractile uncoupling. The mechanism by which induction of mitochondrial MMP-9 leads to mitochondrial permeability is the subject of study. According to our unpublished data, it is suggested that HCY induces mitochondrial MMP-9 and causes degradation of mitochondrial connexin43, eventually leading to contractile uncoupling.

## 7.4 $H_2S$ Hypothesis of Cardioprotection in HHCY

Thiols (e.g., GSH, CGSH, and HCY) are reductants per se that defend organisms from ROS attack and regulate the redox state of cells. In cultured endothelial cells, HCY is toxic, and this action has been related to increased production of hydrogen peroxide [10, 11]. Until now, nitric oxide and carbon monoxide were in focus for therapeutics in cardiovascular diseases. Very recently, a third endogenously produced gaseous signaling molecule, hydrogen sulfide ($H_2S$), has emerged as a potentially important mediator of CV homeostasis and cytoprotection [12]. $H_2S$ is a byproduct during the metabolism of HCY. Cystathione-beta-synthase (CBS) is the major enzyme involved in $H_2S$ production, but its expression is confined to brain, kidney, and lung. This enzyme is absent in cardiac tissue [13]. The other enzyme which has a role in $H_2S$ production is cystathione-gammalyase (CSE). CSE is the enzyme which produces $H_2S$ via cysteine metabolism. $H_2S$ is produced in the vasculature by CSE, where it mediates smooth muscle relaxation and subsequent vasodilation [14]. Vasculoprotective and antihypertensive effects of dietary garlic (Allium sativum) are mediated in large part via the generation of $H_2S$ [15]. Garlic-derived organic polysulfides are converted by red blood cells (RBCs) into hydrogen sulfide gas. $H_2S$ produced from garlic leads to vasorelaxation via vascular smooth muscle KATP channel-mediated hyperpolarization [14].

$H_2S$ has been shown to relax vascular smooth muscle, induce vasodilation of isolated blood vessels, and reduce blood pressure [16]. Recent studies have provided insights into the protective actions of $H_2S$ in the setting of myocardial I/R injury [17, 18]. Pretreatment with the $H_2S$ donor (NaHS) reduces arrhythmias in isolated hearts subjected to global cardiac I/R and improves myocyte survival [18]. Kimura and colleagues have shown that $H_2S$ protects neurons against oxidative damage. Lefer and colleagues on the other hand have shown that an optimum dose of $H_2S$ limits the damage from heart attacks and other coronary diseases. During cardiovascular disease, imbalance occurs in the MMP/TIMP (Tissue Inhibitor of Metalloproteinase) ratio. We have recently observed that $H_2S$ downregulates the expression of TIMP-3, MMP-2, and MMP-9 [19]. TIMP-3 is an important enzyme that promotes apoptosis. Since apoptosis in the heart is detrimental, therefore it is clear that $H_2S$ acts as a protector, particularly for heart tissue. It is clear that $H_2S$ may have a vast role in cardioprotection and further studies are needed to explore this novel gas. We have summarized the putative role of $H_2S$ in cardioprotection in the setting of HHCY in Fig. 7.4.

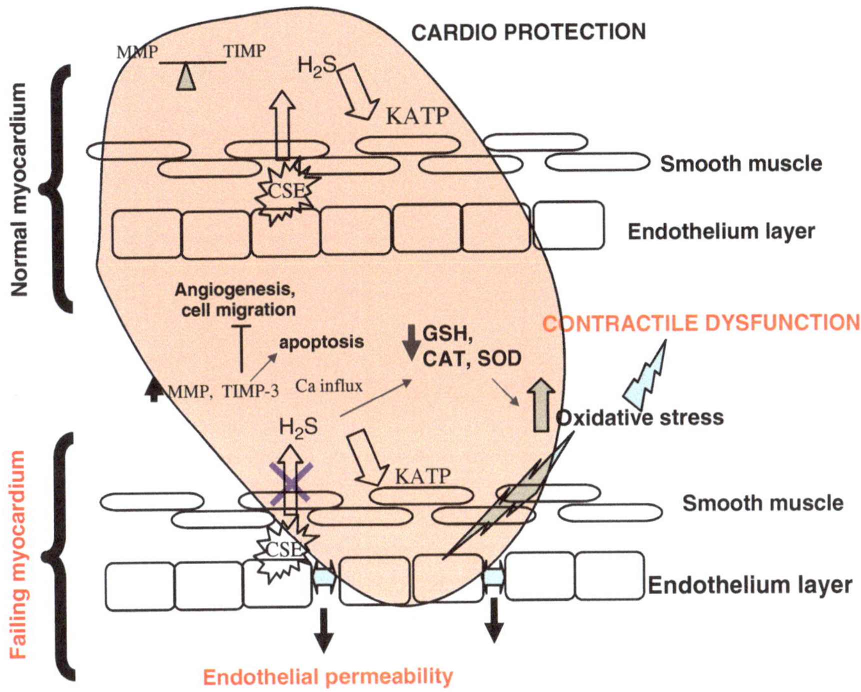


**Fig. 7.4** Suggestive mechanism of cardioprotection by $H_2S$. In normal myocardium $H_2S$ is generated from cythathionine gammalyase in the process of homocysteine metabolism. $H_2S$ activates $K_{ATP}$ channels and imparts relaxation of smooth muscle. $H_2S$ increases glutathione availability, and acts as an antioxidant. MMP/TIMP ratio is maintained in equilibrium. In failing myocardium, the oxidative stress is increased, leading to imbalance in MMP/TIMP ratio. Nonavailability of $H_2S$ leads to inactivation of $K_{ATP}$ channels, leading to dysfunction in contraction and relaxation of smooth muscle. Thus endothelial permeability is increased. In the process, cell migration and angiogenesis are blocked and apoptosis increases in cardiac smooth muscles. We here hypothesize that external supply of $H_2S$ may ameliorate these changes and protect the failing myocardium

## 7.5 Proliferation and Maintenance of Resident Cardiac Stem Cell, MMP/TIMP Levels, and FoxO Transcription Factor

The statistical analyses of mortality by heart failure in America reveal that it is increasing every year [20], demanding better therapies for the disease. Hitherto, the invasive treatment for heart failure deals with improving the life of patients without reviving the damaged cardiac cells, rendering short-term therapy for cardiac function. The long-term therapy requires a source for replenishment of those cells that are lost because of acute myocardial infarction. Stem cells have the potential of regenerating myocardial tissue and have attracted major attention to fulfill this goal through cell transplantation.

Embryonic stem cells (ESCs), bone marrow-derived mesenchymal and hematopoietic stem cells, fetal myocytes, skeletal myoblasts, and endothelial progenitor cells have been transplanted into the postinfarcted myocardium of experimental animals to test their potential for regenerating myocytes and/or vascular structure for tissue repair [21]. The major drawbacks of the use of ESCs for therapy are their tendency to produce teratomas, the ethical issues surrounding their future use, and the necessity of immunosuppression due to their heterologous nature [22]. Bone marrow–derived stems cells may also have myogenic potential [23–26], but this has been questioned by some investigators [22, 27, 28]. In this gloomy scenario, the only avenue to replace lost myocardial cells is by exogenous cell transplantation.

Before 2003, the heart was thought to be a terminally differentiated postmitotic organ with no regenerative potential. However, the finding that the postmitotic organs such as the brain also have regenerative potential, because they harbor tissue-specific adult stem cells [29], tempted investigators to look for adult cardiac stem cells. A combined effort of Anversa's and Nadal's group first identified and characterized a distinct population of resident cardiac stem-progenitor cells (CSCs) in rodents [30]. The resident populations of progenitor cells were also confirmed by several other investigators [31–35]. These cells have three important criteria: they are self-renewing, clonogenic, and multipotent; and they give rise to a minimum of three cardiogenic cell lineages: myocytes, smooth muscle cells, and endothelial cells [36]. It is established that when CSCs are injected into mice, they generate new myocytes.

The key point in stem cell therapy is to understand the molecular mechanism that regulates the choice between self-renewal and differentiation of CSCs. It is reported

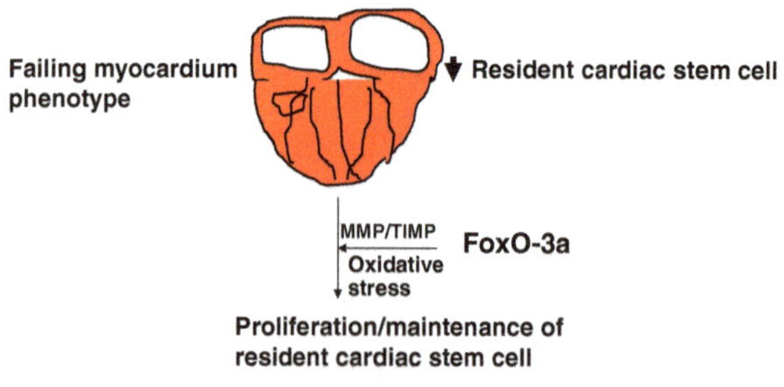


**Fig. 7.5** Putative role of FoxO transcription factor in cardiac stem cell therapy. In the failing myocardium there is a decrease in resident cardiac stem cells. It is anticipated that in the failing myocardium, the increase in oxidative stress modulates the activation levels of FoxO-3a, causing an imbalance in MMP/TIMP levels and leading to decline in the cardiac stem cells

that the stem cell niche plays a fundamental role in the regulation of quiescence, self-renewal, and commitment to differentiation of CSCs [37]. Unfortunately, no concrete knowledge about the cellular and cytokine/growth factor environment of the CSC niche is available [37]. In the normal myocardium, MMP and TIMP levels are maintained in equilibrium by the Fox O-3a transcription factor. A report suggested that Forkhead box transcription factor might be involved in the proliferation and maintenance of resident cardiac stem cells. We anticipate that the increase in oxidative stress may be involved in the FoxO-3a posttranscriptional modifications (phosphorylation/dephosphorylatio; acetylation/deacetylation) that lead to the imbalance of MMP and TIMP levels, which in turn may decrease resident cardiac stem cells, leading to heart failure (Fig. 7.5). A deep insight into the signaling mechanism regulating CSC self-renewal and differentiation will provide a better standing for therapeutic endeavors.

## References

1. Finkelstein JD, Martin JJ (1986) Methionine metabolism in mammals: adaptation to methionine excess. J Biol Chem 261:1582–1587
2. Jacobsen DW (1998) Homocysteine and vitamins in cardiovascular disease. Clin Chem 44:1833–1843
3. Bautista LE, Arenas IA, Penuela A et al (2002) Total plasma homocysteine level and risk of cardiovascular disease: a meta-analysis of prospective cohort studies. J Clin Epidemiol 55:882–887
4. Graham IM, Daly LE, Refsum HM et al (1997) Plasma homocysteine as a risk factor for vascular disease: the European concerted action project. J Am Med Assoc 277:1775–1781
5. Weiss N, Heydrick SJ, Postea O et al (2003) Influence of hyperhomocysteinemia on the cellular redox state: impact on homocysteine-induced endothelial dysfunction. Clin Chem Lab Med 41:1455–1461
6. Hayden MR, Tyagi SC (2002) Intimal redox stress: accelerated atherosclerosis in metabolic syndrome and type 2 diabetes mellitus:atheroscleropathy. Cardiovasc Diabetol 1:9
7. Upchurch GR Jr, Welch GN, Fabian AJ et al (1997) Homocyst(e)ine decreases bioavailable nitric oxide by a mechanism involving glutathione peroxidase. J Biol Chem 272:17012–17017
8. Austin RC, Sood SK, Dorward AM et al (1998) Homocysteine-dependent alterations in mitochondrial gene expression, function and structure: homocysteine and $H_2O_2$ act synergistically to enhance mitochondrial damage. J Biol Chem 273:30808–30817
9. Perez-de-Arce K, Foncea R, Leighton F (2005) Reactive oxygen species mediates homocysteine-induced mitochondrial biogenesis in human endothelial cells: modulation by antioxidants. Biochem Biophys Res Commun 338:1103–1109
10. Sen U, Tyagi N, Kumar M et al. Cystathionine-beta-synthase gene transfer and 3-deazaadenosine ameliorate inflammatory response in endothelial cells. Am J Physiol Cell Physiol 2007; 293: C1779–C1787.
11. Gaucher C, Menu P (2008) How to evaluate blood substitutes for endothelial cell toxicity. Antioxid Redox Signal 10:1153–1162
12. Szabo C (2007) Hydrogen sulphide and its therapeutic potential. Nat Rev Drug Discov 6: 917–935
13. Jhee KH, Kruger WD (2005) The role of cystathionine beta-synthase in homocysteine metabolism. Antioxid Redox Signal 7:813–822
14. Kimura Y, Kimura H (2004) Hydrogen sulfide protects neurons from oxidative stress. FASEB J 18:1165–1167

15. Benavides GA, Squadrito GL, Mills RW et al. Hydrogen sulfide mediates the vasoactivity of garlic. Proc Natl Acad Sci U S A 2007; 13:17977–17982
16. Lefer DJ. A new gaseous signaling molecule emerges: cardioprotective role of hydrogen sulfide. Proc Natl Acad Sci U S A 2007;13: 17907–17908
17. Qu K, Lee SW, Bian JS et al (2008) Hydrogen sulfide: neurochemistry and neurobiology. Neurochem Int 52:155–165
18. Pan TT, Feng ZN, Lee SW et al (2006) Endogenous hydrogen sulfide contributes to the cardioprotection by metabolic inhibition preconditioning in the rat ventricular myocytes. J Mol Cell Cardiol 40:119–130
19. Kumar M, Tyagi N, Moshal KS et al (2008) $GABA_A$ receptor agonist mitigates homocysteine – induced cerbrovascular remodeling in knockout mice. Brain Res 1221:147–153
20. American Heart Association (2005) Heart disease and stroke statistics: 2005 update. American Heart Association, Dallas, TX
21. Dimmeler S, Zeiher AM, Schneider MD (2005) Unchain my heart: the scientific foundations of cardiac repair. J Clin Investg 115:572–583
22. Balsam LB, Wagers AJ, Christensen JL et al (2004) Haematopoietic stem cells adopt mature haematopoietic fates in ischaemic myocardium. Nature 428:668–673
23. Yoon Y, Wecker A, Heyd L et al (2005) Clonally expanded novel multipotent stem cells from human bone marrow regenerate myocardium after myocardial infarction. J Clin Invest 115:326–338
24. Fernandez-Aviles F, San Roman JA, Garcia-Frade J et al (2004) Experimental and clinical regenerative capability of human bone marrow cells after myocardial infarction. Circ Res 95:742–748
25. Eisenberg LM, Burns L, Eisenberg CA (2003) Hematopoietic cells from bone marrow have the potential to differentiate into cardiomyocytes in vitro. Anat Rec A Discov Mol Cell Evol Biol 274:870–882
26. Eisenberg CA, Burch JB, Eisenberg LM (2006) Bone marrow cells transdifferentiate to cardiomyocytes when introduced into the embryonic heart. Stem Cells 24:1236–1245
27. Chien KR (2004) Stem cell: lost in translation. Nature 428:607–608
28. Murry CE, Soonpaa MH, Reinecke H (2004) Haematopoietic stem cells do not transdifferentialte into cardiac myocytes in myocardial infarcts. Nature 428:664–668
29. Korbling M, Extrov Z (2003) Adult stem cells for tissue repair – a new therapeutic concept? N. Engl J Med 349:570–582
30. Beltrami AP, Barlucchi L,Torella D et al (2003) Adult cardiac stem cells are multipotent and support myocardial regeneration. Cell 114:763–776
31. Oh H, Bradfute SB, Gallardo TD et al (2003) Cardiac progenitor cells from adult myocardium: homing, differentiation, and fusion after infarction. Proc Natl Acad Sci USA 100: 12313–12318
32. Matsuura K, Nagai T, Nishigaki N et al (2004) Adult cardiac stem Sca-1 positive cells differentiate into beating cardiomyocytes. J Biol Chem 279:11384–11391
33. Martin CM, Meeson AP, Robertson SM et al (2004) Persistent expression of the ATP-binding cassette transporter, Abcg2, identifies cardiac SP cells in the developing and adult heart. Dev Biol 265:262–275
34. Pfister O, Mouquet F, Jain M et al (2005) CD31- but Not CD31+ cardiac side population cells exhibit functional cardiomyogenic differentiation. Circ Res 97:52–61
35. Tomita Y, Matsumura K, Ogawa S et al (2005) Cardiac neural crest cells contribute to the dormant multipotent stem cell in the mammalian heart. J Cell Biol 170:1135–1146
36. Torella D, Ellison GM, Karakikes I et al (2007) Resident cardiac stem cells. Cell Mol Life Sci 64:661–673
37. Fuchs E, Tumbar T, Guasch G (2004) Socializing with the neighbors: stem cells and their niche. Cell 116:769–778

# Chapter 8
# Functional Studies of NADPH Oxidases in Human Vasculature

**Tomasz J. Guzik**

**Abstract** Oxidative stress in the vascular wall is a characteristic feature of vascular disease states, associated with atherosclerosis in humans. The earliest detectable changes in vascular disease states are abnormalities of the endothelium, resulting in an imbalance between nitric oxide and superoxide production and loss of normal homeostatic functions such as the normal tonic inhibition of inflammation and thrombosis. The current review describes functional studies of NADPH oxidase expression and activity and its role in the regulation of endothelial dysfunction. NADPH oxidases have been indicated as a main source of $O_2^{\bullet-}$ in human vasculature. In humans, it has been confirmed that NADPH oxidase activity is inversely correlated with endothelial function. This relationship exists even when corrected for other major risk factors for atherosclerosis, including diabetes and hypercholesterolemia. Diabetes and hypercholesterolemia are most strongly associated with NADPH oxidase activity in human vessels, while hypertension also plays an important role. Numerous Nox homologs are expressed in human vessels in correlation with risk factors or the presence of atherosclerosis and coronary artery disease. These include predominantly p22phox, Nox2, Nox4, and Nox5. Expression of these homologs is regulated by systemic factors and is correlated in arteries and veins. More extensive translational approaches need to be applied to confirm major findings regarding regulation of NADPH oxidases in human vasculature, in order to assess whether Nox enzymes are realistic therapeutic targets once specific small molecule inhibitors have been developed.

**Keywords** Nox family NADPH oxidases · Nox2 · Superoxide dismutase · Endothelial dysfunction · Nitric oxide bioavailability · Arterial hypertension · Atherosclerosis

---

T.J. Guzik (✉)
Translational Medicine Laboratory, Department of Internal and Agricultural Medicine and Department of Pharmacology Jagiellonian, University School of Medicine, Cracow 31-121, Poland
e-mail: tguzik@cm-uj.krakow.pl

H. Sauer et al. (eds.), *Studies on Cardiovascular Disorders*,
Oxidative Stress in Applied Basic Research and Clinical Practice,
DOI 10.1007/978-1-60761-600-9_8, © Springer Science+Business Media, LLC 2010

## 8.1 Introduction

Oxidative stress in the vascular wall is a characteristic feature of vascular disease states, associated with atherosclerosis in humans. Conditions such as coronary artery disease, cerebrovascular, and peripheral vascular diseases are the largest cause of mortality and morbidity in industrialized countries. Therefore, understanding the mechanisms of these conditions and initial steps of the development of vascular dysfunction and disease is of critical importance. The earliest detectable changes in vascular disease states are abnormalities of the endothelium, resulting in loss of its homeostatic functions that normally act to inhibit disease-related processes such as inflammation and thrombosis. In particular, nitric oxide (NO) produced by NO synthase (eNOS) in the vascular endothelium modulates blood flow and pressure and has important antiatherogenic effects on platelets, vascular smooth muscle, and endothelial cells. In humans, NO-mediated endothelial function inversely correlates with risk factor profile, being deficient in athero-prone conditions such as hypercholesterolemia, diabetes mellitus, hypertension, and smoking. More importantly, prospective studies identify deficient NO-mediated endothelial function as a quantitative, independent predictor of adverse cardiac events. Numerous animal model studies demonstrate the important role of NO in vascular disease pathogenesis. Targeted deletion of the eNOS gene in mice results in hypertension and impaired vascular remodeling. Accordingly, augmenting NO by local gene delivery of NOS improves endothelial function, limits neointimal proliferation, and induces regression of atherosclerotic lesions.

The majority of studies which defined these mechanisms was conducted in endothelial cells in culture or in animal models. Recent observations emphasize a clear need for a "translational" approach when defining mechanisms of diseases, as events described in animal models do not always translate into the clinical setting. This is even more important in relation to the role of NADPH oxidases, since in spite of many years of meticulous studies performed in numerous laboratories and a vast understanding of how NADPH oxidases function, only limited practical use of this knowledge has been applied in searching for new treatments of diverse diseases in which we learned over the years that NADPH oxidases are involved [1]. This chapter will discuss translational studies, which have been performed to characterize the functionality of NADPH oxidases in human blood vessels.

## 8.2 Functional Studies of Oxidative Stress in Human Vasculature

### 8.2.1 *Ex Vivo Studies*

The role of vascular oxidative stress can be most directly assessed ex vivo in isolated vascular rings. Several laboratories have employed this technique, including our group, to study human vessels, including: internal mammary arteries, radial arteries, saphenous veins, and particularly coronary arteries (Fig. 8.1). This approach,

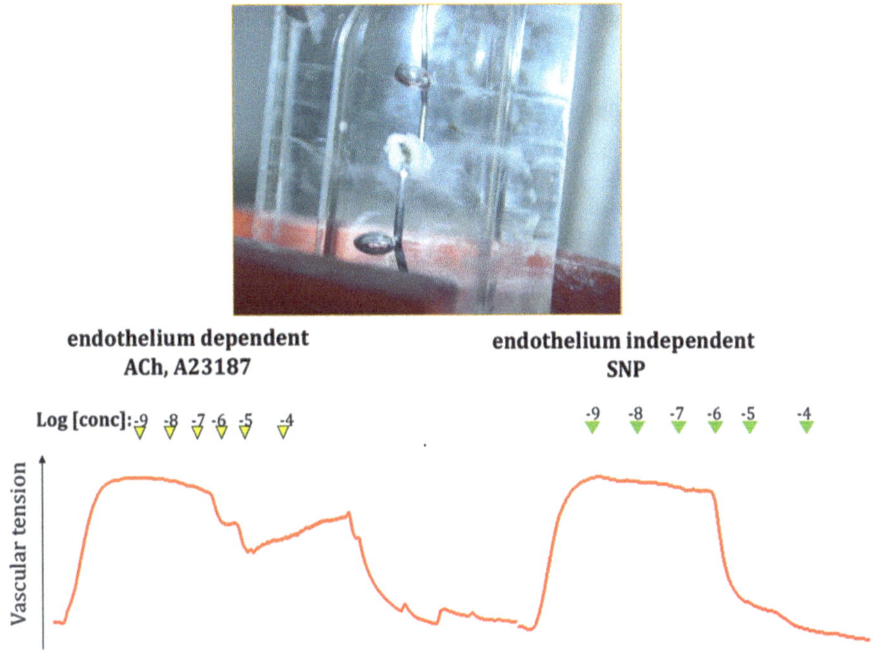


**Fig. 8.1** Human internal mammary artery studied in vitro in an organ chamber experiment, which allows for determination of the contribution of superoxide towards endothelial function measured as nitric oxide bioavailability. Following precontraction, increasing concentrations of vasorelaxants are administered. Experiments may be performed in the presence and absence of superoxide scavengers and oxidase inhibitors. Ach, acetylcholine; SNP, sodium nitroprusside

and in particular the use of human conduit vessels, which are most readily available to study, may give important insight into the pathogenesis of coronary artery disease, since both oxidative stress and endothelial dysfunction are systemic phenomena and are correlated among distinct compartments of circulation. In a recent study, we showed that NADPH oxidase activity and NO bioavailability are closely correlated between human arteries and veins from diverse areas of circulation [2]. These relationships were also observed at the molecular level of expression of NADPH oxidase subunits [2]. While this approach has been criticized by some for its ex vivo character or impossibility of studying healthy vessels, it is much closer to human pathophysiology than animal models or even cells in culture. Moreover, it has proven very useful to demonstrate functional relationships between oxidative stress and clinical factors to which blood vessels are exposed in patients, such as risk factors for atherosclerosis or inflammatory conditions [3, 4]. This approach allows one to address many questions regarding mechanisms of oxidative stress in human vasculature, but is limited to patients with relatively severe atherosclerosis or vascular disease, because they require surgical interventions. This approach also permits an organ culture type of experiments, which consist of exposing human blood vessels to treatments in a similar fashion to that used in cell culture. Oxidative stress and

NADPH oxidases have also been studied and characterized using these techniques in varicose vein disease [5].

An important practical consideration for such studies is that vessels need to be treated with extreme caution, as their endothelial layer is very susceptible to further damage. Therefore, vessels should be obtained prior to surgical preparation and distention. Moreover, close cooperation between the scientist and the surgeon is critical for success of these studies.

Some studies attempt to analyze oxidative stress in postmortem samples. However, this approach may be difficult to use in functional assays, since functional changes related to the pathology occurring during death may be pronounced. However, in spite of the great value of human conduit vessels such as IMA or HSV as a model system, it must be remembered that these vessels do not develop typical atherosclerosis, although endothelial dysfunction and intimal hyperplasia are often observed. Therefore, critical and careful interpretation of results is warranted when extrapolating these observations to human coronary circulation. At the same time, bypass graft disease is a common cause of morbidity in post-CABG patients, and using conduit vessels in functional studies allows one to address these important issues.

### *8.2.2 In Vivo Studies*

In vivo approaches are indirect, but may also be very valuable to provide translation of basic science findings into clinical medicine. A particularly valuable strategy has been implemented by Heitzer et al., who assessed vascular oxidative stress by measuring how vascular function changes in response to intra-arterial administration of high concentrations of superoxide scavengers such as vitamin C [6]. These studies allow for elegant determination of vascular ROS production in humans not undergoing surgical interventions, thus enabling the study of certain clinical and prognostic correlates [6].

## 8.3 Role of Reactive Oxygen Species in the Regulation of Endothelial Function in Human Vasculature

The mechanisms that result in loss of NO bioactivity in vascular disease have remained unclear. Data from experimental models suggest that loss of eNOS protein or activity is not a primary cause in the early states of atherosclerosis and point towards the increased production of reactive oxygen species in this process. The role of oxygen radicals such as superoxide anion in impairing NO-mediated endothelial function has been widely demonstrated. The effects of oxygen radicals in regulating endothelial function is an independent prospective indicator of adverse cardiovascular risk [6]. The reaction between NO and superoxide occurs at almost diffusion-limited rates, six times greater than the removal of superoxide by copper-zinc superoxide dismutase (Cu/Zn SOD). Mutual interaction between NO and superoxide in vascular tissues has been initially described in vitro and in the animal models of hypertension and atherosclerosis [7]. More recently, it was also shown in human arteries and to a lesser extent in veins [8]. Interestingly, under basal conditions, greater superoxide production in arteries than in veins is balanced by increased NO production, resulting in higher arterial peroxynitrite formation [8]. This is also critical for the interpretation of studies investigating vascular NO and superoxide production, as it shows that both nitric oxide and superoxide roles in blood vessel biology need to be considered together, rather than individually. Shapira and colleagues [9] observed greater NO production in radial arteries, as opposed to internal mammary arteries, and both were greater than in saphenous veins. These differences can be attributed to differential activation by signaling pathways, rather than to intrinsic differences in total eNOS activity, since both acetylcholine-induced vasorelaxation (receptor-mediated Ca++-dependent eNOS activation), and contractile response to NOS inhibition (shear stress–like, calcium-independent eNOS activation by isometric tension) were similar among the distinct types of vessel [10].

Similarly to nitric oxide production, superoxide production is also greater in human arteries than veins [11]. However, in our own studies, these differences were found only after NOS blockade (using either L-NAME or L-NMMA) and were abolished at basal state, most likely as the result of superoxide scavenging by NO [8]. Our experiments further confirmed this interaction in blood vessels using superoxide scavengers to inhibit the NO-superoxide interaction or enhancing it through the use of NO donors [8]. Furthermore, corresponding measurements of peroxynitrite formation revealed the converse changes in response to NOS antagonists or NO donors. These experiments clearly showed that NO-superoxide interaction is much more pronounced in human arteries than in veins. Interestingly, Berry et al. [11], as well as Schmalfuss et al. [12], have shown differences in basal superoxide release without NOS blockade, which we did not observe. These differences could be due to distinct clinical or biological characteristics of the studied patients or to different assay conditions. These studies may also allow one to estimate the balance between different radical generation in various blood vessel types or in different layers of the same vessel. An increase in superoxide release after endothelial removal suggests that the endothelium produces a net excess of NO, rather than superoxide, which in turn influences other vessel wall layers. However, inhibition of NOS, but not removal of endothelium, resulted in an even greater increase in superoxide release, showing that the endothelium itself is indeed a significant source of superoxide in human blood vessels, as suggested by DHE fluorescence. While in the endothelium NO seems to be greater than superoxide generation, when the whole blood vessel is taken into account the generation of superoxide at basal state seems to exceed the capacity of endothelium to generate NO in these ex vivo conditions [13], since we found that an NO donor alone added to vascular rings caused a significant increase in peroxynitrite formation [8]. Moreover, incubation of vessel rings with NADPH greatly stimulates superoxide production (>tenfold) [14], but has no effect on peroxynitrite formation (T. Guzik, KM Channon, unpublished data); whereas concomitant application of an

NO donor in addition to NADPH increases peroxynitrite formation. As discussed above, these studies of ex vivo vessels have limitations related to the fact that the NO-superoxide balance may be different in in vivo vessels, due to tonic activation of endothelial NO production by shear stress, and due to other sources of NO such as nitrosothiols. It is also likely that the relative balance between NO and superoxide varies not only with blood vessel type, as we observed, but also under different biological conditions. Vessels from subjects with overt risk factors for atherosclerosis show higher superoxide production and lower NO production than healthy vessels [10], and thus the interaction between superoxide and NO release may be different. A parallel situation occurs in experimental models, e.g., in normocholesterolemic rabbit aorta or nonhypertensive rat, in contrast to aortas from animals with hypercholesterolemia, or angiotensin II infusion.

These findings functionally characterize nitric oxide–superoxide interaction in human vasculature, showing clearly an increased NO-superoxide interaction in human arteries as compared with that in veins [8]. Interestingly, despite higher superoxide and peroxynitrite formation in arteries, endothelium-dependent vasorelaxation is also higher in arteries than in veins. It may be that superoxide or peroxynitrite, in addition to NO, could act as mediators of endothelium-dependent relaxation in these vessels. Higher peroxynitrite formation in arteries may also favor the development of disorders more specific for arteries, such as hypertension or atherosclerosis; whereas veins are not susceptible to classical atherosclerosis, despite exposure to all systemic risk factors except arterial blood pressure [8]. These functional studies in human vessels support the emerging concept that NO-mediated vascular signaling needs to be considered as one part of more complex pathways closely networking with those mediated by superoxide and peroxynitrite, as well as their respective targets. That is, the coordinated regulation of NO, superoxide production, and peroxynitrite formation is biologically more important than the bioactivity of any of these species individually [8].

## 8.4 Vessel Wall Layers Contributing to Total Vascular Superoxide

A most important question to be addressed is the relative contribution towards superoxide production from the intima, media, or adventitia. While all of these vascular layers express NADPH oxidases and other enzymes and are capable of producing superoxide, it is very difficult, based on either cell culture studies or animal models, to address this issue. The implementation of dihydroethidium fluorescent staining (DHE) by Miller et al. has created an opportunity to indirectly address this question [15], although it must be emphasized that in spite of some researchers using DHE to quantify superoxide production, this method has limitations and remains semiquantitative. DHE reveals that indeed all layers of both human arteries and veins produce superoxide. However, while in veins the differences in the relative contribution of

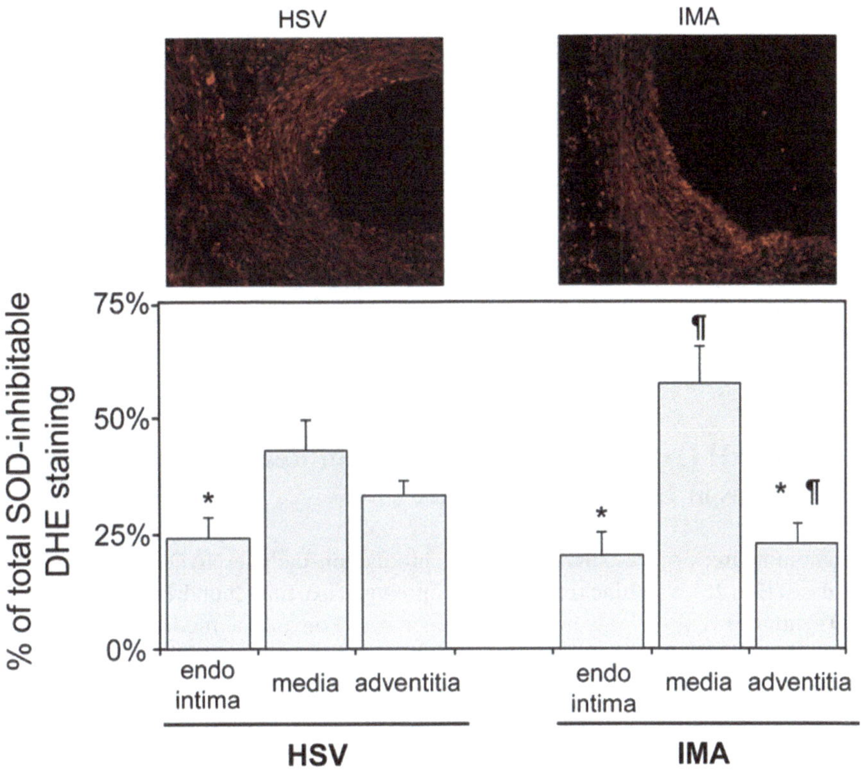


**Fig. 8.2** Relative contribution of individual vessel wall layers towards total superoxide production (based on Guzik et al. [2])

individual layers are not significant, in arteries unquestionably the greatest contribution comes from media, with both endothelium and adventitia contributing to less than half of that production (Fig. 8.2) [2]. The cellular sources of vascular superoxide may differ in distinct vessel types or among diverse species. In aortas from hypertensive angiotensin II–infused rats and mice, the adventitia seems the major source of superoxide [16–18], while the endothelium appears to be a more important contributor in the hypercholesterolemic rabbit aorta or DOCA-salt hypertensive animals [19, 20].

## 8.5 New "Functional Hypothesis" of Oxidative Stress

Traditionally, oxidative stress has been defined as an imbalance between pro-oxidant and antioxidant mechanisms (e.g., between superoxide-producing systems and superoxide dismutases, glutathione peroxidases, thioredoxins, etc.) [21, 22]. Recent studies have suggested that, while profound changes of such balance are observed in advanced stages of disease, in the early stages the alterations occur predominantly within individual cellular compartments or even within individual enzymes, without modification of total cellular redox status. Such changes of electron flow in certain enzyme systems result in the formation of a "pathological," or excessive amount of a correct product leading to disturbances of redox signaling [23]. Endothelial NO synthase uncoupling, leading to a switch towards production of superoxide instead of NO by this enzyme, or superoxide "leakage" from the mitochondrial chain may be examples of such local changes. Therapeutic interventions on the level of total cellular redox status (like antioxidant vitamins) may not be sufficient to correct these disturbances. Novel strategies should instead focus on "repairing" the function of the individual molecules involved in ROS generation [24].

## 8.6 NADPH Oxidases as Main Sources of Reactive Oxygen Species in Human Vasculature

Potential sources of vascular superoxide production include NAD(P)H-dependent oxidases [25, 26], xanthine oxidase [27], lipoxygenase, mitochondrial oxidases, and NO synthases [28]. A large body of evidence based on animal models suggests that a membrane-bound NADPH oxidase, xanthine oxidase, and dysfunctional eNOS are the major sources of superoxide anion in various pre-atherosclerotic conditions [25, 29]. These have recently been confirmed in human peripheral conduit arteries and veins [2] as well as in human coronary arteries [30–32]. Indeed, studies using various oxidase inhibitors in human veins and arteries have shown that inhibition of NADPH oxidase [3, 11] causes the greatest inhibition of superoxide production in human veins and also in arteries (although in both conduit and coronary arteries the sources of ROS are more complex), which will be discussed below.

Isoforms of the phagocytic NADPH oxidase, referred to as Nox enzymes, have been found in all vascular cells, including endothelial cells, VSMCs, and fibroblasts [33]. At least five protein components compose the classical NADPH oxidase complex: the catalytic membrane-bound Nox, which in the phagocyte is refered to as cytochrome $b_{558}$ ($p22^{phox}$ and $gp91^{phox}$ heterodimer) and regulatory cytosolic proteins – $p47^{phox}$, $p67^{phox}$, as well as low molecular weight G protein Rac. There are at least 7 Nox isoforms, characterized by distinct structures of their catalytic subunits, as well as by distinct patterns of association with regulatory subunits (reviewed in detail by Lambeth [34] and in other chapters of this publication). These catalytic subunits possess flavin and heme-binding regions and generate $O_2^{\bullet-}$ via one-electron transfer from NADPH to oxygen. An exception to this is Nox4, which seems to produce predominantly $H_2O_2$, although full understanding of this issue is still ongoing. Of the various Nox isoforms, Nox1, Nox2, and Nox4 are the most important in vascular cells. Except for Nox5, all Nox isoforms require $p22^{phox}$ as a docking subunit. Nox4 is constitutively active and does not require cytosolic subunits. Interestingly, Nox homologs may be differentially associated with various vascular disease phenotypes. Changes in Nox1 expression directly alter cell proliferation [35], and treatment of VSMC with angiotensin II or PDGF upregulates

Nox1, while downregulating Nox4 [36]. Vascular injury increases expression of Nox1, Nox2, and p22$^{phox}$, while Nox4 increases later [37], coinciding with a reduction in the rate of VSMC proliferation. These findings suggest that while Nox1 and Nox2 are involved in acute response to injury or to angiotensin II stimulation, Nox4 is involved in maintaining the quiescent phenotype [37].

Comparisons of human veins and arteries using various potential oxidase substrates and inhibitors have demonstrated that the NADPH oxidases are quantitatively and proportionally a greater source of superoxide in veins, whereas xanthine oxidase (and in some cases uncoupled endothelial nitric oxide synthases) appear to additionally contribute substantially to superoxide production in arteries [2]. Molecular studies have identified that increased vascular NADPH oxidase activity in human vessels is associated particularly with increased protein levels of p22phox, p47phox, and p67phox, and increases in at least p22phox and Nox2 (gp91phox) mRNA expression. The NADPH oxidase is predominantly Nox2-based in saphenous veins, whereas a Nox4-based oxidase appears proportionately more important in mammary arteries [2]. Interestingly, in spite of differences in the expression profiles or different mRNA levels of individual Nox homologs in human vessels, the expression of both p22phox and Nox4 mRNA are strikingly correlated in arteries and veins [2]. The marked differences in the Nox2/Nox4 ratio could at least in part account for the distinct susceptibility of veins to smooth muscle intimal hyperplasia, leading to adverse remodeling and accelerated atherosclerosis in vein grafts [38], whereas mammary artery grafts are much less susceptible to atherosclerosis. The differences in NADPH oxidase subunit molecular composition between arteries and veins could also reflect the differences in the vascular cells that contribute to superoxide production [39], as discussed above. In veins, the predominance of Nox2 expression suggests major contributions from the endothelium and adventitia, as these contain Nox2-based oxidases [40]. Indeed, adventitial superoxide production from a Nox2-containing NADPH oxidase directly contributes to endothelial dysfunction by NO scavenging [41]. In human arteries, our observation of increased Nox4 expression suggests that smooth muscle cells could play a critical role [30, 37]. However, recent data show that human microvascular smooth muscle cells express Nox2 in response to angiotensin II stimulation, mediated by a c-Src pathway [42]. This clearly illustrates that Nox isoform expression in human vascular cells is regulated in a complex manner that can vary with cell type in different vessels and in response to varied pathophysiologic stimuli.

Human functional studies of vascular NADPH oxidases add further insights into the relationships between Nox isoform expression and increased oxidative stress in atherosclerosis. First identifications of p22phox-based NADPH oxidases in human coronary artery plaques by Azumi et al. [43] were then complemented by Sorescu et al., who demonstrated that Nox2 and p22phox are greatly increased with the progression of human coronary artery atherosclerotic plaques, in part because of inflammatory cell infiltration, while Nox4 is increased in early lesions and decreased in very severe lesions [30]. A recent study by Hwang et al. has additionally shown that Nox4 expression is increased by oscillatory versus pulsatile flow, which may be particularly relevant to the development of atherosclerotic plaques in regions of turbulent flow. Importantly, Nox4 expression in this model coincided with increased oxidative stress and LDL oxidation [44]. Indeed, coronary artery disease (CAD) is associated with increased coronary artery $O_2^{\bullet-}$ production [45]. Studies with pharmacological inhibitors and substrate stimulation indicated that NADPH oxidases are the major sources of superoxide production, and that xanthine oxidase also contributes modestly to $O_2^{\bullet-}$ production in CAD [45]. CAD is associated with increased NADPH oxidase subunit expression, mainly p22phox and Nox2, related in part to higher monocyte/macrophage infiltration. In turn, Nox4 is most strongly related to the observed increase in NADPH oxidase activity in CAD [45]. $O_2^{\bullet-}$ production is increased in segments of coronary arteries from CAD patients even when morphologic features of atherosclerosis are absent [45]. The importance of NADPH oxidase for plaque stability is further emphasized by the finding that the shoulder region of plaques is a particularly intense area of ROS production, in association with p22phox and Nox2 expression [30].

The cellular sources of superoxide in coronary arteries appear to be complex. Inflammatory cell infiltrates are likely to be a quantitatively important source [30, 46, 47]. However, dihydroethidium staining, presented here, shows that superoxide production is evenly increased in cells within endothelium, media, and adventitia in CAD, showing that inflammatory cells may not be the sole source of $O_2^{\bullet-}$ in CAD [47]. It is important to emphasize that while NADPH oxidases are the major source of superoxide in human coronary and conduit arteries, in certain clinical conditions, such as CAD or diabetes, xanthine oxidase [45] or uncoupled nitric oxide synthase [48, 49], may also play a quantitatively lesser, but potentially very important role. It is vital to emphasize that enzymatic sources of $O_2^{\bullet-}$ in coronary arteries are similar to those of peripheral arteries [11, 49].

In a recent study, we identified that Nox5 protein and mRNA expression are markedly increased in the vessels of patients with CAD compared to non-CAD subjects [50]. This was associated with increased calcium-dependent membrane NADPH oxidase activity, characteristic of Nox5. The localization of Nox5 is of interest in human vessels. Immunostaining confirmed the increase in Nox5 protein and shows that its presence seems to vary depending on the stage of atherosclerosis. In early lesions, Nox5 seemed in many cases to co-localize with endothelial cells. It is conceivable that the cytokine milieu present in early atherosclerosis could promote Nox5 expression in endothelial cells. In moderately advanced lesions, endothelial staining was less evident; however, a large amount of Nox5 co-localized with vascular smooth muscle cells in subintimal regions. Complex regions show extensive Nox5 staining in the area of plaque. The presence of Nox5 in early lesions and its loss in advanced lesions is reminiscent of the expression of the endothelial nitric oxide synthase in these settings, as it is present in early lesions and is lost in endothelial cells overlying advanced plaques [51]. Since other enzymes expressed and important in human CAD, such as xanthine oxidase, Nox4, and Nox2, are unlikely to be activated by calcium, the relative contribution of Nox5 to coronary artery superoxide production appears to be important, considering that the calcium-dependent activity was similar to the calcium-independent activity in CAD vessels, while in non-CAD membranes its contribution was much less [50]. In CAD, calcium-independent NADPH oxidase activity is increased approximately 2.5-fold, while the increase in calcium-dependent activity is sevenfold. This would indicate that the atherosclerotic milieu provides a potent stimulus for Nox5 in human blood vessels.

In summary, p22$^{phox}$, Nox2, Nox4, Nox5, and to a lesser extent Nox1, are expressed in human vessels [2, 30]. Nox2 and p22$^{phox}$, as well as the recently identified Nox5, are greatly increased with the progression of human atherosclerosis in CAD [45], while Nox4 is increased in early lesions and decreased in severe lesions [30]. A direct spatial relationship between NADPH oxidase-generated ROS and oxidized LDL was demonstrated in carotid plaques, while $O_2^{\bullet-}$ production was increased in lesions associated with unstable angina [46].

### *8.6.1 Regulation of NADPH Oxidases in Human Vasculature*

The activity and expression of NADPH oxidase may be regulated by cytokines (TNF-α, TGF-β, PDGF), and agonists like angiotensin II and thrombin [33]. Activity of these agents is often related to protein kinase C signaling, as suggested by studies in human vessels using the PKC inhibitor chelerethrine regarding regulation of superoxide production in both veins and arteries. Hemodynamic forces, such as oscillatory shear stress and stretch, can also activate the NADPH oxidase and increase Nox expression. Importantly, angiotensin II is one of the most potent stimuli of vascular NADPH oxidase activity and expression [11, 33]. This property clearly links ROS production with activation of the renin-angiotensin system in hypertension, in early stages of atherosclerosis [52] and in heart failure [53]. In rats made hypertensive by chronic angiotensin II infusion, the expression of Nox1 and p22$^{phox}$ mRNA is elevated [54]. Moreover, the increase in blood pressure caused by angiotensin II is markedly reduced in p47$^{phox-/-}$ mice [55].

Human studies have also shown the role of angiotensin II in regulating NADPH oxidase activity and oxidative stress in conduit vessels and coronary arteries [2, 11, 45]. Human coronary artery NADPH oxidase is activated by angiotensin II in vessels of subjects with coronary artery disease and this enzyme is regulated by protein kinase C [45]. Moreover, incubation of native coronary artery disease vessels with angiotensin II receptor blocker (losartan) leads to significant inhibition of total superoxide production even in the absence of exogenous angiotensin II [45]. A direct spatial relationship between NADPH oxidase-generated ROS and oxidized LDL was demonstrated in carotid plaques, indicating a potential role for ox-LDL in regulating human vescular NADPH oxidases [46]. Indeed, ox-LDL has been shown to regulate NADPH oxidase subunit expression in human vessels [56]. Findings of systemic correlations between NADPH oxidase homolog expression and characteristics also add to our understanding of the regulation of vascular oxidases in humans. This emphasizes that while sources of superoxide anion in arteries and veins are not identical, they are clearly both subject to systemic factors, such as diabetes, hypercholesterolemia, or angiotensin II [42, 57–59]. Nox4 expression was most closely correlated between arteries and veins [2]. Although Nox2 was expressed in both veins and arteries, there was no correlation between the two. Nox1, in turn, was expressed only at low levels in a small number of patients, but when detected was present in both arteries and veins. These data suggest that Nox4 might be the principal subunit responsible for variability of NADPH oxidase activity related to systemic factors in human atherosclerosis [2]. In line with these findings, the level of Nox4 expression in human coronary arteries is most strongly correlated with increased NADPH oxidase activity in human coronary arteries, independent of the presence of CAD [45]. These suggestions, however, require further confirmation and consolidation with other studies.

Despite the systemic nature of oxidative stress and endothelial dysfunction discussed above, certain areas of the coronary tree are more prone to atherosclerosis [60, 61]. In particular, increased endothelial dysfunction and plaque burden have been found at bifurcations [60]. We show that superoxide production is almost doubled at bifurcations. This likely is because of disturbed flow at these sites, leading to increased NADPH oxidase expression and activity [62]. Indeed, we have demonstrated that superoxide production and Nox expression is highest at sites of bifurcation of the human coronary arteries in either patients with or without coronary artery disease [45].

### *8.6.2 Central Role of NADPH Oxidases in Regulating Oxidative Stress*

Apart from a critical contribution from NADPH oxidases to human vascular superoxide production, other oxidase systems are also involved, in particular, xanthine oxidase and endothelial nitric oxide synthase, as discussed so far. Interestingly, recent years have brought the surprising discovery that superoxide produced by NADPH oxidases may be critical for the regulation of other vascular oxidases [63]. Data obtained using NADPH oxidase $p47^{phox}$ knockout mice and cells have shown that NADPH oxidase is critical to the regulation of superoxide production from both eNOS [20] and xanthine oxidase [64]. Peroxynitrite, the product of NO-superoxide interaction, oxidizes $H_4B$ and leads to eNOS uncoupling [20]. In endothelial cells lacking $p47^{phox}$, $O_2^{\bullet -}$ production was depressed and XO protein and activity were minimal, while $p47^{phox}$ transfection restored XO protein levels [64]. Similarly, eNOS uncoupling was prevented in the absence of NADPH oxidase [20]. These findings make NADPH oxidases particularly important drug targets for specific "antioxidation," which would modulate ROS production from other sources as well.

## 8.7 Risk Factors for Atherosclerosis and Vascular NADPH Oxidases

Data discussed so far clearly establish that the net basal $O_2^{\bullet -}$ production is greater in human coronary arteries in the setting of coronary artery disease and associated risk factors [45]. Moreover, vascular oxidative stress is generalized in the coronary circulation and not limited to sites of overt atherosclerosis. In line with these observations, our previous findings indicate that superoxide production and endothelial dysfunction are also present in the peripheral systemic circulation in atherosclerotic individuals and actually precede the development of overt lesion formation in experimental animals with either hypertension or hypercholesterolemia [3, 49].

Many common risk factors for vascular disease, such as hypertension and diabetes, remain prevalent in Western and other populations, suggesting that vascular disease will continue to impose a substantial burden on health care resources throughout the next generation. Studies of the relationships between human atherosclerosis risk factors and the activity of NADPH oxidases have demonstrated that NADPH oxidase activity is directly correlated with the number of major risk factors for atherosclerosis present in a given patient [3]. Multivariate analysis has shown that hypercholesterolemia and diabetes mellitus exhibited the strongest independent association with increased NADPH oxidase activity and expression in human vessels from patients undergoing bypass graft surgery [3]. Moreover, the use of statins or AT1-receptor blockers were also found to correlate with vascular NADPH oxidase activity and expression in a randomized clinical trial [65]. Finally, exercise interventions may also decrease expression and activity of NADPH oxidases in human conduit vessels [66].

## 8.8 NADPH Oxidases in Bypass Graft Conduit Vessel Disease and Dysfunction

Bypass graft surgery remains a cornerstone of surgical treatment for multivessel coronary artery disease. Saphenous veins are still used as bypass conduits, but are prone to failure because of accelerated atherosclerosis and occlusion, occurring at rates of 50% or more at 10 years following implantation [67]. Vein grafts are characterized by intimal hyperplasia resulting from vascular smooth muscle cell migration and proliferation, which develops as a consequence of early graft injury [68] and is critical in the subsequent development of accelerated atherosclerosis [69]. In contrast, arterial bypasses such as the internal mammary artery provide a conduit that is rarely affected by atherosclerosis, has excellent long term patency, and confers an independent prognostic benefit in patients undergoing coronary bypass surgery [67]. Differences in the vascular biology of venous vs. arterial conduits may underlie their contrasting susceptibility to atherosclerosis. In particular, mammary arteries have markedly higher levels of endothelial-derived nitric oxide bioactivity [8], that exerts functionally important effects because NO is a pivotal mediator of key vascular functions, including regulation of smooth muscle tone and blood pressure, platelet activation, and vascular cell signaling [70]. Moreover, these functional studies may be critical for understanding and preventing the particularly high sensitivity of human radial artery conduit vessels to spasm, which seems related to increased superoxide production and NADPH oxidase expression [71].

## 8.9 Functional Studies of Genetic Regulation of NADPH Oxidases

Genetic regulation of vascular NADPH oxidase activity is very likely to be an important factor in the regulation of oxidative stress. Several polymorphisms within the p22phox CYBA gene have been described, while genetic variations of other Nox genes remain essentially unknown regarding their importance for vascular oxidative stress. Functional polymorphisms within the p22phox gene include A-930G, C242T, and in some studies also A640G in 3′UTR [72–74]. In an initial report, Inoue et al. found that the T allele of the C242T polymorphism is less common in subjects with MI [73]. This trend was confirmed in a large-scale study, however only in the male population, not in the whole group [75]. Another large study of patients with hypercholesterolemia found an opposite trend with respect to progression of atherosclerosis [76]. Some smaller studies have found a particular association with conditions related to high oxidative stress, e.g., diabetes and its complications [77–79], while others failed to find such correlations [80, 81]. Functional studies of such relationships between genetic polymorphisms within the p22phox gene CYBA and vascular superoxide production or NADPH oxidase activity are less common. Initially, we demonstrated that the presence of the T allele is associated with lower superoxide production and NADPH oxidase activity in human conduit vessels. Following this report, Wyche et al., in an elegant EPR-based study, found that while C242T polymorphism had significant functional effects on neutrophil oxidase activity, this was not observed in relation to promoter A-930G polymorphism [82]. At the same time, Zalba and colleagues studied the role of polymorphisms of the promoter of the CYBA gene in the regulation of gene expression [83, 84] and the regulation of the oxidative burst in neutrophils [85]. However, these studies have not used modern comprehensive genetic approaches and remain inconclusive. The genetics of other NADPH oxidase subunits or homologs remains an unopened chapter, which is surprising, considering the explosion of population genetics studies over the past 10 years [1].

## 8.10 Conclusions

In summary, the NADPH oxidases have been ascribed a major role as a main source of $O_2^{\bullet -}$ in the human vasculature. In humans, it has been confirmed that NADPH oxidase activity is inversely correlated with endothelial function. This relationship exists even when corrected for other major risk factors for atherosclerosis, including diabetes and hypercholesterolemia [86]. Diabetes and hypercholesterolemia are most strongly associated with NADPH oxidase activity in human vessels, while hypertension also plays an important role. Numerous Nox homologs are expressed in human vessels in correlation with risk factors and/or the presence of atherosclerosis or coronary artery disease. These include predominantly p22phox, Nox2, Nox4, and Nox5. Expression of these homologs is regulated by systemic factors and is correlated in arteries and veins. More extensive translational approaches need to be applied to confirm major findings regarding the regulation of NADPH oxidases in human vasculature, in order to assess whether Nox enzymes are realistic therapeutic targets once specific small molecule inhibitors have been developed [1]. Hopefully, the functional characterization of NADPH oxidases in human vessels will eventually lead to the discovery of safe and specific inhibitors of individual homologs of NADPH oxidases, which will move our understanding of NADPH oxidases to a new level of clinical application.

## References


1. Guzik TJ, Griendling K (2009) NADPH oxidases – molecular understanding finally reaching the clinical level? Antioxid Redox Signal 11(10):2365–2370
2. Guzik T, Sadowski J, Kapelak B et al (2004) Systemic regulation of vascular NAD(P)H oxidase activity and Nox isoform expression in human arteries and veins. Arterioscler Thromb Vasc Biol 24:1614–1620
3. Guzik TJ, West NEJ, Black E et al (2000) Vascular superoxide production by NAD(P)H oxidase: association with endothelial dysfunction and clinical risk factors. Circ Res 86: e85–e90
4. Guzik TJ, West NEJ, Black E et al (2000) Functional effect of the C242T polymorphism in the NAD(P)H oxidase p22phox subunit on vascular superoxide production in atherosclerosis. Circulation 102:1744–1747
5. Wali MA, Suleiman SA, Kadoumi OF et al (2002) Superoxide radical concentration and superoxide dismutase (SOD) enzyme activity in varicose veins. Ann Thorac Cardiovasc Surg 8:286–290
6. Heitzer T, Baldus S, von Kodolitsch Y et al (2005) Systemic endothelial dysfunction as an early predictor of adverse outcome in heart failure. Arterioscler Thromb Vasc Biol 25:1174–1179
7. Kojda G, Harrison D (1999) Interactions between NO and reactive oxygen species: pathophysiological importance in atherosclerosis, hypertension, diabetes and heart failure. Cardiovasc Res 43:562–571
8. Guzik TJ, West N, Pillai R et al (2002) Nitric oxide modulates superoxide release and peroxynitrite formation in human blood vessels. Hypertension 39:1088–1094
9. Shapira OM, Xu A, Aldea GS et al (1999) Enhanced nitric oxide-mediated vascular relaxation in radial artery compared with internal mammary artery or saphenous vein. Circulation 100:II322–II327
10. Huraux C, Makita T, Kurz S et al (1999) Superoxide production, risk factors, and endothelium-dependent relaxations in human internal mammary arteries. Circulation 99:53–59
11. Berry C, Hamilton CA, Brosnan MJ et al (2000) Investigation into the sources of superoxide in human blood vessels: angiotensin II increases superoxide production in human internal mammary arteries. Circulation 101:2206–2212
12. Schmalfuss CM, Chen LY, Bott JN et al (1999) Superoxide anion generation, superoxide dismutase activity, and nitric oxide release in human internal mammary artery and saphenous vein segments. J Cardiovasc Pharmacol Ther 4:249–257
13. Fleming I, Bauersachs J, Schafer A et al (1999) Isometric contraction induces the Ca2+-independent activation of the endothelial nitric oxide synthase. Proc Natl Acad Sci USA 96:1123–1128
14. Judkins CP, Sobey CG, Dang TT et al (2006) NADPH-induced contractions of mouse aorta do not involve NADPH oxidase: a role for P2X receptors. J Pharmacol Exp Ther 317:644–650

15. Miller FJ Jr., Gutterman DD, Rios CD et al (1998) Superoxide production in vascular smooth muscle contributes to oxidative stress and impaired relaxation in atherosclerosis. Circ Res 82:1298–1305
16. Pagano PJ, Chanock SJ, Siwik DA et al (1998) Angiotensin II induces p67phox mRNA expression and NADPH oxidase superoxide generation in rabbit aortic adventitial fibroblasts. Hypertension 32:331–337
17. Wang HD, Pagano PJ, Du Y et al (1998) Superoxide anion from the adventitia of the rat thoracic aorta inactivates nitric oxide. Circ Res 82:810–818
18. Pagano PJ, Griswold MC, Najibi S et al (1999) Resistance of endothelium-dependent relaxation to elevation of $O_2^{\bullet-}$ levels in rabbit carotid artery. Am J Physiol 277:H2109–H2114
19. Somers MJ, Mavromatis K, Galis ZS et al (2000) Vascular superoxide production and vasomotor function in hypertension induced by deoxycorticosterone acetate-salt. Circulation 101:1722–1728
20. Landmesser U, Dikalov S, Price SR et al (2003) Oxidation of tetrahydrobiopterin leads to uncoupling of endothelial cell nitric oxide synthase in hypertension. J Clin Invest 111:1201–1209
21. Harrison DG (1995) Alterations of vasomotor regulation in atherosclerosis. Cardiovasc Drugs Ther 9(Suppl 1):55–63
22. Harrison DG (1998) Cellular and molecular mechanisms of endothelial dysfunction. J Clin Invest 100:2153–2157
23. Harrison DG, Gongora MC, Guzik TJ et al (2007) Oxidative stress and hypertension. J Am Soc Hypertens 1:30–44
24. Guzik TJ, Harrison DG (2006) Vascular NADPH oxidases as drug targets for novel antioxidant strategies. Drug Discov Today 11:524–533
25. Rajagopalan S, Kurz S, Munzel T et al (1996) Angiotensin II-mediated hypertension in the rat increases vascular superoxide production via membrane NAD/NADPH oxidase activation. J Clin Invest 97:1916–1923
26. Ushio-Fukai M, Zafari AM, Fukui T et al (1996) p22phox is a critical component of the superoxide-generating NADH/NADPH oxidase system and regulates angiotensin II-induced hypertrophy in vascular smooth muscle cells. J Biol Chem 271:23317–23321
27. White CR, Darley-Usmar V, Berrington WR et al (1996) Circulating plasma xanthine oxidase contributes to vascular dysfunction in hypercholesterolemic rabbits. Proc Natl Acad Sci USA 93:8745–8749
28. Vasquez-Vivar J, Kalyanaraman B, Martasek P et al (1998) Superoxide generation by endothelial nitric oxide synthase: the influence of cofactors. Proc Natl Acad Sci USA 95:9220–9225
29. Ohara Y, Peterson TE, Harrison DG (1993) Hypercholesterolemia increases endothelial superoxide anion production. J Clin Invest 91:2546–2551
30. Sorescu D, Weiss D, Lassegue B et al (2002) Superoxide production and expression of Nox family proteins in human atherosclerosis. Circulation 105:1429–1435
31. Spiekermann S, Landmesser U, Dikalov S et al (2003) Electron Spin Resonance Characterization of Vascular Xanthine and NAD(P)H Oxidase Activity in Patients With Coronary Artery Disease: Relation to Endothelium-Dependent Vasodilation. Circulation 107:1383–1389
32. Landmesser U, Spiekermann S, Dikalov S et al (2002) Vascular oxidative stress and endothelial dysfunction in patients with chronic heart failure: role of xanthine-oxidase and extracellular superoxide dismutase. Circulation 106:3073–3078
33. Griendling KK, Sorescu D, Ushio-Fukai M (2000) NAD(P)H oxidase: role in cardiovascular biology and disease. Circ Res 86:494–501
34. Lambeth JD (2004) NOX enzymes and the biology of reactive oxygen. Nat Rev Immunol 4:181–189
35. Suh YA, Arnold RS, Lassegue B et al (1999) Cell transformation by the superoxide-generating oxidase Mox1. Nature 401:79–82

36. Lassegue B, Sorescu D, Szocs K et al (2001) Novel gp91(phox) homologues in vascular smooth muscle cells: Nox1 mediates angiotensin II-induced superoxide formation and redox-sensitive signaling pathways. Circ Res 88:888–894
37. Szocs K, Lassegue B, Sorescu D et al (2002) Upregulation of Nox-based NAD(P)H oxidases in restenosis after carotid injury. Arterioscler Thromb Vasc Biol 22:21–27
38. West NEJ, Guzik TJ, Black E et al (2001) Enhanced superoxide production in experimental venous bypass graft intimal hyperplasia: role of NAD(P)H oxidase. Atheroscler Thromb Vasc Biol 21:189–194
39. Wang HD, Hope S, Du Y et al (1999) Paracrine role of adventitial superoxide anion in mediating spontaneous tone of the isolated rat aorta in angiotensin II-induced hypertension. Hypertension 33:1225–1232
40. Bayraktutan U, Blayney L, Shah AM (2000) Molecular characterization and localization of the NAD(P)H oxidase components gp91-phox and p22-phox in endothelial cells. Arterioscler Thromb Vasc Biol 20:1903–1911
41. Rey FE, Li XC, Carretero OA et al (2002) Perivascular superoxide anion contributes to impairment of endothelium-dependent relaxation: role of gp91(phox). Circulation 106: 2497–2502
42. Touyz RM, Chen X, Tabet F et al (2002) Expression of a functionally active gp91phox-containing neutrophil-type NAD(P)H oxidase in smooth muscle cells from human resistance arteries: regulation by angiotensin II. Circ Res 90:1205–1213
43. Azumi H, Inoue N, Takeshita S et al (1999) Expression of NADH/NADPH oxidase p22phox in human coronary arteries. Circulation 100:1494–1498
44. Hwang J, Ing MH, Salazar A et al (2003) Pulsatile Versus Oscillatory Shear Stress Regulates NADPH Oxidase Subunit Expression. Implication for Native LDL Oxidation. Circ Res 93:1225–1232
45. Guzik TJ, Sadowski J, Guzik B et al (2006) Coronary artery superoxide production and Nox isoform expression in human coronary artery disease. Arterioscler Thromb Vasc Biol 26: 333–339
46. Azumi H, Inoue N, Ohashi Y et al (2002) Superoxide generation in directional coronary atherectomy specimens of patients with angina pectoris: important role of NAD(P)H oxidase. Arterioscler Thromb Vasc Biol 22:1838–1844
47. Cathcart MK (2004) Regulation of superoxide anion production by NADPH oxidase in monocytes/macrophages: contributions to atherosclerosis. Arterioscler Thromb Vasc Biol 24:23–28
48. Antoniades C, Shirodaria C, Crabtree M et al (2007) Altered plasma versus vascular biopterins in human atherosclerosis reveal relationships between endothelial nitric oxide synthase coupling, endothelial function, and inflammation. Circulation 116:2851–2859
49. Guzik TJ, Mussa S, Gastaldi D et al (2002) Mechanisms of increased vascular superoxide production in human diabetes mellitus: Role of NAD(P)H oxidase and endothelial nitric oxide synthase. Circulation 105:1656–1662
50. Guzik TJ, Chen W, Gongora MC et al (2008) Calcium-dependent NOX5 nicotinamide adenine dinucleotide phosphate oxidase contributes to vascular oxidative stress in human coronary artery disease. J Am Coll Cardiol 52:1803–1809
51. Wilcox JN, Subramanian RR, Sundell CL et al (1997) Expression of multiple isoforms of nitric oxide synthase in normal and atherosclerotic vessels. Arterioscler Thromb Vasc Biol 17:2479–2488
52. Warnholtz A, Nickenig G, Schulz E et al (1999) Increased NADH-oxidase-mediated superoxide production in the early stages of atherosclerosis: evidence for involvement of the renin-angiotensin system. Circulation 99:2027–2033
53. Widder J, Behr T, Fraccarollo D et al (2004) Vascular endothelial dysfunction and superoxide anion production in heart failure are p38 MAP kinase-dependent. Cardiovasc Res 63:161–167
54. Fukui T, Ishizaka N, Rajagopalan S et al (1997) p22phox mRNA expression and NADPH oxidase activity are increased in aortas from hypertensive rats. Circ Res 80:45–51

55. Landmesser U, Cai H, Dikalov S et al (2002) Role of p47(phox) in vascular oxidative stress and hypertension caused by angiotensin II. Hypertension 40:511–515
56. Rueckschloss U, Galle J, Holtz J et al (2001) Induction of NAD(P)H oxidase by oxidized low-density lipoprotein in human endothelial cells: antioxidative potential of hydroxymethylglutaryl coenzyme A reductase inhibitor therapy. Circulation 104:1767–1772
57. Bonetti PO, Lerman LO, Lerman A (2003) Endothelial dysfunction: a marker of atherosclerotic risk. Arterioscler Thromb Vasc Biol 23:168–175
58. Anderson TJ, Uehata A, Gerhard MD et al (1995) Close relation of endothelial function in the human coronary and peripheral circulations. J Am Coll Cardiol 26:1235–1241
59. Neunteufl T, Katzenschlager R, Hassan A et al (1997) Systemic endothelial dysfunction is related to the extent and severity of coronary artery disease. Atherosclerosis 129: 111–118
60. McLenachan JM, Vita J, Fish DR et al (1990) Early evidence of endothelial vasodilator dysfunction at coronary branch points. Circulation 82:1169–1173
61. Ungvari Z, Csiszar A, Edwards JG et al (2003) Increased superoxide production in coronary arteries in hyperhomocysteinemia: role of tumor necrosis factor-alpha, NAD(P)H oxidase, and inducible nitric oxide synthase. Arterioscler Thromb Vasc Biol 23:418–424
62. De Keulenaer GW, Chappell DC, Ishizaka N et al (1998) Oscillatory and steady laminar shear stress differentially affect human endothelial redox state: role of a superoxide-producing NADH oxidase. Circ Res 82:1094–1101
63. Mueller CF, Laude K, McNally JS et al (2005) ATVB in focus: redox mechanisms in blood vessels. Arterioscler Thromb Vasc Biol 25:274–278
64. McNally JS, Davis ME, Giddens DP et al (2003) Role of xanthine oxidoreductase and NAD(P)H oxidase in endothelial superoxide production in response to oscillatory shear stress. Am J Physiol Heart Circ Physiol 285:H2290–H2297
65. Morawietz H, Erbs S, Holtz J et al (2006) Endothelial protection, AT1 blockade and cholesterol-dependent oxidative stress: the EPAS trial. Circulation 114:I296–I301
66. Adams V, Linke A, Krankel N et al (2005) Impact of regular physical activity on the NAD(P)H oxidase and angiotensin receptor system in patients with coronary artery disease. Circulation 111:555–562
67. Campeau L, Enjalbert M, Lesperance J et al (1983) Atherosclerosis and the late closure of aortocoronary saphenous vein grafts: sequential studies at 2 weeks, 1 year, 5–7 years and 10–12 years after surgery. Circulation 68(Suppl II):1–7
68. Ip JH, Fuster V, Badimon L et al (1990) Syndromes of accelerated atherosclerosis: role of vascular injury and smooth muscle cell proliferation. J Am Coll Cardiol 15:1667–1687
69. Schwartz SM, deBlois D, O'Brien ERM (1995) The intima: soil for atherosclerosis and restenosis. Circ Res 77:445–465
70. Cai H, Harrison DG (2000) Endothelial dysfunction in cardiovascular diseases: the role of oxidant stress. Circ Res 87:840–844
71. Ryszawa N, Rudzinski P, Piatek J et al (2007) Different reactivity of the proximal and distal segments of the radial artery to vasoconstrictors in patients undergoing coronary artery bypass grafting. Kardiol Pol 65:1313–1319;discussion 1320
72. Gardemann A, Mages P, Katz N et al (1999) The p22 phox A640G gene polymorphism but not the C242T gene variation is associated with coronary heart disease in younger individuals. Atherosclerosis 145:315–323
73. Inoue N, Kawashima S, Kanazawa K et al (1998) Polymorphism of the NADH/NADPH oxidase p22 phox gene in patients with coronary artery disease. Circulation 97:135–137
74. Cai H, Duarte N, Wilcken DE et al (1999) NADH/NADPH oxidase p22 phox C242T polymorphism and coronary artery disease in the Australian population. Eur J Clin Invest 29:744–748
75. Yamada Y, Izawa H, Ichihara S et al (2002) Prediction of the risk of myocardial infarction from polymorphisms in candidate genes. N Engl J Med 347:1916–1923

76. Cahilly C, Ballantyne CM, Lim DS et al (2000) A variant of p22(phox), involved in generation of reactive oxygen species in the vessel wall, is associated with progression of coronary atherosclerosis. Circ Res 86:391–395
77. Matsunaga-Irie S, Maruyama T, Yamamoto Y et al (2004) Relation between development of nephropathy and the p22phox C242T and receptor for advanced glycation end product G1704T gene polymorphisms in type 2 diabetic patients. Diabetes Care 27:303–307
78. Nakano T, Matsunaga S, Nagata A et al (2003) NAD(P)H oxidase p22phox Gene C242T polymorphism and lipoprotein oxidation. Clin Chim Acta 335:101–107
79. Hayaishi-Okano R, Yamasaki Y, Kajimoto Y et al (2003) Association of NAD(P)H oxidase p22 phox gene variation with advanced carotid atherosclerosis in Japanese type 2 diabetes. Diabetes Care 26:458–463
80. Renner W, Schallmoser K, Gallippi P et al (2000) C242T polymorphism of the p22 phox gene is not associated with peripheral arterial occlusive disease. Atherosclerosis 152:175–179
81. Zafari AM, Davidoff MN, Austin H et al (2002) The A640G and C242T p22(phox) polymorphisms in patients with coronary artery disease. Antioxid Redox Signal 4:675–680
82. Wyche KE, Wang SS, Griendling KK et al (2004) C242T CYBA polymorphism of the NADPH oxidase is associated with reduced respiratory burst in human neutrophils. Hypertension 43:1246–1251
83. Zalba G, San Jose G, Beaumont FJ et al (2001) Polymorphisms and promoter overactivity of the p22(phox) gene in vascular smooth muscle cells from spontaneously hypertensive rats. Circ Res 88:217–222
84. Moreno MU, San Jose G, Orbe J et al (2003) Preliminary characterisation of the promoter of the human p22(phox) gene: identification of a new polymorphism associated with hypertension. FEBS Lett 542:27–31
85. San Jose G, Moreno MU, Olivan S et al (2004) Functional effect of the p22phox-930A/G polymorphism on p22phox expression and NADPH oxidase activity in hypertension. Hypertension 44:163–169
86. Channon KM, Guzik TJ (2002) Mechanisms of superoxide production in human blood vessels: relationship to endothelial dysfunction, clinical and genetic risk factors. J Physiol Pharmacol 53:515–524

# Chapter 9
# Relationship of the *CYBA* Gene Polymorphisms with Oxidative Stress and Cardiovascular Risk

**Guillermo Zalba and Javier Díez**

**Abstract** Oxidative stress plays a key role in the pathophysiology of several major cardiovascular diseases, including atherosclerosis, hypertension, heart failure, stroke, and diabetes. Reactive oxygen species (ROS) induce cardiovascular alterations by modulating cell contraction/dilatation, migration, growth/apoptosis, and extracellular matrix protein turnover, which contribute to vascular and cardiac remodeling. Of the several sources of ROS within the cardiovascular system, the family of the multisubunit NADPH oxidases appears to be a predominant contributor generating superoxide anions. Recent data suggest a significant role of the genetic background in NADPH oxidase regulation. Common genetic polymorphisms within *CYBA*, the gene that encodes the $p22^{phox}$ subunit of the NADPH oxidase, have been characterized in the context of cardiovascular diseases. This chapter aims to present the current state of research into these polymorphisms in their relationship to cardiovascular diseases.

**Keywords** $p22^{phox}$ Gene · Polymorphism · NADPH oxidases · Atherosclerosis

**Abbreviations**

| | |
|---|---|
| ROS | Reactive oxygen species |
| $\bullet O_2^-$ | Superoxide |
| CGD | Chronic granulomatous disease |
| SHR | Spontaneously hypertensive rat |
| CAD | Coronary artery disease |
| C/EBP | CCAAT enhancer-binding protein |
| HIF-1α | Hypoxia-inducible factor-1alpha |
| LD | Linkage disequilibrium |

G. Zalba (✉)
Center for Applied Medical Research, 31008 Pamplona, Spain
e-mail: gzalba@unav.es

H. Sauer et al. (eds.), *Studies on Cardiovascular Disorders*,
Oxidative Stress in Applied Basic Research and Clinical Practice,
DOI 10.1007/978-1-60761-600-9_9, © Springer Science+Business Media, LLC 2010

## 9.1 Introduction

Considerable progress has been made in determining the genetic basis of human diseases since the human genome was sequenced. The discovery of genomic regions and genetic polymorphisms that influence susceptibility to common complex diseases is increasing. Nevertheless, several challenges exist for the identification of the genetic determinants of common, complex diseases, such as cardiovascular diseases [1]. These challenges include (a) phenotypic variability, given that cardiovascular alterations manifest as several clinical phenotypes; (b) genetic heterogeneity, given the multiple causal pathways that may lead to cardiovascular diseases; (c) gene-gene and gene-environment interactions, which requires the performance of studies with large sample sizes and adequate computational resources; and (d) the etiologic spectrum ranges, from common genetic variants with small effects (a single genetic variant that can constitute only a small proportion—1%—of the disease phenotype), which requires the performance of studies in large sample sizes in order to recognize this small effect, to rare genetic variants with large effects on cardiovascular alterations. With these limitations, it is necessary to perform well-designed studies to define clinically relevant phenotypes, identify genes, and define environmental contributions to oxidative stress-mediated cardiovascular complications.

## 9.2 The NADPH Oxidase System

Reactive oxygen species (ROS) are produced by a wide variety of enzymatic sources (mitochondrial transport chain, NADPH oxidase, xanthine oxidase, cyclooxygenases, lipoxygenases and uncoupled nitric oxide synthase). However, evidence over recent years supports the NADPH oxidase family as the predominant cellular source of ROS in the cardiovascular system.

The NADPH oxidase was first described and characterized in phagocytes. It was originally thought that the enzyme was used solely for host defense [2]. The membrane-associated proteins are a large subunit, $gp91^{phox}$ (*phox* from *ph*agocytic *ox*idase), and a small one, $p22^{phox}$, known as a flavocytochrome, $b_{558}$, which contains the entire electron transport apparatus of the phagocyte NADPH oxidase and thus may act as a physical conduit for electron transport across the membrane. In addition, there are up to three cytosolic subunits ($p47^{phox}$, $p67^{phox}$, and $p40^{phox}$) and a low-molecular-weight G protein (rac2). In a resting phagocyte, the NADPH oxidase complex is disassembled, with its components segregated into different parts of the cell. Upon stimulation, the cytosolic components of the complex translocate to the phagosome or plasma membrane and assemble with the integral membrane proteins to form a multisubunit enzyme complex that begins to generate superoxide ($\bullet O_2^-$) [3, 4]. In the last years, similar NADPH oxidases have been described in a wide variety of nonphagocytic cells and tissues [5–7]. In this sense, several homologues of the $gp91^{phox}$ (also defined as Nox2) have been identified, which are known as Nox1, Nox3, Nox4, and Nox5 [5]. In addition, there are also homologues to the cytoplasmic components p47$^{phox}$ and p67$^{phox}$, which are termed NoxO1 and NoxA1, respectively. The small GTPase Rac2 is restricted to leukocytes, but the Rac1 is expressed ubiquitously and can be a substitute for Rac2 [8].

A key role of NADPH oxidases has been demonstrated both in experimental and human cardiovascular diseases such as metabolic syndrome [9, 10], hypertension [11–13], diabetes [14, 15], left ventricular hypertrophy and heart failure [5, 16, 17], renal disease [18–20], atherosclerosis [21–23], and cerebrovascular disease [24]. It is important to note that in cardiovascular diseases, not only the vascular oxidase, but also the phagocytic NADPH oxidase plays an important role in $\bullet O_2^-$ production, because monocytes and lymphocytes can infiltrate cardiovascular tissues and facilitate structural and functional alterations [25–27].

The NADPH oxidases are sensitively regulated by a wide range of (patho)physiologically relevant factors, which include humoral factors [12, 26, 28–31] and mechanical factors [32]. In addition, genetic variants might regulate NADPH oxidase-driven $\bullet O_2^-$ production [33, 34]. In this sense, genetic defects in the genes encoding four of the phox proteins (gp91$^{phox}$, p22$^{phox}$, p47$^{phox}$, and p67$^{phox}$) are known to cause chronic granulomatous disease (CGD), which is a rare inherited disorder of the innate immune system [35]. In spite of NADPH oxidase being involved in the regulation of the vascular wall, it has not been reported whether CGD patients exhibit cardiovascular abnormalities. A possibility for the absence of cardiovascular abnormalities in CGD patients may be because of the compensatory effects of other Nox homologues. While the incidence of CGD is estimated to be 1 in 200,000–250,000 individuals, cardiovascular diseases are a substantial public health problem with >35% of the adult population being affected. In this sense, the study of the genetic variants of the NADPH oxidase subunits and their potential consequences in cardiovascular diseases has gained priority in the last decade.

## 9.3 p22$^{phox}$ Genetic Variants and Cardiovascular Disease

The p22$^{phox}$ protein is an essential subunit for the functionality of most of the known oxidases [36–38]. p22$^{phox}$ may interact with Nox1-4 [7] (Fig. 9.1), and several works support the notion that binding of p22$^{phox}$ to Nox proteins leads to flavocytochrome stabilization. The underlying concept is that Nox proteins and the p22$^{phox}$ protein are stable only as a heterodimer, while monomers are degraded by the proteosome. In line with this concept, the importance of the p22$^{phox}$ subunit for the phagocyte NADPH oxidase was revealed with the identification of p22$^{phox}$-deficient CGD patients, who did not have detectable Nox2 protein [39, 40]. In vitro studies revealed the essential role of p22$^{phox}$ on Nox1-4 stability. Small-interfering RNA-mediated p22$^{phox}$ downregulation leads to decreased function of Nox1-4 [41, 42].

With this background, the scientific community has shown in recent years a particular interest in the gene encoding the p22$^{phox}$ protein. Some experimental data support a pathophysiological role for p22$^{phox}$ gene polymorphisms, some of which

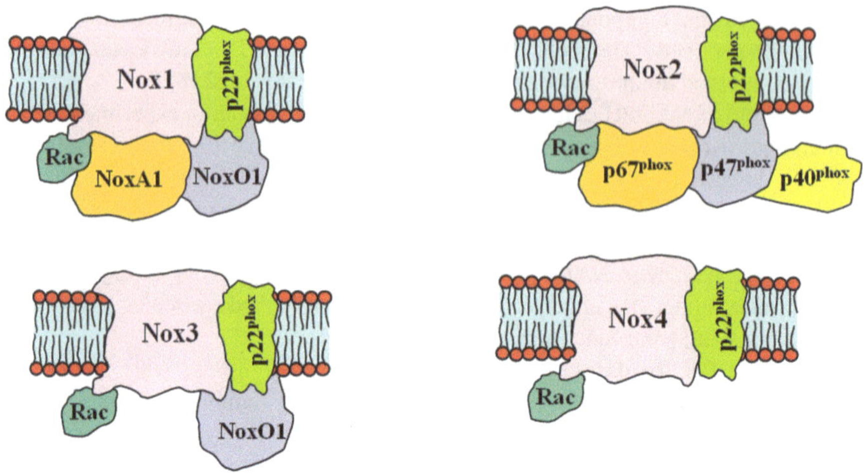


**Fig. 9.1** Comparative structure of NADPH oxidases. Nox family members display a similar structure and enzymatic function, although they differ in their mechanism of activation. The p22$^{phox}$ protein constitutes an essential subunit for the functionality of Nox1, Nox2, Nox3, and Nox4

are able to influence NADPH oxidase gene expression and activity in the context of cardiovascular diseases [43, 44]. For instance, in the animal model of genetic hypertension, the spontaneously hypertensive rat (SHR), both the aortic NADPH oxidase activity and p22$^{phox}$ mRNA expression are enhanced in hypertensive compared to normotensive rats [11]. It has been suggested that this upregulated p22$^{phox}$ mRNA expression might be due to genetic variants in the p22$^{phox}$ gene. In accordance with this possibility, the existence of five genetic variants in the p22$^{phox}$ gene promoter of SHR has been reported, which result in an increase of the transcriptional activity of this gene [45].

The human p22$^{phox}$ is encoded by the *CYBA* gene, located on the long arm of chromosome 16, at position 24. It spans 8.5 kb and is composed of six exons and five introns that encode an open reading frame of approximately 600 bp [39]. A significant number of genetic polymorphisms have been reported within the promoter and exonic sequences of the p22$^{phox}$ gene, some of which are able to influence gene expression and NADPH oxidase activation, leading to significant functional variation between individuals in oxidative stress levels (Table 9.1). Moreover, some of these polymorphisms have been associated with diverse cardiovascular diseases such as hypertension, coronary artery disease (CAD), myocardial infarction, cerebrovascular disease, and diabetic and nondiabetic nephropathy.

### *9.3.1 C242T Polymorphism*

This genetic variant is located in exon 4, at position 214 from ATG codon [39], and encodes a CAC-TAC codon change, which results in a nonconservative substitution

**Table 9.1** *CYBA* polymorphisms in human cardiovascular disease

| Polymorphism | *rs* Number | Location | Position from ATG | Effect | Ref. |
|---|---|---|---|---|---|
| C242T | rs4673 | Exon 4 | +214 | 72 His→Tyr | [39] |
| A640G | rs1049255 | Exon 6 (3′ UTR) | +612 | Not described | [46] |
| C549T | rs1049254 | Exon 6 | +521 | 174 Ala→Val | [39] |
| $-930^{A/G}$ | rs9932581 | Promoter | −930 | Activation by C/EBP | [47] |
| $-675^{A/T}$ | Not assigned | Promoter | −675 | Activation by HIF-1α | [48] |
| $-852^{C/G}$ | rs16966671 | Promoter | −852 | Not described | [49, 48] |
| $-536^{C/T}$ | rs13306296 | Promoter | −536 | Not described | [48] |

of histidine-72 by a tyrosine residue, an alteration that may impair the heme-binding site of the $p22^{phox}$ protein. This replacement might lead to a loss of oxidative function and to a reduced production of ROS and oxidative stress in the vasculature [50].

The association of the *CYBA* C242T polymorphism with *atherosclerosis* has been extensively studied in the past decade, although the results have been conflicting in Asian and Caucasian populations (Table 9.2). Inoue et al. found that the T allele conferred protection against atherosclerosis in a Japanese population [51]. In contrast, other studies did not find any association between the C242T polymorphism and the severity of CAD detected by coronary angiography [58–62], or peripheral arterial occlusive disease [63]. Finally, some studies even showed the opposite effect, that the T allele is significantly associated with progression of CAD [55–57] and cerebrovascular disease [68–70]. In this regard, no differences in coronary epicardial or microvascular responses to acetylcholine or sodium nitroprusiate according to genotypes of this polymorphism were reported [62, 66]. But other studies showed that carriers of the CC genotype exhibited a blunted endothelium-dependent dilator response, which was independent of other risk factors or atherosclerosis [64]. In the context of *diabetes*, the available data are also diverse (Table 9.2). Thus it seems that the prevalence of the CC genotype is higher in type-2 diabetic patients [72] but not in type-1 diabetic patients [73]. Hodgkinson et al. found that the T allele was associated with susceptibility to diabetic nephropathy in patients with type-1 diabetes [74]. In another study, this polymorphism was not associated with diabetic nephropathy secondary to type-2 diabetes mellitus in 612 subjects from a Chinese population [75]. Interestingly, Doi et al. reported a protective effect of the T allele against end-stage renal disease, but only in the nondiabetic group [49]. In agreement with this, Hayaishi-Okano et al. found that type-2 diabetic patients with the TC/TT genotype displayed a significantly lower average intima-media thickness than those with the CC genotype, despite the fact that there were no differences in the risk factors [76]. In patients in stage 5 *chronic kidney disease*, individuals with CC genotype had increased cardiovascular disease mortality compared to CT and TT patients [79]. In acute renal failure, the patients carrying the T allele were associated with biomarkers

**Table 9.2** Association studies of C242T polymorphism

| Association studies with cardiovascular disease | |
|---|---|
| *Atherosclerosis* | |
| T allele protects from atherosclerosis | [51, 52–54] |
| T allele favours progression of atherosclerosis | [55–57] |
| No association of the polymorphism with atherosclerosis | [58–63] |
| T allele association with increased endothelium-dependent dilator response | [64, 65] |
| No association of the polymorphism with the vasodilator response | [66, 67] |
| *Cerebrovascular disease* | |
| T allele favours progression of cerebrovascular disease | [68–70] |
| T allele protects against cardioembolic brain infarction | [71] |
| *Diabetes* | |
| T allele protects from diabetes | [72] |
| No association of the polymorphism with diabetes | [73] |
| T allele favours nephropathy in diabetes | [74] |
| No association of the polymorphism with nephropathy | [75] |
| T allele protects from renal disease | [49] |
| T allele protects from cardiovascular complications | [72, 76] |
| *Hypertension* | |
| T allele protects from essential hypertension | [77] |
| No association of the polymorphism with preeclampsia | [78] |
| *Renal disease* | |
| T allele protects from renal disease | [49] |
| T allele protects from cardiovascular mortality | [79] |
| *Association studies with functional effects* | |
| T allele associates with reduced NADPH oxidase activity in control subjects | [80] |
| T allele associates with reduced NADPH oxidase activity in hypertensive patients | [77] |
| T allele associates with reduced NADPH oxidase activity in atherosclerotic patients | [81, 82] |

of oxidative stress and adverse outcomes [83]. Regarding *hypertension*, few studies have been performed (Table 9.2). Raijmakers et al. reported a lack of association between the C242T polymorphism and preeclampsia [78]. Moreno et al. recently reported a significant association of the C242T polymorphism with essential hypertension in a Caucasian population [77]. In this study, hypertensive patients with the CC genotype exhibited higher plasma levels of von Willebrand factor, a marker of endothelial damage, than those carrying the TT genotype.

Such conflicting data on whether this polymorphism contributes to cardiovascular disease are likely to be due to multiple long-standing risk factors and cardiovascular burdens confounding the possible effect of this polymorphism on polygenic diseases such as diabetes and atherosclerosis. In addition, in many of these studies the size of the studied population was really low. For instance, no association of the C242T polymorphism with vasodilatation in forearm circulation was reported in 90 Caucasian subjects with hypercholesterolemia [67]. In a similar way, in patients with proven CAD and assigned to placebo treatment (n = 152), the risk of recurrence of cardiovascular events after 2.5 years of follow-up was similar across C242T genotypes [55]. On the contrary, in a study performed in a large Japanese population (2,541 subjects with hypercholesterolemia vs. 2,707 subjects without hypercholesterolemia), Shimokata et al. have found that the T allele of the C242T polymorphism was protective against CAD in men with hypercholesterolemia, while the protective effect was less apparent in those without this condition [65]. Similarly, in a study performed in a Japanese population (1,055 patients with brain infarction and 1,055 control subjects), Kuroda et al. have reported a protective effect of the T allele against cardioembolic infarction, but not related to lacunar and atherothrombotic infarction [71].

On the other hand, different studies have recently obtained consistent results in *high-risk populations*. Fan et al. have reported a protective effect of T allele against development of CAD in a high-risk Finnish Caucasian population [52]. Likewise, Corsetti et al. have found a significant association of C242T with the risk of coronary events in a high-risk subgroup of postinfarcted patients defined by inflammation and hypercholesterolemia over a 26-month follow-up [53]. In this study, CC individuals exhibited a 3.14 higher hazard ratio for recurrent coronary events than CT/TT subjects. Finally, a recent study reported by Arca et al. [54] evaluated the incidence of major adverse cardiovascular events in 237 patients with angiographic evidence of coronary stenosis during a median follow-up of 7.8 years. Patients with CT/TT genotypes showed reduced recurrence of cardiovascular deaths, nonfatal MI, and renovascularization procedures compared with homozygous carriers of the C allele. In addition, the presence of the 242T allele was associated with significantly reduced plasma concentrations of 8-hydroxy-2-deoxy-2-deoxyguanosine, a marker of oxidative stress. In the same study the authors found that only the risk of revascularization procedures remained significantly associated with the T allele after controlling for classical risk factors, which allowed them to propose that the protection conferred by the T allele becomes apparent only in the background of a high-risk status.

The C242T polymorphism exerts a functional effect on NADPH oxidase activity in physiological and pathophysiological conditions (Table 9.2). Firstly, Guzik et al. demonstrated that 242T allele was associated with reduced vascular NADPH oxidase activity in saphenous veins of CAD patients, independently of other clinical risk factors [81]. In agreement with this, Wyche et al. showed that the neutrophil NADPH oxidase-dependent respiratory burst activity in homozygous healthy American individuals with the T allele is significantly lower than that of wild-type carriers and heterozygous healthy individuals [80]. Likewise, the NADPH oxidase activity in blood mononuclear cells (lymphocytes and monocytes) from hypertensive patients with the T allele is significantly lower than that of patients carrying the CC genotype [77]. In these studies, the enhanced NADPH oxidase activity was not associated with increased $p22^{phox}$ levels in CC individuals, thus supporting the notion that the polymorphism is in fact reducing the ability of $p22^{phox}$ to anchor $gp91^{phox}$, and subsequently altering NADPH oxidase activity. Finally, Delles et al. [82] have reported an association of C242T polymorphism with vascular $\bullet O_2^-$ generation in pooled analysis of patients with CAD and controls. Although only a borderline level of significance was reached in association of the 242C allele with $\bullet O_2^-$ generation, the number of 242C alleles was significantly correlated with vascular $\bullet O_2^-$ generation in an additive genetic model.

### 9.3.2 A640G Polymorphism

The A640G polymorphism is located in the 3′ untranslated region of *CYBA* [46], with no frank amino acid substitution, and may affect transcriptional rate by means of modifying mRNA processing and stability. However, there is controversy about the functionality of this polymorphism. For example, Gardemann et al. analyzed 2,205 Caucasian males whose coronary anatomy was defined by coronary angiography, and reported that the G allele was more frequent in subjects without coronary disease, which suggested a protective effect exerted by the G allele [58]. Similar findings were observed in a more recent case-control study in which the prevalence of $p22^{phox}$ genotypes was analyzed in 619 Spanish subjects in order to explore the contribution of A640G to coronary heart disease risk [84]. A significant risk was found associated with GG homozygosity. On the other hand, Park et al. found that exercise training in 59 North Americans with high cardiovascular disease risk factors decreased thiobarbituric acid–reactive substances in 16% [85]. In this study the A allele carriers showed greater reduction in thiobarbituric acid–reactive substances than noncarriers at the A640G locus, indicating the A allele as the protector.

Other studies indicate no association of the A640G with cardiovascular diseases. For example, Zafari et al. found no significant frequency distribution of the genotypes among patients with or without angiographically significant coronary disease in a study performed in 216 North American subjects [59]. Inoue et al. found the prevalence of the genotypes of the A640G polymorphism did not differ between 201 controls and 201 patients with CAD of a Japanese population [51]. Other studies have also shown the lack of association of the A640G polymorphism with cardiovascular disease [74, 86]. According to these studies, the $\bullet O_2^-$ production by human neutrophils in 90 healthy North Americans subjects was not altered by the A640G polymorphism [80]. Finally, in a recent study performed in 81 healthy Caucasian volunteers, the $\bullet O_2^-$ production by granulocytes was lower in GG than in AA individuals [87].

### 9.3.3 -930$^{A/G}$ Polymorphism

The -930$^{A/G}$ polymorphism is located in the promoter region of *CYBA* at position -930 from the ATG codon [47]. An in silico analysis of the promoter sequence shows that this polymorphism lies on a potential binding site for CCAAT enhancer-binding protein (C/EBP) transcription factors, and it has been speculated that it might modulate *CYBA* transcriptional activity [88].

The -930$^{A/G}$ polymorphism was reported to be associated with essential hypertension in a Spanish population of 156 subjects [47]. This association with hypertension was confirmed in a greater population of 623 subjects [88]. In particular, a significant increase in the G allele frequency was found in hypertensives. The association of this polymorphism with hypertension was confirmed by Kokubo et al. [89], in a population of 3,652 subjects from Japan, who found that the GG genotype was also associated with hypertension in the male population. In a Brazilian population, this polymorphism was not associated with end-organ damage in hypertensive patients [90]. Likewise, this polymorphism was not associated with diabetic nephropathy secondary to type-2 diabetes mellitus in 612 subjects from a Chinese population [75].

Although there is some controversy about the association of this polymorphism with cardiovascular diseases, the relevance of the functionality of this polymorphism in hypertension has been underlined recently by a study demonstrating that hypertensive subjects with the GG genotype exhibited significantly increased phagocytic p22$^{phox}$ mRNA and protein levels and enhanced NADPH oxidase activity (Fig. 9.2) [88]. In contrast, no differences were found in the NADPH oxidase expression and activity between genotypes within the normotensive group. In accordance with this, in a healthy population, no functional impact of this polymorphism on neutrophil NADPH oxidase-dependent $\bullet O_2^-$ production was found [80]. Finally, mutagenesis experiments have demonstrated a functional role of this polymorphism on the *CYBA* promoter activity [47]. The overexpression of C/EBPδ is able to induce a greater

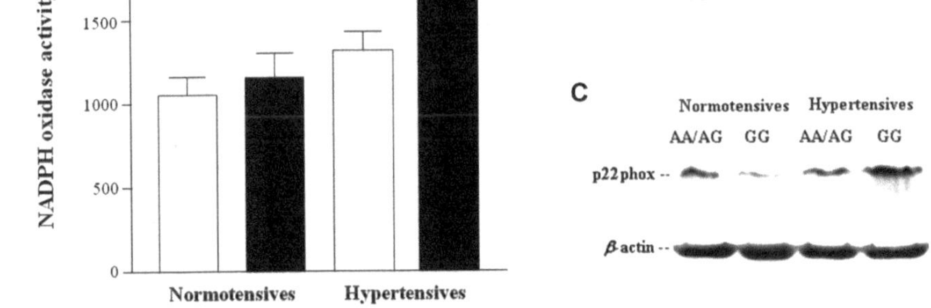


**Fig. 9.2** Association of -930$^{A/G}$ polymorphism with (**a**) NADPH oxidase activity, (**b**) p22$^{phox}$ mRNA levels, and (**c**) p22$^{phox}$ protein expression. Determinations were performed in phagocytic cells from normotensive subjects and hypertensive patients. *Bars* show mean+SEM. **P*<0.05 compared with AA/AG hypertensives and with AA/AG and GG normotensives (modified from reference [88])

effect on the transcriptional activity of the G than the A allelic *CYBA* promoter constructs in vascular smooth muscle cells, suggesting a potential involvement of C/EBPδ in the NADPH oxidase activation present in those patients carrying the G allele [88]. The relevance of these findings is underlined by a study reporting increased C/EBPd levels in experimental hypertension [91].

### 9.3.4 -675$^{A/T}$ Polymorphism

This new genetic variation is located in the promoter region of the *CYBA* gene, at position -675 from the ATG codon [48]. The change of the T allele by the A allele seems to be involved in the elimination of a potential target sequence for the binding of the hypoxia-inducible factor-1alpha (HIF-1α) transcription factor.

The -675$^{A/T}$ polymorphism is associated with essential hypertension [48]. The prevalence of the TT genotype is higher in hypertensive than in normotensive subjects. What is more, hypertensive patients carrying the TT genotype exhibited higher blood pressure values and carotid intima-media thickness than TA/AA patients.

This polymorphism is also associated with the NADPH oxidase activation in phagocytic cells because increased NADPH oxidase-dependent $\bullet O_2^-$ production has been reported in TT subjects compared with TA/AA subjects [48]. Site-directed mutagenesis experiments demonstrated a functional role of this polymorphism on the *CYBA* promoter activity (Fig. 9.3). The overexpression of HIF-1α was therefore able to induce a positive effect on the transcriptional activity of the T but not the A allelic *CYBA* promoter constructs in vascular smooth muscle cells, thus suggesting a potential involvement of HIF-1α in the NADPH oxidase activation present in those patients carrying the T allele.

### 9.3.5 Other CYBA Polymorphisms

The involvement of other *CYBA* allelic variants modulating the expression and activity of the p22$^{phox}$ subunit can not be discarded. Interestingly, two other genetics changes, the -852$^{C/G}$ polymorphism and -536$^{C/T}$ polymorphism, have been characterized in the promoter of *CYBA*, although they were not associated with hypertension in a study performed in a Caucasian population from the south of Europe (865 cases and 496 controls) [48]. Likewise, Doi et al. found no association of the -852$^{C/G}$ polymorphism with end-stage renal disease [49]. In 1990, Dinauer et al. identified another polymorphism (C549T) located in exon 6, at position 521 from the ATG codon [39]. The C549T polymorphism encodes a C-T substitution, which predicts the conservative replacement of alanine with valine at position 174. Further studies will be necessary to analyze the implication of these polymorphisms in other cardiovascular diseases.

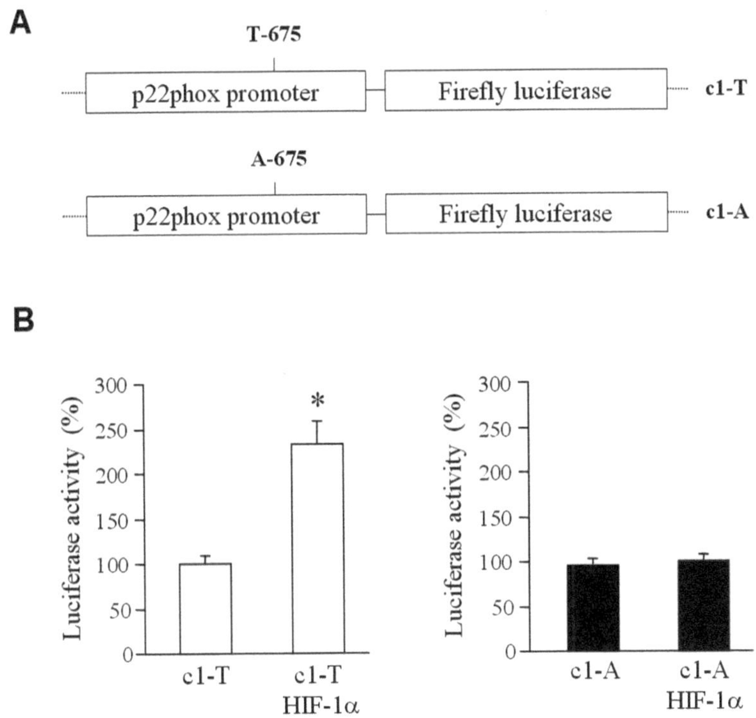


**Fig. 9.3** Functional effect of -675$^{A/T}$ polymorphism on *CYBA* promoter activity. *Transient reporter gene expression assays with constructs containing full-length CYBA promoter.* (**a**) Schematic presentation of reporter gene constructs containing the human p22$^{phox}$ gene promoter, with the only difference between the two constructs being either a T or an A base at the -675 polymorphic site. (**b**) Luciferase levels in A7r5 vascular smooth muscle cells transfected with reporter gene constructs containing the T (c1-T, *open bars*) or A (c1-A, *closed bars*) allelic *CYBA* gene and in cotransfection with HIF-1α. Data shown are mean+SEM. *$P<0.05$ compared with c1-T (modified from reference [48])

## 9.4 Genetic and Environmental Interactions

Linkage disequilibrium (LD) analyses suggest that the *CYBA* polymorphisms are in a low/moderate LD [48]. Overall, the association of the -930$^{A/G}$ polymorphism with other *CYBA* polymorphisms is almost absent (Fig. 9.4). However, the C242T, -675$^{A/T}$, and A640G polymorphisms seem to be in a moderate LD [58, 48]. Thus, the relationships among *CYBA* polymorphisms may add new insights to the genetic studies of *CYBA* and NADPH oxidase-mediated $\cdot O_2^-$ production. In this sense, data suggest that a synergistic effect between the -930$^{A/G}$ and the C242T polymorphisms is able to modulate NADPH oxidase activity. The concurrence of -930GG and 242CC genotypes results in a higher NADPH oxidase activity in hypertensive patients. In an interesting study, Doi et al. have identified a risk haplotype for non

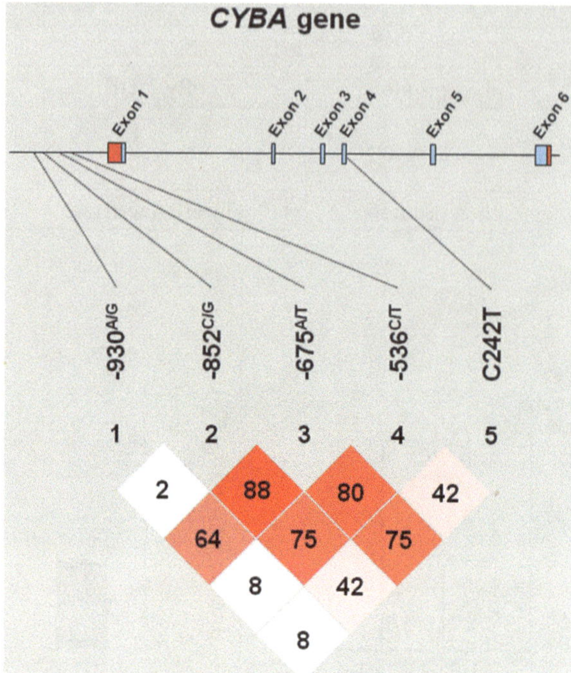


**Fig. 9.4** Linkage disequilibrium (LD) analysis among *CYBA* polymorphisms. LD graphic was obtained from a population of 1,153 subjects with the software Haploview 4.1, and shows the standardized LD coefficients (*D'*). The -852$^{C/G}$, -675$^{A/T}$, and -536$^{C/T}$ polymorphisms were found to be in high LD, within a block that exhibits a mild LD with C242T polymorphism. Interestingly, the LD of -930$^{A/G}$ polymorphism with the other *CYBA* polymorphisms is negligible; although *D'* values between the -930$^{A/G}$ and the -675$^{A/T}$ variants are high enough, these values might be a confounding result as a consequence of the asymmetric distribution of genotypes for the -675$^{A/T}$ polymorphism

diabetic end-stage renal disease in the *CYBA* gene [49]. The CC242-AA640 haplotype was associated with end-stage renal disease after adjusting for confounding factors. Likewise, Park et al. found systemic oxidative stress was decreased to a greater extent in the C242/A640 haplotype carriers compared with the noncarriers with aerobic exercise training [85].

In a recent study, He et al. described a relevant association of the C242T polymorphism with the risk of coronary heart disease [92]. The authors reported that compared with nonsmokers with the CC genotype, nonsmokers with the CT/TT genotype had a decreased risk of heart disease. On the contrary, smokers with the CT/TT genotype had an increased risk compared with smokers with the CC genotype, thus suggesting a potential interaction among smoking and *CYBA* C242T polymorphism in relation to heart disease risk. Likewise, Niemiec et al. reported the risk of CAD associated with the presence of cigarette smoking and hypercholesterolemia was stronger in CT/TT carriers than in CC homozygotes [93].

## 9.5 Summary and Conclusions

In most cases, association studies have revealed significant associations between *CYBA* gene polymorphisms and cardiovascular diseases, although the results have not always been consistent across studies. For example, results of association in mixed populations may actually be due to differences in the frequencies of alleles in different ethnic groups. In most studies, association analyses have been carried out with insufficiently stringent statistical thresholds. Finally, the enzymatic and structural characteristics of the emerging new members of this oxidase family have not been clearly defined.

On the other hand, it may be time to consider that the synergistic effect of *CYBA* variants may identify a particular high-oxidative stress risk subgroup within the healthy population and the population exhibiting cardiovascular risk factors, including hypertension and diabetes. This may therefore help to explain the contradictory results regarding the association of the *CYBA* polymorphism with cardiovascular diseases, and may point toward the importance of multiple polymorphism assessment in functional and association studies of complex diseases, including hypertension, diabetes, and atherosclerosis. We cannot forget that the environmental component (including common cardiovascular risk factors) is relevantly involved in the final phenotype of the cardiovascular diseases. In this sense, a paradigm shift is taking place in human disease diagnosis and care, driven by the capability to perform routine genetic analysis of individuals.

*CYBA* haplotypes for C242T and A640G polymorphisms exhibited differential changes in systemic oxidative stress in response to aerobic exercise training, known to be the most effective nonpharmacological intervention to alleviate oxidative stress [85]. In recent years, the importance that genetic variations have in predicting efficacy of drug therapy has been demonstrated:—*pharmacogenomics*, which is refining individualized approaches to care in cardiovascular patients in view of their genetic susceptibility. Interestingly, several genetic variants of NADPH oxidase may modulate the risk of developing anthracycline-induced cardiotoxicity [94]. Cardiovascular therapy, including antihypertensive drugs, statins, and thiazolidinediones, besides reducing blood pressure, lowering LDL cholesterol, and increasing insulin sensitivity, respectively, exerts other pleiotropic properties, including antioxidant effects, by reducing NADPH oxidase activity [95–99]. Thus, the stratification of cardiovascular patients on the basis of individual *CYBA* risk haplotypes may be useful for developing novel therapeutic approaches.

**Acknowledgments** This work was supported by the agreement between the Foundation for Applied Medical Research and "UTE project CIMA"; European Union (InGenious HyperCare, LSHM-CT-2006-037093); Red Temática de Investigación Cooperativa en Enfermedades Cardiovasculares from the Instituto de Salud Carlos III, Ministry of Health (RD06/0014/0008) and Ministry of Science and Culture (SAF-2007-62553) of Spain.

## References


1. Kullo IJ, Ding K (2007) Mechanisms of disease: the genetic basis of coronary heart disease. Nat Clin Pract Card Medi 4:558–569
2. Bokoch GM, Knaus UG (2003) NADPH oxidases: not just for leukocytes anymore! Trends Biochem Sci 28:502–508
3. Quinn MT, Gauss KA (2004) Structure and regulation of the neutrophil respiratory burst oxidase: comparison with nonphagocyte oxidases. J Leukoc Biol 76:760–781
4. Cross AR, Segal AW (2004) The NADPH oxidase of professional phagocytes–prototype of the NOX electron transport chain systems. Biochim Biophys Acta 1657:1–22
5. Ray R, Shah AM (2005) NADPH oxidase and endothelial cell function. Clin Sci 109:217–226
6. Brandes RP, Kreuzer J (2005) Vascular NADPH oxidases: molecular mechanisms of activation. Cardiovasc Res 65:16–27
7. Bedard K, Krause KH (2007) The NOX family of ROS-generating NADPH oxidases: physiology and pathophysiology. Physiol Rev 87:245–313
8. Quinn MT, Ammons MC, Deleo FR (2006) The expanding role of NADPH oxidases in health and disease: no longer just agents of death and destruction. Clin Sci 111:1–20
9. Inoguchi T, Nawata H (2005) NAD(P)H oxidase activation: a potential target mechanism for diabetic vascular complications, progressive beta-cell dysfunction and metabolic syndrome. Curr Drug Targets 6:495–501
10. Fortuño A, San José G, Moreno MU et al (2006) Phagocytic NADPH oxidase overactivity underlies oxidative stress in metabolic syndrome. Diabetes 55:209–215
11. Zalba G, Beaumont FJ, San José G et al (2000) Vascular NADH/NADPH oxidase is involved in enhanced superoxide production in spontaneously hypertensive rats. Hypertension 35:1055–1061
12. Fortuño A, Oliván S, Beloqui O et al (2004) Association of increased phagocytic NADPH oxidase-dependent superoxide production with diminished nitric oxide generation in essential hypertension. J Hypertens 22:2169–2175
13. Lassegue B, Griendling KK (2004) Reactive oxygen species in hypertension; An update. Am J Hypertens 17:852–860
14. Guzik TJ, Mussa S, Gastaldi D et al (2002) Mechanisms of increased vascular superoxide production in human diabetes mellitus: role of NAD(P)H oxidase and endothelial nitric oxide synthase. Circulation 105:1656–1662
15. Inoguchi T, Tsubouchi H, Etoh T et al (2003) A possible target of antioxidative therapy for diabetic vascular complications-vascular NAD(P)H oxidase. Curr Med Chem 10:1759–1764
16. Heymes C, Bendall JK, Ratajczak P et al (2003) Increased myocardial NADPH oxidase activity in human heart failure. J Am Coll Cardiol 41:2164–2171
17. Cave A, Grieve D, Johar S et al (2005) NADPH oxidase-derived reactive oxygen species in cardiac pathophysiology. Philos Trans R Soc Lond B Biol Sci 360:2327–2334
18. Fortuño A, Beloqui O, San José G et al (2005) Increased phagocytic nicotinamide adenine dinucleotide phosphate oxidase-dependent superoxide production in patients with early chronic kidney disease. Kidney Int Suppl 22:S71–S75
19. Zalba G, Fortuño A, Díez J (2006) Oxidative stress and atherosclerosis in early chronic kidney disease. Nephrol Dial Transplant 21:2686–2690
20. Gill PS, Wilcox CS (2006) NADPH oxidases in the kidney. Antioxid Redox Signal 8: 1597–1607
21. Sorescu D, Weiss D, Lassegue B et al (2002) Superoxide production and expression of nox family proteins in human atherosclerosis. Circulation 105:1429–1435
22. Zalba G, Beloqui O, San José G et al (2005) NADPH oxidase-dependent superoxide production is associated with carotid intima-media thickness in subjects free of clinical atherosclerotic disease. Arterioscler Thromb Vasc Biol 25:1452–1457
23. Zalba G, Fortuño A, Orbe J et al (2007) Phagocytic NADPH oxidase-dependent superoxide production stimulates matrix metalloproteinase-9: implications for human atherosclerosis. Arterioscler Thromb Vasc Biol 27:587–593

24. Miller AA, Drummond GR, Sobey CG (2006) Novel isoforms of NADPH-oxidase in cerebral vascular control. Pharmacol Ther 111:928–948
25. Kalinina N, Agrotis A, Tararak E et al (2002) Cytochrome b558-dependent NAD(P)H oxidase-phox units in smooth muscle and macrophages of atherosclerotic lesions. Arterioscler Thromb Vasc Biol 22:2037–2043
26. Liu J, Yang F, Yang XP et al (2003) NAD(P)H oxidase mediates angiotensin II-induced vascular macrophage infiltration and medial hypertrophy. Arterioscler Thromb Vasc Biol 23:776–782
27. Cifuentes ME, Pagano PJ (2006) Targeting reactive oxygen species in hypertension. Curr Opin Nephrol Hypertens 15:179–186
28. Rajagopalan S, Kurz S, Munzel T et al (1996) Angiotensin II-mediated hypertension in the rat increases vascular superoxide production via membrane NADH/NADPH oxidase activation. Contribution to alterations of vasomotor tone. J Clin Invest 97:1916–1923
29. De Keulenaer GW, Alexander RW, Ushio-Fukai M et al (1998) Tumour necrosis factor alpha activates a p22phox-based NADH oxidase in vascular smooth muscle. Biochem J 329: 653–657
30. Pu Q, Neves MF, Virdis A et al (2003) Endothelin antagonism on aldosterone-induced oxidative stress and vascular remodeling. Hypertension 42:49–55
31. Dorffel Y, Franz S, Pruss A et al (2001) Preactivated monocytes from hypertensive patients as a factor for atherosclerosis? Atherosclerosis 157:151–160
32. Virdis A, Neves MF, Amiri F et al (2002) Spironolactone improves angiotensin-induced vascular changes and oxidative stress. Hypertension 40:504–510
33. Zalba G, San José G, Moreno MU et al (2001) Oxidative stress in arterial hypertension: role of NAD(P)H oxidase. Hypertension 38:1395–1399
34. San José G, Fortuño A, Beloqui O et al (2008) NADPH oxidase CYBA polymorphisms, oxidative stress and cardiovascular diseases. Clin Sci 114:173–182
35. Roos D (1994) The genetic basis of chronic granulomatous disease. Immunol Rev 138: 121–157
36. Ushio-Fukai M, Zafari AM, Fukui T et al (1996) p22phox is a critical component of the superoxide-generating NADH/NADPH oxidase system and regulates angiotensin II-induced hypertrophy in vascular smooth muscle cells. J Biol Chem 271:23317–23321
37. Ushio-Fukai M, Tang Y, Fukai T et al (2002) Novel role of gp91(phox)-containing NAD(P)H oxidase in vascular endothelial growth factor-induced signaling and angiogenesis. Circ Res 91:1160–1167
38. Lassegue B, Clempus RE (2003) Vascular NAD(P)H oxidases: specific features, expression, and regulation. Am J Physiol Regul Integr Comp Physiol 285:R277–R297
39. Dinauer MC, Pierce EA, Bruns GA et al (1990) Human neutrophil cytochrome b light chain (p22-phox). Gene structure, chromosomal location, and mutations in cytochrome-negative autosomal recessive chronic granulomatous disease. J Clin Invest 86:1729–1737
40. Parkos CA, Dinauer MC, Jesaitis AJ et al (1989) Absence of both the 91kD and 22kD subunits of human neutrophil cytochrome b in two genetic forms of chronic granulomatous disease. Blood 73:1416–1420
41. Kawahara T, Ritsick D, Cheng G et al (2005) Point mutations in the proline-rich region of p22phox are dominant inhibitors of Nox1- and Nox2-dependent reactive oxygen generation. J Biol Chem 280:31859–31869
42. Martyn KD, Frederick LM, von Loehneysen K et al (2006) Functional analysis of Nox4 reveals unique characteristics compared to other NADPH oxidases. Cell Signal 18:69–82
43. Zalba G, San José G, Moreno MU et al (2005) NADPH oxidase-mediated oxidative stress: genetic studies of the p22(phox) gene in hypertension. Antioxid Redox Signal 7:1327–1336
44. Soccio M, Toniato E, Evangelista V et al (2005) Oxidative stress and cardiovascular risk: the role of vascular NAD(P)H oxidase and its genetic variants. Eur J Clin Invest 35:305–314
45. Zalba G, San José G, Beaumont FJ et al (2001) Polymorphisms and promoter overactivity of the p22(phox) gene in vascular smooth muscle cells from spontaneously hypertensive rats. Circ Res 88:217–222

46. de Boer M, de Klein A, Hossle JP et al (1992) Cytochrome b558-negative, autosomal recessive chronic granulomatous disease: two new mutations in the cytochrome b558 light chain of the NADPH oxidase (p22-phox). Am J Hum Genet 51:1127–1135
47. Moreno MU, San José G, Orbe J et al (2003) Preliminary characterisation of the promoter of the human p22(phox) gene: identification of a new polymorphism associated with hypertension. FEBS Lett 542:27–31
48. Moreno MU, San José G, Fortuño A et al (2007) A novel *CYBA* variant, the -675A/T polymorphism, is associated with essential hypertension. J Hypertens 25:1620–1626
49. Doi K, Noiri E, Nakao A et al (2005) Haplotype analysis of NAD(P)H oxidase p22 phox polymorphisms in end-stage renal disease. J Hum Genet 50:641–647
50. Whitehead AS, FitzGerald GA (2001) Twenty-first century phox: not yet ready for widespread screening. Circulation 103:7–9
51. Inoue N, Kawashima S, Kanazawa K et al (1998) Polymorphism of the NADH/NADPH oxidase p22 phox gene in patients with coronary artery disease. Circulation 97:135–137
52. Fan M, Kahonen M, Rontu R et al (2006) The p22phox C242T gene polymorphism is associated with a reduced risk of angiographically verified coronary artery disease in a high-risk Finnish Caucasian population. The Finnish Cardiovascular Study. Am Heart J 152:538–542
53. Corsetti JP, Ryan D, Moss AJ et al (2008) NAD(P)H oxidase polymorphism (C242T) and high HDL cholesterol associate with recurrent coronary events in postinfarction patients. Atherosclerosis 196:461–468
54. Arca M, Conti B, Montali A et al (2008) C242T polymorphism of NADPH oxidase p22phox and recurrence of cardiovascular events in coronary artery disease. Arterioscler Thromb Vasc Biol 28:752–757
55. Cahilly C, Ballantyne CM, Lim DS et al (2000) A variant of p22(phox), involved in generation of reactive oxygen species in the vessel wall, is associated with progression of coronary atherosclerosis. Circ Res 86:391–395
56. Nasti S, Spallarossa P, Altieri P et al (2006) C242T polymorphism in CYBA gene (p22phox) and risk of coronary artery disease in a population of Caucasian Italians. Dis Markers 22: 167–173
57. Spence MS, McGlinchey PG, Patterson CC et al (2003) Investigation of the C242T polymorphism of NAD(P)H oxidase p22 phox gene and ischaemic heart disease using family-based association methods. Clin Sci 105:677–682
58. Gardemann A, Mages P, Katz N et al (1999) The p22 phox A640G gene polymorphism but not the C242T gene variation is associated with coronary heart disease in younger individuals. Atherosclerosis 145:315–323
59. Zafari AM, Davidoff MN, Austin H et al (2002) The A640G and C242T p22(phox) polymorphisms in patients with coronary artery disease. Antioxid Redox Signal 4:675–680
60. Saha N, Sanghera DK, Kamboh MI (1999) The p22 phox polymorphism C242T is not associated with CHD risk in Asian Indians and Chinese. Eur J Clin Invest 29:999–1002
61. Cai H, Duarte N, Wilcken DE (1999) NADH/NADPH oxidase p22 phox C242T polymorphism and coronary artery disease in the Australian population. Eur J Clin Invest 29:744–748
62. Li A, Prasad A, Mincemoyer R et al (1999) Relationship of the C242T p22phox gene polymorphism to angiographic coronary artery disease and endothelial function. Am J Med Genet 86:57–61
63. Renner W, Schallmoser K, Gallippi P et al (2000) C242T polymorphism of the p22 phox gene is not associated with peripheral arterial occlusive disease. Atherosclerosis 152:175–179
64. Schachinger V, Britten MB, Dimmeler S (2001) NADH/NADPH oxidase p22 phox gene polymorphism is associated with improved coronary endothelial vasodilator function. Eur Heart J 22:96–101
65. Shimokata K, Yamada Y, Kondo T et al (2004) Association of gene polymorphisms with coronary artery disease in individuals with or without nonfamilial hypercholesterolemia. Atherosclerosis 172:167–173

66. Fricker R, Hesse C, Weiss J et al (2004) Endothelial venodilator response in carriers of genetic polymorphisms involved in NO synthesis and degradation. Br J Clin Pharmacol 58:169–177
67. Schneider MP, Hilgers KF, Huang Y et al (2003) The C242T p22phox polymorphism and endothelium-dependent vasodilation in subjects with hypercholesterolaemia. Clin Sci 105: 97–103
68. Ito D, Murata M, Watanabe K et al (2000) C242T polymorphism of NADPH oxidase p22 PHOX gene and ischemic cerebrovascular disease in the Japanese population. Stroke 31: 936–939
69. Krex D, Ziegler A, Konig IR et al (2003) Polymorphisms of the NADPH oxidase P22PHOX gene in a Caucasian population with intracranial aneurysms. Cerebrovasc Dis 16:363–368
70. Genius J, Grau AJ, Lichy C (2008) The C242T polymorphism of the NAD(P)H oxidase p22phox subunit is associated with an enhanced risk for cerebrovascular disease at young age. Cerebrovasc Dis 26:430–433
71. Kuroda J, Kitazono T, Ago T et al (2007) NAD(P)H oxidase p22phox C242T polymorphism and ischemic stroke in Japan: the Fukuoka Stroke Registry and the Hisayama study. Eur J Neurol 14:1091–1097
72. Matsunaga-Irie S, Maruyama T, Yamamoto Y et al (2004) Relation between development of nephropathy and the p22phox C242T and receptor for advanced glycation end product G1704T gene polymorphisms in type 2 diabetic patients. Diabetes Care 27:303–307
73. Matsunaga S, Maruyama T, Yamada S et al (2003) Nicotinamide adenine dinucleotide phosphate oxidase (NADPH oxidase) P22 Phox C242T gene polymorphism in type 1 diabetes. Ann N Y Acad Sci 1005:324–327
74. Hodgkinson AD, Millward BA, Demaine AG (2003) Association of the p22phox component of NAD(P)H oxidase with susceptibility to diabetic nephropathy in patients with type 1 diabetes. Diabetes Care 26:3111–3115
75. Lim SC, Goh SK, Lai YR et al (2006) Relationship between common functional polymorphisms of the p22phox gene (-930A > G and +242C > T) and nephropathy as a result of Type 2 diabetes in a Chinese population. Diabet Med 23:1037–1041
76. Hayaishi-Okano R, Yamasaki Y, Kajimoto Y et al (2003) Association of NAD(P)H oxidase p22 phox gene variation with advanced carotid atherosclerosis in Japanese type 2 diabetes. Diabetes Care 26:458–463
77. Moreno MU, San José G, Fortuño A et al (2006) The C242T CYBA polymorphism of NADPH oxidase is associated with essential hypertension. J Hypertens 24:1299–1306
78. Raijmakers MT, Roes EM, Steegers EA et al (2002) The C242T-polymorphism of the NADPH/NADH oxidase gene p22phox subunit is not associated with pre-eclampsia. J Hum Hypertens 16:423–425
79. Grahl DA, Axelsson J, Nordfors L et al (2007) Associations between the CYBA 242C/T and the MPO -463G/A polymorphisms, oxidative stress and cardiovascular disease in chronic kidney disease patients. Blood Purif 25:210–218
80. Wyche KE, Wang SS, Griendling KK et al (2004) C242T CYBA polymorphism of the NADPH oxidase is associated with reduced respiratory burst in human neutrophils. Hypertension 43:1246–1251
81. Guzik TJ, West NE, Black E et al (2000) Functional effect of the C242T polymorphism in the NAD(P)H oxidase p22phox gene on vascular superoxide production in atherosclerosis. Circulation 102:1744–1747
82. Delles C, Zimmerli LU, McGrane DJ et al (2008) Vascular stiffness is related to superoxide generation in the vessel wall. J Hypertens 26:946–955
83. Perianayagam MC, Liangos O, Kolyada AY et al (2007) NADPH oxidase p22phox and catalase gene variants are associated with biomarkers of oxidative stress and adverse outcomes in acute renal failure. J Am Soc Nephrol 18:255–263
84. Macías-Reyes A, Rodríguez-Esparragón F, Caballero-Hidalgo A et al (2008) Insight into the role of CYBA A640G and C242T gene variants and coronary heart disease risk. A case-control study. Free Rad Biol Med 42:82–92

85. Park JY, Ferrell RE, Park JJ et al (2005) NADPH oxidase p22phox gene variants are associated with systemic oxidative stress biomarker responses to exercise training. J Appl Physiol 99:1905–1911
86. Mashiba J, Koike G, Kamiunten H et al (2005) Vasospastic angina and microvascular angina are differentially influenced by PON1 A632G polymorphism in the Japanese. Circ J 69: 1466–1471
87. Schirmer M, Hoffmann M, Kaya E et al (2008) Genetic polymorphisms of NAD(P)H oxidase: variation in subunit expression and enzyme activity. Pharmacogenomics J 8:297–304
88. San José G, Moreno MU, Olivan S et al (2004) Functional effect of the p22phox -930A/G polymorphism on p22phox expression and NADPH oxidase activity in hypertension. Hypertension 44:163–169
89. Kokubo Y, Iwai N, Tago N et al (2005) Association analysis between hypertension and CYBA, CLCNKB, and KCNMB1 functional polymorphisms in the Japanese population–the Suita Study. Circ J 69:138–142
90. Sales ML, Ferreira MC, Leme CA Jr et al (2007) Non-effect of p22-phox -930A/G polymorphism on end-organ damage in Brazilian hypertensive patients. J Hum Hypertens 2:504–506
91. Kitami Y, Fukuoka T, Hiwada K et al (1999) A high level of CCAAT-enhancer binding protein-delta expression is a major determinant for markedly elevated differential gene expression of the platelet-derived growth factor-alpha receptor in vascular smooth muscle cells of genetically hypertensive rats. Circ Res 84:64–73
92. He MA, Cheng LX, Jiang CZ et al (2007) Associations of polymorphism of P22(phox) C242T, plasma levels of vitamin E, and smoking with coronary heart disease in China. Am Heart J 153:640.e1–640.e6
93. Niemiec P, Zak I, Wita K (2007) The 242T variant of the CYBA gene polymorphism increases the risk of coronary artery disease associated with cigarette smoking and hypercholesterolemia. Coron Artery Dis 18:339–346
94. Wojnowski L, Kulle B, Schirmer M et al (2005) NAD(P)H oxidase and multidrug resistance protein genetic polymorphisms are associated with doxorubicin-induced cardiotoxicity. Circulation 112:3754–3762
95. Rueckschloss U, Quinn MT, Holtz J et al (2002) Dose-dependent regulation of NAD(P)H oxidase expression by angiotensin II in human endothelial cells: protective effect of angiotensin II type 1 receptor blockade in patients with coronary artery disease. Arterioscler Thromb Vasc Biol 22:1845–1851
96. van der Giet M, Erinola M, Zidek W et al (2002) Captopril and quinapril reduce reactive oxygen species. Eur J Clin Invest 32:732–737
97. Maack C, Kartes T, Kilter H et al (2003) Oxygen free radical release in human failing myocardium is associated with increased activity of rac1-GTPase and represents a target for statin treatment. Circulation 108:1567–1574
98. Hwang J, Kleinhenz DJ, Lassegue B et al (2005) Peroxisome proliferator-activated receptor-gamma ligands regulate endothelial membrane superoxide production. Am J Physiol Cell Physiol 288:C899–C905
99. Morawietz H, Erbs S, Holtz J et al (2006) Endothelial Protection, AT1 blockade and Cholesterol-Dependent Oxidative Stress: the EPAS trial. Circulation 114:I296–I301

# Chapter 10
# Redox-Related Genetic Markers of Cardiovascular Diseases

**Christian Delles and Anna F. Dominiczak**

**Abstract** Multiple factors contribute to the development of cardiovascular diseases, with oxidative stress being one of the most important pathogenetic mechanisms. Redox-related genes are therefore attractive candidate genes for cardiovascular diseases. There is compelling evidence that polymorphisms of genes that are related to production of and defences against reactive oxygen species are associated with levels of free radicals and intermediate cardiovascular phenotypes. Less robust data are available for the relationship between variants of redox-related genes and advanced cardiovascular diseases such as coronary artery disease. Reasons for these negative findings are the complexity of the disease and insufficient characterisation of the phenotype and environmental factors. Large-scale genome-wide association studies are expected to deliver results on the role of redox-related genes in the pathogenesis of cardiovascular diseases, but future strategies will also involve more systematic and integrative approaches, including transcriptomic, proteomic, and metabolomic strategies.

**Keywords** Single nucleotide polymorphisms · Redox signalling genes · Candidate genes · Genome-wide association studies · Free radicals and intermediate cardiac phenotypes

## 10.1 Introduction

Cardiovascular diseases are caused by complex interactions between genetic and environmental factors. Heritability of cardiovascular diseases can be estimated from pedigree and twin studies. For example, in hypertension, 25–65% of blood pressure variation is attributable to genetic factors [1–4]. In advanced cardiovascular diseases

C. Delles (✉)
BHF Glasgow Cardiovascular Research Centre, University of Glasgow, Glasgow G12 8TA, Scotland, UK
e-mail: c.delles@clinmed.gla.ac.uk

H. Sauer et al. (eds.), *Studies on Cardiovascular Disorders*,
Oxidative Stress in Applied Basic Research and Clinical Practice,
DOI 10.1007/978-1-60761-600-9_10, 

such as coronary artery disease and myocardial infarction, it is difficult to dissect the causative genetic factors. Some factors may contribute directly to changes within the coronary arteries, whereas other factors affect blood pressure, lipid levels, or body weight, and thereby contribute indirectly to the genetic risk of coronary artery disease. The concept of the cardiovascular continuum, which proposes a progression of cardiovascular diseases from risk factors and early functional changes to structural vascular changes and established disease, is useful in this context [5–7]. Unravelling the mechanisms of basic vascular pathology such as endothelial dysfunction and inflammation will thereby lead to a better understanding of the causes of more complex cardiovascular diseases.

Oxidative stress is an important pathogenetic mechanism in the development of cardiovascular diseases because of its involvement in every step in the cardiovascular continuum. Oxidative stress has been found to be associated with hypertension [8], endothelial dysfunction [9], increased vascular stiffness [10, 11], and established coronary artery disease [12], to name a few. Redox-related genes are therefore attractive but at the same time particularly difficult candidates by which to explain the genetic background of cardiovascular diseases. They are challenging for a number of reasons. First, because of the involvement of oxidative stress in many processes of vascular pathology, the links between variants of redox-related genes and specific cardiovascular phenotypes are not always tight. Second, especially in advanced cardiovascular diseases, oxidative stress may not only be the cause but also the consequence of the disease. Third, a large number of protein and enzyme systems are involved in regulation of the redox state, and because of compensation by other systems, changes in one or a few of these systems may not always lead to measurable changes in the production of free radicals. Finally, apart from systems that directly produce or interact with free radicals, other systems contribute indirectly to oxidative stress. For example, because of its eminent role in vascular tone and salt and water balance, angiotensin II would not be regarded as a pro-oxidant in the first instance, although increased superoxide production by $AT_1$-receptor–mediated stimulation of NADPH oxidase is a major mechanism of its action [13].

To cover the wide range of the role of redox-related genetic markers in cardiovascular diseases we will therefore focus on some of the most important aspects: human cardiovascular disease, as opposed to ex vivo or animal work, to immediately highlight clinical relevance; hypertension and coronary artery disease as examples, because of less robust genetic studies in other cardiovascular diseases, including stroke and peripheral vessel disease; and reactive oxygen species (ROS) and specifically the superoxide anion, as opposed to other free radicals, because of the prominent role of superoxide in vascular biology.

## 10.2 Phenotypic Quality

Accurate assessment of the phenotype is a crucial factor in genotype-phenotype association studies [1, 6, 14]. The closer the phenotype is related to the gene under examination, the better are the chances of detecting a causative link. Changes in expression and function of redox-related genes will in the first instance affect the levels of ROS. Technical difficulties in measuring ROS translate into difficulties in unraveling the genetics of redox-related pathways in cardiovascular diseases. We highlight a few challenges of phenotyping that apply to cardiovascular genetic studies in general, and to studies into redox-related genetic markers of cardiovascular diseases in particular.

Human studies into the role of oxidative stress in the development of cardiovascular diseases commonly rely on single markers or a combination of markers of oxidative stress, rather than on direct assessment of ROS in vascular or cardiac tissue [8, 15]. These markers of oxidative stress integrate the effects of a number of pro-oxidants and antioxidants and are therefore not particularly well suited to specifically examine genotype-phenotype associations. More direct and precise measurements are complicated and not always possible in human studies. However, some of the most relevant studies into redox-related genetic markers of cardiovascular diseases are based on such precise phenotyping, including the assessment of superoxide production in vascular tissue [10, 16–18] and circulating blood cells [19, 20].

Cardiovascular diseases are continuous traits. These quantitative traits are transformed into qualitative traits for clinical purposes and for genetic studies. Cut-offs that are to some extent arbitrary are being used to define cases and controls in genetic case control studies and family-based studies. In terms of the power of a study, this is problematic if phenotypes of cases and controls are close together on a quantitative scale. It is less problematic if cases and controls are sampled from the extremes on either side of the threshold [1]. This concept has been successfully utilised in the British Genetics of Hypertension (BRIGHT) study where hypertensive cases were derived from the top 5% of the blood pressure distribution in the UK and therefore had clearly different blood pressures compared to control subjects [21]. Genetic studies into continuous cardiovascular traits are possible especially when the trait is accurately measurable. This is, for example, the case with lipid or blood glucose levels, but is already more difficult for blood pressure. The extent of coronary artery disease or carotid plaques can only be described by semiquantitative scores, which are important clinical tools but which have not been validated as high fidelity phenotypes in genetic studies. As mentioned above, there is considerable overlap between cardiovascular disease entities, and accordingly substantial overlap in the genetic variants contributing to these diseases can be expected.

The quality of phenotyping and standardisation of phenotyping especially in multicentre studies has a major impact on the outcome of genotype-phenotype association studies. There is a fine balance between an option of less accurate phenotyping of larger cohorts, an alternative strategy of precise phenotyping of smaller cohorts, and a combination of both approaches. Most importantly, the same accuracy of phenotyping that is used to characterise cases should be applied to controls in order to avoid an unexpected "caseness" of controls [1].

Given the technical difficulties to assess vascular ROS precisely in clinical studies, a pragmatic strategy is to examine the association of variants of redox-related genes with clinically measurable cardiovascular phenotypes such as hypertension and coronary artery disease in large cohorts. These studies have to be powered to detect relatively small effect sizes. Once the role of redox-related genes in a particular cardiovascular trait is established, it is very likely that the functional mechanisms can be shown in specifically designed experiments. It is a general rule that a genetic effect that stands out despite the complexity of clinical cardiovascular traits is likely to be seen in earlier and intermediate cardiovascular phenotypes where its precise pathophysiology can be studied [22].

## 10.3 Strategies to Unravel the Genetics of Redox-Related Diseases

### *10.3.1 Candidate Gene Approach*

There are a number of strategies for the discovery of disease-related genes in cardiovascular diseases. The classic approach is to select plausible candidate genes based on knowledge of their role in pathophysiology. Whilst positive findings from such studies support the role of particular genes in the disease process, negative findings do not necessarily exclude such a role. Historically, candidate gene studies were the only affordable and technically feasible genetic studies and were commonly performed in study cohorts that would nowadays not fulfil standards of genetic association studies [23]. However, candidate gene studies are still one of the most important genetic tools when specific pathways such as redox-related mechanisms are to be examined. We will discuss two examples in more detail, one of a gene involved in the production of ROS, and one of a gene involved in the defences against ROS.

#### 10.3.1.1 NADPH Oxidase

NADPH oxidase was originally described in phagocytes but similar NADPH oxidases are present in a variety of tissues including the vascular endothelium [24, 25]. NADPH oxidase, amongst uncoupled endothelial nitric oxide synthase, xanthine oxidase, lipoxygenases, cyclooxygenases, and the mitochondrial respiratory chain, is the major source of vascular superoxide [26–28]. The role of NADPH oxidase-derived superoxide in the pathogenesis of cardiovascular diseases is well established [25, 27, 28]. NADPH oxidases consist of homologues of $gp91^{phox}$ (Nox1 to 5), the $p22^{phox}$ subunit, and cytosolic subunits such as $p47^{phox}$, $p67^{phox}$, $p40^{phox}$, and/or the G-proteins Rac1 or Rac2 [25]. The $p22^{phox}$ subunit is present in NADPH oxidases containing Nox1 to 4 and is crucial for protein stability. The *CYBA* gene encoding human $p22^{phox}$ protein is located on chromosome 16q24. A large number of polymorphisms within this gene have been described, with a major focus on the C242T polymorphism, that leads to substitution of $His^{72}$ by tyrosine [29]. Other common polymorphisms include the A640G and –930$^{A/G}$ variants. The former is located in the 3′ untranslated region of *CYBA*, the latter is located in the promoter region. The role of genetic variants of *CYBA* has been reviewed in detail by San José et al. [25].

Generally it appears that the T-allele of C242T causes reduced superoxide generation from NADPH oxidases. Superoxide production from neutrophils and the neutrophil respiratory burst were found to be reduced in T-allele carriers [20]. Similar results were seen in mononuclear cells [30]. Interestingly, the effect of this genetic variant on mononuclear cell superoxide production is modulated by the presence or absence of hypertension. No difference across the genotypes was found in normotensive subjects, suggesting additional regulatory mechanisms that are present in normotension but lost in hypertension, thereby unmasking the polymorphism effect [30]. The crucial role of the p22$^{phox}$ subunit and its presence, in particular, in vascular and phagocytic NADPH oxidases, allows drawing conclusions from these studies in white blood cells to other tissues, including the vasculature. Indeed, studies confirmed that the T-allele is associated with reduced direct vascular superoxide production [10, 18].

Consequently, associations of the *CYBA* C242T variant with intermediate cardiovascular phenotypes that are closely related to vascular superoxide production have been studied. The T-allele has been found to be associated with improved endothelial function [31, 32], reduced vascular stiffness [10], and protection from hypertension [30] and coronary artery disease [33]. However, this picture is blurred by the presence of data which did not show significant effects of the C242T polymorphism on endothelial function [34, 35], or which even demonstrated opposite effects, e.g., that the T-allele favours development or progression of coronary artery disease [36, 37]. The recent genome-wide association study (GWAS) from the Wellcome Trust Case Control Consortium (WTCCC) did not find significant associations of markers in 16q24 with hypertension or coronary artery disease where the closest match was moderate evidence of association for a marker in 16q23 for coronary artery disease [38].

The conflicting data are likely attributable to the multifactorial nature of cardiovascular diseases and complex cardiovascular traits with contributions of several pathways to the disease phenotype. A small change in the gene encoding one member of one pathway is unlikely to lead to easily detectable changes in the complex phenotype, whereas earlier and immediately related phenotypes such as superoxide production may be affected more clearly. Furthermore, the *CYBA* C242T polymorphism has also been found to modulate the cardiovascular effects of risk factors such as hypercholesterolaemia [32] and to be associated with other cardiovascular diseases, including cerebrovascular disease [39] and renal disease [40, 41], again however with conflicting results. It is impossible in small cohort studies to dissect whether the effects of this genetic variant on coronary artery disease are direct or if they are indirect by primarily affecting the above cardiovascular risk factors or related cardiovascular traits. Current large-scale collaborative genetic studies that are adequately powered to detect even small effects will give definite answers.

The other polymorphisms of the *CYBA* gene are less well examined. The A640G polymorphism does not affect neutrophil superoxide production [20], and consequently there are conflicting data on an association of the A640G variant with coronary artery disease [33, 42]. The -930$^{A/G}$ polymorphism has been found to be associated with hypertension [19], although the immediate functional relevance of these variants, particularly on superoxide production, is less clear [33, 19]. Even fewer data are available for other polymorphisms such as -852$^{C/G}$, -536$^{C/T}$, and C549T (for review see [25]). However, these studies show that more than one genetic variant with the potential to affect a phenotype may exist within a gene and that these additional variants have to be taken into account individually and by examining haplotypes. In fact, functional interaction between the -930$^{A/G}$ and C242T variants on mononuclear cell NADPH oxidase activity has been demonstrated [19].

#### 10.3.1.2 Superoxide Dismutase

Superoxide dismutases (SODs) are the major defence mechanism against superoxides. In humans three isoenzymes are present. Intracellular Cu-Zn-SOD, mitochondrial Mn-SOD, and extracellular Cu-Zn-SOD are encoded by the *SOD*1, *SOD*2, and *SOD*3 genes located on chromosomes 21q22, 6q25, and 4p15, respectively. The presence not only of isoforms but different distribution of SODs within the cellular, subcellular, and extracellular space adds another level of complexity to the genetics of redox-related enzymes. For example, it has been shown that in vivo gene transfer of extracellular Cu-Zn-SOD but not of intracellular Cu-Zn-SOD or mitochondrial Mn-SOD improves endothelial function in the stroke-prone spontaneously hypertensive rat (SHRSP) [43, 44].

In African-Americans, a genome scan using microsatellite markers showed linkage with hypertension at 6q24 in close proximity to the *SOD*2 gene [45], although this region did not feature prominently for hypertension in the WTCCC study [38]. A substitution of Ala$^{16}$ by valine (A16V) in the signal sequence of Mn-SOD has been functionally characterised. The amino acid exchange in the premature protein affects mitochondrial targeting of the enzyme [46]. The alanine variant has been found to increase Mn-SOD activity and to protect macrophages against oxLDL-induced apoptosis and the risk of coronary artery disease and myocardial infarction [47]. This polymorphism has also been found to be associated with the degree of carotid atherosclerosis in a cohort of hypertensive and normotensive subjects, explaining however only 1.3% of the overall variation [48]. The sample size was suitable for performing additional analyses into sexual dimorphism and effects of the polymorphisms on another cardiovascular risk factor, LDL-cholesterol. A significant interaction between LDL-cholesterol and the A16V variant was found in women. The results of this study should not be overinterpreted, but it shows features of the genetics of redox-related markers of cardiovascular diseases that have already been highlighted above, namely overall minor contribution to phenotypic variability and interaction with other risk factors. Such results may guide us in the design and analysis of other datasets where sexual dimorphism and environmental factors should not only be addressed in post hoc analysis but be considered a priori as part of the study design and power calculations.

A number of other polymorphisms have been described in the *SOD* genes, including a variant of *SOD2* in intron 3 which leads to alternative splicing of exon 4 [49]. This polymorphism is an interesting candidate for further research into its effects on cardiovascular traits, but generalisation of results will be limited by significantly different minor allele frequencies between subjects with African-American and Caucasian backgrounds [49].

Polymorphisms of *SOD3* have also been found to be associated with hypertension [50] and coronary atherosclerosis [51], but these data still need to be confirmed in independent cohorts. A substitution of $Arg^{213}$ by glycine (R213G) in the heparin-binding domain of extracellular SOD leads to reduced affinity to heparin without affecting enzymatic activity and has been found to be associated with increased plasma concentration of the enzyme [52, 53]. The polymorphism affects the risk of ischaemic heart disease [54] and cardiovascular death in patients with end-stage renal failure and diabetes [55]. In a study examining the effect of R213G in spontaneously hypertensive rats (SHR) and Wistar-Kyoto (WKY) rats using a recombinant adenovirus vector, the R213G variant had no significant effect on vascular (aortic) superoxide production and consequently no effect on vasomotor function and blood pressure [56].

#### 10.3.1.3 Other Redox-Related Candidate Genes

Glutathione peroxidases are a family of enzymes that reduce the levels of lipid hydroperoxides and free hydrogen peroxide and thereby contribute crucially to the defences against oxidative damage. Glutathione peroxidase 1 is the most abundant isoenzyme in mammalian cells and its activity has been shown to predict cardiovascular events in patients with coronary artery disease [57]. The gene encoding human glutathione peroxidase 1 (*GPX1*) is located on chromosome 3p21. Polymorphisms of *GPX1* have been examined in the context of cardiovascular diseases and were found to be associated with increased carotid intima-media thickness [58], risk [59, 60] and severity [61] of coronary artery disease, and restenosis after coronary stenting [62]. A *GPX1* variant has also been shown to protect against thoracic aortic aneurysms in patients with hypertension [63], whereas no association was found between a *GPX1* variant and erythrocyte glutathione peroxidase activity and stroke [64]. These studies generate interesting hypotheses which have to be confirmed in future investigations before any definite conclusions can be drawn.

Some data exist on polymorphisms of the gene encoding catalase (*CAT*; located on chromosome 11p13) and their association with cardiovascular disease. Jiang et al. [65] examined promoter polymorphisms of *CAT* and found an association between a $-844^{C/T}$ variant and hypertension. Recently, the $-262^{C/T}$ variant of *CAT* (and also the C242T variant of *CYBA*) has been found to be associated with adverse outcomes in patients with acute renal failure [41]. The latter study is not immediately related to cardiovascular diseases, but shows another principle of the genetics of oxidative stress-related genes. Some of these variants may be only functionally relevant in a scenario of additional stressors for the oxidant-antioxidant balance. In other words, in situations that are characterised by particularly high levels of ROS, such as acute renal failure, cancer, or cytotoxic drug therapy, subtle changes in redox-related genes may have more pronounced effects than in conditions characterised by low oxidative stress.

There is also evidence for an antiatherogenic role of paraoxonases, most likely because of attenuation of oxidative modification of lipoproteins. Studies of associations between genetic variants of paraoxonase and atherosclerosis have been reviewed by Ng et al. [66] and will not be discussed in more detail.

### *10.3.2 Rodent Models and Translational Approaches*

Rodent models are invaluable tools to examine the pathophysiology of human cardiovascular diseases. Inbred rat models are of particular interest, as they mimic the complexity of human cardiovascular diseases. Recent advances in molecular and genetic tools and the availability of sequence data in the rat have led to an acceleration of gene discovery and characterisation of molecular pathways affected by genetic variants [67, 68]. The SHRSP is a well-characterised experimental model for human essential hypertension and displays multiple gene-gene and gene-environment interactions [69]. The SHRSP is characterised by increased vascular superoxide production and associated phenotypes including endothelial dysfunction, cardiac hypertrophy, and stroke [70–73]. A strategy to discover cardiovascular disease-related genes in rodent models such as the SHRSP and then to translate these findings into the human situation appears most promising, although only limited data have been produced so far. We will describe an example of this strategy as a model for future research.

Genome-wide linkage studies have successfully localised quantitative trait loci (QTL) for blood pressure regulation in the SHRSP [69]. Two blood pressure QTL have been mapped to rat chromosome 2 [74] and were subject to selective breeding of congenic animals by introducing various segments of chromosome 2 from the WKY rat into the SHRSP rat background and vice versa [75, 76]. Introgression of the region of chromosome 2 of the WKY rat into the SHRSP rat background led to a significant reduction in blood pressure [76]. Gene expression profiling by microarray technique in the parental strains and in SP.WKYGla2a and SP.WKYGla2c* congenic strains identified the glutathione S-transferase mu type 1 gene (*Gstm1*) as a positional candidate for hypertension in the SHRSP [76]. This gene is also a functional candidate for hypertension, as glutathione S-transferases are involved in the detoxification of ROS [77]. Subsequent studies have confirmed the gene expression data [78, 79] and demonstrated reduced Gstm1 protein expression in the renal tissue, specifically in the collecting ducts, of the SHRSP [80].

The role of variants in genes encoding human glutathione S-transferases have been examined extensively in human cancer for three reasons. First, as outlined above, cancer is a condition characterised by increased oxidative stress, and it is more likely to detect a functional role of a redox-related gene in this context. Second, the chemotherapy of cancer leads to additional oxidative stress, and the role of glutathione S-transferases in chemotherapy-related oxidative damage and also in detoxification of these drugs is of immediate clinical relevance. Third, the deletion genotypes of two members of the *GST* gene family (*GSTM*1 and *GSTT*1) are relatively easy to genotype and are due to the complete absence of a protein likely to have greater effects on a phenotype than subtle changes in single nucleotides. Based on these considerations, and with the characteristics of many patients with cancer enrolled in clinical studies, clinical databases, and national registers, a wealth of studies into *GST* polymorphisms and cancer are available. It generally seems that deletions of *GSTM*1 and/or *GSTT*1, but also other polymorphisms of *GST* genes, are associated with increased susceptibility to cancer, especially in the presence of other carcinogens such as cigarette smoking [81–88]. Similar results have also been found for other conditions with increased oxidative and inflammatory stress, including asthma and allergies [89, 90].

*GSTM* genes were subject to association studies in human cardiovascular diseases as well, but the results are far less consistent compared to studies in cancer. For example, there has been evidence of increased risk of hypertension in carriers of the *GSTM*1 deletion if they also had the *GSTA*1 *B allele [91]; whereas another study reports a protective role of the *GSTT*1 deletion and no effect of the *GSTM*1 deletion on hypertension status [92]. In coronary artery disease, there are also conflicting results as to whether the presence or the absence of a *GST* gene is associated with increased risk [93, 94]. This inconsistency indicates that genetic association studies have to adhere to strict quality criteria. There is certainly no doubt about the quality of the studies mentioned above; but with today's knowledge of genetic statistics they may be considered underpowered and, in particular, characterised by the absence of replication in an independent population. Such quality criteria for genetic association studies have, for example, been published by the National Cancer Institute and the National Human Genome Research Institute Working Group on Replication in Association Studies [23]. Replication and sufficient power are cornerstones of these criteria, as well as the demonstration of immediate functional consequences of a polymorphism, along with data on gene and protein expression and cross-species analysis.

We have therefore performed a definitive association study of *GSTM* genes in hypertension [95]. Resequencing of *GSTM* genes, genotyping for the *GSTM*1 deletion, and studies in three independent cohorts using both case control and family-based designs have excluded an association of *GSTM* genes with human essential hypertension. However, even this study was not able to adjust for all environmental factors that are able to affect the genotype-phenotype association and may not be the last word. It still provides an example and template for future studies of candidate genes for cardiovascular diseases.

### *10.3.3 Genome-Wide Association Studies*

In contrast to candidate gene studies, GWAS are in the first instance hypothesis-free and examine the association of a large number of genetic variants with a specific phenotype. Until recently, genome-wide coverage was achieved by genotyping of a relatively small number of genetic markers (for example, about 400 in the BRIGHT study [21]), which are in turn highly polymorphic. With advances in genotyping technology, modern GWAS genotype for a large number of genetic markers (for example, about 500,000 in the WTCCC study [38]), which are in turn less polymorphic (usually one or the other allele). The better coverage of the genome by the larger number of genetic markers in modern GWAS is at first glance an advantage, but even in this era, many of the markers are not located within genes or are of unknown functional relevance, with subsequent challenges to explain even the most significant results. This includes, for example, the locus on chromosome 9, which has been repeatedly found to be associated with coronary artery disease [38, 96]. GWAS rely on large numbers of cases and controls or large numbers of families or parts of families, and can only detect associations between phenotypes and genetic variants with a certain effect size. High quality phenotyping improves the signal-to-noise ratio, and one has to be careful not to forget about the value of smaller functional genetic studies in well phenotyped cohorts as opposed to large-scale collaborative GWAS in cohorts phenotyped in multiple centres according to different protocols. Whereas positive findings in GWAS are certainly pointing to the most promising candidate regions in the human genome for a given disease, a negative finding does not exclude the functional relevance of a gene in a particular genomic region. Such genes can be implicated in the pathogenesis of cardiovascular diseases at heritabilities below the limit of detection of a given GWAS and might be detectable in association studies of candidate genes.

Without questioning the enormous implications of recent GWAS in the discovery of novel genetic markers of cardiovascular disease, one should not be surprised that there is no immediate evidence of a gene directly related to the redox state that has been identified by such studies into cardiovascular phenotypes. Nevertheless, it is very likely that we will see such results in the near future. First, the increasing sample sizes of collaborative genotyping enterprises will lead to the detection of genes with smaller effect sizes, which will apply to the majority of cardiovascular disease-related genes, including genes related to redox pathways. For example, in the WTCCC study [38], approximately 2,000 patients with coronary artery disease and 2,000 patients with hypertension were genotyped at approximately 500,000 loci, leading to a power (averaged across single nucleotide polymorphisms with minor allele frequencies above 5%) of 43% for alleles with a relative risk of 1.3, and of 80% for a relative risk of 1.5, for a *P*-value threshold of $5 \times 10^{-7}$. These effect sizes may be well above the expected relative risks of some genes involved in the pathogenesis of cardiovascular diseases. Second, a number of large-scale genotyping projects are currently under way which will benefit from excellent phenotyping, not least also with regards to markers of oxidative stress and vascular function. One of these projects is a European collaborative study into the genetics of hypertension and hypertension-related diseases (InGenious HyperCare; www.hypercare.eu).

### 10.3.4 Mitochondria

The mitochondrial electron transport system is formed by protein complexes I–IV and two individual molecules, coenzyme Q and cytochrome c. This system creates a gradient of protons across the inner mitochondrial membrane, which is then used by the ATPase complex to generate ATP. Mitochondria are one of the major sources of ROS and the role of mitochondria in the pathogenesis of cardiovascular disease has been reviewed in detail [27, 97, 98]. For example, mitochondria-derived oxidative stress is present in human essential hypertension and can be reduced by antihypertensive treatment [99]. In the SHRSP, a mitochondria-targeted antioxidant reduces blood pressure and hypertension-related target organ damage [100].

Mitochondria contain their own DNA (mtDNA). Human mtDNA is a circular molecule containing 37 genes encoding for 13 proteins of the electron transport system, and 2 rRNA and 22 tRNA molecules [101]. However, the majority of mitochondrial proteins are encoded by the nuclear genome. Therefore, this major player in cellular oxidative stress is characterised by an assembly of proteins using genetic code from two different DNAs. Traditionally there is a focus on nuclear genomics, but recently large-scale genotyping of mtDNA variants has become feasible. Also, statistical methods to examine family-based studies for evidence of mitochondria-based transmission are available [102].

The mitochondrial genome is characterised by substantial differences from the nuclear genome not only with regard to DNA structure. Embryonic mitochondria derive virtually exclusively from the oocyte, i.e., from the mother, thereby leading to a non-Mendelian but maternal pattern of transmission. The mutation rate of mtDNA is significantly higher than that of nuclear DNA because of the lack of histones, less developed repair mechanisms, and the immediate contact of mtDNA with mitochondria-derived ROS. The high mutation rate and the large number of mitochondria within the cells can lead to genetic heterogeneity between mitochondria in any given cell or organism; and even major genetic defects in the genomes of some mitochondria can be compensated by the presence of other mitochondria that function normally.

There are some examples showing that genes that encode mitochondrial proteins are candidate genes for cardiovascular diseases. $p66^{Shc}$ is a growth factor adaptor protein that is involved in mitochondrial ROS production and mediation of ROS-induced apoptosis [103]. It is one of three splice products of the *SHC*1 gene product. *SHC*1 is located on human chromosome 1 [104]. Targeted $p66^{Shc}$ mutations in the mouse are associated with reduced production of ROS and prolonged lifespan [105]. $p66^{Shc}$ –/– mice also exhibit increased lifespan and reduced oxidative stress, and are protected against age-related endothelial dysfunction [105, 106]. A number of studies suggest associations between *SHC*1 gene variants and longevity [107, 108], diabetes [109], and premature cardiovascular disease in humans [110].

The potential of an mtDNA variant directly affecting cardiovascular phenotypes is illustrated by the syndrome of mitochondrial myopathy, encephalopathy, lactic acidosis, and stroke-like episodes (MELAS). It is caused by the A3243G variant of the mitochondrial tRNALeu(UUR) gene. However, this variant has also been associated with other mitochondrial disorders, including maternally inherited diabetes [111]. Moreover, a mutation in the mitochondrial tRNALys gene that has been found to be related to diabetes and myoclonus epilepsy with ragged red fibres (MERRF) (A8296G) has also been found to be associated with MELAS [112]. MERRF in turn has been found to be associated with another mtDNA variant in the tRNALys gene (A8344G) [113]. These gene products are not immediately related to redox processes within the mitochondria, but they affect mitochondrial function in general. Detailed studies of functional consequences on mitochondrial and cellular oxidative stress will be challenging because of the rarity of mitochondrial disorders.

Finally, dilated cardiomyopathy has been found to be associated with variants in mtDNA. Remes et al. [114] described the presence of mtDNA deletions in patients with idiopathic dilated cardiomyopathy. A number of other mtDNA mutations have been found to be associated with cardiomyopathy [115–118]. It will be interesting to characterise mitochondrial ROS production in patients with mitochondrial disorders, but also to examine whether more subtle changes in mitochondrial function in combination with other genetic and/or environmental factors are also involved in the pathogenesis of cardiovascular diseases.

## 10.4 Interactions Between Genes and Environment

### *10.4.1 Antioxidant Therapy*

Antioxidant therapy is a potential confounder of studies into the genetics of redox-related markers, as it may affect cardiovascular phenotypes such as endothelial dysfunction. As with any environmental factor, careful assessment and phenotyping of study participants is required, and should also include intake of antioxidants and dietary habits to facilitate statistical adjustment for these factors. However, the lack of immediately detectable effects of antioxidants on cardiovascular outcomes [119, 120] may suggest that antioxidant use is a less important confounder in studies of more advanced phenotypes such as coronary artery disease.

Independent of genetic effects, there is also "inheritance" of maternal characteristics through a process of in utero programming [102, 121]. In population studies into inheritance of cardiovascular and other phenotypes, it is difficult to differentiate in utero programming from maternal transmission of mitochondrial disorders. Genotyping of mtDNA can help to resolve any uncertainties. We are not aware of any immediate consequences of in utero programming on current studies into the genetics of redox-related markers. However, it has been reported that in cholesterol-fed New Zealand white rabbits, maternal antioxidant treatment reduces atherosclerosis in offspring [122]. This is an example of a complex environment-phenotype interaction that again demonstrates the need for careful assessment of environmental factors which could even include maternal antioxidant intake.

### *10.4.2 Smoking*

Cigarette smoking is associated with the production of ROS and may significantly affect studies of oxidative stress-related cardiovascular phenotypes. It is possible that an additional stressor such as cigarette smoking enhances the otherwise rather subtle effects of genetic variants of redox-related genes. This has been demonstrated in studies of the association of genetic variants of *GSTM* genes with cancer, in which cigarette smoking has been found to interact with the *GSTM* genotype [81, 87]. Similar results have been obtained in studies of *GST* genes in cardiovascular diseases [123, 124]. Particularly, studies into the genetics of redox-related markers should therefore strictly control and adjust for cigarette smoking.

The same applies to other environmental stressors such as air pollution. Interactions between air pollution, and specifically diesel exhaust, and *GSTM* genotype have been shown in studies of asthma [125], and there is evidence that the genotype-phenotype relationship of *GSTM*1 and cardiovascular traits is also affected by particulate air pollution [126].

### *10.4.3 Medication and Pharmacogenetics*

Redox-related genes are attractive candidates for pharmacogenetic studies. For example, *GSTM*1 and *GSTT*1 genotypes have been found to be associated with responses to chemotherapy in cancer [127]. The *GST* gene family shows that enzyme systems involved in detoxification and metabolism of drugs can be involved in production of or defences against oxidative stress. Genetic variants of these redox-related genes may lead to different drug levels and thereby affect the efficacy of and adverse reactions to drugs. Changes in the production of ROS are then only one effect of a genetic variant; and the main effect of a genetic variant can be different from that on ROS levels.

A wealth of datasets from clinical trials is available for pharmacogenetic studies, and promising results have been published recently. Of particular interest in this context are studies examining the effects of variants of redox-related genes on drug response. For example, genetic polymorphisms of *CYP2C*19 have been found to modulate the pharmacokinetics and pharmacodynamics of clopidogrel [128, 129]. Very recently, two studies demonstrated an association of *CYP2C*19 loss-of-function genotype with response to clopidogrel following myocardial infarction [130, 131]. CYP2C19 has arachidonate monooxygenase activity [132] and is involved in the oxidative activation of clopidogrel. This pathway is an interesting example of how a redox-related enzyme, by affecting drug metabolism, is related to the outcome of cardiovascular disease.

More data are expected to be derived from large-scale pharmacogenetic studies, particularly from secondary analysis of GWAS. Response to pharmacotherapy can be used as a qualitative and quantitative trait, and this strategy has been successfully employed in the analysis of responses to hypertensive therapy in the BRIGHT study [133]. Other GWAS offer opportunities to examine the pharmacogenomics of drug response, provided that detailed data on drug therapy are available for the study participants. Another useful example is the recent discovery that common variants in *SLCO1B1* encoding the organic anion-transporting polypeptide are strongly associated with an increased risk of statin-induced myopathy [134]. It will be exciting to see if these strategies demonstrate significant effects of redox-related genes.

## 10.5 Regulation of Transcription

Finally, it should be noted that apart from genetic mutations in redox-related genes causing changes in ROS levels, there is also an opposite effect of changes in ROS leading to changes in the transcription of genes. Cellular oxidative stress regulates, directly or indirectly, the structure, subcellular localisation, and activity of transcription factors [135]. For example, the expression of *GSTP1* and *CYP3A4* in a human erythroleukemia cell line has been found to be upregulated by hydrogen peroxide [136]. These oxidative stress-related changes in enzyme systems that are critically involved in drug metabolism may then affect responses to cardiovascular drugs and ultimately affect cardiovascular outcomes.

Another feature in this area of research has been the recent discovery of a new functional class of noncoding small RNA molecules (microRNAs) that act as translational inhibitors or target the specific mRNA molecule for degradation [137, 138]. miRNAs target multiple mRNAs [139] and may provide an integrated view of transcriptional regulation. This suggests the potential for miRNA expression profiling in other complex diseases, including cardiovascular diseases [140].

## 10.6 Summary and Conclusions

There are very good reasons to hypothesise that redox-related genes are involved in the pathogenesis of cardiovascular diseases. This hypothesis is supported by a plethora of functional genetic studies using candidate gene approaches, but is still awaiting further support from large-scale genome scans and independent replication studies. The latter are not specific to redox-related genes, but in fact apply to virtually all "classic" cardiovascular candidate genes, including members of the renin-angiotensin-aldosterone system and adrenergic receptor genes, which also did not stand out significantly in recent GWAS. We are confident that further improvements in phenotyping, but also in the study design of GWAS, allowing detection of smaller effect sizes and advances in bioinformatics, will lead to confirmation of the importance of many redox-related genes in the pathogenesis of cardiovascular diseases.

Nevertheless, more comprehensive answers can be expected from approaches that go beyond genetic and genomic studies. Genetic analysis of pathways of multiple pro-oxidant and antioxidant enzyme systems across several pathogenetically related diseases, complemented by transcriptomic, proteomic, and metabolomic strategies, will explain the role of redox-related genes and their involvement in pathophysiological pathways.

## References


1. Padmanabhan S, Melander O, Hastie C, Menni C, Delles C, Connell JM, Dominiczak AF (2008) Hypertension and genome-wide association studies: combining high fidelity phenotyping and hypercontrols. J Hypertens 26:1275–1281
2. Staessen JA, Wang J, Bianchi G, Birkenhager WH (2003) Essential hypertension. Lancet 361:1629–1641
3. Mongeau JG, Biron P, Sing CF (1986) The influence of genetics and household environment upon the variability of normal blood pressure: the Montreal Adoption Survey. Clin Exp Hypertens A 8:653–660
4. Feinleib M, Garrison RJ, Fabsitz R, Christian JC, Hrubec Z, Borhani NO, Kannel WB, Rosenman R, Schwartz JT, Wagner JO (1977) The NHLBI twin study of cardiovascular disease risk factors: methodology and summary of results. Am J Epidemiol 106:284–285
5. Dzau V, Braunwald E (1991) Resolved and unresolved issues in the prevention and treatment of coronary artery disease: a workshop consensus statement. Am Heart J 121:1244–1263
6. Delles C, Moreno MU, Padmanabhan S, Graham D, McBride MW, Dominiczak AF (2007) Functional genomics of the oxidative stress pathway. Curr Hypertens Rev 3:156–165
7. Delles C, Dominiczak AF (2006) Vascular failure or sick vessel syndrome: The cardiovascular continuum is a useful concept for clinical research. J Hypertens 24:2147–2148
8. Red6n J, Oliva MR, Tormos C, Giner V, Chaves J, Iradi A, Sáez GT (2003) Antioxidant activities and oxidative stress byproducts in human hypertension. Hypertension 41: 1096–1101
9. Taddei S, Virdis A, Ghiadoni L, Magagna A, Salvetti A (1998) Vitamin C improves endothelium-dependent vasodilation by restoring nitric oxide activity in essential hypertension. Circulation 97:2222–2229
10. Delles C, Zimmerli LU, McGrane DJ, Koh-Tan CH, Pathi VL, McKay AJ, Steedman T, Dargie HJ, Hamilton CA, Dominiczak AF (2008) Vascular stiffness is related to superoxide generation in the vessel wall. J Hypertens 26:946–955
11. Wilkinson IB, Megson IL, MacCallum H, Sogo N, Cockcroft JR, Webb DJ (1999) Oral vitamin C reduces arterial stiffness and platelet aggregation in humans. J Cardiovasc Pharmacol 34:690–693
12. Heitzer T, Schlinzig T, Krohn K, Meinertz T, Münzel T (2001) Endothelial dysfunction, oxidative stress, and risk of cardiovascular events in patients with coronary artery disease. Circulation 104:2673–2678
13. Kaschina E, Unger T (2003) Angiotensin AT1/AT2 receptors: regulation, signalling and function. Blood Press 12:70–88
14. McBride MW, Graham D, Delles C, Dominiczak AF (2006) Functional genomics in hypertension. Curr Opin Nephrol Hypertens 15:145–151
15. Himmelfarb J, Stenvinkel P, Ikizler TA, Hakim RM (2002) The elephant in uremia: oxidant stress as a unifying concept of cardiovascular disease in uremia. Kidney Int 62:1524–1538

16. Antoniades C, Shirodaria C, Van Assche T, Cunnington C, Tegeder I, Lötsch J, Guzik TJ, Leeson P, Diesch J, Tousoulis D, Stefanadis C, Costigan M, Woolf CJ, Alp NJ, Channon KM (2008) GCH1 haplotype determines vascular and plasma biopterin availability in coronary artery disease effects on vascular superoxide production and endothelial function. J Am Coll Cardiol 52:158–165
17. Guzik TJ, Sadowski J, Guzik B, Jopek A, Kapelak B, Przybylowski P, Wierzbicki K, Korbut R, Harrison DG, Channon KM (2006) Coronary artery superoxide production and nox isoform expression in human coronary artery disease. Arterioscler Thromb Vasc Biol 26:333–339
18. Guzik TJ, West NE, Black E, McDonald D, Ratnatunga C, Pillai R, Channon KM (2000) Functional effect of the C242T polymorphism in the NAD(P)H oxidase p22phox gene on vascular superoxide production in atherosclerosis. Circulation 102:1744–7174
19. San José G, Moreno MU, Oliván S, Beloqui O, Fortuño A, Díez J, Zalba G (2004) Functional effect of the p22phox -930A/G polymorphism on p22phox expression and NADPH oxidase activity in hypertension. Hypertension 44:163–169
20. Wyche KE, Wang SS, Griendling KK, Dikalov SI, Austin H, Rao S, Fink B, Harrison DG, Zafari AM (2004) C242T CYBA polymorphism of the NADPH oxidase is associated with reduced respiratory burst in human neutrophils. Hypertension 43:1246–1251
21. Caulfield M, Munroe P, Pembroke J, Samani N, Dominiczak A, Brown M, Benjamin N, Webster J, Ratcliffe P, O'Shea S, Papp J, Taylor E, Dobson R, Knight J, Newhouse S, Hooper J, Lee W, Brain N, Clayton D, Lathrop GM, Farrall M, Connell J; MRC (2003) British Genetics of Hypertension Study. Genome-wide mapping of human loci for essential hypertension. Lancet 361:2118–2123
22. Bianchi G (2005) Genetic variations of tubular sodium reabsorption leading to 'primary' hypertension: from gene polymorphism to clinical symptoms. Am J Physiol Regul Integr Comp Physiol 289:R1536–R1549
23. NCI-NHGRI Working Group on Replication in Association Studies (2007) Replicating genotype-phenotype associations. Nature 447:655–660
24. Bokoch GM, Knaus UG (2003) NADPH oxidases: not just for leukocytes anymore! Trends Biochem Sci 28:502–508
25. San José G, Fortuño A, Beloqui O, Díez J, Zalba G (2008) NADPH oxidase CYBA polymorphisms, oxidative stress and cardiovascular diseases. Clin Sci 114:173–182
26. Hamilton CA, Brosnan MJ, Al-Benna S, Berg G, Dominiczak AF (2002) NAD(P)H oxidase inhibition improves endothelial function in rat and human blood vessels. Hypertension 40:755–762
27. Hamilton CA, Miller WH, Al-Benna S, Brosnan MJ, Drummond RD, McBride MW, Dominiczak AF (2004) Strategies to reduce oxidative stress in cardiovascular disease. Clin Sci 106:219–234
28. Mueller CF, Laude K, McNally JS, Harrison DG (2005) ATVB in focus: redox mechanisms in blood vessels. Arterioscler Thromb Vasc Biol 25:274–278
29. Dinauer MC, Pierce EA, Bruns GA, Curnutte JT, Orkin SH (1990) Human neutrophil cytochrome b light chain (p22-phox). Gene structure, chromosomal location, and mutations in cytochrome-negative autosomal recessive chronic granulomatous disease. J Clin Invest 86:1729–1737
30. Moreno MU, San José G, Fortuño A, Beloqui O, Díez J, Zalba G (2006) The C242T CYBA polymorphism of NADPH oxidase is associated with essential hypertension. J Hypertension 24:1299–1306
31. Schächinger V, Britten MB, Dimmeler S, Zeiher AM (2001) NADH/NADPH oxidase p22 phox gene polymorphism is associated with improved coronary endothelial vasodilator function. Eur Heart J 22:96–101
32. Shimokata K, Yamada Y, Kondo T, Ichihara S, Izawa H, Nagata K, Murohara T, Ohno M, Yokota M (2004) Association of gene polymorphisms with coronary artery disease in individuals with or without nonfamilial hypercholesterolemia. Atherosclerosis 172:167–173

33. Inoue N, Kawashima S, Kanazawa K, Yamada S, Akita H, Yokoyama M (1998) Polymorphism of the NADH/NADPH oxidase p22 phox gene in patients with coronary artery disease. Circulation 97:135–137
34. Schneider MP, Hilgers KF, Huang Y, Delles C, John S, Oehmer S, Schmieder RE (2003) The C242T p22phox polymorphism and endothelium-dependent vasodilation in subjects with hypercholesterolaemia. Clin Sci 105:97–103
35. Li A, Prasad A, Mincemoyer R, Satorius C, Epstein N, Finkel T, Quyyumi AA (1999) Relationship of the C242T p22phox gene polymorphism to angiographic coronary artery disease and endothelial function. Am J Med Genet 86:57–61
36. Cahilly C, Ballantyne CM, Lim DS, Gotto A, Marian AJ (2000) A variant of p22(phox), involved in generation of reactive oxygen species in the vessel wall, is associated with progression of coronary atherosclerosis. Circ Res 86: 391–395
37. Cai H, Duarte N, Wilcken DE, Wang XL (1999) NADH/NADPH oxidase p22 phox C242T polymorphism and coronary artery disease in the Australian population. Eur J Clin Invest 29:744–874
38. Wellcome Trust Case Control Consortium (2007) Genome-wide association study of 14,000 cases of seven common diseases and 3,000 shared controls. Nature 447: 661–678
39. Ito D, Murata M, Watanabe K, Yoshida T, Saito I, Tanahashi N, Fukuuchi Y (2000) C242T polymorphism of NADPH oxidase p22 PHOX gene and ischemic cerebrovascular disease in the Japanese population. Stroke 31:936–939
40. Matsunaga-Irie S, Maruyama T, Yamamoto Y, Motohashi Y, Hirose H, Shimada A, Murata M, Saruta T (2004) Relation between development of nephropathy and the p22phox C242T and receptor for advanced glycation end product G1704T gene polymorphisms in type 2 diabetic patients. Diabetes Care 27:303–307
41. Perianayagam MC, Liangos O, Kolyada AY, Wald R, MacKinnon RW, Li L, Rao M, Balakrishnan VS, Bonventre JV, Pereira BJ, Jaber BL (2007) NADPH oxidase p22phox and catalase gene variants are associated with biomarkers of oxidative stress and adverse outcomes in acute renal failure. J Am Soc Nephrol 18:255–263
42. Gardemann A, Mages P, Katz N, Tillmanns H, Haberbosch W (1999) The p22 phox A640G gene polymorphism but not the C242T gene variation is associated with coronary heart disease in younger individuals. Atherosclerosis 145:315–323
43. Fennell JP, Brosnan MJ, Frater AJ, Hamilton CA, Alexander MY, Nicklin SA, Heistad DD, Baker AH, Dominiczak AF (2002) Adenovirus-mediated overexpression of extracellular superoxide dismutase improves endothelial dysfunction in a rat model of hypertension. Gene Ther 9:110–117
44. Alexander MY, Brosnan MJ, Hamilton CA, Fennell JP, Beattie EC, Jardine E, Heistad DD, Dominiczak AF (2000) Gene transfer of endothelial nitric oxide synthase but not Cu/Zn superoxide dismutase restores nitric oxide availability in the SHRSP. Cardiovasc Res 47:609–617
45. Zhu X, Luke A, Cooper RS, Quertermous T, Hanis C, Mosley T, Gu CC, Tang H, Rao DC, Risch N, Weder A (2005) Admixture mapping for hypertension loci with genome-scan markers. Nat Genet 37:177–181
46. Shimoda-Matsubayashi S, Matsumine H, Kobayashi T, Nakagawa-Hattori Y, Shimizu Y, Mizuno Y (1996) Structural dimorphism in the mitochondrial targeting sequence in the human manganese superoxide dismutase gene. A predictive evidence for conformational change to influence mitochondrial transport and a study of allelic association in Parkinson's disease. Biochem Biophys Res Commun 226:561–565
47. Fujimoto H, Taguchi J, Imai Y, Ayabe S, Hashimoto H, Kobayashi H, Ogasawara K, Aizawa T, Yamakado M, Nagai R, Ohno M (2008) Manganese superoxide dismutase polymorphism affects the oxidized low-density lipoprotein-induced apoptosis of macrophages and coronary artery disease. Eur Heart J 29:1267–1274

48. Kakko S, Päivänsalo M, Koistinen P, Kesäniemi YA, Kinnula VL, Savolainen MJ (2003) The signal sequence polymorphism of the MnSOD gene is associated with the degree of carotid atherosclerosis. Atherosclerosis 168:147–152
49. Shao J, Chen L, Marrs B, Lee L, Huang H, Manton KG, Martin GM, Oshima J (2007) SOD2 polymorphisms: unmasking the effect of polymorphism on splicing. BMC Med Genet 8:7
50. Naganuma T, Nakayama T, Sato N, Fu Z, Soma M, Aoi N, Usami R (2008) A haplotype-based case-control study examining human extracellular superoxide dismutase gene and essential hypertension. Hypertens Res 31:1533–1540
51. Samoila OC, Carter AM, Futers ST, Otiman G, Anghel A, Tamas L, Seclaman E (2008) Polymorphic variants of extracellular superoxide dismutase gene in a Romanian population with atheroma. Biochem Genet 46:634–643
52. Sandström J, Nilsson P, Karlsson K, Marklund SL (1994) Ten-fold increase in human plasma extracellular superoxide dismutase content caused by a mutation in heparin-binding domain. J Biol Chem 269:19163–19166
53. Adachi T, Yamada H, Yamada Y, Morihara N, Yamazaki N, Murakami T, Futenma A, Kato K, Hirano K (1996) Substitution of glycine for arginine-213 in extracellular-superoxide dismutase impairs affinity for heparin and endothelial cell surface. Biochem J 313: 235–239
54. Juul K, Tybjaerg-Hansen A, Marklund S, Heegaard NH, Steffensen R, Sillesen H, Jensen G, Nordestgaard BG (2004) Genetically reduced antioxidative protection and increased ischemic heart disease risk: The Copenhagen City Heart Study. Circulation 109:59–65
55. Yamada H, Yamada Y, Adachi T, Fukatsu A, Sakuma M, Futenma A, Kakumu S (2000) Protective role of extracellular superoxide dismutase in hemodialysis patients. Nephron 84:218–223
56. Chu Y, Alwahdani A, Iida S, Lund DD, Faraci FM, Heistad DD (2005) Vascular effects of the human extracellular superoxide dismutase R213G variant. Circulation 112:1047–1053
57. Blankenberg S, Rupprecht HJ, Bickel C, Torzewski M, Hafner G, Tiret L, Smieja M, Cambien F, Meyer J, Lackner KJ (2003) AtheroGene Investigators. Glutathione peroxidase 1 activity and cardiovascular events in patients with coronary artery disease. N Engl J Med 349:1605–1613
58. Hamanishi T, Furuta H, Kato H, Doi A, Tamai M, Shimomura H, Sakagashira S, Nishi M, Sasaki H, Sanke T, Nanjo K (2004) Functional variants in the glutathione peroxidase-1 (GPx-1) gene are associated with increased intima-media thickness of carotid arteries and risk of macrovascular diseases in Japanese type 2 diabetic patients. Diabetes 53: 2455–2460
59. Tang NP, Wang LS, Yang L, Gu HJ, Sun QM, Cong RH, Zhou B, Zhu HJ, Wang B (2008) Genetic variant in glutathione peroxidase 1 gene is associated with an increased risk of coronary artery disease in a Chinese population. Clin Chim Acta 395:89–93
60. Winter JP, Gong Y, Grant PJ, Wild CP (2003) Glutathione peroxidase 1 genotype is associated with an increased risk of coronary artery disease. Coron Artery Dis 14:149–153
61. Nemoto M, Nishimura R, Sasaki T, Hiki Y, Miyashita Y, Nishioka M, Fujimoto K, Sakuma T, Ohashi T, Fukuda K, Eto Y, Tajima N (2007) Genetic association of glutathione peroxidase-1 with coronary artery calcification in type 2 diabetes: a case control study with multi-slice computed tomography. Cardiovasc Diabetol 6:23
62. Oguri M, Kato K, Hibino T, Yokoi K, Segawa T, Matsuo H, Watanabe S, Nozawa Y, Murohara T, Yamada Y (2007) Genetic risk for restenosis after coronary stenting. Atherosclerosis 194:e172–e178
63. Kato K, Oguri M, Kato N, Hibino T, Yajima K, Yoshida T, Metoki N, Yoshida H, Satoh K, Watanabe S, Yokoi K, Murohara T, Yamada Y (2008) Assessment of genetic risk factors for thoracic aortic aneurysm in hypertensive patients. Am J Hypertens 21:1023–1027
64. Forsberg L, de Faire U, Marklund SL, Andersson PM, Stegmayr B, Morgenstern R (2000) Phenotype determination of a common Pro-Leu polymorphism in human glutathione peroxidase 1. Blood Cells Mol Dis 26:423–426

65. Jiang Z, Akey JM, Shi J, Xiong M, Wang Y, Shen Y, Xu X, Chen H, Wu H, Xiao J, Lu D, Huang W, Jin L (2001) A polymorphism in the promoter region of catalase is associated with blood pressure levels. Hum Genet 109:95–98
66. Ng CJ, Shih DM, Hama SY, Villa N, Navab M, Reddy ST (2005) The paraoxonase gene family and atherosclerosis. Free Radic Biol Med 38:153–163
67. Delles C, McBride MW, Padmanabhan S, Dominiczak AF (2008) The genetics of cardiovascular disease. Trends Endocrinol Metab 199:309–316
68. Aitman TJ, Critser JK, Cuppen E, Dominiczak A, Fernandez-Suarez XM, Flint J, Gauguier D, Geurts AM, Gould M, Harris PC, Holmdahl R, Hubner N, Izsvák Z, Jacob HJ, Kuramoto T, Kwitek AE, Marrone A, Mashimo T, Moreno C, Mullins J, Mullins L, Olsson T, Pravenec M, Riley L, Saar K, Serikawa T, Shull JD, Szpirer C, Twigger SN, Voigt B, Worley K (2008) Progress and prospects in rat genetics: a community view. Nat Genet 40: 516–522
69. Rapp JP (2000) Genetic analysis of inherited hypertension in the rat. Physiol Ver 80:135–172
70. Dominiczak AF, Devlin AM, Lee WK, Anderson NH, Bohr DF, Reid JL (1996) Vascular smooth muscle polyploidy and cardiac hypertrophy in genetic hypertension. Hypertension 27:752–759
71. Jeffs B, Clark JS, Anderson NH, Gratton J, Brosnan MJ, Gauguier D, Reid JL, Macrae IM, Dominiczak AF (1997) Sensitivity to cerebral ischaemic insult in a rat model of stroke is determined by a single genetic locus. Nat Genet 16:364–367
72. Kerr S, Brosnan MJ, McIntyre M, Reid JL, Dominiczak AF, Hamilton CA (1999) Superoxide anion production is increased in a model of genetic hypertension: role of the endothelium. Hypertension 33:1353–1358
73. Hamilton CA, Brosnan MJ, McIntyre M, Graham D, Dominiczak AF (2001) Superoxide excess in hypertension and aging: a common cause of endothelial dysfunction. Hypertension 37:529–534
74. Clark JS, Jeffs B, Davidson AO, Lee WK, Anderson NH, Bihoreau MT, Brosnan MJ, Devlin AM, Kelman AW, Lindpaintner K, Dominiczak AF (1996) Quantitative trait loci in genetically hypertensive rats. Possible sex specificity. Hypertension 28:898–906
75. Jeffs B, Negrin CD, Graham D, Clark JS, Anderson NH, Gauguier D, Dominiczak AF (2000) Applicability of a "speed" congenic strategy to dissect blood pressure quantitative trait loci on rat chromosome 2. Hypertension 35:179–187
76. McBride MW, Carr FJ, Graham D, Anderson NH, Clark JS, Lee WK, Charchar FJ, Brosnan MJ, Dominiczak AF (2003) Microarray analysis of rat chromosome 2 congenic strains. Hypertension 41:847–853
77. Hayes JD, Strange RC (2000) Glutathione S-transferase polymorphisms and their biological consequences. Pharmacology 61:154–166
78. Friese RS, Mahboubi P, Mahapatra NR, Mahata SK, Schork NJ, Schmid-Schönbein GW, O'Connor DT (2005) Common genetic mechanisms of blood pressure elevation in two independent rodent models of human essential hypertension. Am J Hypertens 18: 633–652
79. Hübner N, Wallace CA, Zimdahl H, Petretto E, Schulz H, Maciver F, Mueller M, Hummel O, Monti J, Zidek V, Musilova A, Kren V, Causton H, Game L, Born G, Schmidt S, Müller A, Cook SA, Kurtz TW, Whittaker J, Pravenec M, Aitman TJ (2005) Integrated transcriptional profiling and linkage analysis for identification of genes underlying disease. Nat Genet 37:243–253
80. McBride MW, Brosnan MJ, Mathers J, McLellan LI, Miller WH, Graham D, Hanlon N, Hamilton CA, Polke JM, Lee WK, Dominiczak AF (2005) Reduction of Gstm1 expression in the stroke-prone spontaneously hypertension rat contributes to increased oxidative stress. Hypertension 45:786–792
81. Gronau S, Koenig-Greger D, Jerg M, Riechelmann H (2003) Gene polymorphisms in detoxification enzymes as susceptibility factor for head and neck cancer? Otolaryngol Head Neck Surg 128:674–680

82. Canbay E, Dokmetas S, Canbay EI, Sen M, Bardakci F (2003) Higher glutathione transferase GSTM1 0/0 genotype frequency in young thyroid carcinoma patients. Curr Med Res Opin 19:102–106
83. Sierra-Torres CH, Au WW, Arrastia CD, Cajas-Salazar N, Robazetti SC, Payne DA, Tyring SK (2003) Polymorphisms for chemical metabolizing genes and risk for cervical neoplasia. Environ Mol Mutagen 41:69–76
84. Zhong S, Wyllie AH, Barnes D, Wolf CR, Spurr NK (1993) Relationship between the GSTM1 genetic polymorphism and susceptibility to bladder, breast and colon cancer. Carcinogenesis 14:1821–1824
85. Liloglou T, Walters M, Maloney P, Youngson J, Field JK (2002) A T2517C polymorphism in the GSTM4 gene is associated with risk of developing lung cancer. Lung Cancer 37:143–146
86. Balta G, Yuksek N, Ozyurek E, Ertem U, Hicsonmez G, Altay C, Gurgey A (2003) Characterization of MTHFR, GSTM1, GSTT1, GSTP1, and CYP1A1 genotypes in childhood acute leukemia. Am J Hematol 73:154–160
87. Bardakci F, Canbay E, Degerli N, Coban L, Canbay EI (2003) Relationship of tobacco smoking with GSTM1 gene polymorphism in laringeal cancer. J Cell Mol Med 7:307–312
88. Lee KM, Park SK, Kim SU, Doll MA, Yoo KY, Ahn SH, Noh DY, Hirvonen A, Hein DW, Kang D (2003) N-acetyltransferase (NAT1, NAT2) and glutathione S-transferase (GSTM1, GSTT1) polymorphisms in breast cancer. Cancer Lett 196:179–186
89. Saadat M, Ansari-Lari M (2007) Genetic polymorphism of glutathione S-transferase T1, M1 and asthma, a meta-analysis of the literature. Pak J Biol Sci 10:4183–4189
90. Gilliland FD, Li YF, Saxon A, Diaz-Sanchez D (2004) Effect of glutathione-S-transferase M1 and P1 genotypes on xenobiotic enhancement of allergic responses: randomised, placebo-controlled crossover study. Lancet 363:119–125
91. Oniki K, Hori M, Takata K, Yokoyama T, Mihara S, Marubayashi T, Nakagawa K (2008) Association between glutathione S-transferase A1, M1 and T1 polymorphisms and hypertension. Pharmacogenet Genomics 18:275–277
92. Marinho C, Alho I, Arduíno D, Falcão LM, Brás-Nogueira J, Bicho M (2007) GST M1/T1 and MTHFR polymorphisms as risk factors for hypertension. Biochem Biophys Res Commun 353:344–350
93. Abu-Amero KK, Al-Boudari OM, Mohamed GH, Dzimiri N (2006) T null and M null genotypes of the glutathione S-transferase gene are risk factor for CAD independent of smoking. BMC Med Genet 7:38
94. Wilson MH, Grant PJ, Hardie LJ, Wild CP (2000) Glutathione S-transferase M1 null genotype is associated with a decreased risk of myocardial infarction. FASEB J 14:791–796
95. Delles C, Padmanabhan S, Lee WK, Miller WH, McBride MW, McClure JD, Brain NJ, Wallace C, Marçano AC, Schmieder RE, Brown MJ, Caulfield MJ, Munroe PB, Farrall M, Webster J, Connell JM, Dominiczak AF (2008) Glutathione S-transferase variants and hypertension. J Hypertens 26:1343–1352
96. Samani NJ, Erdmann J, Hall AS, Hengstenberg C, Mangino M, Mayer B, Dixon RJ, Meitinger T, Braund P, Wichmann HE, Barrett JH, König IR, Stevens SE, Szymczak S, Tregouet DA, Iles MM, Pahlke F, Pollard H, Lieb W, Cambien F, Fischer M, Ouwehand W, Blankenberg S, Balmforth AJ, Baessler A, Ball SG, Strom TM, Braenne I, Gieger C, Deloukas P, Tobin MD, Ziegler A, Thompson JR, Schunkert H (2007) WTCCC and the Cardiogenics Consortium. Genomewide association analysis of coronary artery disease. N Engl J Med 357:443–453
97. Madamanchi NR, Runge MS (2007) Mitochondrial dysfunction in atherosclerosis. Circ Res 100:460–473
98. Delles C, Miller WH, Dominiczak AF (2008) Targeting reactive oxygen species in hypertension. Antioxid Redox Signal 10:1061–1077
99. Espinosa O, Jimenez-Almazan J, Chaves FJ, Tormos MC, Clapes S, Iradi A, Salvador A, Fandos M, Redón J, Sáez GT (2007) Urinary 8-oxo-7,8-dihydro-2′-deoxyguanosine (8-oxo-dG), a reliable oxidative stress marker in hypertension. Free Radic Res 1:546–554

100. Graham D, Spiers A, Beattie E, Taylor K, Murphy MP, Hamilton C, Miller WH, Dominiczak AF (2007) Mitochondrial-targeted antioxidant treatment improves cardiovascular function. J Hypertens 25(Suppl 2):S145; (abstract)
101. Anderson S, Bankier AT, Barrell BG, de Bruijn MH, Coulson AR, Drouin J, Eperon IC, Nierlich DP, Roe BA, Sanger F, Schreier PH, Smith AJ, Staden R, Young IG (1981) Sequence and organization of the human mitochondrial genome. Nature 290: 457–465
102. Delles C (2007) Mitochondria, maternal transmission and hypertension. J Hypertens 25:2001–2003
103. Bonfini L, Migliaccio E, Pelicci G, Lanfrancone L, Pelicci PG (1996) Not all Shc's roads lead to Ras. Trends Biochem Sci 21:257–261
104. Huebner K, Kastury K, Druck T, Salcini AE, Lanfrancone L, Pelicci G, Lowenstein E, Li W, Park SH, Cannizzaro L, Pelicci PG, Schlessinger J (1994) Chromosome locations of genes encoding human signal transduction adapter proteins, Nck (NCK), Shc (SHC1), and Grb2 (GRB2. Genomics 22:281–287
105. Migliaccio E, Giorgio M, Mele S, Pelicci G, Reboldi P, Pandolfi PP, Lanfrancone L, Pelicci PG (1999) The p66shc adaptor protein controls oxidative stress response and life span in mammals. Nature 402:309–313
106. Francia P, Delli Gatti C, Bachschmid M, Martin-Padura I, Savoia C, Migliaccio E, Pelicci PG, Schiavoni M, Lüscher TF, Volpe M, Cosentino F (2004) Deletion of p66shc gene protects against age-related endothelial dysfunction. Circulation 110:2889–2895
107. Kamei H, Adati N, Arai Y, Yamamura K, Takayama M, Nakazawa S, Ebihara Y, Gondo Y, Akechi M, Noguchi T, Hirose N, Sakaki Y, Kojima T (2003) Association analysis of the SHC1 gene locus with longevity in the Japanese population. J Mol Med 81:724–728
108. Mooijaart SP, van Heemst D, Schreuder J, van Gerwen S, Beekman M, Brandt BW, Eline Slagboom P, Westendorp RG (2004) 'Long Life' Study Group. Variation in the SHC1 gene and longevity in humans. Exp Gerontol 39:263–268
109. Almind K, Ahlgren MG, Hansen T, Urhammer SA, Clausen JO, Pedersen O (1999) Discovery of a Met300Val variant in Shc and studies of its relationship to birth weight and length, impaired insulin secretion, insulin resistance, and type 2 diabetes mellitus. J Clin Endocrinol Metab 84:2241–2244
110. Sentinelli F, Romeo S, Barbetti F, Berni A, Filippi E, Fanelli M, Fallarino M, Baroni MG (2006) Search for genetic variants in the p66Shc longevity gene by PCR-single strand conformational polymorphism in patients with early-onset cardiovascular disease. BMC Genet 7:14
111. Finsterer J (2007) Genetic, pathogenetic, and phenotypic implications of the mitochondrial A3243G tRNALeu(UUR) mutation. Acta Neurol Scand 116:1–14
112. Sakuta R, Honzawa S, Murakami N, Goto Y, Nagai T (2002) Atypical MELAS associated with mitochondrial tRNA(Lys) gene A8296G mutation. Pediatr Neurol 27:397–400
113. Remes AM, Kärppä M, Moilanen JS, Rusanen H, Hassinen IE, Majamaa K, Uimonen S, Sorri M, Salmela PI, Karvonen SL, Karvonen SL (2003) Epidemiology of the mitochondrial DNA 8344A>G mutation for the myoclonus epilepsy and ragged red fibres (MERRF) syndrome. J Neurol Neurosurg Psychiatr 74:1158–1159
114. Remes AM, Hassinen IE, Ikäheimo MJ, Herva R, Hirvonen J, Peuhkurinen KJ (1994) Mitochondrial DNA deletions in dilated cardiomyopathy: a clinical study employing endomyocardial sampling. J Am Coll Cardiol 23:935–942
115. Ruppert V, Nolte D, Aschenbrenner T, Pankuweit S, Funck R, Maisch B (2004) Novel point mutations in the mitochondrial DNA detected in patients with dilated cardiomyopathy by screening the whole mitochondrial genome. Biochem Biophys Res Commun 318: 535–543
116. Grasso M, Diegoli M, Brega A, Campana C, Tavazzi L, Arbustini E (2001) The mitochondrial DNA mutation T12297C affects a highly conserved nucleotide of tRNA(Leu(CUN)) and is associated with dilated cardiomyopathy. Eur J Hum Genet 9:311–315

117. Shin WS, Tanaka M, Suzuki J, Hemmi C, Toyo-oka T (2000) A novel homoplasmic mutation in mtDNA with a single evolutionary origin as a risk factor for cardiomyopathy. Am J Hum Genet 67:1617–1620
118. Arbustini E, Diegoli M, Fasani R, Grasso M, Morbini P, Banchieri N, Bellini O, Dal Bello B, Pilotto A, Magrini G, Campana C, Fortina P, Gavazzi A, Narula J, Viganò M (1998) Mitochondrial DNA mutations and mitochondrial abnormalities in dilated cardiomyopathy. Am J Pathol 153:1501–1510
119. Heart Protection Study Collaborative Group (2002) MRC/BHF Heart Protection Study of antioxidant vitamin supplementation in 20,536 high-risk individuals: a randomised placebo-controlled trial. Lancet 360:23–33
120. Yusuf S, Dagenais G, Pogue J, Bosch J, Sleight P (2000) Vitamin E supplementation and cardiovascular events in high-risk patients. The Heart Outcomes Prevention Evaluation Study Investigators. N Engl J Med 342:154–160
121. Alexander BT (2006) Fetal programming of hypertension. Am J Physiol Regul Integr Comp Physiol 290:R1–R10
122. Napoli C, Witztum JL, Calara F, de Nigris F, Palinski W (2000) Maternal hypercholesterolemia enhances atherogenesis in normocholesterolemic rabbits, which is inhibited by antioxidant or lipid-lowering intervention during pregnancy: an experimental model of atherogenic mechanisms in human fetuses. Circ Res 87:946–952
123. Probst-Hensch NM, Imboden M, Felber Dietrich D, Barthélemy JC, Ackermann-Liebrich U, Berger W, Gaspoz JM, Schwartz J (2008) Glutathione S-transferase polymorphisms, passive smoking, obesity, and heart rate variability in nonsmokers. Environ Health Perspect 116:1494–1499
124. Wang LS, Tang JJ, Tang NP, Wang MW, Yan JJ, Wang QM, Yang ZJ, Wang B (2008) Association of GSTM1 and GSTT1 gene polymorphisms with coronary artery disease in relation to tobacco smoking. Clin Chem Lab Med 46:1720–1725
125. Melén E, Nyberg F, Lindgren CM, Berglind N, Zucchelli M, Nordling E, Hallberg J, Svartengren M, Morgenstern R, Kere J, Bellander T, Wickman M, Pershagen G (2008) Interactions between glutathione S-transferase P1, tumor necrosis factor, and traffic-related air pollution for development of childhood allergic disease. Environ Health Perspect 116:1077–1084
126. Chahine T, Baccarelli A, Litonjua A, Wright RO, Suh H, Gold DR, Sparrow D, Vokonas P, Schwartz J (2007) Particulate air pollution, oxidative stress genes, and heart rate variability in an elderly cohort. Environ Health Perspect 115:1617–1622
127. Medeiros R, Pereira D, Afonso N, Palmeira C, Faleiro C, Afonso-Lopes C, Freitas-Silva M, Vasconcelos A, Costa S, Osório T, Lopes C (2003) Platinum/paclitaxel-based chemotherapy in advanced ovarian carcinoma: glutathione S-transferase genetic polymorphisms as predictive biomarkers of disease outcome. Int J Clin Oncol 8:156–161
128. Brandt JT, Close SL, Iturria SJ, Payne CD, Farid NA, Ernest CS 2nd, Lachno DR, Salazar D, Winters KJ (2007) Common polymorphisms of CYP2C19 and CYP2C9 affect the pharmacokinetic and pharmacodynamic response to clopidogrel but not prasugrel. J Thromb Haemost 5:2429–2436
129. Hulot JS, Bura A, Villard E, Azizi M, Remones V, Goyenvalle C, Aiach M, Lechat P, Gaussem P (2006) Cytochrome P450 2C19 loss-of-function polymorphism is a major determinant of clopidogrel responsiveness in healthy subjects. Blood 108:2244–2247
130. Simon T, Verstuyft C, Mary-Krause M, Quteineh L, Drouet E, Méneveau N, Steg PG, Ferrières J, Danchin N, Becquemont L (2009) the French Registry of Acute ST-Elevation and Non–ST-Elevation Myocardial Infarction (FAST-MI) Investigators. Genetic determinants of response to clopidogrel and cardiovascular events. N Engl J Med 360(4): 363–375
131. Mega JL, Close SL, Wiviott SD, Shen L, Hockett RD, Brandt JT, Walker JR, Antman EM, Macias W, Braunwald E, Sabatine MS (2009) Cytochrome P-450 polymorphisms and response to clopidogrel. N Engl J Med 360(4):354–362

132. Keeney DS, Skinner C, Travers JB, Capdevila JH, Nanney LB, King LE Jr, Waterman MR (1998) Differentiating keratinocytes express a novel cytochrome P450 enzyme, CYP2B19, having arachidonate monooxygenase activity. J Biol Chem 273:32071–32079
133. Padmanabhan S, Wallace C, Munroe PB, Dobson R, Brown M, Samani N, Clayton D, Farrall M, Webster J, Lathrop M, Caulfield M, Dominiczak AF, Connell JM (2006) Chromosome 2p shows significant linkage to antihypertensive response in the British Genetics of Hypertension Study. Hypertension 47:603–608
134. SEARCH Collaborative Group; Link E, Parish S, Armitage J, Bowman L, Heath S, Matsuda F, Gut I, Lathrop M, Collins R (2008) SLCO1B1 variants and statin-induced myopathy–a genomewide study. N Engl J Med 359:789–799
135. Liu H, Colavitti R, Rovira II, Finkel T (2005) Redox-dependent transcriptional regulation. Circ Res 97:967–974
136. Nagai F, Kato E, Tamura HO (2004) Oxidative stress induces GSTP1 and CYP3A4 expression in the human erythroleukemia cell line, K562. Biol Pharm Bull 27:492–495
137. Meltzer PS (2005) Cancer genomics: small RNAs with big impacts. Nature 435:745–746
138. Bartel DP, Chen CZ (2004) Micromanagers of gene expression: the potentially widespread influence of metazoan microRNAs. Nat Rev Genet 5:396–400
139. Lim LP, Lau NC, Garrett-Engele P, Grimson A, Schelter JM, Castle J, Bartel DP, Linsley PS, Johnson JM (2005) Microarray analysis shows that some microRNAs downregulate large numbers of target mRNAs. Nature 433:769–773
140. Hatfield SD, Shcherbata HR, Fischer KA, Nakahara K, Carthew RW, Ruohola-Baker H (2005) Stem cell division is regulated by the microRNA pathway. Nature 435:974–978

# Chapter 11
# NADPH Oxidases and Blood-Brain Barrier Dysfunction in Stroke

**Timo Kahles, Sabine Heumüller, and Ralf P. Brandes**

**Abstract** The opening of the blood-brain barrier, brain edema, intracerebral hemorrhage, and loss of neuronal tissue are consequences of ischemic stroke. The phase of ischemia is often followed by reperfusion, either occurring spontaneously or as a result of recanalization therapies such as thrombolysis or mechanical devices. An increased formation of reactive oxygen species (ROS) occurs during ischemia/hypoxia, but most importantly during reperfusion/reoxygenation, and contributes to tissue injury. Although mitochondria have long been considered the primary source of ROS during ischemic stroke and reperfusion, recently NADPH oxidases have been shown to be of great importance for the process, too. All cells in the brain express functional NADPH oxidases, and genetic deletion of $p47^{phox}$ or Nox2 reduces the brain infarct size in mice. The opening of the blood-brain barrier (BBB) within minutes after reperfusion is an early sign of vascular dysfunction after stroke. The initial phase of this process, which involves contraction of endothelial cells, is mediated by NADPH oxidases, and genetic deletion or pharmacological inhibition of the oxidases prevents the BBB opening. In this chapter, the contribution of ROS and the NADPH oxidase in particular for tissue injury and BBB dysfunction in ischemic stroke will be reviewed.

**Keywords** Stroke · Reactive oxygen species · Ischemia · Reperfusion · Hypoxia · Reoxygenation · Brain · Endothelium

## 11.1 Introduction

Reactive oxygen species (ROS) have been linked to a large number of pathophysiological processes and diseases of the central nervous system (CNS). Although there is ample evidence that low amounts of ROS in the brain also contribute

T. Kahles (✉)
Institut für Kardiovaskuläre Physiologie, Fachbereich Medizin der Goethe-Universität, 60596 Frankfurt am Main, Germany
e-mail: kahles@zphys1.uni-frankfurt.de

H. Sauer et al. (eds.), *Studies on Cardiovascular Disorders*, Oxidative Stress in Applied Basic Research and Clinical Practice,
DOI 10.1007/978-1-60761-600-9_11, 

to physiological signaling [1], the classic oxidative stress model in which ROS are direct mediators of tissue damage predominates [2]. As with all other organs, depending on the disease situation and the model studied, ROS formation has been attributed to different enzymatic sources. Traditionally, pathology-associated ROS were believed to be produced by mitochondria, cyclooxygenase, monoaminooxygenase, and xanthine oxidase [3]. Just recently, with the growing interest in vascular NADPH oxidases [4], the role of this family of enzymes for pathology and physiological cell function in the brain has gained considerable interest [5]. Nowadays, growing evidence points to a major role for NADPH oxidases as the key enzyme in a broad spectrum of diseases of the CNS, such as Alzheimer disease [6], amyotrophic lateral sclerosis [7], Parkinson disease [8], subarachnoidal hemorrhage [9], intracerebral hemorrhage [10], and ischemic stroke [11].

Particularly in the latter condition, and if followed by reperfusion, ROS formation has been demonstrated to be of pathophysiological significance. Nevertheless, relatively little is known about the role of NADPH oxidases in this context. Furthermore, our knowledge of the expression, the activation, and the signaling of the individual NADPH oxidases in cerebral tissue is still limited because of the complexity of the cellular composition and the heterogeneity of the brain.

The focus of this chapter is on the contribution of NADPH oxidases to reperfusion injury after cerebral ischemia, with special regard to their actions on the blood-brain barrier.

## 11.2 The Clinical Setting of Stroke

Stroke is the third leading cause of death in the industrialized nations. The World Health Organization estimates that about 15 million people suffer from stroke each year (www.who.int). One third will not survive the first year after the event, and another third will be permanently disabled. Ischemic stroke is a consequence of a thrombotic or embolic occlusion of a brain-supplying vessel, and accounts for more than 80% of strokes, compared to 20% which are of hemorrhagic nature. Early restoration of cerebral blood flow is mandatory in preventing persistent brain damage in ischemic stroke. Delayed reperfusion results in the loss of neuronal cells and in blood-brain barrier disruption. The consequences of these processes are potentially life-threatening brain edema and intracerebral hemorrhage. Ischemia followed by reperfusion has been studied extensively in animal experiments. The situation in patients, however, is very different. Although reperfusion therapy for myocardial infarction was established a long time ago, the same strategy for the brain is still rarely used. The reason for this is the potential of devastating intracerebral hemorrhages and the small therapeutic window in which to rescue the neuronal tissue. Indeed, as neurons are more sensitive to ischemia, a beneficial effect of medical reperfusion therapy is restricted to the first 4.5 h after the onset of symptoms [12].

## 11.3 Reactive Oxygen Species in Ischemic Brain Injury

There is a strong rationale for increased ROS formation during ischemia and reperfusion [13]. In animal experiments, signs of oxidative stress, such as a reduction in glutathione [14] and vitamins, have been observed in cerebral tissue; and this process was accompanied by an increase in lipid peroxidation products [15]. In several studies, increased levels of different ROS species in infarcted tissue were observed, with a particular increase in ROS formation directly after reperfusion [16–19]. Also, oxidative stress in patients as measured by lipid peroxidation products in the plasma has been linked to stroke [20, 21].

As a consequence, a large number of studies have determined the effects of antioxidants and enzyme inhibitors on experimental stroke in animal models, and the vast majority of the publications have reported a beneficial effect with this approach [13].

Numerous enzymatic sources could account for ROS formation in stroke. Among them, mitochondria are considered of particular importance [22] but, because of the critical function of these organella, mitochondria-dependent ROS formation is impossible to approach with inhibitors in vivo [23, 24].

In addition to mitochondria, other enzyme systems have been reported to be of importance: Cyclooxygenase can promote ROS formation by oxidative modifications of arachidonic acid, which leads to superoxide anion formation. This process, however, which is mediated by COX 2, requires induction of the enzyme and thus is not involved in ROS formation in the first 6 h after transient ischemia [25].

In the heart, xanthine oxidase is thought to be a main source of ROS during ischemia/reperfusion. Whether the enzyme is also important for ROS formation in the brain is still controversial. Although xanthine oxidase inhibition has yielded some beneficial effects [13], the formation of the enyzme in stroke has not been observed [26].

Some studies suggest a role for leukocytes in ROS formation in stroke: Neutropenia induced by chemotherapeutics and anti-Cd18 antibodies reduced the cerebral ROS formation after stroke [27, 28]. The Nox2-containing NADPH oxidase is the main source of ROS in leukocytes, but the enzyme is also expressed in other cells of the brain.

## 11.4 NADPH Oxidases in the Central Nervous System

The Nox homologues of the NADPH oxidase family are diversely expressed across different brain regions and cell types. The coexistence of multiple Nox isoforms in a single cell was demonstrated, and it is assumed that individual Nox proteins have different effects. This homologue-specific action is most probably a consequence of a different subcellular localization, expression intensity, and mode of activation of the individual Nox proteins [29, 30].

The vast majority of publications address the expression of Nox2 or its subunits. Serrano et al. [31] showed expression of subunits of the Nox2-containing NADPH oxidase in neurons distributed over the whole brain of the adult mouse, including the hippocampus, using immunohistochemistry. Interestingly, although resident microglia are involved in host defense, and the corresponding circulating leukocytes are a prominent source of NADPH-mediated ROS, no Nox2 protein expression could be detected in microglia of the resting mouse brain. Under pathological conditions, Nox2 expression in these cells, however, increases to detectable levels [32], and pronounced Nox2 mRNA expression can be observed in cultured microglia. The high expression of Nox2 in the hippocampus points towards a role for this NADPH oxidase in long-term potentiation and memory. To our knowledge, however, whether the memory and learning function of mice is affected by genetic deletion of the protein has not been studied. Nox2 expression was also observed in other regions of the brain, in particular the medulla and the spinal cord [33]. Although the physiological function of the protein in these regions is still unclear, Nox2 is potentially involved in the control of sympathetic tone and the central responses to angiotensin II and salt [34, 35].

Only very few studies have addressed the cerebral expression of Nox1, 3, and 4. Vallet et al. [36] investigated the role of Nox4 in cerebral ischemia and could also show a wide distribution of the NADPH oxidase in neurons of various brain regions in mice using in situ hybridization and immunohistochemical techniques. Also, Nox4 mRNA was verified in total human brain by RT-PCR. During permanent cerebral ischemia Nox4 mRNA significantly increased in neurons of the ischemic hemisphere within 1–3 days, peaking at days 7–15, and declining till day 30. This peak paralleled capillary formation and reorganization processes in the ischemic area, suggesting a role for Nox4 in brain repair. The cellular sources of Nox4 are, however, uncertain. In cultured cells from the mouse brain, a high expression of Nox4 was observed in endothelial cells and neurons, whereas microglia expressed only a very low mRNA level (Heumüller, unpublished observations, Fig. 11.1). Cats and mice lack the gene for Nox5, but mRNA expression of this NADPH oxidase was reported for the human brain [37].

Besides Nox2 and 4, some studies also suggested the expression of other isoforms such as Nox1 and 3 [38]. Nox3, however, appears to be restricted to the inner ear, compared to other brain regions, where it is involved in otoconia formation [39].

A more quantitative approach was performed by Infanger et al. [40]. That study compared the expression levels of Nox1, 2, and 4 at the mRNA level in various murine brain regions, thereby circumventing the lack of sufficient specific antibodies. Nox2 and 4 were dominant in the fore-, mid-, and hindbrain. Nox1, in contrast, was only slightly expressed. The relative expression levels of Nox2 and 4 homologues differed depending on the brain region. Nox2 was prevalent in the forebrain, Nox4 in the midbrain, and almost similar levels could be found in the hindbrain.

Although these studies already draw a complex picture, the conclusions presented here have not been reproduced by all observers, and partially conflicting results have been reported [11, 41, 42]. Finally, very little is yet known regarding the expression of the individual Nox proteins in cerebral disease conditions, and even less regarding the individual functions of Nox proteins in the brain.

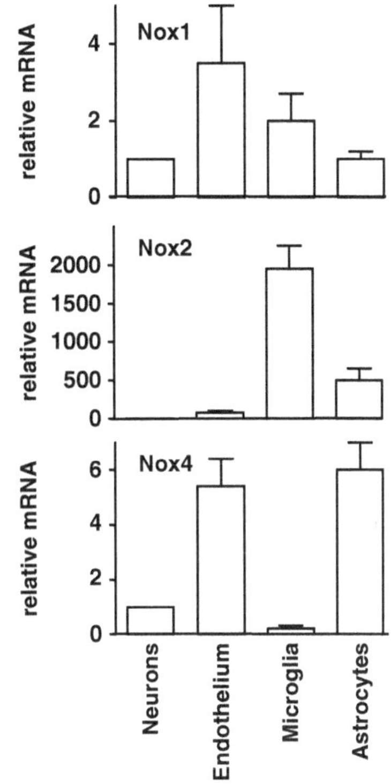


**Fig. 11.1** Expression of Nox mRNA in cerebral cells. Cell populations of neurons, endothelial cells, microglia, and astrocytes were isolated from the neonatal mouse brain and cultured separately. mRNA expression of individual Nox homologues was determined by qRT-PCR standardized to elongation factors using the delta CT method. Numbers shown are normalized to the expression observed in neurons

The brain is a heterocellular tissue. Thus, the different distribution of Nox homologues might also be a consequence of a different cellular composition. Isolation of individual cells from the brain has been carried out. For endothelial cells, these studies have confirmed observations from the systemic circulation and report an expression of Nox1, 2, and 4 [42–46]. We observed an abundant expression of Nox2 mRNA in microglia, and also substantial expression of the mRNA for Nox2 in astrocytes and endothelial cells, but only low expression in neurons. In contrast, Nox1 was predominantly found in murine cerebral endothelial cells, and in smaller quantities in microglia, neurons, and astrocytes (Kahles, unpublished observation) (Fig. 11.1).

For a more detailed view of the expression profile of NADPH oxidases in cerebral tissue, see [47].

## 11.5 The Role of NADPH Oxidases in Ischemic Stroke

### *11.5.1 NADPH Oxidases in Ischemia/Reperfusion Outside of the Brain*

As mentioned above, mitochondria are considered an important source of ROS in ischemia/reperfusion. Several facts, however, also argue in favor of a contribution by

NADPH oxidases: The oxidases are involved in the process of ROS-triggered ROS release, as $H_2O_2$ has been shown to activate them [48]. Moreover, the upstream signaling cascades, which are involved in NADPH oxidase assembly, are activated during ischemia/reperfusion; as a consequence of the inhibition of lipid phosphatases and activation of the PI3-kinase, Rac1 is activated. In addition, release of arachidonic acid and phosphorylation of $p47^{phox}$ are required for NADPH oxidase assembly. Increases in intracellular calcium should facilitate sufficient activation of calcium-dependent phospholipase A2 isoforms to generate arachidonate [49]. $P47^{phox}$ can be phosphorylated by several kinases, including protein kinase B/Akt, p21-activated protein kinase (p21-PAK), as well as protein kinase C (PKC) isoforms. Akt and PKCs are both activated during hypoxia/reoxygenation [50], and p21-PAK is a direct downstream target of Rac1.

Indeed, there is experimental evidence that NADPH oxidases are activated during ischemia/reperfusion in several cells of the brain. In cultured endothelial cells, the hypoxia-induced increase in ROS formation was blocked by p22phox antisense oligo-nucleotides [51]. Also the inhibition of Rac1 prevented the hypoxia/reoxygenation-induced increase in several cell culture models.

There are also in vivo data for a role of NADPH oxidase in ischemia/reperfusion–induced ROS formation: In the isolated lung, ischemia/reperfusion failed to increase ROS formation in Nox2-deficient animals [52]. Also in Nox2-deficient animals as well, ischemia/reperfusion-induced liver injury was attenuated after inhibition of Rac1 [53].

## *11.5.2 Cerebral NADPH Oxidases and Ischemic Brain Injury*

With the aid of neurons obtained from Nox2 –/– mice, it was shown that this NADPH oxidase was the main source of ROS after reoxygenation in a cell culture model [54]. In vivo studies on cerebral ROS formation are rare and difficult because the distribution of the tracers is affected by the altered perfusion in the setting of ischemia. Thus, measurements require confirmation by specific inhibitors or genetic deletion of NADPH oxidases. In rats, the ROS formation in the penumbra, but not the necrotic core, was increased 2 h after the onset of reperfusion and this was associated with increased NADPH oxidase activity. Importantly, pretreatment with HMG-CoA reductase inhibitors, that prevent the activation of Rac1 and thus the NADPH oxidase, reduced ROS formation [55]. A direct involvement of the NADPH oxidase for cerebral ROS formation after stroke was documented with the aid of $p47^{phox}$ –/– mice 3 h after the onset of reperfusion [56]. Importantly, in both cases, reduced NADPH oxidase–dependent ROS formation was associated with attenuated neuronal injury. This observation is in accordance with data obtained in Nox2 –/– mice: The infarct volume after transient middle cerebral artery occlusion for 2 h [57], and even after permanent occlusion [58], was significantly smaller in Nox2–/– mice (Fig. 11.2). Several studies have used the compound apocynin to determine

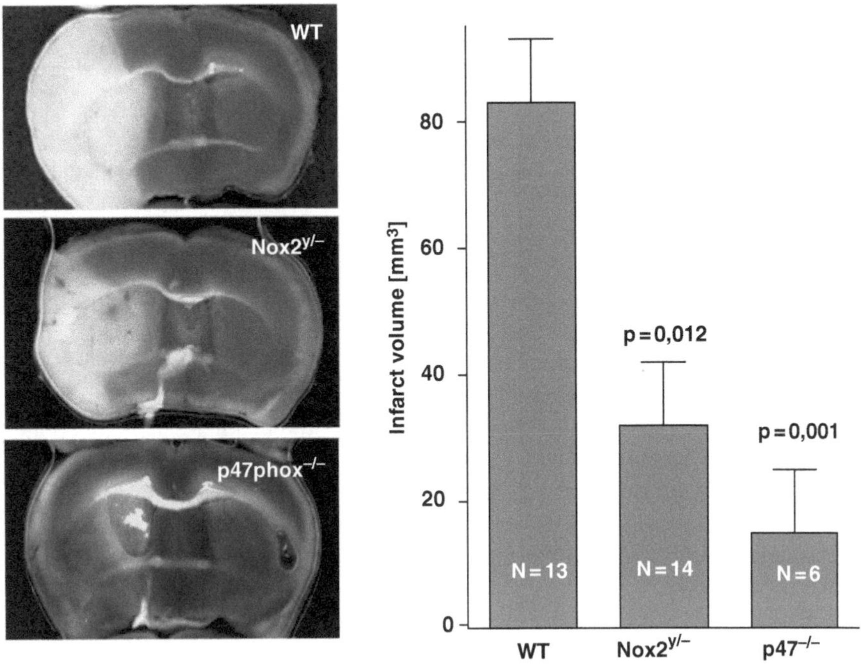


**Fig. 11.2** Genetic deletion of NADPH oxidase subunits lowers the infarct size of the murine brain. Mice were subjected to 2 h of transient ischemia using the middle cerebral artery filament occlusion technique followed by 22 h of reperfusion. Shown are the exemplary scans and statistical analysis of the infarct size as determined by tetrazolium staining

the effect of NADPH oxidase inhibition on cerebral ROS formation after experimental stroke. Indeed, it was observed that apocynin reduced ROS formation in ischemic murine brain [57, 59, 60] and rat brain [61] and in gerbil hypocampus [62]. This effect was accompanied with a reduction in brain lesion volume [60, 61] and neuronal death [62]. The interpretation of data obtained with apocynin is complicated by the fact that the compound is an exquisite antioxidant [63]. The recent observation that the protective effect of apocynin is not observed in Nox2-deficient mice, however, indicates that indeed Nox2 is a major source of ROS during cerebral ischemia/reperfusion [59].

## 11.6 The Blood-Brain Barrier

### *11.6.1 Structural Components of the Blood-Brain Barrier*

The blood-brain barrier (BBB) is part of the neurovascular unit consisting of neurons, glia, and microvessels. It represents the physical and metabolic interface between the systemic circulation and the CNS, and maintains a tightly regulated microenvironment for neuronal function. The BBB is formed by endothelial cells, pericytes, the basal lamina, and the adjacent astrocytic end-feet. In contrast to the endothelial layers in the systemic vasculature, the cerebral endothelium is not fenestrated, and endothelial cells are connected by adherens and tight junctions, which lead to a very restrictive barrier with high transendothelial electrical resistance and low permeability. Endothelial junctions are characterized by specific adhesion molecules. Vascular endothelial cadherin (VE-cadherin) is the most important contributor to adherens junctions, which are predominantly linked to the cytoskeleton by catenins. Tight junctions (TJ) mainly consist of occludin and claudins 1, 3, and 5, as well as the junctional adhesion molecules JAM A, B, and C, and endothelial selective adhesion molecules (ESAM). At the cytoplasmatic side of the TJ, large protein complexes are formed, including Zonula occludentes (ZO 1–3), membrane-associated guanylate kinases with inverted orientation of protein-protein interaction domains (MAGI 1–3), calcium-dependent serine protein kinase (CASK), partitioning defective proteins (PAR 3 and 6), and junction associated coiled protein (JACOP) which are connected to the actin-cytoskeleton. The complex architecture of the endothelial junctions not only guarantees a firm and tight connection of endothelial cells but also facilitates strain-induced signal transduction and dynamic reorganization [64].

Although the specialized endothelium is considered the basis for the tightness of the BBB, pericytes, astrocytes, and the basal lamina are important for the structural integrity of the barrier, and alterations in any of these components result in blood-brain barrier dysfunction. Pericytes are contractile cells and thus can modulate capillary blood flow and vascular perfusion pressure [65, 66]. Astrocytes, with the aid of their end-feet, are in control of BBB maturation and of maintenance of barrier properties [67] as well as in control of cerebral blood flow [68, 69]. Finally, the basal lamina (BL), composed of different matrix proteins like laminin, fibronectin, collagen IV, and heparan sulphate proteoglycans, provides the scaffold for the otherwise very plastic structures. The BL is linked to the luminal endothelial cells by endothelial integrins, and to the abluminal astrocytic end-feet through the dystrophin-dystroglycan-complex [67]. Disruption of the basal lamina results in intracerebral hemorrhage, which is a typical late feature of BBB dysfunction [70, 71]

## *11.6.2 In Vivo Regulation of the Blood-Brain Barrier*

Basically all afflictions of the CNS are accompanied by an increase in BBB permeability, which is exploited diagnostically by the determination of the penetration and retention of contrast agents into the tissue, increased interstitial water content, or the presence of proteins in the cerebral spinal fluid. The underlying pathomechanisms may vary according to the underlying diseases. In inflammation, locally produced cytokines activate the endothelium [72, 73]; whereas in cerebral malignancies, either vascular leakiness in the course of de novo vessel formation or hypoxic cellular dysfunction are involved. The latter mechanism, which also mediates BBB dysfunction after stroke, is rather complex, because energy depletion and subsequent cellular calcium overload, rigor, and hydrops are accompanied by the release of a multitude of signaling molecules and alterations in gene expression and a localized stress reaction which resembles an inflammatory phenotype [74]. Hypoxia-induced changes in gene expression also contribute to high altitude brain edema, which is an exclusive consequence of local changes in brain perfusion and BBB permeability [75]. Also, structural alterations may contribute to BBB dysfunction, such as the degradation of the collagen in the basal lamina after reperfusion or thrombolytic therapy, which is a consequence of an activation of matrix metallo-proteinases [76, 77].

The fact that oxidative stress and increased BBB permeability are observed in almost all disease conditions of the CNS gives rise to the question whether both conditions are just indicators of some undefined "dysfunction" or if alterations in BBB function might result from oxidative stress. Indeed, several trials utilizing nonspecific antioxidants reported beneficial effects on BBB permeability: In animal experiments several antioxidants revealed a neuroprotective effect. Clinical trials tested the efficacy of Ebselen, a glutathione peroxidase mimetic, and the free radical scavengers Trialazid, Edaravone, and NXY-059 in ischemic stroke. Ebselen proved to be effective in two studies [78, 79] regarding clinical outcome at 1 and 3 months, especially if treatment starts within 24 h of stroke onset. Trialazid on the other hand failed to be effective in stroke treatment [80, 81]. The antioxidant Edaravone initially showed a beneficial outcome in the treatment of acute ischemic stroke [82]. It was introduced to the Japanese market in 2001. However, further stroke trials, either conducted as single therapy or in combination with the thrombin inhibitor Argatroban, did not show a significant benefit for the drug [83]. NXY-059 appeared to be a very promising free radical scavenger after completing the tolerability studies and a phase III trial [84]; unfortunately, the latest clinical trial [85] was ineffective for the treatment of acute ischemic stroke within 6 h after the onset of symptoms. This failure might be due to several causes, such as indiscriminate inclusion of all stroke subtypes, the possibility of inactivity of the drug because of the known instability of free radical trapping agents, or even preclinical drug study problems [86].

### *11.6.3 Blood-Brain Barrier Dysfunction in Stroke*

Brain edema is a major factor contributing to tissue damage in ischemic stroke [87]. The local edema-mediated increase in interstitial pressures impedes tissue perfusion and thus extends the ischemic area beyond the region directly affected by the vascular occlusion. The goal of current treatment strategies for ischemic stroke is the protection of this region of risk belonging to the penumbra [88]. Although cellular swelling is certainly the main factor for edema formation in acute stroke, increased BBB permeability also contributes to this process [89]. The opening of the BBB in ischemic stroke follows a biphasic pattern [90], which refers to a process called vasogenic edema. Endothelial cell contraction early after the onset of ischemia leads to a functional opening of the BBB, which results in increased water extravasation. Subsequently, the vascular matrix is degraded by proteases activated in the ischemic tissue and vascular integrity is disrupted. Reperfusion and thrombolytic therapy have both shown to promote these processes, providing some explanation for the increased rate of intracerebral hemorrhages associated with this reperfusion therapy [91].

The pathophysiology of reperfusion, either induced experimentally or by recanalization therapy such as thrombolysis, is complex. Because a substantial part of vital but critically ischemic tissue dies after initiation of reperfusion, much effort has been devoted to minimizing reperfusion injury by different treatment regimes [92, 93].

Numerous studies have provided a link between ROS and the opening of the BBB: A main hypothesis is that mitochondrial damage from calcium overload and energy depletion manifests just after restoration of the oxygen supply [94]. Indeed, dysfunctional mitochondria are able to release massive amounts of ROS leaking from the respiratory chain. These ROS further aggravate the tissue damage and further promote mitochondrial dysfunction. This process also contributes to endothelial damage and BBB opening (for review, see [95]). Based on this concept, a large number of animal experiments and even studies in patients have been conducted to test the effects of antioxidants on brain edema formation; and indeed, antioxidants attenuated vascular leakage following ischemia in vivo [96, 97].

## *11.6.4 Mechanisms of Blood-Brain Barrier Opening*

From the before mentioned in vivo data, it becomes obvious that the tight monolayer of endothelial cells in the brain is not a static structure but responds to a broad panel of stimuli by changing transport rates, adhesion molecule expression, autacoid secretion, and most importantly alterations in permeability [98]. With respect to the molecular mechanisms involved, the latter aspect has been extensively studied in different cell culture models. From these experiments, it is evident that the opening of the BBB represents a complex cellular reaction pattern which not only involves contraction of the endothelial cells as a consequence of the polymerization of the actin-cytoskeleton but also the active disassembly of the cellular junctions.

The contractile state of the endothelium of the BBB depends to a large extent on actin polymerization. Resting endothelial cells are characterized by coronary actin structures stabilizing the cell, whereas after activation these structures are degraded and stress fibers are formed which contract the cell. Actin dynamics are highly complex and virtually hundreds of proteins modulate the actin turnover. Although it is certainly beyond the scope of this article to preview this reorganization process, it should be mentioned that in particular small GTPases of the Rho family, which include RhoA, cdc42, and Rac1, control the process. Rac1 on the other hand is also involved in the activation of NADPH oxidases [99], and importantly, activation of

Rho-GTPases including RhoA und Rac1 have all been reported to be increased by ROS [100].

The effects of pro-inflammatory stimuli as well as hypoxia/reoxygenation on endothelial monolayer permeability have been extensively studied; and nowadays the signal transduction cascades of tumor necrosis factor α (TNFα) or VEGF are well characterized and offer options for therapeutic intervention. Interestingly, both compounds are known to increase the endothelial formation of ROS [101], and permeability changes in response to TNFα [102] and VEGF [103] are dependent on NADPH oxidases. For hypoxia/reoxygenation the situation is more complex, in particular because this important in vivo condition can only be incompletely recapitulated by cell culture systems. Monocultures of endothelial cells have a low oxygen demand and metabolic turnover, whereas cocultures are often too complex for a detailed analysis. Ischemia/reperfusion in vivo involves not only hypoxia but also asphyxia, and different degrees of glucose deprivation and acidosis.

During reperfusion in vivo, normal nutrient supply to the tissue is quickly achieved, but hyperperfusion and cellular activation and dysfunction persist and render the tissue highly susceptible to ROS-mediated damages. It is therefore not surprising that depending on the observer and the experimental conditions, hypoxia/reoxygenation of endothelial cells results in a variety of effects ranging from no reaction to massive apoptosis. Given these limitations and the problems arising from the determination of ROS formation in situations with changing redox potential, direct determination of the effects of ROS on cellular function has become popular.

Indeed, exposure of endothelial cells to $H_2O_2$ promotes the opening of adherens junctions and stress fiber formation in cultured endothelial cells [104, 105]. Pharmacological analysis of this process, which directly increases permeability in cultured porcine endothelial cells, suggested the involvement of an increase in intracellular calcium concentration, as well as activation of PI3 kinase and rho kinase [104, 105]. Moreover, several tyrosine kinases, such as Src, are activated in response to $H_2O_2$ [106]. Recently it has been shown that Src phosphorylates the Rac1-GEF Tiam1, which leads to the disassembly of junctions [107]. The cellular reactions in response to extracellular ROS, however, are complex and not necessarily identical to the reactions of endogenous ROS. Extracellular stimulation with $H_2O_2$ will activate matrix and protease-dependent pathways such as the release of cyclophilins [108] of heparin-binding EGF, which, in addition to the direct redox-mediated effect on signal transduction, have prominent effects themselves. Among these indirect effects, $H_2O_2$ has even been demonstrated to activate the NADPH oxidase [48].

## 11.7 The Role of NADPH Oxidases in Blood-Brain Barrier Dysfunction

Very little is known about the role of NADPH oxidases in early BBB disruption. We have recently shown that BBB permeability was reduced in Nox2-deficient mice [57]. Moreover, inhibition of the activation of NADPH oxidases by the

Rac1-inhibitory clostrium-toxin lethal toxin A also prevented BBB opening. That the effect could also be exploited pharmacologically was demonstrated by the fact that not only apocynin but also atorvastatin prevented the early BBB opening [57] (Fig. 11.3). Atorvastatin, like all HMG-CoA reductase inhibitors, inhibits NADPH oxidase by preventing the geranylgeranylation of Rac-1 [109].

The mechanisms of how the NADPH oxidase promotes BBB opening are incompletely understood. We observed that ROS increase the actin polymerization of cerebral endothelial cells leading to cellular contraction by a mechanism sensitive to PI3-kinase and Rho kinase inhibition [57] (Fig. 11.4). Kuhlmann et al.

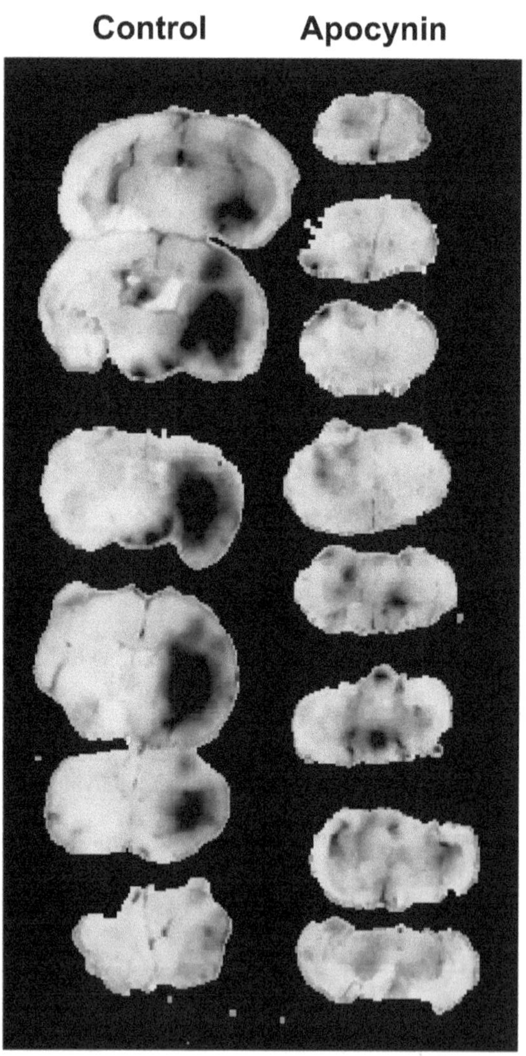


**Fig. 11.3** Effect of apocynin on the early increase in blood-brain barrier permeability. Mice were treated with apocynin (40 mg/kg) and subsequently subjected to transient middle cerebral artery occlusion for 2 h. Directly after the onset of reperfusion, Evans blue (150 μl of a 2% solution) was injected and brains were harvested 1 h later. Evans blue extravasation was visualized using an infrared laser fluorescent scanner (Odyssey, Licor)

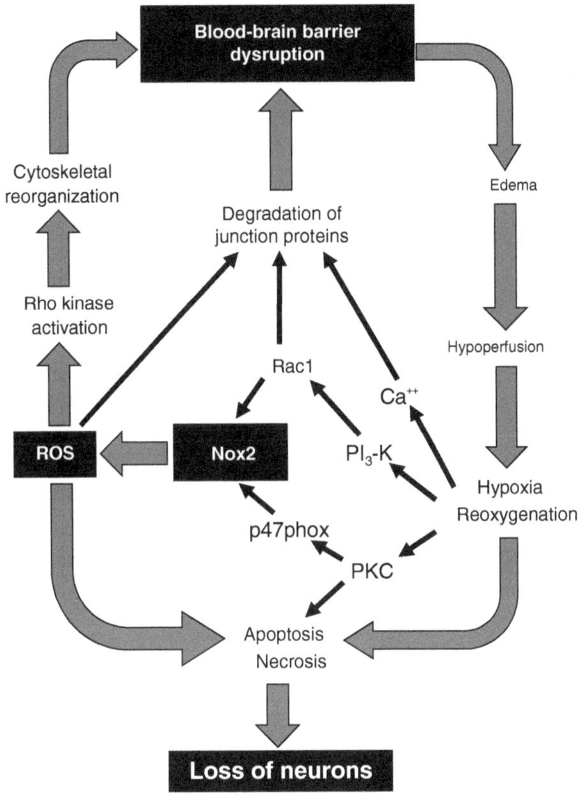


**Fig. 11.4** Model of the role of the NADPH oxidase Nox2 in cerebral ischemia/reperfusion. Hypoxia/reoxygenation directly result in neuronal loss by apoptosis and necrosis. Hypoxia/reoxygenation however also activate the Nox2-containing NADPH oxidase through a pathway involving protein kinase C (PKC)–mediated phosphorylation of $p47^{phox}$ and phosphatidylinositol-tris-phosphate-kinase–mediated activation of Rac1 (PI3-K). Reactive oxygen species (ROS) generated by Nox2 elicit direct neurotoxic effects. ROS also activate the Rho kinase which leads to cytoskeletal reorganization and endothelial cell contraction, a process which is also promoted by increases in intracellular calcium and activation of Rac1. ROS and the latter two mediators induce the degradation of junction proteins and this together with the cellular contraction results in blood-brain barrier disruption. The subsequent increase in vascular permeability promotes edema formation with the consequence of cerebral hypoperfusion

[110] have shown the activation of the endothelial contractile machinery with a subsequent increase in barrier permeability following a hypoxia-induced increase in intracellular calcium levels, which itself enhanced ROS generation by NADPH oxidases, leading to the activation of the myosin light chain kinase and phosphorylation of its substrate in vitro and in vivo. Moreover, it was recently observed that $H_2O_2$ increases the paracellular permeability by activating p44/42 MAP kinases and increasing calcium [105].

There is also a potential for a role of NADPH oxidase in the delayed BBB opening.

Gasche et al. [111] showed that oxidative stress is involved in mediating BBB disruption during the first 3–6 h of reperfusion after experimental ischemia through MMP-9 activation. The gelatinase MMP-9 in turn degrades the basal lamina and leads to an increase in BBB permeability and subsequently intracerebral hemorrhage. Liu et al. [112] demonstrated that treatment of transient focal cerebral ischemia with normobaric oxygen reduces NADPH oxidase-mediated MMP-9 induction.

Aquaporin 4 (AQP4), a glial membrane water channel which is abundantly expressed at the BBB towards perivascular [113] and endothelial cells, also potentially contributes to delayed brain edema formation. In the filament occlusion model of the middle cerebral artery, brain edema at 72 h was significantly larger in SOD2 –/– mice compared to their wild-type littermates. This increase was paralleled by an increase in AQP4-expression [93]. In a previous study, Manley et al. [114] demonstrated that genetic deletion of AQP4 in mice attenuates cerebral edema following ischemic stroke. It is still a matter of debate whether this increase in AQP4-expression after ROS-mediated reperfusion injury contributes to brain edema formation or if it is just a clearance mechanism to remove an excess of water.

## 11.8 Summary and Conclusion

Oxidative stress aggravates reperfusion injury following transient focal cerebral ischemia. NADPH oxidase-derived ROS contribute to BBB opening in this context. Genetic deletion or pharmacological inhibition of NADPH oxidases significantly attenuates postischemic damage. Thus potentially life-threatening complications of stroke such as space-occupying cerebral edema or intracerebral hemorrhage, especially after restoration of cerebral blood flow, could potentially be avoided by NADPH oxidase inhibitors.

For a better understanding of the contribution of the different Nox homologues to stroke pathophysiology, and their interaction among each other, as well as the underlying mechanisms, specific inhibitors and antibodies will have to be developed.

## References

1. D'Autreaux B, Toledano MB (2007) ROS as signalling molecules: mechanisms that generate specificity in ROS homeostasis. Nat Rev Mol Cell Biol 8:813–824
2. Fisher M (2008) Injuries to the vascular endothelium: vascular wall and endothelial dysfunction. Rev Neurol Dis 5(Suppl 1):S4–S11
3. Traystman RJ, Kirsch JR, Koehler RC (1991) Oxygen radical mechanisms of brain injury following ischemia and reperfusion. J Appl Physiol 71:1185–1195
4. Cai H, Griendling KK, Harrison DG (2003) The vascular NAD(P)H oxidases as therapeutic targets in cardiovascular diseases. Trends Pharmacol Sci 24:471–478

5. Chrissobolis S, Faraci FM (2008) The role of oxidative stress and NADPH oxidase in cerebrovascular disease. Trends Mol Med 14:495–502
6. Block ML (2008) NADPH oxidase as a therapeutic target in Alzheimer's disease. BMC Neurosci 9(Suppl 2):S8
7. Wu DC, Re DB, Nagai M, Ischiropoulos H, Przedborski S (2006) The inflammatory NADPH oxidase enzyme modulates motor neuron degeneration in amyotrophic lateral sclerosis mice. Proc Natl Acad Sci USA 103:12132–12137
8. Wu DC, Teismann P, Tieu K, Vila M, Jackson-Lewis V, Ischiropoulos H, Przedborski S (2003) NADPH oxidase mediates oxidative stress in the 1-methyl-4-phenyl-1,2,3,6-tetrahydropyridine model of Parkinson's disease. Proc Natl Acad Sci USA 100: 6145–6150
9. Zheng JS, Zhan RY, Zheng SS, Zhou YQ, Tong Y, Wan S (2005) Inhibition of NADPH oxidase attenuates vasospasm after experimental subarachnoid hemorrhage in rats. Stroke 36:1059–1064
10. Tang J, Liu J, Zhou C, Ostanin D, Grisham MB, Neil GD, Zhang JH (2005) Role of NADPH oxidase in the brain injury of intracerebral hemorrhage. J Neurochem 94:1342–1350
11. Paravicini TM, Drummond GR, Sobey CG (2004) Reactive oxygen species in the cerebral circulation: physiological roles and therapeutic implications for hypertension and stroke. Drugs 64:2143–2157
12. Hacke W, Kaste M, Bluhmki E, Brozman M, Davalos A, Guidetti D, Larrue V, Lees KR, Medeghri Z, Machnig T, Schneider D, von KR, Wahlgren N, Toni D (2008) Thrombolysis with alteplase 3 to 4.5 hours after acute ischemic stroke. N Engl J Med 359:1317–1329
13. Margaill I, Plotkine M, Lerouet D (2005) Antioxidant strategies in the treatment of stroke. Free Radic Biol Med 39:429–443
14. Uemura Y, Miller JM, Matson WR, Beal MF (1991) Neurochemical analysis of focal ischemia in rats. Stroke 22:1548–1553
15. Serteser M, Ozben T, Gumuslu S, Balkan S, Balkan E (2002) Lipid peroxidation in rat brain during focal cerebral ischemia: prevention of malondialdehyde and lipid conjugated diene production by a novel antiepileptic, lamotrigine. Neurotoxicology 23:111–119
16. Lancelot E, Callebert J, Revaud ML, Boulu RG, Plotkine M (1995) Detection of hydroxyl radicals in rat striatum during transient focal cerebral ischemia: possible implication in tissue damage. Neurosci Lett 197:85–88
17. Kato N, Yanaka K, Nagase S, Hirayama A, Nose T (2003) The antioxidant EPC-K1 ameliorates brain injury by inhibiting lipid peroxidation in a rat model of transient focal cerebral ischaemia. Acta Neurochir (Wien) 145:489–493
18. Mizui T, Kinouchi H, Chan PH (1992) Depletion of brain glutathione by buthionine sulfoximine enhances cerebral ischemic injury in rats. Am J Physiol 262:H313–H317
19. Choi-Kwon S, Park KA, Lee HJ, Park MS, Lee JH, Jeon SE, Choe MA, Park KC (2004) Temporal changes in cerebral antioxidant enzyme activities after ischemia and reperfusion in a rat focal brain ischemia model: effect of dietary fish oil. Brain Res Dev Brain Res 152:11–18
20. Alexandrova M, Bochev P, Markova V, Bechev B, Popova M, Danovska M, Simeonova V (2004) Dynamics of free radical processes in acute ischemic stroke: influence on neurological status and outcome. J Clin Neurosci 11:501–506
21. Demirkaya S, Topcuoglu MA, Aydin A, Ulas UH, Isimer AI, Vural O (2001) Malondialdehyde, glutathione peroxidase and superoxide dismutase in peripheral blood erythrocytes of patients with acute cerebral ischemia. Eur J Neurol 8:43–51
22. Moro MA, Almeida A, Bolanos JP, Lizasoain I (2005) Mitochondrial respiratory chain and free radical generation in stroke. Free Radic Biol Med 39:1291–1304
23. Chan PH (2001) Reactive oxygen radicals in signaling and damage in the ischemic brain. J Cereb Blood Flow Metab 21:2–14
24. Warner DS, Sheng H, Batinic-Haberle I (2004) Oxidants, antioxidants and the ischemic brain. J Exp Biol 207:3221–3231

25. Nogawa S, Zhang F, Ross ME, Iadecola C (1997) Cyclo-oxygenase-2 gene expression in neurons contributes to ischemic brain damage. J Neurosci 17:2746–2755
26. Betz AL, Randall J, Martz D (1991) Xanthine oxidase is not a major source of free radicals in focal cerebral ischemia. Am J Physiol 260:H563–H568
27. Beray-Berthat V, Croci N, Plotkine M, Margaill I (2003) Polymorphonuclear neutrophils contribute to infarction and oxidative stress in the cortex but not in the striatum after ischemia-reperfusion in rats. Brain Res 987:32–38
28. Fabian RH, Kent TA (1999) Superoxide anion production during reperfusion is reduced by an antineutrophil antibody after prolonged cerebral ischemia. Free Radic Biol Med 26: 355–361
29. Bedard K, Krause KH (2007) The NOX family of ROS-generating NADPH oxidases: physiology and pathophysiology. Physiol Rev 87:245–313
30. Lambeth JDNOX (2004) enzymes and the biology of reactive oxygen. Nat Rev Immunol 4:181–189
31. Serrano F, Kolluri NS, Wientjes FB, Card JP, Klann E (2003) NADPH oxidase immunoreactivity in the mouse brain. Brain Res 988:193–198
32. Bianca VD, Dusi S, Bianchini E, Dal Pra I, Rossi F (1999) Beta-amyloid activates the O-2 forming NADPH oxidase in microglia, monocytes, and neutrophils. A possible inflammatory mechanism of neuronal damage in Alzheimer's disease. J Biol Chem 274:15493–15499
33. Green SP, Cairns B, Rae J, Errett-Baroncini C, Hongo JA, Erickson RW, Curnutte JT (2001) Induction of gp91-phox, a component of the phagocyte NADPH oxidase, in microglial cells during central nervous system inflammation. J Cereb Blood Flow Metab 21:374–384
34. Mayorov DN (2007) Brain superoxide as a key regulator of the cardiovascular response to emotional stress in rabbits. Exp Physiol 92:471–479
35. Peterson JR, Sharma RV, Davisson RL (2006) Reactive oxygen species in the neuropathogenesis of hypertension. Curr Hypertens Rep 8:232–241
36. Vallet P, Charnay Y, Steger K, Ogier-Denis E, Kovari E, Herrmann F, Michel JP, Szanto I (2005) Neuronal expression of the NADPH oxidase NOX4, and its regulation in mouse experimental brain ischemia. Neuroscience 132:233–238
37. Kawahara T, Quinn MT, Lambeth JD (2007) Molecular evolution of the reactive oxygen-generating NADPH oxidase (Nox/Duox) family of enzymes. BMC Evol Biol 7:109
38. de la Monte SM, Wands JR (2006) Molecular indices of oxidative stress and mitochondrial dysfunction occur early and often progress with severity of Alzheimer's disease. J Alzheimers Dis 9:167–181
39. Banfi B, Malgrange B, Knisz J, Steger K, Dubois-Dauphin M, Krause KH (2004) NOX3, a superoxide-generating NADPH oxidase of the inner ear. J Biol Chem 279:46065–46072
40. Infanger DW, Sharma RV, Davisson RL (2006) NADPH oxidases of the brain: distribution, regulation, and function. Antioxid Redox Signal 8:1583–1596
41. Lassegue B, Clempus RE (2003) Vascular NAD(P)H oxidases: specific features, expression, and regulation. Am J Physiol Regul Integr Comp Physiol 285:R277–R297
42. Miller AA, Drummond GR, Schmidt HH, Sobey CG (2005) NADPH oxidase activity and function are profoundly greater in cerebral versus systemic arteries. Circ Res 97:1055–1062
43. Paravicini TM, Chrissobolis S, Drummond GR, Sobey CG (2004) Increased NADPH-oxidase activity and Nox4 expression during chronic hypertension is associated with enhanced cerebral vasodilatation to NADPH in vivo. Stroke 35:584–589
44. Ago T, Kitazono T, Kuroda J, Kumai Y, Kamouchi M, Ooboshi H, Wakisaka M, Kawahara T, Rokutan K, Ibayashi S, Iida M (2005) NAD(P)H oxidases in rat basilar arterial endothelial cells. Stroke 36:1040–1046
45. Miller AA, Drummond GR, Mast AE, Schmidt HH, Sobey CG (2007) Effect of gender on NADPH-oxidase activity, expression, and function in the cerebral circulation: role of estrogen. Stroke 38:2142–2149
46. Miller AA, Drummond GR, De Silva TM, Mast AE, Hickey H, Williams JP, Broughton BR, Sobey CG (2009) NADPH oxidase activity is higher in cerebral versus systemic arteries of four animal species: role of Nox2. Am J Physiol Heart Circ Physiol 296:H220–H225

47. Sorce S, Krause KH (2009) NOX enzymes in the central nervous system: from signaling to disease. Antioxid Redox Signal 11(10):2365–2370
48. Li WG, Miller FJ Jr, Zhang HJ, Spitz DR, Oberley LW, Weintraub NL (2001) H(2)O(2)-induced O(2) production by a non-phagocytic NAD(P)H oxidase causes oxidant injury. J Biol Chem 276:29251–29256
49. Sun GY, Horrocks LA, Farooqui AA (2007) The roles of NADPH oxidase and phospholipases A2 in oxidative and inflammatory responses in neurodegenerative diseases. J Neurochem 103:1–16
50. Barnett ME, Madgwick DK, Takemoto DJ (2007) Protein kinase C as a stress sensor. Cell Signal 19:1820–1829
51. Schafer M, Schafer C, Ewald N, Piper HM, Noll T (2003) Role of redox signaling in the autonomous proliferative response of endothelial cells to hypoxia. Circ Res 92:1010–1015
52. Al-Mehdi AB, Zhao G, Dodia C, Tozawa K, Costa K, Muzykantov V, Ross C, Blecha F, Dinauer M, Fisher AB (1998) Endothelial NADPH oxidase as the source of oxidants in lungs exposed to ischemia or high K+. Circ Res 83:730–737
53. Ozaki M, Deshpande SS, Angkeow P, Bellan J, Lowenstein CJ, Dinauer MC, Goldschmidt-Clermont PJ, Irani K (2000) Inhibition of the Rac1 GTPase protects against nonlethal ischemia/reperfusion-induced necrosis and apoptosis in vivo. FASEB J 14:418–429
54. Abramov AY, Scorziello A, Duchen MR (2007) Three distinct mechanisms generate oxygen free radicals in neurons and contribute to cell death during anoxia and reoxygenation. J Neurosci 27:1129–1138
55. Hong H, Zeng JS, Kreulen DL, Kaufman DI, Chen AF (2006) Atorvastatin protects against cerebral infarction via inhibition of NADPH oxidase-derived superoxide in ischemic stroke. Am J Physiol Heart Circ Physiol 291:H2210–H2215
56. Suh SW, Shin BS, Ma H, Van HM, Brennan AM, Yenari MA, Swanson RA (2008) Glucose and NADPH oxidase drive neuronal superoxide formation in stroke. Ann Neurol 64: 654–663
57. Kahles T, Luedike P, Endres M, Galla HJ, Steinmetz H, Busse R, Neumann-Haefelin T, Brandes RP (2007) NADPH oxidase plays a central role in blood-brain barrier damage in experimental stroke. Stroke 38:3000–3006
58. Walder CE, Green SP, Darbonne WC, Mathias J, Rae J, Dinauer MC, Curnutte JT, Thomas GR (1997) Ischemic stroke injury is reduced in mice lacking a functional NADPH oxidase. Stroke 28:2252–2258
59. Jackman KA, Miller AA, De Silva TM, Crack PJ, Drummond GR, Sobey CG (2009) Reduction of cerebral infarct volume by apocynin requires pretreatment and is absent in Nox2-deficient mice. Br J Pharmacol 156:680–688
60. Tang XN, Cairns B, Cairns N, Yenari MA (2008) Apocynin improves outcome in experimental stroke with a narrow dose range. Neuroscience 154:556–562
61. Tang LL, Ye K, Yang XF, Zheng JS (2007) Apocynin attenuates cerebral infarction after transient focal ischaemia in rats. J Int Med Res 35:517–522
62. Wang Q, Tompkins KD, Simonyi A, Korthuis RJ, Sun AY, Sun GY (2006) Apocynin protects against global cerebral ischemia-reperfusion-induced oxidative stress and injury in the gerbil hippocampus. Brain Res 1090:182–189
63. Heumuller S, Wind S, Barbosa-Sicard E, Schmidt HH, Busse R, Schroder K, Brandes RP (2008) Apocynin is not an inhibitor of vascular NADPH oxidases but an antioxidant. Hypertension 51:211–217
64. Lehoux S, Castier Y, Tedgui A (2006) Molecular mechanisms of the vascular responses to haemodynamic forces. J Intern Med 259:381–392
65. Shepro D, Morel NM (1993) Pericyte physiology. FASEB J 7:1031–1038
66. Fisher M (2009) Pericyte signaling in the neurovascular unit. Stroke 40:S13–S15
67. Wolburg H, Noell S, Mack A, Wolburg-Buchholz K, Fallier-Becker P (2009) Brain endothelial cells and the glio-vascular complex. Cell Tissue Res 335:75–96
68. Takano T, Tian GF, Peng W, Lou N, Libionka W, Han X, Nedergaard M (2006) Astrocyte-mediated control of cerebral blood flow. Nat Neurosci 9:260–267

69. Gordon GR, Mulligan SJ, MacVicar BA (2007) Astrocyte control of the cerebrovasculature. Glia 55:1214–1221
70. Hamann GF, del Zoppo GJ, Von KR (1999) Hemorrhagic transformation of cerebral infarction–possible mechanisms. Thromb Haemost 82(Suppl 1):92–94
71. Rosell A, Cuadrado E, Ortega-Aznar A, Hernandez-Guillamon M, Lo EH, Montaner J (2008) MMP-9-positive neutrophil infiltration is associated to blood-brain barrier breakdown and basal lamina type IV collagen degradation during hemorrhagic transformation after human ischemic stroke. Stroke 39:1121–1126
72. Amantea D, Nappi G, Bernardi G, Bagetta G, Corasaniti MT (2009) Post-ischemic brain damage: pathophysiology and role of inflammatory mediators. FEBS J 276: 13–26
73. Persidsky Y, Ramirez SH, Haorah J, Kanmogne GD (2006) Blood-brain barrier: structural components and function under physiologic and pathologic conditions. J Neuroimmune Pharmacol 1:223–236
74. Dirnagl U, Iadecola C, Moskowitz MA (1999) Pathobiology of ischaemic stroke: an integrated view. Trends Neurosci 22:391–397
75. Hackett PH (1999) High altitude cerebral edema and acute mountain sickness. A pathophysiology update. Adv Exp Med Biol 474:23–45
76. Zhao BQ, Tejima E, Lo EH (2007) Neurovascular proteases in brain injury, hemorrhage and remodeling after stroke. Stroke 38:748–752
77. Heo JH, Lucero J, Abumiya T, Koziol JA, Copeland BR, del Zoppo GJ (1999) Matrix metalloproteinases increase very early during experimental focal cerebral ischemia. J Cereb Blood Flow Metab 19:624–633
78. Yamaguchi T, Sano K, Takakura K, Saito I, Shinohara Y, Asano T, Yasuhara H (1998) Ebselen in acute ischemic stroke: a placebo-controlled, double-blind clinical trial. Ebselen Study Group. Stroke 29:12–17
79. Ogawa A, Yoshimoto T, Kikuchi H, Sano K, Saito I, Yamaguchi T, Yasuhara H (1999) Ebselen in acute middle cerebral artery occlusion: a placebo-controlled, double-blind clinical trial. Cerebrovasc Dis 9:112–118
80. Tirilazad International Steering Committee (2000) Tirilazad mesylate in acute ischemic stroke: A systematic review. Stroke 31:2257–2265
81. van der Worp HB, Kappelle LJ, Algra A, Bar PR, Orgogozo JM, Ringelstein EB, Bath PM, van GJ (2002) The effect of tirilazad mesylate on infarct volume of patients with acute ischemic stroke. Neurology 58:133–135
82. Effect of a novel free radical scavenger, edaravone (MCI-186), on acute brain infarction (2003) Randomized, placebo-controlled, double-blind study at multicenters. Cerebrovasc Dis 15:222–229
83. Toyoda K, Fujii K, Kamouchi M, Nakane H, Arihiro S, Okada Y, Ibayashi S, Iida M (2004) Free radical scavenger, edaravone, in stroke with internal carotid artery occlusion. J Neurol Sci 221:11–17
84. Lees KR, Zivin JA, Ashwood T, Davalos A, Davis SM, Diener HC, Grotta J, Lyden P, Shuaib A, Hardemark HG, Wasiewski WW (2006) NXY-059 for acute ischemic stroke. N Engl J Med 354:588–600
85. Shuaib A, Lees KR, Lyden P, Grotta J, Davalos A, Davis SM, Diener HC, Ashwood T, Wasiewski WW, Emeribe U (2007) NXY-059 for the treatment of acute ischemic stroke. N Engl J Med 357:562–571
86. Savitz SI, Schabitz WRA (2008) Critique of SAINT II: wishful thinking, dashed hopes, and the future of neuroprotection for acute stroke. Stroke 39:1389–1391
87. Rosenberg GA, Yang Y (2007) Vasogenic edema due to tight junction disruption by matrix metalloproteinases in cerebral ischemia. Neurosurg Focus 22:E4
88. Baron JC (1999) Mapping the ischaemic penumbra with PET: implications for acute stroke treatment. Cerebrovasc Dis 9:193–201
89. Ayata C, Ropper AH (2002) Ischaemic brain oedema. J Clin Neurosci 9:113–124

90. Kuroiwa T, Ting P, Martinez H, Klatzo I (1985) The biphasic opening of the blood-brain barrier to proteins following temporary middle cerebral artery occlusion. Acta Neuropathol 68:122–129
91. Wang X, Tsuji K, Lee SR, Ning M, Furie KL, Buchan AM, Lo EH (2004) Mechanisms of hemorrhagic transformation after tissue plasminogen activator reperfusion therapy for ischemic stroke. Stroke 35:2726–2730
92. Hamann GF, Okada Y, Fitridge R, del Zoppo GJ (1995) Microvascular basal lamina antigens disappear during cerebral ischemia and reperfusion. Stroke 26:2120–2126
93. Maier CM, Hsieh L, Crandall T, Narasimhan P, Chan PH (2006) Evaluating therapeutic targets for reperfusion-related brain hemorrhage. Ann Neurol 59:929–938
94. Liu S, Liu M, Peterson S, Miyake M, Vallyathan V, Liu KJ (2003) Hydroxyl radical formation is greater in striatal core than in penumbra in a rat model of ischemic stroke. J Neurosci Res 71:882–888
95. Heo JH, Han SW, Lee SK (2005) Free radicals as triggers of brain edema formation after stroke. Free Radic Biol Med 39:51–70
96. Kato N, Yanaka K, Hyodo K, Homma K, Nagase S, Nose T (2003) Stable nitroxide Tempol ameliorates brain injury by inhibiting lipid peroxidation in a rat model of transient focal cerebral ischemia. Brain Res 979:188–193
97. Kim GW, Lewen A, Copin J, Watson BD, Chan PH (2001) The cytosolic antioxidant, copper/zinc superoxide dismutase, attenuates blood-brain barrier disruption and oxidative cellular injury after photothrombotic cortical ischemia in mice. Neuroscience 105: 1007–1018
98. Abbott NJ (2005) Dynamics of CNS barriers: evolution, differentiation, and modulation. Cell Mol Neurobiol 25:5–23
99. Diekmann D, Abo A, Johnston C, Segal AW, Hall A (1994) Interaction of Rac with p67phox and regulation of phagocytic NADPH oxidase activity. Science 265:531–533
100. Jin L, Ying Z, Webb RC (2004) Activation of Rho/Rho kinase signaling pathway by reactive oxygen species in rat aorta. Am J Physiol Heart Circ Physiol 287:H1495–H1500
101. Ushio-Fukai M (2007) VEGF signaling through NADPH oxidase-derived ROS. Antioxid Redox Signal 9:731–739
102. Gertzberg N, Neumann P, Rizzo V, Johnson A (2004) NAD(P)H oxidase mediates the endothelial barrier dysfunction induced by TNF-alpha. Am J Physiol Lung Cell Mol Physiol 286:L37–L48
103. Al-Shabrawey M, Rojas M, Sanders T, Behzadian A, El-Remessy A, Bartoli M, Parpia AK, Liou G, Caldwell RB (2008) Role of NADPH oxidase in retinal vascular inflammation. Invest Ophthalmol Vis Sci 49:3239–3244
104. Schreibelt G, Kooij G, Reijerkerk A, van DR, Gringhuis SI, van der PS, Weksler BB, Romero IA, Couraud PO, Piontek J, Blasig IE, Dijkstra CD, Ronken E, De Vries HE (2007) Reactive oxygen species alter brain endothelial tight junction dynamics via RhoA, PI3 kinase, and PKB signaling. FASEB J 21:3666–3676
105. Fischer S, Wiesnet M, Renz D, Schaper W (2005) $H_2O_2$ induces paracellular permeability of porcine brain-derived microvascular endothelial cells by activation of the p44/42 MAP kinase pathway. Eur J Cell Biol 84:687–697
106. Haorah J, Ramirez SH, Schall K, Smith D, Pandya R, Persidsky Y (2007) Oxidative stress activates protein tyrosine kinase and matrix metalloproteinases leading to blood-brain barrier dysfunction. J Neurochem 101:566–576
107. Woodcock SA, Rooney C, Liontos M, Connolly Y, Zoumpourlis V, Whetton AD, Gorgoulis VG, Malliri A (2009) SRC-induced disassembly of adherens junctions requires localized phosphorylation and degradation of the rac activator tiam1. Mol Cell 33: 639–653
108. Jin ZG, Melaragno MG, Liao DF, Yan C, Haendeler J, Suh YA, Lambeth JD, Berk BC, Cyclophilin A (2000) is a secreted growth factor induced by oxidative stress. Circ Res 87:789–796

109. Pyne-Geithman GJ, Morgan CJ, Wagner K, Dulaney EM, Carrozzella J, Kanter DS, Zuccarello M, Clark JF (2005) Bilirubin production and oxidation in CSF of patients with cerebral vasospasm after subarachnoid hemorrhage. J Cereb Blood Flow Metab 25:1070–1077
110. Kuhlmann CR, Tamaki R, Gamerdinger M, Lessmann V, Behl C, Kempski OS, Luhmann HJ (2007) Inhibition of the myosin light chain kinase prevents hypoxia-induced blood-brain barrier disruption. J Neurochem 102:501–507
111. Gasche Y, Copin JC, Sugawara T, Fujimura M, Chan PH (2001) Matrix metalloproteinase inhibition prevents oxidative stress-associated blood-brain barrier disruption after transient focal cerebral ischemia. J Cereb Blood Flow Metab 21:1393–1400
112. Liu W, Sood R, Chen Q, Sakoglu U, Hendren J, Cetin O, Miyake M, Liu KJ (2008) Normobaric hyperoxia inhibits NADPH oxidase-mediated matrix metalloproteinase-9 induction in cerebral microvessels in experimental stroke. J Neurochem 107:1196–1205
113. Amiry-Moghaddam M, Ottersen OP (2003) The molecular basis of water transport in the brain. Nat Rev Neurosci 4:991–1001
114. Manley GT, Fujimura M, Ma T, Noshita N, Filiz F, Bollen AW, Chan P, Verkman AS (2000) Aquaporin-4 deletion in mice reduces brain edema after acute water intoxication and ischemic stroke. Nat Med 6:159–163

# Chapter 12
# Smoking-Induced Oxidative Stress in the Pathogenesis of Cardiovascular Diseases

David Bernhard

**Abstract** Cigarette smoke, the prototypical form of oxidative stressor for the vascular system, is a complex mix of thousands of oxidative and nonoxidative compounds. Although it is still not clear which smoke chemicals are responsible for CVD initiation and progression, some important chemicals have been identified. Importantly, many cigarette smoke chemicals increase the oxidative burden for the vasculature directly, but also indirectly, e.g., by causing cellular damage and inflammation. The processes underlying the impact of smoking on CVD initiation and progression is discussed in this chapter.

**Keywords** Cigarette smoke · Smoking · Oxidants · Radicals · Endothelial · Atherosclerosis · Arteriosclerosis · Metals · Cd · Superoxide · NO · Inflammation · Necrosis · Thrombosis · Pathophysiology · Smooth muscle cells · Plaque · Macrophages · Lipids · Oxidised LDL · Lymphocytes

## Abbreviations

| | |
|---|---|
| WHO | World Health Organisation |
| CVDs | cardiovascular diseases |
| CS | cigarette smoke |
| LPS | lipopolysaccharides |
| FMD | flow-mediated dilation |
| ECM | extracellular matrix |

D. Bernhard (✉)
Cardiac Surgery – Research Laboratories, Department of Surgery, Medical University of Vienna/AKH, Ebene 8, G09/07, Währinger Gürtel 18-20, A-1090 Vienna, Austria
e-mail: david.bernhard@meduniwien.ac.at

H. Sauer et al. (eds.), *Studies on Cardiovascular Disorders*,
Oxidative Stress in Applied Basic Research and Clinical Practice,
DOI 10.1007/978-1-60761-600-9_12, © Springer Science+Business Media, LLC 2010

## 12.1 Introduction

The prototypic source for oxidant delivery to the human body, smoking, reduces the life span of humans by 6 years and disease-free life by 13 years. By the year 2020 tobacco consumption will kill over 8 million humans annually worldwide, making it the number one single cause of death in the world. The number of smokers and smoking-related deaths, which are mainly due to cancers, chronic obstructive pulmonary disease, and CVDs, still raises dramatically, and smoking is becoming more and more a problem of the developing world. The WHO estimates that by 2030 80% of all worldwide tobacco-related deaths will occur in developing countries. Accordingly, the role of smoking as a risk factor for CVD—which is often referred to as the most easily preventable risk factor—continues to gain relevance not only as a health burden for individuals, but also as a burden for health care systems and our societies.

Despite intensive, mainly clinical studies clearly correlating the risk for CVDs with the smoking status of humans, the identity of the CVD-causing agents in CS is still not known. Although some CVD-causing and -promoting CS chemicals have been identified, the data available so far are sufficient to make clear that CVD initiation and development cannot be ascribed to a single CS component, but rather to a mix of various smoke chemicals. A crucial role of radicals and oxidants in the CVD-causing and -promoting activity of CS is assured.

In this chapter the interaction between the classical pro-oxidative CVD risk factor, smoking, which accounts for up to 30% of all CVDs, and the pathophysiological processes in CVD initiation and progression will be discussed.

## 12.2 Smoking as a Source for Oxidative Stress in the Cardiovascular System

### *12.2.1 Generation of Oxidants and Radicals by Combustion of Cigarette Constituents*

Upon lighting a cigarette, a combustion process is initiated in the cigarette. The cigarette components that are burned and later inhaled by the smoker range from cigarette paper, adhesives, print colour, and tobacco, to several hundred different tobacco additives (~10% of the cigarette filler). The temperature in the glow of a cigarette is about 500°C and increases to about 900°C during puffing, generating a large amount of different redox active and inactive chemicals, both of which are inhaled by the smoker. These chemicals can roughly be split into chemicals of the tar fraction and those of the volatile phase. Both fractions can be separated by a Cambridge filter, trapping the particulate matter (i.e., the tar fraction; particle size > 1 μm), and letting the volatile gas fraction pass. The mainstream smoke (i.e., the smoke that is actively inhaled by drawing the smoke through the lit cigarette) comprises 8% particulate matter (tar fraction) and 92% gaseous chemicals (volatile fraction) [1]. The tar fraction precipitates mainly on the epithelia of the respiratory tract, and only very small particles (< 2.5 μm) and volatile cigarette smoke constituents reach the alveoli. Filtered by the lung, volatile, mainly hydrophilic compounds, reach the circulation [2]. Nevertheless, the tar fraction contains a range of hydrophobic compounds such as polycyclic aromatic hydrocarbons (PAHs), which can pass through cellular membranes, enabling them also to enter the circulation [3]. It is estimated that cigarette smoke contains about 4800 different compounds, but this number is basically defined by the lower limit of detection of the analysis applied, and is therefore probably much higher. Because of incomplete combustion, but also as a result of ionisation processes during smoking, a large number of different oxidants and radicals are generated and inhaled by the smoker, ranging from small volatile molecules (mainly small carbon- and oxygen-centred radicals [4, 5]) to large long living organic radicals (e.g., quinone/hydroquinone complex type radicals that reduce molecular oxygen to form superoxide). Pryor et al. reported that 1 g of the volatile fraction of cigarette smoke contains approximately $10^{15}$ short-lived highly reactive radicals and that 1 g of the tar fraction of cigarette smoke contains approximately $10^{17}$ long-lived radicals, indicating the massive additional oxidative burden for the human body of cigarette smoking.

## *12.2.2 Secondary Generation of Oxidants and Radicals by Cigarette Smoke in the Cardiovascular System*

Apart from the direct delivery of oxidants and radicals to the human body, cigarette smoke also increases the oxidative burden for the vascular system indirectly by causing necrotic cell death and inflammation, by reducing oxidant defence enzymes, and by upregulating pro-oxidative enzymes.

### 12.2.2.1 Secondary Oxidative Stress in the Vessel Wall by Smoking-Caused Inflammation

Direct and indirect generation of oxidants and radicals, but also other CS-related processes (e.g., adduct formation) lead to the modification of macromolecules (lipids, sugars, nucleic acids, and proteins). Many of these modifications alter the structure and function of these molecules, which leads to malfunctions in signal transduction or changes in the structure and steady state of cells. At a certain amount of cellular malfunction or damage, cells activate repair processes, and in case the cellular repair does not suffice to deal with the damage, cells activate cell death programs [6, 7]. In the case of cigarette smoke it was shown, also by our group, that cell demise occurs via apoptosis [8], but also via necrosis [6]. In contrast to classical programmed cell death (apoptosis), necrotic cell death is characterised by a rupture of the plasma membrane and a leakage of cytosolic constituents into the environment. As a result of this leakage, macrophages and dendritic cells are attracted and activated [9, 10], and consequently initiate an inflammatory cascade in the vessel wall. In the course of this inflammation, hydrogen peroxide and hydroxyl radicals are produced that are meant to kill infectious agents. However, in the absence of pathogens, a vicious circle of cell damage and inflammation is initiated, leading to an enduring local and systemic pro-oxidative state in the smokers' circulatory system.

In addition, CS was also demonstrated to directly activate endothelial cells, leading to the release of a number of factors like IL-6 and Weibel-Palade body–contained vonWillebrand factor, P-selectin, and IL-8, mediators known to contribute to inflammation and chemoattraction of the above cell types. CS chemicals were shown to provoke the above-mentioned response in vitro and in vivo, which crucially contributes to infiltration of the vessel wall by cells of the immune system, inflammation, and oxidative stress [7]. These events cause further damage to cells of the vasculature and thereby again contribute to a vicious circle of damage, inflammation, and oxidative stress, which has been shown to endure for decades, even if smokers quit smoking.

Further, cigarette smoke contains considerable amounts of fungal and bacterial constituents like LPS [11]. LPS are a part of bacterial cell walls which activate macrophages and endothelial cells by binding to TOLL-like receptors (TLR). Upon activation, TLRs induce translocation of NFkappaB into the nucleus, resulting in enhanced transcription and the secretion of pro-inflammatory cytokines and chemokines. This process contributes to inflammation of the lung in the course of smoking. It is not clear whether sufficient amounts of LPS can enter the circulation; but a systemic inflammatory status, initiated by deposition of LPS in the lung, contributes to increased oxidative burden in the circulation, being per se a risk factor for atherosclerosis.

The-above mentioned modification of macromolecules by oxidants and radicals from cigarette smoking contributes to yet another source of inflammation and an oxidative state in the vessel wall. Modified macromolecules represent foreign antigens for the immune system. These modified self-antigens, also termed "alter self" molecules, are attacked by the immune system; and this process again causes cellular damage, inflammation, and oxidative stress.

#### 12.2.2.2 Oxidative Stress by Smoking-Caused Reduction of Physiological Antioxidants

Aerobic organisms, like humans, reduce oxygen to derive most of their energy. On the other hand, these organisms are also susceptible to the damaging effects of $O_2^\bullet$, $OH^\bullet$, and $H_2O_2$ that are produced during oxygen metabolism. The human body has developed safeguard mechanisms, i.e., antioxidants, to scavenge these ROS. However, the delicate balance between radical-generating and radical-scavenging systems is crucial in maintaining organismal and cellular homeostasis, and even a small shift or imbalance between the two systems is associated with many diseases. It has been reported that cigarette smoke disturbs this balance not only by generating ROS (see above), but also by keeping antioxidant systems busy and even by reducing the activity and amount of antioxidants and antioxidative enzymes. Accordingly, the adaptation and basal capacity of endogenous antioxidants and detoxifying enzymes plays a key role in the final biological outcome of smoking. Among the systems that crucially define an individual's susceptibility to cigarette smoke–caused damage, the balance between GSH and GSSG plays a major role. Cigarette smoking has been shown to cause a shift from GSH to GSSG, reducing the buffering capacity of this important antioxidative system. Further, amounts of the elements, like selenium and zinc, known to play a role in the active centre of antioxidative enzymes are reduced in the circulation of smokers [12]. In most cases the interaction of cigarette smoking and homeostasis of elements is not understood, but it may be linked to secondary changes in the physiology of various organs (e.g., kidneys) by cigarette smoking.

#### 12.2.2.3 Smoking-Mediated Modulation of Gene Expression as a Source for Cardiovascular Oxidative Stress

Genetic factors not only have been linked to the initiation and progression of CVDs, but could also explain the dramatically increased risk of certain individuals for CVDs. Based on the DNA sequences' variability between humans, the different sensitivities of individuals to the effects of cigarette smoke chemicals, including oxidants and radicals, are defined by their genetic backgrounds. These variations modify the otherwise linear relationship between cigarette smoking and disease conditions. On the other hand, cigarette smoke also modifies genes. Cigarette smoking affects genes both quantitatively (gene expression) and qualitatively (mutations). Both will not only directly cause biological effects such as CVDs, but also define how an individual will respond to cigarette smoking. The balance between smoke exposure and defence against oxidative smoke chemicals determines the variety in responses and outcomes among individuals. It is believed that these differences are largely determined by the above-mentioned genetic variation affecting the structure, function, and/or level of expression of candidate genes, which, in turn, differentially affect a variety of cellular and systemic processes. Further, it is anticipated that individual responses to cigarette smoking are determined by multiple genotype-environment interactions. This intricate smoke-gene interaction hypothesis is supported by numerous population-based studies, animal models, and in vitro experiments [13]. To counteract cigarette smoke–induced toxicity, the human body utilises detoxification systems such as the cytochrome P450 enzyme family and antioxidants, which may be more relevant to atherogenesis [14]; and antioxidant systems, including superoxide dismutase (SOD), cysteine, methionine, urate, and glutathione. SOD, by scavenging $O_2^{\bullet}$, plays a central role in preserving biological NO. This capacity can be adversely affected by factors such as hypercholesterolemia and/or cigarette smoking [15] (see also below). Glutathione reductase maintains high levels of reduced glutathione in the cytosol. Reduced glutathione is widely distributed intracellularly and is essential for the inhibition of oxidation [16]. There are several isozymes of glutathione peroxidases distributed in various cells and plasma which protect cells and other enzymes from oxidative damage by catalysing the reduction of hydrogen peroxide, lipid hydroperoxides, and organic hydroperoxide

[17]. Paraoxonase associated with HDL particles is another important antioxidant that combats oxidative stress [18]. Further smoke chemicals (e.g., acrolein) have been shown to modulate the expression of oxidant/radical–generating enzymes like NADPH oxidase—generating superoxide.

#### 12.2.2.4 Cigarette Smoke–Contained Metals, a Source for Chronic Oxidative Stress in the Vessel Wall

Tobacco plants have evolved mechanisms that allow for the uptake of trace elements from soil. In addition to essential elements, however, these mechanisms also allow for the uptake of toxic elements. For this reason tobacco plants are utilised to, e.g., detoxify metal-contaminated soil. In the case of tobacco farming, tobacco plants also take up toxic metals like Cd, which are then deposited in the leaves. Because of that, smoking is the most important source for cadmium uptake by humans [19]; and other metals like aluminium, arsenic, chromium, copper, iron, lead, nickel, and vanadium are also transferred to the human body by smoking [20]. After having entered the body, metals follow different routes for elimination, but are also deposited in various organs. Cd, for example, deposits in the liver, testis, and kidneys. Another system that accumulates high amounts of cadmium is the vasculature. Although the pathophysiological role of the metals delivered by smoking is not well understood, it is clear that a number of metals function as oxidation reaction catalysts. Copper and iron catalyse the formation of hydroxyl radicals via the Fenton reaction. Elements like cadmium exhaust glutathione pools and cause lipid peroxidation. An excellent review on oxidative mechanisms in metal toxicity was given by Stohs and Bagchi [21]. An important principle of metal delivery to the human body by smoking is that metals accumulate in the vasculature slowly over time. We hypothesised previously that metals which accumulate in the vessel wall may catalyse oxidation reactions locally, and thereby contribute to the initiation and progression of CVDs. Other phenomena that secondarily affect the circulatory system to cause CVDs, e.g., by causing kidney damage and failure, may also be at play.

## 12.3 Smoking-Caused Oxidative Stress as a Pathophysiological Factor in Cardiovascular Disease Initiation and Progression

### *12.3.1 The Role of Smoking-Caused Oxidative Stress in CVD Initiation*

#### 12.3.1.1 Impact of Smoking on Endothelial Function

Endothelial function plays a particularly important role in processes like the regulation of vascular tension, or the control of compound and cell transfer from the bloodstream into the vessel wall. In more or less all modern concepts regarding the initiation of atherogenesis, endothelial dysfunction plays a central role, or is even considered to be the first step on the way from a healthy vessel wall towards a diseased one. Cigarette smoking interferes with endothelial function in many ways.

Vascular tension is controlled by different endothelial-dependent and -independent systems. Endothelium-based vasorelaxation is induced by NO-mediated signalling. Cigarette smoke oxidants (also generated secondarily by metals catalysis), in particular $H_2O_2$ (hydrogen peroxide), $OH^-$, $OH^\bullet$ (hydroxyl radical), lipid radicals, and peroxides [5, 22, 23], interact with each other to form new oxidants like superoxide. Superoxide interacts with NO to form $ONOO^-$ (peroxynitrite) [24]. $ONOO^-$ formation–mediated reduction in NO-bioavailability may contribute to smoking-related hypertension and reduced FMD, a well know indicator of early atherosclerosis.

Apart from reducing NO availability, cigarette smoke oxidants and radicals directly cause endothelial cell damage and death (see above), and thereby cause a shift from a closed towards a more open endothelium. Impaired endothelial barrier function will allow for a less controlled transfer of compound and cells from the blood stream into the vessel wall, and may thereby contribute to enhanced lipid deposition in the vessel wall. Reduced barrier function will also allow monocytes and lymphocytes to more easily infiltrate into the vessel wall. Increased infiltration of these cell types will promote lipid deposition, fatty streak formation, and inflammation.

#### 12.3.1.2 Smoking and the Autoimmune Hypothesis of Atherosclerosis

It is generally accepted that atherosclerosis is characterised not just by deposition of lipids and hardening and thickening of the vessel wall, but that in itself, it is an inflammatory disease of the vessel wall. Inflammation is relevant throughout all stages of the disease, and modern concepts see a critical role for the immune system also at incipient stages of atherosclerosis [25]. It is thought that endothelial cell stress by hypercholesterolemia or hypertension, but also by smoking, causes the translocation of the mitochondrial heat shock protein 60 to the surface of endothelial cells, where cross-reactive antibacterial heat shock protein antibodies, generated during bacterial infections in childhood, bind to the surface of stressed endothelial cells, and induce cell damage, activation of T-cells, and again inflammation and oxidative stress. Recently we observed that CS extracts also induce the translocation of heat shock protein 60 to the surface of endothelial cells, and antibody-dependent cytotoxicity could be initiated by treating cells with CS extracts (unpublished observations). It is speculated that oxidative modification of endothelial cell surface proteins by CS alters their antigenic properties, resulting in immunologically relevant altered-self molecules on the surface of endothelial cells. A mechanism similar to heat shock protein–mediated cell damage would be the consequence, also resulting in endothelial damage and dysfunction.

#### 12.3.1.3 Smoking and Lipid Oxidation

Fatty streak formation in the vessel wall is the first macroscopically visible sign of beginning atherosclerosis. Fatty streaks indicate lipid deposits which have been proven to be one of the most crucial factors, if not the most crucial one, in atherosclerosis initiation. These lipid deposits are the result of the uptake of lipids by macrophages, which leads to their transformation into foam cells. The uptake of lipids into macrophages is normally a controlled process; however, upon oxidation of lipids (oxidised LDL, and minimally modified LDL), an uncontrolled uptake is initiated, resulting in foam cell formation, lipid overload, and eventually also in cell death. Cell death of lipid-laden macrophages is the basis for extracellular lipid stores in the vessel wall as a part of atherosclerotic plaques. In vivo studies could show that smokers have increased levels of antioxidised LDL antibodies [26], indicative of the increased presence of these modifications of LDL in smokers, and in vitro analyses could confirm the LDL oxidising activity of cigarette smoke. Further studies could demonstrate that in the course of foam cell formation, cells become activated and start secreting pro-inflammatory cytokines, which lead to further infiltration of the vessel wall by T-cells, and again to inflammation, and consequently to an increased oxidative burden in the vessel wall, and acceleration of lipid oxidation. Another smoking-induced vicious circle of oxidation and inflammation is initiated.

## *12.3.2 The Role of Smoking-Caused Oxidative Stress in CVD Progression*

### 12.3.2.1 Smoking-Mediated Oxidative Stress and Inflammation in CVD Progression

Primary but also secondary oxidative stress (e.g., by inflammation) plays a crucial role throughout CVD progression. Like the diseases themselves, with phases of no or slow progression and enhanced progression, the load of oxidative stress will also change over time. In smokers an enduring inflammatory status can be observed even years after cessation; and hand in hand with inflammation, the systemic oxidative burden will also be present in the human body. The origin or the location of the inflammatory focus is not entirely clear, but it is likely to involve the lung and the vascular wall. Lung inflammation from smoking is due to direct tissue damage, as well as the ongoing burden for the lung and the immune system to deal with particulate matter, and toxic as well as foreign compounds. A systemic inflammatory status per se, e.g., indicated by increased levels of CRP, is a risk factor for CVDs, as is reduced lung function. Both phenomena contribute to smoking-caused CVD progression.

In the progression phase of CVDs, inflammation and chronic oxidative burden are constantly present, particularly in the active phases of CVDs. The deposition of metals in the vessel wall and their accumulation over years will lead to local catalysis of oxidation reactions favouring and enhancing modification of proteins, cellular damage, reduced NO availability, and lipid oxidation. As a consequence, the balance of the vascular system will be permanently shifted in favour of a pro-atherogenic state. In line with this, and apart from the direct damaging activity of excess levels of oxidants, these chemicals also have pivotal roles as signalling molecules. Typical consequences of oxidative stress are smooth muscle cell proliferation and migration

[27, 28], which constitute another hallmark not only in the initiation of CVDs but also in their progression. Increased proliferation of SMCs and migration into the intima results in further thickening of the vessel wall, tissue remodelling processes, and ultimately the genesis of advanced lesions and vulnerable plaques.

#### 12.3.2.2 Smoking-Induced Vascular Aging as a CVD-Promoting Factor

As discussed above, cigarette smoke–contained oxidants and radicals modify genes not only quantitatively, but also qualitatively by causing mutations. The accumulation of mutations not only results in the transformation of cells causing cancer, but is also an integral part of the aging process. If aging is defined as the process of deterioration of a system over time, smoking clearly accelerates aging of the human body, including the vasculature [29]. Apart from specific processes associated with aging, such as the shortening of telomeres [30], and the accumulation of mutations in the mitochondrial and nuclear genome, smoking promotes processes such as the formation of advanced glycation end-products and the aging of ECM [31], the generation of malfunctioning damaged proteins [32], alterations in cellular structures [6, 7], and the induction of inflammation and oxidative stress (see above). However, the exact mechanisms by which smoking contributes to the aging of the vascular system, which in turn contributes to CVD, still remain to be defined.

#### 12.3.2.3 Smoking, Oxidative Stress, and Thrombogenesis

Cigarette smoking is a well-known risk factor for acute cardiovascular events such as myocardial infarction. This observation is in part explained by the finding that smoking promotes thrombogenesis and alters fibrinolysis [33]. The exact pathophysiologies are not well understood, but smokers' platelets exhibit hyperaggregability, produce less NO, and are less responsive to exogenous NO [34–36]. Furthermore, in smokers, the equilibrium between antithrombotic and prothrombotic factors is disturbed in favour of the latter. Similarly, basal and stimulated t-PA release is reduced in cultured endothelial cells exposed to smokers' serum, suggesting a dysfunction in fibrinolysis [37]. The relevance of this observation for smoking-related CVD is underlined by the fact that smokers have increased circulating fibrinogen levels, which per se constitute a risk factor for acute cardiovascular events [38]. Importantly, inflammation and oxidative stress are linked to both phenomena [39].

### *12.3.3 Oxidative Stress–Independent Mechanisms in CVD Initiation and Progression*

#### 12.3.3.1 Nonoxidative Smoke Chemicals and CVD Initiation

Apart from endothelial damage by oxidants and radicals, nonoxidative cigarette smoke constituents have also been reported to interfere with endothelial homeostasis and CVD initiation. The best studied constituent of cigarette smoke, nicotine, was shown to cause endothelial damage in vitro and also to cause reduced FMD in vivo. Further nicotine-dependent reduction in NO levels was reported. Still, the mechanism of action of nicotine is not clear; but apart from nonoxidative effects, oxidative and carcinogenic phenomena have also been described.

In addition to nicotine, metals like Cd and lead also contribute to atherosclerosis initiation. Although some metals have clearly been shown to catalyse oxidation reactions directly, other metals may exert their atherogenic activity via mechanisms other than oxidation. Published results are not always free of contradictions, making clear conclusions on the precise activity of metals difficult; but apart from pro-oxidative activities, metals also contribute to nonoxidant pathways in CVD initiation. Polycyclic aromatic hydrocarbons, many of which belong to the group of long-living radicals, were also shown to exert atherogenic activity, apart from functioning as radicals. Via the binding to and activation of aryl hydrocarbon receptors, transcriptional cascades are activated that, for instance, also lead to the secretion of pro-inflammatory cytokines. Based on these findings, it can be concluded that oxidants and radicals, apart from causing direct effects due to their chemical nature, are involved in nonoxidative processes too; and also that nonoxidative compounds in cigarette smoke may secondarily contribute to an oxidative environment in the vessel wall. Oxidation, particularly cigarette smoke–caused oxidation, plays without doubt a central role in atherogenesis and CVD progression; but nonoxidation-based factors are also crucially involved in these processes.

#### 12.3.3.2 Smoking, Collagen Synthesis, and Plaque Stability

Atherosclerosis progression is classified as fatty streak, fatty plaque, atheroma, fibroatheroma, and complicated lesion—or AHA grades 0–6 [40], but these grades do not necessarily correlate with clinical severity. Advanced atherosclerotic lesions are characterised by a lipid core, which is surrounded by pools of extracellular lipids and foam cells, macrophages, lymphocytes, and vascular smooth muscle cells (VSMC). Extracellular matrix molecules can also be found in the space between cells. An atheroma is covered by a thin or thick fibrous cap, which can be symmetric or eccentric. Acute coronary syndromes (unstable angina and acute myocardial infarction) play a central role in clinical cardiology, and are the most deadly clinical manifestation of coronary artery disease. The basis for acute coronary syndromes is often acute thrombosis on ruptured or eroded plaques, even under conditions of low luminal stenosis [41]. The most important constituent of a fibrous cap is collagen. Especially at vulnerable sites of human atheroma, impaired collagen synthesis or increased collagen degradation can be observed [42]. Despite the clinical relevance of plaque rupture, the underlying pathophysiological processes are still unclear. Although the role of smoking and smoke chemicals in these processes has up to now not been properly studied, clinically phenomena are associated with smoking and may be explained by the effects of smoking on collagen production or degradation. Evidence that indicates this interaction includes the following:

1. Smokers have an increased risk for acute coronary syndromes [43] and are prone to develop thromboses [44]. In this case local processes, i.e., the inhibition of collagen synthesis, may play a role, and would lead to a thin fibrous cap that may rupture.
2. The smokers paradox describes the phenomenon that smokers have a reduced risk to develop restenosis after percutaneous coronary angioplasty [45, 46]. If CS inhibits collagen production in the restenosis site, it can then slow down the process of restenosis.
3. Smokers have an increased risk for arterial aneurysm [47]. Disorders in collagen production are thought to significantly contribute to the occurrence of aneurysm formation.
4. The wound healing properties in smokers are reduced [48]. It was previously shown that smoking reduces collagen production by switching off P4Hα1 expression [49].

## 12.4 Summary and Conclusions

Cigarette smoke with its ~4800 different constituents is probably the most complex risk factor for CVD. The contribution of cigarette smoking to CVD was confirmed by a very large number of clinical studies; and second-hand smoke has clearly been linked to CVDs in large studies. The number of potentially CVD-relevant compounds in cigarette smoke, and the larger number of interference sites in the initiation and progression of CVDs, has so far hampered the identification of the relevant CS chemicals. Still, as summarized in this chapter, many different oxidation processes are involved in the pathophysiology of CVDs. Accordingly, the major conclusion of this chapter is that oxidants/radicals play a central role in smoking-caused CVDs, but also that a lot of work is still needed to identify the major chemicals and pathways via which smoking causes CVDs.

**Acknowledgments** This project was supported by the Austrian National Bank (Project # 12697).

## References

1. Ambrose JA, Barua RS (2004) The pathophysiology of cigarette smoking and cardiovascular disease: an update. J Am Coll Cardiol 43:1731–1737
2. Burns DM (1991) Cigarettes and cigarette smoking. Clin Chest Med 12:631–642
3. Ding YS, Yan XJ, Jain RB, Lopp E, Tavakoli A, Polzin GM, Stanfill SB, Ashley DL, Watson CH (2006) Determination of 14 polycyclic aromatic hydrocarbons in mainstream smoke from U.S. brand and non-U.S. brand cigarettes. Environ Sci Technol 40:1133–1138
4. Church DF, Pryor WA (1985) Free-radical chemistry of cigarette smoke and its toxicological implications. Environ Health Perspect 64:111–126
5. Pryor WA, Church DF, Evans MD, Rice WY Jr, Hayes JR (1990) A comparison of the free radical chemistry of tobacco-burning cigarettes and cigarettes that only heat tobacco. Free Radic Biol Med 8:275–279

6. Bernhard D, Pfister G, Huck CW, Kind M, Salvenmoser W, Bonn GK, Wick G (2003) Disruption of vascular endothelial homeostasis by tobacco smoke: impact on atherosclerosis. FASEB J 17:2302–2304
7. Bernhard D, Csordas A, Henderson B, Rossmann A, Kind M, Wick G (2005) Cigarette smoke metal-catalyzed protein oxidation leads to vascular endothelial cell contraction by depolymerization of microtubules. FASEB J 19:1096–1107
8. Wang J, Wilcken DE, Wang XL (2001) Cigarette smoke activates caspase-3 to induce apoptosis of human umbilical venous endothelial cells. Mol Genet Metab 72:82–88
9. Basu S, Binder RJ, Suto R, Anderson KM, Srivastava PK (2000) Necrotic but not apoptotic cell death releases heat shock proteins, which deliver a partial maturation signal to dendritic cells and activate the NF-kappa B pathway. Int Immunol 12:1539–1546
10. Cocco RE, Ucker DS (2001) Distinct modes of macrophage recognition for apoptotic and necrotic cells are not specified exclusively by phosphatidylserine exposure. Mol Biol Cell 12:919–930
11. Hasday JD, Bascom R, Costa JJ, Fitzgerald T, Dubin W (1999) Bacterial endotoxin is an active component of cigarette smoke. Chest 115:829–835
12. Kafai MR, Ganji V (2003) Sex, age, geographical location, smoking, and alcohol consumption influence serum selenium concentrations in the USA: third National Health and Nutrition Examination Survey, 1988-1994. J Trace Elem Med Biol 17:13–18
13. Wang J, Dudley D, Wang XL (2002) Haplotype-specific effects on endothelial NO synthase promoter efficiency: modifiable by cigarette smoking. Arterioscler Thromb Vasc Biol 22:1–4
14. Bennett MR (2001) Reactive oxygen species and death: oxidative DNA damage in atherosclerosis. Circ Res 88:648–650
15. Pryor WA, Stone K (1993) Oxidants in cigarette smoke. Radicals, hydrogen peroxide, peroxynitrate, and peroxynitrite. Ann N Y Acad Sci 686:12–27
16. Arteel GE, Briviba K, Sies H (1999) Protection against peroxynitrite. FEBS Lett 445:226–230
17. Chu FF (1994) The human glutathione peroxidase genes GPX2, GPX3, and GPX4 map to chromosomes 14, 5, and 19, respectively. Cytogenet Cell Genet 66:96–98
18. Heinecke JW, Lusis AJ (1998) Paraoxonase-gene polymorphisms associated with coronary heart disease: support for the oxidative damage hypothesis? Am J Hum Genet 62:20–24
19. Bernhard D, Rossmann A, Henderson B, Kind M, Seubert A, Wick G (2006) Increased serum cadmium and strontium levels in young smokers: effects on arterial endothelial cell gene transcription. Arterioscler Thromb Vasc Biol 26:833–838
20. Bernhard D, Rossmann A, Wick G (2005) Metals in cigarette smoke. IUBMB Life 57: 805–809
21. Stohs SJ, Bagchi D (1995) Oxidative mechanisms in the toxicity of metal ions. Free Radic Biol Med 18:321–336
22. Burke A, Fitzgerald GA (2003) Oxidative stress and smoking-induced vascular injury. Prog Cardiovasc Dis 46:79–90
23. Church DF, Pryor WA (1985) Free-radical chemistry of cigarette smoke and its toxicological implications. Environ Health Perspect 64:111–126
24. Kojda G, Harrison D (1999) Interactions between NO and reactive oxygen species: pathophysiological importance in atherosclerosis, hypertension, diabetes and heart failure. Cardiovasc Res 43:562–571
25. Wick G, Knoflach M, Xu Q (2004) Autoimmune and inflammatory mechanisms in atherosclerosis. Annu Rev Immunol 22:361–403
26. Heitzer T, Yla-Herttuala S, Luoma J, Kurz S, Munzel T, Just H, Olschewski M, Drexler H (1996) Cigarette smoking potentiates endothelial dysfunction of forearm resistance vessels in patients with hypercholesterolemia. Role of oxidized LDL. Circulation 93:1346–1353
27. Herbert JM, Bono F, Savi P (1996) The mitogenic effect of H2O2 for vascular smooth muscle cells is mediated by an increase of the affinity of basic fibroblast growth factor for its receptor. FEBS Lett 395:43–47

28. Wang Z, Castresana MR, Newman WH (2001) Reactive oxygen and NF-kappaB in VEGF-induced migration of human vascular smooth muscle cells. Biochem Biophys Res Commun 285:669–674
29. Bernhard D, Moser C, Backovic A, Wick G (2007) Cigarette smoke—an aging accelerator? Exp Gerontol 42:160–165
30. Morla M, Busquets X, Pons J, Sauleda J, MacNee W, Agusti AG (2006) Telomere shortening in smokers with and without COPD. Eur Respir J 27:525–528
31. Vlassara H, Palace MR (2003) Glycoxidation: the menace of diabetes and aging. Mt Sinai J Med 70:232–241
32. Sun L, Luo C, Long J, Wei D, Liu J (2006) Acrolein is a mitochondrial toxin: Effects on respiratory function and enzyme activities in isolated rat liver mitochondria. Mitochondrion 6(3):136–142
33. Burke A, Fitzgerald GA (2003) Oxidative stress and smoking-induced vascular injury. Prog Cardiovasc Dis 46:79–90
34. Blache D (1995) Involvement of hydrogen and lipid peroxides in acute tobacco smoking-induced platelet hyperactivity. Am J Physiol 268:H679–H685
35. Ichiki K, Ikeda H, Haramaki N, Ueno T, Imaizumi T (1996) Long-term smoking impairs platelet-derived nitric oxide release. Circulation 94:3109–3114
36. Sawada M, Kishi Y, Numano F, Isobe M (2002) Smokers lack morning increase in platelet sensitivity to nitric oxide. J Cardiovasc Pharmacol 40:571–576
37. Newby DE, Wright RA, Labinjoh C, Ludlam CA, Fox KA, Boon NA, Webb DJ (1999) Endothelial dysfunction, impaired endogenous fibrinolysis, and cigarette smoking: a mechanism for arterial thrombosis and myocardial infarction. Circulation 99:1411–1415
38. Tuut M, Hense HW (2001) Smoking, other risk factors and fibrinogen levels. evidence of effect modification. Ann Epidemiol 11:232–238
39. Shebuski RJ, Kilgore KS (2002) Role of inflammatory mediators in thrombogenesis. J Pharmacol Exp Ther 300:729–735
40. McGill HC Jr, McMahan CA, Zieske AW, Tracy RE, Malcom GT, Herderick EE, Strong JP (2000) Association of coronary heart disease risk factors with microscopic qualities of coronary atherosclerosis in youth. Circulation 102:374–379
41. Libby P (2001) Current concepts of the pathogenesis of the acute coronary syndromes. Circulation 104:365–372
42. Libby P (1995) Molecular bases of the acute coronary syndromes. Circulation 91:2844–2850
43. Kennon S, Suliman A, MacCallum PK, Ranjadayalan K, Wilkinson P, Timmis AD (1998) Clinical characteristics determining the mode of presentation in patients with acute coronary syndromes. J Am Coll Cardiol 32:2018–2022
44. Burke AP, Farb A, Malcom GT, Liang YH, Smialek J, Virmani R (1997) Coronary risk factors and plaque morphology in men with coronary disease who died suddenly. N Engl J Med 336:1276–1282
45. Cohen DJ, Doucet M, Cutlip DE, Ho KK, Popma JJ, Kuntz RE (2001) Impact of smoking on clinical and angiographic restenosis after percutaneous coronary intervention: another smoker's paradox? Circulation 104:773–778
46. Hasdai D, Rihal CS, Lerman A, Grill DE, Holmes DR Jr (1997) Smokers undergoing percutaneous coronary revascularization present with fewer narrowings in the target coronary artery. Am J Cardiol 80:1212–1214
47. Ellamushi HE, Grieve JP, Jager HR, Kitchen ND (2001) Risk factors for the formation of multiple intracranial aneurysms. J Neurosurg 94:728–732
48. Krueger JK, Rohrich RJ (2001) Clearing the smoke: the scientific rationale for tobacco abstention with plastic surgery. Plast Reconstr Surg 108:1063–1073
49. Raveendran M, Senthil D, Utama B, Shen Y, Dudley D, Wang J, Zhang Y, Wang XL (2004) Cigarette suppresses the expression of P4Halpha and vascular collagen production. Biochem Biophys Res Commun 323:592–598

# Chapter 13
# Oxidative Stress in Vascular Aging

**Anna Csiszar and Zoltan Ungvari**

**Abstract** The free radical theory of aging postulates that macromolecular damage due to the increased production of reactive oxygen species (ROS) drives the aging process. While many of the original assumptions of the theory are currently debated, there is convincing evidence for the crucial role of ROS in the development of many age-associated diseases. Advancing age is one of the most significant risk factors for the development of atherosclerosis in the Western world. The theme that emerges from this overview is that aging is associated with both NAD(P)H oxidase- and mitochondria-derived ROS overproduction, which promotes inflammatory phenotypic alterations in the vascular wall, facilitating the development of atherosclerosis. We also discuss some of the possible therapeutic strategies by which age-related vascular oxidative stress and inflammation can be delayed or reversed, improving cardiovascular health in the elderly.

**Keywords** Aging · Oxidative stress · Free radical · Senescence · Cardiovascular disease

## 13.1 Introduction

The population in the Western world is aging. By 2030 one fourth of the European population will be 65 years of age or older and the majority of them will suffer from age-associated cardiovascular disease [1–3]. There is increasing evidence that in the absence of other risk factors, aging per se promotes development of atherosclerosis and increases the morbidity and mortality of myocardial infarction and stroke. Understanding the critical mechanisms underlying cardiovascular aging and age-related arterial pathophysiological alterations may hold promise in developing novel

A. Csiszar (✉)
Reynolds Oklahoma Center on Aging, Department of Geriatric Medicine, University of Oklahoma Health Sciences Center, 975 NE 10th Street, BRC-1313, Oklahoma City, OK 73104
e-mail: anna-csiszar@ouhsc.edu

H. Sauer et al. (eds.), *Studies on Cardiovascular Disorders*,
Oxidative Stress in Applied Basic Research and Clinical Practice,
DOI 10.1007/978-1-60761-600-9_13, 

interventional treatments for promoting cardiovascular health and extending the productive lifespan of older persons.

The free radical theory of aging, originally proposed by Harman [4] half a century ago, postulates that free radical reactions underlie aging. According to this theory, increased production of ROS results in a variety of macromolecular oxidative modifications with age, and the accumulation of such oxidative damage of proteins, lipids, and DNA is the primary causal factor in the aging process. There is considerable evidence that aging in mammals is associated with oxidative stress, and that oxidative macromolecular damage accrues with age in virtually every tissue studied [5–13]. The oxidative stress theory of aging predicts that longer-lived species should produce less ROS and/or exhibit superior resistance to the adverse effects of oxidative stress. In support of this theory, successfully aging species, including the naked mole rat (*Heterocephalus glaber*; maximum lifespan: >28 years), the white-footed mouse (*Peromyscus leucopus*; maximum lifespan: >8 years), and the little brown bat (*Myotis lucifugus*; maximum lifespan: >30 years), exhibit significantly lower arterial ROS production and/or superior cellular resistance to oxidative stress than shorter-living species, such as the house mouse (*Mus musculus*; maximum lifespan: ~3.5 years) [14–17]. Antioxidants neutralize ROS, and thereby may attenuate damage accrual. In lower organisms, overexpression of antioxidant enzymes and/or treatment with antioxidants seems to extend lifespan [18], which accords with the predictions of the free radical theory of aging. Experimental testing of the free radical theory of aging in mammals yielded mixed results [19–24]. For example, Schriner et al. found that mice that overexpress human catalase targeted to mitochondria exhibited increased life span [22]. Yet in other studies transgenic mice overexpressing other antioxidant enzymes do not exhibit an extended longevity phenotype [23, 24]. Furthermore, dietary supplementation with antioxidants does not appear to increase lifespan in mammals. Possible explanations for these observations include the compartmentalization of ROS production and ROS signaling and the species-specificity of the cause of death (the leading cause of mortality in mice is cancer, which may or may not be influenced by antioxidants).

The general concept that oxidative stress is involved in many age-related diseases, including development of coronary artery disease, cataract formation, and Alzheimer disease, appears robust. This overview focuses on emerging evidence that reactive oxygen species (ROS) play a central role in cardiovascular aging [1–3, 25, 26], and discusses the role of caloric restriction and treatment with the caloric restriction mimetic resveratrol in modulation of the endothelial oxidative stress response and prevention of cardiovascular disease during aging.

## 13.2 Oxidative Stress in Vascular Aging: Role of NAD(P)H Oxidases

There is strong evidence that oxidative stress develops with age in the arterial system both in humans [27–31] and in laboratory animals [8, 10–13]. An important consequence of increased oxidative stress in aging is a functional inactivation of endothelium-derived NO by high concentrations of $O_2^{\bullet -}$ [8, 11, 13, 27, 30, 32]. It is known that severe impairment of NO bioavailability decreases vasodilator capacity, thereby limiting tissue blood supply [9, 33]. It has been suggested that age-related decline in eNOS expression [8, 34–37] and/or a decreased intracellular L-arginine accessibility [38] can further aggravate the already impaired NO bioavailability.

One of the major sources of elevated $O_2^{\bullet -}$ production in aging is an increased activity of NAD(P)H oxidases [8, 12, 32, 39, 40]. Inhibition of NAD(P)H oxidases was shown to improve endothelial function in aged vessels from various vascular beds [8, 10, 32, 41]. NAD(P)H oxidase can be induced by inflammatory cytokines, and there is data suggesting that upregulation of TNFα in the aged vascular wall contributes to the increased NAD(P)H oxidase activation in aged vessels [41, 42] (see below). NAD(P)H oxidase activation was also suggested to underlie age-related alterations of cerebrovascular regulation [43]. Importantly, amyloid β peptide, which is a key factor in the pathogenesis of Alzheimer's disease, can also activate the vascular $gp91^{phox}$-containing NAD(P)H oxidase, and oxidative stress and cerebrovascular dysfunction do not occur in transgenic mice overexpressing the amyloid precursor protein but lacking $gp91^{phox}$ [44].

Many of the adverse consequences of oxidative stress are not directly due to $O_2^{\bullet -}$ itself but are mediated via production of highly reactive oxidant peroxynitrite, the reaction product of NO and superoxide [45]. There is solid evidence for a substantially enhanced cardiovascular $ONOO^-$ formation in aging [8, 11, 13, 32]. There are many downstream targets of peroxynitrite-induced cytotoxicity [45]. Peroxynitrite readily reacts with enzymes, macromolecules, and lipid membranes, which leads to cellular dysfunction. For example, tyrosine nitration may lead to dysfunction of nitrated proteins, as has been shown in the case of Mn-superoxide dismutase (MnSOD). Peroxynitrite may also inhibit superoxide dismutase, glutaredoxin and other antioxidant systems, which leads to positive feedback cycles of intracellular oxidant generation and oxidative injury [46, 47]. A recent study analyzing protein nitration in cardiac tissue from old rats using proteomics identified several enzymes of the glycolytic machinery (α-enolase-1, α-aldolase, and GAPDH) as targets for protein nitration [48]. Mitochondrial proteins, including aconitase and ATP synthase and other proteins involved in electron transfer, appear to be especially sensitive to aging-related nitration [48]. In addition, peroxynitrite-modified cellular proteins are subject to accelerated degradation via the proteosome.

## 13.3 Role of Mitochondrial Oxidative Stress in Arterial Aging

Mitochondria are responsible for ~90% of cellular oxygen consumption, and there is strong evidence that mitochondrial ROS production increases with age in most tissues from a variety of species. There is increasing evidence that mitochondria are also a major source of ROS in aged blood vessels [49, 50]. The pathophysiological consequences of mitochondrial oxidative stress are likely multifaceted and involve both mitochondrial oxidative decline affecting cellular energetics and the signaling role of ROS.

The mitochondrial theory of aging, first proposed in 1972 by Harman [51], postulates that a vicious cycle exists, in which free radical–induced mutation of mtDNA impairs respiratory chain function, enhancing the production of more DNA-damaging oxygen radicals. According to the theory, a bioenergetic crisis finally ensues, which leads to tissue dysfunction and degeneration. The mitochondrial theory of aging is supported by circumstantial evidence. While the nuclear DNA is protected by histones and various repair enzymes, the mitochondria lack histones and efficient DNA repair systems to offer protection from free radical–mediated damage. Studies on various laboratory species and humans suggest that old age in many tissues is associated with oxidative mitochondrial decay, mtDNA damage, and/or impaired cytochrome C oxidase (COX) activity (of note, 3 of 13 proteins of complex IV are encoded for by mtDNA). Importantly, increased mitochondrial ROS production is also associated with a significant decline in COX activity in aged rodent arteries [50]. The role of mitochondrial oxidative stress in vascular aging is clearly demonstrated by the findings that in aged $MnSOD^{+/-}$ mice, high levels of mitochondrial ROS formation lead to a severe impairment of endothelial function associated with significant mtDNA damage [52, 53]. Mice that overexpress human catalase targeted to mitochondria exhibited increased life span and delayed age-related cardiac alterations relative to control wild-type mice, suggesting that improved antioxidant defenses in mitochondria promote mitochondrial and organismal health [22]. Whether attenuation of mitochondrial oxidative stress per se would delay vascular aging in this model is yet to be determined. Further evidence for an intimate link between mitochondrial oxidative stress in aging and endothelial dysfunction came from studies of $p66^{Shc-}$null mice [13]. The mitochondrial enzyme $p66^{Shc}$ is an adaptor protein, which plays an important role in the regulation of mitochondrial ROS production and programmed cell death [54, 55]. Genetic deletion of $p66^{Shc}$ results in reduced production of mitochondrial ROS and extended longevity in mice, associated with increased endothelial bioavailability of NO and improved endothelial function [54, 55].

Despite the aforementioned findings, recent studies suggest that not every mouse model of extended longevity is characterized by a reduced mitochondrial ROS production and endothelial protection. It is well documented that plasma growth hormone (GH) levels decline with age in humans and in experimental animals, and there are a number of studies extant linking GH deficiency to age-related pathological conditions, such as cardiac and microvascular dysfunction, cognitive decline, sarcopenia, and frailty [56–59]. However, during the last decade, studies in *Caenorhabditis elegans* created a controversy regarding the role of GH/insulin-like growth factor (IGF) pathway in the aging process, showing that reduced insulin-like signaling may actually promote longevity in lower organisms by altering oxidative stress resistance and metabolism [60]. The observation that mice with hereditary dwarfism (Ames dwarf) exhibit a significant extension of life span (over 40%) [61] raised the possibility that insulin-like signals also play a role in the regulation of mammalian longevity. Ames dwarf mice are deficient in GH, prolactin, and thyroid stimulating hormone because of a mutation in *Prop*-1, a factor required for differentiation of the pituitary gland during development [62]. Since this original observation, it has been documented that phenotypically identical Snell dwarf mice [63] and GH receptor/binding protein gene knockout mice [64, 65] also exhibit a longevity phenotype. All of these GH-impaired mutant mice have very low circulating IGF-I levels. A central role for defective IGF signaling in the longevity phenotype is suggested by the finding that female Igf1r$^{+/-}$ mice also live significantly longer than their wild-type counterparts. We have recently found that mitochondrial ROS generation is increased in the arteries and the heart of GH/IGF-1–deficient Ames dwarf mice [66]. Also, administration of IGF-1 or GH in a dose-dependent manner upregulates MnSOD and attenuates mitochondrial ROS production in cultured endothelial cells and cardiac myocytes [66]. These studies suggest that the GH/IGF-1 axis exerts primarily vasoprotective functions by attenuating vascular oxidative stress; they also raise the possibility that the age-related decline in GH and IGF-1 levels may aggravate mitochondrial oxidative stress in aged arteries.

Animal studies have clearly shown that aging is associated with substantial changes in substrate metabolism in the heart. Importantly, the capacity to oxidize fatty acids significantly declines with advanced age [67]. Vascular endothelial and smooth muscle cells have been shown to use fatty acids as substrates for oxidative phosphorylation [68–70], and there is reason to believe that vascular mitochondria show age-related decline [49, 50, 71] similar to that of cardiac mitochondria. Impaired mitochondrial energy metabolism in aging vessels is likely to contribute to vascular dysfunction in aging [71]. This view is supported by the observation that mimicking the decline in mitochondrial energy metabolism in aging by pharmacological inhibition of oxidative phosphorylation by rotenone (which inhibits electron transport at the level of flavin mononucleotide) results in marked impairment of endothelium-dependent relaxation of vascular preparations from various species [71–74]. Similar findings were reported with antimycin A (which inhibits electron transport at the level of cytochrome b-$c_1$) and oligomycin (which inhibits mitochondrial $F_1$-ATPase) as well [74, 75], suggesting that alterations of mitochondrial energy metabolism have a direct influence on endothelial NO mediation. Rotenone does not seem to affect vascular relaxations induced by either NO donors [71, 73] or endothelium-independent vasorelaxants [72].

Mitochondria-derived ROS, in addition to causing oxidative mtDNA damage, play important signaling roles. The findings that inhibition of mitochondrial ROS production or scavenging of $H_2O_2$ attenuate NF-κB activation and NF-κB–dependent gene expression in aged vessels [49] suggest that mitochondrial $H_2O_2$ production is involved in the regulation of endothelial NF-κB activity. In contrast, mitochondria-derived $O_2^{\bullet -}$ is likely to play a lesser signaling role. First, $O_2^{\bullet -}$ is membrane-impermeable (except in the protonated perhydroxyl radical form, which represents only a small fraction of the total $O_2^{\bullet -}$ produced); whereas $H_2O_2$ easily penetrates the mitochondrial membranes. Second, because high levels of SOD in mitochondria (MnSOD in the matrix and on the inner membrane and Cu,Zn-SOD in the intermembrane space) efficiently scavenge $O_2^{\bullet -}$, it is likely that mitochondria-derived $H_2O_2$ is a major factor in initiating inflammatory signaling processes in endothelial cells. Furthermore, exogenous $H_2O_2$ significantly increases NF-κB activation in the arteries of young rats, mimicking the aging phenotype [49].

## 13.4 Low-Grade Vascular Inflammation During Aging: Role of Oxidative Stress

Chronic low-grade inflammation is a well-known corollary of the vascular aging process [76] and is believed to significantly contribute to morbidity and mortality of age-associated diseases. Inflammation is considered to be a critical initial step in the development of atherosclerosis during aging. There is abundant evidence that arterial aging, even in the absence of traditional risk factors for atherosclerosis (hypertension, diabetes, smoking, etc.), is associated with a proinflammatory shift in gene expression profile [8, 9, 25, 77, 78]. Proinflammatory changes in endothelial phenotype during aging, termed "endothelial activation," involve induction of cellular adhesion molecules, an increase in endothelial-leukocyte interactions, as well as alterations in the secretion of autocrine/paracrine factors, which are pivotal to inflammatory responses.

This intrinsic low-grade inflammatory state in aging is in part due to cell-autonomous mechanisms and in part mediated by paracrine factors produced in the vascular wall. As noted above, the available evidence suggest that activation of NF-κB, a redox-sensitive transcription factor, plays a central role in endothelial activation in aging [27, 49, 79, 80]. Accordingly, recent studies showed that transcriptional activity of NF-κB increases during aging [27, 49, 81], and is likely responsible for the increased expression of adhesion molecules, iNOS, and many paracrine mediators found in aged vessels [8, 49, 82]. Chronic activation of NF-κB leads to a proinflammatory microenvironment in the vascular wall, which predisposes arteries to atherosclerosis [83]. Disruption of NF-κB–regulated inflammatory processes has the potential to confer vasoprotection. Indeed, pharmacological inhibition of NF-κB attenuates endothelial activation, decreasing monocyte adhesiveness to endothelial cells of aged arteries [49, 79, 80]. NF-κB activation and chronic inflammation seem to be a generalized phenomenon during aging, since increases in NF-κB activity have been observed in aged rat skeletal muscle, liver, brain, and cardiac muscle [81, 84–86]. Recent studies suggest that multiple pathways can regulate NF-κB activation, promoting arterial inflammation during aging [76]. In arterial cells, NF-κB is present as an inactive, IκB-bound complex in the cytoplasm. Upon stimulation, NF-κB translocates to the nucleus and initiates inflammatory gene expression. Cellular signal transduction pathways that lead to the activation of NF-κB converge on oxygen free radical–dependent activation of a high molecular weight complex that contains an IκB kinase (IKK). Activation of IKK complex leads to the phosphorylation and degradation of IκB, consequently unmasking NF-κB. ROS-mediated pathways that converge on NF-κB, contributing to endothelial activation during aging, likely include mitochondrial

ROS-induced pathways, TNFα signaling, and the local renin-angiotensin system (RAS) and pathways associated with innate immunity (recently reviewed elsewhere [76]). Among these mechanisms, induction of NF-κB by mitochondrial ROS represents a cell-autonomous effect. This concept is supported by the findings that NF-κB transcriptional activity is increased in cultured arterial cells derived from aged primates (Csiszar, Lakatta, and Ungvari, unpublished observation) and rodents [87] irrespective of the presence of other cell types or the in vivo context.

The inflammatory transcriptomes of blood vessels from both aged rodents and primates change with marked similarity, including inflammatory cytokines [41, 77, 78]. Functional genomic analysis of these genes suggests that TNFα was involved in the paracrine regulation of endothelial function [41]. An increased TNFα production has been demonstrated in the aged coronary arteries, carotid arteries, aorta, and heart [40, 41, 88, 89]. Because an NF-κB binding site is present on the promoter region of the TNFα gene [90], the possibility that NF-κB activation induced by mitochondria-derived ROS promotes TNFα expression in the arterial wall cannot be ruled out. We have previously demonstrated that arterial, endothelial, and smooth muscle express the TNFα converting enzyme (TACE/ADAM17) [78], suggesting the presence of an autocrine/paracrine TNFα-dependent regulatory pathway in the arterial wall. Plasma levels of TNFα also increase in aging [91–95]. Previous studies showed that TNFα induces oxidative stress in endothelial and smooth muscle cells by upregulating/activating NAD(P)H oxidase [41, 96]; and recent clinical and experimental studies have linked TNFα to endothelial impairment, atherosclerosis, and heart failure [97, 98]. Etanercept (Enbrel) is an FDA-approved drug (composed of the extracellular ligand-binding portion of human TNF receptor 2) which binds and inactivates circulating TNFα. It is significant that chronic anti-TNFα treatment with etanercept exerts multifaceted vasculoprotective effects in aged rats [41, 42, 99]. Among these, etanercept treatment significantly improves endothelial function and decreases vascular NAD(P)H oxidase activity and expression [41, 99]. There is solid evidence that TNFα-induced NAD(P)H oxidase–dependent ROS generation contributes to the activation of NF-κB [41, 100]. Accordingly, in endothelial cells [100], TNFα treatment results in NF-κB–dependent upregulation of proatherogenic inflammatory mediators, which can, in turn, be attenuated by NAD(P)H oxidase inhibitors. Neutralization of TNFα by chronic etanercept treatment was shown to attenuate expression of adhesion molecules in arteries of aged rats [41]. Previous studies also suggest that increased endothelial apoptosis is a feature of advanced aging [8, 17, 41, 78]. Chronic etanercept treatment [17] decreased apoptotic cell death in aged vessels, suggesting that increased TNFα levels also promote programmed endothelial cell death, which may likely contribute to age-related cardiovascular pathophysiology [25].

## 13.5 Caloric Restriction Attenuates Vascular Oxidative Stress in Aging

The dietary regimen known as caloric restriction can delay aging and extend lifespan in evolutionary distant organisms (including the invertebrate *C. elegans*, *D.*

*melanogaster*, and the bowl and doily spider, *Frontinella pyramitela* [101, 102], as well as laboratory rodents) [103–115]. Caloric restriction also slows the functional decline associated with aging in various organ systems, such as skeletal muscle, brain, heart, and the immune system, and delays early onset of age-related diseases (e.g., cancer, sarcopenia, osteoporosis, and cataract formation) in mammals [33, 40, 116]. The available evidence suggests that caloric restriction also exerts vasoprotective effects, which may prevent/delay development of cardiovascular disease (reviewed recently elsewhere [33]). Sohal and Weindruch put forward the original hypothesis that antiaging action of caloric restriction is derived from the ability of cells to attenuate oxidative stress associated with aging [117]. We have recently found that lifelong caloric restriction in aged F344 rats significantly attenuates oxidative stress, decreases NAD(P)H oxidase activity, and improves endothelial function in the aorta [118]. Previous studies also showed that mitochondria isolated from caloric-restricted animals produce significantly less ROS than those from ad libitum–fed controls [119]. The reduction of vascular ROS production is also associated with downregulation of inflammatory markers and a decreased NF-κB activity [33, 120]. The mechanisms underlying the antioxidative effect of caloric restriction are likely multifaceted, involving both cell-autonomous effects (e.g., changes in mitochondrial function), changes in paracrine regulation (altered secretome), and effects mediated by circulating neuroendocrine factors [33]. In addition, caloric restriction may also attenuate vascular oxidative stress by improving plasma lipid profile, normalizing glucose levels, and decreasing blood pressure. Previously Cabo et al. [121] demonstrated that in vitro treatment of cultured hepatocytes with sera from caloric-restricted animals mimics phenotypic effects observed in vivo during caloric restriction. Our recent data showed that circulating factors within the plasma of caloric-restricted animals significantly attenuate ROS production in primary coronary arterial endothelial cells in culture [118]. These findings support the view that neuroendocrine factors mediate, at least in part, the antioxidant vascular effects of caloric restriction.

Multiple lines of evidence indicate that the sirtuin family of NAD+-dependent deacetylases and ADP ribosyltransferases mediates the lifespan extension by caloric restriction in lower organisms [122–131]. In mammals SIRT1 (a homologue to the *Saccharomyces cerevisiae* Sir2 protein) is also inducible by caloric restriction [126], suggesting a central role for this enzyme in mammalian physiology and stress response as well. SIRT1 is expressed in the cardiovascular system [132, 133] and is induced by caloric restriction [133]. We recently demonstrated that knockdown of SIRT1 diminishes the reduction of ROS production in cultured endothelial cells elicited by treatment with sera from caloric-restricted rats [118]. This finding suggests that SIRT1 activation contributes to the antioxidative vasoprotective effects of caloric restriction. Because serum from caloric-restricted humans also induces SIRT1 in detector cells [134], it is logical to assume that caloric restriction–induced SIRT1 activation confers similar protective effects in humans as well.

## 13.6 Attenuation of Age-Related Vascular Oxidative Stress by the Caloric Restriction Mimetic Resveratrol

Although caloric restriction exerts potent antioxidative effects during aging, such a diet is unlikely to be widely adopted by the elderly because of compliance issues. As an alternative, current research focuses on the development of caloric restriction mimetic compounds that provide some of the benefits of caloric restriction without a reduction in caloric intake. A group of polyphenolic SIRT1-activating compounds has been identified recently [124, 135]. One of the most potent natural SIRT1 activating compounds—based on in vitro and now in vivo studies in rodents—is resveratrol (3,5,4′-trihydroxystilbene), a polyphenol that lowers the $K_m$ of SIRT1 for the acetylated substrate and for $NAD^+$ [123, 124, 128, 136, 137]. The evidence that activation of SIRT1 by resveratrol and other SIRT1 activating compounds is physiologically relevant appears quite strong [128, 138, 139]. As with caloric restriction, resveratrol has extended the lifespan of very distantly related species, including *S. cerevisiae* [124], *Caenorhabditis elegans* [140], *Drosophila melanogaster* [123, 141], and the vertebrate fish *Nothobranchius furzeri* [142]. In the first three species, lifespan extension is dependent on a SIRT1 homologue. Resveratrol was also shown to improve a number of health parameters and extend lifespan in obese mice [40, 143].

There is accumulating evidence that resveratrol can exert vasoprotective effects and attenuate vascular oxidative stress in aging [40]. Chronic treatment of aged mice with resveratrol significantly decreased expression and activity of NAD(P)H oxidase and normalized endothelial function [40]. Antioxidative effects of resveratrol were associated with a significant attenuation of vascular inflammation in aging [40]. Diabetes mellitus is associated with accelerated vascular aging characterized by oxidative stress and inflammation. Recent studies suggest that resveratrol can effectively attenuate vascular oxidative stress and protect endothelial function in diabetes [40]. Using an animal model of exogenous oxidative stress and accelerated vascular aging (cigarette smoke exposure in rats), we have shown that resveratrol treatment effectively decreases vascular oxidative stress induced by exogenous activation of NAD(P)H oxidases [132, 144]. Importantly, in vitro treatment with cigarette smoke extract also increased ROS production in rat arteries and cultured coronary arterial endothelial cells, which was attenuated by resveratrol treatment [132, 144]. The aforementioned protective effects of resveratrol were abolished by knockdown of SIRT1, whereas overexpression of SIRT1 mimicked the effects of resveratrol [132].

Oxidative stress and the resulting vascular inflammation during aging are associated with endothelial apoptosis [77]. Importantly, chronic resveratrol treatment of aged mice significantly attenuates the rate of endothelial apoptosis [40]. Similar findings were demonstrated in animal models of type 2 diabetes [40] and cigarette smoking [132] as well. Previously we found that in cultured endothelial cells and in aorta segments maintained in organoid culture resveratrol treatment prevents induction of apoptosis by oxidative stressors (oxidized LDL, TNFα, or exposure to $UV_{240\,nm}$) [145]. Resveratrol treatment upregulated the expression of glutathione peroxidase, catalase, and heme oxygenase-1 in cultured endothelial cells and arterial segments [145]. The protective effect of resveratrol was attenuated by inhibition of glutathione peroxidase and heme oxygenase-1, suggesting a role for antioxidant systems in the antiapoptotic action of resveratrol [145].

## 13.7 Conclusions

In conclusion, aging is associated with oxidative/nitrosative stress and inflammatory changes in the vascular transcriptome and secretome. Whether conventional treatments with antioxidant and antiinflammatory properties (e.g., a combination of antioxidant vitamins, statins, nonsteroidal antiinflammatory drugs, and ACE-inhibitors) are able to reverse or delay the aging-induced considerable functional decline of the cardiovascular system remains a subject of current debate. Overall, we can expect that recent advances in our understanding of the role of cellular stress response and prosurvival pathways underlying cardiovascular aging will, in the not so distant future, yield novel antiaging therapeutic approaches that will be exploited for the benefit of elderly patients.

**Acknowledgments** This work was supported by grants from the American Diabetes Association (to ZU), the American Federation for Aging Research (to AC) and the NIH (HL077256 and HL43023 to ZU and AC).

## References

1. Lakatta EG (2003) Arterial and cardiac aging: major shareholders in cardiovascular disease enterprises: Part III: cellular and molecular clues to heart and arterial aging. Circulation 107:490–497
2. Lakatta EG, Levy D (2003) Arterial and cardiac aging: major shareholders in cardiovascular disease enterprises: Part II: the aging heart in health: links to heart disease. Circulation 107:346–354
3. Lakatta EG, Levy D (2003) Arterial and cardiac aging: major shareholders in cardiovascular disease enterprises: Part I: aging arteries: a "set up" for vascular disease. Circulation 107:139–146
4. Harman D, Aging: A (1956) Theory based on free radical and radiation chemistry. J Gerontol 11:298–300
5. Van Remmen H, Richardson A (2001) Oxidative damage to mitochondria and aging. Exp Gerontol 36:957–968
6. Van Remmen H, Hamilton ML, Richardson A (2003) Oxidative damage to DNA and aging. Exerc Sport Sci Rev 31:149–153
7. Hamilton ML, Van Remmen H, Drake JA et al (2001) Does oxidative damage to DNA increase with age? Proc Natl Acad Sci USA 98:10469–10474
8. Csiszar A, Ungvari Z, Edwards JG et al (2002) Aging-induced phenotypic changes and oxidative stress impair coronary arteriolar function. Circ Res 90:1159–1166
9. Csiszar A, Pacher P, Kaley G et al (2005) Role of oxidative and nitrosative stress, longevity genes and poly(ADP-ribose) polymerase in cardiovascular dysfunction associated with aging. Curr Vasc Pharmacol 3:285–291
10. Hamilton CA, Brosnan MJ, McIntyre M et al (2001) Superoxide excess in hypertension and aging: a common cause of endothelial dysfunction. Hypertension 37:529–534

11. Sun D, Huang A, Yan EH et al (2004) Reduced release of nitric oxide to shear stress in mesenteric arteries of aged rats. Am J Physiol Heart Circ Physiol 286:H2249–H2256
12. van der Loo B, Labugger R, Skepper JN et al (2000) Enhanced peroxynitrite formation is associated with vascular aging. J Exp Med 192:1731–1744
13. Francia P, delli Gatti C, Bachschmid M et al (2004) Deletion of p66shc gene protects against age-related endothelial dysfunction. Circulation 110:2889–2895
14. Ungvari Z, Buffenstein R, Austad SN et al (2008) Oxidative stress in vascular senescence: lessons from successfully aging species. Front Biosci 13:5056–5070
15. Ungvari Z, Krasnikov BF, Csiszar A et al (2008) Testing hypotheses of aging in long-lived mice of the genus Peromyscus: association between longevity and mitochondrial stress resistance, ROS detoxification pathways and DNA repair efficiency. Age 30:121–133
16. Labinskyy N, Csiszar A, Orosz Z et al (2006) Comparison of endothelial function, $O_2^{\bullet -}$ and $H_2O_2$ production, and vascular oxidative stress resistance between the longest-living rodent, the naked mole rat, and mice. Am J Physiol 291:H2698–H2704
17. Csiszar A, Labinskyy N, Orosz Z et al (2007) Vascular aging in the longest-living rodent, the naked mole rat. Am J Physiol 293:H919–H927
18. Sampayo JN, Olsen A, Lithgow GJ (2003) Oxidative stress in Caenorhabditis elegans: protective effects of superoxide dismutase/catalase mimetics. Aging Cell 2:319–326
19. Sentman ML, Granstrom M, Jakobson H et al (2006) Phenotypes of mice lacking extracellular superoxide dismutase and copper- and zinc-containing superoxide dismutase. J Biol Chem 281:6904–6909
20. Mansouri A, Muller FL, Liu Y et al (2006) Alterations in mitochondrial function, hydrogen peroxide release and oxidative damage in mouse hind-limb skeletal muscle during aging. Mech Ageing Dev 127:298–306
21. Van Remmen H, Ikeno Y, Hamilton M et al (2003) Life-long reduction in MnSOD activity results in increased DNA damage and higher incidence of cancer but does not accelerate aging. Physiol Genomics 16:29–37
22. Schriner SE, Linford NJ, Martin GM et al (2005) Extension of murine life span by overexpression of catalase targeted to mitochondria. Science 308:1909–1911
23. Mele J, Van Remmen H, Vijg J et al (2006) Characterization of transgenic mice that overexpress both copper zinc superoxide dismutase and catalase. Antioxid Redox Signal 8:628–638
24. Huang TT, Carlson EJ, Gillespie AM et al (2000) Ubiquitous overexpression of CuZn superoxide dismutase does not extend life span in mice. J Gerontol 55:B5–B9
25. Ungvari Z, Csiszar A, Kaley G (2004) Vascular inflammation in aging. Herz 29:733–740
26. Labinskyy N, Csiszar A, Veress G et al (2006) Vascular dysfunction in aging: potential effects of resveratrol, an anti-inflammatory phytoestrogen. Curr Med Chem 13: 989–996
27. Donato AJ, Eskurza I, Silver AE et al (2007) Direct evidence of endothelial oxidative stress with aging in humans: relation to impaired endothelium-dependent dilation and upregulation of nuclear factor-kappaB. Circ Res 100:1659–1666
28. Eskurza I, Kahn ZD, Seals DR (2006) Xanthine oxidase does not contribute to impaired peripheral conduit artery endothelium-dependent dilatation with ageing. J Physiol (Lond) 571:661–668
29. Eskurza I, Monahan KD, Robinson JA et al (2004) Effect of acute and chronic ascorbic acid on flow-mediated dilatation with sedentary and physically active human ageing. J Physiol (Lond) 556:315–324
30. Jablonski KL, Seals DR, Eskurza I et al (2007) High-dose ascorbic acid infusion abolishes chronic vasoconstriction and restores resting leg blood flow in healthy older men. J Appl Physiol 103:1715–1721
31. Gates PE, Boucher ML, Silver AE et al (2006) Impaired flow-mediated dilation with age is not explained by L-arginine bioavailability or endothelial asymmetric dimethylarginine protein expression. J Appl Physiol 291:H985–H1002

32. Adler A, Messina E, Sherman B et al (2003) NAD(P)H oxidase-generated superoxide anion accounts for reduced control of myocardial $O_2$ consumption by NO in old Fischer 344 rats. Am J Physiol Heart Circ Physiol 285:H1015–H1022
33. Ungvari Z, Parrado-Fernandez C, Csiszar A et al (2008) Mechanisms underlying caloric restriction and lifespan regulation: implications for vascular aging. Circ Res 102: 519–528
34. Tanabe T, Maeda S, Miyauchi T et al (2003) Exercise training improves ageing-induced decrease in eNOS expression of the aorta. Acta Physiol Scand 178:3–10
35. Woodman CR, Price EM, Laughlin MH (2002) Aging induces muscle-specific impairment of endothelium-dependent dilation in skeletal muscle feed arteries. J Appl Physiol 93: 1685–1690
36. Matsushita H, Chang E, Glassford AJ et al (2001) eNOS activity is reduced in senescent human endothelial cells: Preservation by hTERT immortalization. Circ Res 89:793–798
37. Hoffmann J, Haendeler J, Aicher A et al (2001) Aging enhances the sensitivity of endothelial cells toward apoptotic stimuli: important role of nitric oxide. Circ Res 89:709–715
38. Berkowitz DE, White R, Li D et al (2003) Arginase reciprocally regulates nitric oxide synthase activity and contributes to endothelial dysfunction in aging blood vessels. Circulation 108:2000–2006
39. Jacobson A, Yan C, Gao Q et al (2007) Aging enhances pressure-induced arterial superoxide formation. Am J Physiol Heart Circ Physiol 293:H1344–H1350
40. Pearson KJ, Baur JA, Lewis KN et al (2008) Resveratrol delays age-related deterioration and mimics transcriptional aspects of dietary restriction without extending life span. Cell Metab 8:157–168
41. Csiszar A, Labinskyy N, Smith K et al (2007) Vasculoprotective effects of anti-tumor necrosis factor-{alpha} treatment in aging. Am J Pathol 170:388–698
42. Arenas IA, Xu Y, Davidge ST (2006) Age-associated impairment in vasorelaxation to fluid shear stress in the female vasculature is improved by TNF-{alpha} antagonism. Am J Physiol Heart Circ Physiol 290:H1259–H1263
43. Park L, Anrather J, Girouard H et al (2007) Nox2-derived reactive oxygen species mediate neurovascular dysregulation in the aging mouse brain. J Cereb Blood Flow Metab 27: 1908–1918
44. Park L, Anrather J, Zhou P et al (2005) NADPH-oxidase-derived reactive oxygen species mediate the cerebrovascular dysfunction induced by the amyloid beta peptide. J Neurosci 25:1769–1777
45. Pacher P, Beckman JS, Liaudet L (2007) Nitric oxide and peroxynitrite in health and disease. Physiol Rev 87:315–424
46. Szabo C (2003) Multiple pathways of peroxynitrite cytotoxicity. Toxicol Lett 140–141: 105–112
47. Turko IV, Murad F (2002) Protein nitration in cardiovascular diseases. Pharmacol Rev 54:619–634
48. Kanski J, Behring A, Pelling J et al (2004) Proteomic identification of 3-nitrotyrosine-containing rat cardiac proteins: effect of biological aging. Am J Physiol Heart Circ Physiol 288:H371–H381
49. Ungvari ZI, Orosz Z, Labinskyy N et al (2007) Increased mitochondrial $H_2O_2$ production promotes endothelial NF-kB activation in aged rat arteries. Am J Physiol Heart Circ Physiol 293:H37–H47
50. Ungvari ZI, Labinskyy N, Gupte SA et al (2008) Dysregulation of mitochondrial biogenesis in vascular endothelial and smooth muscle cells of aged rats. Am J Physiol Heart Circ Physiol 294:H2121–H2128
51. Harman D (1972) The biologic clock: the mitochondria? J Am Geriatr Soc 20:145–147
52. Wenzel P, Schuhmacher S, Kienhofer J et al (2008) Manganese superoxide dismutase and aldehyde dehydrogenase deficiency increase mitochondrial oxidative stress and aggravate age-dependent vascular dysfunction. Cardiovasc Res 80:280–289

53. Brown KA, Didion SP, Andresen JJ et al (2007) Effect of aging, MnSOD deficiency, and genetic background on endothelial function: evidence for MnSOD haploinsufficiency. Arterioscler Thromb Vasc Biol 27:1941–1946
54. Camici GG, Cosentino F, Tanner FC et al (2008) The role of p66Shc deletion in age-associated arterial dysfunction and disease states. J Appl Physiol 105:1628–1631
55. Cosentino F, Francia P, Camici GG et al (2008) Final common molecular pathways of aging and cardiovascular disease: role of the p66Shc protein. Arterioscler Thromb Vasc Biol 28:622–628
56. Ceda GP, Dall'Aglio E, Maggio M et al (2005) Clinical implications of the reduced activity of the GH-IGF-I axis in older men. J Endocrinol Invest 28:96–100
57. Sonntag WE, Lynch CD, Cooney PT et al (1997) Decreases in cerebral microvasculature with age are associated with the decline in growth hormone and insulin-like growth factor 1. Endocrinology 138:3515–3520
58. Groban L, Pailes NA, Bennett CD et al (2006) Growth hormone replacement attenuates diastolic dysfunction and cardiac angiotensin II expression in senescent rats. J Gerontol A Biol Sci Med Sci 61:28–35
59. Wannenburg T, Khan AS, Sane DC et al (2001) Growth hormone reverses age-related cardiac myofilament dysfunction in rats. Am J Physiol Heart Circ Physiol 281:H915–H922
60. Tatar M, Bartke A, Antebi A (2003) The endocrine regulation of aging by insulin-like signals. Science 299:1346–1351
61. Brown-Borg HM, Borg KE, Meliska CJ et al (1996) Dwarf mice and the ageing process. Nature 384:33
62. Sornson MW, Wu W, Dasen JS et al (1996) Pituitary lineage determination by the Prophet of Pit-1 homeodomain factor defective in Ames dwarfism. Nature 384:327–333
63. Flurkey K, Papaconstantinou J, Miller RA et al (2001) Lifespan extension and delayed immune and collagen aging in mutant mice with defects in growth hormone production. Proc Natl Acad Sci USA 98:6736–6741
64. Hauck SJ, Aaron JM, Wright C et al (2002) Antioxidant enzymes, free-radical damage, and response to paraquat in liver and kidney of long-living growth hormone receptor/binding protein gene-disrupted mice. Horm Metab Res 34:481–486
65. Al-Regaiey KA, Masternak MM, Bonkowski M et al (2005) Long-lived growth hormone receptor knockout mice: interaction of reduced insulin-like growth factor i/insulin signaling and caloric restriction. Endocrinology 146:851–860
66. Csiszar A, Labinskyy N, Perez V et al (2008) Endothelial function and vascular oxidative stress in long-lived GH/IGF-deficient Ames dwarf mice. Am J Physiol Heart Circ Physiol 295:H1882–H1894
67. Abu-Erreish GM, Neely JR, Whitmer JT et al (1977) Fatty acid oxidation by isolated perfused working hearts of aged rats. Am J Physiol 232:E258–E262
68. Hulsmann WC, Dubelaar ML (1992) Carnitine requirement of vascular endothelial and smooth muscle cells in imminent ischemia. Mol Cell Biochem 116:125–129
69. Imesch E, Nef P, Giacobino JP (1984) Study in pig coronary smooth muscle cell subcellular fractions of the activity of various enzymes involved in lipid metabolism and of the beta-receptor adenylate cyclase couple. Comp Biochem Physiol B 77:501–506
70. Gillies PJ, Bell FP (1979) Carnitine palmitoyltransferase activity in mitochondrial fractions isolated from aortas of rabbits fed cholesterol-supplemented diets. Atherosclerosis 34:25–34
71. Csiszar A, Labinskyy N, Orosz Z et al (2006) Altered mitochondrial energy metabolism may play a role in vascular aging. Med Hypotheses 67:904–908
72. Weir CJ, Gibson IF, Martin W (1991) Effects of metabolic inhibitors on endothelium-dependent and endothelium-independent vasodilatation of rat and rabbit aorta. Br J Pharmacol 102:162–166
73. Rodman DM, Mallet J, McMurtry IF (1991) Difference in effect of inhibitors of energy metabolism on endothelium-dependent relaxation of rat pulmonary artery and aorta. Am J Respir Cell Mol Biol 4:237–242

74. Griffith TM, Edwards DH, Newby AC et al (1986) Production of endothelium derived relaxant factor is dependent on oxidative phosphorylation and extracellular calcium. Cardiovasc Res 20:7–12
75. Dionisi O, Galeotti T, Terranova T et al (1975) Superoxide radicals and hydrogen peroxide formation in mitochondria from normal and neoplastic tissues. Biochim Biophys Acta 403:292–300
76. Csiszar A, Wang M, Lakatta EG et al (2008) Inflammation and endothelial dysfunction during aging: role of NF-{kappa}B. J Appl Physiol 105:1333–1341
77. Csiszar A, Ungvari Z, Koller A et al (2003) Aging-induced proinflammatory shift in cytokine expression profile in rat coronary arteries. FASEB J 17:1183–1185
78. Csiszar A, Ungvari Z, Koller A et al (2004) Proinflammatory phenotype of coronary arteries promotes endothelial apoptosis in aging. Physiol Genomics 17:21–30
79. Sung B, Park S, Yu BP et al (2006) Amelioration of age-related inflammation and oxidative stress by PPARgamma activator: suppression of NF-kappaB by 2,4-thiazolidinedione. Exp Gerontol 41:590–599
80. Chung HY, Sung B, Jung KJ et al (2006) The molecular inflammatory process in aging. Antioxid Redox Signal 8:572–581
81. Helenius M, Hanninen M, Lehtinen SK et al (1996) Aging-induced up-regulation of nuclear binding activities of oxidative stress responsive NF-kB transcription factor in mouse cardiac muscle. J Mol Cell Cardiol 28:487–498
82. Cernadas MR, Sanchez de Miguel L, Garcia-Duran M et al (1998) Expression of constitutive and inducible nitric oxide synthases in the vascular wall of young and aging rats. Circ Res 83:279–286
83. Hajra L, Evans AI, Chen M et al (2000) The NF-kappa B signal transduction pathway in aortic endothelial cells is primed for activation in regions predisposed to atherosclerotic lesion formation. Proc Natl Acad Sci USA 97:9052–9057
84. Zhang J, Dai J, Lu Y et al (2004) In vivo visualization of aging-associated gene transcription: evidence for free radical theory of aging. Exp Gerontol 39:239–247
85. Korhonen P, Helenius M, Salminen A (1997) Age-related changes in the regulation of transcription factor NF-kappa B in rat brain. Neurosci Lett 225:61–64
86. Radak Z, Chung HY, Naito H et al (2004) Age-associated increase in oxidative stress and nuclear factor kappaB activation are attenuated in rat liver by regular exercise. FASEB J 18:749–750
87. Yan ZQ, Sirsjo A, Bochaton-Piallat ML et al (1999) Augmented expression of inducible NO synthase in vascular smooth muscle cells during aging is associated with enhanced NF-kappaB activation. Arterioscler Thromb Vasc Biol 19:2854–2862
88. Lee CK, Allison DB, Brand J et al (2002) Transcriptional profiles associated with aging and middle age-onset caloric restriction in mouse hearts. Proc Natl Acad Sci USA 99: 14988–14993
89. Belmin J, Bernard C, Corman B et al (1995) Increased production of tumor necrosis factor and interleukin-6 by arterial wall of aged rats. Am J Physiol 268:H2288–H2293
90. Pelletier C, Varin-Blank N, Rivera J et al (1998) Fc epsilonRI-mediated induction of TNF-alpha gene expression in the RBL- 2H3 mast cell line: regulation by a novel NF-kappaB-like nuclear binding complex. J Immunol 161:4768–4776
91. Bruunsgaard H, Skinhoj P, Pedersen AN et al (2000) Ageing, tumour necrosis factor-alpha (TNF-alpha) and atherosclerosis. Clin Exp Immunol 121:255–260
92. Schulz S, Schagdarsurengin U, Suss T et al (2004) Relation between the tumor necrosis factor-alpha (TNF-alpha) gene and protein expression, and clinical, biochemical, and genetic markers: age, body mass index and uric acid are independent predictors for an elevated TNF-alpha plasma level in a complex risk model. Eur Cytokine Netw 15: 105–111
93. Yamamoto K, Shimokawa T, Yi H et al (2002) Aging and obesity augment the stress-induced expression of tissue factor gene in the mouse. Blood 100:4011–4018

94. Spaulding CC, Walford RL, Effros RB (1997) Calorie restriction inhibits the age-related dysregulation of the cytokines TNF-alpha and IL-6 in C3B10RF1 mice. Mech Ageing Dev 93:87–94
95. Harris TB, Ferrucci L, Tracy RP et al (1999) Associations of elevated interleukin-6 and C-reactive protein levels with mortality in the elderly. Am J Med 106:506–512
96. Ungvari Z, Csiszar A, Edwards JG et al (2003) Increased superoxide production in coronary arteries in hyperhomocysteinemia: role of tumor necrosis factor-alpha, NAD(P)H oxidase, and inducible nitric oxide synthase. Arterioscler Thromb Vasc Biol 23: 418–424
97. Bozkurt B, Kribbs SB, Clubb FJ Jr et al (1998) Pathophysiologically relevant concentrations of tumor necrosis factor-alpha promote progressive left ventricular dysfunction and remodeling in rats. Circulation 97:1382–1391
98. Bozkurt B, Torre-Amione G, Warren MS et al (2001) Results of targeted anti-tumor necrosis factor therapy with etanercept (ENBREL) in patients with advanced heart failure. Circulation 103:1044–1047
99. Arenas IA, Armstrong SJ, Xu Y et al (2005) Chronic tumor necrosis factor-alpha inhibition enhances NO modulation of vascular function in estrogen-deficient rats. Hypertension 46:76–81
100. Csiszar A, Smith K, Labinskyy N et al (2006) Resveratrol attenuates TNF-{alpha}-induced activation of coronary arterial endothelial cells: role of NF-{kappa}B inhibition. Am J Physiol 291:H1694–H1699
101. Austad SN (1989) Life extension by dietary restriction in the bowl and doily spider, Frontinella pyramitela. Exp Gerontol 24:83–92
102. Spencer RP (1990) Relationship of reproductive success and median longevity to food intake, in the captive female spider Frontinella pyramitela. Mech Ageing Dev 55:9–13
103. Weindruch R, Kayo T, Lee CK et al (2002) Gene expression profiling of aging using DNA microarrays. Mech Ageing Dev 123:177–193
104. Kayo T, Allison DB, Weindruch R et al (2001) Influences of aging and caloric restriction on the transcriptional profile of skeletal muscle from rhesus monkeys. Proc Natl Acad Sci USA 98:5093–5098
105. Weindruch R, Kayo T, Lee CK et al (2001) Microarray profiling of gene expression in aging and its alteration by caloric restriction in mice. J Nutr 131:918S–923S
106. Zainal TA, Oberley TD, Allison DB et al (2000) Caloric restriction of rhesus monkeys lowers oxidative damage in skeletal muscle. FASEB J 14:1825–1836
107. Lee CK, Weindruch R, Prolla TA (2000) Gene-expression profile of the ageing brain in mice. Nat Genet 25:294–297
108. Lee CM, Aspnes LE, Chung SS et al (1998) Influences of caloric restriction on age-associated skeletal muscle fiber characteristics and mitochondrial changes in rats and mice. Ann N Y Acad Sci 854:182–191
109. Lass A, Sohal BH, Weindruch R et al (1998) Caloric restriction prevents age-associated accrual of oxidative damage to mouse skeletal muscle mitochondria. Free Radic Biol Med 25:1089–1097
110. Edwards IJ, Rudel LL, Terry JG et al (1998) Caloric restriction in rhesus monkeys reduces low density lipoprotein interaction with arterial proteoglycans. J Gerontol A Biol Sci Med Sci 53:B443–B448
111. Moore WA, Davey VA, Weindruch R et al (1995) The effect of caloric restriction on lipofuscin accumulation in mouse brain with age. Gerontology 41(Suppl 2):173–185
112. Feuers RJ, Weindruch R, Hart RW (1993) Caloric restriction, aging, and antioxidant enzymes. Mutat Res 295:191–200
113. Weindruch R (1992) Effect of caloric restriction on age-associated cancers. Exp Gerontol 27:575–581
114. Harper ME, Bevilacqua L, Hagopian K et al (2004) Ageing, oxidative stress, and mitochondrial uncoupling. Acta Physiol Scand 182:321–331

115. Higami Y, Barger JL, Page GP et al (2006) Energy restriction lowers the expression of genes linked to inflammation, the cytoskeleton, the extracellular matrix, and angiogenesis in mouse adipose tissue. J Nutr 136:343–352
116. Pearson KJ, Lewis KN, Price NL et al (2008) Nrf2 mediates cancer protection but not prolongevity induced by caloric restriction. Proc Natl Acad Sci USA 105:2325–2330
117. Sohal RS, Weindruch R (1996) Oxidative stress, caloric restriction, and aging. Science 273:59–63
118. Csiszar A, Labinskyy N, Jimenez R et al (2009) Anti-oxidative and anti-inflammatory vasoprotective effects of caloric restriction in aging: role of circulating factors and SIRT1. Mech Ageing Dev 130:518–527
119. Lambert AJ, Merry BJ (2004) Effect of caloric restriction on mitochondrial reactive oxygen species production and bioenergetics: reversal by insulin. Am J Physiol Regul Integr Comp Physiol 286:R71–R79
120. Zou Y, Yoon S, Jung KJ et al (2006) Upregulation of aortic adhesion molecules during aging. J Gerontol 61:232–244
121. de Cabo R, Furer-Galban S, Anson RM et al (2003) An in vitro model of caloric restriction. Exp Gerontol 38:631–639
122. Kaeberlein M, McVey M, Guarente L (1999) The SIR2/3/4 complex and SIR2 alone promote longevity in Saccharomyces cerevisiae by two different mechanisms. Genes Dev 13: 2570–2580
123. Wood JG, Rogina B, Lavu S et al (2004) Sirtuin activators mimic caloric restriction and delay ageing in metazoans. Nature 430:686–689
124. Howitz KT, Bitterman KJ, Cohen HY et al (2003) Small molecule activators of sirtuins extend Saccharomyces cerevisiae lifespan. Nature 425:191–196
125. Brunet A, Sweeney LB, Sturgill JF et al (2004) Stress-dependent regulation of FOXO transcription factors by the SIRT1 deacetylase. Science 303:2011–2015
126. Cohen HY, Miller C, Bitterman KJ et al (2004) Calorie restriction promotes mammalian cell survival by inducing the SIRT1 deacetylase. Science 305:390–392
127. Anderson RM, Bitterman KJ, Wood JG et al (2003) Nicotinamide and PNC1 govern lifespan extension by calorie restriction in Saccharomyces cerevisiae. Nature 423:181–185
128. Milne JC, Lambert PD, Schenk S et al (2007) Small molecule activators of SIRT1 as therapeutics for the treatment of type 2 diabetes. Nature 450:712–716
129. Tissenbaum HA, Guarente L (2001) Increased dosage of a sir-2 gene extends lifespan in Caenorhabditis elegans. Nature 410:227–230
130. Rogina B, Helfand SL (2004) Sir2 mediates longevity in the fly through a pathway related to calorie restriction. Proc Natl Acad Sci USA 101:15998–16003
131. Lin SJ, Defossez PA, Guarente L (2000) Requirement of NAD and SIR2 for life-span extension by calorie restriction in Saccharomyces cerevisiae. Science 289:2126–2128
132. Csiszar A, Labinskyy N, Podlutsky A et al (2008) Vasoprotective effects of resveratrol and SIRT1: attenuation of cigarette smoke-induced oxidative stress and proinflammatory phenotypic alterations. Am J Physiol Heart Circ Physiol 294:H2721–H2735
133. Shinmura K, Tamaki K, Bolli R (2008) Impact of 6-mo caloric restriction on myocardial ischemic tolerance: possible involvement of nitric oxide-dependent increase in nuclear Sirt1. Am J Physiol Heart Circ Physiol 295:H2348–H2355
134. Allard JS, Heilbronn LK, Smith C et al (2008) In vitro cellular adaptations of indicators of longevity in response to treatment with serum collected from humans on calorie restricted diets. PLoS ONE 3:e3211
135. Porcu M, Chiarugi A (2005) The emerging therapeutic potential of sirtuin-interacting drugs: from cell death to lifespan extension. Trends Pharmacol Sci 26:94–103
136. Baur JA, Sinclair DA (2006) Therapeutic potential of resveratrol: the in vivo evidence. Nat Rev 5:493–506
137. Yang H, Baur JA, Chen A et al (2007) Design and synthesis of compounds that extend yeast replicative lifespan. Aging Cell 6:35–43

138. Mai A, Massa S, Lavu S et al (2005) Design, synthesis, and biological evaluation of sirtinol analogues as class III histone/protein deacetylase (Sirtuin) inhibitors. J Med Chem 48: 7789–7795
139. Firestein R, Blander G, Michan S et al (2008) The SIRT1 deacetylase suppresses intestinal tumorigenesis and colon cancer growth. PLoS ONE 3:e2020
140. Viswanathan M, Kim SK, Berdichevsky A et al (2005) A role for SIR-2.1 regulation of ER stress response genes in determining C. elegans life span. Dev Cell 9:605–615
141. Bauer JH, Goupil S, Garber GB et al (2004) An accelerated assay for the identification of lifespan-extending interventions in Drosophila melanogaster. Proc Natl Acad Sci USA 101:12980–12985
142. Valenzano DR, Terzibasi E, Genade T et al (2006) Resveratrol prolongs lifespan and retards the onset of age-related markers in a short-lived vertebrate. Curr Biol 16:296–300
143. Baur JA, Pearson KJ, Price NL et al (2006) Resveratrol improves health and survival of mice on a high-calorie diet. Nature 444:337–342
144. Orosz Z, Csiszar A, Labinskyy N et al (2007) Cigarette smoke-induced proinflammatory alterations in the endothelial phenotype: role of NAD(P)H oxidase activation. Am J Physiol 292:H130–H139
145. Ungvari Z, Orosz Z, Rivera A et al (2007) Resveratrol increases vascular oxidative stress resistance. Am J Physiol 292:H2417–H2424

# Chapter 14
# Oxidative Stress and Cardiovascular Disease in Diabetes Mellitus

**Divya Gupta, Kathy K. Griendling, and W. Robert Taylor**

**Abstract** Diabetes has a profound impact on the cardiovascular system, and oxidative stress is likely an important mechanism through which diabetes adversely affects that system. Numerous animal and some human studies support the role of oxidative stress as a unifying hypothesis linking hyperglycemia to distinct cardiovascular pathophysiologic processes. The ultimate mechanism of excess production of ROS in diabetes likely involves multiple enzymatic sources of ROS that are both convergent upon common cellular and molecular targets and interrelated with positive feedback loops occurring between these different enzymatic systems. Particular roles of mitochondrial electron transport chain, Nox family NADPH oxidases, and uncoupled NO synthase(s) have been documented. While the experimental data linking oxidative stress to the cardiovascular complications of diabetes are quite extensive, as is the case in many other settings, there is a lack of convincing data in humans demonstrating a protective effect of antioxidants on diabetic cardiovascular disease. Nonetheless, strategies to reduce disrupted redox cell signaling and oxidative stress may find applicability regarding the treatment and prevention of the cardiovascular complications of diabetes.

**Keywords** NADPH oxidases · Mitochondrial electron transport · Redox signaling · Glucose intolerance · Insulin resistance · Uncoupled NO synthases · Advanced glycatino end products · Polyol pathway

## 14.1 Introduction

Diabetes has an enormous impact on global health with an estimated prevalence over 250 million people worldwide, while the numbers are predicted to increase to 380 million by 2026 [1]. Currently, diabetes affects 5.9% of the world's adult

D. Gupta (✉)
Departments of Medicine, The Atlanta VA Medical Center, Emory University School of Medicine, Atlanta, GA, USA
e-mail: dgupta2@emory.edu

H. Sauer et al. (eds.), *Studies on Cardiovascular Disorders*,
Oxidative Stress in Applied Basic Research and Clinical Practice,
DOI 10.1007/978-1-60761-600-9_14, © Springer Science+Business Media, LLC 2010

population, 80% of whom are in developing countries [1]. Two forms of diabetes exist, insulin-dependent, or type I diabetes, and non-insulin–dependent, or type II diabetes. Type I diabetes occurs because of a presumed autoimmune destruction of the pancreatic beta-islet cells, with a resultant reduction in insulin production and subsequent decrease in glucose uptake and metabolism. These patients make up just 5–10% of the total diabetes population [2]. The remaining 90–95% of patients with diabetes fall under the category of non-insulin–dependent diabetes [2] which is characterized by an initial insulin resistance—an inability of the bodies' adipose and skeletal muscle cells to appropriately respond to the available insulin, leading to hyperglycemia. Diabetes adversely affects virtually every organ system in the body. However, the major cause of increased morbidity and mortality in diabetics is because of the effects of diabetes on the cardiovascular system.

Patients with diabetes have a twofold to eightfold increase in cardiovascular disease (CVD), primarily through increases in atherosclerosis, as well as thrombosis. However, diabetes can have a broad range of cardiovascular effects, including altered endothelial function, decreased vascular compliance, microvascular disease, and development of cardiomyopathy [3, 4]. The adverse cardiovascular effects of diabetes begin very early on in the disease process, as evidenced by the observation that patients without frank diabetes but the presence of an abnormal 2-h oral glucose tolerance test have a twofold increased risk of macrovascular disease [5].

## 14.2 Enzymatic Sources of Reactive Oxygen Species in Diabetes

While there are many potential mechanisms through which diabetes causes cardiovascular disease, oxidative stress, mainly as a result of increased levels of reactive oxygen species (ROS), has been proposed to play a pivotal role in virtually all aspects of increased cardiovascular dysfunction. The pathological mechanisms involved in the increase in ROS levels in diabetes are characterized by the convergence of multiple sources of ROS and classic positive feedback mechanisms that ultimately result in their overactivity. There are at least three major sources of reactive oxygen species in diabetes that negatively impact the cardiovascular system. They are the NADPH oxidase, the mitochondria [6], and the endothelial nitric oxide synthase (eNOS or NOS III). In addition, it is likely that other enzymatic systems may also play a role. The interactions between these sources and their downstream targets are likely responsible for the enormous impact of diabetes on cardiovascular disease.

### *14.2.1 DAG-PKC Activation*

Several studies have shown that increased ROS production can occur simply because of hyperglycemia. One of the mechanisms involves the diacylglycerol (DAG)-protein kinase C (PKC) pathway [7]. PKC activation and DAG accumulation are increased in hyperglycemic states. The elevated DAG levels are present chronically, thereby contributing to the chronic sequelae known to be characteristic of diabetes [8]. Increased PKC activity and DAG accumulation have tissue-specific responses that are due in part to the presence of specific PKC isoforms [9]. In vascular cells, PKC has been shown to increase the activity and expression of eNOS and perhaps even contribute to eNOS uncoupling (see below). PKC has also been implicated in increased Nox family NADPH oxidase activity as well as expression of its subunits [9].

In normal physiologic states, eNOS, found in endothelial cells, works to produce endogenous NO via conversion of L-arginine to L-citrulline [10]. Increased levels of DAG activate PKC [11], which phosphorylates nitric oxide synthase (eNOS or NOS III), increasing its expression, most likely as a means of increasing NO production to counterbalance some effects of superoxide [7]. However, instead of increased NO effects, decreased NO production or bioavailability has been noted [7, 12]. It has been shown that chronic exposure to glucose increases endothelial cell production of superoxide about threefold [13]. The complex interactions between ROS and eNOS are detailed later in this chapter. In an analogous fashion, activation of PKC by hyperglycemia can increase the expression and activity of NADPH oxidase in virtually all cell types within the vascular wall, and potentially the myocardium as well. Thus, PKC activation leads directly to an increase in oxidative stress within the cardiovascular system through its effects on both eNOS and the NADPH oxidase system. It is important to note that several effects of PKC activation may be suppressed by alterations in ROS fluxes (Fig. 14.1).

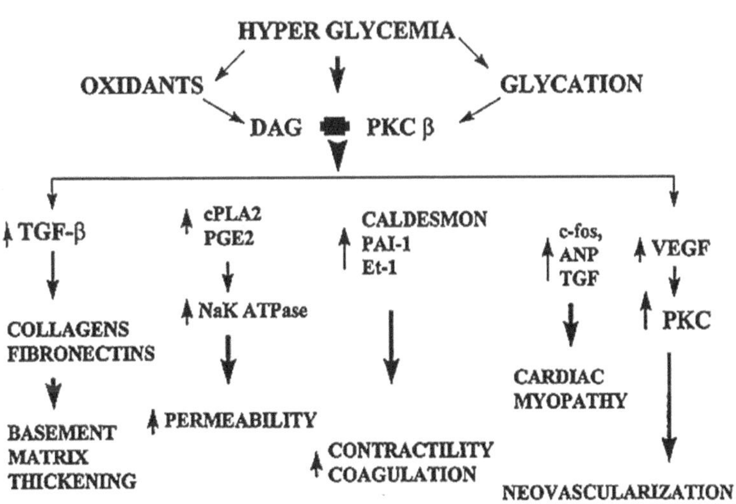


**Fig. 14.1** Adverse effects of DAG-PKC activation due to hyperglycemia. *PAI*-1, Plasminogen activator inhibitor-1; *ET*-1, endothelin-1; *ANP*, atrial naturetic peptide [11]

### *14.2.2 NADPH Oxidase*

As indicated above, increased PKC levels also work to activate superoxide-producing enzymes, such as the nicotinamide adenine dinucleotide phosphate oxidase (NADPH oxidase) [10]. Studies have shown that hyperglycemia results in increased mRNA expression of NADPH oxidase subunits in endothelial cells, accompanied by greater NADPH oxidase activity [7]. PKC involvement in this process was confirmed via loss-of-function experiments with chelerythrine, a PKC inhibitor [7]. Many cardiovascular cells, such as vascular smooth muscle and endothelial cells, adventitial and cardiac fibroblasts, and cardiomyocytes have a continuous low level of NADPH (or to a much lower extent, NADH-dependent ROS-generating activity) and this ROS production can increase when presented with the appropriate stimuli (ROS, cytokines, oxidized LDL, hyperglycemia, AGEs, angiotensin II, etc.); this effect is blunted by specific inhibitors [14].

NADPH oxidase was first recognized in phagocytes as necessary for killing ingested pathogens. Its importance in this process was clearly established with the understanding of chronic granulomatous disease, in which a genetic defect causes the oxidase complex to be nonfunctional, leaving the patient predisposed to recurrent infections [15]. Further studies deemed two oxidase subunits to be critical to its functioning, $p22^{phox}$ and $gp91^{phox}$. Without these subunits, electron transfer from NADPH to molecular oxygen is not possible and superoxide does not form [14]. Although $p22^{phox}$ can be found in almost all cell types, $gp91^{phox}$ is not uniformly present. However, isoforms of this catalytic subunit, now termed Nox (1–5) or Duox (1 or 2), are present [14].

While NADPH oxidase directly produces ROS as a dedicated enzyme complex, it may also potentiate the production of reactive oxygen species through other enzymatic systems (Fig. 14.2). Thus, NADPH oxidase-triggered ROS can trigger the production of much larger ROS amounts through the cell.

### *14.2.3 Cellular Respiration*

Mitochondrial electron transport also plays a part in increased superoxide production in the setting of diabetes. The amount of superoxide produced by the mitochondrial electron transport chain (Fig. 14.3) is increased when cultured endothelial cells are exposed to a hyperglycemic environment [17]. This effect is blunted by a superoxide dismutase mimetic [18, 19] or inhibitors of oxidative phosophorylation [18, 16]. Increased superoxide production by this mechanism may also lead to increased polyol pathway activity (see below) and PKC activity, thus increasing ROS production even further.

### *14.2.4 Oxidative Stress and Advanced Glycation End Products*

In the setting of excess glucose, there is an irreversible, nonenzymatic protein glycosylation, the Maillard Reaction. This leads to Amadori products and Schiff bases.

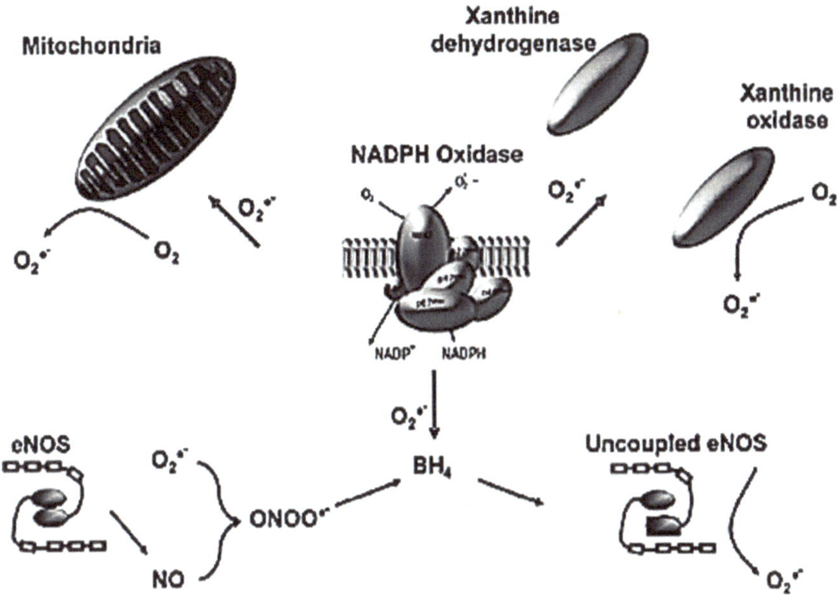


**Fig. 14.2** Interplay between NADPH oxidase and other sources of ROS [14]

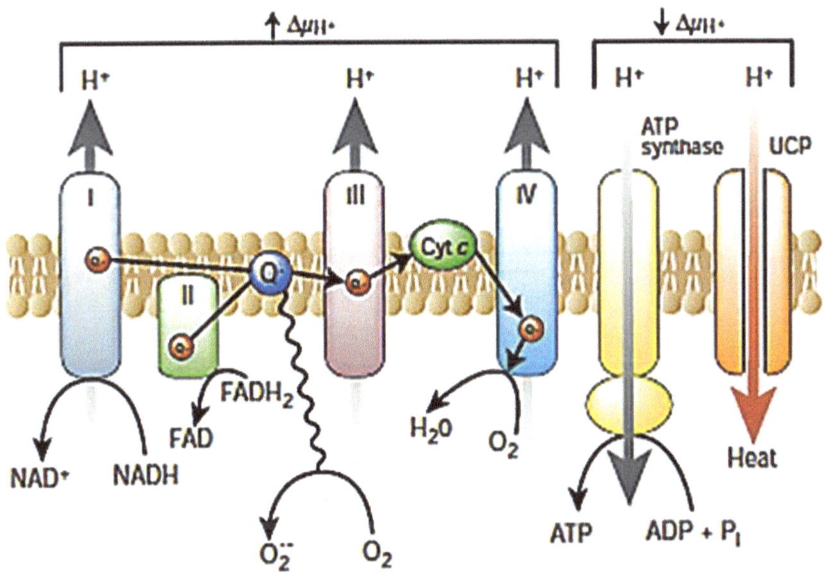


**Fig. 14.3** Production of superoxide by the mitochondrial electron transport chain. Increased hyperglycemia-derived electron donors from the TCA cycle (NADH and FADH2) generate a high mitochondrial membrane potential (DmH+) by pumping protons across the mitochondrial inner membrane. This inhibits electron transport at complex III, increasing the half-life of free radical intermediates of coenzyme Q (ubiquinone), which reduces $O_2$ to superoxide [16]

The Amadori products undergo oxidative degradation, leading to the formation of advanced glycation end products (AGEs). Oxidative stress can increase via AGEs alone, and/or through AGE interaction with the receptor for AGEs (RAGE) potentially participating in many cardiovascular complications of diabetes. As the ligands for RAGE, including AGEs, accumulate, RAGE is upregulated, further amplifying the effects of AGEs [20].

AGEs have also been shown to increase NADPH oxidase activation and the resultant superoxide production [21]. The increased superoxide production in turn leads to increased AGE formation, creating a cyclical pattern with positive feedback [22]. AGEs have also been shown to increase mitochondrial ROS generation. However, the mechanism of this effect is not well understood [23].

### *14.2.5 Oxidative Stress and the Polyol Pathway*

In euglycemic states, aldose reductase (AR) has a low affinity for glucose, accounting for the lack of activation of the polyol pathway. However, as glucose levels increase, its conversion to sorbitol and fructose through the polyol pathway increases as well [24] (Fig. 14.4). With the help of cytosolic NADPH, AR, a cytosolic enzyme, reduces glucose to sorbitol, which is then converted to fructose with the help of a second enzyme, sorbitol dehydrogenase (SDH) and $NAD^+$ [25]. This reduction of $NAD^+$ to NADH leads to the cytosolic accumulation of NADH [25]. Increased superoxide also leads to increased polyol pathway flux [22]. Activation of this pathway leads to increased NADPH consumption which is necessary for

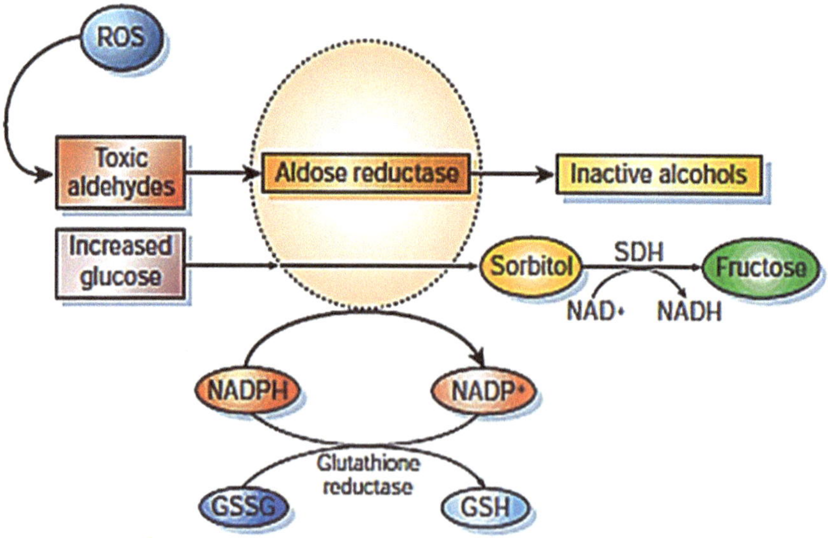


**Fig. 14.4** The polyol pathway [16]

GSH generation, an intracellular thiol buffer [3]. Also, conversion of sorbitol to fructose causes $NAD^+$ to be reduced to NADH. This has been postulated as a mechanism to generate ROS via NADH oxidase [26]. However, the kinetics and substrate specificity of NADPH oxidase make this latter mechanism less likely.

Other sources of superoxide production are also increased in hyperglycemia leading to increased oxidative stress [27] (see Figs. 14.5 and 14.6). The mechanisms described above are currently thought to be the principal processes for superoxide production in the setting of diabetes. Other enzymatic systems such as xanthine oxidase and arachidonic acid metabolism are likely to be involved as well. However, data in support of these other enzymatic sources of ROS in diabetics are presently less convincing.

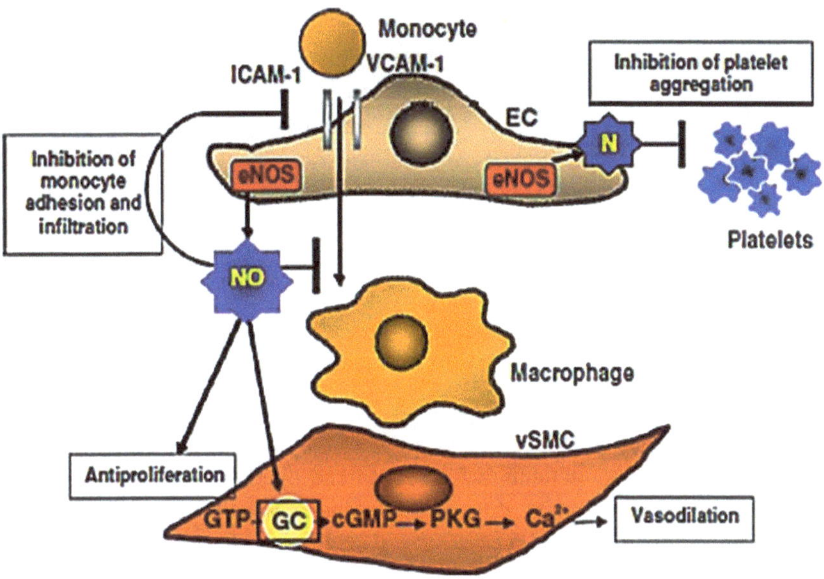


**Fig. 14.5** Normal endothelial function [6]

## 14.3 Role of Reactive Oxygen Species in the Cardiovascular Consequences of Diabetes

Diabetes affects virtually every aspect of cardiovascular disease with wide ranging effects on the vasculature and myocardium. Indeed, virtually every pathologic state within the vasculature has been postulated to have a link to alterations in oxidative stress. As diabetes affects this most fundamental process through a variety of cellular and molecular mechanisms, it is not surprising that the end result is a widespread effect involving multiple cells with a divergent set of pathological events.

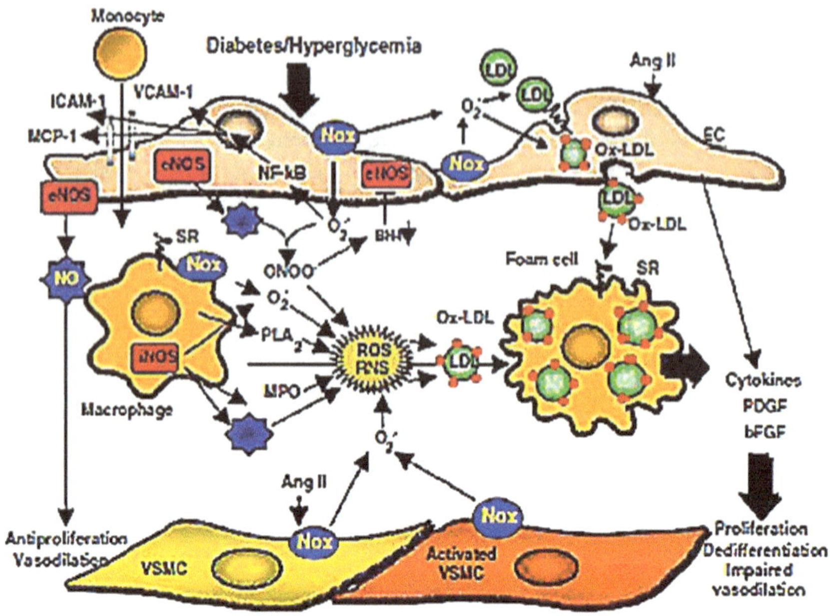


**Fig. 14.6** Simplified schematic of endothelial dysfunction in the face of diabetes/hyperglycemia [6]

## 14.3.1 Endothelial Dysfunction

Type II diabetics bear most of the vascular complications of diabetes. Poorly controlled diabetes in these patients, which is marked by elevated glucose levels and insulin resistance, exerts many of its effects on the endothelium, which is an essential factor in maintaining normal function and health. Many of the body's tissues are able to protect themselves from the deleterious effects of this disease by maintaining near normal intracellular glucose levels through decreased intracellular glucose transport. Unfortunately, the endothelium appears to be particularly vulnerable to elevated glucose [3]. Increased oxidative stress creates an imbalance in this system, leading to elevated blood pressure and increased vascular proliferation, by exerting its effects on the endothelium itself, as well as the vascular smooth muscle cells that surround the endothelium.

There are three forms of NO synthase (NOS), each one coded by a distinct gene: endothelial (eNOS or NOS III), inducible (iNOS or NOS II), and neuronal (nNOS or NOS I). In this section, we will be focusing on eNOS and its changing function in the face of hyperglycemia.

Adequate levels of bioavailable nitric oxide (NO) are essential for optimal endothelial function and vascular health. Vascular relaxation is dependent on NO bioavailability, which is decreased through its scavenging by superoxide radicals [28] and/or decreased NO production [14]. In the simplest form, increased superoxide decreases the availability of NO through its chemical conversion into peroxynitrite ($ONOO^-$) and derived oxidant species. Interestingly, although there is a decrease in NO in poorly controlled diabetics, there is also an increase in eNOS protein expression. This paradox is presumably the result of an ineffective compensatory mechanism that may attempt to increase NO production. Increased eNOS expression is ineffective in increasing bioavailable NO for at least two reasons. First, superoxide depletes NO through its conversion to peroxynitrite, as described above. Second, the functionality of eNOS is altered because of what is termed uncoupling, during which excess oxidative stress oxidizes tetrahydrobiopterin (a critical cofactor for eNOS activity), resulting in increased electron leakage towards molecular oxygen and the consequent production of superoxide by eNOS at the expense of NO (see Fig. 14.7 for details). This mechanism accounts for a positive feedback loop in which additional superoxide further impairs the ability of eNOS to generate NO and also further depletes tetrahydrobiopterin, resulting in the generation of additional superoxide. It is important to note that peroxynitrite is a potent oxidant which also causes direct oxidative damage to cells [30, 29]. Finally, eNOS expression is

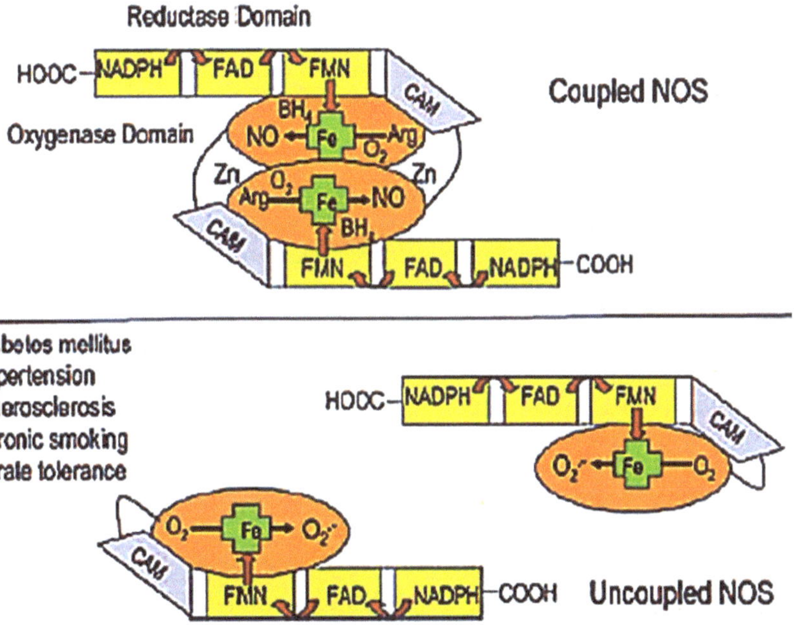


**Fig. 14.7** Coupled vs. uncoupled eNOS. Electron flow starts from NADPH to flavins FAD and FMN of the reductase domain, which delivers the electrons to the iron of the heme (oxygenase domain) and to the $BH3^-$ radical generated as an intermediate in the catalytic cycle. BH4 is essential to donate an electron and proton to versatile intermediates in the reaction cycle of L-arginine/$O_2$ to L-citrullin/NO. Calmodulin (CAM) controls electron flow in eNOS. Zinc ions(Zn) bound to NOS are required for dimer formation and stability. Monomeric eNOS or BH4/*L-arginine – deficient* eNOS is uncoupled and produces $O_2^-$ [29]

increased by hydrogen peroxide. Hydrogen peroxide levels have been documented to be elevated in states of hyperglycemia. This is due not only to the abundance of superoxide, which undergoes dismutation, but also to increased amounts and activity of Cu/Zn superoxide dismutase (SOD). Excess CuZn SOD might promote hydrogen peroxide accumulation in a situation in which superoxide would otherwise be diverted to a product other than hydrogen peroxide, namely peroxynitrite [7]. In summary, this system of increased expression of the synthetic enzyme in the setting of a diminished cofactor and excessive superoxide converts a potentially protective enzymatic mechanism into one that is pro-oxidant and potentially deleterious to the cardiovascular system.

### *14.3.2 Diabetes and Hypertension*

Hypertension is very often diagnosed in patients with type II diabetes, and the combination portends a particularly poor prognosis in terms of cardiovascular disease. While this association is common, the causality of diabetes in terms of increasing the risk of hypertension and vice versa remains a subject of debate. In the context of oxidative stress, it is clear that both diseases share some common pathological mechanisms. NO from the endothelium normally diffuses to the vascular smooth muscle cells, activating guanylate cyclase (GC) which induces vascular smooth muscle relaxation [6]. With decreased endothelial relaxation in diabetes because of decreased NO bioavailability (as described in the previous section), there is a resultant decrease in bioavailable NO to smooth muscle cells and increased vascular tone and elevated blood pressure. The structural effects of diabetes on the vasculature wall may also lead to an increase in the prevalence of hypertension. As described above, AGEs induce cross-linking of extracellular matrix proteins, which leads to a decrease in the compliance of the arterial wall. This stiffening of the arterial wall can cause an increase in pulse pressure, which is clinically translated into so-called systolic hypertension. Finally, the linkage between diabetes and angiotensin II raises the possible involvement of the renin-angiotensin system. The effects of angiotensin II, a potent vasoconstrictor and cause of hypertension, are modified by angiotensin II type I (AT1) receptor overexpression in the face of hyperinsulinemia, hyperglycemia, and oxidative stress [31, 32], as is seen in type II diabetes. Increased AT1 receptor expression is also linked to increased activation of NADPH oxidase [31] and increased ROS production, creating a cyclical process.

Thus, while the clinical association between hypertension and diabetes and the need to aggressively treat hypertension in diabetic patients are both clear, the cellular and molecular links between these two disease processes remain poorly defined. However, it is clear that these two diseases share the involvement of ROS as upstream and downstream mediators of their cardiovascular complications. This may likely explain the synergy between hypertension and diabetes in terms of cardiovascular disease.

### 14.3.3 Diabetes and Atherosclerosis

Many of the previously discussed factors are major contributors to increased atherosclerosis in patients with diabetes. Increased oxidative stress leads to the inactivation of nitric oxide, DNA and protein modification, and the activation of redox-sensitive gene expression (including adhesion molecules, proinflammatory cytokines, and matrix metalloproteinases), which are key to the initiation and progression of atherosclerosis [33]. In addition, several critical elements that are dysregulated in the setting of diabetes have been identified to be involved in determining plaque vulnerability, including macrophage infiltration and coronary calcification [22] (see Figs. 14.5 and 14.6). Both macrophage infiltration and coronary calcification were shown to be especially elevated in diabetic patients with poor glycemic control. The progression of coronary calcification is related to the length of time a patient has had diabetes, independently of other risk factors and descriptors of the diabetic state [22, 34].

Diabetes also has profound effects on vascular smooth muscle cells within the vascular wall. Vascular smooth muscle cells exist along a continuum of two states or phenotypes. The quiescent state, which is the dominant phenotype in healthy vessels, is characterized by a contractile phenotype. The synthetic or proliferative phenotype is more prevalent in areas of remodeling such as in neointimal proliferation in early atherosclerotic lesions [35]. In diabetes, vascular smooth muscle cells are more likely to exhibit characteristics of the proliferative phenotype in vivo [36]. In diabetic atherosclerotic animals, smooth muscle proliferation within atherosclerotic lesions is increased [37]. In contrast, studies using cultured smooth muscle cells or freshly isolated smooth muscle cells from normal and diabetic animals have provided conflicting results in terms of the effects of hyperglycemia on smooth muscle cell proliferation. However, it is possible to conclude that at least in the in vivo setting, diabetes is associated with an increase in smooth muscle cell proliferation. This is likely a consequence of the complex milieu of the smooth muscle cells in vivo, where they are impacted by other cell types and inflammatory cytokines.

ROS obviously also have direct effects on the oxidation of LDL through lipooxidation, leading to increased atherosclerosis. In vitro studies have shown that LDL alone is not strongly atherogenic; however, in its modified—and particularly oxidized—state, it becomes proatherogenic [38]. Lipid oxidation occurs when pathophysiologic levels of glucose are available, causing protein oxidation via peroxidation of polyunsaturated fatty acids [39]. This process is increased in the presence of oxidative stress, as evidenced by in vitro studies in which there is a decrease in LDL oxidation in the presence of antioxidants [40, 41]. OxLDL leads to increased transformation of monocytes and macrophages into lipid-laden foam cells, accounting for the evolution of atherosclerosis [41].

In addition, there can be indirect effects, such as induction of the leptin-like oxLDL receptor (LOX-1). It has been proposed that LOX-1 activation can be a proximal signal in the inflammatory cascade, leading to increased expression of adhesion molecules and cytokines in a redox-sensitive manner [42].

### 14.3.4 Diabetes and Thrombosis

Thrombosis is a key factor leading to vascular occlusion and is affected by multiple factors. In normal individuals, maintenance of vascular homeostasis decreases the likelihood that detrimental clotting will lead to a myocardial infarction, vascular insufficiency, cerebral vascular accidents, etc. However, with diabetes and its associated oxidative stress, the balance tilts towards that of increased thrombosis, leading to many of the vascular complications. Endothelial dysfunction, as discussed in the earlier section, plays a critical role in increased thrombogenesis through increased platelet adhesion, procoagulant activity, and impaired fibrinolysis [6]. Increased ROS works to increase the formation of the tissue factor complex, which generates thrombin. Increased thrombin, in turn, activates vascular NADPH oxidase through yet another positive feedback loop, further increasing ROS production and enhancing the prothrombotic state [43]. Another mechanism of increased thrombosis occurs via oxidized LDL, which causes increased platelet adhesion, as well as decreased tissue-type plasminogen activator (tPA), and increased plasminogen activator inhibitor-1 (PAI-1), leading to increased clot formation via platelet adhesion [6]. Oxidative stress induces vascular injury and modulates these key regulators of thrombosis in a way that increases the prothrombotic state in diabetics, likely contributing to vascular complications.

### 14.3.5 Diabetic Cardiomyopathy

Data from the Framingham studies have shown that diabetic men are twice as likely to develop congestive heart failure, and that diabetic women are five times more likely to develop congestive heart failure, when compared with age-matched controls [44]. Both systolic and diastolic dysfunction are prevalent in diabetics [45]. Echocardiographic studies have shown that in diabetic patients without coronary atherosclerosis, decreased diastolic filling, increased atrial filling, and increased isovolumetric relaxation are all present [46]. A causal role for oxidative stress in diabetic cardiomyopathy has not been definitively ascertained. However, the heart is likely to be particularly susceptible to oxidative stress, at least in part considering that, relative to other tissues, cardiac tissues have decreased levels of antioxidant enzymes [47]. Furthermore, in diabetics, the antioxidant capacity of the myocardium is further diminished, thus increasing the likelihood of ROS inducing myocyte dysfunction [48].

In the setting of diabetic cardiomyopathy, it is likely that most of the enzymatic sources of ROS discussed previously are also involved. Mitochondrial damage in diabetic cardiomyocytes has been well established and likely contributes significantly to the local production of ROS. However, there is also evidence for nonmitochondrial sources of ROS in the diabetic myocardium. Several studies have suggested that this is an indirect mechanism involving local production of angiotensin II, which may increase oxidative stress via NADPH oxidase.

ROS production in diabetes, as discussed earlier, has been shown to lead to apoptotic and damaged myocytes. Oxidative stress also increases the formation of excessive and abnormal extracellular matrix, leading to increased fibrosis, which may be relevant to both systolic and diastolic dysfunction [49]. Also, increased ROS can enhance apoptosis, as evidenced by increased TUNEL staining and caspase-3 activation, as well as increased DNA damage and impaired DNA repair, all associated with abnormal cardiac remodeling [50]. Nonmitochondrial sources of ROS can also lead to an increase in mitochondrial ROS via delivery of reducing equivalents to the electron transport chain, leading to mitochondrial uncoupling [4]. Many studies have indeed demonstrated mitochondrial uncoupling and dysfunction [4], in addition to indirect evidences, e.g., that overexpression of mitochondrial superoxide dismutase (SOD2) in the hearts of diabetic mice reverses the maladaptive changes in mitochondria and preserves cardiomyocyte function [51]. Thus, oxidative stress leads to a variety of changes in the myocardium that are interrelated and result in myocyte death as well as alterations in the extracellular matrix.

## 14.3.6 Arrhythmia

Approximately 50% of deaths in patients with cardiomyopathy are sudden, and the vast majority are due to ventricular arrhythmias [52]. Delayed cardiac action potential repolarization is believed to be the etiology of most cardiomyopathy-associated dysrrhythmias, attributed to a significant decrease in the calcium-independent, transient outward current $It_O$, one of the four major $K^+$ currents [53]; and oxidative stress is believed to be the culprit leading to this downregulation (Fig. 14.8). In myocytes, it is the ratio of reduced glutathione (GSH) to oxidized glutathione (GSSG) that

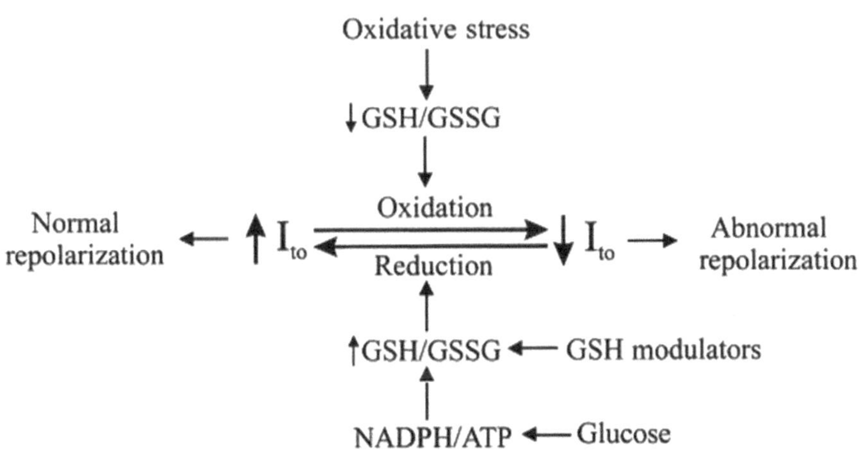


**Fig. 14.8** Proposed redox mechanism of transient outward current ($It_O$) downregulation in the diseased ventricle. GSH, reduced glutathione; GSSG, oxidized glutathione; NADPH, reduced nicotinamide adenine dinucleotide phosphate [54]

is key in determining the amount of oxidative stress [55]. As that ratio decreases, as it does in uncontrolled diabetes, it is believed that the density of $It_0$ decreases. Diabetic rat cardiomyocytes treated with GSH exhibit increased $It_0$ when compared to diabetic rat myocytes that were not treated with GSH [56]. It has also been shown that $It_0$ increases toward near normal levels in cardiomyocytes from insulin-treated diabetic rats [57], showing the link between insulin responsiveness, oxidative stress, and the crucial $It_0$ current.

## 14.4 Summary

Diabetes has a profound impact on the cardiovascular system, and oxidative stress is likely an important mechanism through which diabetes adversely affects that system. Numerous animal and some human studies support this as a unifying hypothesis linking hyperglycemia to multiple distinct cardiovascular pathophysiologic processes. The ultimate mechanism of excess production of ROS in diabetes likely involves multiple enzymatic sources of ROS that are both convergent upon common cellular and molecular targets and interrelated with positive feedback loops occurring between distinct enzymatic systems. While the experimental data linking oxidative stress to the cardiovascular complications of diabetes are quite extensive, as is the case in many other settings, there is a lack of convincing data in humans demonstrating a protective effect of antioxidants on diabetic cardiovascular disease. Nonetheless, therapeutic strategies that reduce reactive oxygen species remain attractive targets for the treatment and prevention of the cardiovascular complications of diabetes.

## References

1. Gan D (ed) (2006) Diabetes epidemic out of control, in diabetes atlas. International Diabetes Federation, Cape Town
2. Diabetes Public Health Resource (2005) National Diabetes Fact Sheet. http://www.cdc.gov/diabetes/pubs/general.htm#what . Accessed 20 Dec 2005
3. Haidara M, Yassin HZ, Rateb M, Ammar H, Zorkani MA (2006) Role of oxidative stress in development of cardiovascular complications in diabetes mellitus. Curr Vasc Pharmacol 4(3):215–227
4. Boudina S, Abel ED (2007) Diabetic cardiomyopathy revisited. Circulation 115(25): 3213–3223
5. Laakso M (1997) Epidemiology of macrovascular disease in diabetes. Diabetes Rev 5: 284–315
6. Pandolfi A, De Filippis EA (2007) Chronic hyperglicemia and nitric oxide bioavailablity play a pivotal role in pro-atherogenic vascular modifications. Genes Nutr 2:195–208
7. Hink U, Li H, Mollnau H, Oelze M, Matheis E, Hartmann M, Skatchkov M, Thaiss F, Stahl RA, Warnholtz A, Meinertz T, Griendling K, Harrison DG, Forstermann U, Munzel T (2001) Mechanisms underlying endothelial dysfunction in diabetes mellitus. Circ Res 88:e14–e22
8. Xia P, Inoguchi T, Kern TS, Engerman RL, Oates PJ, King GL (1994) Characterization of the mechanism for the chronic activation of diacylglycerol-protein kinase C pathway in diabetes and hypergalactosemia. Diabetes 43:1122–1129

9. Rask-Madsen C, King GL (2005) Proatherosclerotic mechanisms involving protein kinase C in diabetes and insulin resistance. Arterioscler Thromb Vasc Biol 25(3):487–496
10. Hadi H, Suwaidi JA (2007) Endothelial dysfunction in diabetes mellitus. Vasc Health Risk Manag 3(6):853–876
11. Ishii H, Koya D, King GL (1998) Protein kinase C activation and its role in the development of vascular complications in diabetes mellitus. J Mol Med 76:21–31
12. Hirata K, Hirata K, Kuroda R, Sakoda T, Katayama M, Inoue N, Suematsu M, Kawashima S, Yokoyama M (1995) Inhibition of endothelial nitric oxide synthase activity by protein kinase C. Hypertension 25:180–185
13. Cosentino F, Hishikawa K, Katusic ZS, Lüscher TF (1997) High glucose increases nitric oxide synthase expression and superoxide anion generation in human aortic endothelial cells. Circulation 96:25–28
14. Cave A, Brewer AC, Narayanapanicker A, Ray R, Grieve DJ, Walker S, Shah AM (2006) NADPH oxidases in cardiovascular health and disease. Antiox Redox Signal 8(5–6): 691–728
15. Dinauer M, Orkin SH (1992) Chronic granulomatous disease. Annu Rev Med 43:117–124
16. Brownlee M (2001) Biochemistry and molecular cell biology of diabetic complications. Nature 414(6865):813–820
17. Du XL, Edelstein D, Rossetti L, Fantus IG, Goldberg H, Ziyadeh F, Wu J, Brownlee M (2000) Hyperglycemia-induced mitochondrial superoxide overproduction activates the hexosamine pathway and induces plasminogen activator inhibitor-1 expression by increasing Sp1 glycosylation. Proc Natl Acad Sci USA 97(22):12222–12226
18. Subramaniam P (2007) Mechanisms for oxidative stress in diabetic cardiovascular disease. Antiox Redox Signal 9(7):955–969
19. Salvemini D, Wang ZQ, Zweier JL, Samouilov A, Macarthur H, Misko TP, Currie MG, Cuzzocrea S, Sikorski JA, Riley DP (1999) A nonpeptidyl mimic of superoxide disutase with therapeutic activity in rats. Science 286:304–306
20. Stern D, Yan SD, Yan SF, Schmidt AM (2002) Receptor for advanced glycation endproducts: a multiligand receptor magnifying cell stress in diverse pathologic settings. Adv Drug Deliv Rev 54:1615–1625
21. Cifuentes ME, Rey FE, Carretero OA, Pagano PJ (2000) Upregulation of $p67^{phox}$ and $gp91^{phox}$ in aortas from angiotensin II-infused mice. Am J Physiol Heart Circ Physiol 279:H2234–H2240
22. Meerarani P, Badimon JJ, Zias E, Fuster V, Moreno PR (2006) Metabolic syndrome and diabetic atherothrombosis: implications in vascular complications. Curr Mol Med 6:501–514
23. Basta G, Lazzerini G, Del Turco S, Ratto GM, Schmidt AM, De Caterina R (2005) At least 2 distinct pathways generating reactive oxygen species mediate vascular cell adhesion molecule-1 induction by advanced glycation end products. Arterioscler Thromb Vasc Biol 25:1401–1407
24. Lee AY, Chung SK, Chung SS (1995) Demonstration that the polyol accumulation is responsible fro diabetic cataract by the use of transgenic mice expressing the aldose reductase gene in the lens. Proc Natl Acad Sci USA 92:2780–2784
25. Wilson DK, Bohren KM, Gabbay KH, Quiocho FA (1992) An unlikely sugar substrate site in the 1.65 A structure of the human aldose reductase holoenzyme implicated in diabetic complications. Science 257:81–84
26. Morré DM, Lenaz G, Morré DJ (2000) Surface oxidase and oxidative stress propagation in aging. J Exp Biol 203:1513–1521
27. Griendling KK, Sorescu D, Lassègue B, Ushio-Fukai M (2000) Modulation of protein kinase activity and gene expression by reactive oxygen species and their role in vascular physiology and pathophysiology. Arterioscler Thromb Vasc Biol 20:2175–2183
28. Schafer A, Bauersachs J (2008) Endothelial dysfunction, impaired endogenous platelet inhibition and platelet activation in diabetes and atherosclerosis. Curr Vasc Pharmacol 6:52–60

29. Munzel T, Daiber A, Ullrich V, Mülsch A (2005) Vascular consequences of endothelial nitric oxide synthase uncoupling for the activity and expression of the soluble guanylyl cyclase and the cGMP-dependent protein kinase. Arterioscler Thromb Vasc Biol 25:1551–1557
30. Smolenski A, Bachmann C, Reinhard K, Hönig-Liedl P, Jarchau T, Hoschuetzky H, Walter U (1998) Analysis and regulation of vasodilator-stimulated phosphoprotein serine 239 phosphorylation in vitro and in intact cells using a phosphospecific monoclonal antibody. J Bio Chem 273:20029–20035
31. Wassman S, Nickenig G (2006) Pathophysiological regulation for the AT1-receptor and implications for vascular disease. J Hypertens Suppl 24:s15–s21
32. Banday AA, Lokhandwala MF (2008) Oxidative stress-induced renal angiotensin AT1 receptor upregulation causes increased stimulation of sodium transporters and hypertension. Am J Physiol Renal Physiol 295:F698–F706
33. Nickenig G, Harrison DG (2002) The AT(1)-type angiotensin receptor in oxidative stress and atherogenesis: part I: oxidative stress and atherogenesis. Circulation 105:393–396
34. Wolfe ML, Iqbal N, Gefter W, Mohler ER 3rd, Rader DJ, Reilly MP (2002) Coronary artery calcification at electron beam computed tomography is increased in asymptomatic type 2 diabetics independent of traditional risk factors. J Cardiovasc Risk 9:369–376
35. Mosse PR, Campbell GR, Wang ZL, Campbell JH (1985) Smooth muscle phenotypic expression in human carotid arteries. I. Comparison of cells with diffuse intial thickenings adjacent to atheromatous plaques with those of the media. Lab Invest 53:556–562
36. Schwartz CJ, Valente AJ, Sprague EA, Kelley JL, Cayatte AJ, Rozek MM (1992) Pathogenesis of the atherosclerotic lesion. Implications for diabetes mellitus. Diabetes Care 15:1156–1167
37. Suzuki LA, Poot M, Gerrity RG, Bornfeldt KE (2001) Diabetes accelerates smooth muscle accumulation in lesions of atherosclerosis: lack of direct growth-promoting effects of high glucose levels. Diabetes 50(4):851–860
38. Witztum JL, Steinberg D (1991) Role of oxidized low density lipoprotein in atherogenesis. J Clin Invest 88:1785–1792
39. Pennathur S, Ido Y, Heller JI, Byun J, Danda R, Pergola P, Williamson JR, Heinecke JW (2005) Reactive carbonyls and polyunsaturated fatty acids produce a hydroxyl radical-like species: a potential pathway for oxidative damage of retinal proteins in diabetes. J Bio Chem 280:22706–22714
40. Steinbrecher U, Parthasarathy S, Leake DS, Witztum JL, Steinberg D (1984) Modification of low density lipoprotein by endothelial cells involves lipid peroxidation and degradation of low density lipoprotein phospholipids. Proc Natl Acad Sci USA 81:3883–3887
41. Steinberg D (1997) Low density lipoprotein oxidation and its pathobiological significance. J Biol Chem 272:20963–20966
42. Mehta JL, Rasouli N, Sinha AK, Molavi B (2006) Oxidative stress in diabetes: a mechanistic overview of its effects on atherogenesis and myocardial dysfunction. Int J Biochem Cell Biol 38(5–6):794–803
43. Herkert O, Djordjevic T, BelAiba RS, Görlach A (2004) Insights into the redox control of blood coagulation: role of vascular NADPH oxidase-derived reactive oxygen species in the thrombogenic cycle. Antiox Redox Signal 6(4):765–776
44. Kannel WB, McGee DL (1979) Diabetes and cardiovascular disease: the Framingham study. J Am Med Assoc 241:2035–2038
45. Liu JE, Palmieri V, Roman MJ, Bella JN, Fabsitz R, Howard BV, Welty TK, Lee ET, Devereux RB (2001) The impact of diabetes on left ventricular filling pattern in normotensive and hypertensive adults: the Strong Heart Study. J Am Coll Cardiol 37:1943–1949
46. Schannwell CM, Schneppenheim M, Perings S, Plehn G, Strauer BE (2002) Left ventricular diastolic dysfuntion as an early manifestation of diabetic cardiomyopathy. Cardiology 98: 33–39
47. Srikanthan P, Hsueh W (2004) Preventing heart failure in patients with diabetes. Med Clin North Am 88:1237–1256

48. Wold LE, Ceylan-Isik AF, Ren J (2005) Oxidative stress and stress signaling: menace of diabetic cardiomyopathy. Acta Pharmacol Sin 26(8):908–917
49. Adeghate E (2004) Molecular and cellular basis of the aetiology and management of diabetic cardiomyopathy: a short review. Mol Cell Biochem 261:187–191
50. Barouch L, Berkowitz DE, Harrison RW, O'Donnell CP, Hare JM (2003) Disruption of leptin signaling contributes to cardiac hypertrophy independently of body weight in mice. Circulation 108:754–759
51. Shen X, Zheng S, Metreveli NS, Epstein PN (2006) Protection of cardiac mitochondria by overexpression of MnSOD reduces diabetic cardiomyopathy. Diabetes 55:798–805
52. Kjekshus J (1990) Arrhythmias and mortality in congestive heart failure. Am J Cardiol 65:I-42–I-48
53. Tomaselli G, Marbán E (1999) Electrophysiological remodeling in hypertrophy and heart failure. Cardiovasc Res 42:270–283
54. Rozanski GJ, Xu Z (2002) A metabolic mechanism for cardiac $K^+$ channel remodelling. Clin Exp Pharmacol Physiol 29:132–137
55. Meister A (1995) Glutathione metabolism. Meth Enzymol 251:3–7
56. Xu Z (1999) Interaction of glucose and glutathione metabolism in regulating K+ channels in diabetic cardiomyocytes. FASEB J 13:A97
57. Xu Z, Patel KP, Rozanski GJ (1996) Metabolic basis of decreased transient outward K+ current in ventricular myocytes from diabetic rats. Am J Physiol 271:H2190–H2196

# Chapter 15
# Reactive Oxygen Species, Oxidative Stress, and Hypertension

**Rhian M. Touyz, Andreia Chignalia, and Mona Sedeek**

**Abstract** Reactive oxygen species (ROS) influence many physiological processes including host defense, hormone biosynthesis, fertilization and cellular signaling. Increased ROS production has been implicated in various chronic diseases, including hypertension, atherosclerosis, diabetes and kidney disease. Oxidative stress may be both a cause and a consequence of hypertension. Although oxidative injury may not be the sole etiology, it amplifies blood pressure elevation in the presence of other prohypertensive factors, such as salt loading, activation of the renin-angiotensin system and sympathetic hyperactivity. Oxidative stress is a multisystem phenomenon in hypertension and involves the heart, kidneys, nervous system, and vessels. Compelling evidence indicates the importance of the vasculature in the pathophysiology of hypertension, and therefore much emphasis has been placed on the (patho)biology of ROS in the vascular system. A major source for cardiovascular and renal ROS is a family of nonphagocytic NAD(P)H oxidases, including the prototypic Nox2 homologue-based NAD(P)H oxidase, as well as other NAD(P)H oxidases, such as Nox1 and Nox4. Other possible sources include mitochondrial electron transport enzymes, xanthine oxidase, cyclooxygenase, lipoxygenase, and uncoupled nitric oxide synthase (NOS). NAD(P)H oxidase-derived ROS is important in regulating endothelial function and vascular tone, and oxidative stress is implicated in endothelial dysfunction, inflammation, hypertrophy, apoptosis, migration, fibrosis, angiogenesis, and rarefaction, important processes involved in vascular remodeling in hypertension. These findings have evoked considerable interest because of the possibilities that therapies targeted against nonphagocytic NAD(P)H oxidase to decrease ROS generation and/or strategies to increase nitric oxide (NO) availability and antioxidants may be useful in minimizing vascular injury and thereby prevent or regress target organ damage associated with hypertension.

R.M. Touyz (✉)
Ottawa Hospital Research Institute, Kidney Research Centre, University of Ottawa, Ottawa, ON K1H 8M5, Canada
e-mail: rtouyz@uottawa.ca

H. Sauer et al. (eds.), *Studies on Cardiovascular Disorders*, Oxidative Stress in Applied Basic Research and Clinical Practice,
DOI 10.1007/978-1-60761-600-9_15, © Springer Science+Business Media, LLC 2010

**Keywords** Renin-angiotensin system · NADPH oxidases · Mitochondrial electron transport · Redox signaling · Uncoupled NO synthases · Vascular endothelium

## 15.1 Introduction

Hypertension affects 30% of adults in the Western world and is the leading cause of morbidity and mortality worldwide [1]. Although the exact etiology still remains largely unknown, with only about 5% of hypertensive patients having an identifiable cause, it is clear that hypertension is due to dynamic and complex interactions involving many systems (heart, kidney, brain, vessels), between genes, physiology, and environment (Fig. 15.1). At the molecular level, multiple factors have been implicated in the pathophysiology of hypertension, including activation of the renin-angiotensin-aldosterone system, inflammation, aberrant G protein-coupled receptor signaling and endothelial dysfunction [2–5]. Common to these processes is oxidative stress due, in large part, to excess production of reactive oxygen species (ROS), to decreased nitric oxide (NO) bioavailability, and to decreased antioxidant capacity in the vessels, heart, brain, and kidneys [6–9].

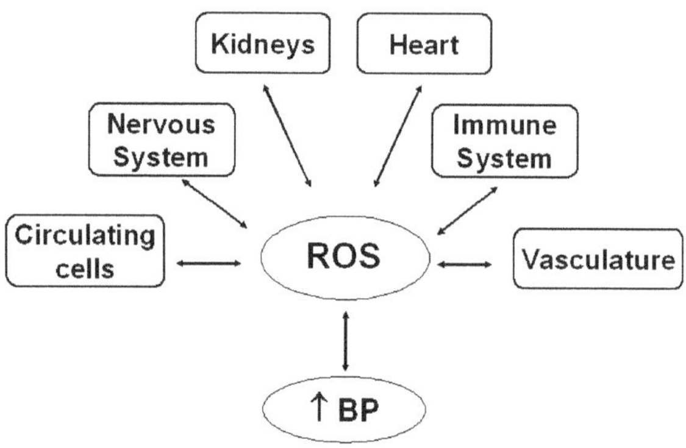


**Fig. 15.1** Generation of ROS in hypertension is a multisystem phenomenon, involving multiple organs. Oxidative stress may be both a cause and a consequence of hypertension

ROS, originally considered to induce negative and injurious cellular effects, such as apoptosis, are now recognized to have important positive actions, such as the induction of host defense genes, activation of transcription factors, and mobilization of ion transport systems [10–13]. In the vascular system ROS play a physiological role in controlling endothelial function and vascular tone, and a pathophysiological role in endothelial dysfunction, inflammation, hypertrophy, proliferation, apoptosis, migration, fibrosis, angiogenesis, and rarefaction, important processes underlying vascular remodeling in hypertension and other cardiovascular diseases. Molecular processes whereby ROS induce cardiovascular injury involve activation of redox-sensitive signaling pathways [14–16]. Superoxide anion and $H_2O_2$ stimulate mitogen-activated protein kinases, tyrosine kinases, and transcription factors (NFB, AP-1, and HIF-1), and inactivate protein tyrosine phosphatases [17–19]. ROS also increase $[Ca^{2+}]i$ and upregulate protooncogene and proinflammatory gene expression and activity [20–22]. These phenomena occur through oxidative modification of proteins by altering key amino acid residues, by inducing protein dimerization, and by interacting with metal complexes such as Fe-S moieties [23, 24]. Changes in the intracellular redox state through glutathione and thioredoxin systems may also influence intracellular signaling events [25, 26].

The association between free radicals and hypertension was suggested as early as 1960 [27]; but it was some 40 years later that this association was investigated in detail, when it was demonstrated that Ang II-mediated hypertension in rats increases vascular superoxide production via membrane NAD(P)H oxidase activation [28]. Almost all models of hypertension display some form of oxidative excess, including genetic forms (SHR, SHRSP), surgically-induced (2K1C, aortic banding), endocrine-induced (Ang II, aldosterone, DOCA), and diet-induced hypertension (salt, fat) [29–33]. Mice deficient in ROS-generating enzymes have lower blood pressure compared with wild-type counterparts, and Ang II infusion fails to induce hypertension in these mice [31, 34]. Since inhibition of ROS-generating enzymes, antioxidants, and ROS scavengers reduce blood pressure, whereas pro-oxidants increase blood pressure, it has been suggested that ROS are causally associated with hypertension, at least in animal models.

In human hypertension, biomarkers of systemic oxidative stress, including levels of plasma thiobarbituric acid-reactive substances and 8-epi-isoprostanes, are elevated [35–37]. Factors contributing to increased oxidative stress in human hypertension include decreased antioxidant activity, reduced levels of ROS scavengers, and activation of ROS-generating enzymes [38–40]. A causal link between ROS and high blood pressure has not yet been unambiguously established in humans. Only a few small clinical studies showed a blood pressure lowering effect of antioxidants [41–43], whereas many large antioxidant clinical trials failed to demonstrate any cardiovascular benefit and blood pressure reduction [44–46]. Nevertheless, what is becoming increasingly evident is that oxidative stress plays a critical role in the molecular mechanisms associated with cardiovascular and renal injury in hypertension, and that hypertension itself can contribute to oxidative stress. A greater understanding of the (patho)biology of ROS may lead to new insights and novel diagnostics and treatments for hypertension.

## 15.2 Biology of ROS

Reactive oxygen species are produced as intermediates in reduction-oxidation (redox) reactions leading from $O_2$ to $H_2O$ [47, 48]. The sequential univalent reduction of $O_2$ is: $O_2 \xrightarrow{e-} \cdot O_2^- \xrightarrow{e-} H_2O_2 \xrightarrow{e-} OH\cdot \xrightarrow{e-} H_2O + O_2$. Of the ROS generated in cardiovascular cells, $O_2^{\bullet -}$ and $H_2O_2$ appear to be particularly important. In biological systems, $O_2^{\bullet -}$ is short-lived owing to its rapid reduction to $H_2O_2$

by superoxide dismutase (SOD), of which there are three mammalian isoforms, copper/zinc SOD (SOD1), mitochondrial SOD (Mn SOD, SOD2), and extracellular SOD (EC-SOD, SOD3) [49–52]. The major vascular SOD is EC-SOD. The charge on the superoxide anion makes it unable to cross cellular membranes except possibly through ion channels. $H_2O_2$ has a longer lifespan than $O_2^{\bullet-}$, is relatively stable, and is easily diffusible within and between cells. The main source of $H_2O_2$ in vascular tissue is the dismutation of $O_2^{\bullet-}$: $2\,O_2^{\bullet-} + 2H^+ \rightarrow H_2O_2 + O_2$. This reaction can be spontaneous or it can be catalyzed by SOD.

The distinct chemical properties between $O_2^{\bullet-}$ and $H_2O_2$ and their different sites of distribution mean that different species of ROS activate diverse signaling pathways, which lead to divergent, and potentially opposing, biological responses. For example, in the vasculature, increased $O_2^{\bullet-}$ levels inactivate the vasodilator NO, leading to endothelial dysfunction and vasoconstriction [53, 54]; whereas $H_2O_2$ acts as a direct vasodilator in some vascular beds, including the cerebral, coronary, and mesenteric arteries [55–57].

## 15.3 Production and Metabolism of ROS in the Cardiovascular System

ROS are produced by all vascular cell types, including endothelial, smooth muscle, and adventitial cells, and can be formed by many enzymes. Enzymatic sources of ROS important in vascular disease and hypertension are xanthine oxidoreductase, uncoupled NO synthase (NOS), mitochondrial respiratory enzymes, and nicotinamide adenine dinucleotide phosphate (NADPH) oxidase [58–61].

### *15.3.1 Xanthine Oxidase*

Xanthine oxidase (XO) and xanthine dehydrogenase (XDH) are interconvertible forms of the same enzyme, known as xanthine oxidoreductase. Physiologically, XO and XDH participate in many biochemical reactions, with the primary role being degradation of purines and the conversion of hypoxanthine to xanthine, and xanthine to uric acid. As a byproduct in the purine degradation pathway, XO oxidizes NADH to form $O_2^{\bullet-}$ and $H_2O_2$. In the vascular wall, XO-derived $O_2^{\bullet-}$ reacts rapidly with NO to form $ONOO^-$, which can lead to a negative feedback of the enzyme [58, 62, 63]. Uric acid, which has antioxidant potential, also acts as a feedback inhibitor of XO. Xanthine oxidase is expressed in vascular cells, it circulates in the plasma, and it binds to endothelial cell extracellular matrix. Although xanthine oxidase-derived $O_2^{\bullet-}$ has been studied mainly in the context of cardiac disease and atherosclerosis, there is evidence suggesting involvement in hypertension. Spontaneously hypertensive rats (SHR) and DOCA-salt hypertensive rats demonstrate elevated levels of endothelial XO and increased ROS production, which are associated with increased arteriolar tone [64]. This may be mediated, in part, through an adrenal pathway, because adrenalectomy reduces XO expression [65]. Endothelial dysfunction in transgenic rats with overexpression of renin and angiotensinogen has also been associated with increased XO activity [66]. In addition to effects on the vasculature, XO may play a role in end-organ damage in hypertension. In experimental models of hypertension, XO activity is increased in the kidney. Long-term inhibition of XO with allopurinol in SHR reduced renal XO activity without lowering blood pressure, indicating that the increased renal ROS production was a consequence of hypertension rather than a contributing factor [67]. The finding that allopurinol can improve cardiac and renal hypertrophy in SHR and slow the progression of renal disease in patients with chronic kidney disease and hypertension [68], whilst having a minimal impact on blood pressure [69], supports a role for XO in hypertensive end-organ damage. This may be mediated through direct vascular effects of XO-produced uric acid [70]. To further support a role for XO in the pathogenesis of hypertension, allopurinol decreased blood pressure in hyperuricemic adolescents with newly diagnosed hypertension [71]. However, it still remains unclear whether $O_2^{\bullet-}$ or uric acid is the primary factor involved in XO-sensitive hypertension.

## 15.3.2 Uncoupled Nitric Oxide Synthase

Under physiological conditions, nitric oxide synthase (NOS), in the presence of cofactors L-arginine and tetrahydrobiopterin ($BH_4$), produces NO. In the absence of these cofactors, because of oxidative destruction or downregulation of GTP cyclohydrolase-1, which is the rate-limiting enzyme in BH4 production, uncoupled NOS produces $O_2^{\bullet-}$ instead of driving electrons towards NO production [72, 73]. All three NOS isoforms are capable of "uncoupling" that leads to the preferential formation of $O_2^{\bullet-}$ [72, 73]. eNOS uncoupling has been demonstrated in DOCA-salt-induced hypertension and in SHR [74, 75], and has been implicated in atherosclerosis and endothelial dysfunction in low-density lipoprotein receptor-deficient mice (LDLR−/−) fed a high salt, high fat diet [76, 77]. Dysfunctional eNOS is also important in cardiac remodeling from pressure overload. In mice subjected to proximal aortic constriction, oral $BH_4$ prevented progressive chamber dilation and dysfunction, reversed fibrosis and hypertrophy, and improved myocyte function and calcium handling [78]. This was associated with eNOS recoupling and reduced oxidative stress. Whether effects of uncoupled NOS are due to increased $O_2^{\bullet-}$ generation or to decreased NO bioavailability still remains unclear [60]. Nevertheless, $BH_4$ has been suggested as a treatment modality for hypertension, endothelial dysfunction, atherosclerosis, diabetes, cardiac hypertrophic remodeling, and heart failure [79–81]. While previously difficult to use clinically because of chemical instability and cost, newer methods to synthesize stable BH4 suggest its novel potential as a therapeutic agent [82]. In fact, some classical antihypertensive drugs, including the beta blocker nebivolol, have been shown to induce effects by preventing eNOS uncoupling [83].

### *15.3.3 Mitochondrial Respiratory Enzymes*

Mitochondrial biogenesis is involved in the control of cell metabolism, signal transduction, and regulation of mitochondrial ROS production. More than 95% of $O_2$ consumed by cells is reduced by four electrons to yield two molecules of $H_2O$ via mitochondrial electron transport chain complexes (I-IV), with up to 1–2% of the electron flow leaking onto $O_2$ to form $O_2^{\bullet-}$ under specific normoxic conditions [84]. Mitochondrial ROS production is modulated by many factors, including mitochondrial electron transport chain efficiency [85], mitochondrial antioxidant content [86], local oxygen [86, 87], NO concentrations [88], availability of metabolic electron donors [89], uncoupling protein (UCP) activity [90], cytokines, and vasoactive agonists [91–94].

Ang II and ET-1 stimulate mitochondrial ROS generation in endothelial and vascular smooth muscle cells and in rat aorta in vivo [91–97]. Mechanisms whereby these vasoactive agents stimulate mitochondrial ROS production are unclear but could involve the opening of mitochondrial potassium channels (mitoKATP) [98] and mitochondrial permeability transition (MPT) [99–101]. Interestingly, Ang II may interact directly with mitochondria, as evidenced by studies demonstrating that labelled $^{125}$I-Ang II is detectable in cardiac, brain, and smooth muscle mitochondria [101].

Alterations in mitochondrial biogenesis are associated with mitochondrial dysfunction and mitochondrial oxidative stress. Impaired activity and/or decreased expression of mitochondrial electron transport chain complexes I, III, and IV have been implicated in vascular aging and cardiovascular disease [102]; and an association between mitochondrial dysfunction and blood pressure has been reported in human and experimental hypertension [103–106]. Ang II-sensitive hypertension is also linked to mitochondrial-derived oxidative stress, since AT1 receptor blockade attenuates $H_2O_2$ production [107] and mitochondrial dysfunction in SHR; and in mice, Ang II infusion is associated with decreased expression of cardiac mitochondrial electron transport genes [108]. In DOCA-salt hypertension, mitochondrial-derived ROS plays an important role in oxidative vascular damage, an effect mediated via ET-1/ETA receptors [109, 110]. Chan and coworkers [111] have provided new evidence that mitochondrial dysfunction and mitochondrial-localized ROS production in the central nervous system is important in cardiovascular function. They demonstrated a relationship between decreased activity of complex I and III and increased ROS production. When electron transport was re-established, ROS formation was decreased, and blood pressure was reduced. Clinically, Yang et al showed that mitochondrial heritability for systolic blood pressure was about 5% and mitochondrial effects may account for 35% of hypertensive pedigrees [112, 113]. In African Americans with hypertension-associated end-stage renal disease, mitochondrial DNA mutations in the kidneys have been identified [114].

### 15.3.4 ROS-Generating Nox Family NAD(P)H Oxidases

NAD(P)H oxidases were originally considered as enzymes expressed only in phagocytic cells involved in host defense and innate immunity. Recent evidence indicates that there is a family of NAD(P)H oxidases, based on the discovery of $gp91^{phox}$ homologues. The new homologues, along with $gp91^{phox}$, are now designated the Nox family of NAD(P)H oxidases [115–117] and are key sources of ROS in the vasculature. The prototypical NAD(P)H oxidase is a multimeric enzyme found in phagocytes and comprises five subunits: $p47^{phox}$ ("phox" stands for phagocyte oxidase), $p67^{phox}$, $p40^{phox}$, $p22^{phox}$, and the catalytic subunit $gp91^{phox}$ (also termed Nox2) [118, 119]. In unstimulated cells $p47^{phox}$, $p67^{phox}$, and $p40^{phox}$ exist in the cytosol, whereas $p22^{phox}$ and $gp91^{phox}$ are in the membrane, where they occur as a heterodimeric flavoprotein (cytochrome b558). Upon stimulation $p47^{phox}$ is phosphorylated and the cytosolic subunits form a complex that translocates to the membrane, where it associates with cytochrome b558 to assemble the active oxidase, which transfers electrons from the substrate to $O_2$ forming $O_2^{\bullet-}$ [120, 121]. Activation also requires participation of Rac 2 (or Rac 1) and Rap 1A.

The mammalian Nox family comprises seven members, characterized primarily by the catalytic subunit that they utilize. These include Nox1, Nox2 (formerly $gp91^{phox}$), Nox3, Nox4, Nox5, Duox1, and Duox2 [122–124]. Nox family NAD(P)H oxidases are expressed in many tissues and mediate diverse biological functions. All Noxes are transmembrane proteins that transport electrons across biological membranes to reduce $O_2$ to $O_2^{\bullet-}$. They have conserved structural properties, including an NADPH-binding site at the COOH terminus, a FAD-binding site in the COOH terminus, six conserved transmembrane domains, and four conserved heme-binding histidines. Nox1, Nox2, Nox4, and Nox5 have been identified in vascular tissue [125]. In vessels, in addition to vascular cells possessing functional Noxes, resident macrophages, neutrophils, and platelets express NAD(P)H oxidase, particularly in pathological states. Accordingly, these cells can also contribute to vascular oxidative stress in disease.

Nox1 is found primarily in colon epithelial cells as well as in other cell types such as endothelial cells and vascular smooth muscle cells, and is involved in host defense and cell growth [126, 127]. Nox1 requires the membrane subunit $p22^{phox}$ for its activity as well as the cytosloic subunits $p47^{phox}$ and $p67^{phox}$. It is regulated by the redox chaperone protein disulfide isomerase (PDI) in vascular smooth muscle cells [128], and has recently been implicated in vascular smooth muscle cell migration, proliferation, and extracellular matrix production, effects mediated by cofilin [129]. Nox2 is the catalytic subunit of the respiratory burst oxidase in phagocytes, but is also expressed in vascular, cardiac, renal, and neural cells [130–134].

Human Nox2 is a highly glycosylated protein that runs with an apparent molecular mass of ~70–90 kDa on SDS-PAGE. Nox2 is unstable without p22$^{phox}$ and requires the cytosolic subunits for its full activation. In neutrophils Nox2 localizes to intracellular and plasma membranes, and in vascular smooth muscle cells it also localizes with the cytoskeleton. The Nox2 gene, located on the X chromosome, is inducible and is highly regulated by Ang II and stretch, and is upregulated in various forms of hypertension [134–136]. Nox3 is found in fetal tissue and the adult inner ear and is involved in vestibular function. It has not been identified in vascular cells and has not been implicated in the pathogenesis of cardiovascular disease.

Nox4, originally termed Renox (renal oxidase) because of its extensive abundance in the kidney, is also found in vascular cells, fibroblasts, and osteoclasts [137–139]. In vascular smooth muscle cells, Nox4 and p22$^{phox}$ colocalize with vinculin in focal adhesions. Nox4 has also been found in the endoplasmic reticulum and nucleus of vascular cells [140–142]. Nox 4 antibodies recognize two bands, one of 75–80 kDa and a second of 65 kDa from both endogenous Nox4 expressing cells and Nox4-transfected cells. Nox4 produces mainly $H_2O_2$, while Nox1 generates mostly $O_2^{\bullet-}$ that is subsequently converted to $H_2O_2$. The difference in the products generated by Nox1 and Nox4 may contribute to distinct roles of these Noxes in cell signaling. Regulation of Nox 4 is controversial. It has been reported that Nox4 forms a heterodimer with p22$^{phox}$ for full activity and stabilization of the enzyme complex [143]. However, forms of p22$^{phox}$ mutated in the proline-rich region (PRR) region inhibited ROS production by Nox1, Nox2, and Nox3, but not for Nox4 [144]. Nox 4 does not seem to require p47$^{phox}$, p67$^{phox}$, p40$^{phox}$, or Rac for its activation; although Nox R1, a Nox 4-binding protein, was recently identified, which may be important for Nox4 regulation [145]. In vascular smooth muscle and endothelial cells, Nox4 localizes to focal adhesions and the endoplasmic reticulum, and has been implicated in cell migration, proliferation, tube formation, angiogenesis, and cell differentiation [146, 147]. In the kidney, Nox4 has been suggested to function as an oxygen sensor that regulates erythropoietin synthesis [148]. Overproduction of renal ROS has important pathophysiological consequences, because it is associated with tissue injury and inflammatory reactions which affect tubular and glomerular cell functions [148, 150].

Nox5 is a $Ca^{2+}$-dependent homologue, found in the testes and lymphoid tissue, but also in vascular cells [151–153]. While all Nox proteins are present in rodents and man, the mouse and rat genome does not contain the nox5 gene. Four splice variants of Nox5, namely Nox5α, Nox5β, Nox5γ, and Nox5δ, have been identified [154, 155]. Unlike other vascular Noxes, Nox5 possesses an amino-terminal calmodulin-like domain with four binding sites for $Ca^{2+}$ (EF hands) and does not require p22$^{phox}$ or other subunits for its activation. Nox5 is directly regulated by intracellular $Ca^{2+}$ ($[Ca^{2+}]i$), the binding of which induces a conformational change leading to enhanced ROS generation [154, 155]. The functional significance of vascular Nox5 is unknown, although it has been implicated in endothelial cell proliferation and angiogenesis, in PDGF-induced proliferation of vascular smooth muscle cells, and in oxidative damage in atherosclerosis [149, 156, 157]. Vascular Nox5 has been shown to be activated by thrombin, PDGF, ionomycin, Ang II and ET-1 [157–159]. Duox1 and duox2 are thyroid Noxes involved in thyroid hormone biosynthesis [160]. Whether they play a role in vascular function is unknown.

#### 15.3.4.1 Distribution of Noxes in the Vascular Wall

The three major cell tyes of the vascular wall, including endothelial cells, smooth muscle cells, and adventitial fibroblasts, all possess functionally active Nox isoforms [122–124]. In pathological conditions associated with vascular injury, such as atherosclerosis, diabetes, and hypertension, macrophages and leukocytes invade the vessel and become resident cells in the vascular media [161]. These cells are rich in NAD(P)H oxidase and may also contribute to vascular ROS generation. Endothelial cells express mRNA and protein for Nox2, Nox4, and associated regulatory proteins $p22^{phox}$, $p47^{phox}$, and $p67^{phox}$ and play a role in endothelial cell biology [162]. Nox2 is the major source of ROS in endothelial cells under basal conditions, and in pathological conditions Nox1 and Nox4 may be upregulated [163, 164]. Nox2, Nox4, and Nox5 appear to localize primarily in the perinuclear area associated with membranes on the endoplasmic reticulum and nucleus, although Nox2 is also found in the plasma membrane within cholesterol-enriched domains, which may serve as signaling platforms for ROS generation in vascular disease [149, 156, 157, 165]. Vascular smooth muscle cells possess Nox2 (in human resistance arteries) and Nox4, which are major sources of ROS. Nox1, present in low concentrations in basal states, is upregulated in disease. Adventitial fibroblasts also possess Noxes (Nox2, Nox4) important in adventitial ROS formation.

#### 15.3.4.2 Regulation of Noxes

How the NAD(P)H oxidase subunits interact in cardiovascular cells and how they generate $O_2^{\bullet -}$ is still unclear. All Noxes, except Nox5, appear to have an obligatory need for $p22^{phox}$ [144, 166, 167]. Whereas Nox2 requires $p47^{phox}$ and $p67^{phox}$ for its activity, Nox1 may interact with homologues of $p47^{phox}$ (NAD(P)H oxidase organizer 1 (NOXO1)) and $p67^{phox}$ (NAD(P)H oxidase activator 1 (NOXA1)) [168, 169]. Oxidase activation involves Rac translocation, phosphorylation of $p47^{phox}$ and its translocation, possibly with $p67^{phox}$, and $p47^{phox}$ association with cytochrome b558. Nox2 and Nox 4 are constitutively active. However, induction of Nox mRNA expression is observed in response to physical stimuli (shear stress, pressure); growth factors (platelet-derived growth factor, epidermal growth factor, and transforming growth factor β); cytokines (tumor necrosis factor-α, interleukin-1, and platelet aggregation factor); mechanical forces (cyclic stretch, laminar and oscillatory shear stress); metabolic factors (hyperglycemia, hyperinsulinemia, free fatty acids, advanced glycation end products (AGE)); and G protein-coupled receptor agonists (serotonin, thrombin, bradykinin, endothelin, and Ang II) [170–175]. Ang II is an important and potent regulator of cardiovascular NAD(P)H oxidase, which activates NAD(P)H oxidase via AT1 receptors through stimulation of signaling pathways involving c-Src p21Ras, PKC, PLD, and PLA2 [176–179]. Ang II also influences NAD(P)H oxidase activation through transcriptional regulation of oxidase subunits.

## 15.4 Protecting Against Oxidative Stress: Antioxidant Defenses

Enzymatic and nonenzymatic systems have evolved to protect against injurious oxidative stress. Major enzymatic antioxidants are SOD, catalase, glutathione peroxidases, thioredoxin, and peroxiredoxin [180–183]. Nonenzymatic antioxidants include ascorbate, tocopherols, glutathione, billirubin, and uric acid; and scavenge OH· and other free radicals [184]. SOD catalyzes the dismutation of $O_2^{\bullet-}$ into $H_2O_2$ and $O_2$. Extracellular SOD, the major vascular SOD, is produced and secreted by vascular smooth muscle cells, binds to glycosaminoglycans in the vascular extracellular matrix, and regulates oxidant status in the vascular interstitium [180, 183]. Reduced glutathione plays a major role in the regulation of the intracellular redox state of vascular cells by providing reducing equivalents for many biochemical pathways [184–186]. Glutathione peroxidase (GPX) reduces $H_2O_2$ and lipid peroxides to water and lipid alcohols, respectively, and in turn oxidizes glutathione to glutathione disulfide [186]. Oxidized glutathione (GSSG) can be recycled by glutathione reductase to reduced GSH, utilizing NADPH as a substrate; or it can be exported from the cell via active transport by the multidrug resistance protein 1 (MRP1) [187, 188]. Hypertension induced by DOCA-salt or Ang II was attenuated in $MRP^{-/-}$ mice, and vascular glutathione flux was blunted in $MRP1^{-/-}$ mice, allowing recycling of GSSG to reduced glutathione and promoting increased intracellular antioxidant capacity [187, 188]. These findings suggest that MRP1 inhibition may protect against oxidant stress by preventing loss of glutathione from vascular cells, thereby improving endothelial function and attenuating development of hypertension. Catalase is an intracellular antioxidant enzyme that is mainly located in cellular peroxisomes and catalyzes the reaction of $H_2O_2$ to water and $O_2$ [189]. Catalase is very effective in high-level oxidative stress and protects cells from $H_2O_2$ produced within the cell. The enzyme is especially important in the case of limited glutathione content or reduced GPX activity. Thioredoxin reductase participates in thiol-dependent cellular reductive processes [190–192]. Low antioxidant bioavailability promotes cellular oxidative stress and has been implicated in cardiovascular and renal oxidative damage associated with hypertension [180]. Activity of SOD, catalase, and GSH peroxidase is lower and the GSSG/GSH is higher in plasma and circulating cells from hypertensive patients than normotensive subjects [193]. In mice deficient in EC-SOD and in rats in which GSH synthesis is inhibited, blood pressure is significantly elevated, demonstrating that reduced antioxidant capacity is associated with elevated blood pressure [51, 194]. Failure to upregulate antioxidant genes and reduced antioxidant capacity are also associated with age-accelerated atherosclerosis [195].

## 15.5 ROS and Vascular (Patho)Biology in Hypertension

ROS have been implicated in the regulation of vascular tone by modulating vasodilation directly ($H_2O_2$ may have vasodilator actions), or indirectly by decreasing NO bioavailability through quenching by $O_2^{\bullet-}$ to form $ONOO^-$ [196, 197]. ROS, through the regulation of hypoxia-inducible factor-1 (HIF-1), are also important in $O_2$ sensing [198], which is essential for maintaining normal $O_2$ homeostasis. In pathological conditions ROS are involved in inflammation, endothelial dysfunction, cell proliferation, migration and activation, extracellular matrix deposition, fibrosis, angiogenesis, and vascular remodeling (Fig. 15.2). These effects are mediated through redox-sensitive regulation of multiple signaling molecules and second messengers, including mitogen-activated protein (MAP) kinases, protein tyrosine phosphatases, tyrosine kinases, proinflammatory genes, ion channels, and $Ca^{2+}$ [199–201, 202] (Fig. 15.3).

Mechanisms by which ROS cause hypertension through changes in vascular function and structure probably relate to reduced vasodilation, increased contraction, and structural remodeling, causing increased peripheral resistance and elevated

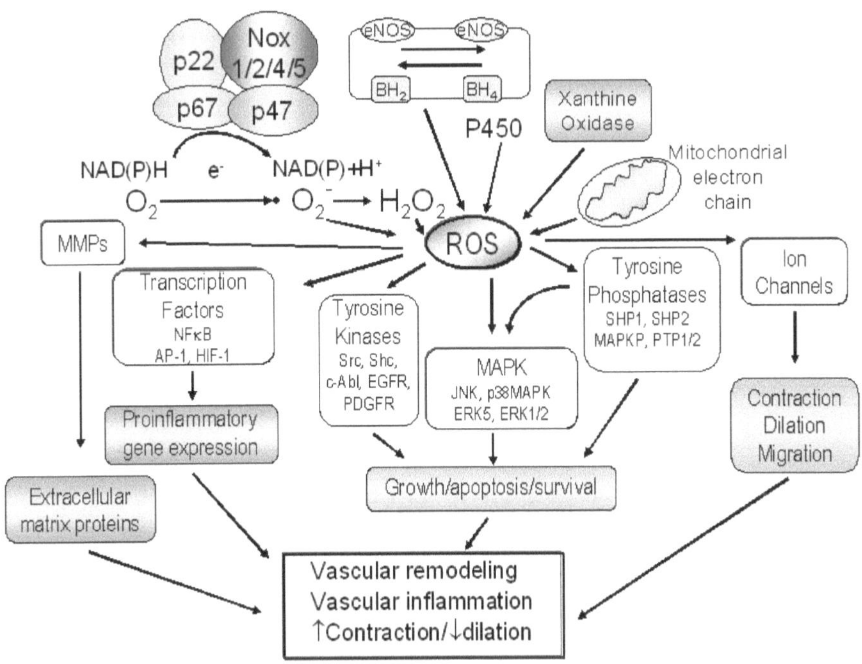


**Fig. 15.2** Activation of ROS-generating enzymes, such as NAD(P)H oxidase, uncoupling of NOS and mitochondrial enzymes in vascular cells results in generation of reactive oxygen species, which in turn influence signaling molecules involved in vascular growth, fibrosis, contraction/dilation and inflammation. These redox-sensitive processes contribute to vascular damage and remodeling in hypertension and other cardiovascular diseases. MAPK, mitogen-activated protein kinases; MMPs, matrix metalloproteinases; BH4, tetrahydrobiopterin

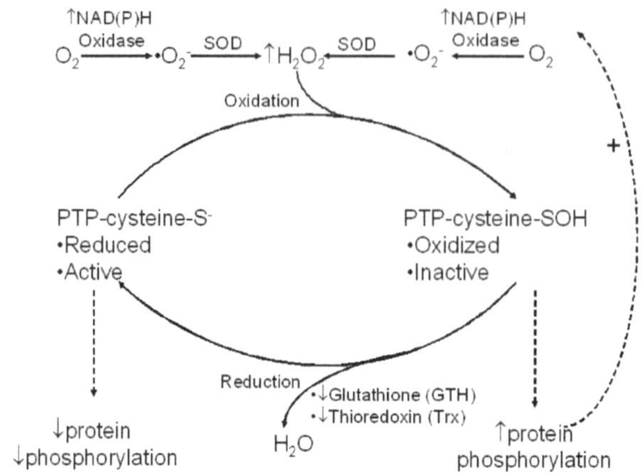


**Fig. 15.3** Molecular targets of ROS in vascular cells. Protein tyrosine phosphatases (PTP) contain highly conserved cysteine residues that are oxidized in the presence of ROS. Increased NAD(P)H oxidase-derived ROS results in oxidation of PTPs, leading to inactivation of PTPs and consequent increased phosphorylation of downstream protein targets. Activated proteins can in turn stimulate activation of NAD(P)H oxidase, which further increases ROS generation Oxidation of PTPs is a reversible process, which in the presence of antioxidants, such as glutathione or thioredoxin, results in reduction of PTPs and consequent activation of phosphatases. +, positive feedback effect

blood pressure [210, 203] (Figs. 15.2 and 15.4). ROS formation in organs other than the vasculature also contributes to hypertension. In animal models, NAD(P)H oxidase activation and ROS generation are increased and antioxidant enzyme expression is reduced in the kidneys [204, 205]. Renal oxidative stress is associated with glomerular damage, proteinuria, sodium and volume retention, and nephron loss, all important in the development of hypertension [206–208]. Centrally produced ROS by NAD(P)H oxidase in the hypothalamic and circumventricular organs are implicated in the central control of hypertension, in part through sympathetic outflow [209–212].

## 15.6 Oxidative Stress in Experimental Hypertension

The relationship between oxidative stress and increased blood pressure has been demonstrated in many models of hypertension. Increased ROS formation precedes development of hypertension in SHR, and is implicated in fetal programming and development of hypertension later in life, supporting the important role of ROS in the genesis and maintenance of hypertension [213, 214]. Markers of oxidative stress, such as TBARS, and F2α-isoprostanes, tissue concentrations of $O_2^{\bullet-}$ and $H_2O_2$, and activation of NAD(P)H oxidase and xanthine oxidase are increased; whereas levels of NO and antioxidant enzymes are reduced in experimental hypertension [215–218].

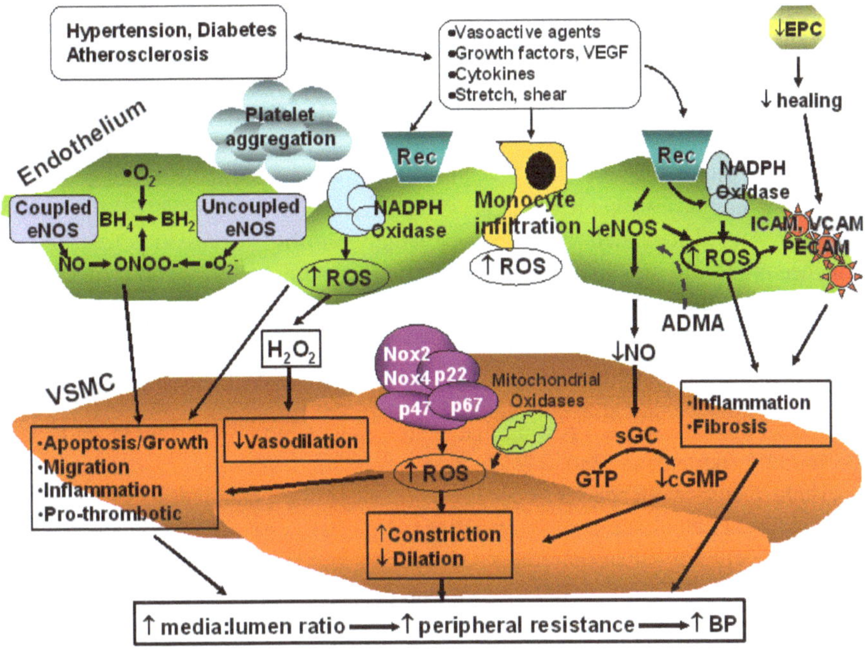


**Fig. 15.4** Putative mechanisms whereby changes in vascular redox status influence blood pressure. Increased oxidative stress results in activation of redox-sensitive signaling molecules, which induce vascular growth, constriction, fibrosis and inflammation. These processes contribute to reduced vasodilation, increased contraction and structural remodeling, causing increased peripheral resistance and elevated blood pressure. BP, blood pressure; ROS, reactive oxygen species; Rec, receptor; EPC, endothelial progenitor cells; ADMA, asymmetric dimethylarginine; NO, nitric oxide

Ang II-dependent hypertension is particularly sensitive to NAD(P)H oxidase-derived ROS. In rats and mice made hypertensive by Ang II infusion, expression of NAD(P)H oxidase subunits (Nox1, Nox2, Nox4, $p22^{phox}$), oxidase activity, and generation of ROS are increased [219, 222]. To support a role for NAD(P)H oxidase-derived ROS generation in the pathogenesis of Ang II-induced hypertension, various mouse models with altered NAD(P)H oxidase subunit expression have been studied [34, 223–225]. In $p47^{phox}$ knockout mice and in $gp91^{phox}$ (Nox2) knockout mice, Ang II infusion fails to induce hypertension, and these animals do not show the same increases in $O_2^{\bullet-}$ production, vascular hypertrophy, and endothelial dysfunction observed in Ang II-infused wild-type mice [226, 227]. In Ang II-infused mice treated with siRNA targeted to renal $p22^{phox}$, renal NAD(P)H oxidase activity was blunted, ROS formation was reduced, and blood pressure elevation was attenuated, suggesting that $p22^{phox}$ is required for Ang II-induced oxidative stress and hypertension [228]. On the other hand, overexpression of vascular $p22^{phox}$ was associated with increased oxidative stress and vascular dysfunction, but no significant increase in blood pressure [229]. Treatment with apocynin or diphenylene iodonium, nonspecific pharmacological inhibitors of NAD(P)H oxidase, or with gp91dstat, a novel specific inhibitor of NAD(P)H oxidase, reduced vascular $O_2^{\bullet-}$ production, prevented cardiovascular remodeling, and attenuated development of hypertension in Ang II-treated mice [230–232]. Nox1-deficient mice have reduced vascular $O_2^{\bullet-}$ production, and blood pressure elevation in response to Ang II is blunted [233, 234]; whereas in transgenic mice in which Nox1 is overexpressed in the vascular wall, Ang II-mediated vascular hypertrophy and blood pressure elevation are enhanced [235]. In most of these models, Ang II was infused for a short time period (1–3 weeks), inducing an acute hypertensive response. In a model of chronic Ang II-dependent hypertension, where we crossed transgenic mice expressing human renin (which exhibit an Ang II-sensitive hypertensive phenotype) with $Nox2^{-/-}$ or $Nox1^{-/-}$ mice, development of hypertension was not prevented even though oxidative stress was reduced, suggesting that Noxes may be more important in acute than in chronic hypertension [236, 237].

There is also evidence for ROS involvement in the pathogenesis of hypertension independent of direct Ang II actions. In SHR, vascular, renal, and cardiac $O_2^{\bullet-}$ production is enhanced compared with normotensive controls [238–240]. In stroke-prone SHR, aortic expression of Nox1 and Nox4 is significantly increased compared with WKY [241]. In DOCA-salt–induced mineralocorticoid hypertension, vascular $O_2^{\bullet-}$ production involving elevated NAD(P)H oxidase activity, uncoupling of endothelial NOS and mitochondrial sources is increased, in part through the endothelin-1 (ET-1)/ETA receptor pathway [110, 242]. Infusion of ET-1 increases NAD(P)H oxidase-dependent $O_2^{\bullet-}$ production; however, preventing such increase in ROS generation does not inhibit the development of hypertension in these animals [245]. Overexpression of human ET-1 in mice also induces vascular remodeling and impairs endothelial function, via activation of NAD(P)H oxidase [246].

To further support a role for oxidative stress in hypertension, many studies have shown that treatment with antioxidant vitamins, the antioxidant compound tempol (4-hydroxy-2,2,6,6-tetramethyl piperidinoxyl), other free radical scavengers, or tetrahydrobiopterin ($BH_4$) attenuate or prevent development of hypertension and associated target organ damage [247–249].

## 15.7 Oxidative Stress and Clinical Hypertension

Although studies in humans have not been as convincing as those in experimental models, there is evidence that oxidative stress is increased in patients with essential hypertension, renovascular hypertension, malignant hypertension, salt-sensitive hypertension, cyclosporine-induced hypertension, and preeclampsia [250–254]. These findings are based, in general, on increased levels of plasma thiobarbituric acid–reactive substances and 8-epi-isoprostanes, which are biomarkers of lipid peroxidation and oxidative stress [253–255]. Polymorphonuclear leukocyte- and platelet-derived $O_2^{\bullet -}$, which also participate in vascular oxidative stress and inflammation, are increased in hypertensive patients [256, 257].

Hypertensive patients exhibit significantly higher circulating plasma levels of $H_2O_2$ than normotensive subjects [258]. Additionally, normotensive subjects with a family history of hypertension have greater $H_2O_2$ production than blood pressure–matched normotensives without a family history of hypertension, suggesting that there may be a genetic component that leads to elevated production of hydrogen peroxide [258, 259]. Lacy et al. determined familial correlations for $H_2O_2$ production as a quantitative trait in a family-based cohort of hypertensive subjects and used these results to estimate the heritability of this trait. Heritability estimates revealed that approximately 20–35% of the observed variance in $H_2O_2$ production could be attributed to genetic factors, suggesting an important heritable component to the overall determination of this trait [259]. Plasma levels of asymmetric dimethylarginine (ADMA) (eNOS inhibitor) and the lipid peroxidation product of linoleic acid, 13-hydroxyoctadecadienoic acid (HODE), a marker of ROS production, were inversely correlated with microvascular emdothelial dysfunction and elevated blood pressure in hypertensive patients [260].

We showed that ROS production is increased in vascular smooth muscle cells from resistance arteries of hypertensive patients and that this is associated with upregulation of vascular NAD(P)H oxidase [261, 262]. The importance of this oxidase in oxidative stress in human cardiovascular disease is supported by studies from Zalba and colleagues, who demonstrated that polymorphisms in NAD(P)H oxidase subunits are associated with increased atherosclerosis and hypertension [263]. In particular, the -930(A/G) polymorphism in the $p22^{(phox)}$ promoter may be a novel genetic marker associated with hypertension [263]. The C242T CYBA polymorphism is associated with essential hypertension; and hypertensive patients carrying the CC genotype of this polymorphism exhibit features of NAD(P)H oxidase-mediated oxidative stress and endothelial damage, and are prone to cerebrovascular disease [264, 265]. In a Japanese population, the G(-930)A polymorphism of CYBA was confirmed to be important in the pathogenesis of hypertension [266]. Polymorphisms -337GA and 565+64CT of the xanthine oxidase gene have been shown to be related to blood pressure and oxidative stress in hypertension, further supporting a role for xanthine oxidase in hypertension.

In addition to excess ROS generation, decreased antioxidant defense mechanisms contribute to oxidative stress in patients with hypertension. Hypertensive patients have reduced activity and decreased content of antioxidant enzymes, including SOD, glutathione peroxidase, and catalase [267–269]. Decreased levels of antioxidant vitamins A, C, and E have been demonstrated in newly diagnosed, untreated hypertensive patients, compared with normotensive controls [269]. Moreover, SOD activity has been demonstrated to correlate inversely with blood pressure in patients with hypertension [269]. Antioxidant vitamins reduced blood pressure and arterial stiffness in patients with diabetes [270], but had no effect in postmenopausal women or in healthy subjects [271]. In patients with white coat hypertension, serum protein carbonyl (PCO, indicating protein oxidation) was increased, and endogenous antioxidant proteins (protein thiol, SOD, glutathione) were decreased compared with normotensive individuals, further supporting a relationship between oxidative stress and hypertension [272].

## 15.8 Antioxidant Therapy and Human Hypertension

The potential of antioxidants in treating conditions associated with oxidative stress is supported by experimental investigations, observational findings, small clinical studies, and epidemiological data [270, 273]. However, findings are inconsistent, and clinical trial data are inconclusive [274, 275]. Many large trials have been published regarding antioxidant vitamin effects on risks of cardiovascular disease, including the Cambridge Heart Antioxidant Study (CHAOS; 2002 patients); the Alpha Tocopherol, Beta-Carotene cancer prevention study (ATBC; 27,271 males); the Gruppo Italiano per lo Studio della Sopravvivenza nell'Infarto Miocardico (GISSI)-Prevenzione trial (3658 patients); the Heart Outcomes Prevention Evaluation (HOPE) study (2545 subjects); the Medical Research Council/British Heart Foundation (MRC/BHF) heart protection study (20,536 adults); the Primary Prevention Project (PPP; 4495 patients); and the Antioxidant Supplementation in Atherosclerosis Prevention (ASAP) study (520 subjects) [274, 275]. In the HOPE-TOO study, which was a follow-up of a subset of the original HOPE trial (Heart Outcomes Prevention Evaluation), patients taking 400 IU vitamin E showed increased incidence of heart failure [276]. Except for the ASAP study, which demonstrated that six-year supplementation of daily vitamin E and slow-release vitamin C reduced progression of carotid atherosclerosis, the other studies failed to demonstrate significant beneficial effects of antioxidants on BP or on cardiovascular end points [274, 275]. Thus, the overall results of clinical trials have been negative.

Unlike the large multicenter trials, smaller clinical studies have shown positive responses in hypertensive patients treated with antioxidants, either in combination (zinc, ascorbic acid, $\alpha$-tocopherol, $\beta$-carotene) or as monotherapy (vitamin C or vitamin E). This has been particularly true for vitamin C. Most studies demonstrated an inverse relationship between plasma ascorbate levels and blood pressure in both normotensive and hypertensive populations [193, 277]. In the SU.VI.MAX study, a decreasing trend was observed with vitamin C levels and risk of hypertension in women but not in men [278]. Vitamin C supplementation is associated with reduced blood pressure in hypertensive patients, with systolic blood pressure falling by 3.6–17.8 mmHg for each 50 $\mu$mol/L increase in plasma ascorbate [39]. However, Ward et al. found that a six-week treatment with vitamin C and grape seed polyphenols was associated with a paradoxical increase in ambulatory blood pressure in hypertensive patients [279]. This was not attributed to increased oxidative stress.

Human studies of vitamin E (400–1,000 IU/day) have demonstrated beneficial effects in improving insulin sensitivity, lowering serum glucose levels, increasing intracellular $Mg^{2+}$, inhibiting thromboxane effects, and reducing vascular resistance [193, 277, 281]. Data from the 1946 British Birth Cohort reported that low vitamin

E intake during childhood and adulthood was a good predictor of hypertension at age 43 years [282]. However, reductions in blood pressure in hypertensive subjects treated with vitamin E have been inconsistent [193, 277]. Similar trends have been observed in preeclampsia, where early studies suggested that vitamins C and E may prevent against preeclampsia in high risk patients [283, 284], whereas more recent evidence indicates that supplementation with vitamins C and E during pregnancy does not reduce the risk of preeclampsia in nulliparous women [285–287]. If vitamin E does in fact have an antihypertensive effect, it is probably small and may be limited to untreated patients or those with vascular disease or other concomitant diseases, such as diabetes [193, 288].

In general, the results of clinical studies investigating antioxidant effects have been disappointing, given the consistent and promising findings from experimental investigations, clinical observations, and epidemiological data. Possible reasons relate to (1) the type of antioxidants used, (2) the patient cohorts included in trials, and (3) the trial design itself. With respect to antioxidants, it is possible that the agents examined were ineffective and nonspecific and that dosing regimens and duration of therapy were insufficient. For example, vitamins C and E may have pro-oxidant properties with harmful and deleterious interactions. It is also possible that orally administered antioxidants may be inaccessible to the source of free radicals, particularly if ROS are generated in intracellular compartments and organelles [289]. Furthermore, antioxidant vitamins do not scavenge $H_2O_2$, which may be more important than $O_2^{\bullet-}$ in cardiovascular disease. Another factor of importance is that antioxidants do not inhibit ROS production. Regarding cohorts included in large trials, most subjects had significant cardiovascular disease, in which case the damaging effects of oxidative stress may be irreversible. Another confounding factor is that most of the enrolled subjects were taking aspirin prophylactically. Since aspirin has intrinsic antioxidant properties [290], additional antioxidant therapy may be ineffective. Moreover, in the patients studied in whom negative results were obtained, it was never proven that these individuals did in fact have increased oxidative stress. To date, there are no large clinical trials in which patients were recruited based on evidence of elevated ROS formation. Also, none of the large clinical trials were designed to examine the effects of antioxidants specifically on blood pressure.

## 15.9 Other Possible Strategies to Reduce Oxidative Stress

Theoretically, agents that reduce oxidant formation should be more efficacious than nonspecific, inefficient antioxidant vitamin scavengers. This is based on experimental evidence in which it has been demonstrated that inhibition of NAD(P)H oxidase–mediated $O_2^{\bullet-}$ generation, using pharmacological and gene-targeted strategies, leads to regression of vascular remodeling, improved endothelial function, and lowering of blood pressure [289–293]. In fact, vascular NAD(P)H oxidase, specifically gp91$^{phox}$ (Nox2) homologues, may be novel therapeutic targets for vascular disease [289, 291–293]. Harrison and colleagues [187, 188] proposed a new strategy to increase antioxidant capacity without the use of exogenous antioxidants. They suggest that drugs that selectively inhibit MRP1 would prevent cellular glutathione loss and thereby protect against oxidative damage, endothelial dysfunction, and hypertension [187, 188]. Another interesting approach is targeting glucose-6-phosphate dehydrogenase (G6PD), which is a source of NADPH, the substrate for NAD(P)H oxidase [294]. Inhibition of G6PD has been shown to ameliorate development of pulmonary hypertension, possibly through decreased oxidative stress. To date only investigational G6PD inhibitors are available.

In view of current data and the lack of evidence to prove the benefits from use of antioxidants to prevent cardiovascular disease [295], antioxidant supplementation is not recommended for the prevention or treatment of hypertension. However, most therapeutic guidelines suggest that the general population consumes a diet emphasizing antioxidant-rich fruits and vegetables and whole grains [296, 297, 298, 299]. Another important lifestyle modification that may have cardiovascular protective and blood pressure lowering effects by reducing oxidative stress is exercise. In experimental models of hypertension and in human patients with coronary artery disease, exercise reduced vascular NAD(P)H oxidase activity and ROS production, ameliorated vascular injury, and reduced blood pressure [300, 301, 302, 303, 304, 305, 306, 307].

Some of the beneficial effects of classical antihypertensive agents such as ß-adrenergic blockers, ACE inhibitors, AT1 receptor antagonists, and $Ca^{2+}$ channel blockers may be mediated, in part, by decreasing vascular oxidative stress [303, 304, 305, 306, 307]. These effects have been attributed to direct inhibition of NAD(P)H oxidase activity and to the intrinsic antioxidant properties of the drugs.

## 15.10 Conclusions

In physiological conditions, ROS play an important role in vascular biology by regulating endothelial function and vascular tone through highly controlled redox-sensitive signaling pathways. Uncontrolled production/degradation of ROS results in oxidative stress, which induces cardiovascular and renal damage with associated increase in blood pressure. Although oxidative damage may not be the sole cause of hypertension, it facilitates and amplifies blood pressure elevation in the presence of other pro-hypertensive factors, such as salt loading, activation of the renin-angiotensin system, and sympathetic hyperactivity. Compelling findings from experimental and animal studies suggest a causative role for oxidative stress in the pathogenesis of hypertension. However, from a clinical viewpoint, current data are less conclusive. This may relate to the heterogeneity of the populations studied, inappropriate or insensitive methodologies to evaluate oxidative state clinically, and the suboptimal antioxidant therapies used. Further research in the field of oxidative stress and human hypertension is warranted. In particular, there is an urgent need for the development of sensitive and specific biomarkers to assess the oxidant status of patients. Also needed are clinical trials designed to specifically address the role of oxidative stress in the development of hypertension. With a better understanding of mechanisms regulating ROS metabolism and identification of processes that promote oxidative excess, it should be possible to target therapies more effectively, so that the detrimental actions of oxygen free radicals can be reduced and the beneficial effects of nitric oxide can be enhanced. Such therapies could have potential in the management of diseases associated with vascular damage, including hypertension.

**Acknowledgments** Work from the author's laboratory was supported by grants 44018 and 57886, both from the Canadian Institutes of Health Research.

## References

1. Kakar P, Lip GY (2006) Towards understanding the aetiology and pathophysiology of human hypertension: where are we now? J Hum Hypertens 20(11):833–836
2. Touyz RM (2005) Molecular and cellular mechanisms in vascular injury in hypertension: role of angiotensin II. Curr Opin Nephrol Hypertens 14(2):125–131
3. Harrison DG, Widder J, Grumbach I, Chen W, Weber M, Searles C (2006) Endothelial mechanotransduction, nitric oxide and vascular inflammation. J Intern Med 259(4):351–363
4. Harrison DG, Guzik TJ, Goronzy J, Weyand C (2008) Is hypertension an immunologic disease? Curr Cardiol Rep 10(6):464–469
5. Harris DM, Cohn HI, Pesant S, Eckhart AD (2008) GPCR signalling in hypertension: role of GRKs. Clin Sci 115(3):79–89
6. Touyz RM (2003) Reactive oxygen species in vascular biology: role in arterial hypertension. Expert Rev Cardiovasc Ther 1:91–106
7. Tain YL, Baylis C (2006) Dissecting the causes of oxidative stress in an in vivo model of hypertension. Hypertension 48(5):828–829
8. Vaziri ND, Rodriguez-Iturbe B (2006) Mechanisms of disease: oxidative stress and inflammation in the pathogenesis of hypertension. Nat Clin Pract Nephrol 2(10):582–593
9. Landmesser U, Harrison DG, Drexler H (2006) Oxidant stress-a major cause of reduced endothelial nitric oxide availability in cardiovascular disease. Eur J Clin Pharmacol 62: 13–19
10. Touyz RM, Schiffrin EL (2004) Reactive oxygen species in vascular biology: implications in hypertension. Histochem Cell Biol 122(4):339–352
11. Droge W (2002) Free radicals in the physiological control of cell function. Physiol Rev 82(1):47–95
12. Mueller CF, Laude K, McNally JS, Harrison DG (2005) ATVB in focus: redox mechanisms in blood vessels. Arterioscler Thromb Vasc Biol 25(2):274–278
13. Pawlak K, Naumnik B, Brzosko S, Pawlak D, Mysliwiec M (2004) Oxidative stress-a link between endothelial injury, coagulation activation, and atherosclerosis in haemodialysis patients. Am J Nephrol 24(1):154–161
14. Ushio-Fukai M, Alexander RW, Akers M, Griendling KK (1998) p38 Mitogen-activated protein kinase is a critical component of the redox-sensitive signaling pathways activated by angiotensin II. Role in vascular smooth muscle cell hypertrophy. J Biol Chem 273: 15022–15029
15. Griendling KK, Sorescu D, Lassegue B, Ushio-Fukai M (2000) Modulation of protein kinase activity and gene expression by reactive oxygen species and their role in vascular physiology and pathophysiology. Arterioscler Thromb Vasc Biol 20:2175–2183
16. Zhang Y, Griendling KK, Dikalova A, Owens GK, Taylor WR (2005) Vascular hypertrophy in angiotensin II-induced hypertension is mediated by vascular smooth muscle cell-derived $H_2O_2$. Hypertension 46:732–737

17. Hool LC, Corry B (2007) Redox control of calcium channels: from mechanisms to therapeutic opportunities. Antioxid Redox Signal 9(4):409–435
18. Touyz RM, Tabet F, Schiffrin EL (2003) Redox-dependent signalling by angiotensin II and vascular remodelling in hypertension. Clin Exp Pharmacol Physiol 30(11):860–866
19. Touyz RM (2005) Reactive oxygen species as mediators of calcium signalling by angiotensin II: implications in vascular physiology and pathophysiology. Antioxid Redox Signal 7(9–10):1302–1314
20. Millar TM, Phan V, Tibbles LA (2007) ROS generation in endothelial hypoxia and reoxygenation stimulates MAP kinase signaling and kinase-dependent neutrophil recruitment. Free Radic Biol Med 42(8):1165–1677
21. Kimura S, Zhang GX, Nishiyama A, Shokoji T, Yao L, Fan YY, Rahman M, Abe Y (2005) Mitochondria-derived reactive oxygen species and vascular MAP kinases: comparison of angiotensin II and diazoxide. Hypertension 45(3):438–444
22. Tabet F, Savoia C, Schiffrin EL, Touyz RM (2004) Differential calcium regulation by hydrogen peroxide and superoxide in vascular smooth muscle cells from spontaneously hypertensive rats. J Cardiovasc Pharmacol 44(2):200–208
23. Gutierrez J, Ballinger SW, Darley-Usmar VM, Landar A (2006) Free radicals, mitochondria, and oxidized lipids: the emerging role in signal transduction in vascular cells. Circ Res 99(9):924–932
24. Usatyuk PV, Parinandi NL, Natarajan V (2006) Redox regulation of 4-hydroxy-2-nonenal-mediated endothelial barrier dysfunction by focal adhesion, adherens, and tight junction proteins. J Biol Chem 281(46):35554–35566
25. Yoshioka J, Schreiter ER, Lee RT (2006) Role of thioredoxin in cell growth through interactions with signaling molecules. Antioxid Redox Signal 8(11–12):2143–2151
26. Anathy V, Aesif SW, Guala AS, Havermans M, Reynaert NL, Ho YS, Budd RC, Janssen-Heininger YM (2009) Redox amplification of apoptosis by caspase-dependent cleavage of glutaredoxin 1 and S-glutathionylation of Fas. J Cell Biol 184(2):241–252
27. Romanowski A, Murray IR, Huston MJ (1960) Effects of hydrogen peroxide on normal and hypertensive rats. Pharm Acta Helv 35:354–357
28. Rajagopalan S, Kurz S, Munzel T, Tarpey M, Freeman BA, Griendling KK et al. (1996) Angiotensin II-mediated hypertension in the rat increases vascular superoxide production via membrane NADH/NAD(P)H oxidase activation. Contribution to alterations of vasomotor tone. J Clin Invest 97:1916–1923
29. Zalba G, Beaumont FJ, San Jose G, Fortuno A, Fortuno MA, Etayo JC et al. (2000) Vascular NADH/NAD(P)H oxidase is involved in enhanced superoxide production in spontaneously hypertensive rats. Hypertension 35:1055–1061
30. Akasaki T, Ohya Y, Kuroda J, Eto K, Abe I, Sumimoto H, Iida M (2006) Increased expression of gp91phox homologues of NAD(P)H oxidase in the aortic media during chronic hypertension: involvement of the renin-angiotensin system. Hypertens Res 29(10): 813–820
31. Jung O, Schreiber JG, Geiger H, Pedrazzini T, Busse R, Brandes RP (2004) gp91phox-containing NAD(P)H oxidase mediates endothelial dysfunction in renovascular hypertension. Circulation 109(14):1795–1801
32. Kagota S, Tada Y, Kubota Y, Nejime N, Yamaguchi Y, Nakamura K, Kunitomo M, Shinozuka K (2007) Peroxynitrite is involved in the dysfunction of vasorelaxation in SHR/NDmcr-cp rats, spontaneously hypertensive obese rats. J Cardiovasc Pharmacol 50(6):677–685
33. Klanke B, Cordasic N, Hartner A, Schmieder RE, Veelken R, Hilgers KF (2008) Blood pressure versus direct mineralocorticoid effects on kidney inflammation and fibrosis in DOCA-salt hypertension. Nephrol Dial Transplant 23(11):3456–3463
34. Landmesser U, Cai H, Dikalov S, McCann L, Hwang J, Jo H et al. (2002) Role of p47(phox) in vascular oxidative stress and hypertension caused by angiotensin II. Hypertension 40: 511–515

35. Lavi S, Yang EH, Prasad A, Mathew V, Barsness GW, Rihal CS, Lerman LO, Lerman A (2008) The interaction between coronary endothelial dysfunction, local oxidative stress, and endogenous nitric oxide in humans. Hypertension 51(1):127–133
36. Franco MC, Kawamoto EM, Gorjão R, Rastelli VM, Curi R, Scavone C, Sawaya AL, Fortes ZB, Sesso R (2007) Biomarkers of oxidative stress and antioxidant status in children born small for gestational age: evidence of lipid peroxidation. Pediatr Res 62(2):204–208
37. Cottone S, Mulè G, Guarneri M, Palermo A, Lorito MC, Riccobene R, Arsena R, Vaccaro F, Vadalà A, Nardi E, Cusimano P, Cerasola G (2009) Endothelin-1 and F2-isoprostane relate to and predict renal dysfunction in hypertensive patients. Nephrol Dial Transplant 24(2): 497–503
38. Mistry HD, Wilson V, Ramsay MM, Symonds ME, Broughton Pipkin F (2008) Reduced selenium concentrations and glutathione peroxidase activity in preeclamptic pregnancies. Hypertension 52(5):881–888
39. Duffy SJ, Gokce N, Holbrook M, Huang A, Frei B, Keaney JF Jr, Vita JA (1999) Treatment of hypertension with ascorbic acid. Lancet 354(9195):2048–2049
40. Duffy SJ, Gokce N, Holbrook M, Hunter LM, Biegelsen ES, Huang A, Keaney JF Jr, Vita JA (2001) Effect of ascorbic acid treatment on conduit vessel endothelial dysfunction in patients with hypertension. Am J Physiol Heart Circ Physiol 280(2):H528–H534
41. Kurl S, Tuomainen TP, Laukkanen JA, Nyyssönen K, Lakka T, Sivenius J, Salonen JT (2002) Plasma vitamin C modifies the association between hypertension and risk of stroke. Stroke 33(6):1568–1573
42. Hajjar IM, George V, Sasse EA, Kochar MS (2002) A randomized, double-blind, controlled trial of vitamin C in the management of hypertension and lipids. Am J Ther 9(4):289–293
43. Svetkey LP, Loria CM (2002) Blood pressure effects of vitamin C: what's the key question? Hypertension 40(6):789–791
44. Darko D, Dornhorst A, Kelly FJ, Ritter JM, Chowienczyk PJ (2002) Lack of effect of oral vitamin C on blood pressure, oxidative stress and endothelial function in Type II diabetes. Clin Sci 103(4):339–344
45. Hatzitolios A, Iliadis F, Katsiki N, Baltatzi M (2008) Is the anti-hypertensive effect of dietary supplements via aldehydes reduction evidence based? A systematic review. Clin Exp Hypertens 30(7):628–639
46. Wray DW, Uberoi A, Lawrenson L, Bailey DM, Richardson RS (2009) Oral antioxidants and cardiovascular health in the exercise-trained and untrained elderly: a radically different outcome. Clin Sci 116(5):433–441
47. Fridovich I (1997) Superoxide anion radical ($O_2^{-\bullet}$), superoxide dismutases, and related matters. J Biol Chem 272(30):18515–18517
48. Johnson F, Giulivi C (2005) Superoxide dismutases and their impact upon human health. Mol Aspects Med 26(4–5):340–352
49. Faraci FM, Didion SP (2004) Vascular protection: superoxide dismutase isoforms in the vessel wall. Arterioscler Thromb Vasc Biol 24(8):1367–1373
50. Mendez JI, Nicholson WJ, Taylor WR (2005) SOD isoforms and signaling in blood vessels: evidence for the importance of ROS compartmentalization. Arterioscler Thromb Vasc Biol 25(5):887–888
51. Welch WJ, Chabrashvili T, Solis G, Chen Y, Gill PS, Aslam S, Wang X, Ji H, Sandberg K, Jose P, Wilcox CS (2006) Role of extracellular superoxide dismutase in the mouse angiotensin slow pressor response. Hypertension 48(5):934–941
52. Jung O, Marklund SL, Xia N, Busse R, Brandes RP (2007) Inactivation of extracellular superoxide dismutase contributes to the development of high-volume hypertension. Arterioscler Thromb Vasc Biol 27(3):470–477
53. Cai H, Harrison DG (2000) Endothelial dysfunction in cardiovascular diseases: the role of oxidant stress. Circ Res 87:840–844
54. Cai H (2005) Hydrogen peroxide regulation of endothelial function: origins, mechanisms, and consequences. Cardiovasc Res 68(1):26–36

55. Paravicini TM, Chrissobolis S, Drummond GR, Sobey CG (2004) Increased NAD(P)H-oxidase activity and Nox4 expression during chronic hypertension is associated with enhanced cerebral vasodilatation to NAD(P)H in vivo. Stroke 35:584–589
56. Liu Y, Zhao H, Li H, Kalyanaraman B, Nicolosi AC, Gutterman DD (2003) Mitochondrial sources of $H_2O_2$ generation play a key role in flow-mediated dilation in human coronary resistance arteries. Circ Res 93:573–580
57. Matoba T, Shimokawa H, Nakashima M, Hirakawa Y, Mukai Y, Hirano K et al. (2000) Hydrogen peroxide is an endothelium-derived hyperpolarizing factor in mice. J Clin Invest 106:1521–1530
58. Nishino T, Okamoto K, Eger BT, Pai EF, Nishino T (2008) Mammalian xanthine oxidoreductase – mechanism of transition from xanthine dehydrogenase to xanthine oxidase. FEBS J 275(13):3278–3289
59. Seshiah PN, Weber DS, Rocic P, Valppu L, Taniyama Y, Griendling KK (2002) Angiotensin II stimulation of NAD(P)H oxidase activity: upstream mediators. Circ Res 91: 406–413
60. Adlam D, Bendall JK, De Bono JP, Alp NJ, Khoo J, Nicoli T, Yokoyama M, Kawashima S, Channon KM (2007) Relationships between nitric oxide-mediated endothelial function, eNOS coupling and blood pressure revealed by eNOS-GTP cyclohydrolase 1 double transgenic mice. Exp Physiol 92(1):119–126
61. Moens AL, Kass DA (2006) Tetrahydrobiopterin and cardiovascular disease. Arterioscler Thromb Vasc Biol 26(11):2439–2444
62. Mukhopadhyay P, Rajesh M, Bátkai S, Kashiwaya Y, Haskó G, Liaudet L, Szabó C, Pacher P (2009) Role of superoxide, nitric oxide, and peroxynitrite in doxorubicin-induced cell death in vivo and in vitro. Am J Physiol Heart Circ Physiol 296(5):H1466–H1483
63. Förstermann U (2008) Oxidative stress in vascular disease: causes, defense mechanisms and potential therapies. Nat Clin Pract Cardiovasc Med 5(6):338–349
64. Suzuki H, DeLano FA, Parks DA, Jamshidi N, Granger DN, Ishii H et al. (1998) Xanthine oxidase activity associated with arterial blood pressure in spontaneously hypertensive rats. Proc Natl Acad Sci USA 95:4754–4759
65. DeLano FA, Parks DA, Ruedi JM, Babior BM, Schmid-Schonbein GW (2006) Microvascular display of xanthine oxidase and NAD(P)H oxidase in the spontaneously hypertensive rat. Microcirculation 13(7):551–566
66. Mervaala EM, Cheng ZJ, Tikkanen I, Lapatto R, Nurminen K, Vapaatalo H et al. (2001) Endothelial dysfunction and xanthine oxidoreductase activity in rats with human renin and angiotensinogen genes. Hypertension 37:414–418
67. Laakso J, Mervaala E, Himberg JJ, Teravainen TL, Karppanen H, Vapaatalo H et al. (1998) Increased kidney xanthine oxidoreductase activity in salt-induced experimental hypertension. Hypertension 32:902–906
68. Siu YP, Leung KT, Tong MK, Kwan TH (2006) Use of allopurinol in slowing the progression of renal disease through its ability to lower serum uric acid level. Am J Kidney Dis 47(1): 51–59
69. Laakso JT, Teravainen TL, Martelin E, Vaskonen T, Lapatto R (2004) Renal xanthine oxidoreductase activity during development of hypertension in spontaneously hypertensive rats. J Hypertens 22:1333–1340
70. Corry DB, Tuck ML (2006) Uric acid and the vasculature. Curr Hypertens Rep 8(2):116–119
71. Feig DI, Soletsky B, Johnson RJ (2008) Effect of allopurinol on blood pressure of adolescents with newly diagnosed essential hypertension: a randomized trial. J Am Med Assoc 300(8):924–932
72. Andrew PJ, Mayer B (1999) Enzymatic function of nitric oxide synthases. Cardiovasc Res 43:521–531
73. Vasquez-Vivar J, Kalyanaraman B, Martasek P, Hogg N, Masters BS, Karoui H et al. (1998) Superoxide generation by endothelial nitric oxide synthase: the influence of cofactors. Proc Natl Acad Sci USA 95:9220–9225

74. Landmesser U, Dikalov S, Price SR, McCann L, Fukai T, Holland SM et al. (2003) Oxidation of tetrahydrobiopterin leads to uncoupling of endothelial cell nitric oxide synthase in hypertension. J Clin Invest 111:1201–1209
75. Hong HJ, Hsiao G, Cheng TH, Yen MH (2001) Supplemention with tetrahydrobiopterin suppresses the development of hypertension in spontaneously hypertensive rats. Hypertension 38:1044–1048
76. Ketonen J, Mervaala E (2008) Effects of dietary sodium on reactive oxygen species formation and endothelial dysfunction in low-density lipoprotein receptor-deficient mice on high-fat diet. Heart Vessels 23(6):420–429
77. Moens AL, Takimoto E, Tocchetti CG, Chakir K, Bedja D, Cormaci G, Ketner EA, Majmudar M, Gabrielson K, Halushka MK, Mitchell JB, Biswal S, Channon KM, Wolin MS, Alp NJ, Paolocci N, Champion HC, Kass DA (2008) Reversal of cardiac hypertrophy and fibrosis from pressure overload by tetrahydrobiopterin: efficacy of recoupling nitric oxide synthase as a therapeutic strategy. Circulation 117(20):2626–2636
78. Moens AL, Kass DA (2007) Therapeutic potential of tetrahydrobiopterin for treating vascular and cardiac disease. J Cardiovasc Pharmacol 50(3):238–246
79. Bauersachs J, Widder JD (2009) Tetrahydrobiopterin, endothelial nitric oxide synthase, and mitochondrial function in the heart. Hypertension 53(6):907–908
80. Katusic ZS, d'Uscio LV, Nath KA (2009) Vascular protection by tetrahydrobiopterin: progress and therapeutic prospects. Trends Pharmacol Sci 30(1):48–54
81. Wang S, Xu J, Song P, Wu Y, Zhang J, Chul Choi H, Zou MH (2008) Acute inhibition of guanosine triphosphate cyclohydrolase 1 uncouples endothelial nitric oxide synthase and elevates blood pressure. Hypertension 52(3):484–490
82. Porkert M, Sher S, Reddy U, Cheema F, Niessner C, Kolm P, Jones DP, Hooper C, Taylor WR, Harrison D, Quyyumi AA (2008) Tetrahydrobiopterin: a novel antihypertensive therapy. J Hum Hypertens 22(6):401–407
83. Oelze M, Daiber A, Brandes RP, Hortmann M, Wenzel P, Hink U, Schulz E, Mollnau H, von Sandersleben A, Kleschyov AL, Mülsch A, Li H, Förstermann U, Münzel T (2006) Nebivolol inhibits superoxide formation by NAD(P)H oxidase and endothelial dysfunction in angiotensin II-treated rats. Hypertension 48(4):677–684
84. Brand MD, Affourtit C, Esteves TC, Green K, Lambert AJ, Miwa S, Pakay JL, Parker N (2004) Mitochondrial superoxide: production, biological effects, and activation of uncoupling proteins. Free Radic Biol Med 37(6):755–767
85. Manna SK, Zhang HJ, Yan T, Oberley LW, Aggarwal BB (1998) Overexpression of manganese superoxide dismutase suppresses tumor necrosis factor-induced apoptosis and activation of nuclear transcription factor-kappaB and activated protein-1. J Biol Chem 273(21):13245–13254
86. Alvarez S, Valdez LB, Zaobornyj T, Boveris A (2003) Oxygen dependence of mitochondrial nitric oxide synthase activity. Biochem Biophys Res Commun 305(3):771–775
87. Boveris A, Chance B (1973) The mitochondrial generation of hydrogen peroxide. General properties and effect of hyperbaric oxygen. Biochem J 134(3):707–716
88. Poderoso JJ, Carreras MC, Lisdero C, Riobo N, Schopfer F, Boveris A (1996) Nitric oxide inhibits electron transfer and increases superoxide radical production in rat heart mitochondria and submitochondrial particles. Arch Biochem Biophys 328(1):85–92
89. Loschen G, Flohe L, Chance B (1971) Respiratory chain linked $H_2O_2$ production in pigeon heart mitochondria. FEBS Lett 18(2):261–264
90. Ceaser EK, Ramachandran A, Levonen AL, Darley-Usmar VM (2003) Oxidized low-density lipoprotein and 15-deoxy-delta 12,14-PGJ2 increase mitochondrial complex I activity in endothelial cells. Am J Physiol Heart Circ Physiol 285(6):H2298–H2308
91. Mari M, Caballero F, Colell A, Morales A, Caballeria J, Fernandez A, Enrich C, Fernandez-Checa JC, Garcia-Ruiz C (2006) Mitochondrial free cholesterol loading sensitizes to TNF- and Fas-mediated steatohepatitis. Cell Metab 4(3):185–198

92. Wosniak J, Santos CX, Kowaltowski AJ, Laurindo FR (2009) Cross-talk between mitochondria and NAD(P)H oxidase: effects of mild mitochondrial dysfunction on angiotensin II-mediated increase in Nox isoform expression and activity in vascular smooth muscle cells. Antioxid Redox Signal 11(6):1265–1278
93. Nozoe M, Hirooka Y, Koga Y, Araki S, Konno S, Kishi T, Ide T, Sunagawa K (2008) Mitochondria-derived reactive oxygen species mediate sympathoexcitation induced by angiotensin II in the rostral ventrolateral medulla. J Hypertens 26(11): 2176–2184
94. De Giusti VC, Correa MV, Villa-Abrille MC, Beltrano C, Yeves AM, de Cingolani GE, Cingolani HE, Aiello EA (2008) The positive inotropic effect of endothelin-1 is mediated by mitochondrial reactive oxygen species. Life Sci 83(7–8):264–271
95. Fernandez-Patron C (2007) Therapeutic potential of the epidermal growth factor receptor transactivation in hypertension: a convergent signaling pathway of vascular tone, oxidative stress, and hypertrophic growth downstream of vasoactive G-protein-coupled receptors? Can J Physiol Pharmacol 85(1):97–104
96. de Cavanagh EM, Ferder L, Toblli JE, Piotrkowski B, Stella I, Fraga CG, Inserra F (2008) Renal mitochondrial impairment is attenuated by AT1 blockade in experimental type I diabetes. Am J Physiol Heart Circ Physiol 294(1):H456–H465
97. Zhang GX, Lu XM, Kimura S, Nishiyama A (2007) Role of mitochondria in angiotensin II-induced reactive oxygen species and mitogen-activated protein kinase activation. Cardiovasc Res 76(2):204–212
98. Gunter TE, Pfeiffer DR (1990) Mechanisms by which mitochondria transport calcium. Am J Physiol 258(5 Pt 1):C755–C786
99. Zoratti M, Szabo I (1995) The mitochondrial permeability transition. Biochim Biophys Acta 1241(2):139–176
100. Kowaltowski AJ, Castilho RF, Vercesi AE (2001) Mitochondrial permeability transition and oxidative stress. FEBS Lett 495(1–2):12–15
101. Koncz P, Szanda G, Rajki A, Spät A (2006) Reactive oxygen species, $Ca^{2+}$ signaling and mitochondrial NAD(P)H level in adrenal glomerulosa cells. Cell Calcium 40(4): 347–357
102. Ungvari Z, Labinskyy N, Gupte S, Chander PN, Edwards JG, Csiszar A (2008) Dysregulation of mitochondrial biogenesis in vascular endothelial and smooth muscle cells of aged rats. Am J Physiol Heart Circ Physiol 294(5):H2121–H2128
103. Postnov YV, Orlov SN, Budnikov YY, Doroschuk AD, Postnov AY (2007) Mitochondrial energy conversion disturbance with decrease in ATP production as a source of systemic arterial hypertension. Pathophysiology 14(3–4):195–204
104. Zhang H, Luo Y, Zhang W, He Y, Dai S, Zhang R, Huang Y, Bernatchez P, Giordano FJ, Shadel G, Sessa WC, Min W (2007) Endothelial-specific expression of mitochondrial thioredoxin improves endothelial cell function and reduces atherosclerotic lesions. Am J Pathol 170(3):1108–1120
105. Rodriguez-Iturbe B, Sepassi L, Quiroz Y, Ni Z, Wallace DC, Vaziri ND (2007) Association of mitochondrial SOD deficiency with salt-sensitive hypertension and accelerated renal senescence. J Appl Physiol 102(1):255–260
106. Zorov DB, Juhaszova M, Sollott SJ (2006) Mitochondrial ROS-induced ROS release: an update and review. Biochim Biophys Acta 1757(5–6):509–517
107. Doughan AK, Harrison DG, Dikalov SI (2008) Molecular mechanisms of angiotensin II-mediated mitochondrial dysfunction: linking mitochondrial oxidative damage and vascular endothelial dysfunction. Circ Res 102(4):488–496
108. de Cavanagh EM, Inserra F, Ferder M, Ferder L (2007) From mitochondria to disease: role of the renin-angiotensin system. Am J Nephrol 27(6):545–553
109. Touyz RM, Yao G, Viel E, Amiri F, Schiffrin EL (2004) Angiotensin II and endothelin1 regulate MAP kinases through different redox-dependent mechanisms in human vascular smooth muscle cells. J Hypertens 22(6):1141–1149

110. Callera GE, Tostes RC, Yogi A, Montezano AC, Touyz RM (2006) Endothelin-1-induced oxidative stress in DOCA-salt hypertension involves NADPH-oxidase-independent mechanisms. Clin Sci 110(2):243–253
111. Chan SH, Wu KL, Chang AY, Tai MH, Chan JY (2009) Oxidative impairment of mitochondrial electron transport chain complexes in rostral ventrolateral medulla contributes to neurogenic hypertension. Hypertension 53(2):217–227
112. Yang Q, Kim SK, Sun F, Cui J, Larson MG, Vasan RS, Levy D, Schwartz F (2007) Maternal influence on blood pressure suggests involvement of mitochondrial DNA in the pathogenesis of hypertension: the Framingham Heart Study. J Hypertens 25(10):2067–2073
113. Rachek LI, Grishko VI, LeDoux SP, Wilson GL (2006) Role of nitric oxide-induced mtDNA damage in mitochondrial dysfunction and apoptosis. Free Radic Biol Med 40(5): 754–762
114. Puddu P, Puddu GM, Cravero E, De PS, Muscari A (2007) The putative role of mitochondrial dysfunction in hypertension. Clin Exp Hypertens 29(7):427–434
115. Miller AA, Drummond GR, Sobey CG (2006) Novel isoforms of NADPH-oxidase in cerebral vascular control. Pharmacol Ther 111(3):928–948
116. Bayraktutan U, Blayney L, Shah AM (2000) Molecular characterization and localization of the NAD(P)H oxidase components gp91-phox and p22-phox in endothelial cells. Arterioscler Thromb Vasc Biol 20:1903–1911
117. Li JM, Shah AM (2002) Intracellular localization and preassembly of the NAD(P)H oxidase complex in cultured endothelial cells. J Biol Chem 277:19952–19960
118. Babior BM (2004) NAD(P)H oxidase. Curr Opin Immunol 16(1):42–47
119. Vignais PV (2002) The superoxide-generating NAD(P)H oxidase: structural aspects and activation mechanism. Cell Mol Life Sci 59(9):1428–1459
120. Taura M, Miyano K, Minakami R, Kamakura S, Takeya R, Sumimoto H (2009) A region N-terminal to the tandem SH3 domain of p47phox plays a crucial role in the activation of the phagocyte NAD(P)H oxidase. Biochem J 419(2):329–338
121. Bokoch GM, Zhao T (2006) Regulation of the phagocyte NAD(P)H oxidase by Rac GTPase. Antioxid Redox Signal 8(9–10):1533–1548
122. Geiszt M (2006) NAD(P)H oxidases: New kids on the block. Cardiovasc Res 71:289–299
123. Cave AC, Brewer AC, Panicker AN, Ray R, Grieve DJ, Walker S, Shah AM (2006) NAD(P)H oxidases in cardiovascular health and disease. Antiox Redox Sig 8:691–727
124. Griendling KK (2006) NAD(P)H oxidases: new regulators of old functions. Antioxid Redox Signal 8(9–10):1443–1445
125. Guzik TJ, Chen W, Gongora MC, Guzik B, Lob HE, Mangalat D, Hoch N, Dikalov S, Rudzinski P, Kapelak B, Sadowski J, Harrison DG (2008) Calcium-dependent NOX5 nicotinamide adenine dinucleotide phosphate oxidase contributes to vascular oxidative stress in human coronary artery disease. J Am Coll Cardiol 52(22):1803–1809
126. Nisimoto Y, Tsubouchi R, Diebold BA, Qiao S, Ogawa H, Ohara T, Tamura M (2008) Activation of NAD(P)H oxidase 1 in tumour colon epithelial cells. Biochem J 415(1): 57–65
127. Muzaffar S, Shukla N, Bond M, Newby AC, Angelini GD, Sparatore A, Del Soldato P, Jeremy JY (2008) Exogenous hydrogen sulfide inhibits superoxide formation, NOX-1 expression and Rac1 activity in human vascular smooth muscle cells. J Vasc Res 45(6):521–528
128. Fernandes DC, Manoel AHO, Wosniak J, Laurindo FR (2009) Protein disulfide isomerise overexpression in vascular smooth muscle cells induces spontaneous preemptive NAD(P)H oxidase activation and Nox1 mRNA expression: effects of nitrosothiol exposure. Arch Biochem Biophys 484(2):197–204
129. Lee MY, San Martin A, Mehta PK, Dikalova AE, Garrido AM, Datla SR, Lyons E, Krause KH, Banfi B, Lambeth JD, Lassègue B, Griendling KK (2009) Mechanisms of vascular smooth muscle NAD(P)H oxidase 1 (Nox1) contribution to injury-induced neointimal formation. Arterioscler Thromb Vasc Biol 29(4):480–487

130. Chabrashvili T, Tojo A, Onozato ML, Kitiyakara C, Quinn MT, Fujita T, Welch WJ, Wilcox CS (2002) Expression and cellular localization of classic NAD(P)H oxidase subunits in the spontaneously hypertensive rat kidney. Hypertension 39(2):269–274
131. De Silva TM, Broughton BR, Drummond GR, Sobey CG, Miller AA (2009) Gender influences cerebral vascular responses to angiotensin II through Nox2-derived reactive oxygen species. Stroke 40(4):1091–1097
132. Touyz RM, Yao G, Schiffrin EL (2003) c-Src induces phosphorylation and translocation of p47phox: role in superoxide generation by angiotensin II in human vascular smooth muscle cells. Arterioscler Thromb Vasc Biol 23:981–987
133. Lassegue B, Clempus RE (2003) Vascular NAD(P)H oxidases: specific features, expression, and regulation. Am J Physiol Regul Integr Comp Physiol 285:R277–R297
134. Touyz RM, Chen X, Tabet F, Yao G, He G, Quinn MT, Pagano PJ, Schiffrin EL (2002) Expression of a functionally active gp91phox-containing neutrophil-type NAD(P)H oxidase in smooth muscle cells from human resistance arteries: regulation by angiotensin II. Circ Res 90(11):1205–1213
135. Goyal P, Weissmann N, Grimminger F, Hegel C, Bader L, Rose F, Fink L, Ghofrani HA, Schermuly RT, Schmidt HH, Seeger W, Hanze J (2004) Upregulation of NAD(P)H oxidase 1 in hypoxia activates hypoxia-inducible factor 1 via increase in reactive oxygen species. Free Radic Biol Med 36(10):1279–1288
136. Gupte SA, Kaminski PM, George S, Kouznestova L, Olson SC, Mathew R, Hintze TH, Wolin MS (2009) Peroxide generation by p47phox-Src activation of Nox2 has a key role in protein kinase C-induced arterial smooth muscle contraction. Am J Physiol Heart Circ Physiol 296(4):H1048–H1057
137. Goettsch C, Goettsch W, Muller G, Seebach J, Schnittler HJ, Morawietz H (2009) Nox4 overexpression activates reactive oxygen species and p38 MAPK in human endothelial cells. Biochem Biophys Res Commun 380(2):355–360
138. Xiao Q, Luo Z, Pepe AE, Margariti A, Zeng L, Xu Q (2009) Embryonic stem cell differentiation into smooth muscle cells is mediated by Nox4-produced $H_2O_2$. Am J Physiol Cell Physiol 296(4):C711–C723
139. Cheng G, Cao Z, Xu X, van Meir EG, Lambeth JD (2001) Homologs of gp91phox: cloning and tissue expression of Nox3, Nox4, and Nox5. Gene 269(1–2):131–140
140. Kuroda J, Nakagawa K, Yamasaki T, Nakamura K, Takeya R, Kuribayashi F, Imajoh-Ohmi S, Igarashi K, Shibata Y, Sueishi K, Sumimoto H (2005) The superoxide-producing NAD(P)H oxidase Nox4 in the nucleus of human vascular endothelial cells. Genes Cells 10(12):1139–1151
141. Chen K, Kirber MT, Xiao H, Yang Y, Keaney JF Jr (2008) Regulation of ROS signal transduction by NAD(P)H oxidase 4 localization. J Cell Biol 181(7):1129–1139
142. Mittal M, Roth M, König P, Hofmann S, Dony E, Goyal P, Selbitz AC, Schermuly RT, Ghofrani HA, Kwapiszewska G, Kummer W, Klepetko W, Hoda MA, Fink L, Hänze J, Seeger W, Grimminger F, Schmidt HH, Weissmann N (2007) Hypoxia-dependent regulation of nonphagocytic NAD(P)H oxidase subunit NOX4 in the pulmonary vasculature. Circ Res 101(3):258–267
143. Kawahara T, Lambeth JD (2007) Molecular evolution of Phox-related regulatory subunits for NAD(P)H oxidase enzymes. BMC Evol Biol 7:178–181
144. Kawahara T, Ritsick D, Cheng G, Lambeth JD (2005) Point mutations in the proline-rich region of p22phox are dominant inhibitors of Nox1- and Nox2-dependent reactive oxygen generation. J Biol Chem 280(36):31859–31869
145. Clempus RE, Sorescu D, Dikalova AE, Pounkova L, Jo P, Sorescu GP, Schmidt HH, Lassègue B, Griendling KK (2007) Nox4 is required for maintenance of the differentiated vascular smooth muscle cell phenotype. Arterioscler Thromb Vasc Biol 27(1):42–48
146. Zhang L, Sheppard OR, Shah AM, Brewer AC (2008) Positive regulation of the NAD(P)H oxidase NOX4 promoter in vascular smooth muscle cells by E2F. Free Radic Biol Med 45(5):679–685

147. Petry A, Djordjevic T, Weitnauer M, Kietzmann T, Hess J, Görlach A (2006) NOX2 and NOX4 mediate proliferative response in endothelial cells. Antioxid Redox Signal 8(9–10):1473–1484
148. Shiose A, Kuroda J, Tsuruya K, Hirai M, Hirakata H, Naito S, Hattori M, Sakaki Y, Sumimoto H (2001) A novel superoxide-producing NAD(P)H oxidase in kidney. J Biol Chem 276(2):1417–1423
149. Jay DB, Papaharalambus CA, Seidel-Rogol B, Dikalova AE, Lassègue B, Griendling KK (2008) Nox5 mediates PDGF-induced proliferation in human aortic smooth muscle cells. Free Radic Biol Med 45(3):329–335
150. Sabeur K, Ball BA (2007) Characterization of NAD(P)H oxidase 5 in equine testis and spermatozoa. Reproduction 134(2):263–270
151. Serrander L, Jaquet V, Bedard K, Plastre O, Hartley O, Arnaudeau S, Demaurex N, Schlegel W, Krause KH (2007) NOX5 is expressed at the plasma membrane and generates superoxide in response to protein kinase C activation. Biochimie 89(9):1159–1167
152. Si J, Fu X, Behar J, Wands J, Beer DG, Souza RF, Spechler SJ, Lambeth D, Cao W (2007) NAD(P)H oxidase NOX5-S mediates acid-induced cyclooxygenase-2 expression via activation of NF-kappaB in Barrett's esophageal adenocarcinoma cells. J Biol Chem 282(22):16244–16255
153. Sumimoto H (2008) Structure, regulation and evolution of Nox-family NAD(P)H oxidases that produce reactive oxygen species. FEBS J 275(13):3249–3277
154. BelAiba RS, Djordjevic T, Petry A, Diemer K, Bonello S, Banfi B, Hess J, Pogrebniak A, Bickel C, Görlach A (2007) NOX5 variants are functionally active in endothelial cells. Free Radic Biol Med 42(4):446–459
155. Jagnandan D, Church JE, Banfi B, Stuehr DJ, Marrero MB, Fulton DJ (2007) Novel mechanism of activation of NAD(P)H oxidase 5. Calcium sensitization via phosphorylation. J Biol Chem 282(9):6494–6507
156. Tirone F, Cox JA (2007) NAD(P)H oxidase 5 (NOX5) interacts with and is regulated by calmodulin. FEBS Lett 581(6):1202–1208
157. Schulz E, Münzel T (2008) NOX5, a new "radical" player in human atherosclerosis? J Am Coll Cardiol 52(22):1810–1812
158. Montezano AC, Paravicini TM, Chignalia AZ, Yusuf H, Almasri M, He Y, He G, Callera GE, Krause K-H, Lambeth D, Touyz RM (2010) Nicotinamide Adenine Dinucleotide Phosphate Reduced Oxidase 5 (Nox5) Regulation by Angiotensin II and Endothelin-1 is Mediated via Calcium/Calmodulin-dependent Pathways in Human Endothelial Cells. Circ Res 106(8):1363–1373
159. El Jamali A, Valente AJ, Lechleiter JD, Gamez MJ, Pearson DW, Nauseef WM, Clark RA (2008) Novel redox-dependent regulation of NOX5 by the tyrosine kinase c-Abl. Free Radic Biol Med 44(5):868–881
160. Milenkovic M, De Deken X, Jin L, De Felice M, Di Lauro R, Dumont JE, Corvilain B, Miot F (2007) Duox expression and related $H_2O_2$ measurement in mouse thyroid: onset in embryonic development and regulation by TSH in adult. J Endocrinol 192(3):615–626
161. Sorescu D, Weiss D, Lassègue B, Clempus RE, Szöcs K, Sorescu GP, Valppu L, Quinn MT, Lambeth JD, Vega JD, Taylor WR, Griendling KK (2002) Superoxide production and expression of nox family proteins in human atherosclerosis. Circulation 105(12):1429–1435
162. Selemidis S, Sobey CG, Wingler K, Schmidt HH, Drummond GR (2008) NAD(P)H oxidases in the vasculature: molecular features, roles in disease and pharmacological inhibition. Pharmacol Ther 120(3):254–291
163. Zhang R, Harding P, Garvin JL, Juncos R, Peterson E, Juncos LA, Liu R (2009) Is forms and functions of NAD(P)H oxidase at the macula densa. Hypertension 53(3):556–563
164. Li S, Tabar SS, Malec V, Eul BG, Klepetko W, Weissmann N, Grimminger F, Seeger W, Rose F, Hänze J (2008) NOX4 regulates ROS levels under normoxic and hypoxic conditions, triggers proliferation, and inhibits apoptosis in pulmonary artery adventitial fibroblasts. Antioxid Redox Signal 10(10):1687–1698

165. Hilenski LL, Clempus RE, Quinn MT, Lambeth JD, Griendling KK (2004) Distinct subcellular localizations of Nox1 and Nox4 in vascular smooth muscle cells. Arterioscler Thromb Vasc Biol 24(4):677–683
166. Takeya R, Sumimoto H (2006) Regulation of novel superoxide-producing NAD(P)H oxidases. Antioxid Redox Signal 8(9–10):1523–1532
167. Ambasta RK, Kumar P, Griendling KK, Schmidt HH, Busse R, Brandes RP (2004) Direct interaction of the novel Nox proteins with p22phox is required for the formation of a functionally active NAD(P)H oxidase. J Biol Chem 279(44):45935–45941
168. Ambasta RK, Schreiber JG, Janiszewski M, Busse R, Brandes RP (2006) Noxa1 is a central component of the smooth muscle NAD(P)H oxidase in mice. Free Radic Biol Med 41(2):193–201
169. Cheng G, Lambeth JD (2005) Alternative mRNA splice forms of NOXO1: differential tissue expression and regulation of Nox1 and Nox3. Gene 356:118–126
170. Ibi M, Matsuno K, Shiba D, Katsuyama M, Iwata K, Kakehi T, Nakagawa T, Sango K, Shirai Y, Yokoyama T, Kaneko S, Saito N, Yabe-Nishimura C (2008) Reactive oxygen species derived from NOX1/NAD(P)H oxidase enhance inflammatory pain. J Neurosci 28(38):9486–9494
171. Peng YJ, Yuan G, Jacono FJ, Kumar GK, Prabhakar NR (2006) 5-HT evokes sensory long-term facilitation of rodent carotid body via activation of NAD(P)H oxidase. J Physiol 576(Pt 1):289–295
172. Gao L, Mann GE (2009) Vascular NAD(P)H oxidase activation in diabetes: a double-edged sword in redox signalling. Cardiovasc Res 82(1):9–20
173. Suh SW, Shin BS, Ma H, Van Hoecke M, Brennan AM, Yenari MA, Swanson RA (2008) Glucose and NAD(P)H oxidase drive neuronal superoxide formation in stroke. Ann Neurol 64(6):654–663
174. Nistala R, Whaley-Connell A, Sowers JR (2008) Redox control of renal function and hypertension. Antioxid Redox Signal 10(12):2047–2089
175. Brandes RP, Schröder K (2008) Differential vascular functions of Nox family NAD(P)H oxidases. Curr Opin Lipidol 19(5):513–518
176. El-Benna J, Dang PM, Gougerot-Pocidalo MA, Marie JC, Braut-Boucher F (2009) p47phox, the phagocyte NAD(P)H oxidase/NOX2 organizer: structure, phosphorylation and implication in diseases. Exp Mol Med 41(4):217–225
177. Maehara Y, Miyano K, Sumimoto H (2009) Role for the first SH3 domain of p67phox in activation of superoxide-producing NAD(P)H oxidases. Biochem Biophys Res Commun 379(2):589–593
178. Montezano AC, Callera GE, Yogi A, He Y, Tostes RC, He G, Schiffrin EL, Touyz RM (2008) Aldosterone and angiotensin II synergistically stimulate migration in vascular smooth muscle cells through c-Src-regulated redox-sensitive RhoA pathways. Arterioscler Thromb Vasc Biol 28(8):1511–1518
179. Block K, Eid A, Griendling KK, Lee DY, Wittrant Y, Gorin Y (2008) Nox4 NAD(P)H oxidase mediates Src-dependent tyrosine phosphorylation of PDK-1 in response to angiotensin II: role in mesangial cell hypertrophy and fibronectin expression. J Biol Chem 283(35):24061–24076
180. Gongora MC, Qin Z, Laude K, Kim HW, McCann L, Folz JR, Dikalov S, Fukai T, Harrison DG (2006) Role of extracellular superoxide dismutase in hypertension. Hypertension 48(3):473–481
181. Sindhu RK, Ehdaie A, Farmand F, Dhaliwal KK, Nguyen T, Zhan CD, Roberts CK, Vaziri ND (2005) Expression of catalase and glutathione peroxidase in renal insufficiency. Biochim Biophys Acta 1743(1–2):86–92
182. Sui H, Wang W, Wang PH, Liu LS (2005) Effect of glutathione peroxidase mimic ebselen (PZ51) on endothelium and vascular structure of stroke-prone spontaneously hypertensive rats. Blood Press 14(6):366–372

183. Wassmann S, Wassmann K, Nickenig G (2004) Modulation of oxidant and antioxidant enzyme expression and function in vascular cells. Hypertension 44(4):381–386
184. Tajima M, Kurashima Y, Sugiyama K, Ogura T, Sakagami H (2009) The redox state of glutathione regulates the hypoxic induction of HIF-1. Eur J Pharmacol 606(1–3):45–49
185. Wong CH, Bozinovski S, Hertzog PJ, Hickey MJ, Crack PJ (2008) Absence of glutathione peroxidase-1 exacerbates cerebral ischemia-reperfusion injury by reducing post-ischemic microvascular perfusion. J Neurochem 107(1):241–252
186. Chung SS, Kim M, Youn BS, Lee NS, Park JW, Lee IK, Lee YS, Kim JB, Cho YM, Lee HK, Park KS (2009) Glutathione peroxidase 3 mediates the antioxidant effect of peroxisome proliferator-activated receptor gamma in human skeletal muscle cells. Mol Cell Biol 29(1):20–30
187. Widder JD, Guzik TJ, Mueller CF, Clempus RE, Schmidt HH, Dikalov SI, Griendling KK, Jones DP, Harrison DG (2007) Role of the multidrug resistance protein-1 in hypertension and vascular dysfunction caused by angiotensin II. Arterioscler Thromb Vasc Biol 27(4): 762–768
188. Mueller CF, Wassmann K, Widder JD, Wassmann S, Chen CH, Keuler B, Kudin A, Kunz WS, Nickenig G (2008) Multidrug resistance protein-1 affects oxidative stress, endothelial dysfunction, and atherogenesis via leukotriene C4 export. Circulation 117(22):2912–2918
189. Zamocky M, Furtmüller PG, Obinger C (2008) Evolution of catalases from bacteria to humans. Antioxid Redox Signal 10(9):1527–1548
190. Wilcox CS, Pearlman A (2008) Chemistry and antihypertensive effects of tempol and other nitroxides. Pharmacol Ver 60(4):418–469
191. Chrissobolis S, Didion SP, Kinzenbaw DA, Schrader LI, Dayal S, Lentz SR, Faraci FM (2008) Glutathione peroxidase-1 plays a major role in protecting against angiotensin II-induced vascular dysfunction. Hypertension 51(4):872–877
192. Ebrahimian T, Touyz RM (2008) Thioredoxin in vascular biology: role in hypertension. Antioxid Redox Signal 10(6):1127–1136
193. Redon J, Oliva MR, Tormos C, Giner V, Chaves J, Iradi A et al. (2003) Antioxidant activities and oxidative stress byproducts in human hypertension. Hypertension 41: 1096–1101
194. Zhou XJ, Vaziri ND, Wang XQ, Silva FG, Laszik Z (2002) Nitric oxide synthase expression in hypertension induced by inhibition of glutathione synthase. J Pharmacol Exp Ther 300(3):762–767
195. Collins AR, Lyon CJ, Xia X, Liu JZ, Tangirala RK, Yin F, Boyadjian R, Bikineyeva A, Praticò D, Harrison DG, Hsueh WA (2009) Age-accelerated atherosclerosis correlates with failure to upregulate antioxidant genes. Circ Res 104(6):e42–e54
196. Cohen RA, Adachi T (2006) Nitric-oxide-induced vasodilatation: regulation by physiologic s-glutathiolation and pathologic oxidation of the sarcoplasmic endoplasmic reticulum calcium ATPase. Trends Cardiovasc Med 16(4):109–114
197. Münzel T, Daiber A, Mülsch A (2005) Explaining the phenomenon of nitrate tolerance. Circ Res 97(7):618–628
198. Chavez A, Miranda LF, Pichiule P, Chavez JC (2008) Mitochondria and hypoxia-induced gene expression mediated by hypoxia-inducible factors. Ann N Y Acad Sci 1147: 312–320
199. Feissner RF, Skalska J, Gaum WE, Sheu SS (2009) Crosstalk signaling between mitochondrial $Ca^{2+}$ and ROS. Front Biosci 14:1197–1218
200. Bashan N, Kovsan J, Kachko I, Ovadia H, Rudich A (2009) Positive and negative regulation of insulin signaling by reactive oxygen and nitrogen species. Physiol Rev 89(1):27–71
201. Monteiro HP, Arai RJ, Travassos LR (2008) Protein tyrosine phosphorylation and protein tyrosine nitration in redox signaling. Antioxid Redox Signal 10(5):843–889
202. Callera GE, Montezano AC, Yogi A, Tostes RC, Touyz RM (2007) Vascular signaling through cholesterol-rich domains: implications in hypertension. Curr Opin Nephrol Hypertens 16(2):90–104

203. Harrison DG, Gongora MC, Guzik TJ, Widder J (2007) Oxidative stress and hypertension. J Am Soc Hypertens 1:30–44
204. Gill PS, Wilcox CS (2006) NAD(P)H oxidases in the kidney. Antioxid Redox Signal 8(9–10):1597–1607
205. Wilcox CS (2005) Oxidative stress and nitric oxide deficiency in the kidney: a critical link to hypertension? Am J Physiol Regul Integr Comp Physiol 289(4):R913–R935
206. Evans RG, Fitzgerald SM (2005) Nitric oxide and superoxide in the renal medulla: a delicate balancing act. Curr Opin Nephrol Hypertens 14(1):9–15
207. Agarwal R, Campbell RC, Warnock DG (2004) Oxidative stresses in hypertension and chronic kidney disease: role of angiotensin II. Semin Nephrol 24(2):101–114
208. Kaysen GA, Eiserich JP (2004) The role of oxidative stress-altered lipoprotein structure and function and microinflammation on cardiovascular risk in patients with minor renal dysfunction. J Am Soc Nephrol 15(3):538–548
209. Hirooka Y (2008) Role of reactive oxygen species in brainstem in neural mechanisms of hypertension. Auton Neurosci 142(1–2):20–24
210. Mayorov DN (2007) Brain superoxide as a key regulator of the cardiovascular response to emotional stress in rabbits. Exp Physiol 92(3):471–479
211. Peterson JR, Sharma RV, Davisson RL (2006) Reactive oxygen species in the neuropathogenesis of hypertension. Curr Hypertens Rep 8(3):232–241
212. Girouard H, Iadecola C (2006) Neurovascular coupling in the normal brain and in hypertension, stroke, and Alzheimer disease. J Appl Physiol 100(1):328–335
213. Nuyt AM (2008) Mechanisms underlying developmental programming of elevated blood pressure and vascular dysfunction: evidence from human studies and experimental animal models. Clin Sci 114(1):1–17
214. Friese RS, Mahboubi P, Mahapatra NR, Mahata SK, Schork NJ, Schmid-Schönbein GW, O'Connor DT (2005) Common genetic mechanisms of blood pressure elevation in two independent rodent models of human essential hypertension. Am J Hypertens 18(5 Pt 1):633–652
215. Török J (2008) Participation of nitric oxide in different models of experimental hypertension. Physiol Res 57(6):813–825
216. Viel EC, Benkirane K, Javeshghani D, Touyz RM, Schiffrin EL (2008) Xanthine oxidase and mitochondria contribute to vascular superoxide anion generation in DOCA-salt hypertensive rats. Am J Physiol Heart Circ Physiol 295(1):H281–H288
217. Puddu P, Puddu GM, Cravero E, Rosati M, Muscari A (2008) The molecular sources of reactive oxygen species in hypertension. Blood Press 17(2):70–77
218. Roghair RD, Segar JL, Volk KA, Chapleau MW, Dallas LM, Sorenson AR, Scholz TD, Lamb FS (2009) Vascular nitric oxide and superoxide anion contribute to sex-specific programmed cardiovascular physiology in mice. Am J Physiol Regul Integr Comp Physiol 296(3):R651–R662
219. Fukai T, Ishizaka N, Rajagopalan S, Laursen JB, Capers QT, Taylor WR et al. (1997) p22phox mRNA expression and NAD(P)H oxidase activity are increased in aortas from hypertensive rats. Circ Res 80:45–51
220. Tornavaca O, Pascual G, Barreiro ML, Grande MT, Carretero A, Riera M, Garcia-Arumi E, Bardaji B, González-Núñez M, Montero MA, López-Novoa JM, Meseguer A (2009) Kidney androgen-regulated protein transgenic mice show hypertension and renal alterations mediated by oxidative stress. Circulation 119(14):1908–1917
221. Hopps E, Lo Presti R, Caimi G (2009) Pathophysiology of polymorphonuclear leukocyte in arterial hypertension. Clin Hemorheol Microcirc 41(3):209–218
222. Carlström M, Persson AE (2009) Important role of NAD(P)H oxidase 2 in the regulation of the tubuloglomerular feedback. Hypertension 53(3):456–457
223. Park YM, Lim BH, Touyz RM, Park JB (2008) Expression of NAD(P)H oxidase subunits and their contribution to cardiovascular damage in aldosterone/salt-induced hypertensive rat. J Korean Med Sci 23(6):1039–1045

224. Inaba S, Iwai M, Furuno M, Tomono Y, Kanno H, Senba I, Okayama H, Mogi M, Higaki J, Horiuchi M (2009) Continuous activation of renin-angiotensin system impairs cognitive function in renin/angiotensinogen transgenic mice. Hypertension 53(2):356–362
225. Haque MZ, Majid DS (2008) Reduced renal responses to nitric oxide synthase inhibition in mice lacking the gene for gp91phox subunit of NAD(P)H oxidase. Am J Physiol Renal Physiol 295(3):F758–F764
226. Nisbet RE, Graves AS, Kleinhenz DJ, Rupnow HL, Reed AL, Fan TH, Mitchell PO, Sutliff RL, Hart CM (2009) The role of NAD(P)H oxidase in chronic intermittent hypoxia-induced pulmonary hypertension in mice. Am J Respir Cell Mol Biol 40(5):601–609
227. Byrne JA, Grieve DJ, Bendall JK, Li JM, Gove C, Lambeth JD, Cave AC, Shah AM (2003) Contrasting roles of NAD(P)H oxidase isoforms in pressure-overload versus angiotensin II-induced cardiac hypertrophy. Circ Res 93(9):802–805
228. Modlinger P, Chabrashvili T, Gill PS, Mendonca M, Harrison DG, Griendling KK, Li M, Raggio J, Wellstein A, Chen Y, Welch WJ, Wilcox CS (2006) RNA silencing in vivo reveals role of p22phox in rat angiotensin slow pressor response. Hypertension 47(2):238–244
229. Laude K, Cai H, Fink B, Hoch N, Weber DS, McCann L, Kojda G, Fukai T, Schmidt HH, Dikalov S, Ramasamy S, Gamez G, Griendling KK, Harrison DG (2005) Hemodynamic and biochemical adaptations to vascular smooth muscle overexpression of p22phox in mice. Am J Physiol Heart Circ Physiol 288(1):H7–H12
230. Virdis A, Neves MF, Amiri F, Touyz RM, Schiffrin EL (2004) Role of NAD(P)H oxidase on vascular alterations in angiotensin II-infused mice. J Hypertens 22:535–542
231. Hu L, Zhang Y, Lim PS, Miao Y, Tan C, McKenzie KU, Schyvens CG, Whitworth JA (2006) Apocynin but not L-arginine prevents and reverses dexamethasone-induced hypertension in the rat. Am J Hypertens 19(4):413–418
232. Rey FE, Cifuentes ME, Kiarash A, Quinn MT, Pagano PJ (2001) Novel competitive inhibitor of NAD(P)H oxidase assembly attenuates vascular $O_2^-$ and systolic blood pressure in mice. Circ Res 89:408–414
233. Gavazzi G, Banfi B, Deffert C, Fiette L, Schappi M, Herrmann F, Krause KH (2006) Decreased bloodpressure in NOX1-deficient mice. FEBS Lett 580(2):497–504
234. Matsuno K, Yamada H, Iwata K, Jin D, Katsuyama M, Matsuki M, Takai S, Yamanishi K, Miyazaki M, Matsubara H, Yabe-Nishimura C (2005) Nox1 is involved in angiotensin II-mediated hypertension: a study in Nox1-deficient mice. Circulation 112(17):2677–2685
235. Dikalova A, Clempus R, Lassegue B, Cheng G, McCoy J, Dikalov S et al. (2005) Nox1 overexpression potentiates angiotensin II-induced hypertension and vascular smooth muscle hypertrophy in transgenic mice. Circulation 112:2668–2676
236. Touyz RM, Mercure C, He Y, Javeshghani D, Yao G, Callera GE, Yogi A, Lochard N, Reudelhuber TL (2005) Angiotensin II-dependent chronic hypertension and cardiac hypertrophy are unaffected by gp91phox-containing NAD(P)H oxidase. Hypertension 45(4):530–537
237. Yogi A, Mercure C, Touyz J, Callera GE, Montezano AC, Aranha AB, Tostes RC, Reudelhuber T, Touyz RM (2008) Renal redox-sensitive signaling, but not blood pressure, is attenuated by Nox1 knockout in angiotensin II-dependent chronic hypertension. Hypertension 51(2):500–506
238. Peixoto EB, Pessoa BS, Biswas SK, Lopes de Faria JB (2009) Antioxidant SOD mimetic prevents NAD(P)H oxidase-induced oxidative stress and renal damage in the early stage of experimental diabetes and hypertension. Am J Nephrol 29(4):309–318
239. García-Redondo AB, Briones AM, Beltrán AE, Alonso MJ, Simonsen U, Salaices M (2009) Hypertension increases contractile responses to hydrogen peroxide in resistance arteries through increased thromboxane A2, $Ca^{2+}$, and superoxide anion levels. J Pharmacol Exp Ther 328(1):19–27
240. Takaki A, Morikawa K, Murayama Y, Yamagishi H, Hosoya M, Ohashi J, Shimokawa H (2008) Roles of endothelial oxidases in endothelium-derived hyperpolarizing factor responses in mice. J Cardiovasc Pharmacol 52(6):510–517

241. Yamamoto E, Tamamaki N, Nakamura T, Kataoka K, Tokutomi Y, Dong YF, Fukuda M, Matsuba S, Ogawa H, Kim-Mitsuyama S (2008) Excess salt causes cerebral neuronal apoptosis and inflammation in stroke-prone hypertensive rats through angiotensin II-induced NAD(P)H oxidase activation. Stroke 39(11):3049–3056
242. Somers MJ, Mavromatis K, Galis ZS, Harrison DG (2000) Vascular superoxide production and vasomotor function in hypertension induced by deoxycorticosterone acetate-salt. Circulation 101:1722–1728
243. Callera GE, Touyz RM, Teixeira SA, Muscara MN, Carvalho MH, Fortes ZB et al. (2003) ETA receptor blockade decreases vascular superoxide generation in DOCA-salt hypertension. Hypertension 42:811–817
244. Park JB, Touyz RM, Chen X, Schiffrin EL (2002) Chronic treatment with a superoxide dismutase mimetic prevents vascular remodeling and progression of hypertension in salt-loaded stroke-prone spontaneously hypertensive rats. Am J Hypertens 15(1 Pt 1):78–84
245. Elmarakby AA, Loomis ED, Pollock JS, Pollock DM (2005) NAD(P)H oxidase inhibition attenuates oxidative stress but not hypertension produced by chronic ET-1. Hypertension 45:283–287
246. Amiri F, Virdis A, Neves MF, Iglarz M, Seidah NG, Touyz RM et al. (2004) Endothelium-restricted overexpression of human endothelin-1 causes vascular remodeling and endothelial dysfunction. Circulation 110:2233–2240
247. Quiroz Y, Ferrebuz A, Vaziri ND, Rodriguez-Iturbe B (2009) Effect of chronic antioxidant therapy with superoxide dismutase-mimetic drug, tempol, on progression of renal disease in rats with renal mass reduction. Nephron Exp Nephrol 112(1):e31–e42
248. Castro MM, Rizzi E, Rodrigues GJ, Ceron CS, Bendhack LM, Gerlach RF, Tanus-Santos JE (2009) Antioxidant treatment reduces matrix metalloproteinase-2-induced vascular changes in renovascular hypertension. Free Radic Biol Med 46(9):1298–1307
249. Chen X, Touyz RM, Park JB, Schiffrin EL (2001) Antioxidant effects of vitamins C and E are associated with altered activation of vascular NAD(P)H oxidase and superoxide dismutase in stroke-prone SHR. Hypertension 38(3 Pt 2):606–611
250. Fortuno A, Olivan S, Beloqui O, San Jose G, Moreno MU, Diez J et al. (2004) Association of increased phagocytic NAD(P)H oxidase-dependent superoxide production with diminished nitric oxide generation in essential hypertension. J Hypertens 22:2169–2175
251. Higashi Y, Sasaki S, Nakagawa K, Matsuura H, Oshima T, Chayama K (2002) Endothelial function and oxidative stress in renovascular hypertension. N Engl J Med 346:1954–1962
252. Lip GY, Edmunds E, Nuttall SL, Landray MJ, Blann AD, Beevers DG (2002) Oxidative stress in malignant and non-malignant phase hypertension. J Hum Hypertens 16:333–336
253. Lee VM, Quinn PA, Jennings SC, Ng LL (2003) Neutrophil activation and production of reactive oxygen species in pre-eclampsia. J Hypertens 21:395–402
254. Ward NC, Hodgson JM, Puddey IB, Mori TA, Beilin LJ, Croft KD (2004) Oxidative stress in human hypertension: association with antihypertensive treatment, gender, nutrition, and lifestyle. Free Radic Biol Med 36:226–232
255. Ide T, Tsutsui H, Ohashi N, Hayashidani S, Suematsu N, Tsuchihashi M, Tamai H, Takeshita A (2002) Greater oxidative stress in healthy young men compared with premenopausal women. Arterioscler Thromb Vasc Biol 22(3):438–442
256. Minuz P, Patrignani P, Gaino S, Seta F, Capone ML, Tacconelli S, Degan M, Faccini G, Fornasiero A, Talamini G, Tommasoli R, Arosio E, Santonastaso CL, Lechi A, Patrono C (2004) Determinants of platelet activation in human essential hypertension. Hypertension 43:64–70
257. Yasunari K, Maeda K, Nakamura M, Yoshikawa J (2002) Oxidative stress in leukocytes is a possible link between blood pressure, blood glucose, and C-reacting protein. Hypertension 39:777–780
258. Lacy F, Kailasam MT, O'Connor DT, Schmid-Schonbein GW, Parmer RJ (2000) Plasma hydrogen peroxide production in human essential hypertension: role of heredity, gender, and ethnicity. Hypertension 36(5):878–884

259. Lacy F, O'Connor DT, Schmid-Schönbein GW (1998) Plasma hydrogen peroxide production in hypertensives and normotensive subjects at genetic risk of hypertension. J Hypertens 16:291–303
260. Wang D, Strandgaard S, Iversen J, Wilcox CS (2009) Asymmetric dimethylarginine, oxidative stress, and vascular nitric oxide synthase in essential hypertension. Am J Physiol Regul Integr Comp Physiol 296(2):R195–R200
261. Touyz RM, Schiffrin EL (2001) Increased generation of superoxide by angiotensin II in smooth muscle cells from resistance arteries of hypertensive patients: role of phospholipase D-dependent NAD(P)H oxidase-sensitive pathways. J Hypertens 19(7):1245–1254
262. Touyz RM, Yao G, Quinn MT, Pagano PJ, Schiffrin EL (2005) p47phox associates with the cytoskeleton through cortactin in human vascular smooth muscle cells: role in NAD(P)H oxidase regulation by angiotensin II. Arterioscler Thromb Vasc Biol 25(3):512–518
263. Zalba G, San Jose G, Moreno MU, Fortuno A, Diez J (2005) NAD(P)H oxidase-mediated oxidative stress: genetic studies of the p22(phox) gene in hypertension. Antioxid Redox Signal 7(9–10):1327–1336
264. Moreno MU, Jose GS, Fortuno A, Beloqui O, Diez J, Zalba G (2006) The C242T CYBA polymorphism of NAD(P)H oxidase is associated with essential hypertension. J Hypertens 24(7):1299–1306
265. Genius J, Grau AJ, Lichy C (2008) The C242T polymorphism of the NAD(P)H oxidase p22phox subunit is associated with an enhanced risk for cerebrovascular disease at a young age. Cerebrovasc Dis 26(4):430–433
266. Kokubo Y, Iwai N, Tago N, Inamoto N, Okayama A, Yamawaki H, Naraba H, Tomoike H (2005) Association analysis between hypertension and CYBA, CLCNKB, and KCNMB1 functional polymorphisms in the Japanese population – the Suita Study. Circ J 69(2): 138–142
267. Pacher P, Nivorozhkin A, Szabó C (2006) Therapeutic effects of xanthine oxidase inhibitors: renaissance half a century after the discovery of allopurinol. Pharmacol Rev 58: 87–114
268. Saez GT, Tormos C, Giner V, Chaves J, Lozano JV, Iradi A, Redon J (2004) Factors related to the impact of antihypertensive treatment in antioxidant activities and oxidative stress by-products in human hypertension. Am J Hypertens 17(9):809–816
269. Simic DV, Mimic-Oka J, Pljesa-Ercegovac M, Savic-Radojcvic A, Opacic M, Matic D, Ivanovic B, Simic T (2006) Byproducts of oxidative protein damage and antioxidant enzyme activities in plasma of patients with different degrees of essential hypertension. J Hum Hypertens 20(2):149–155
270. Mullan BA, Young IS, Fee H, McCance DR (2002) Ascorbic acid reduces blood pressure and arterial stiffness in type 2 diabetes. Hypertension 40(6):804–809
271. Zureik M, Galan P, Bertrais S, Mennen L, Czernichow S, Blacher J, Ducimetière P, Hercberg S (2004) Effects of long-term daily low-dose supplementation with antioxidant vitamins and minerals on structure and function of large arteries. Arterioscler Thromb Vasc Biol 24(8):1485–1491
272. Caner M, Karter Y, Uzun H, Curgunlu A, Vehid S, Balci H, Yucel R, Güner I, Kutlu A, Yaldiran A, Oztürk E (2006) Oxidative stress in human sustained and white coat hypertension. Int J Clin Pract 60(12):1565–1571
273. Chen J, He J, Hamm L, Batuman V, Whelton PK (2002) Serum antioxidant vitamins and blood pressure in the United States population. Hypertension 40:810–816
274. Hasnain BI, Mooradian AD (2004) Recent trials of antioxidant therapy: what should we be telling our patients? Cleve Clin J Med 71:327–334
275. Jialal I, Devaraj S (2003) Antioxidants and atherosclerosis: don't throw out the baby with the bath water. Circulation 107:926–928
276. Bosch J, Lonn E, Pogue J, Arnold JM, Dagenais GR, Yusuf S (2005) Long-term effects of ramipril on cardiovascular events and on diabetes: results of the HOPE study extension. HOPE/HOPE-TOO Study Investigators. Circulation 112(9):1339–1346

277. Houston MC (2005) Nutraceuticals, vitamins, antioxidants, and minerals in the prevention and treatment of hypertension. Prog Cardiovasc Dis 47(6):396–449
278. Czernichow S, Bertrais S, Blacher J, Galan P, Briancon S, Favier A, Safar M, Hercberg S (2005) Effect of supplementation with antioxidants upon long-term risk of hypertension in the SU.VI.MAX study: association with plasma antioxidant levels. J Hypertens 23(11):2013–2018
279. Bates CJ, Walmsley CM, Prentice A, Finch S (1998) Does vitamin C reduce blood pressure? Results of a large study of people aged 65 or older. J Hypertens 16(7):925–932
280. Ward NC, Hodgson JM, Croft KD, Burke V, Beilin LJ, Puddey IB (2005) The combination of vitamin C and grape seed polyphenols increases blood pressure: a randomized, double-blind, placebo-controlled trial. J Hypertens 23(2):427–434
281. Barbagallo M, Dominguez LJ, Tagliamonte MR, Resnick LM, Paolisso G (1999) Effects of vitamin E and glutathione on glucose metabolism: role of magnesium. Hypertension 34(4 Pt 2):1002–1006
282. Mishra GD, Malik NS, Paul AA, Wadsworth ME, Bolton-Smith C (2003) Childhood and adult dietary vitamin E intake and cardiovascular risk factors in mid-life in the 1946 British Birth Cohort. Eur J Clin Nutr 57(11):1418–1425
283. Poston L, Raijmakers M, Kelly F (2004) Vitamin E in preeclampsia. Ann N Y Acad Sci 1031:242–248
284. Rumiris D, Purwosunu Y, Wibowo N, Farina A, Sekizawa A (2006) Lower rate of preeclampsia after antioxidant supplementation in pregnant women with low antioxidant status. Hypertens Pregnancy 25(3):241–253
285. Kelly RP, Poo Yeo K, Isaac HB, Lee CY, Huang SH, Teng L, Halliwell B, Wise SD (2008) Lack of effect of acute oral ingestion of vitamin C on oxidative stress, arterial stiffness or blood pressure in healthy subjects. Free Radic Res 42(5):514–522
286. Rumbold AR, Crowther CA, Haslam RR, Dekker GA, Robinson JS (2006) ACTS Study Group. Vitamins C and E and the risks of preeclampsia and perinatal complications. N Engl J Med 354(17):1796–1806
287. Beazley D, Ahokas R, Livingston J, Griggs M, Sibai BM (2005) Vitamin C and E supplementation in women at high risk for preeclampsia: a double-blind, placebo-controlled trial. Am J Obstet Gynecol 192(2):520–521
288. Skyrme-Jones RA, O'Brien RC, Berry KL, Meredith IT (2000) Vitamin E supplementation improves endothelial function in type I diabetes mellitus: a randomized, placebo-controlled study. J Am Coll Cardiol 36(1):94–102
289. Cai H, Griendling KK, Harrison DG (2003) The vascular NAD(P)H oxidases as therapeutic targets in cardiovascular diseases. Trends Pharmacol Sci 24:471–478
290. Wu R, Lamontagne D, de Champlain J (2002) Antioxidative properties of acetylsalicylic acid on vascular tissues from normotensive and spontaneously hypertensive rats. Circulation 105:387–392
291. Dulak J, Zagorska A, Wegiel B, Loboda A, Jozkowicz A (2006) New strategies for cardiovascular gene therapy: regulatable pre-emptive expression of pro-angiogenic and antioxidant genes. Cell Biochem Biophys 44(1):31–42
292. Cave A (2009) Selective targeting of NAD(P)H oxidase for cardiovascular protection. Curr Opin Pharmacol 9(2):208–213
293. Fang J, Seki T, Maeda H (2009) Therapeutic strategies by modulating oxygen stress in cancer and inflammation. Adv Drug Deliv Rev 61(4):290–302
294. Gupte SA (2008) Glucose-6-phosphate dehydrogenase: a novel therapeutic target in cardiovascular diseases. Curr Opin Investig Drugs 9(9):993–1000
295. Huang HY, Caballero B, Chang S, Alberg AJ, Semba RD, Schneyer CR, Wilson RF, Cheng TY, Vassy J, Prokopowicz G, Barnes GJ 2nd, Bass EB (2006) The efficacy and safety of multivitamin and mineral supplement use to prevent cancer and chronic disease in adults: a systematic review for a National Institutes of Health state-of-the-science conference. Ann Intern Med 145(5):372–385

296. Tribble DL (1999) Antioxidant consumption and risk of coronary heart disease: emphasis on vitamin C, vitamin E and β-carotene. A statement for the healthcare professionals from the American Heart Association. Circulation 99:591–595
297. Touyz RM, Campbell N, Logan A, Gledhill N, Petrella R, Padwal R (2004) Canadian Hypertension Education Program. The 2004 Canadian recommendations for the management of hypertension: Part III-Lifestyle modifications to prevent and control hypertension. Can J Cardiol 20:55–83
298. Lopes HF, Martin KL, Nashar K, Morrow JD, Goodfriend TL, Egan BM (2003) DASH diet lowers blood pressure and lipid-induced oxidative stress in obesity. Hypertension 41(3): 422–430
299. John JH, Ziebland S, Yudkin P, Roe LS, Neil HAW (2002) Effects of fruit and vegetable consumption on plasma antioxidant concentrations and blood pressure: a randomized controlled trial. Lancet 359:1969–1973
300. Wang JS, Lee T, Chow SE (2006) Role of exercise intensities in oxidized low-density lipoprotein-mediated redox status of monocyte in men. J Appl Physiol 101(3):740–744
301. Adams V, Linke A, Krankel N, Erbs S, Gielen S, Mobius-Winkler S, Gummert JF, Mohr FW, Schuler G, Hambrecht R (2005) Impact of regular physical activity on the NAD(P)H oxidase and angiotensin receptor system in patients with coronary artery disease. Circulation 111(5):555–562
302. Pan YX, Gao L, Wang WZ, Zheng H, Liu D, Patel KP, Zucker IH, Wang W (2007) Exercise training prevents arterial baroreflex dysfunction in rats treated with central angiotensin. Hypertension 49(3):519–527
303. Chen S, Ge Y, Si J, Rifai A, Dworkin LD, Gong R (2008) Candesartan suppresses chronic renal inflammation by a novel antioxidant action independent of AT1R blockade. Kidney Int 74(9):1128–1138
304. Oliveira PJ, Goncalves L, Monteiro P, Providencia LA, Moreno AJ (2005) Are the antioxidant properties of carvedilol important for the protection of cardiac mitochondria? Curr Vasc Pharmacol 3(2):147–158
305. Cifuentes ME, Pagano PJ (2006) Targeting reactive oxygen species in hypertension. Curr Opin Nephrol Hypertens 15(2):179–186
306. Berk BC (2007) Novel approaches to treat oxidative stress and cardiovascular diseases. Trans Am Clin Climatol Assoc 118:209–214
307. Sugiura T, Kondo T, Kureishi-Bando Y, Numaguchi Y, Yoshida O, Dohi Y, Kimura G, Ueda R, Rabelink TJ, Murohara T (2008) Nifedipine improves endothelial function: role of endothelial progenitor cells. Hypertension 52(3):491–498

# Chapter 16
# Peripartum Cardiomyopathy: Role of STAT-3 and Reactive Oxygen Species

**Denise Hilfiker-Kleiner, Arash Haghikia, and Andres Hilfiker**

**Abstract** Enhanced oxidative stress related to high metabolic turnover and elevated tissue oxygen requirements are the characteristic physiological state in pregnancy. In women with noneventful pregnancy and peripartum periods, this process appears to be paralleled by an increase in systemic antioxidant capacity. While these biochemical changes may not have pathophysiological consequences in healthy women, they may sensitize women with additional risk factors in late pregnancy and the early postpartum period to cardiovascular diseases such as preeclampsia and peripartum cardiomyopathy (PPCM). PPCM is a serious, potentially life-threatening heart disease of uncertain etiology in previously healthy women. Recent experimental findings associate unbalanced peripartum oxidative stress with the generation of a potent angiostatic, pro-apoptotic and proinflammatory factor, 16-kDa prolactin. Consistent with this notion, enhancing antioxidative capacity or pharmacological inhibition of prolactin secretion prevents PPCM in experimental models and seems to be promising in initial clinical approaches. Thus, unbalanced oxidative stress and high prolactin levels in combination seem to be key factors in PPCM and may therefore represent novel specific therapeutic targets to treat PPCM. The present article summarizes the current knowledge on peripartum oxidative stress mechanisms and associated cardiovascular disease forms and reports on potential pathomechanisms and novel treatment options for PPCM.

**Keywords** Peripartum cardiomyopathy · Preeclampsia · Oxidative stress · Prolactin · STAT3

---

D. Hilfiker-Kleiner (✉)
Department of Cardiology and Angiology, Department of Cardiac, Thoracic, Transplantation, and Vascular Surgery, Hannover Medical School, 30625 Hannover, Germany
e-mail: hilfiker.denise@mh-hannover.de

H. Sauer et al. (eds.), *Studies on Cardiovascular Disorders*,
Oxidative Stress in Applied Basic Research and Clinical Practice,
DOI 10.1007/978-1-60761-600-9_16, 

## 16.1 Introduction

Pregnancy is a physiological state associated with enhanced oxidative stress related to high metabolic turnover and elevated tissue oxygen requirements. The placenta has been identified as an important source of lipid peroxides because of its high polyunsaturated fatty acid content [1]. Levels of peroxidation markers, such as lipid hydroperoxide and malondialdehyde, are higher in pregnant than in nonpregnant women [2]. Lipid peroxidation is enhanced in the second trimester, tapers off later in gestation, and decreases after delivery. The placenta is also a source of antioxidative enzymes controlling placental lipid peroxidation during uncomplicated pregnancy. All the major antioxidative defense systems, including SOD, catalase, GPx, glutathione, vitamin C, and vitamin E, are found in the placenta and may suffice for control of lipid peroxidation in normal pregnancies [3, 4].

Peripartum cardiomyopathy (PPCM) is a rare but potentially life-threatening disorder of unknown etiology and pathophysiology. Because of its rare incidence, the geographical differences, and its heterogeneous presentation, PPCM continues to be incompletely characterized and understood. Diagnosis of PPCM is based on four primary diagnostic criteria, as outlined by the workshop recommendations of the National Heart Lung and Blood Institute and the Office of Rare Diseases [5]. These are: (A) development of the disease in the last month of pregnancy or within five months of delivery; (B) absence of an identifiable cause of heart failure; (C) absence of recognizable heart disease prior to the last month of pregnancy; and (D) LV systolic dysfunction demonstrated by classical echocardiographic criteria. At present, PPCM is listed as a form of dilated cardiomyopathy and is treated according to the guidelines for dilated cardiomyopathy with no other specific therapy [6]. The prognosis of affected women is poor, with reported mortality rates of 15% and full recovery in only 23% of PPCM patients, while continuous deterioration is reported in up to 50% of cases despite optimal medical treatment [6–11]. In the context of PPCM, risk factors such as age >30 years, preeclampsia, African origin, tocolytic therapy, and twin pregnancy are discussed but have not been confirmed in recent prospective studies [6].

Little is known about the pathophysiology of peripartum-induced cardiomyopathy. There have been speculations about the involvement of inflammation, myocarditis, autoimmune reactions, and apoptosis [10, 12–14]. More recently, a mouse model of PPCM has suggested an involvement of cardioprotective signaling pathways (i.e., signal transducer and activator of transcription-3 [STAT3] signaling), impaired oxidant defense, and subsequent enhanced oxidative stress in conjunction with an unfavorable cleavage of the nursing hormone prolactin into its detrimental 16 kD form [15].

The present article summarizes oxidative stress–related mechanisms in normal and disease states of pregnancy and postpartum, and highlights oxidative stress–mediated pathomechanisms and their potential influence on the development of PPCM, as well as potential novel treatment strategies.

## 16.2 Oxidative Stress and Antioxidative Defense During Pregnancy and Postpartum

### *16.2.1 Oxidative Stress Factors*

Lipid metabolism is altered during pregnancy and is characterized by normal or even low cholesterol during early pregnancy and hypertriglyceridaemia in late pregnancy [16]. It is assumed that the anabolic phase of early pregnancy produces metabolic changes that encourage lipogenesis and fat storage in preparation for the catabolic phase of late pregnancy, in which there is rapid fetal growth [17]. The insulin resistance of pregnancy increases lipolysis in adipose tissue, leading to an enhanced flux of fatty acids to the liver. This promotes the synthesis of very low density lipoproteins (VLDL) and, as a result, increased triglyceride concentrations. In addition, insulin resistance reduces the activity of lipoprotein lipase, an insulin-dependent enzyme that is responsible for VLDL clearance from plasma. Therefore, VLDL remains in the plasma longer and ultimately leads to accumulation of low-density lipoprotein (LDL) [17]. When LDL is oxidized (oxLDL), it produces endothelial dysfunction and inflammation, as is described in atherosclerotic lesions, thereby initiating vascular occlusion and endothelial dysfunction [18].

The placenta has been identified as an important source of lipid peroxides because of its high polyunsaturated fatty acid content [1]; and lipid hydroperoxides and malondialdehyde are higher in pregnant than in nonpregnant women [2]. Analysis of plasma lipid hydroperoxide (LHP) levels, as a direct marker for oxidative stress, showed no significant difference between LHP concentrations during the first trimester of pregnancy and nonpregnant healthy controls [16]. LHP significantly increased during the second trimester, but without exceeding the upper limit of controls [16]. In the third trimester, LHP concentrations increased further, to values well above the normal range and comparable to high-risk populations, such as diabetics with vascular disease [16, 19]. In the early postpartum period, LHP concentrations decreased substantially, but did not reach values similar to those observed in nonpregnant controls [16]. Thus, during pregnancy, there are marked changes in serum cholesterol, triglycerides, and LDL subfractions [16]. Such alterations are normally associated with an increased risk for coronary artery disease [20].

### *16.2.2 Antioxidant Capacity*

As outlined elsewhere in this book, antioxidants may be broadly classified into enzymatic (superoxide dismutases [SODs], catalase, glutathione peroxidase (GPx), and glutathione or their precursors); or nonenzymatic components (vitamins: A, E, C, co-enzyme Q, β-carotene; reducing agents: glutathione, cysteine, thioredoxin; binding proteins: albumin, ceruloplasmin, lactoferrin, transferrin; constituents of enzymes: uric acid, copper, zinc, selenium; and others: bilirubin, erythropoietin). The sum of these components has been determined as the total antioxidant capacity [1, 16]. Interestingly, the placenta has not only been determined as a source of lipid peroxides, but also as a source of antioxidative enzymes controlling placental lipid peroxidation [1, 16]. In fact, all the major antioxidative defense systems, including SOD, catalase, GPx, glutathione, vitamin C and vitamin E, are found in the placenta and may suffice for control of lipid peroxidation in normal pregnancies [1, 16].

During pregnancy there are substantial changes in the total antioxidant capacity of the circulation. It appears that the serum total antioxidant capacity is decreased in the first trimester of pregnancy, compared to nonpregnant controls [16]. During the second and third trimester of pregnancy the total antioxidant capacity increases but remains slightly below normal levels [16]. During the early postpartum period, the total antioxidant capacity increases further to values well within the normal range or even above that of healthy adults [16]. It is assumed that alterations in the total antioxidant capacity during pregnancy mainly reflect alterations in uric acid because, once this is removed, total antioxidant capacity does not appear to change. Uric acid concentrations are reduced in early pregnancy because of increased renal clearance, while the end of pregnancy is characterized by a significant increase in uric acid concentrations because of an increased rate of catabolism and a raised uric acid pool [21], suggesting an important antioxidant value of serum uric acid in late pregnancy. However, preeclampsia is strongly associated with hyperuricemia, and in some studies the increase in serum uric acid has been found to correlate with both maternal and fetal morbidity [22].

Other studies have shown an early fall in vitamin C, and that vitamin E increases progressively during normal pregnancy [1]. However, the results of two large randomized controlled trials evaluating the supplementation of pregnant women with high dosages of oral vitamin C and vitamin E for preventing preeclampsia revealed no significant differences between the vitamin and placebo groups for the occurrence of preeclampsia, death, or serious outcomes in the infant, or for having an infant with low birth weight [23, 24], pointing to a minor role for vitamins in total antioxidant capacity during pregnancy. In a case-control study, significantly lower levels of SODs and of GPx were found in placentas from preeclampsia patients than in control placentas, pointing to decreased enzymatic antioxidant capacity in the placental tissue of women suffering from preeclampsia [25]. Taken together, the total antioxidant capacity undergoes substantial changes during pregnancy, but its precise regulation, its source, and the role of different antioxidative systems are not fully understood.

### *16.2.3 Summary*

In women with noneventful pregnancy and peripartum periods, naturally increased oxidative stress appears to be paralleled by an increase in systemic antioxidant capacity to ensure that pregnancy-associated biochemical changes have no pathophysiological consequences. However, pregnancy-induced alterations of lipid metabolism (cholesterol, high-density lipoprotein (HDL)-cholesterol, triglycerides, and LDL) may reflect a particularly sensitive period in late pregnancy and early postpartum where additional oxidative stress–promoting factors, i.e., smoking, obesity, hypertension, or diabetes, may tip the balance towards a pathophysiological state. Such a scenario would be similar to atherosclerosis, where LDL is oxidized, thereby promoting endothelial inflammation, vascular occlusion, and endothelial dysfunction [16]. As a consequence, unbalanced oxidative stress could contribute to disorders during pregnancy, childbirth, and the postnatal period, such as preeclampsia and PPCM [15, 26].

## 16.3 Peripartum Cardiomyopathy (PPCM)

PPCM is a distinct entity of a dilated cardiomyopathy that occurs in women between one month antepartum and six months postdelivery [5, 6]. PPCM can be distinguished from other forms of postinfectious and idiopathic cardiomyopathies by virtue of the fact that it develops in the context of pregnancy relatively rapidly during the six-month period beginning in the late third trimester antepartum to five months postpartum in women without preexisting cardiac disease [6, 27]. The diagnosis of PPCM is based on diagnostic criteria outlined by the workshop recommendations of the National Heart Lung and Blood Institute and the Office of Rare Diseases [5], as discussed earlier.

The incidence of PPCM is largely unknown, and estimates vary among different geographic regions. The current roughly estimated incidence rate in western countries, largely based on retrospective analyses, is 1:3000–1:4000 [5]. Higher incidences of PPCM are reported for South Africa with 1:1000, for Haiti with 1:300, and in certain sub-Saharan zones with 1:100 pregnancies [6, 28, 29]. Yet no prospective data are available.

Because of its rare incidence, the geographical differences, and its heterogeneous presentation, the mechanisms leading to PPCM are unclear and the pathophysiology of PPCM continues to be incompletely characterized and understood. Nevertheless, a number of mechanisms have been proposed as potential contributing factors, including preeclampsia, nutritional deficiencies, genetic disorders, viral or autoimmune etiologies, hormonal problems, volume overload, alcohol, the physiologic stress of pregnancy, or the unmasking of latent idiopathic dilated cardiomyopathy [6].

## 16.4 Potential Risk Factors for PPCM

### *16.4.1 Infectious Agents*

Some authors suggest a potential role of infectious agents in PPCM because selected studies have found the presence of viral transcripts in cardiac tissues of patients with PPCM [14, 30]. A retrospective review of endomyocardial biopsy specimens from 34 PPCM patients showed a comparable incidence of myocarditis (8.8%) to that found in age- and sex-matched patients undergoing transplantation for idiopathic dilated cardiomyopathy (IDC: 9.1%) [31], indicating that the frequency of viral infections is not higher in PPCM then in IDC. Also, the presence of HIV infection seemed not to have an additional adverse effect on PPCM patients [29]. Interestingly, experimental data with encephalomyocarditis virus in mice suggested that viral infection increases the severity of myocardial damage in postpartum mice compared with nonpregnant control mice [32]. Thus, myocardial viral infections may not be a very common factor to trigger or drive PPCM, but the peripartum physiology may accelerate damage to the heart induced by some types of viruses.

### *16.4.2 Autoimmune Responses*

Autoimmune responses as potential risk factors for PPCM have also been discussed and are supported by experimental observations that serum derived from PPCM patients affects in vitro maturation of dendritic cells differently compared with serum from healthy postpartum women [33]. Whether these alterations are causally connected to PPCM remains to be defined. There may also be an increased risk for PPCM in patients with lupus; several case reports on such conditions have been published [34, 35].

### *16.4.3 Preeclampsia*

A history of preeclampsia during pregnancy appears frequently in reports of patients with PPCM [6, 36, 37]. Preeclampsia is a characteristic hypertensive disorder of human pregnancy and a leading cause of maternal and fetal mortality and morbidity worldwide. Preeclampsia and eclampsia occur in 6–8% of all pregnancies [38]. Although the progression of preeclampsia to eclampsia and HELLP syndrome (hemolysis, elevated liver enzymes, low platelet count) is potentially fatal, preeclampsia itself can be asymptomatic. Current research suggests a two-stage model of the pathophysiology of preeclampsia, with the first stage being marked by reduced placental perfusion, which then translates into the multisystemic maternal syndrome of preeclampsia [39]. The notion that reduced placental perfusion results in preeclampsia only in some women implies that the development of preeclampsia results from the interaction of pregnancy-specific physiological changes, e.g., metabolic alteration and increased inflammatory response, with maternal constitutional factors, such as obesity, diabetes, hypertension, hyperhomocysteinemia, and African origin. Obviously, these maternal factors predispose to cardiovascular disease postpartum and in later life as supported by follow-up studies [40]. In this regard, many patients with a postpartum cardiomyopathy have experienced preeclampsia during pregnancy [41], suggesting at least some common pathophysiological conditions between these two diseases.

As stated earlier, pregnancy is accompanied by substantial metabolic and physiological alterations, such as insulin resistance, hypertriglyceridemia [42], and enhanced immune function [43]. These alterations contribute to a decreased threshold for endothelial functional abnormality and sensitize the endothelium to insults, which is derailed by abnormal placental perfusion and maternal predisposing factors leading to preeclampsia-associated complications; whereas under normal conditions, after completion of pregnancy, these factors are resolved [44]. Most pathophysiological features of preeclampsia either contribute to the generation of oxidative stress or are stimulated by oxidative stress; some of these features are illustrated in the following paragraphs.

#### 16.4.3.1 Oxidative Modification of Lipids

In preeclampsia a dyslipidemia, already recognized in normal pregnancy, is more prominently present. This state is marked by reduced HDL, increased triglycerides, and very low LDL. Under conditions of enhanced oxidative stress, the formation of oxidized LDL (oxLDL) is accelerated [45]. This is evident for preeclamptic women with increased plasma and tissue concentrations of markers of oxidative stress, and elevated antibodies to oxLDL [46]. OxLDL in turn impairs local endothelial function and promotes the activation of selectins, resulting in augmented recruitment of monocytes to the endothelial surface. Involvement of oxidative stress in the genesis of preeclampsia-related endothelial dysfuntion would indicate that therapeutic reduction of oxidative stress by means of antioxidants could prevent or attenuate the maternal preeclampsia syndrome [39]. Indeed, one small trial evaluating the effect of antioxidant therapy with vitamins C and E showed promising results in terms of reducing the incidence of preeclampsia [47]. However, larger studies failed to provide similar results, indicating that vitamins are not efficient to serve for antioxidant therapy [23, 24].

#### 16.4.3.2 Activation of the Immune System by Oxidative Stress Mechanisms

Oxidized lipids produced in human placenta are potent activators of leucocytes, in particular of monocytes and neutrophils [48, 49]. In women with preeclampsia, placental production of oxidized lipids is significantly higher than in women with normal pregnancies. It is speculated that activation of neutrophils in preeclamptic women probably occurs as the neutrophils circulate through the intervillous space and are directly exposed to oxidized lipids released by the placenta. Indeed, leukocytes from preeclamptic patients release more reactive oxygen species [50]. Additionally, the placenta produces proinflammatory cytokines in response to hypoxia, activating monocytes and neutrophils. As the activated neutrophils return to the maternal circulation, they could relay the oxidative stress of the placenta to the maternal circulation by releasing toxic compounds, such as ROS. If the activated neutrophils were to adhere to the vascular endothelium, they could cause maternal vascular oxidative stress and inflammation. Activated monocytes move through the endothelium to release ROS, to uptake oxLDL, and to form foam cells, further contributing to endothelial injury. One of the consequences of endothelial cell oxidation is that the integrity of the endothelium is compromised, allowing proteins to leak out of the circulation. This event can result in edema in the maternal systemic circulation and proteinuria in the kidney. Endothelial oxidation could, therefore, explain edema and proteinuria, two of the major clinical symptoms of preeclampsia (reviewed by Wash et al. [26]).

#### 16.4.3.3 Asymmetric Dimethylarginine (ADMA)

Besides oxLDL, urinary prolactin and asymmetric dimethylarginine (ADMA) levels have been mentioned as potential prognostic markers for the outcome of preeclampsia [51]. ADMA as an endogenous inhibitor of nitric oxide synthases (NOS) is involved in the regulation of the cellular redox state [52] and has aroused interest in pregnancy-related disease research. The accumulation of cytosolic ADMA depends on the rate of protein turnover when methylated arginine residues are released upon protein degradation. ADMA is mainly metabolized by the catalytic activity of dimethylarginine demethylaminohydrolase (DDAH), rather than excreted. In normal pregnancy ADMA levels have been demonstrated to fall, while women with preeclampsia reveal increased ADMA levels [53]. Furthermore, a clear correlation between increased ADMA levels and endothelial dysfunction has been shown only for women with high ADMA levels in the early phase of pregnancy, who subsequently suffered from preeclampsia; whereas this correlation was absent in women who were devoid of endothelial dysfunction [54]. This indicates that ADMA-associated cardiovascular complications of pregnancy are linked to increased susceptibility of ADMA-induced effects on the vasculature. For women who develop PPCM, correlations to general risk factors of cardiovascular events have been demonstrated, e.g., hypertension, hypercholesterolemia, and diabetes [6, 37]—all cardiovascular risk states in which increased ADMA has been detected [55]. However, the relevance of ADMA for the pathogenesis and clinical course of PPCM awaits future investigations.

All the pathophysiological features of preeclampsia listed above contribute to endothelial dysfunction and subsequent reduced maternal systemic organ perfusion [56]. Additional factors, including soluble fms-like tyrosine kinase-1 [57], angiotensin II type 1 receptor autoantibodies, and cytokines such as tumor necrosis factor-alpha, which generate widespread dysfunction of the maternal vascular endothelium, are discussed as contributors to preeclampsia and to PPCM [6, 58]. In preeclampsia and PPCM, blood flow can be further compromised by activation of the coagulation cascade and the formation of microthrombi [44]. Profoundly reduced perfusion causes glomerular and mesangial structural changes that ultimately lead to impairment of the glomerular ultrafiltration capacity, thereby explaining edema and proteinuria, two of the major clinical symptoms of preeclampsia [39] and PPCM [6].

A recent study described a correlation between urinary prolactin, its cleaved 16 kDa derivative, the disease severity, and the occurrence of adverse outcomes in patients with preeclampsia [51]. In this regard, the authors suggest that cleavage of circulating prolactin by cathepsin-D is rather a local process, and thus the antiangiogenic effects of 16 kDa prolactin fragment are exerted directly on the glomerular endothelium, contributing to deranged ultrafiltration properties. This feature points to further similarities between preeclampsia and PPCM, since elevated serum prolactin and cathepsin-D–mediated cleavage of prolactin in its 16 kDa form were also described in patients with PPCM [15]. Thus it is tempting to speculate about similar pathomechanisms in these two diseases, suggesting that findings from one disease may provide insights into the other, and vice versa.

## 16.5 Mechanistic Insights into the Pathophysiology of Peripartum Cardiomyopathy

### *16.5.1 The Estrogen-PI3-Akt Connection*

During pregnancy, the heart undergoes homeostatically regulated remodeling, including hypertrophy paralleled by a proportional growth of the capillary network without cardiac fibrosis and changes in classical markers of pathological hypertrophy (e.g., myosin heavy chains [alpha and beta], atrial natriuretic peptide, phospholamban, and sarcoplasmic reticulum $Ca^{2+}$-ATPase) to accommodate increased pregnancy-related hemodynamic volume overload and to maintain normal maternal-fetal health [15, 59, 60]. Plasma estrogen levels are known to be elevated during pregnancy with a sharp decline postpartum, and estrogen promotes activation of cardioprotective Akt signaling in cardiomyocytes [61]. Increased serum estrogen levels in late pregnancy seem to induce stretch-activated c-Src-kinase (c-Src), and subsequently Akt signaling in the maternal heart [15, 59, 60]. Since estrogen promotes the activation of cardioprotective c-Src-Akt signaling in cardiomyocytes [61], it is conceivable that estrogen also promotes cardioprotection during pregnancy. The delivery of the placenta results in a sudden drop in estrogen, which is associated with a decrease in cardiac Akt signaling in postpartum mice [15]. Estrogen-mediated cardioprotection during pregnancy may explain why the maternal heart seems to be less sensitive to pathological effects during preeclampsia and why PPCM patients, who after an episode of PPCM become pregnant again, tolerate pregnancy quite well but show a severe recurrence of cardiac failure after delivery [62].

### *16.5.2 STAT3, the Guardian of Postpartum Hearts*

We recently reported that mice with a cardiomyocyte-specific deletion of signal transducer and activator of transcription-3 (STAT3-KO) develop a cardiomyopathy phenotype quite similar to that observed in PPCM patients [15]. STAT3-KO mice show normal pregnancy-mediated Akt activation, hypertrophy, and vessel growth, and never develop symptoms during pregnancy, when hemodynamic load culminates [15]. However, STAT3-KO mice invariably develop a PPCM with systolic dysfunction and a high mortality rate after giving birth to their litters [15]. Prominent features of postpartum STAT3-KO hearts were a rapid loss of myocardial capillaries, increased apoptosis, extensive fibrosis, and ventricular dilatation, implicating an important role for STAT3 in postpartum cardioprotection [15]. Indeed, myocardial activation of STAT3 was noted in wild-type but not in STAT3-KO mice late in pregnancy and postpartum [15]. The nursing hormone prolactin is known to activate STAT3 via its specific receptors (the short and long forms of prolactin receptor) in various cell types, including cardiomyocytes in vitro and the heart in vivo [15, 63]. Therefore, prolactin might at least in part be responsible for postpartum activation of cardioprotective STAT3 signaling.

Thus, different signaling pathways seem to be required for protection of the maternal heart during different phases of reproduction: c-Src-Akt is likely to exert protection against pregnancy-mediated stress, while STAT3 appears necessary to protect from postpartum-mediated stress.

### *16.5.3 STAT3 and Antioxidant Pathways in the Postpartum Heart: An Important Role for MnSOD*

It has been shown that STAT3 mediates protection from oxidative stress in the heart and in cardiomyocytes in part by upregulating antioxidant enzymes such as manganese sodium dismutase (MnSOD), a powerful ROS scavenging enzyme located in mitochondria [64]. Indeed, we observed an upregulation of MnSOD in postpartum hearts from wild-type, but not from STAT3-KO mice [15], indicating that STAT3 promotes cardiac MnSOD expression postpartum. In line with a lower antioxidative defense, enhanced levels of reactive oxygen species (ROS) were noted in postpartum STAT3-KO hearts [15]. Moreover, while reduction of MnSOD protein levels are not sufficient to induce cardiomyopathy and heart failure in nonpregnant mice [65], the addition of pregnancy/postpartum stress to MnSOD heterozygous females resulted in severe nonreversible hypertrophic cardiomyopathy, implying that a reduction by 50% of this protein is sufficient to impair postpartum cardioprotection [15]. Further evidence for an important role of MnSOD in postpartum protection derives from experiments with tetrakis (4-benzoic acid) porphyrin (MnTBAP), one of the so-called MnSOD mimetics, that has catalytic activities similar to MnSOD, and acts as a powerful pharmacological suppressor of ROS [66]. Treatment with MnTBAP attenuated ROS generation, preserved cardiac function, and prevented postpartum-related mortality in STAT3-KO female mice, but had no effect on left ventricular dilation [15].

Thus, STAT3 via MnSOD plays an important role for the antioxidant defense in the postpartum heart [15]. The observation that mice with genetically reduced MnSOD protein levels do not develop the typical dilated cardiac phenotype of PPCM but rather a hypertrophic cardiomyopathy, together with the finding that

MnTBAP mediated only a partial rescue from PPCM in STAT3-KO mice, suggest that MnSOD, even though important, is not the only antioxidant factor needed for cardioprotection postpartum.

### *16.5.4 Oxidative Stress and High Prolactin Levels: A Detrimental Combination*

As mentioned previously, STAT3-KO mice display a profound loss of capillaries and a rapid dilatation of all ventricles in the early postpartum phase [15]. Indeed there is a connection between the nursing hormone prolactin, the loss in cardiac capillaries, adverse left ventricular remodeling, and cardiac failure in STAT3-KO mice, because the PPCM phenotype in terms of adverse remodeling, cardiac function, and postpartum mortality was completely prevented by pharmacological blockade of prolactin with bromocriptine [15]. Bromocriptine is a dopamine-D2-receptor agonist, known to block prolactin release from the piturary gland efficiently in humans [67] and mice [68]. Interestingly, bromocriptine did not affect increased ROS production in the immediate postpartum phase in STAT3-KO mice, supporting the notion that enhanced oxidative stress alone is not triggering PPCM [15].

Prolactin has been hypothesized as a potential factor in the pathogenesis of PPCM previously [69]. Interestingly, recent work showed that prolactin is a hormone that can either stimulate or inhibit various stages of vessel formation and remodeling. This potential to exert opposing effects on angiogenesis resides in the proteolytic processing of the proangiogenic full-length 23-kDa prolactin by the protease cathepsin D or by various metalloproteinases (MMP) into an antiangiogenic 16-kDa form, which is known to induce endothelial cell dissociation and apoptosis [70–72].

Oxidative stress, which is clearly increased in STAT3-KO females, is a potent stimulus for the activation of cathepsin D, because it triggers its release from lysosomes in cardiomyocytes [70, 73]. In fact, increasing systemic oxidative stress, for example, by a single injection of the anthracycline doxorubicin [74], is sufficient to increase the expression and activation of cathepsin D in many organs, including the heart [15]. While enhanced oxidative stress and activated cathepsin D after a single low dose of doxorubicin infusion in nonpregnant mice had no adverse effects, the addition of high levels of circulating prolactin to this setting provoked a high mortality rate because of multiorgan failure in these mice, further confirming the detrimental effects of the combination of oxidative stress, cathepsin D, and prolactin [15].

Oxidative stress also promotes the activation of MMP-2 [75], another enzyme able to generate the 16-kDa form from the 23-kDa prolactin. While cathepsin D works best under acidic conditions [70], we showed that active cathepsin D can be released from cardiomyocytes into the cell culture supernatant in vitro, where it is able to generate 16-kDa prolactin from recombinant 23-kDa prolactin even under physiological conditions [15]. Furthermore, we presented evidence that prolactin is processed in its 16-kDa form in postpartum STAT3-KO hearts [15].

Prolactin production is not restricted to the pituitary gland. In fact, various cell types, including fibroblasts, are able to produce prolactin [68]. Since PPCM is often associated with a high degree of cardiac fibrosis [15, 76, 77], locally produced prolactin may, in addition to circulating pituitary prolactin, contribute to the generation of 16-kDa prolactin. Locally produced 16-kDa prolactin may enhance cardiac damage even when serum prolactin is already diminished. Interestingly, inhibitors of prolactin release, such as bromocriptine, block prolactin secretion by fibroblasts and decrease at the same time the release of prolactin-cleaving MMPs from these cells [68], suggesting that bromocriptine may exert more direct cardiac protection by interfering with cardiac prolactin metabolism in PPCM.

Thus, the coincidence of unbalanced oxidative stress, prolactin-cleaving enzymes (cathepsin D and/or MMPs), and high prolactin levels (piturary and cardiac) appears to be causative for PPCM in STAT3-KO mice.

### *16.5.5 Impact of the 16-kDa Prolactin on the Cardiovascular System*

From the physiological point of view, it is unlikely that the full-length 23-kDa prolactin, which induces lactation and activates cardioprotective STAT3 signaling, is responsible for PPCM. Indeed, systemic infusion of 23-kDa prolactin in wild-type and STAT3-KO mice had no adverse effects on the heart [15]; and patients with prolactinomas who experience high prolactin serum levels are not known for a high incidence of heart failure. In contrast, high expression of 16-kDa prolactin, even in the absence of the postpartum physiology, destroyed the cardiac microvasculature, lowered cardiac function, and promoted ventricular dilatation. Furthermore, it affected cardiomyocyte metabolism and contractility in vitro [15]. The detrimental effect of 16-kDa prolactin on the cardiac microvasculature is consistent with recent observations in tumor biology, where 16-kDa prolactin induces apoptosis and dissociation of endothelial cells and prevents their proliferation and migration [71, 78]. Moreover, 16-kDa prolactin promotes vasoconstriction [79]. Interestingly, 16-kDa prolactin does not act via the known prolactin receptors [80].

There might be a connection between IFN-gamma, prolactin, and chronic inflammation, since its upregulation correlates with oxLDL and prolactin during the progression of PPCM [81]. IFN-gamma is an important mediator of inflammation and innate immune response, and 16-kDa prolactin strongly enhances adhesion of inflammatory cells to the endothelium [80]. Furthermore, 16-kDa prolactin stimulates the expression of IFN-gamma–responsive genes such as interferon-stimulated protein (28 and 15 kDa) and interferon-responsive factor [80, 82]. But also the full-length prolactin may promote proinflammatory immune responses, since it causes an increase in the binding activity of the intracellular transcription factors nuclear factor-kappaB (NFkappaB) and interferon regulatory factor-1 (IRF-1), which are known to promote secretion of proinflammatory cytokines such as TNF-$\alpha$ and IL-12 [83]. Vice versa, inflammatory cytokines could promote a “prolactin-cytokine positive feedback loop" by stimulating the release of pituitary prolactin [84]. Thus, in the early stage of PPCM, prolactin (mainly its 16 kDa form) may induce a strong inflammatory reaction, preferentially in the endothelium, by upregulating IFN-gamma and related proinflammatory signaling pathways. Accelerated inflammation may in turn lead into a positive prolactin-cytokine feedback loop, further increasing oxidative stress, subsequent prolactin cleavage, and cardiovascular damage.

This collection of adverse effects derived from prolactin (and mainly its cleaved 16-kDa form) on the cardiovascular system suggests that oxidative stress and 16-kDa prolactin are key factors in the pathophysiology of PPCM.

## 16.6 How Relevant is the STAT3–Oxidative Stress–Prolactin Hypothesis for Human PPCM?

### *16.6.1 Gene Polymorphisms and Dysregulation of STAT3 Signaling Pathways in Human PPCM*

The higher incidence of PPCM in certain geographic areas, i.e., the sub-Saharan region of Africa, South Africa, and Haiti, emphasize the involvement of genetic factors [6]. This feature is further supported by reports of PPCM in a mother and her daughter by J. Fett in Haiti [85], and our personal observation of PPCM in sisters in South Africa and in Germany [37]. However, so far, no gene polymorphism has been associated with an increased risk for PPCM.

Polymorphisms in the STAT3 gene have been associated with cardiovascular diseases in dialysis patients [86] and with differences in responses to IFN-alpha therapy [87]. Various SNPs were detected in the coding region of the STAT3 gene PPCM patients from South Africa, but none has been associated with a higher risk for PPCM so far [15].

While the STAT3-KO mouse model developed PPCM because of the genetic deletion of STAT3 in cardiomyocytes, it is conceivable that additional genes, either upstream or downstream of STAT3, might be affected. In fact, various polymorphisms have been described for JAK2, the major upstream protein of STAT3. However, no associations of common SNPs or the JAK2 V617F mutation have been reported for pregnancy-associated disease yet [88].

In line with a potential downregulation of cardiac STAT3 expression in PPCM by a still unknown mechanism, STAT3 protein expression is largely decreased in end-stage failing hearts from patients with PPCM [15]. However, similar observations were made in end-stage failing hearts from patients with other types of heart disease [89]. Therefore, downregulation of cardiac STAT3 expression may not be specifically related to PPCM, but may rather be secondary to heart failure in PPCM patients.

Gene polymorphisms have also been described for MnSOD [90], a downstream target of STAT3 for oxidative protection, but no association has been described yet for pregnancy-associated heart disease or PPCM.

Thus, no direct evidence for genetic alterations within the JAK/STAT signaling cascade or MnSOD exists so far in association with PPCM; and future studies are needed to evaluate potential genetic risk factors for this disease.

## *16.6.2 Evidence for the Oxidative Stress–Prolactin Hypothesis in Human PPCM*

### 16.6.2.1 Oxidative Stress and Inflammation

In a recent study it was observed that patients with acute onset of PPCM displayed significantly higher baseline levels of oxLDL (indicative for increased oxidative stress) than normal age- and pregnancy-matched women [15, 81]. Along the same lines, preeclampsia, a condition during pregnancy associated with higher oxidative stress as outlined above, is frequently reported in patients who develop PPCM in the peripartum period [91]. A subsequent analysis showed that high serum levels of oxLDL are not only present in patients with acute PPCM, but are also persistently high in patients unable to recover from the disease [15, 81]. Furthermore, persistently high oxLDL levels positively correlate with high serum levels of interferon-$\gamma$ (IFN$\gamma$) in PPCM patients who did not recover from PPCM [81]. Thus, these observations emphasize the major pathophysiological role of enhanced oxidative stress in PPCM and suggest that oxidative stress and inflammation may be interconnected in the initiation and during progression of PPCM.

### 16.6.2.2 Cathepsin D, Prolactin Cleavage, and Bromocriptine

A unique aspect of pregnancy, labor, and birth is profound hormonal change. In this regard, prolactin, a dominant hormone during pregnancy and early postpartum, has been hypothesized as a potential factor in the pathogenesis of PPCM [92]. Interestingly, baseline serum prolactin levels are significantly higher among PPCM patients compared with postpartum controls [81]. Furthermore, prolactin levels decrease significantly during recovery in PPCM patients, while no significant decrease was observed in patients who were unable to recover from PPCM [81]. While it is unlikely that the uncleaved 23-kDa nursing hormone alone is harmful in PPCM patients, there is evidence for enhanced prolactin cleavage in PPCM patients compared to healthy nursing women. Indeed, higher levels of activated cathepsin D together with higher levels of the angiostatic and proapoptotic 16-kDa prolactin were found in sera from PPCM patients compared with pregnancy-matched healthy controls [15, 81]. These observations strongly suggest the presence of a systemically activated oxidative stress–cathepsin D-16-kDa prolactin cascade in human PPCM [15]. It is therefore likely that activation of this cascade is a key feature of PPCM in humans. This notion is further supported by observations from small pilot studies and healing attempts in which prolactin was pharmacologically blocked with bromocriptine in PPCM patients. In this regard, patients who had suffered from PPCM in a previous pregnancy and presented with a subsequent pregnancy, are at a high risk for developing the disease again [62]. Six patients with subsequent pregnancies obtained bromocriptine in addition to standard therapy for heart failure, and all of them had an uneventful postpregnancy follow-up. In contrast, six patients with similar conditions who obtained only standard therapy for heart failure suffered from recurrence of PPCM, three patients died subsequently, and the surviving three patients remained in heart failure [15]. Thus, it seems that the preventive effect of bromocriptine in patients with a high risk to develop the disease can be recapitulated. Meanwhile, there are also some case reports on recovery from acute PPCM after the addition of bromocriptine to the standard therapy of heart failure without any further complications [36, 93, 94], suggesting that prolactin blockade by bromocriptine may be efficient in acute PPCM. Controlled and randomized studies are awaited to prove this promising novel PPCM-specific therapy approach.

#### 16.6.2.3 16-kDa Prolactin in Prepartum Cardiovascular Disease

As outlined in previous chapters, preeclampsia appears to be a potential risk factor for the development of PPCM. A recent study found that the presence of prolactin and its angiostatic 16-kDa form in urinary samples of pregnant women were more frequently detected in women with severe forms of preeclampsia, eclampsia, and HELLP syndrome, and were also frequently found in women who developed placental abruption, acute renal failure, or pulmonary edema [51]. This observation extends the potential detrimental roles of the angiostatic 16-kDa to the prepartum phase. Moreover, it points to a potential value of 16-kDa prolactin as a prognostic marker in pregnancy and postpartum for cardiovascular complications.

#### 16.6.2.4 Summary

Taken together, there is strong evidence for the presence of 16-kDa prolactin in patients with PPCM and in patients with severe forms of preeclampsia, supporting the notion that an oxidative stress–cathepsin D-16-kDa prolactin cascade could be a central pathophysiological process in PPCM and severe preeclampsia. It has to be noted that the patient numbers are too small to be conclusive at this time, especially since spontaneous recovery from PPCM is reported in 25–30% of patients [6]. Randomized and controlled studies are currently being performed in South Africa and in Europe to test the efficacy of bromocriptine in the treatment of PPCM patients.

### *16.6.3 Prolactin, Bromocriptine, and the Risk for Thrombosis*

Concerning the safety of bromocriptine in pregnancy, a survey of more than 1400 pregnant women who took bromocriptine primarily during the first few weeks of pregnancy found no evidence of increased rates of abortion or congenital malformations [95].

However, in early postpartum women, there are some case reports on myocardial infarction, which occurred in association with taking bromocriptine [96]. It should be noted that there is in general an increased risk for myocardial infarction in peripartum women because of changes in coagulation activity in the maternal blood characterized by elevation of factors VII, X, VIII, fibrinogen, and von Willebrand factor, which is maximal around term [97]. Indeed, the risk for thrombotic complications and aortic dissection is increased in postpartum women independently from the use of bromocriptine [98–100]. This feature may have evolved to protect women from the bleeding challenges of miscarriage and childbirth. Thrombin, a central protease in the coagulation cascade, can generate a C-terminal 16-kDa fragment of human prolactin at a physiological pH that is not angiostatic and retains little mitogenic activity [101]. Accordingly, prolactin may modulate the availability of thrombin in the coagulation cascade. As a consequence, bromocriptine may only be used in conjugation with anticoagulation therapy such as low-molecular heparin, a substance that is given to patients with heart failure anyway. An additional interesting feature in terms of heparin therapy in PPCM patients comes from a report showing that full-length prolactin and its cleaved angiostatic 16-kDa fragment are bound by heparin [102], suggesting that heparin may lead to the depletion of both prolactin forms from the circulation.

Thus, interfering with the prolactin system may indeed alter coagulation activity in postpartum women. Therefore, we recommend that bromocriptine treatment should always be conducted with anticoagulation, i.e., heparin, to keep coagulation under control at the same time.

## 16.7 Summary and Conclusions

In summary, it is likely that multiple independent factors may trigger PPCM, but it appears that factors associated with increased oxidative stress are quite likely to play a central role for initiation and progression of PPCM. In this regard, we postulate that: (1) a powerful antioxidant defense is needed to prevent pathophysiological processes in pregnancy and postpartum; (2) the cardiovascular system is especially vulnerable to unbalanced oxidative stress during pregnancy and postpartum; and (3) that oxidative stress may be the common intersecting pathway leading to clinical manifestation of preeclampsia and PPCM.

With the recent discovery of an oxidative stress–cathepsin D-16-kDa prolactin cascade in experimental and human PPCM, a specific pathophysiological mechanism for PPCM has emerged which may provide a rational basis for a specific therapeutic intervention. Bromocriptine, a drug blocking the release of prolactin systemically and locally, which has been used for many years in women to stop lactation, should now be tested in randomized trials for its efficacy in the treatment of acute PPCM. Moreover, systematic collection of data prospectively is required, as well as international cardiac registries to study the etiology and different pathogenic mechanisms of PPCM, including potential genetic and lifestyle aspects.

**Acknowledgments** The original work reported here was supported by the Deutsche Forschungsgemeinschaft and the Jean Leducq Foundation.

## References

1. Shoji H, Koletzko B (2007) Oxidative stress and antioxidant protection in the perinatal period. Curr Opin Clin Nutr Metab Care 10:324–328
2. Morris JM, Gopaul NK, Endresen MJ et al (1998) Circulating markers of oxidative stress are raised in normal pregnancy and pre-eclampsia. Br J Obstet Gynaecol 105:1195–1199
3. Gitto E, Reiter RJ, Karbownik M et al (2002) Causes of oxidative stress in the pre- and perinatal period. Biol Neonate 81:146–157
4. Mueller A, Koebnick C, Binder H et al (2005) Placental defence is considered sufficient to control lipid peroxidation in pregnancy. Med Hypotheses 64:553–557
5. Pearson GD, Veille JC, Rahimtoola S et al (2000) Peripartum cardiomyopathy: National Heart, Lung, and Blood Institute and Office of Rare Diseases (National Institutes of Health) workshop recommendations and review. J Am Med Assoc 283:1183–1188
6. Sliwa K, Fett J, Elkayam U (2006) Peripartum cardiomyopathy. Lancet 368:687–693
7. Brar SS, Khan SS, Sandhu GK et al (2007) Incidence, mortality, and racial differences in peripartum cardiomyopathy. Am J Cardiol 100:302–304
8. Elkayam U, Akhter MW, Singh H et al (2005) Pregnancy-associated cardiomyopathy: clinical characteristics and a comparison between early and late presentation. Circulation 111:2050–2055
9. Reimold SC, Rutherford JD (2001) Peripartum cardiomyopathy. N Engl J Med 344: 1629–1630
10. Sliwa K, Skudicky D, Bergemann A et al (2000) Peripartum cardiomyopathy: analysis of clinical outcome, left ventricular function, plasma levels of cytokines and Fas/APO-1. J Am Coll Cardiol 35:701–705
11. Sliwa K, Skudicky D, Candy G et al (2002) The addition of pentoxifylline to conventional therapy improves outcome in patients with peripartum cardiomyopathy. Eur J Heart Fail 4:305–309
12. Hayakawa Y, Chandra M, Miao W et al (2003) Inhibition of cardiac myocyte apoptosis improves cardiac function and abolishes mortality in the peripartum cardiomyopathy of Galpha(q) transgenic mice. Circulation 108:3036–3041
13. Sliwa K, Forster O, Libhaber E et al (2006) Peripartum cardiomyopathy: inflammatory markers as predictors of outcome in 100 prospectively studied patients. Eur Heart J 27:441–446
14. Fett JD (2007) Viral infection as a possible trigger for the development of peripartum cardiomyopathy. Int J Gynaecol Obstet 97:149–150
15. Hilfiker-Kleiner D, Kaminski K, Podewski E et al (2007) A cathepsin D-cleaved 16 kDa form of prolactin mediates postpartum cardiomyopathy. Cell 128:589–600
16. Toescu V, Nuttall SL, Martin U et al (2002) Oxidative stress and normal pregnancy. Clin Endocrinol (Oxf) 57:609–613
17. Kaaja R (1998) Insulin resistance syndrome in preeclampsia. Semin Reprod Endocrinol 16:41–46
18. Steinberg D, Carew TE, Fielding C et al (1989) Lipoproteins and the pathogenesis of atherosclerosis. Circulation 80:719–723
19. Nuttall SL, Dunne F, Kendall MJ et al (1999) Age-independent oxidative stress in elderly patients with non-insulin-dependent diabetes mellitus. QJM 92:33–38
20. Belo L, Santos-Silva A, Quintanilha A et al (2008) Similarities between pre-eclampsia and atherosclerosis: a protective effect of physical exercise? Curr Med Chem 15:2223–2229
21. Merviel P, Ba R, Beaufils M et al (1998) Lone hyperuricemia during pregnancy: maternal and fetal outcomes. Eur J Obstet Gynecol Reprod Biol 77:145–150

22. Kang DH, Finch J, Nakagawa T et al (2004) Uric acid, endothelial dysfunction and pre-eclampsia: searching for a pathogenetic link. J Hypertens 22:229–235
23. Rumbold AR, Crowther CA, Haslam RR et al (2006) Vitamins C and E and the risks of preeclampsia and perinatal complications. N Engl J Med 354:1796–1806
24. Poston L, Briley AL, Seed PT et al (2006) Vitamin C and vitamin E in pregnant women at risk for pre-eclampsia (VIP trial): randomised placebo-controlled trial. Lancet 367: 1145–1154
25. Vanderlelie J, Venardos K, Clifton VL et al (2005) Increased biological oxidation and reduced anti-oxidant enzyme activity in pre-eclamptic placentae. Placenta 26:53–58
26. Walsh SW (2007) Obesity: a risk factor for preeclampsia. Trends Endocrinol Metab 18: 365–370
27. Fett JD (2002) Peripartum cardiomyopathy. Insights from Haiti regarding a disease of unknown etiology. Minn Med 85:46–48
28. Fett JD, Christie LG, Carraway RD et al (2005) Five-year prospective study of the incidence and prognosis of peripartum cardiomyopathy at a single institution. Mayo Clin Proc 80:1602–1606
29. Mayosi BM (2007) Contemporary trends in the epidemiology and management of cardiomyopathy and pericarditis in sub-Saharan Africa. Heart 93:1176–1183
30. Cenac A, Djibo A, Djangnikpo L (1993) Peripartum dilated cardiomyopathy. A model of multifactor disease? Rev Med Interne 14:1033
31. Rizeq MN, Rickenbacher PR, Fowler MB et al (1994) Incidence of myocarditis in peripartum cardiomyopathy. Am J Cardiol 74:474–477
32. Yokoyama T, Kanda T, Suzuki T et al (1993) Enhancement of myocardial damage and alteration of lymphocyte subsets in murine model of postpartum myocarditis. Am J Cardiovasc Pathol 4:343–351
33. Ellis JE, Ansari AA, Fett JD et al (2005) Inhibition of progenitor dendritic cell maturation by plasma from patients with peripartum cardiomyopathy: role in pregnancy-associated heart disease. Clin Dev Immunol 12:265–273
34. Raskin RJ, Haddock JB, Lawless OJ (1983) Systemic lupus erythematosus with myocarditis complicating pregnancy. South Med J 76:258–260
35. Seo A, Sakamoto H, Tanaka Y et al (2006) Peripartum cardiomyopathy with antiphospholipid antibody: a case report. J Cardiol 47:261–266
36. Hilfiker-Kleiner D, Meyer GP, Schieffer E et al (2007) Recovery from postpartum cardiomyopathy in 2 patients by blocking prolactin release with bromocriptine. J Am Coll Cardiol 50:2354–2355
37. Hilfiker-Kleiner D, Sliwa K, Drexler H (2008) Peripartum cardiomyopathy: recent insights in its pathophysiology. Trends Cardiovasc Med 18:173–179
38. Munjuluri N, Lipman M, Valentine A et al (2005) Postpartum eclampsia of late onset. BMJ 331:1070–1071
39. Roberts JM, Gammill HS (2005) Preeclampsia: recent insights. Hypertension 46:1243–1249
40. Sibai BM, El-Nazer A, Gonzalez-Ruiz A (1986) Severe preeclampsia-eclampsia in young primigravid women: subsequent pregnancy outcome and remote prognosis. Am J Obstet Gynecol 155:1011–1016
41. Mehta NJ, Mehta RN, Khan IA (2001) Peripartum cardiomyopathy: clinical and therapeutic aspects. Angiology 52:759–762
42. Hubel CA, McLaughlin MK, Evans RW et al (1996) Fasting serum triglycerides, free fatty acids, and malondialdehyde are increased in preeclampsia, are positively correlated, and decrease within 48 hours post partum. Am J Obstet Gynecol 174:975–982
43. Sacks GP, Studena K, Sargent K et al (1998) Normal pregnancy and preeclampsia both produce inflammatory changes in peripheral blood leukocytes akin to those of sepsis. Am J Obstet Gynecol 179:80–86
44. Roberts JM, Lain KY (2002) Recent insights into the pathogenesis of pre-eclampsia. Placenta 23:359–372

45. Steinberg D (1995) Role of oxidized LDL and antioxidants in atherosclerosis. Adv Exp Med Biol 369:39–48
46. Uotila J, Solakivi T, Jaakkola O et al (1998) Antibodies against copper-oxidised and malondialdehyde-modified low density lipoproteins in pre-eclampsia pregnancies. Br J Obstet Gynaecol 105:1113–1117
47. Chappell LC, Seed PT, Briley AL et al (1999) Effect of antioxidants on the occurrence of pre-eclampsia in women at increased risk: a randomised trial. Lancet 354: 810–816
48. Collot-Teixeira S, Martin J, McDermott-Roe C et al (2007) CD36 and macrophages in atherosclerosis. Cardiovasc Res 75:468–477
49. Kopprasch S, Pietzsch J, Graessler J (2004) The protective effects of HDL and its constituents against neutrophil respiratory burst activation by hypochlorite-oxidized LDL. Mol Cell Biochem 258:121–127
50. Tsukimori K, Maeda H, Ishida K et al (1993) The superoxide generation of neutrophils in normal and preeclamptic pregnancies. Obstet Gynecol 81:536–540
51. Leanos-Miranda A, Marquez-Acosta J, Cardenas-Mondragon GM et al (2008) Urinary prolactin as a reliable marker for preeclampsia, its severity, and the occurrence of adverse pregnancy outcomes. J Clin Endocrinol Metab 93:2492–2499
52. Scalera F, Borlak J, Beckmann B et al (2004) Endogenous nitric oxide synthesis inhibitor asymmetric dimethyl L-arginine accelerates endothelial cell senescence. Arterioscler Thromb Vasc Biol 24:1816–1822
53. Holden DP, Fickling SA, Whitley GS et al (1998) Plasma concentrations of asymmetric dimethylarginine, a natural inhibitor of nitric oxide synthase, in normal pregnancy and preeclampsia. Am J Obstet Gynecol 178:551–556
54. Savvidou MD, Hingorani AD, Tsikas D et al (2003) Endothelial dysfunction and raised plasma concentrations of asymmetric dimethylarginine in pregnant women who subsequently develop pre-eclampsia. Lancet 361:1511–1517
55. Palm F, Onozato ML, Luo Z et al (2007) Dimethylarginine dimethylaminohydrolase (DDAH): expression, regulation, and function in the cardiovascular and renal systems. Am J Physiol Heart Circ Physiol 293:H3227–H3245
56. Friedman SA, Taylor RN, Roberts JM (1991) Pathophysiology of preeclampsia. Clin Perinatol 18:661–682
57. Maynard SE, Min JY, Merchan J et al (2003) Excess placental soluble fms-like tyrosine kinase 1 (sFlt1) may contribute to endothelial dysfunction, hypertension, and proteinuria in preeclampsia. J Clin Invest 111:649–658
58. Irani RA, Xia Y (2008) The functional role of the renin-angiotensin system in pregnancy and preeclampsia. Placenta 29:763–771
59. Eghbali M, Deva R, Alioua A et al (2005) Molecular and functional signature of heart hypertrophy during pregnancy. Circ Res 96:1208–1216
60. Eghbali M, Wang Y, Toro L et al (2006) Heart hypertrophy during pregnancy: a better functioning heart? Trends Cardiovasc Med 16:285–291
61. Patten RD, Pourati I, Aronovitz MJ et al (2004) 17beta-estradiol reduces cardiomyocyte apoptosis in vivo and in vitro via activation of phospho-inositide-3 kinase/Akt signaling. Circ Res 95:692–699
62. Sliwa K, Forster O, Zhanje F et al (2004) Outcome of subsequent pregnancy in patients with documented peripartum cardiomyopathy. Am J Cardiol 93:1441–1443, A1410
63. Cataldo L, Chen NY, Yuan Q et al (2000) Inhibition of oncogene STAT3 phosphorylation by a prolactin antagonist, hPRL-G129R, in T-47D human breast cancer cells. Int J Oncol 17:1179–1185
64. Negoro S, Kunisada K, Fujio Y et al (2001) Activation of signal transducer and activator of transcription 3 protects cardiomyocytes from hypoxia/reoxygenation-induced oxidative stress through the upregulation of manganese superoxide dismutase. Circulation 104:979–981

65. Van Remmen H, Williams MD, Guo Z et al (2001) Knockout mice heterozygous for Sod2 show alterations in cardiac mitochondrial function and apoptosis. Am J Physiol Heart Circ Physiol 281:H1422–H1432
66. Houstis N, Rosen ED, Lander ES (2006) Reactive oxygen species have a causal role in multiple forms of insulin resistance. Nature 440:944–948
67. Harrison RG (1979) Suppression of lactation. Semin Perinatol 3:287–297
68. Nagafuchi H, Suzuki N, Kaneko A et al (1999) Prolactin locally produced by synovium infiltrating T lymphocytes induces excessive synovial cell functions in patients with rheumatoid arthritis. J Rheumatol 26:1890–1900
69. Kothari SS (1997) Aetiopathogenesis of peripartum cardiomyopathy: prolactin-selenium interaction? Int J Cardiol 60:111–114
70. Corbacho AM, Martinez De La Escalera G, Clapp C (2002) Roles of prolactin and related members of the prolactin/growth hormone/placental lactogen family in angiogenesis. J Endocrinol 173:219–238
71. Tabruyn SP, Sorlet CM, Rentier-Delrue F et al (2003) The antiangiogenic factor 16 K human prolactin induces caspase-dependent apoptosis by a mechanism that requires activation of nuclear factor-kappaB. Mol Endocrinol 17:1815–1823
72. Macotela Y, Aguilar MB, Guzman-Morales J et al (2006) Matrix metalloproteases from chondrocytes generate an antiangiogenic 16 kDa prolactin. J Cell Sci 119:1790–1800
73. Roberg K, Ollinger K (1998) Oxidative stress causes relocation of the lysosomal enzyme cathepsin D with ensuing apoptosis in neonatal rat cardiomyocytes. Am J Pathol 152: 1151–1156
74. Ferreira AL, Matsubara LS, Matsubara BB (2008) Anthracycline-induced cardiotoxicity. Cardiovasc Hematol Agents Med Chem 6:278–281
75. Schulz R (2007) Intracellular targets of matrix metalloproteinase-2 in cardiac disease: rationale and therapeutic approaches. Annu Rev Pharmacol Toxicol 47:211–242
76. Cole P, Cook F, Plappert T et al (1987) Longitudinal changes in left ventricular architecture and function in peripartum cardiomyopathy. Am J Cardiol 60:871–876
77. Diwan A, Wansapura J, Syed FM et al (2008) Nix-mediated apoptosis links myocardial fibrosis, cardiac remodeling, and hypertrophy decompensation. Circulation 117:396–404
78. Lee SH, Kunz J, Lin SH et al (2007) 16-kDa prolactin inhibits endothelial cell migration by down-regulating the Ras-Tiam1-Rac1-Pak1 signaling pathway. Cancer Res 67:11045–11053
79. Gonzalez C, Corbacho AM, Eiserich JP et al (2004) 16 K-prolactin inhibits activation of endothelial nitric oxide synthase, intracellular calcium mobilization, and endothelium-dependent vasorelaxation. Endocrinology 145:5714–5722
80. Tabruyn SP, Sabatel C, Nguyen NQ et al (2007) The angiostatic 16 K human prolactin overcomes endothelial cell anergy and promotes leukocyte infiltration via NF-{kappa}B activation. Mol Endocrinol 21(6):1422–1429
81. Forster O, Hilfiker-Kleiner D, Ansari AA et al (2008) Reversal of IFN-gamma, oxLDL and prolactin serum levels correlate with clinical improvement in patients with peripartum cardiomyopathy. Eur J Heart Fail 10:861–868
82. Gira AK, Casper KA, Otto KB et al (2003) Induction of interferon regulatory factor 1 expression in human dermal endothelial cells by interferon-gamma and tumor necrosis factor-alpha is transcriptionally regulated and requires iron. J Invest Dermatol 121:1191–1196
83. Brand JM, Frohn C, Cziupka K et al (2004) Prolactin triggers pro-inflammatory immune responses in peripheral immune cells. Eur Cytokine Netw 15:99–104
84. McMurray RW (2001) Estrogen, prolactin, and autoimmunity: actions and interactions. Int Immunopharmacol 1:995–1008
85. Fett JD, Sundstrom BJ, Etta King M et al (2002) Mother-daughter peripartum cardiomyopathy. Int J Cardiol 86:331–332
86. Zhang L, Kao WH, Berthier-Schaad Y et al (2006) Haplotype of signal transducer and activator of transcription 3 gene predicts cardiovascular disease in dialysis patients. J Am Soc Nephrol 17:2285–2292

87. Ito N, Eto M, Nakamura E et al (2007) STAT3 polymorphism predicts interferon-alfa response in patients with metastatic renal cell carcinoma. J Clin Oncol 25:2785–2791
88. Ge D, Gooljar SB, Kyriakou T et al (2008) Association of common JAK2 variants with body fat, insulin sensitivity and lipid profile. Obes 16:492–496
89. Podewski EK, Hilfiker-Kleiner D, Hilfiker A et al (2003) Alterations in Janus kinase (JAK)-signal transducers and activators of transcription (STAT) signaling in patients with end-stage dilated cardiomyopathy. Circulation 107:798–802
90. Kim YJ, Park HS, Park MH et al (2005) Oxidative stress-related gene polymorphism and the risk of preeclampsia. Eur J Obstet Gynecol Reprod Biol 119:42–46
91. de Beus E, van Mook WN, Ramsay G et al (2003) Peripartum cardiomyopathy: a condition intensivists should be aware of. Intensive Care Med 29:167–174
92. Ntusi NB, Mayosi BM (2008) Aetiology and risk factors of peripartum cardiomyopathy: A systematic review. Int J Cardiol 131(2):168–179
93. Habedank D, Kuhnle Y, Elgeti T et al (2008) Recovery from peripartum cardiomyopathy after treatment with bromocriptine. Eur J Heart Fail 10(11):1149–1151
94. Jahns BG, Stein W, Hilfiker-Kleiner D et al (2008) Peripartum cardiomyopathy – a new treatment option by inhibition of prolactin secretion. Am J Obstet Gynecol 199:e
95. Turkalj I, Braun P, Krupp P (1982) Surveillance of bromocriptine in pregnancy. J Am Med Assoc 247:1589–1591
96. Hopp L, Haider B, Iffy L (1996) Myocardial infarction postpartum in patients taking bromocriptine for the prevention of breast engorgement. Int J Cardiol 57:227–232
97. Brenner B (2004) Haemostatic changes in pregnancy. Thromb Res 114:409–414
98. James AH, Brancazio LR, Ortel TL (2005) Thrombosis, thrombophilia, and thromboprophylaxis in pregnancy. Clin Adv Hematol Oncol 3:187–197
99. Patti G, Nasso G, D'Ambrosio A et al (1999) Myocardial infarction during pregnancy and postpartum: a review. G Ital Cardiol 29:333–338
100. Goland S, Schwarz ER, Siegel RJ et al (2007) Pregnancy-associated spontaneous coronary artery dissection. Am J Obstet Gynecol 197:e11–e13
101. Khurana S, Liby K, Buckley AR et al (1999) Proteolysis of human prolactin: resistance to cathepsin D and formation of a nonangiostatic, C-terminal 16 K fragment by thrombin. Endocrinology 140:4127–4132
102. Khurana S, Kuns R, Ben-Jonathan N (1999) Heparin-binding property of human prolactin: a novel aspect of prolactin biology. Endocrinology 140:1026–1029

# Chapter 17
# Oxidative Stress and Inflammation after Coronary Angiography

**Raymond Farah**

**Abstract** Percutaneous coronary intervention (PCI) as an invasive procedure includes inflation of a balloon and/or implantation of an endovascular prosthesis (stent) in an atherosclerotic coronary vessel at a level where the plaque narrows its cross-sectional area by more than 75%. Various reports have demonstrated that balloon inflation or stent implantation triggers inflammation and subsequent growth of smooth muscle cells. Both oxidative stress (OS) and inflammation parameters worsen, increasing the risk of complications. The polymorphonuclear leukocyte (PMNL) is one of the inflammatory cells releasing reactive oxygen species contributing to OS, inflammation, and endothelial injury (Tardif, Cardiol Rounds 7(9), 2003).

**Keywords** Oxidative stress · Percutaneous coronary intervension · Inflammation · Atherosclerosis · Polymorphonuclear leukocyte

## 17.1 Introduction

Atherosclerotic disease remains a leading cause of death in Western societies, and a major contributor to loss of disability-adjusted life-years worldwide. There has been about a 28% elevation in death from cardiovascular disease in developing countries during the last 5 years [1, 2]. Atherosclerosis is a disease characterized by chronic inflammation-related oxidative stress (OS) resulting in complications that include ischemia, acute coronary syndromes, and stroke. OS plays a critical role in the formation of plaques, and along with inherent vascular inflammation, may be a strong predictor of atherosclerosis.

Thus, understanding the atherogenesis, behavior, diagnosis, and treatment of coronary heart disease has become of high priority among clinical and laboratory researchers.

---

R. Farah (✉)
Department of Internal Medicine B, Ziv Medical Center, Safed, Israel
e-mail: raymond.f@ziv.health.gov.il

H. Sauer et al. (eds.), *Studies on Cardiovascular Disorders*,
Oxidative Stress in Applied Basic Research and Clinical Practice,
DOI 10.1007/978-1-60761-600-9_17, © Springer Science+Business Media, LLC 2010

## 17.2 Oxidative Stress During Percutaneous Coronary Intervention

Percutaneous coronary intervention (PCI) as an invasive procedure includes inflation of a balloon and/or implantation of an endovascular prosthesis (stent) in an atherosclerotic coronary vessel at a level where the plaque narrows the blood vessel's cross-sectional area by more than 75% [2]. Various reports have demonstrated that the barometric trauma to the vessel wall by balloon inflation or stent implantation triggers inflammation and the subsequent growth of smooth muscle cells. This process leads in 20–40% of cases to a significant narrowing of the previously treated vessel (restenosis). In this setting, both OS and inflammation parameters are worsened, increasing the risk of complications. Different reports have demonstrated the significant elevation of inflammatory markers after the PCI procedure and coronary angiography. PCI in patients with stable angina without any inflammatory disorders caused an elevation in C-reactive protein (CRP). Patients with high levels of CRP frequently need another revascularization 6 months later [3, 4]. High serum levels of high-sensitivity C-reactive protein (hs-CRP), interleukin-6, and tumor necrosis factor-alpha have been shown to be predictors of adverse outcomes in patients with coronary artery disease (CAD) [4]. One study clearly supports the role of inflammation in restenosis after PCI, as measured by statistically higher levels of Lp(a) and fibrinogen in patients with major adverse clinical events (repeat PCI, CABG, myocardial infarction, and death) and CRP in patients with repeat angina [5]. Recent studies showed that uncomplicated diagnostic coronary angiography triggers a systemic inflammatory response in patients with stable angina, and should be considered in interpreting the significance of the systemic inflammatory response observed after PCI [6]. Polymorphonuclear leukocytes (PMNL) are among the inflammatory cells releasing reactive oxygen species contributing to OS, inflammation, and endothelial injury [7]. Activated PMNLs damage the surrounding tissue by releasing reactive oxygen species (ROS) and proteolytic enzymes before self-necrosis. OS and inflammation will result in endothelial damage and atherosclerosis in the long run [8–12].

## 17.3 Antioxidant Approaches in Clinical Practice?

Normally the body maintains a balance between its antioxidant defenses and free radicals. But an imbalance can be dangerous. Biochemical processes in the body generate reactive oxygen species that are normally mopped up by antioxidant defense mechanisms. Under certain conditions, an imbalance can develop between the antioxidant defenses and the formation of ROS. The resulting accumulation of ROS, called oxidative stress, enables them to interact with physiological mediators in the body. Such an interaction inactivates those mediators and can result in the formation of toxic products. An example of this is nitric oxide (NO), a blood vessel dilator and antithrombotic agent generated in the lining of blood vessels, which reacts with superoxide anion ($O_2^-$). This interaction inactivates NO, leading to a condition in which the blood vessels fail to respond normally to the beneficial stimuli of the blood vessel dilators. This condition is predictive of cardiovascular disease and occurs in subjects with risk factors but no overt symptoms of disease. The reaction between NO and $O_2^-$ also leads to the formation of peroxynitrite, a powerful oxidant species that has been implicated in conditions such as hypercholesterolaemia, diabetes, and coronary artery disease. Another example is prostacyclin, the generation of which is decreased by lipid peroxides produced by the interaction between normal lipids in the body and ROS.

Experimental and clinical studies suggest that oxidative stress contributes to the development and progression of cardiovascular disease. However, clinical trials with classic vitamin antioxidants failed to demonstrate any benefit in cardiovascular outcomes. Recent advances in our understanding of mechanisms involved in free radical generation reinstate that a more comprehensive approach targeting the prevention of reactive oxygen species (ROS) formation early in the disease process may prove beneficial. Before a potential role for antioxidants in the treatment of CVD is eliminated, more carefully designed studies with classic as well as new antioxidants in well-defined patient populations are warranted to provide a definitive answer [13].

Several key unanswered questions in relation to oxidative stress and atherosclerosis are raised, and proposed as fruitful areas of research [14].

There is emerging evidence for genetic components from genome-wide gene expression studies and from systematic evaluation of candidate genes within the oxidative stress pathways. In both cases it can be concluded that the restoration of vascular reactive oxygen species to normal is an important but frequently neglected therapeutic target [15].

### *17.3.1 Myeloperoxidase (MPO) as a Biomarker of Oxidative Stress in Cardiovascular Disease*

Oxidative stress and inflammation play important roles in the pathogenesis of destabilization of coronary artery disease (CAD) leading to acute coronary syndromes (ACS). Infiltrating macrophages and neutrophils participate in the transformation of stable coronary artery plaques to unstable lesions [16, 17]. Recently, there has been a renewed interest in myeloperoxidase (MPO), a proinflammatory enzyme that is abundant in ruptured plaque [18] and can be measured in peripheral blood. MPO is a hemoprotein that is stored in azurophilic granules of polymorphonuclear neutrophils and macrophages. MPO catalyzes the conversion of chloride and hydrogen peroxide to hypochlorite and is secreted during inflammatory conditions. It has been implicated in the oxidation of lipids contained within LDL cholesterol. In addition, MPO consumes endothelial-derived NO, thereby reducing NO bioavailability and impairing its vasodilating and anti-inflammatory properties.

Major evidence for MPO as an enzymatic catalyst for oxidative modification of lipoproteins in the artery wall has been suggested in a number of studies performed with low-density lipoprotein [19]. In contrast to low-density lipoprotein, plasma levels of high-density lipoprotein (HDL)-cholesterol and apoAI, the major apolipoprotein of HDL, inversely correlate with the risk of developing coronary artery disease. There is now strong evidence that HDL is a selective in vivo target for MPO-catalyzed oxidation, that may represent a specific molecular mechanism for converting the cardioprotective lipoprotein into a dysfunctional form, raising the possibility that the enzyme represents a potential therapeutic target for preventing vascular disease in humans [20]. Zhou et al. [21] showed that atorvastatin reduced serum MPO and CRP concentrations in patients with ACS.

MPO activity can be measured in blood and tissues by spectrophotometric assays using hydrogen peroxide and o-dianisidine dihydrochloride as substrates. In addition, MPO content can be measured in neutrophils as an index of degranulation with the Coulter counter, and flow cytometry and circulating MPO by ELISA. Very recently, commercial methods allowing low-cost and high-volume measurements have been proposed. The introduction of these methods of measurement might make MPO a new and useful cardiac biomarker.

There have been a few but important clinical studies examining the role of MPO as a marker of risk for CAD. Using an enzyme assay, Zhang et al. [22] showed that blood and leukocyte MPO activity were higher in patients with CAD than angiographically verified normal controls, and that this increased activity was significantly associated with the presence of CAD (odds ratio, 11.9; 95% confidence interval (CI), 5.5–25.5). Results were independent of the patient's age, sex, hypertension, smoking, diabetes status, LDL concentration, leukocyte count, and Framingham global risk score. More recently, Meuwese et al. [23], in the European Prospective Investigation into Cancer and Nutrition (EPIC)-Norfolk prospective population study, have evaluated the association of MPO levels with the risk of future CAD in apparently healthy individuals. MPO was measured in baseline samples of a case-control study nested in the prospective EPIC-Norfolk population study: case subjects ($n = 1{,}138$) were apparently healthy men and women who developed CAD during eight years of follow-up; control subjects ($n = 2{,}237$) matched for age, gender, and enrollment time, remained free of CAD. The MPO levels were significantly higher in case subjects than in control subjects, and correlated with C-reactive protein (CRP) and white blood cell count. Risk of future CAD increased in consecutive quartiles of MPO concentration, with an odds ratio (OR) of 1.49 in the top vs. bottom quartile. After adjustment for traditional risk factors, the OR in the top quartile remained significant at 1.36 (95% CI 1.07–1.73). Of interest in this study, serum MPO levels were associated with the risk of future development of CAD in apparently healthy individuals, but the association was weaker than that of traditional risk factors and CRP. However MPO, at variance with CRP, was largely independent of classical risk factors.

In ACS, MPO has been consistently found to be associated with the presence of instability and the risk of future events in the studies that have explored these topics. Biasucci et al. [24] first observed that circulating neutrophils in patients with acute myocardial infarction (AMI) and unstable angina (UA) have a low MPO content, and therefore high MPO levels in the circulation, as compared with those with chronic stable angina and variant angina. This is indicative of a significant release of MPO from neutrophils related to their activation. The lack of neutrophil activation in patients with variant angina, and after stress tests, suggests that this phenomenon may occur independently of ischemic episodes. Therefore, MPO is prevalently a marker of instability and not simply a marker of oxidative stress and damage. Furthermore, in this study MPO did not correlate with CK-MB and troponin T release; this observation is clinically important because an extremely sensitive and specific marker of damage already exists (troponin), but no definite markers of instability exist so far. In this study, MPO content was determined on the Coulter counter, which measures the neutrophil count by flow cytometry and subsequently calculates the mean MPO content in that population.

Using the same method, Buffon et al. [25] studied 65 patients who underwent cardiac catheterization with coronary sinus sampling. The MPO content of the leukocytes collected from the arterial circulation and the coronary sinus effluent were compared. The authors found a gradient of MPO across the coronary circulation in patients with ACS; and this gradient was present even when the culprit lesion involved with the ACS was in the distribution of the right coronary artery, which does not drain into the coronary sinus. In this study, as in the previous one, a significant correlation was found between systemic levels of C-reactive protein and either the aortic or coronary sinus neutrophil MPO.

The potential usefulness for risk stratification of blood concentrations of MPO was examined in two recent studies. In the CAPTURE trial [26], MPO mass concentration was measured in 1,090 patients with ACS. Rates of death and myocardial infarction (MI) were determined at six months of follow-up. An MPO cutoff of 350 μg/L was associated with an adjusted hazard ratio of 2.25 (95% CI, 1.32–3.82). The effects were particularly impressive in patients with undetectable cardiac troponin T (cTnT < 0.01 μg/L), in whom the hazard ratio was 7.48 (95% CI, 1.98–28.29). Interestingly, the increase in risk was already evident after 72 h, increasing only slightly thereafter. This observation is in keeping with the data by Biasucci et al. [24], who had shown return of MPO to baseline levels in all patients, including those with myocardial infarction, within one week. This point is important, as it suggests a peculiar characteristic of MPO, at variance with other inflammatory markers commonly used (like CRP or fibrinogen) and with other proposed inflammatory markers that remain elevated for a relatively long time or have an extremely short and unreliable half-life (such as interleukins). The predictive value of MPO was independent of C-reactive protein; and high MPO serum levels indicated increased cardiac risk, both in patients with medium C-reactive protein serum levels (20.0 vs. 5.9%; $P < .001$) and in those with low C-reactive protein serum levels (17.8 vs. 0%; $P < .001$), suggesting that recruitment and degranulation of neutrophils is a primary event and is followed by release of other systemic mediators and acute-phase proteins such as C-reactive protein. At variance from CRP levels, levels of MPO were not influenced by troponin, suggesting a prognostic role of MPO independent from troponin and confirming that inflammation is a primary phenomenon in ACS [27].

More recently, several studies have also investigated the value of MPO in predicting long-term outcomes. Recently, Li et al. [28] studied 176 consecutive patients who underwent coronary angiography. The patients were divided into four groups according to the quartile of MPO level. They found: (1) ACS rate (36.2%) in the fourth quartile group of MPO level was six times higher than that in the first quartile group of MPO level, $P < .01$. (2) The Gensini score in the fourth quartile group of MPO level was significantly higher than that in the first quartile group ($P < .01$). WBC in the fourth quartile group was also significantly higher than that in the first quartile group, $P < .05$. In addition, Kaplan-Meier event rate curves showed that there was a significant difference in outcome (death, AMI, revascularization) between the groups with MPO $\geq$ 62.9 AUU/L and those with $\leq$62.9 AUU/L of MPO serum level at the six-month follow-up visit (chi (2) = 13.5, $P = .01$). Furthermore, Cavusoglu et al. [29] have investigated the long-term prognostic significance of baseline MPO levels in a well-characterized cohort of 193 men with ACS. All patients were followed prospectively for the development of death and MI, and follow-up data were available for all patients at 24 months. Using the median MPO value of the entire cohort of patients (20.34 ng/mL) as a prespecified cutoff, the MI-free survival at 24 months for the group with MPO values below cutoff were significantly lower than in those with values above cutoff.

Mocatta et al. have investigated the relationship between plasma MPO and clinical outcome after AMI [30]. They studied 512 AMI patients at hospital admission and measured their plasma MPO; they found a significant association of MPO with follow-up events. Importantly, MPO was of incremental prognostic value on the top of ejection fraction and BNP, a finding observed also by Khan in a similar population of patients with STEMI [31–34].

### *17.3.2 Role of PMNLs*

The PMNL is one of the inflammatory cells releasing ROS and thus contributing to OS, inflammation, and endothelial injury [7]. Priming of PMNLs, resulting in OS, increased self-necrosis as well as releasing various chemotactic agents and cytokines, and recruiting more PMNLs [11, 12]. Necrosis starts a chain of inflammatory reactions, inducing cell recruitment and, in the long run, OS and endothelial dysfunction. Understanding the contribution of PMNLs to OS and inflammation may illuminate new mechanisms through which endothelial dysfunction evolves and causes angiopathy and atherosclerosis. We have previously shown the contribution of PMNLs to other clinical situations known to be associated with endothelial dysfunction and accelerated atherosclerosis and coronary heart disease such as uremia, type 2 diabetes, hypertension, and cigarette smoking [8–10, 12]. In patients undergoing a PCI procedure, compared to those undergoing diagnostic coronary angiography, we divided the tested patients into two groups by OS and inflammation parameters. Already at time "0," these parameters were higher in the PCI group of patients, compared to those undergoing only diagnostic coronary angiography. OS parameters decreased significantly following the PCI procedure, compared to those undergoing diagnostic coronary angiography, in whom the parameters did not change (based on the level of primed PMNLs and the rate of superoxide release and CD11b level). Thus we concluded that PCI itself induces increased OS and inflammation. Patients who underwent invasive procedures such as PCI (balloon inflation or stent insertion) had a significant decrease in PMNLs activation one month later, compared to patients who underwent only a diagnostic angiography. Despite this decrement of PMNLs activation after one month, the cells were still primed, probably because of the chronic diseases of these patients [8–10, 12]. Other reports have previously described similar results: Diaz-Araya et al. showed an increased OS in patients with acute myocardial infarction who underwent PCI with either balloon inflation or stent insertion. Increasing OS was expressed by the increment of lipid peroxidasation and reduction in enzymatic antioxidant reserve determined by superoxide dismutase, glutathione peroxidase, and catalase erythrocyte activities at admission and at 0.5 and 24 h after angioplasty [35]. As with our study, Berg et al. demonstrated an elevated OS following angioplasty or stent implantation, in comparison to patients following coronary angiographies only [36, 37]. An activation of PMNLs 1 h after PCI was also demonstrated as well, accelerating their rate of superoxide release. Serdar et al. demonstrated that increased lipid and protein oxidation products and decreased antioxidant enzymes and vitamins contribute to increased OS, which in turn is related to the severity of the disease [38]. It is important to note that we found a significant positive linear regression between cardiac parameters, such as serum CPK, and the rate of superoxide release from separated PMNLs, indicating a positive correlation between PMNL priming and the severity of cardiac disease [39–41].

Other inflammatory markers related to primed PMNLs, e.g., WBC and PMNL counts, did not differ between either group of patients before and after angiography, even though some transient elevation in PMNL counts could be shown. Other systemic inflammation parameters, such as fibrinogen and CRP, showed a significant decrease in the PCI group after the procedure. It has to be emphasized that the change in fibrinogen indicates that it is a more sensitive marker than CRP. It is important to highlight that all these markers are usually elevated in patients with risk factors such as hypertension, hyperlipidemia, diabetes, smoking, and coronary heart disease. We have previously shown that in other clinical situations known to be associated with endothelial dysfunction and accelerated atherosclerosis, such as uremia, diabetes, hypertension, and smoking, PMN are primed with an elevated number, contributing to OS and inflammation, compared to healthy persons [8–10, 12].

## 17.4 Summary

In conclusion, the PMNL contribution to OS and inflammation is lower in patients not treated with PCI but undergoing only diagnostic coronary angiography, compared to the PCI group undergoing balloon inflation or stent implantation. This article adds new facets to the evaluation of cardiac patients—whether they will undergo future intervention with PCI procedure or only diagnostic coronary angiography. This implication might predict which patient will undergo PCI, and thus recommend the optimal preventive treatment for eventual complications, including antioxidant treatment. We need other research with more time periods and samples in order to check the relation between the OS index and other inflammatory markers, and between cardiac parameters during diagnostic angiography, PCI, and thereafter.

## References


1. Tardif JC (2003) Oxidative stress and coronary heart disease. Cardiol Rounds 7(9):1–6
2. Tardif JC, Gregoire J, L'Allier PL (2002) Prevention of restenosis with antioxidants: mechanisms and implications. Am J Cardiovasc Drugs 2:323–324
3. Azar RR, McGay RG, Kiernan FJ, Seecharran B, Feng YJ, Fram DB, Wu AHB, Walters DD: (1997) Coronary angioplasty induces a systemic inflammatory response. Am J Cardiol 80:1476–1478
4. Sukhija R, Fahdi I, Garza L, Fink L, Scott M, Aude W, Pacheco R, Bursac Z, Grant A, Mehta JL (2007) Inflammatory markers, angiographic severity of coronary artery disease, and patient outcome. Am J Cardiol 99:879–884
5. Rahel BM, Visseren FL, Suttorp MJ, Plokker TH, Kelder JC, de Jong BM, Bouter KP, Diepersloot RJ: (2003) Preprocedural serum levels of acute-phase reactants and prognosis after percutaneous coronary intervention. Cardiovasc Res 60:136–140
6. Goldberg A, Zinder O, Zdorovyak A, Diamond E, Lischinsky S, Gruberg L, Markiewicz W, Beyar R, Aronson D: (2003) Diagnostic coronary angiography induces a systemic inflammatory response in patients with stable angina. Am Heart J 146:819–823
7. Smedly LA, Tonnesen MG, Sandhaus RA: (1986) Neutrophil-mediated injury to endothelial cells. J Clin Invest 77:1233–1243
8. Kristal B, Shurtz-Swirski R, Shapiro G, Chezar J, Manaster J, Shasha SM, Sela S: (1998) Participation of peripheral polymorphonuclear leukocytes in the oxidative stress and inflammation in patients with essential hypertension. Am J Hypertens 11:921–928
9. Kristal B, Shurtz-Swirski R, Shasha SM, Manaster J, Shapiro G, Furmanov M, Hassan K, Weissman I, Sela S: (1999) Interaction between erythropoietin and peripheral polymorphonuclear leukocytes in hemodialysis patients. Nephron 81:406–413
10. Shurtz-Swirski R, Sela S, Shapiro G, Nasser L, Shasha SM, Kristal B: (2001) Involvement of polymorphonuclear leukocytes in oxidative stress and inflammation in type 2 diabetes. Diabetes Care 24:104–110
11. Weiss SJ: (1989) Tissue destruction by neutrophils. N Engl J Med 320:365–376
12. Sela S, Shurtz-Swirski R, Awad J, Shapiro G, Nasser L, Shasha SM, Kristal B: (2002) The involvement of peripheral polymorphonuclear leukocytes in oxidative stress and inflammation in cigarette smokers. Isr Med Assoc 4:1015–1019
13. Cantor EJ, Mancini EV, Seth R, Yao XH, Netticadan T (2003) Oxidative stress and heart disease: cardiac dysfunction, nutrition, and gene therapy. Curr Hypertens Rep 5(3):215–221
14. Heistad DD (2006) Oxidative stress and vascular disease: 2005 Duff lecture. Arterioscler Thromb Vasc Biol 26(4):689–695; Epub 2006 Jan 12
15. Dominiczaks AF (2005) Oxidative stress and cardiovascular disease. Endocr Abstr 9:S23
16. Takahiko N, Makiko U, Kazuo H et al (2002) Neutrophil infiltration of culprit lesions in acute coronary syndromes. Circulation 106(23): 2894–2900
17. Sugiyama S, Okada Y, Sukhova GK, Virmani R, Heinecke JW, Libby P (2001) Macrophage myeloperoxidase regulation by granulocyte macrophage colony-stimulating factor in human atherosclerosis and implications in acute coronary syndromes. Am J Pathol 158(3): 879–891

18. Mullane KM, Kraemer R, Smith B (1985) Myelopeoxidase activity as a quantitative assessment of neutrophil infiltration into ischemic myocardium. J Pharmacol Methods 14(3):157–167
19. Holvoet P (1998) Oxidative modification of low-density lipoproteins in atherothrombosis. Acta Cardiol 53 (5):253–260
20. Shao B, Oda MN, Oram JF, Heinecke JW (2006) Myeloperoxidase: an inflammatory enzyme for generating dysfunctional high density lipoprotein. Curr Opin Cardiol 21(4):322–328
21. Zhou T, Zhou S-H, Qi S-S, Shen X-Q, Zeng G-F, Zhou H-N (2006) The effect of atrovastatin on serum myeloperoxidase and CRP levels in patients with acute coronary syndrome. Clin Chim Acta 368(1–2):168–172
22. Zhang R, Brennan M-L, Fu X et al (2001) Association between myeloperoxidase levels and risk of coronary artery disease. J Am Med Assoc 286(17):2136–2142
23. Meuwese MC, Stroes ESG, Hazen SL et al (2007) Serum myeloperoxidase levels are associated with the future risk of coronary artery disease in apparently healthy individuals. The EPIC-Norfolk prospective population study. J Am Coll Cardiol 50(2):159–165
24. Biasucci LM, D'Onofrio G, Liuzzo G et al (1996) Intracellular neutrophil myeloperoxidase is reduced in unstable angina and acute myocardial infarction, but its reduction is not related to ischemia. J Am Coll Cardiol 27(3):611–616
25. Buffon A, Biasucci LM, Liuzzo G, D'Onofrio G, Crea F, Maseri A (2002) Widespread coronary inflammation in unstable angina. N Engl J Med 347(1):5–12
26. Baldus S, Heeschen C, Meinertz T et al (2003) Myeloperoxidase serum levels predict risk in patients with acute coronary syndrome. Circulation 108(12):1440–1445
27. Brennan M-L, Penn MS, Van Lente F et al (2003) Prognostic value of myeloperoxidase in patients with chest pain. N Engl J Med 349(17):1595–1604
28. Li S-H, Xing Y-W, Li Z-Z, Bai S-G, Wang J (2007) Clinical implications of relationship between myeloperoxidase and acute coronary syndromes. Zhonghua Xin Xue Guan Bing Za Zhi 35(3):241–244
29. Cavusoglu E, Ruwende C, Eng C et al (2007) Usefulness of baseline plasma myeloperoxidase levels as an independent predictor of myocardial infarction at two years in patients presenting with acute coronary syndrome. Am J Cardiol 99(10):1364–1368
30. Mocatta TJ, Pilbrow AP, Cameron VA et al (2007) Plasma concentrations of myeloperoxidase predict mortality after myocardial infarction. J Am Coll Cardiol 49(20):1993–2000
31. Morrow. DA (2007) Appraisal of myeloperoxidase for evaluation of patients with suspected acute coronary syndrome. J Am Coll Cardiol 49(20):2001–2002
32. Khan SQ, Kelly D, Quinn P, Davies JE, Ng. LL (2007) Myeloperoxidase aids prognostication together with N-terminal pro-B–type natriuretic peptide in high-risk patients with acute ST elevation myocardial infarction. Heart 93(7):826–831
33. Loria V, Dato I, De Maria GL, Biasucci LM (2008 Oct) Markers of acute coronary syndrome in emergency room. Minerva Med 99(5):497–517
34. Loria V, Dato I, Graziani F, Biasucci LM (2008) Myeloperoxidase: a new biomarker of inflammation in ischemic heart disease and acute coronary syndromes. Mediators Inflamm. 2008:135625
35. Diaz-Araya G, Nettle D, Castro P, Miranda F, Greig D, Campos X, Chiong M, Nazzal C, Corbalan R, Lavandero S: (2002) Oxidative stress after reperfusion with primary coronary angioplasty: lack of effect of glucose-insulin-potassium infusion. Crit Care Med 30:417–421
36. Berg K, Wiseth R, Bjerve K, Brurok H, Gunnes S, Skarra S, Jynge P, Basu S (2004) Oxidative stress and myocardial damage during elective percutaneous coronary interventions and coronary angiography. A comparison of blood-borne isoprostane and troponin release. Free Radic Res 38:517–525
37. Berg K, Jynge P, Bjerve K, Skarra S, Basu S, Wiseth R: (2005) Oxidative stress and inflammatory response during and following coronary interventions for acute myocardial infarction. Free Radic Res 39:629–636

38. Serdar Z, Aslan K, Dirican M, Sarandol E, Yesilbursa D, Serdar A: (2006) Lipid and protein oxidation and antioxidant status in patients with angiographically proven coronary artery disease. Clin Biochem 39:794–803
39. Sela S, Shurtz-Swirski R, Cohen-Mazor M, Mazor R, Chezar J, Shapiro G, Hassan K, Shkolnik G, Geron R, Kristal B: (2005) The primed peripheral polymorphonuclear leukocyte – a culprit underlying chronic low-grade inflammation and systemic oxidative stress in chronic kidney disease. J Am Soc Nephrol 16:2431–2438
40. Yoon JW, Pahl MV, Vaziri ND: (2007) Spontaneous leukocyte activation and oxygen-free radical generation in end-stage renal disease. Kidney Int 71:167–172
41. Swain SD, Rohn TT, Quinn MT: (2002) Neutrophil priming in host defense: role of oxidants as priming agents. Antioxid Redox Signal 4:69–83

# Chapter 18
# Oxidative Stress in Cardiac Transplantation

**Galen M. Pieper and Ashwani K. Khanna**

**Abstract** Early experimental studies which measured antioxidant defenses and accumulation of lipid peroxidation products suggested that oxidative stress plays a role in cardiac transplant rejection. This led to studies designed to test the efficacy of intervention with antioxidants in experimental and clinical cardiac transplantation. This chapter focuses on the critical evidence (both indirect and direct) for oxidative stress. The review considers how oxidative stress may be increased as a consequence of the various facets of transplantation. These include the contributions of organ preservation, ischemia-reperfusion injury, rejection, and the superimposition of immunosuppressant therapy. It is acknowledged that each of these components may combine to contribute to the overall concept of increased oxidative stress in cardiac transplant recipients.

**Keywords** Transplantation · Rejection · Organ preservation · Immunosuppressant · Apoptosis

## 18.1 Introduction

For over a decade, there has been ongoing interest in the concept that oxidative stress plays a role in cardiac transplantation, and that antioxidants might be useful strategies to counteract complications of cardiac transplantation [1]. Most of the early evidence of increased production of reactive oxygen species and oxidative stress arose out of studies examining the role of ischemia-reperfusion on post-transplant function and injury. The present review examines the role of oxidative stress in various facets of cardiac transplantation, including ischemia-reperfusion injury, acute and chronic cardiac rejection, chronic transplant vasculopathy, and the superimposed effects of concurrent immunosuppressant therapy (Fig. 18.1).

---

G.M. Pieper (✉)
Division of Transplant Surgery, Department of Surgery, Medical College of Wisconsin, Cardiovascular Research Center and the Free Radical Research Center, Milwaukee, WI, USA
e-mail: gmpieper@mcw.edu

H. Sauer et al. (eds.), *Studies on Cardiovascular Disorders*,
Oxidative Stress in Applied Basic Research and Clinical Practice,
DOI 10.1007/978-1-60761-600-9_18, 

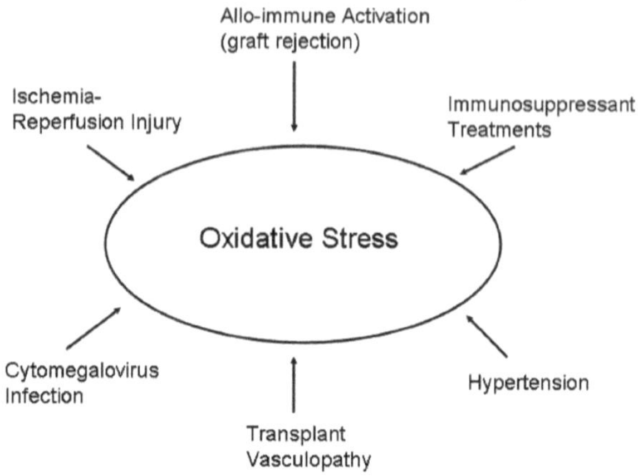


**Fig. 18.1** Sources of oxidative stress in cardiac transplantation

The issue of immunosuppressant therapy is potentially important in this consideration, since certain immunosuppressant medications which offset rejection elicit hypertension as a side effect. In this regard, hypertension is a known risk factor for increased oxidative stress. The incidence of hypertension in the case of pediatric heart transplant recipients is in the range of 47% and 63% at 1 and 5 years following transplantation, respectively [2]. For adults, the most recent data is that the incidence is nearly 75% in the first year but increases to 93.8% of all adult cardiac transplant recipients by 5 years [3].

The fact that many of these stages have demonstrable inflammatory components supports the presumed notion that oxidative stress may play a role at each of these individual stages. However, a closer examination of the literature reveals a remarkable lack of both experimental and human studies that document measurements of reactive oxygen production or good biomarkers of oxidative stress in cardiac transplantation. It should be noted that most of the evidence for transplantation-induced oxidative stress arises from studies in renal transplantation.

## 18.2 Oxidative Stress in Human Cardiac Transplantation

Increased levels of plasma lipid peroxides in cardiac transplant recipients were seen in several studies suggesting increased oxidative stress [4–9]. In these studies, levels of thiobarbituric acid–reactive products were used as markers of evidence of increased lipid peroxidation. A caveat of these traditional assays of lipid peroxidation is that this technique is prone to artifact and is rapidly losing favor among free radical researchers as a specific, reliable, and accurate determinant of lipid peroxidation. Measurement of hydroperoxides of cholesterol esters and triglycerides can bypass some of the limitations of the thiobarbituric acid assay for lipid peroxidation. Clinically, these alternative end points have been used to show increased oxidative stress in heart and kidney transplant recipients, albeit with concurrent immunosuppression-induced hypertension [10]. Alternatives such as determination of 4-hydroxynonenal or the "gold standard," $F_2$-isoprostanes, have since emerged [11]. Measures of $F_2$-isoprostanes have been performed in patients with liver transplantation. In these patients, it is interesting to note that the level of isoprostanes was not correlated with acute cellular rejection, graft failure, or hepatic C infection of the graft [12]. Thus the authors concluded that oxidative stress is unlikely to be a significant mechanism for these transplant-related events. A concern is that this conclusion was based upon an analysis that was performed on a small population ($n = 20$) of patients. In contrast, serum and urine $F_2$-isoprostane levels were increased in patients with chronic kidney allograft nephropathy, suggesting a role for oxidative stress in transplant nephropathy [13]. To our knowledge, there are no published clinical studies of $F_2$-isoprostane levels in human cardiac transplant recipients.

An alternative, noninvasive measurement of oxidative stress in patients is the test for breath alkane levels, which is determined by gas chromatography and mass spectroscopy. This analysis was used in a multicenter study of 539 cardiac transplant patients known as the Heart Allograft Rejection: Detection with Breath Alkanes in Low Levels (HARDBALL) study [14, 15]. In a follow-up study to the initial HARDBALL study, breath alkane levels were shown to be 100% sensitive for detection of grade 3 rejection in this large cohort of cardiac transplant recipients [16]. In a smaller clinical study using a univariate analysis, it was found that blood glutathione levels were the only independent predictor of another complication of cardiac transplantation, notably cardiac allograft vasculopathy [17].

The impact of oxidative stress on coronary artery disease in cardiac transplant recipients was indicated in a study of 99 patients with baseline and follow-up angiography. The study showed that baseline levels of oxidized LDL were a strong predictor ($\chi^2 = 16$, $P < 0.0001$) of the development of cardiac transplant vasculopathy [18]. In contrast, in another study, cytomegalovirus infection and rejection were not correlated with the development of vasculopathy, suggesting that oxidative stress may not be an independent risk to the development of this transplant-associated pathology. A subsequent study of 36 patients within one year posttransplantation showed that circulating levels of antibodies to oxidized LDL were a stronger determinant of endothelial dysfunction than native LDL [19]. Thus certain, but not all, aspects of transplantation may be correlated with oxidative stress.

Oxidative stress could also arise from changes in antioxidant defense. In this context, decreases in levels of superoxide dismutase (SOD) [4] or decreases in nonenzymatic levels of $\alpha$-tocopherol in erythrocytes [4] have been reported in transplant patients exhibiting rejection, compared to transplant recipient patients not displaying rejection. Biopsies of cardiac tissue have also demonstrated decreased SOD in cardiac transplant recipients [20]. This clinical data agrees with findings in our laboratory of decreased manganese SOD (MnSOD) activity and protein levels in a rodent model of acute cardiac transplant rejection [21, 22]. Collectively, these findings suggest a possible link between oxidative stress and cardiac transplant rejection.

## 18.3 Rationale for Antioxidant/Vitamin Intervention

These scattered reports lend credence to the hypothesis that interventions such as antioxidant vitamin supplementation may be utilized to counteract the impact of oxidative stress on transplant-associated complications. Several studies in experimental models have shown that antioxidants are protective in acute cardiac transplant rejection. This protection was based upon evidence of decreasing histological rejection scores, prolonging graft survival or improving graft function, given either alone or concurrently with cyclosporine A. These interventions include the antioxidant vitamins C and E [23–25], dimethylthiourea [26], iron chelators [27, 28], riboflavin [29], a free radical spin-trapping agent [30], superoxide dismutase mimetics [31, 32], or a peroxynitrite decomposition catalyst [33].

Experimental studies and epidemiological data had suggested that vitamin supplementation might prevent cancer and cardiovascular events [34]. In general, several clinical trials have failed to confirm this hypothesis, possibly due to the short duration of treatment. However, in a long-term clinical trial, no differences were observed in major cardiovascular events; but there was, unexpectedly, a higher risk of heart failure and hospitalizations due to heart failure, as shown in the HOPE and HOPE-TOO trials [35]. This study complemented other studies showing that high doses of vitamin E increased the risk of death. These findings raised new concerns about the appropriate mode of intake, dosage, and formulation of vitamin E in trials and the rationale of a single antioxidant strategy. So some individuals have suggested that a combination vitamin strategy should be pursued.

The first clinical trial to demonstrate the role of antioxidants on the development of endothelial dysfunction in cardiac transplant recipients was reported in the year 2002. The authors found in this double-blind, randomized trial of 40 patients that supplementation with vitamins C and E did not alter the frequency of acute parenchymal rejection, but did retard the development of coronary arteriosclerosis [36]. With the availability and use of statin drugs, a follow-up study of 40 patients by the same group showed that vitamins C and E also inhibited the progression of coronary arteriosclerosis in patients receiving pravastatin concurrent with immunosuppressant therapy [37]. In another study, vitamin C alone given via intracoronary administration was shown to acutely improve responses to serotonin [38].

A consideration in evaluating the effects of antioxidants in transplant recipients is the impact of such intervention on therapeutic levels of immunosuppressant agents. In a retrospective study of 29 patients, it was demonstrated that oral administration of vitamins C and E in cardiac transplant recipients decreased the trough levels of cyclosporine by 30%, but had no effect on trough level of another immunosuppressant agent, tacrolimus. The decrease in trough levels of cyclosporine held for both Sandimmune® and Neoral® formulations [39]. A similar finding of a 25% decrease in trough levels of cyclosporine was reported with renal transplant patients treated with vitamins E and C [40].

These studies draw attention to a thorough understanding of the interaction of water-soluble vs. lipid-soluble vitamins in the context of changes in absorption of agents such as cyclosporine. Cyclosporine is highly lipophilic and requires bile acid for oral absorption. It was found in human subjects that water-soluble derivatives of vitamin E such as Liqui-E® or D-α-tocopheryl poly(ethylene glycol 1,000) succinate actually improved oral absorption of cyclosporine due to the surfactant nature of the poly(ethylene glycol 1,000) forming micelles that allow lipid-soluble drugs such as cyclosporine to be taken up more readily [41]. Water-soluble derivatives apparently have no impact on oral absorption of immmunosuppressant agents such as sirolimus (Rapamune®) [42]. In the context of these findings, it would be interesting to understand the precise formulation of the vitamins used in clinical studies and whether the outcomes of interventions using lipid-soluble antioxidants affect therapeutic levels of immunosuppressant agents used in the transplant setting.

Many of the clinical trials using vitamin E intervention in transplant recipients used a dose of 400 IU/day. Implicit in all of the experimental designs is that the dose utilized actually was a dose that could decrease oxidative stress in humans. A recent study of hypercholesterolemic patients treated with vitamin E revealed that $F_2$-isoprostane levels were decreased, but only after 16 weeks of treatment and only using a much higher dose (i.e., 1,600 and 3,200 IU/day) than the dose of vitamin E previously used in the clinical trials of cardiac transplant recipients mentioned above [43]. This finding might explain the conflicting data arising out of the earlier trials.

## 18.4 Donor Heart Preservation, Ischemia-Reperfusion Injury, and Oxidative Stress

For decades it has been recognized that reactive oxygen is released during reperfusion following warm ischemia. Thus, cardioplegic preservation solutions have been developed that facilitate the cold storage protection of donor hearts for transplantation purposes. Many of the common solutions used clinically antagonize free radicals as one mechanism of protection [44, 45]. However, prolonged cold storage of hearts results in injury following revascularization. Experimentally, this is manifested by a decrease in tissue levels of reduced glutathione (GSH) and accumulation of oxidized glutathione (GSSG) [46, 47] and lipid peroxides [29, 47, 48]. In 2005, Renner and colleagues [49] were the first to show the presence of 4-hydroxy-2-nonenal protein adducts following experimental ischemia in transplanted rat hearts, thereby implicating oxidative modification of donor heart proteins.

Previous studies have shown that intervention with a lipophilic antioxidant, probucol, inhibited ischemia-reperfusion–mediated left ventricular dysfunction in heterotopically-transplanted rat isografts [50, 51]. The role of superoxide anion radical formation in cold ischemia-reperfusion injury in cardiac transplants was revealed by findings that adenoviral transfection to overexpress MnSOD resulted in improved ventricular function following transplantation [52].

## 18.5 Specific Role of Superoxide

### 18.5.1 Superoxide in Cardiac Transplantation

Experimental evidence suggests an important role of superoxide anion radical generation in ischemia-reperfusion injury of cardiac grafts, as well as acute and chronic rejection and chronic transplant vasculopathy. The findings arise out of experimental studies that utilized SOD mimetics or SOD overexpression strategies.

Two studies evaluated the effects of superoxide production arising in cardiac transplants following short-term (1 h or less) cold ischemia. Treatment of donor hearts with polyethylene glycol-conjugated SOD was able to inhibit increased expression of matrix metalloproteinase-9, protein nitration, and vascular endothelial dysfunction at 24 h posttransplantation without any effect on hemodynamic parameters [53]. Using M40401, a cell-permeable SOD mimetic, investigators were able to show a significant decrease in early posttransplant increase in myeloperoxidase activity, TNFα levels, and expression of cell adhesion molecules ICAM-1, VCAM-1, and ELAM-1 after short-term (40 min) cold ischemia [31]. Long-term in this model, the presence of the SOD mimetic in the preservation solution significantly attenuated coronary artery luminal narrowing and the intima-media ratio by posttransplant day 90 in animals treated with cyclosporine for the first 9 days following transplantation. This study was important in showing that superoxide released during ischemia-reperfusion of cardiac grafts can play a long-term role in the development of graft coronary artery disease in cardiac transplants.

In one study, authors evaluated the effects of long-term (6 h) cold ischemia using UW organ-preserving solution in cardiac transplants of wild-type rats and rats overexpressing MnSOD by adenoviral transfection. In this case, the SOD transgene provided improved cardiac function following reperfusion [52]. In another related study, overexpression of CuZnSOD transgene in mice was studied in syngeneic allografts in a model of short-term (20 min) cold ischemia in saline solution (i.e., without preservation solution) [54]. After 4 h of reperfusion, the protective actions of the SOD transgene were demonstrated by reduced superoxide levels and decreased caspase 3 and caspase 9 (but not caspase 8) activity. Caspase activity was used as a marker of apoptosis. The presence of the SOD transgene also significantly decreased the levels of myeloperoxidase, suggesting an impact on neutrophil infiltration. Similarly, decreases were seen in TNFα and MCP-1/CCL2, indicating a protective action on the pro-inflammatory component of ischemia-reperfusion injury of cardiac grafts. Protection against injury was confirmed by the decreased number of apoptotic cardiomyocytes and the decreased release of myocardial CPK into serum.

### 18.5.2 Superoxide in Cardiac Rejection

The role of superoxide in the rejection phase of cardiac transplant recipients has been demonstrated in two studies of experimental cardiac transplant rejection in rodents. In acute cardiac transplant rejection, we have shown that treatment with a cell-permeable SOD mimetic known as manganese (III) tetrakis (1-methyl-4-pyridyl)porphyrin pentachloride (abbreviated as MnTmPyP) inhibited caspase 3 activity and improved graft function, as demonstrated by measurements of diastolic and systolic segment length and fractional shortening using in situ sonomicrometry [32]. The conclusion is that the SOD mimetic interfered with cardiac dysfunction by inhibiting apoptosis of cardiomyocytes. Another group examined coronary artery disease in major histocompatibility complex class II mismatched mouse strains in wild-type and CuZnSOD transgenic mice [54]. In this case, SOD transgene resulted in improved graft function, less intimal thickening, less luminal narrowing, and lower intima-to-media ratio. This suggests that superoxide production plays a key role in the development of graft coronary artery disease independent of the effects of superoxide on the contribution of ischemia-reperfusion injury to long-term graft vasculopathy.

### *18.5.3 Direct Measures of Superoxide in Cardiac Grafts*

As before, superoxide production can potentially impact cardiac transplantation at different stages, including organ storage (ischemia), early reperfusion injury, and acute and chronic rejection. Increases in superoxide production could potentially arise from decreased SOD activity or by increased superoxide production by NAD(P)H oxidases. Nevertheless, direct experimental evidence that superoxide is, in fact, increased in experimental cardiac transplantation is quite rare. Furthermore, the limited amount of published data requires further validation and confirmation. The studies to date arise primarily from our own studies and those of another group at Stanford University.

The effect of ischemia-reperfusion injury on posttransplant production of superoxide was detailed in a rat heart transplant model using dihydroethidine fluorescence of tissue sections [53]. This procedure is based upon the principle that hydroethidine reacts with superoxide to form a fluorescent product. Investigators were able to show that treatment with polyethylene glycol-conjugated SOD diminished fluorescence, suggesting increased superoxide production. In addition, we utilized lucigenin-enhanced chemiluminescence of graft homogenates and dihydroethidine fluorescence of tissue sections of cardiac allografts to show that superoxide levels increased acute cardiac allograft rejection in the absence of immunosuppressant treatments [32]. That signals were specific for superoxide was initially implied by the ability of the cell-permeable SOD mimetic, MnTmPyP, to quench or substantially eliminate the alloimmune-induced increase in signal.

*Limitations.* The dihydroethidine fluorescence assay for superoxide production has been widely used in vascular and cardiac tissue in many disease models. Most studies have used this technique in a fluorescence microscopy modality. Traditionally, the theory was that the probe reacted with superoxide to form the ethidium cation ($E^+$), and the red fluorescence appearing in cells was attributed to

$E^+$. Recently, using an HPLC adaptation for the detection of fluorescence products, it has been learned that the reaction product with superoxide is actually the 2-hydroxyethidium cation (2-OH-$E^+$), another fluorescent product [55, 56]. Furthermore, no $E^+$ was shown to be produced by the reaction of hydroethidine and superoxide.

The finding that the spectra of $E^+$ overlaps that of 2-OH-$E^+$ has raised new questions whether the fluorescence microscopic technique can be used as a reliable method to detect superoxide. Furthermore, porphyrinic SOD mimetics have been traditionally used in a variety of superoxide detection assays to document specificity for detection of superoxide. It was not considered until recently that the SOD mimetics may interact directly with hydroethidine. Indeed, it was shown that the decrease in 2-OH-$E^+$ shown in their presence instead resulted from interaction with and depletion of the precursor, hydroethidine [57]. Thus it is now shown that manganese-derived porphyrinic SOD mimetics cannot be used to unequivocally validate the specificity of dihydroethidine fluorescence for superoxide detection, because any reduction in signal for 2-OH-$^+$ could be due solely to depletion of the precursor for the reaction. Because of these new findings, many new precautions have been advised in the HPLC-detection of superoxide in biological tissue [58, 59]. Although the HPLC method has proven to be successful as a tool to detect superoxide in cells and some tissues, it remains to be proven that it can be successful in the detection of superoxide in cardiac tissue.

Lucigenin-enhanced chemiluminescence is another technique that has been widely used for detection of superoxide in a variety of vascular and cardiac diseases, including ischemia-reperfusion [60], atherosclerosis, diabetes mellitus [61, 62], sepsis [63], hypertension, and hypertrophy [64]. Using the lucigenin-dependent chemiluminescence technique, we were able to show that NADH and NADPH-dependent chemiluminescence was increased in cardiac allografts compared to isografts as controls, suggesting multiple sources of superoxide (unpublished observations). The NADH-dependent production of superoxide was also inhibited by diphenyleneiodonium (DPI), which has typically been interpreted as suggesting the role of flavoprotein-dependent superoxide production. The finding of NADH-dependent superoxide production suggests the possibility that xanthine oxidase, a NADH-dependent generator of superoxide production, or mitochondrial production of superoxide via a NADH:ubiquinoine oxidoreductase or specifically complex I may be candidate sources. However, future studies are necessary to validate this hypothesis.

*Limitations.* The use of lucigenin-enhanced chemiluminescence for superoxide detection in biological tissue has some caveats. These limitations are relevant to detection in cardiac transplants. It has been shown that redox cycling of lucigenin in cells can itself serve as a source of superoxide production [65]. In this process, redox cycling involves the auto-oxidation of lucigenin cation-radical, leading to superoxide production. So it is possible that lucigenin may overestimate NAD(P)H oxidase-driven chemiluminescence detection of superoxide in various biological systems [66]. The use of low concentrations (e.g., 5 $\mu$M) of lucigenin has been advised in order to minimize contributions of such artifacts. So we acknowledge that our original attempts to measure superoxide production in cardiac allograft rejection may, in fact, overestimate alloimmune-induced superoxide production in cardiac allografts.

One group of investigators has used an alternative approach for detection of superoxide production in rat heart transplantation [67, 68]. These researchers used an electron paramagnetic resonance (EPR) analysis of superoxide production using the spin probe, 4-amino-2,2,6,9-tetramethylpiperidine-1-oxyl, or tempamine. The theory behind this is that tempamine traps and stabilizes the radical. For tempamine, the molecule interacts with other radicals and loses its paramagnetic character, and thus the EPR signal for the parent compound, tempamine, is lost. Researchers found increased superoxide production in cold ischemic hearts transplanted into recipients which was dependent on the duration of ischemia [68]. Studies indicated that the superoxide signal at 4 h after reperfusion was increased 1.3-fold after 120 min of cold ischemia vs. 30 min of cold ischemia; but that the signal was not influenced by treatment with a protein kinase C inhibitor [67].

EPR techniques to detect radicals have been used in various experimental vascular and in a few cardiac disease models. Yet, to our knowledge, the report above is the only one to have utilized EPR techniques in the setting of cardiac transplantation, albeit for ischemic-reperfusion injury in cardiac transplants. A limitation of that study is that the investigators did not determine that the loss of EPR signal necessarily arose from superoxide generation as compared to some other radical. In a mouse model of cardiac transplantation from the same group [54], it was demonstrated that superoxide production was increased in transplanted hearts relative to native hearts of graft recipients, showing that the increase was specific to transplantation.

*Limitations.* The specificity for superoxide detection using EPR technology appears not to have been validated to date using SOD mimetics during the analysis procedure. Interestingly, when compared to mice overexpressing SOD1, the presence of the human SOD1 transgene had no effect on the rate of degradation of tempamine EPR signal generated in native heart [54]. Furthermore, there was at most only a 30% lower level of signal degradation following transplantation of hearts in SOD1 transgenic mice compared to wild-type mice [54]. This finding suggests that a majority of the loss of tempamine EPR signal in this model arises from a superoxide-independent mechanism.

Although an advanced biophysical tool, the use of EPR suffers from some of the same issues as lucigenin-enhanced chemiluminescence for detection of superoxide in biological systems; but there are other factors too. Here, tempamine shares properties of piperidine nitroxides, which in the native state are EPR-visible. The rapid reduction of nitroxide to its EPR-invisible, hydroxylamine form limits its application in certain biological settings. The oxoammonium cation species produced by interaction of the superoxide anion with the nitroxide spin probe is susceptible to a two-electron reduction to the hydroxylamine by NAD(P)H and thiol levels [69]. Thus, the intracellular levels of ascorbate and glutathione can significantly alter interpretation of EPR data derived from biological systems [69]. Ubiquinol in mitochondria also convert mitochondrial-targeted nitroxide to hydroxylamine [70].

Given that the myocardium is one tissue that has abundant levels of mitochondria, this could be a serious limitation to its use in superoxide detection in cardiac tissue.

In summary, despite the use of advanced, state-of-the art technology, there appears to be no unequivocal proof that superoxide anion levels are increased in cardiac transplantation.

## 18.6 NADPH Oxidase in Cardiac Transplantation

Upregulation of expression of NADPH oxidases have been reported in the heart due to remodeling and hypertrophy [71, 72]. However, there is a paucity of information on NADPH oxidases as a potential source of superoxide production in cardiac transplantation. Upregulation of phagocytic NAPDH oxidases would be intuitive because of graft infiltration of immune cells (e.g., T-lymphocytes, macrophages, and neutrophils). The information to date in experimental animals and clinically is complicated by the contributions of calcineurin inhibitor-induced hypertension (see later section). One group reported increases in thiobarbituric acid products, an index of lipid peroxidation, and upregulation of protein expression for NADPH oxidase subunits $p22^{phox}$ and $p67^{phox}$ in rat cardiac allografts vs. isografts without immunosuppression [73]. Likewise, we have found increased protein levels of $gp91^{phox}$ in acute rat cardiac rejection (unpublished observations). The increase was seen in allograft hearts but not native hearts of allograft recipients, and was decreased by treatment with cyclosporine. This indicated that NAPDH oxidase upregulation was due to alloimmune activation. To our knowledge, there has been no published report of upregulation of NAPDH oxidases due either to the ischemia-reperfusion injury component of cardiac transplantation or to chronic rejection.

## 18.7 Apoptosis and Oxidative Stress in Cardiac Transplantation

### *18.7.1 Role of Ischemia-Reperfusion–Induced Apoptosis*

It has been considered that prolonged ischemia time could be a risk factor for the development of apoptosis in cardiac transplantation. TUNEL staining and activities of caspases (2, 3, 8, and 9) increased with the duration of cold ischemia following 4 h of reperfusion in a murine model of cardiac transplantation [68]. Activities of caspases 2, 3, and 9, but not caspase 8, could be inhibited in longer-term cold ischemia via a combination of protein kinase C$\epsilon$ activator and protein kinase C$\delta$ inhibitor peptides [67]. TUNEL staining and activities of caspase 3 and 9, but not caspase 8, were significantly inhibited 4 h after transplantation of donor hearts overexpressing CuZnSOD [54]. This suggests that superoxide plays a pivotal role in apoptosis because of the ischemia-reperfusion injury component of posttransplantation.

### *18.7.2 Role of Intrinsic vs. Extrinsic Pathways of Ischemia-Reperfusion–Induced Apoptosis*

Caspase 3 activation is a convergent point for apoptotic signaling via intrinsic vs. extrinsic pathways. Conventionally, activation of caspase 8 via activation of CD95 antigen (Fas) with its ligand CD95L (FasL), and of interaction with tumor necrosis factor with its receptor constitutes extrinsic activation. In contrast, superoxide-mediated mitochondrial activation of caspase 9 mediates the intrinsic apoptotic cascade signaling system. From some studies, the mitochondrial, intrinsic pathway plays a prominent role in cardiomyocyte death after ischemia-reperfusion [74]. This conclusion arises from studies using a genetic inactivation of Fas receptor (C57Bl/6-*lpr* mouse) and examination of mitochondrial permeability transition and its blockade by cyclosporine, an inhibitor of the mitochondrial permeability transition pore opening. However, other studies show activation of both caspase 8 and caspase 9 following short-term reperfusion, suggesting a redundant system of pathways for apoptosis in ischemia-reperfusion injury in transplanted hearts [67, 68]. Studies using the SOD transgenic mouse suggest that the bulk of inhibition of caspase 3 activation and TUNEL staining for apoptosis by SOD transgene likely arises via inhibition of the intrinsic, superoxide-dependent pathway [54].

### *18.7.3 Relationship of Oxidative Stress and Apoptosis in Cardiac Rejection*

Szabolcs et al. [75] were among the first to show apoptosis of cardiac myocytes in acute cardiac allograft rejection in rats. The apoptosis was related to expression of inducible nitric oxide synthase (iNOS). Similar conclusions were made by the same group in human cardiac rejection [76]. Experimentally, a linkage of iNOS expression to apoptosis in acute cardiac allograft rejection has been demonstrated using iNOS knockout mice [77, 78]. These studies are consistent with a variety of studies in many cell types that show that high levels of nitric oxide (NO) can produce apoptosis of cells. However, these studies do not directly implicate NO as the nascent signaling molecule causing apoptosis, but rather implicate iNOS or the possibility of another iNOS-mediated pathway. It could not be excluded that iNOS could possibly be causing apoptosis through another mechanism. For example, iNOS could be producing peroxynitrite, a potent inducer of apoptosis.

Peroxynitrite can be formed by the chemical interaction of NO and superoxide. It is theoretically possible that iNOS via peroxynitrite generation could also mediate apoptosis in cardiac rejection. With this concept in mind, we were able to show that both a specific iNOS inhibitor and a SOD mimetic were individually capable of decreasing lipid peroxidation, apoptosis, and histological rejection in acute cardiac allograft rejection [24]. This suggested that apoptosis could be occurring via oxidative or nitrative stress, or both. These findings were consistent with earlier studies showing that intervention with a metalloporphyrinic peroxynitrite decomposition catalyst inhibited apoptosis, as shown by decreased poly(ADP-ribose) polymerase (PARP) activation in rat cardiac allograft rejection [33]. Furthermore, the attenuation of apoptosis was associated with prolongation of graft survival used either alone or combined with treatment with the immunosuppressant agent cyclosporine.

## 18.8 Reactive Oxygen Species and Immune Suppression

### *18.8.1 Cyclosporine-Induced Production of Superoxide*

The first antirejection drug used in organ transplantation was cyclosporine. Since its development, there has been ongoing concern regarding the issue of cyclosporine-induced side effects, which include nephrotoxicity, hepatotoxicity, neurotoxicity, and hypertension. Several papers have alluded to the role of reactive oxygen species in these side effects owing to the amelioration or reduction of side effects by coadministration of antioxidants (see review [79]). Such studies give indirect evidence that immunosuppressant agents, in particular calcineurin inhibitors, may be a source of oxidative stress in cardiac transplant recipients.

It has been demonstrated that cyclosporine can stimulate the production of superoxide in various cell types, including aortic endothelial cells [80, 81], aortic smooth muscle cells [82], kidney proximal tubule cells [83], and HeLa cells [84]. In vivo utilization of adenoviral transfer of superoxide dismutase and utilization of spin-trapping and electron spin resonance spectroscopy has unequivocally shown that in vivo administration of cyclosporine produces superoxide anion [85]. Xanthine oxidase has been ruled out as an enzymatic source of superoxide because of findings that inhibitors of xanthine oxidase do not alter superoxide production [82, 83]. Other studies indicate that superoxide is derived by the metabolism of cyclosporine by the cytochrome P450 system [82, 83, 86]. This finding was used to explain the lack of increase in oxidized glutathione in heart vs. kidney and liver in rodents treated for 4 weeks with a high dose of cyclosporine [87]. Early studies using the flavoprotein inhibitor diphenyliodonium ruled out NAPDH oxidase as a source of superoxide, at least in rat aortic smooth muscle cells [82].

The possibility that superoxide production could result from NADPH oxidase arose from important clinical studies reported by Calò et al. [88]. These authors showed increased mRNA levels for the $p22^{phox}$ subunit of NAPDH oxidase in monocytes of hypertensive, renal transplant patients receiving cyclosporine or tacrolimus. The finding that treatment with ramipril, an inhibitor of angiotensin converting enzyme (ACE) reversed these changes suggested the role of angiotensin II in cyclosporine-induced superoxide production via NAPDH oxidase in humans. It would be interesting to know whether the same increase in NAPDH oxidase could be demonstrated in patients without hypertension, considering that hypertension alone is a well-known risk factor to increased NADPH oxidase activity in various experimental models [89]. In experimental renal transplantation, we have also shown that both mRNA and protein for NOX-1 and $p22^{phox}$ were upregulated in rats treated with tacrolimus [90]. The increase in NOX-1 suggests a nonphagocytic NAPDH oxidase as a potential significant source of superoxide production due to immunosuppressant therapy. This was shown by us in cultured proximal tube cells in which both cyclosporine and tacrolimus individually could upregulate NOX-1 in these cells (unpublished observations). In this case, phagocytic cells were absent, indicating that NADPH oxidases in parenchymal cells of organ transplant recipients might be activated directly as a consequence of the immunosuppressant therapy.

The consequence of increased superoxide production by cyclosporine on oxidative stress per se has been shown in in vivo nontransplant models. In this case, chronic treatment with cyclosporine, but not rapamycin, resulted in increased aortic levels of 8-isoprostanes [91]. Increased oxidative stress due to cyclosporine was also shown in a nontransplant model by the increase in urinary 8-hydroxy-2′-deoxyguanosine [92]. An increase in urinary $F_2$-isoprostanes in uninephrectomized normal rats as a consequence of treatment with cyclosporine was prevented by concurrent treatment with vitamin E [93, 94]. Collectively, these studies were crucial in showing that oxidative stress could be induced directly by cyclosporine, independent of inflammatory responses that might be occurring in the setting of alloimmune-induced transplant rejection.

### *18.8.2 Cyclosporine-Induced Oxidative Stress*

As reported above, one study indicated that treatment of normal rats without transplantation with 30 mg/kg cyclosporine did not increase oxidized glutathione levels in the heart but did so in the kidney and liver [87], suggesting that the heart may be resistant to the oxidative stress elicited by cyclosporine. However, another study showed that treatment of normal rats for 21 days with a lower concentration of 15 mg/kg reduced glutathione levels and increased lipid peroxides while concurrently decreasing antioxidant enzymes in heart tissue [95]. The latter agreed with earlier findings of increased reactive oxygen species production as determined by dichlorofluorescein fluorescence in normal rats treated for 21 days with 15 mg/kg cyclosporine [96]. It is not clear from these studies if the oxidative stress in the heart is related secondarily to cyclosporine-induced hypertension arising from cyclosporine-induced effects on renal function. Studies conducted in an embryonic rat heart myoblast-derived cell line (H9c2) show that cyclosporine can elicit a concentration-dependent increase in reactive oxygen production in cardiac cells [97]. Similarly, cyclosporine A increased lipid peroxidation in isolated rat cardiac myocytes [98].

### *18.8.3 Reactive Oxygen and Other Immunosuppressant Agents*

Much of the evidence that immunosuppressant agents may be a source of oxidative stress through production of reactive oxygen species comes from studies examining the calcineurin inhibitor cyclosporine. Much less is known about other immunosuppressant agents. Transplant patients often receive a combination of drugs including cyclosporine, glucocorticoids, tacrolimus, and sirolimus, depending on conditions and the specific organ that is transplanted. However, clinically it is very difficult to dissect the precise mechanism of oxidative stress because of the complexities of multiple drug regimens.

Experimental studies have shown that glucocorticoids may enhance oxidative stress. Indeed, microvascular oxidative stress has been implicated in glucocorticoid-induced hypertension in rats [99]. The addition of low concentrations of dexamethasone directly to human umbilical vein endothelial cells in culture resulted in increases in reactive oxygen species as determined by dichlorofluoroscein and confirmed by EPR spectroscopy [100]. The source of reactive oxygen species production was determined to be via the mitochondrial electron transport chain, NAPDH oxidase, and xanthine oxidase. These findings indicate that glucocorticoids can produce reactive oxygen species independent of vascular disease and could be a potential mechanism of oxidative stress in cardiac transplant recipients.

In vivo studies in rodents confirm the presence of oxidative stress in dexamethasone-induced hypertension by increases in $F_2$-isoprostantes [101, 102]. To our knowledge, however, there is no convincing evidence that glucocorticoids increase reactive oxygen species production in humans nor whether they contribute to oxidative stress in transplant recipients.

The role of increased ROS production from treatment with tacrolimus, rapamycin, or sirolimus is less certain. The possibility for increased production was demonstrated in Wistar rats with sirolimus-releasing stents [103]. Increased reactive oxygen species production was demonstrated by chemiluminescence and increased mRNA expression of NADPH oxidase subunits, NOX-1, and membrane translocation of cytosolic $p67^{phox}$. Participation of mitochondria as an additional source of reactive oxygen species was indicated, but reactive oxygen species production arising from eNOS uncoupling was excluded. As before, there is no clinical evidence to support that these agents, as opposed to cyclosporine, contribute to oxidative stress in human cardiac transplantation.

### *18.8.4 Immunosuppression, Cytomegalovirus, and Oxidative Stress*

The survival of organ transplants requires the use of immunosuppressant agents. Because of this strategy, opportunistic infections are a side effect. Cytomegalovirus (CMV) infection found in transplant recipients is a significant risk factor for transplant-associated coronary arteriosclerosis. A few studies have implicated CMV infection with increased production of reactive oxygen species. Exposure of human coronary smooth muscle cells or endothelial cells to CMV causes increases in reactive oxygen species production [104, 105]. NADPH oxidase was recently implicated for the first time in studies showing that CMV also increases NADPH oxidase activity in coronary artery smooth muscle cells [106]. Future studies are necessary to determine the type of NADPH oxidase implicated in CMV infection-induced production of reactive oxygen species. There are no studies that have determined the relative difference in oxidant stress in CMV seropositive transplant recipients vs. recipients without CMV infection.

## 18.9 The Triad of the Renin-Angiotensin System, Tgf-β, and Oxidative Stress in Transplantation

### *18.9.1 Effect of Angiotensin II on the Heart*

Activation of the renin-angiotensin system in the kidney due to immunosuppressant agents can feed back to regulate actions directly on the heart. These actions include contractile dysfunction [107], hypertrophy [108], fibrosis [109], and cardiomyocyte apoptosis [110, 111]. The use of angiotensin receptor II–type blockers or angiotensin converting enzyme inhibitors have been proven to protect against chronic transplant-associated vasculopathy in experimental rodent models of cardiac transplantation [112, 113]. ACE inhibitor treatment has been shown in vitro to inhibit oxidative stress in endothelial cells [114]. Clinically, the use of ACE inhibitors has been shown to decrease cardiac allograft vasculopathy [114, 115]. The cardiomyocyte dysfunction and apoptosis caused by angiotensin II was antagonized by antibodies specific to TGF-β1, suggesting a role for TGF-β1 in signaling [107, 111].

### *18.9.2 TGF-β, Oxidative Stress, and Cardiac Transplantation*

TGF-β is widely recognized as a generator of reactive oxygen species and as a fibrogenic cytokine (Fig. 18.2). Clinically, expression of TGF-β positively correlated with left ventricular dysfunction in human cardiac transplant recipients [116, 117]. This study indicated an important role of TGF-β in clinical cases of cardiac transplantation.

The renal fibrogenic effects of cyclosporine treatment were early shown to be blocked by anti-TGF-β antibody, suggesting that the detrimental actions of cyclosporine are mediated by TGF-β [118]. This was corroborated in later studies [119]. Low doses, but not high doses, of anti-TGF-β antibody were also shown to inhibit cyclosporine-induced nephrotoxicity in a rat model of cardiac transplantation without altering its immunosuppressant activities [120]. This was the first study to show the potential efficacy of anti-TGF-β strategies to provide renal protection in the setting of cardiac transplantation.

TGF-β has been known to elicit production of reactive oxygen species in different cell types [121–124]. More recent studies suggest the role of NADPH oxidase in cellular signaling, in cardiac fibroblast differentiation into myofibroblasts [125].

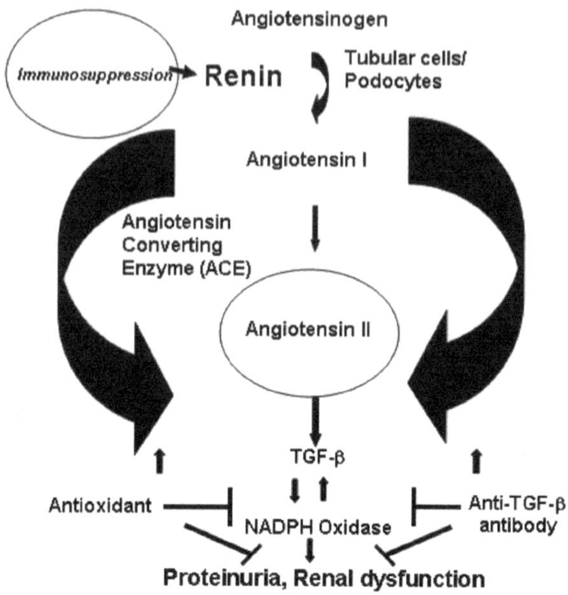


**Fig. 18.2** Diagram showing the triad of interaction of the renin-angiotensin system, TGF-β, and oxidative stress in cardiac transplantation

This was one of the first studies to implicate TGF-β-mediated signaling of reactive oxygen species generation via NADPH oxidase.

Previously it was shown that both tacrolimus and cyclosporine upregulated TGF-β and fibrogenic gene expression in renal transplant recipients [126]. Experimentally, we have shown in a rat transplant model that tacrolimus increased NADPH oxidase subunit expression [90]. In addition to increased NADPH oxidase, tacrolimus also decreased endogenous antioxidant defense systems (i.e., superoxide dismutase and thioredoxin). This indicates that increased oxidative stress might be mediated both by increased reactive oxygen production and by decreased removal of reactive oxygen species. These findings give added support to the concept that antioxidants may be potential therapeutic agents used to control immunosuppression-induced oxidative stress in transplant recipients.

A role of TGF-β in the effects of tacrolimus was shown by findings that anti-TGF-β antibody, but not control antibody, ameliorated the increase in expression of NADPH oxidase as well as the decrease in expression of endogenous antioxidant defenses. Recent studies have suggested that reactive oxygen species production caused by cyclosporine and tacrolimus may act upstream of TGF-β signaling by augmenting the activation of TGF-β release from its latent form [127]. Our studies illustrate that increases in TGF-β as a consequence of immunosuppression may act as an upstream regulator of NADPH oxidase activity as a source of oxidative stress. Finally, our finding that anti-TGF-β antibody prevented nephrotoxicity associated with treatment with tacrolimus [90] illustrates the central role of TGF-β in linking immunosuppression, renal dysfunction, and oxidative stress in transplant recipients.

**Acknowledgments** This research was supported, in part, by National Institutes of Health, Heart and Lung Institute Grant #HL078937.

# References

1. Lehr HA, Messmer K (1996) Rationale for the use of antioxidant vitamins in clinical organ transplantation. Transplantation 62:1197–1199
2. Boucek M, Waltz D, Edwards L et al (2006) Registry of the International Society for Heart and Lung Transplantation: Ninth official pediatric heart transplant report-2006. J Heart Lung Transplant 25:893–903
3. Taylor DO, Edwards LB, Aurora P et al (2008) Registry of the International Society for Heart and Lung Transplantation: Twenty-fifth official adult heart transplant report-2005. J Heart Lung Transplant 27:943–956
4. Cristol JP, Vela C, Maggi MF et al (1998) Oxidative stress and lipid abnormalities in renal transplant recipients with or without chronic rejection. Transplantation 65:1322–1328
5. De Lorgeril M, Richard MJ, Arnaud J et al (1993) Lipid peroxides and antioxidant defenses in accelerated transplantation-associated coronary arteriosclerosis. Am Heart J 125:974–980
6. Heidland A, Šebeková K, Frangiosa A et al (2004) Paradox of circulating advanced glycation end product concentrations in patients with congestive heart failure and after heart transplantation. Heart J 11:1269–1274
7. Kucharská J, Gvozdjáková A, Mizera S et al (1998) Participation of coenzyme $Q_{10}$ in the rejection development of the transplanted heart: a clinical study. Physiol Rev 47:399–404
8. Pechán I, Danová K, Olejarová I et al (2003) Oxidative stress and antioxidant defense systems in patients after heart transplantation. Wien Klin Wochenschr 115:648–651
9. Pérez O, Castro P, Diaz-Araya G et al (2002) Persistence of oxidative stress after heart transplantation: a comparative study of patients with heart transplant versus chronic stable heart failure. Rev Esp Cardiol 55:831–837
10. Calò L, Semplicini A, Davis PA et al (2000) Cyclosporin-induced endothelial dysfunction and hypertension: are nitric oxide system abnormality and oxidative stress involved? Transpl Int 13(suppl 1):S413–S418
11. Roberts LJII, Morrow JD (2000) Measurement of $F_2$-isoprostanes as an index of oxidative stress in vivo. Free Radical Biol Med 28:505–513
12. Burke A, FitzGerald GA, Lucey MR (2002) A prospective analysis of oxidative stress and liver transplantation. Transplantation 74:217–221
13. Djamali A, Sadowski EA, Muehrer RJ et al (2007) BOLD-MRI assessment of intrarenal oxygenation and oxidative stress in patients with chronic kidney allograft dysfunction. Am J Phyiol Renal Physiol 292:F513–F222
14. Phillips M, Boehmer J, Cataneo R et al (2002) Heart allograft rejection: detection with breath alkanes in low levels (the HARDBALL study). J Am Coll Cardiol 1:12–13
15. Phillips M, Boehmer J, Cataneo R et al (2004) Heart allograft rejection: detection with breath alkanes in low levels (the HARDBALL study). J Heart Lung Transplant 23:701–704
16. Phillips M, Boehmer J, Cataneo R et al (2004) Prediction of heart transplant rejection with a breath test for markers of oxidative stress. Am J Cardiol 94:1593–1594
17. De Chiara B, Bigi R, Campolo J et al (2005) Blood glutathione as a marker of cardiac allograft vasculopathy in heart transplant recipients. Clin Transplant 19:367–371
18. Holvoet P, Van Cleemput J, Collen D et al (2000) Oxidized low density lipoprotein is a prognostic marker of transplant-associated coronary artery disease. Arterioscler Thromb Vasc Biol 20:698–702
19. Fang JC, Finlay S, Behrendt D et al (2002) Circulating autoantibodies to oxidized LDL correlated with impaired coronary endothelial function after cardiac transplantation. Arterioscler Thromb Vasc Biol 22:2044–2048

20. Schimke I, Schikora M, Meyer R et al (2000) Oxidative stress in the human heart is associated with changes in the antioxidative defense as shown after heart transplantation. Mol Cell Biochem 204:89–96
21. Nilakantan V, Halligan NLN, Nguyen TK et al (2005) Post-translational modification of manganese superoxide dismutase in acutely rejecting cardiac transplants: role of inducible nitric oxide synthase. J Heart Lung Transplant 24:1591–1599
22. Roza AM, Pieper GM, Moore-Hilton G et al (1994) Free radicals in pancreatic and cardiac allograft rejection. Transplant Proc 26:544–545
23. Nguyen TK, Nilakantan V, Felix CC et al (2006) Beneficial effect of α-tocopheryl succinate in rat cardiac transplants. J Heart Lung Transplant 25:707–715
24. Pieper GM, Nilakantan V, Nguyen NK et al (2008) Reactive oxygen and reactive nitrogen as signaling molecules for caspase 3 activation in acute cardiac transplant rejection. Antiox Redox Signal 10:1031–1039
25. Slakey DP, Roza AM, Pieper GM et al (1993) Delayed cardiac allograft rejection due to combined cyclosporine and antioxidant therapy. Transplantation 56:1305–1309
26. Pieper GM, Olds C, Hilton G et al (2001) Antioxidant treatment inhibits activation of myocardial nuclear factor κB and inhibits nitrosylation of myocardial heme protein in cardiac transplant rejection. Antioxid Redox Signal 3:81–88
27. Cooper M, Lindholm P, Pieper G et al (1998) Myocardial nuclear factor-κB activity and nitric oxide production in rejecting cardiac allografts. Transplantation 66:838–844
28. Pieper GM, Nilakantan V, Hilton G et al (2002) Mechanisms of the protective action of diethyldithiocarbamate-iron complex on acute cardiac allograft rejection. Am J Physiol Heart Circ Physiol 284:H1542–H1551
29. Iwanaga K, Hasegawa T, Hultquist DE et al (2007) Riboflavin-mediated reduction of oxidant injury, rejection, and vaculopathy after cardiac allotransplantation. Transplantation 83: 747–753
30. Pieper GM, Nilakantan V, Zhou X et al (2005) Treatment with α-phenyl-*N*-*tert*-butylnitrone, a free radical-spin trapping agent, abrogates inflammatory cytokine gene expression during alloimmune activation in rat cardiac allografts. J Pharmacol Exp Thera 312: 774–779
31. Murata S, Miniati DN, Kown MH et al (2004) Superoxide dismutase mimetic M40401 reduces ischemia-reperfusion injury and graft coronary artery disease in rodent cardiac allografts. Transplantation 78:1166–1171
32. Nilakantan V, Zhou X, Hilton G et al (2006) Antagonizing reactive oxygen by treatment with a manganese (III) metalloporphyrin-based superoxide dismutase mimetic in cardiac transplants. J Thorac Cardiovasc Surg 131:898–906
33. Pieper GM, Nilakantan V, Chen M et al (2005) Protective mechanisms of a metalloporphyrinic peroxynitrite decomposition catalyst, WW85, in rat cardiac transplants. J Pharmacol Exp Thera 314:53–60
34. Stanner SA, Hughes J, Kelly CNM et al (2003) A review of the epidemiological evidence for the 'antioxidant hypothesis'. Public Health Nutr 7:407–422
35. HOPE and HOPE-TOO Trial Investigators (2005) Effects of long-term vitamin E supplementation on cardiovascular events and cancer: A randomized control trial. J Am Med Assoc 293:1338–1347
36. Fang JC, Kinlay S, Beltrame J et al (2002) Effect of vitamin C and E on progression of transplant-associated arteriosclerosis: a randomized trial. Lancet 359: 1108–1113
37. Behrendt D, Beltrame J, Hikiti H et al (2006) Impact of coronary endothelial dysfunction on the progression of cardiac transplant-associated arteriosclerosis: effect of anti-oxidant vitamins C and E. J Heart Lung Transplant 25:426–433
38. Berkenboom G, Preumont N, Pradier O et al (2006) Relation of coronary hypersensitivity to serotonin in cardiac transplant recipients to vessel wall morphology and effect of vitamin C. Am J Cardiol 97:561–566

39. Lake KD, Aaronson KD, Gorman LE et al (2005) Effect of oral vitamin E and C therapy on calcineurin inhibitor levels in heart transplant recipients. J Heart Lung Transplant 24: 990–994
40. Blackhall ML, Fassett RG, Sharman JE et al (2005) Effects of antioxidant supplementation on blood cyclosporine A and glomerular filtration rate in renal transplant recipients. Nephrol Dial Transplant 20:1970–1975
41. Chang T, Benet LZ, Hebert MF (1996) The effect of water-soluble vitamin E on cyclosporine pharmacokinetics in healthy volunteers. Clin Pharmacol Thera 59:297–303
42. Wacher VJ, Silverman JA, Wong S et al (2002) Sirolimus oral absorption in rats is increased by ketoconazole but is not affected by D-α-tocopheryl poly(ethylene glycol 1000) succinate. J Pharmacol Exp Thera 303:308–313
43. Roberts LJ II, Oates JA, Linton MF et al (2007) The relationship between dose of vitamin E and suppression of oxidative stress in humans. Free Radical Biol Med 43:1388–1393
44. Jahania MS, Sanchez JA, Narayan P et al (1999) Heart preservation for transplantation: principles and strategies. Ann Thorac Surg 68:1983–1987
45. Maathuis MHJ, Leuvenink HGD, Ploeg RJ (2007) Perspectives in organ preservation. Transplantation 83:1289–1298
46. Cargnoni A, Ceconi C, Bernocchi P et al (1999) Changes in oxidative stress and cellular redox potential during myocardial storage for transplantation: experimental studies. J Heart Lung Transplant 18:478–487
47. Renner A, Sagstetter MR, Götz ME et al (2004) Heterotopic rat heart transplantation: severe loss of glutathione in 8-hour ischemic hearts. J Heart Lung Transplant 23:1093–1102
48. Nakao A, Neto JS, Kanna S et al (2004) Protection against ischemia/reperfusion injury in cardiac and renal transplantation with carbon monoxide, biliverdin and both. Am J Transplant 5:282–291
49. Renner A, Sagstetter MR, Harms H (2005) Formation of 4-hydroxy-2-nonenal protein adducts in the ischemic rat heart after transplantation. J Heart Lung Transplant 24: 730–736
50. Rabkin DG, Jia CX, Spotnitz HM (1999) Attenuation of reperfusion injury with probucol in the heterotopic rat cardiac isograft. J Heart Hung Transplant 18:775–780
51. Rabkin DG, Weinberg AD, Spotnitz HM (2003) Optimizing probucol administration to preserve left ventricular compliance after reperfusion injury in the heterotopic rat heart isograft. J Heart Lung Transplant 22:959–966
52. Abunasra H, Smolenski RT, Yap J et al (2006) Comparison of two gene transfer models for attenuation of myocardial ischemia-reperfusion injury following preservation for cardiac transplantation. Eur J Cardio-Thorac Surg 29:772–778
53. Egi K, Conrad NE, Kwan J et al (2004) Inhibition of inducible nitric oxide synthase and superoxide production reduces matrix metalloproteinase-9 activity and restores coronary vasomotor function in rat cardiac allografts. Eur J Cardio-Thorac Surg 26:262–269
54. Tanaka M, Mokhtari GK, Terry RD et al (2004) Overexpression of human copper/zinc superoxide dismutase (SOD1) suppresses ischemia-reperfusion injury and subsequent development of graft coronary artery disease in murine cardiac grafts. Circulation 110(Suppl II):II-200–II-206
55. Zhao H, Joseph J, Fales HM et al (2005) Detection and characterization of the product of hydroethidine and intracellular superoxide by HPLC and limitations of fluorescence. Proc Natl Acad Sci USA 102:5727–5732
56. Zhao H, Kalivendi S, Zhang H et al (2003) Superoxide reacts with hydroethidine but forms a fluorescent product that is distinctly different from ethidium: potential implications in intracellular fluorescence detection of superoxide. Free Radical Biol Med 34: 1359–1368
57. Zielonka J, Vasquez-Vivar J, Kalyanaraman B (2006) The confounding effects of light, sonication, and Mn(III)TBAP on quantitation of superoxide using hydroethidine. Free Radical Biol Med 41:1050–1057

58. Laurindo FRM, Fernandes DC, Santos CXC (2008) Assessment of superoxide production and NADPH oxidase activity by HPLC analysis of dihydroethidium oxidation products. Meth Enzymol 441:237–260
59. Zielonka J, Vasquez-Vivar J, Kalyanaraman B (2008) Detection of 2-hydroxyethidium in cellular systems: a unique marker product of superoxide and hydroethidine. Nature Protocols 3:8–21
60. Bassenge E, Sommer O, Schwemmer M et al (2000) Antioxidant pyruvate inhibits cardiac formation of reactive oxygen species through changes in redox state. Am J Physiol Heart Circ Physiol 279:H2431–H2438
61. Guo Z, Zia X, Jiang J et al (2007) Downregulation of NADPH oxidase, antioxidant enzymes, and inflammatory markers in the heart of streptozotocin-induce diabetic rats by N-aceyl-L-cysteine. Am J Physiol Heart Circ Physiol 292:H1728–H1736
62. Wendt MC, Daiber A, Kleschyov AL et al (2005) Differential effects of diabetes on the expression of the gp91$^{phox}$ homologues nox1 and nox4. Free Radical Biol Med 39(381–391):2005
63. Khadour FH, Panas D, Ferdinandy P et al (2002) Enhanced NO and superoxide generation in dysfunctional hearts from endotoxemic rats. Am J Physiol Heart Circ Physiol 283: H1108–H1115
64. Li JM, Gall NP, Grieve DJ et al (2002) Activation of NADPH oxidase during progression of cardiac hypertrophy to failure. Hypertension 40:477–484
65. Vásquez-Vivar J, Hogg N, Pritchard KA Jr et al (1997) Superoxide anion formation from lucigenin: an electron spin resonance spin-trapping study. FEBS Let 403:127–130
66. Janiszewksi M, Souza HP, Liu X et al (2002) Overestimation of NADH-driven vascular oxidase activity due to lucigenin artifacts. Free Radical Biol Med 32:446–453
67. Tanaka M, Gunawan F, Terry RD et al (2005) Inhibition of heart transplant injury and graft coronary artery disease after prolonged organ ischemia by selective protein kinase C regulators. J Thorac Cardiovasc Surg 129:1160–1167
68. Tanaka M, Mokhtari GK, Terry RD et al (2005) Prolonged cold ischemia in rat cardiac allografts promotes ischemia-reperfusion injury and the development of graft coronary artery disease in a linear fashion. J Heart Lung Transplant 24:1906–1914
69. Bobko AA, Kirilyuk IA, Grigorév IA et al (2007) Reversible reduction of nitroxides to hydroxylamines: roles for ascorbate and glutathione. Free Radical Biol Med 42:404–412
70. Trnka J, Blaikie FH, Smith RAJ et al (2008) A mitochondrial-targeted niroxide is reduced to its hydroxylamine by ubiquinol in mitochondria. Free Radical Biol Med 44:1406–1419
71. Dworakowski R, Anilkumar N, Zhang M et al (2006) Redox signaling involving NADPH oxidase-derived reactive oxygen species. Biochem Soc Trans 34:960–963
72. Sirker A, Zhang M, Murdoch C et al (2007) Involvement of NAPDH oxidases in cardiac remodeling and heart failure. Am J Nephrol 27:649–660
73. Vela C, Thomsen M, Delbosc S et al (2007) Lipid and oxidative stress disorders in a rat model of chronic rejection. Transplant Proc 39:2617–2619
74. Gomez L, Chavanis N, Argaud L et al (2005) Fas-independent mitochondrial damage triggers cardiomyocyte death after ischemia-reperfusion. Am J Physiol Heart Circ Physiol 289:H2153–H2158
75. Szabolcs M, Michler RE, Yang W et al (1996) Apoptosis of cardiac myocytes during cardiac allograft rejection: Relation to induction of nitric oxide synthase. Circulation 94:1665–1673
76. Szabolcs MJ, Ravalli S, Minanov O et al (1998) Apoptosis and increased expression of inducible nitric oxide synthase in human allograft rejection. Transplantation 65:804–812
77. Koglin J, Granville DJ, Glysing-Jensen T et al (1999) Attenuated acute cardiac rejection in NOS2–/– recipients correlates with reduced apoptosis. Circulation 99:836–842
78. Szabolcs MJ, Ninsheng M, Athan E et al (2001) Acute cardiac allograft rejection in nitric oxide synthase-2$^{-/-}$ and nitric oxide synthase-2$^{+/+}$ mice: Effect of cellular chimeras on myocardial inflammation and cardiomyocyte damage and apoptosis. Circulation 103: 2514–2520

79. Rezzani R (2006) Exploring cyclosporine A side effects and the protective role played by antioxidants: the morphological and immunohistochemical studies. Histol Histopathol 21:302–316
80. López-Ongil S, Hernández-Perara O, Navarro-Antolín J et al (1998) Role of reactive oxygen species in signaling cascade of cyclosporine A-mediated up-regulation of eNOS in vascular endothelial cells. Brit J Pharmacol 124:447–454
81. Navarro-Antolín J, López-Muñoz MJ, Soria J et al (2002) Superoxide limits cyclosporine-induced formation of peroxynitrite in endothelial cells. Free Radical Biol Med 32:702–711
82. Krauskopf A, Lhote P, Petermann O et al (2005) Cyclosporin A generates superoxide in smooth muscle cells. Free Radical Res 39:913–919
83. Vetter M, Chen ZJ, Chang GD et al (2003) Cyclosporin A disrupts bradykinin signaling through superoxide. Hypertension 41:1136–1142
84. Horáková K, Sovčíková A, Šeemannová Z et al (2001) Detection of drug-induced, superoxide-mediated cell damage and its prevention by antioxidants. Free Radical Biol Med 30:650–664
85. Zhong Z, Connor HD, Yin M et al (2001) Viral delivery of superoxide dismutase gene reduces cyclosporine A-induced nephrotoxicity. Kidney Int 59:1397–1404
86. Ahmed SS, Napoli KL, Strobel HW (1995) Oxygen radical formation during cytochrome P450-catalyzed cyclosporine metabolism in rat and human liver microsomes at varying hydrogen ion concentrations. Mol Cell Biochem 151:131–140
87. Haberland A, Henk W, Grune T et al. (1997) Differential response of oxygen radical metabolism in rat heart, liver and kidney to cyclosporine A treatment. Inflamm Res 46:452–454
88. Calò LA, Davis PA, Giacon B et al (2002) Oxidative stress in kidney transplant patients with calcineurin inhibitor-induced hypertension: effect of ramipril. J Cardiovasc Pharmacol 40:625–631
89. Cave AC, Brewer AC, Narayanapanicker A et al (2006) Comprehensive invited review NADPH oxidases in cardiovascular health and disease. Antiox Redox Signal 8:691–728
90. Khanna AK, Pieper GM (2007) NADPH oxidase subunits (NOX-1, $p22^{phox}$, Rac-1) and tacrolimus-induced nephrotoxicity in a rat renal transplant model. Nephrol Dial Transplant 22:376–385
91. Ramzy D, Rao V, Tumiati LC et al (2006) Role of endothelin-1 and nitric oxide bioavailability in transplant-related vascular injury. Comparative effects of rapamycin and cyclosporine. Circulation 114 (suppl I):I–214–I–219
92. Ghee JY, Han DH, Song HK et al (2008) The role of macrophage in the pathogensisis of chronic cyclosporine-induced nephropathy. Nephrol Dial Transplant. doi:10.1093/ndt/gfn388
93. Kanji VK, Wang C, Salahudeen AK (1999) Vitamin E suppresses cyclosporine A-induced increased in the urinary excretion of arachidonic acid metabolites including $F_2$-isoprostanes in the rat model. Transplant Proc 21:1724–1728
94. Lungu AO, Zin ZG, Yamawaki H et al (2004) Cyclosporin A inhibits flow-mediated activation of endothelial nitric-oxide synthase by altering cholesterol content in caveolae. J Biol Chem 279:48794–48800
95. Yüce A, Ateşşahin A, Ceribaşı AO (2008) Amelioration of cyclosporine A-induced renal, hepatic and cardiac damages by ellagic acid in rats. Basic Clin Pharmacol Toxicol 103: 186–191
96. Rezzani R, Giugno L, Buffoli B et al (2005) The protective effect of caffeic acid phenethyl ester against cyclosporine A-induced cardiotoxicity in rats. Toxicology 212:155–164
97. Chen HW, Chien CT, Yu SL et al (2002) Cyclosporine A regulate oxidative stress-induced apoptosis in cardiomyocytes: mechanisms via ROS generation, iNOS and Hsp70. Brit J Pharmacol 137:771–781
98. Florio S, Ciarcia R, Crispino L et al (2003) Hydrocortisone has a protective effect on cyclosporine A-induced cardiotoxicity. J Cell Physiol 195:21–26

99. Suzuki H, Swei A, Zweifach BW et al (1995) In vivo evidence for microvascular oxidative stress in spontaneously hypertensive rats: hydroethidine microfluorography. Hypetension 25:1083–1089
100. Iuchi T, Akaike M, Mitsui T et al (2003) Glucocorticoid excess induces superoxide production in vascular endothelial cells and elicits vascular endothelial dysfunction. Circ Res 92:81–87
101. Miao Y, Zhang Y, Lim PS et al (2007) Folic acid prevents and partially reverses glucocorticoid-induced hypertension in the rat. Am J Hypertens 20:304–310
102. Zhang K, Fujii S, Igarashi J et al (2004) Effects of thiol antioxidants on reduced nicotinamide adenine dinucleotide phosphate oxidase in hypertensive Dahl salt-sensitive rats. Free Radical Biol Med 37:1813–1820
103. Jabs A, Göbel S, Wenzel P et al (2008) Sirolimus-induced vascular dysfunction increased mitochondrial and nicotinamide adenosine-dinucleotide phosphate oxidase-dependent superoxide production and decreased vascular nitric oxide formation. J Am Coll Cardiol 51:2130–2138
104. Speier E, Yu ZX, Takeda K et al (2000) Antioxidant effect of estrogen on cytomegalovirus-induced gene expression in coronary artery smooth muscle cells. Circulation 102:2990–2996
105. Weis M, Kledal TN, Lin KY et al (2004) Cytomegalovirus infection impairs the nitric oxide synthase pathway: Role of asymmetric dimethylarginine in transplant arteriosclerosis. Circulation 109:500–505
106. Dhaunsi GS, Kaur J, Turner RB (2003) Role of NADPH oxidase in cytomegalovirus-induced proliferation of human coronary artery smooth muscle cells. J Biomed Sci 10:505–509
107. Mufti S, Wenzel S, Euler G et al (2008) Angiotensin II-dependent loss of cardiac function: mechanisms and pharmacological targets attenuating this effect. J Cell Physiol 217:242–249
108. Crowley SD, Gurley SB, Herrera MJ et al (2006) Angiotensin II causes hypertension and cardiac hypertrophy through its receptors in the kidney. Proc Natl Acad Sci USA 103: 17985–17990
109. Johar S, Cave AC, Narayanapanicker A et al (2006) Aldosterone mediates angiotensin II-induced interstitial cardiac fibrosis via a Nox2-containing NAPDH oxidase. FASEB J 20:E846–E854
110. Grishko V, Pastukh V, Solodushko V et al (2003) Apoptotic cascade initiated by angiotensin II in neonatal cardiomyocytes: role of DNA damage. Am J Physiol Heart Circ Physiol 285:H2364–H2372
111. Schröder D, Heger J, Piper HM et al (2006) Angiotensin II stimulates apoptosis via TGF-β1 signaling in ventricular cardiomyocytes of rat. J Mol Med 84:975–983
112. Richter MHC, Richter HR, Olbrich HG et al (2003) Two good reasons for an angiotensin-II type 1 receptor blockade with losartan after cardiac transplantation: reduction of incidence and severity of transplant vasculopathy. Transplant Int 16:26–32
113. Richter M, Skupin M, Grabs R et al (2000) New approach in the therapy of chronic rejection? ACE- and AT1-blocker reduce the development of chronic rejection after cardiac transplantation in a rat model. J Heart Lung Transplant 19:1047–1055
114. Steinhauff S, Pehlivanli S, Bakovic-Alt R et al (2004) Beneficial effects of quinaprilat on coronary vasomotor function, endothelial oxidative stress, and endothelin activation after human heart transplantation. Transplantation 77:1859–1865
115. Bae JH, Rihal CS, Edwards BS et al (2006) Association of angiotensin-converting enzyme inhibitors and serum lipids with plaque regression in cardiac allograft vasculopathy. Transplantation 82:1108–1111
116. Aziz TM, Burgess MI, Haselton PS et al (2003) Transforming growth factor β and diastolic left ventricular dysfunction after heart transplantation: echocardiographic and histologic evaluation. J Heart Lung Transplant 22:663–673
117. Aziz T, Saad RA, Burgess M et al (2003) Transforming growth factor beta and myocardial dysfunction following heart transplantation. Eur J Cardio-thorac Surg 20:177–186

118. Khanna AK, Cairns VR, Becker CG et al (1999) Transforming growth factor (TGF)-β mimics and anti-TGF-β antibody abrogates the in vivo effects of cyclosporine. Transplantation 67:882–889
119. Ling H, Li X, Jha S et al (2003) Therapeutic role of TGF-β-neutralizing antibody in mouse cyclosporine A nephropathy: morphologic improvement associated with functional preservation. J Am Soc Nephrol 14:377–388
120. Khanna AK, Plummer MS, Hilton G et al (2004) Anti-transforming growth factor antibody at low but not high doses limits cyclosporine-mediated nephrotoxicity without altering rat cardiac allograft survival: Potential of therapeutic applications. Circulation 110:3822–3829
121. Hertig I, Hassoun P, Zuleta J et al (1993) Mechanism of basal and transforming growth factor β1 stimulated $H_2O_2$ release by endothelial cells. Trans Assoc Am Phys 106:179–186
122. Junn E, Lee KN, Ju HR et al (2000) Requirement of hydrogen peroxide generation in TGF-β1 signal transduction in human lung fibroblast cells: involvement of hydrogen peroxide and $Ca^{2+}$ in TGF-β1-induced IL-6 expression. J Immunol 165:2190–2197
123. Li PF, Dietz R, von Hardsorf R (1999) Superoxide induces apoptosis in cardiomyocytes, but proliferation and expression of transforming growth factor-beta1 in cardiac fibroblasts. FEBS Lett 448:206–210
124. Thannickal V, Day R, Klinz S et al (2000) Ras-dependent and -independent regulation of reactive oxygen species by mitogenic growth factors and TGF-β. FASEB J 14:1741–1748
125. Cucoranu I, Clempus R, Dikalova A et al (2005) NAD(P)H oxidase 4 mediates transforming growth factor-β1-induced differentiation of cardiac fibroblasts into myofibroblasts. Circ Res 97:900–907
126. Khanna AK, Plummer M, Bromberek C et al (2002) Expression of TGF-β and fibrogenic genes in transplant recipients with tacrolimus and cyclosporine nephrotoxicity. Kidney Int 62:2257–2263
127. Akool E-S, Doller A, Babelova A et al (2008) Molecular mechanisms of TGFβ receptor-triggered signaling cascades rapidly induced by the calcineurin inhibitors cyclosporine A and FK506. J Immunol 181:2831–2845

# Chapter 19
# Oxidative Stress and Atrial Fibrillation

**Ali A. Sovari and Samuel C. Dudley**

**Abstract** The pathological processes involved in initiation and perpetuation of atrial fibrillation (AF) are still unclear. AF is associated with systemic and cardiac oxidative stress and inflammation. Many risk factors for AF, such as aging and diabetes, are associated with an increased level of reactive oxygen species. In addition, oxidative stress has been shown at both cellular and tissue levels to be arrhythmogenic. Mechanisms of oxidative stress–induced arrhythmia involve a wide range of biological processes and signaling pathways, mainly resulting in abnormal $Na^+$ current and intracellular $Ca^{2+}$ handling. These lead to early and delayed afterdepolarizations as well as effects on conduction velocity through gap junctional remodeling. Oxidative stress is likely to participate with other central mechanisms of arrhythmia, particularly with inflammation and myocardial fibrosis, to promote AF. Understanding these mechanisms should provide better potential therapeutic targets for treatment of the arrhythmia and its complications. In this chapter, we summarize the role of oxidative stress in AF and some potential therapeutic strategies.

**Keywords** Arrhythmia · Atrial fibrillation · Oxidative stress · Afterdepolarization · Electrical remodeling · Inflammation

## 19.1 Introduction

Atrial fibrillation (AF), which affects approximately 2.5 million patients in the United States, is the most common arrhythmia requiring medical attention. The hallmark of AF is rapid, disorganized, and irregular atrial activity (Fig. 19.1). The ventricular response, which depends on the atrioventricular (AV) conduction rate, is usually irregular and rapid, with a rate of 100–160 bpm. AF is clinically classified as "paroxysmal" if it terminates spontaneously in less than 7 days, "persistent" if it fails to terminate spontaneously in less than 7 days, and "permanent" if it persists

S.C. Dudley (✉)
Section of Cardiology, University of Illinois at Chicago, Chicago, IL 60612, USA
e-mail: scdudley@uic.edu

H. Sauer et al. (eds.), *Studies on Cardiovascular Disorders*,
Oxidative Stress in Applied Basic Research and Clinical Practice,
DOI 10.1007/978-1-60761-600-9_19, 

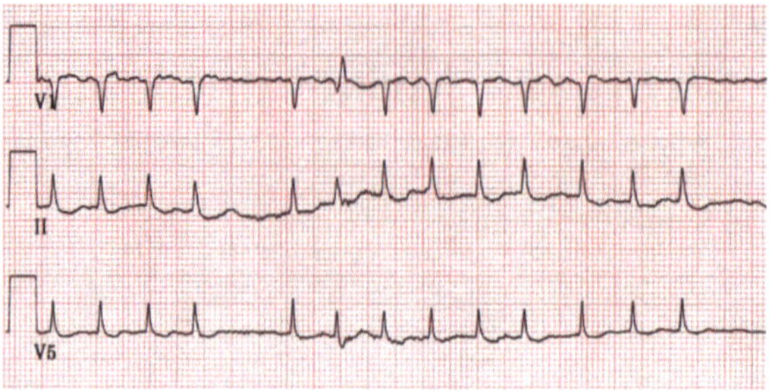


**Fig. 19.1** ECG of a 69-year-old patient presented to the emergency department complaining of shortness of breath. Leads V1, II, and V5 show narrow complex tachycardia with irregular RR intervals. The completely irregular atrial activity can be detected as fibrillation of low amplitude p waves in lead V1

for more than 1 year. Lone AF, which occurs in the absence of identifiable structural abnormalities of the heart, usually develops in people younger than 60 years.

The incidence of AF increases significantly with advancing age and is more common in men. In a study by Krahn and colleagues, the incidence of AF increased with age from less than 0.5 per 1,000 person-years in individuals younger than 50 years to approximately 10 per 1,000 person-years in those older than 70 years [1]. An analysis from the Framingham Heart Study showed that the lifetime risk for the development of AF was 1 in 4 in men and women 40 years or older and that this risk was slightly greater in men [2]. Table 19.1 shows the main clinical risk factors of AF and their association with oxidative stress. An elevated level of ROS has been linked with increasing age, heart failure [6, 7], diabetes mellitus [18], coronary artery disease [4, 5], obesity [21], and alcohol intoxication [20], all conditions associated with AF.

Symptoms caused by AF vary greatly. The prevalence of AF may be underestimated because many patients with AF are asymptomatic. Others may experience mild-to-severe palpitations and/or dizziness. The loss of effective atrial contractility in people with AF, particularly those with heart failure or another pathologic cardiac condition, can cause severe pulmonary congestion, hypotension, and angina.

The most serious adverse effect of AF is thrombus formation in the left atrium (LA) and the left atrial appendage (LAA), which causes stroke and peripheral embolization. The risk of subsequent stroke after a diagnosis of AF increases from 1.3% to 5.1% per person-year as age increases from 50 to 90 years [22]. Stroke or peripheral embolization may be the initial manifestation of AF. The increased risk of thrombus formation in AF patients is linked to the loss of LA and LAA contractility and associated stasis, the development of a hypercoagulable state, and endocardial dysfunction in AF. Heppel et al. showed that thrombus formation in AF is not only associated with reduced peak LAA velocity but also is associated with increased plasma markers of platelet activation (i.e., beta thromboglobulin and platelet factor 4) and of thrombogenesis (i.e., thrombin-antithrombin complexes and D-dimers) [23]. They showed that von Willebrand factor was also increased and

**Table 19.1** The main clinical risk factors of atrial fibrillation and their association with oxidative stress

| Risk factors of atrial fibrillation | Association with ROS elevation |
|---|---|
| *Cardiac diseases* | |
| Hypertension | Yes [3] |
| Coronary artery disease | Yes [4, 5] |
| Cardiomyopathies and heart failure | Yes [6, 7] |
| Valvular diseases | Yes [8, 9] |
| *Pulmonary diseases* | |
| Pulmonary embolism | Yes [10] |
| Chronic obstructive pulmonary disease | Yes [11] |
| Obstructive sleep apnea | Yes [12] |
| Pneumonia | Yes [13] |
| *Surgeries* | |
| CABG and valve surgeries | Yes [14] |
| Cardiac transplantation | Yes [15] |
| Other surgeries | Maybe |
| *Other diseases and conditions* | |
| Aging | Yes [16] |
| Hyperthyroidism | Yes [17] |
| Diabetes Mellitus and obesity | Yes [18] |
| Autonomic dysfunction | Maybe [19] |
| Alcohol | Yes [20] |

that LAA velocity, beta thromboglobulin, and the von Willebrand factor levels are independently associated with LAA thrombosis, more strongly than the presence of spontaneous echo contrast [23]. To prevent stroke and peripheral embolism, most patients with AF should be treated with an antiplatelet (e.g., aspirin or clopidogrel) or anticoagulant (e.g., warfarin) agent.

Most clinical studies have not shown that rhythm control confers a greater survival advantage than does a ventricular rate control strategy using atrioventricular nodal blocking agents. Nevertheless, in symptomatic patients, a rhythm control strategy is desirable. For prevention of the arrhythmia, Vaughan Williams class IC (e.g., flecainide, propafenone) and class III (e.g., amiodarone, sotalol) antiarrhythmic agents are often used [24, 25]. Newer antiarrhythmic drugs (e.g., RSD 1235, AVE 0118, and AZD 7009) [26–28] that target atrial selective ion channels and agents such as dronedarone [29] that produce fewer noncardiac adverse effects are under investigation for the treatment of AF. In addition, catheter ablation [30] and various surgical procedures [31] that are designed to disconnect triggers from the substrate and/or modify the substrate of AF are nonmedical options. Mapping and radiofrequency (RF) ablation of AF is one of the most complex ablation procedures, and a number of approaches can be used, depending on the expertise of the cardiac electrophysiologist and the characteristics of the AF. Nevertheless, most of these strategies have limited success, in part, because the pathogenesis of the disease is unclear.

## 19.2 The Electrical Basis of AF

The human atrial myocyte action potential depends on the orchestrated activation and inactivation of several ion channels. In brief, the inward $Na^+$ current in phase 0 depolarizes the atrial myocyte from its resting potential, which is about –80 mV, to a potential of about +20 mV (Fig. 19.2). Then two outward currents through the potassium channels $I_{to}$ and $IK_{UR}$ make a notch on phase 1 of the action potential. While the $I_{Ca}$ current contributes to the plateau of phase 2, the outward current of $IK_{UR}$ continues to take the myocyte into phase 3, where IKr and IKs delayed rectifier currents are the major outward currents, after which the myocyte returns to its resting potential. There is regional heterogeneity in the action potential duration (APD), and the APD is shorter in the LA than in the right atrium (RA) [32]. In addition, there is also heterogeneity in the APD within each atrial chamber. For example, it is thought that the APD is even shorter in myocytes of pulmonary veins than in myocytes of other regions of the LA [33]. This suggests that there is variation in ion channel expression as a function of location.

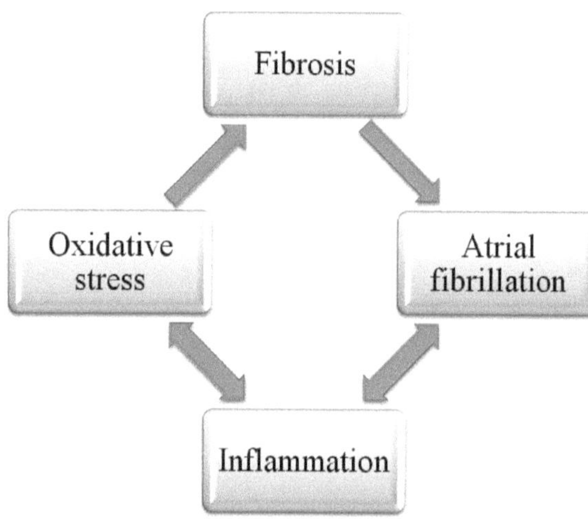


**Fig. 19.2** A schematic figure to show how oxidative stress may interact with inflammation and fibrosis to promote atrial fibrillation

Myocytes in the normal heart are well coupled to each other through transmembrane channels known as gap junctions. This coupling facilitates the orderly spread of the electrical excitation wave and allows for synchronous contractions [34]. In addition to the abnormalities in membrane properties and myocyte ion channels, a pathologic gap junctional connection between myocytes may also result in the abnormal conduction of the excitation wave and in arrhythmia. Furthermore, the specific 3-dimensional anatomy of the atria with respect to veins (such as the pulmonary veins and the vena cava) and atrioventricular valves provides anatomic obstacles that facilitate the formation of electrical reentrant circuits, one basis of AF.

Reentrant circuits and rapidly firing focal ectopic activities are thought to be the primary framework for our understanding of mechanisms of AF [35]. During reentry, impulses propagate around an anatomic or functional electrical block. Initially, one part of the circuit is not excitable, but when the wave front returns to that area, sufficient time has elapsed to allow recovery from the refractory period, and the initially unexcitable tissue responds to stimulation. As a result, the wave front is able to become self-sustaining and can propagate to other parts of the heart to produce the tachyarrhythmia. In AF, either multiple reentrant circuits exist or the conduction of a single reentrant circuit is fibrillatory, giving rise to the irregular surface electrical activity that is the hallmark of this condition. Because the minimum path length for reentry is determined by the product of the effective refractory period (ERP) and the conduction velocity (CV), short ERPs and slower CVs contribute to reentry by reducing the minimum required circuit size and allowing the atria to accommodate more circuits. Similarly, larger atrial size provides more space for circuits. Increasing the ERP or slowing CV are the mechanisms of action for most antiarrhythmic drugs prescribed to manage AF. Most ablation procedures are designed to reduce the tissue area available to support the arrhythmia or to isolate anatomic obstructions and automatic foci facilitating AF.

Although single-circuit or multiple-circuit reentry may have a major role in maintaining AF, the arrhythmia is usually thought to be initiated by focal activities. Haissaguerre and colleagues showed that pulmonary veins are the major source of the rapid focal activities that initiate paroxysmal AF and that RF ablation of those foci terminates AF [36]. Rapid focal activity is thought to result from either enhanced automaticity or triggered activity. In enhanced automaticity, the myocyte reaches the depolarization threshold in phase 4 of its action potential earlier than normal. This causes an increased automatic rate of firing. Triggered activity may arise from early afterdepolarizations (EADs) or delayed afterdepolarizations (DADs). EADs result primarily from reactivation of L-type calcium channels during repolarization secondary to the prolongation of AP and a reduced repolarization reserve [37]. DADs, which emerge primarily during diastole, are caused by spontaneous $Ca^{2+}$ release from an overloaded sarcoplasmic reticulum, which in turn activates other mechanisms such as the $Na^{+}/Ca^{2+}$ exchanger (NCX) that triggers the depolarization of the membrane [38]. Rapid multifocal activities or a single rapid focal activity with irregular conduction may initiate AF and contribute to its persistence.

## 19.3 The Central Role of Myocardial Fibrosis, Inflammation, and Oxidative Stress in AF

A variety of etiologic factors, including channelopathies and genetic defects [39], abnormalities in autonomic tone [40, 41], nerve sprouting [42], chaos, and chaos synchronization [43], have been suggested to play a role in the pathogenesis of AF. Among the etiologic factors, myocardial fibrosis, inflammation, and oxidative stress interrelate to play a central role in the initiation and perpetuation of AF.

Two important characteristics of myocardial fibrosis are increased collagen deposition and fibroblast proliferation. Fibroblasts also undergo phenotypic changes. They express α-smooth muscle actin, form gap junctional connections, and become myofibroblasts [44]. Myocardial fibrosis alters atrial conduction properties as a result of insulating barriers created by collagen deposition in the interstitial matrix, which could lead to reentry [45]. In addition, it has been shown that fibrosis promotes ectopic activity because of fibroblast-myocyte gap junctional coupling, which reduces the myocyte repolarization reserve caused by the lower resting membrane potential of fibroblasts [46]. Reactive oxygen species (ROS) have been shown to enhance fibroblast proliferation and type I collagen gene expression [47], thus promoting fibrosis. Antioxidant therapy has been shown to reduce fibrosis by decreasing the level of transforming growth factor (TGF) [48].

A strong case for the role of inflammation in AF has been made by the demonstration of increased serum levels of the inflammatory marker C-reactive protein in patients with AF [49], and by the predictive ability of inflammatory markers in the onset of AF in patients after cardiac surgery [50] or cardioversion [51]. Inflammation and oxidative stress are interrelated. It has been shown that activated neutrophils, eosinophils, monocytes, and macrophages produce ROS and lysosomal hydrolytic enzymes at sites of inflammation [52]. Also, it is known that ROS can enhance the inflammatory response via the activation of signaling events that mediate the expression of inflammatory genes [53] and by the regulation of a critical element, nuclear factor-kappa B (NF-κB) [54]. NF-κB has been shown to be activated by various stimuli, including ROS, as well as after myocardial infarction, in ischemic states, and during reperfusion [55, 56]. Recently, we have shown that ROS elevation by angiotensin II can activate NF-κB, which in turn transcriptionally downregulates $Na^+$ currents [57]. Therefore, in addition to the direct role of ROS in the promotion of AF, ROS also promotes AF indirectly by enhancing myocardial fibrosis and inflammation.

## 19.4 Biomarkers and Cellular Mechanisms of Oxidative Stress in AF

ROS thought to be involved in cardiac arrhythmogenesis include the superoxide anion ($O_2^{\bullet -}$), $H_2O_2$, the hydroxyl radical ($^{\bullet}OH$), peroxynitrite ($ONOO^-$), and nitric oxide (NO). Many ROS are highly reactive because they contain unpaired electrons in their outer shells, allowing them to donate this electron and oxidize other species. Small amounts of ROS are formed as natural byproducts of the normal metabolism of oxygen and may function as signaling molecules in diverse physiologic processes. Nevertheless, excessive ROS is injurious in a number of ways.

Recent technical advances have allowed the measurement of ROS in biological systems. Animal studies suggest an increase in cardiac oxidation during AF. Multiple studies have shown an increase in markers of oxidative stress in humans with AF. A common source of tissue for this class of experiments has been the human right atrial appendage (RAA) excised during cardiac surgery. Mihm and colleagues used the activity of myofibrillar creatine kinase (MM-CK), which is redox sensitive, as an indirect marker of oxidative stress [58]. Those authors showed that there was less MM-CK activity and an increase in the level of immunodetectable 3-nitrotyrosine, a marker for the presence of peroxynitrite, in RAA taken from patients with AF as opposed to patients with a normal sinus rhythm. Moreover, coronary artery bypass surgery, a procedure that causes AF in a substantial percentage of patients, is associated with an increase in oxidized glutathione and lipid peroxidation [59]. We have reported a statistically significant association between the level of oxidative markers and AF [60]. We used derivatives of reactive oxidative metabolites (DROMs) and ratios of oxidized to reduced glutathione [E(h) GSH] and cysteine [E(h) CySH] to quantify oxidative stress in patients with or without persistent or permanent AF [60]. The increase in the odds ratios for AF for E(h) GSH, E(h) CySH, and DROMs were 6.1 (95% CI, 1.3-28.3; $P = 0.02$), 13.6 (95% CI, 2.5-74.1; $P = 0.01$), and 15.9 (95% CI, 1.7-153.9; $P = 0.02$), respectively, and in multivariate analysis, corrected for AF risk factors that differed among the groups, the association of AF and oxidative stress remained statistically significant [60].

In addition, genomic studies have shown that the gene expression pattern of atrial tissue in patients with AF is associated with oxidative stress. Kim and colleagues used radioactive complementary DNA microarrays to evaluate changes in the expression of 1,152 known genes in atrial tissue of patients undergoing the Maze procedure and those in normal sinus rhythm undergoing coronary artery bypass graft surgery [61]. They showed a significant reduction in the gene expression of antioxidant genes as well as a significant increase in the gene expression of five genes related to ROS in AF patients, representing a clear shift toward the pro-oxidation state in AF [61]. In Kim and colleagues' study, the gene expression of glutathione peroxidase-1 and heme oxygenase-2 decreased; while the gene expression of flavin containing monooxygenase-1, monoamine oxidase-B, uniquin specific protease-8, tyrosine-related protein-1, and tyrosine 3-monooxygenase increased in the atrial tissue of AF patients [61].

There are many plausible ways in which oxidative stress may contribute to arrhythmic risk (Fig. 19.3). A direct cardiac arrhythmogenic effect of ROS has been demonstrated by Berecewicz and Horackova in rat and guinea pig myocytes [62]. Those authors showed that $H_2O_2$ prolongs the APD and induces triggered activities by EAD and DAD [62]. Song et al. showed that $H_2O_2$-induced APD prolongation and EADs are at least partially because of enhanced late $Na^+$ current [63]. Also, we showed that $H_2O_2$ and angiotensin II treatment enhances late sodium current in isolated myocytes [57]. In addition to its effects on the $Na^+$ current, $H_2O_2$ stimulates the L-type calcium current. This results in abnormal intracellular calcium cycling in myocytes, which in turn facilitates EADs [64].

Other alterations in $Ca^{2+}$ handling proteins with ROS have been noted. Hinata and colleagues demonstrated the increased activity of the $Na^+/Ca^{2+}$ exchanger (NCX) upon exposure to $H_2O_2$ by the activation of a mitogen-activated protein kinase signaling pathway and a Src family tyrosine kinase [65]. NCX current is

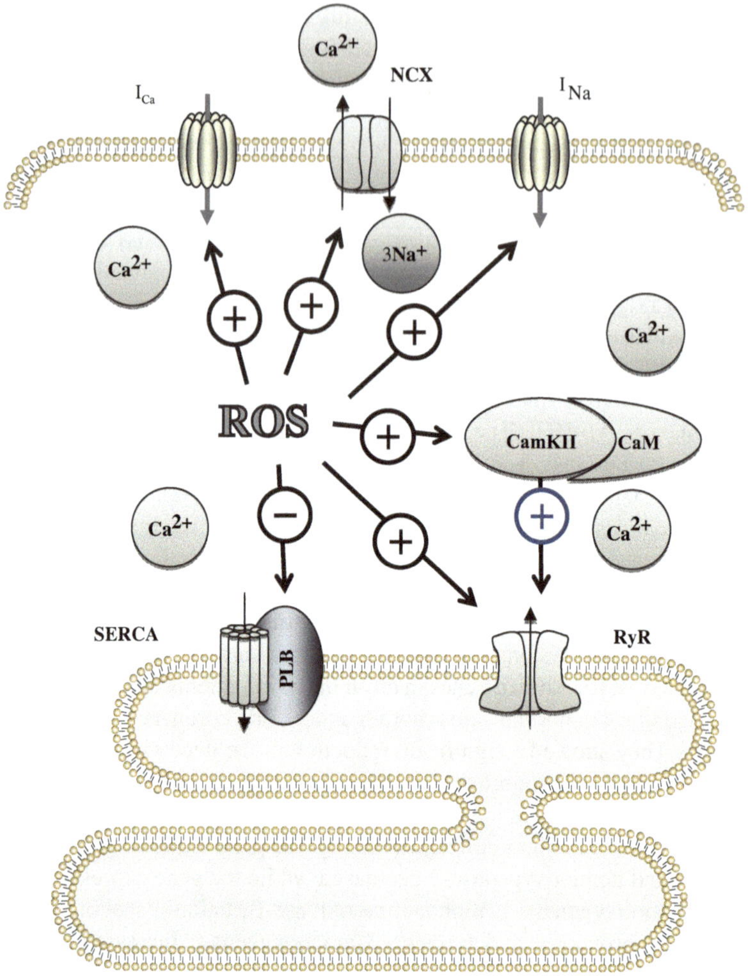


**Fig. 19.3** Schematic review of ROS-mediated subcellular ionic changes that may result in afterdepolarization. ROS-mediated inhibition of SERCA, enhanced $Ca^{2+}$ release from RyR, and enhanced inward L-type $Ca^{2+}$ current increase intracellular $Ca^{2+}$ level. In addition, it has recently been shown that ROS, directly and independently of a $Ca^{2+}$–mediated mechanism, can activate CaMKII; this in turn can increase $Ca^{2+}$ release from the RyR. ROS also activate NCX and $I_{Na}$, that reduce repolarization reserve, which along with increased intracellular $Ca^{2+}$ level may result in emergence of afterdepolarizations and triggered activity

a depolarizing current that contributes to DAD and EAD formation and reduction in repolarization reserve. Even a brief exposure to hydroxyl radicals markedly decreases the sarcoplasmic reticulum (SR) $Ca^{2+}$ uptake [66] that will increase diastolic $Ca^{2+}$ level in myocytes. Also, calmodulin could increase the SR $Ca^{2+}$ uptake in myocytes exposed to $H_2O_2$ [66]. It has recently been shown that calcium/calmodulin

($Ca^{2+}$/CaM)-dependent protein kinase II activity is enhanced by prooxidant conditions [67]. Activated $Ca^{2+}$/calmodulin-dependent protein kinase II (CaMKII) can phosphorylate a variety of proteins in the heart, including the ryanodine receptor (RyR), which is an ion channel primarily responsible for mediating the release of $Ca^{2+}$ from the SR [68]. It appears that CaMKII increases RyR open probability and promotes SR $Ca^{2+}$ leak, which results in EADs and DADs [68]. Although the physiological effect of CaMKII on RyR is controversial, the effect of $H_2O_2$ on increasing the open probability of RyR and increasing the $Ca^{2+}$ release from the SR have been shown in other studies [69], regardless of the potential role of CaMKII phosphorylation.

The combined effects of $H_2O_2$ on cardiac ion channels and calcium ion transports may lead to an increase in intracellular diastolic $Ca^{2+}$ coupled with a reduced repolarization reserve that increases the propensity for EAD and DAD formation. EAD-mediated triggered activity as a cause of AF has been demonstrated in intact hearts by using combined single-cell microelectrode recordings and high-resolution optical mapping [70]. Therefore, there are several plausible mechanisms whereby oxidative stress may be mechanistically related to AF induction or maintenance.

## 19.5 Oxidative Stress and Thromboembolism in AF

Virchow's triad of blood stasis, endothelial dysfunction, and a hypercoagulable state summarize the main etiologic factors thought to determine thrombus formation. The presence of left atrial spontaneous echo contrast and chamber enlargement, both of which are evidence of blood flow stasis in patients with AF, are strongly associated with an increased risk for cerebral ischemic events in this condition [71, 23]. The proclivity for thrombus formation in the LAA, which is a blind pouch containing the lowest velocity blood flow, emphasizes the central role of stasis in thromboembolism associated with AF.

Aside from blood stasis, oxidative stress may contribute to procoagulant changes in the LAA. Oxidative stress can increase the risk of thromboembolism in AF by deteriorating the contractile dysfunction in the left atrium. In patients with AF, however, several clinical risk factors (male sex, older age, hypertension, and diabetes) for thromboembolism do not affect hemodynamics [72–75], but are associated with evidence of systemic oxidative stress [76, 77]. We have shown that AF is associated with the downregulation of NO and the upregulation of $O_2^{\bullet -}$ production in the left atrium [78], which shifts the endocardial balance toward thrombogenicity with increased platelet aggregation and adhesion [79, 80], overexpression of the prothrombotic protein plasminogen activator inhibitor-1 [81, 82], and increased tissue factor with the exposure of adhesion molecules on the endocardial surface. These changes may help explain the proclivity of thrombus to form in this chamber and may represent a future therapeutic target for the prevention of strokes associated with AF.

## 19.6 Therapeutic Implications of Increased ROS in AF

Carnes and colleagues tested the impact of supplemental ascorbate as an antioxidant (extended release, 2 g the evening before surgery, with 1 g/day through the first 5 days after surgery) with respect to the incidence of AF in the perioperative period in a small nonrandomized clinical trial [83]. They found postoperative AF occurred in 16.3% of 43 treated patients vs. 34.9% of 43 concomitant controls [83]. Nevertheless, this beneficial effect of ascorbate could not be validated in larger randomized and blinded trials, and the investigators attributed the failure to the short half-life of ascorbate [84]. ROS molecules tend to react with proteins and lipids adjacent to their point of production, so compartmentalization may provide an explanation for the disappointing results of some of the clinical studies on antioxidant therapies (i.e., vitamins C and E) that directly target ROS. It is not clear why these compounds failed to affect AF, but without evidence that they served as effective antioxidants in the relevant areas of the heart, the potential of antioxidants should not be dismissed out of hand.

On the other hand, it may be more effective to target the sources of oxidative stress in AF. Of the several atrial oxidative systems, the NADPH oxidase has been demonstrated to play a central role in ROS production, causing oxidative stress-mediated AF. Activation of other oxidative systems such as xanthine oxidase may also be partially dependent on NADPH oxidase activity [85]. The NADPH oxidase is thought to play an important role in redox signaling in a variety of pathologic cardiac conditions such as hypertension [86] and both nonischemic [87] and ischemic [88] cardiomyopathy. We have shown that increased $O_2^{\bullet -}$ production in the LAA is caused by increased NADPH oxidase and xanthine oxidase activity in pig hearts after 1 week of AF induced by rapid atrial pacing [89]. Kim and colleagues [90] demonstrated a statistically significant increase in NADPH oxidase activity in 170 postoperative patients with AF who underwent coronary artery bypass surgery.

Interestingly, lipid-lowering drugs, particularly statins and fibrates that possess anti-inflammatory and antioxidant properties have been shown to prevent AF in heart failure patients independent of their effect on the lipid profiles [91]. Angiotensin-converting enzyme inhibitors and angiotensin II receptor blockers have also been shown to be effective against atrial fibrillation in patients with heart failure and hypertension [92, 93]. Since angiotensin II is implicated in increased NADPH oxidase activity, it is tempting to speculate that the antifibrillation benefit of angiotensin II blockers may, in part, be because of the inhibition of NADPH oxidase activity. Targeting excess activity of NADPH oxidase may offer therapeutic effect in the management of AF. Other sources of ROS that may be targeted in the future include mitochondria, xanthine oxidases, uncoupled NO synthases, cytochrome P450 reductases, and cyclooxygenases [56].

## 19.7 Conclusion

Risk factors for AF are associated with oxidative stress. AF is associated with oxidative stress in animals and humans. ROS have a myriad of effects that can contribute to arrhythmic risk, including promotion of interstitial fibrosis and cardiac structural and electrical remodeling. Mitigation of oxidative stress may explain the antiarrhythmic effects of drugs without ion channel altering properties, such as the angiotensin receptor blockers and angiotensin-converting enzyme inhibitors. Identifying the sources of oxidative stress may prove a valuable therapeutic strategy to address this arrhythmia and its most deadly consequence, LAA thrombus formation and stroke.

**Acknowledgments** Funding: R01 HL085520, R01 HL085558, R01 HL073753, an American Heart Association Established Investigator Award 0440164 N, and a Veterans Affairs MERIT grant.

**Author Disclosure Statement** SCD holds a patent entitled: Oxidative Stress Markers Predict Atrial Fibrillation 60/835,074. SCD is a recipient of a grant from Pfizer, Inc. to run at trial, Statins for the prevention of atrial fibrillation (StoP-AF; NCT00252967).

## References

1. Krahn AD, Manfreda J, Tate RB, Mathewson FA, Cuddy TE (1995 May) The natural history of atrial fibrillation: incidence, risk factors, and prognosis in the Manitoba Follow-Up Study. Am J Med 98(5):476–484
2. Lloyd-Jones DM, Wang TJ, Leip EP et al (2004) Lifetime risk for development of atrial fibrillation: the Framingham Heart Study. Circulation 110(9):1042–1046
3. De Champlain J, Wu R, Girouard H et al (2004 Oct–Nov) Oxidative stress in hypertension. Clin Exp Hypertens 26(7–8):593–601
4. Vassalle C, Petrozzi L, Botto N, Andreassi M, Zucchelli GC (2004 Oct) Oxidative stress and its association with coronary artery disease and different atherogenic risk factors. J Intern Med 256(4):308–315
5. Madamanchi NR, Vendrov A, Runge MS (2005 Jan) Oxidative stress and vascular disease. Arterioscler Thromb Vasc Biol 25(1):29–38
6. Ide T, Tsutsui H, Kinugawa S et al (2000) Direct evidence for increased hydroxyl radicals originating from superoxide in the failing myocardium. Circ Res 86:152–157
7. Sam F, Kerstetter DL, Pimental DR et al (2005) Increased reactive oxygen species production and functional alterations in antioxidant enzymes in human failing myocardium. J Card Fail 11:473–480
8. Miller JD, Chu Y, Brooks RM, Richenbacher WE, Peña-Silva R, Heistad DD (2008) Dysregulation of antioxidant mechanisms contributes to increased oxidative stress in calcific aortic valvular stenosis in humans. J Am Coll Cardiol 52(10):843–850
9. Liberman M, Bassi E, Martinatti MK et al (2008 Mar) Oxidant generation predominates around calcifying foci and enhances progression of aortic valve calcification. Arterioscler Thromb Vasc Biol 28(3):463–470
10. Ovechkin AV, Lominadze D, Sedoris KC, Robinson TW, Tyagi SC, Roberts AM (2007) Lung ischemia-reperfusion injury: implications of oxidative stress and platelet-arteriolar wall interactions. Arch Physiol Biochem 113(1):1–12
11. Repine JE, Bast A, Lankhorst I et al (1997) Oxidative Stress in Chronic Obstructive Pulmonary Disease. Am J Respir Crit Care Med Volume 156:341–357

12. Yamauchi M, Nakano H, Maekawa J et al (2005) Oxidative stress in obstructive sleep apnea. Chest 127(5):1674–1679
13. Duflo F, Debon R, Goudable J, Chassard D, Allaouchiche B (2002) Alveolar and serum oxidative stress in ventilator-associated pneumonia. Br J Anaesth 89:231–236
14. Milei J, Ferreira R, Grana DR, Boveris A (2001) Oxidative stress and mitochondrial damage in coronary artery bypass graft surgery: effects of antioxidant treatments. Compr Ther 27(2):108–116
15. Kofler S, Petrakopoulou P, Nickel T, Weis M (2006) Cardiac allograft endothelial dysfunction. Eur J Clin Pharmacol 62:79–82
16. Kregel KC, Zhang HJ (2007) An integrated view of oxidative stress in aging: basic mechanisms, functional effects, and pathological considerations. Am J Physiol Regul Integr Comp Physiol 292(1):R18-R36
17. Civelek S, Seymen O, Seven A, Yigit G, Hatemi H, Burçak G (2001) Oxidative stress in heart tissue of hyperthyroid and iron supplemented rats. J Toxicol Environ Health A 64:499–506
18. Li N, Frigerio F, Maechler P (2008) The sensitivity of pancreatic β-cells to mitochondrial injuries triggered by lipotoxicity and oxidative stress. Biochem Soc Trans 36(Pt 5):930–934
19. Serap D, Tülay K, Cigdem G et al (2005) Autonomic dysfunction is associated with oxidative stress in non-diabetic hemodialysis patients. Dial Transplant vol. 34:74–87
20. Cederbaum AI (2001) Introduction—Serial review: Alcohol, oxidative stress, and cell injury. *Free Radic Biol Med* 31:1524–1526
21. Furukawa S, Fujita T, Shimabukuro M et al (2004) Increased oxidative stress in obesity and its impact on metabolic syndrome. J Clin Invest 114:1752–1761
22. Frost L, Engholm G, Johnsen S, Moller H, Husted S (2000) Incident stroke after discharge from the hospital with a diagnosis of atrial fibrillation. Am J Med 108:36–40
23. Heppell RM, Berkin KE, McLenachan JM, Davies JA (1997) Haemostatic and haemodynamic abnormalities associated with left atrial thrombosis in non-rheumatic atrial fibrillation. Heart 77:407–411
24. Wyse DG, Waldo AL, DiMarco JP et al (2002) A comparison of rate control and rhythm control in patients with atrial fibrillation. N Engl J Med 347:1825–1833
25. Van Gelder IC, Hagens VE, Bosker HA et al (2002) A comparison of rate control and rhythm control in patients with recurrent persistent atrial fibrillation. N Engl J Med 347:1834–1840
26. Naccarelli GV, Wolbrette DL, Samii S et al (2008) Vernakalant: pharmacology, electrophysiology, safety and efficacy. Drugs Today (Barc) 44:325–329
27. Christ T, Wettwer E, Voigt N et al (2008) Pathology-specific effects of the IKur/Ito/IK,ACh blocker AVE0118 on ion channels in human chronic atrial fibrillation. Br J Pharmacol 154:1619–1630
28. Crijns HJ, Van Gelder IC, Walfridsson H et al (2006) Safe and effective conversion of persistent atrial fibrillation to sinus rhythm by intravenous AZD7009. Heart Rhythm 3:1321–1331
29. Singh BN, Connolly SJ, Crijns HJ et al (2007) Dronedarone for maintenance of sinus rhythm in atrial fibrillation or flutter. N Engl J Med 357:987–999
30. Callahan TD 4th, Natale A (2008) Catheter ablation of atrial fibrillation. Med Clin North Am 92:179–201
31. Gillinov AM, Saltman AE (2008) Surgical approaches for atrial fibrillation. Med Clin North Am 92:203–215
32. Li D, Zhang L, Kneller J, Shi H, Nattel S (2001) Ionic mechanism of repolarization differences between canine right and left atrium. Circ Res 88:1168–1175
33. Ehrlich J, Cha TJ, Chartier D, Zhang L, Hohnloser S, Nattel S (2002) Ionic basis of unique pulmonary vein cardiomyocyte action potential properties. Circulation 106(Suppl II):II-179; (abstract)
34. Severs NJ, Dupont E, Thomas N, Kaba R, Rothery S, Jain R et al (2006) Alterations in cardiac connexin expression in cardiomyopathies. Adv Cardiol 42:228–242

35. Nattel S, Shiroshita-Takeshita A, Brundel BJ, Rivard L (2005) Mechanisms of atrial fibrillation: lessons from animal models. Prog Cardiovasc Dis 48:9–28
36. Haïssaguerre M, Jaïs P, Shah DC et al (1998) Spontaneous initiation of atrial fibrillation by ectopic beats originating in the pulmonary veins. N Engl J Med 339:659–666
37. January CT, Moscucci A (1992) Cellular mechanisms of early afterdepolarizations. Ann N Y Acad Sci 644:23–32
38. Pogwizd SM, Schlotthauer K, Li L, Yuan W, Bers DM (2001) Arrhythmogenesis and contractile dysfunction in heart failure: Roles of sodium-calcium exchange, inward rectifier potassium current, and residual beta-adrenergic responsiveness. Circ Res 88:1159–1167
39. Tsai CT, Lai LP, Hwang JJ, Lin JL, Chiang FT (2008) Molecular genetics of atrial fibrillation. J Am Coll Cardiol 52:241–250; Review
40. Takahashi Y, Jaïs P, Hocini M et al (2006) Shortening of fibrillatory cycle length in the pulmonary vein during vagal excitation. J Am Coll Cardiol 47:774–780
41. Bertaglia E, Zoppo F, Bonanno C, Pellizzari N, Frigato N, Pascotto P (2005) Autonomic modulation of the sinus node following electrical cardioversion of persistent atrial fibrillation: relation with early recurrence. Int J Cardiol 102(2):219–223
42. Tan AY, Zhou S, Ogawa M et al (2008 Aug 26) Neural mechanisms of paroxysmal atrial fibrillation and paroxysmal atrial tachycardia in ambulatory canines. Circulation 118(9): 916–925; Epub 2008 Aug 12. PMID: 18697820
43. Sato D, Xie LH, Sovari AA, Tran DX, Morita N, Xie F, Karagueuzian H, Garfinkel A, Weiss JN, Qu Z (2009) Synchronization of chaotic early afterdepolarizations in the genesis of cardiac arrhythmias. Proc Natl Acad Sci USA 106:2983–2988
44. Miragoli M, Gaudesius G, Rohr S (2006) Electrotonic modulation of cardiac impulse conduction by myofibroblasts. Circ Res 98:801–810
45. Spach MS, Dolber PC (1986) Relating extracellular potentials and their derivatives to anisotropic propagation at a microscopic level in human cardiac muscle. Evidence for electrical uncoupling of side-to-side fiber connections with increasing age. Circ Res 58(3):356–371
46. Miragoli M, Salvarani N, Rohr S (2007) Myofibroblasts induce ectopic activity in cardiac tissue. Circ Res 101(8):755–758
47. Murrell GA, Francis MJ, Bromley L (1990) Modulation of fibroblast proliferation by oxygen free radicals. Biochem J 265:659–665
48. Lee KS, Buck M, Houglum K, Chojkier M (1995) Activation of hepatic stellate cells by TGF alpha and collagen type I is mediated by oxidative stress through c-myb expression. J Clin Invest 96:2461–2468
49. Issac TT, Dokainish H, Lakkis NM (2007) Role of inflammation in initiation and perpetuation of atrial fibrillation: a systematic review of the published data. J Am Coll Cardiol 50: 2021–2028
50. Lo B, Fijnheer R, Nierich AP, Bruins P, Kalkman CJC (2005) Reactive protein is a risk indicator for atrial fibrillation after myocardial revascularization. Ann Thorac Surg 79: 1530–1535
51. Watanabe E, Arakawa T, Uchiyama T, Kodama I, Hishida H (2006) Highsensitivity C-Reactive protein is predictive of successful cardioversion for atrial fibrillation and maintenance of sinus rhythm after conversion. Int J Cardiol 108:346–353
52. Morel F, Doussers J, Vignais PV (1991) The superoxide-generating oxidase of phagocytic cells. Physiological, molecular and pathological aspects. Eur J Biochem 201:523–546
53. Suzuki YJ, Forman HJ, Sevanian A (1997) Oxidants as stimulators of signal transduction. Free Radic Biol Med 22:269–285
54. Kabe Y, Ando K, Hirao S et al (2005) Redox regulation of NF-κB activation: distinct redox regulation between the cytoplasm and the nucleus. Antiox Redox Signal 7:395–403
55. Lu L, Chen SS, Zhang JQ et al (2004) Activation of nuclear factor κB and its proinflammatory mediator cascade in the infarcted rat heart. Biochem Biophys Res Commun 321:879–885
56. Seddon M, Looi YH, Shah AM (2007) Oxidative stress and redox signalling in cardiac hypertrophy and heart failure. Heart 93:903–907

57. Shang LL, Sanyal S, Pfahnl AE et al (2008) NF-κB-dependent transcriptional regulation of the cardiac scn5a sodium channel by angiotensin II. Am J Physiol Cell Physiol 294: C372–C379
58. Mihm MJ, Yu F, Carnes CA et al (2001) Impaired myofibrillar energetics and oxidative injury during human atrial fibrillation. Circulation 104:174–180
59. De Vecchi E, Pala MG, Di Credico G et al (1998) Relation between left ventricular function and oxidative stress in patients undergoing bypass surgery. Heart 79:242–247
60. Neuman RB, Bloom HL, Shukrullah I et al (2007) Oxidative stress markers are associated with persistent atrial fibrillation. Clin Chem 53(9):1652–1657
61. Kim YH, Lim DS, Lee JH et al (2003) Gene expression profiling of oxidative stress on atrial fibrillation in humans. Exp Mol Med 35:336–349
62. Beresewicz A, Horackova M (1991) Alterations in electrical and contractile behavior of isolated cardiomyocytes by hydrogen peroxide: possible ionic mechanisms. J Mol Cell Cardiol 23:899–918
63. Song Y, Shryock JC, Wagner S, Maier LS, Belardinelli L (2006) Blocking late sodium current reduces hydrogen peroxide-induced arrhythmogenic activity and contractile dysfunction. J Pharmacol Exp Ther 318(1):214–222
64. Thomas GP, Sims SM, Cook MA, Karmazyn M (1998) Hydrogen peroxide-induced stimulation of L-type calcium current in guinea pig ventricular myocytes and its inhibition by adenosine A1 receptor activation. J Pharmacol Exp Ther 286:1208–1214
65. Hinata M, Matsuoka I, Iwamoto T, Watanabe Y, Kimura J (2007) Mechanism of $Na^+/Ca^{2+}$ exchanger activation by hydrogen peroxide in guinea-pig ventricular myocytes. J Pharmacol Sci 103:283–292
66. Morris TE, Sulakhe PV (1997) Sarcoplasmic reticulum $Ca^{2+}$-pump dysfunction in rat cardiomyocytes briefly exposed to hydroxyl radicals. Free Radic Biol Med 22:37–47
67. Erickson JR, Joiner ML, Guan X et al (2008) Dynamic pathway for calcium-independent activation of CaMKII by methionine oxidation. Cell 133:462–474
68. Erickson JR, Anderson ME (2008) CaMKII and its role in cardiac arrhythmia. J Cardiovasc Electrophysiol 19:1332–1336
69. Anzai K, Ogawa K, Kuniyasu A, Ozawa T, Yamamoto H, Nakayama H (1998) Effects of hydroxyl radical and sulfhydryl reagents on the open probability of the purified cardiac ryanodine receptor channel incorporated into planar lipid bilayers. Biochem Biophys Res Commun 249:938–942
70. Ono N, Hayashi H, Kawase A et al (2007) Spontaneous atrial fibrillation initiated by triggered activity near the pulmonary veins in aged rats subjected to glycolytic inhibition. Am J Physiol 292:H639–H648
71. Jones EF, Calafiore P, McNeil JJ, Tonkin AM, Donnan GA (1996) Atrial fibrillation with left atrial spontaneous contrast detected by transesophageal echocardiography is a potent risk factor for stroke. Am J Cardiol 78:425–429
72. Celermajer DS, Sorensen KE, Spiegelhalter DJ et al (1994) Aging is associated with endothelial dysfunction in healthy men years before the age-related decline in women. J Am Coll Cardiol 24:471–476
73. Wang X, Desai K, Juurlink BH et al (2006) Gender-related differences in advanced glycation endproducts, oxidative stress markers and nitric oxide synthases in rats. Kidney Int 69: 281–287
74. (1994) Risk factors for stroke and efficacy of antithrombotic therapy in atrial fibrillation. Analysis of pooled data from five randomized controlled trials. Arch Intern Med 154:1449–1457
75. Ostgren CJ, Merlo J, Rastam L et al (2004) Atrial fibrillation and its association with type 2 diabetes and hypertension in a Swedish community. Diabetes Obes Metab 6:367–374
76. Keaney JF Jr, Larson MG, Vasan RS et al (2003) Obesity and systemic oxidative stress: clinical correlates of oxidative stress in the Framingham Study. Arterioscler Thromb Vasc Biol 2:434–439

77. Watanabe H, Tanabe N,Watanabe T et al (2008) Metabolic syndrome and risk of development of atrial fibrillation. The Niigata Preventive Medicine Study. Circulation 117:1249–1251
78. Cai H, Li Z, Goette A, Mera F, Honeycutt C, Feterik K, Wilcox JN, Dudley SC Jr, Harrison DG, Langberg JJ (2002) Downregulation of endocardial nitric oxide synthase expression and nitric oxide production in atrial fibrillation: potential mechanisms for atrial thrombosis and stroke. Circulation 106:2854–2858
79. Radomski MW, Palmer RM, Moncada S (1987) Comparative pharmacology of endothelium-derived relaxing factor, nitric oxide and prostacyclin in platelets. Br J Pharmacol 92:181–187
80. Freedman JE, Loscalzo J, Barnard MR et al (1997) Nitric oxide released from activated platelets inhibits platelet recruitment. J Clin Invest 100:350–356
81. Bouchie JL, Hansen H, Feener EP (1998) Natriuretic factors and nitric oxide suppress plasminogen activator inhibitor-1 expression in vascular smooth muscle cells: role of cGMP in the regulation of the plasminogen system. Arterioscler Thromb Vasc Biol 18:1771–1779
82. Swiatkowska M, Cierniewska-Cieslak A, Pawlowska Z et al (2000) Dual regulatory effects of nitric oxide on plasminogen activator inhibitor type 1 expression in endothelial cells. Eur J Biochem 267:1001–1007
83. Carnes CA, Chung MK, Nakayama T et al (2001) Ascorbate attenuates atrial pacing-induced peroxynitrite formation and electrical remodeling and decreases the incidence of postoperative atrial fibrillation. Circ Res. 89:E32–E38
84. Van Wagoner DR (2008) Oxidative stress and inflammation in atrial fibrillation: role in pathogenesis and potential as a therapeutic target. J Cardiovasc Pharmacol 52:306–313
85. Doerries C, Grote K, Hilfiker-Kleiner D et al (2007) Critical role of the NAD(P)H oxidase subunit p47phox for left ventricular remodeling/dysfunction and survival after myocardial infarction. Circ Res 100:894–903
86. Williams HC, Griendling KK (2007) NADPH oxidase inhibitors: new antihypertensive agents? J Cardiovasc Pharmacol 50:9–16
87. Nakagami H, Takemoto M, Liao JK (2003) NADPH xidase-derived superoxide anion mediates angiotensin II-induced cardiac hypertrophy. J Mol Cell Cardiol 35:851–859
88. Sun Y (2007) Oxidative stress and cardiac repair/remodeling following infarction. Am J Med Sci 334:197–205
89. Dudley SC Jr, Hoch NE, McCann LF et al (2005) Atrial fibrillation increases production of superoxide by the left atrium and left atrial appendage: role of the NADPH and xanthine oxidases. Circulation 112:1266–1273
90. Kim YM, Kattach H, Ratnatunga C, Pillai R, Channon KM, Casadei B (2008) Association of atrial nicotinamide adenine dinucleotide phosphate oxidase activity with the development of atrial fibrillation after cardiac surgery. J Am Coll Cardiol 51:68–74
91. Hanna IR, Heeke B, Bush H et al (2006) Lipid-lowering drug use is associated with reduced prevalence of atrial fibrillation in patients with left ventricular systolic dysfunction. Heart Rhythm 3:881–886
92. Maggioni AP, Latini R, Carson PE et al (2005) Valsartan reduces the incidence of atrial fibrillation in patients with heart failure: results from the Valsartan Heart Failure Trial (Val-HeFT). Am Heart J 149:548–557
93. Wachtell K, Lehto M, Gerdts E et al (2005) Angiotensin II receptor blockade reduces new-onset atrial fibrillation and subsequent stroke compared to atenolol: the Losartan Intervention For End Point Reduction in Hypertension (LIFE) study. J Am Coll Cardiol 45:712–719

# Chapter 20
# Oxidative Stress and the Antioxidative Capacity in Myocardial Infarction

**Yao Sun**

**Abstract** Extensive cardiac remodeling after myocardial infarction (MI) contributes significantly to ventricular dysfunction. Factors regulating left ventricular remodeling at different stages following MI are under investigation. There is growing recognition and experimental evidence that oxidative stress mediated by reactive oxygen species (ROS) plays a role in the pathogeneses of myocardial repair/remodeling in various cardiac diseases. After acute MI, oxidative stress is developed in both infarcted and noninfarcted myocardium. Accumulating evidence has demonstrated that oxidative stress participates in several aspects of cardiac repair/remodeling following infarction that includes cardiomyocyte apoptosis, inflammatory/fibrogenic responses, and hypertrophy. The exact pathways of ROS-mediated myocardial remodeling are under investigation. The therapeutic potential of oxidative stress–directed drugs in myocardial remodeling following infarction has not been fully realized. This chapter will address oxidative stress, antioxidative capacity, and their effect on cardiac repair/remodeling and dysfunction following MI.

**Keywords** Myocardial infarction · Oxidative stress · Cardiac repair · Myocardial remodeling · Heart failure

## 20.1 Introduction

Heart failure has emerged as a major health problem during the past two decades. It appears most commonly in patients with previous MI. Myocardial remodeling, which occurs in both infarcted and noninfarcted myocardium, contributes significantly to the development of heart failure [1–3]. Following MI, cardiac structural

Y. Sun (✉)
Division of Cardiovascular Diseases, Department of Medicine, University of Tennessee, Health Science Center, Memphis, TN 38163, USA
e-mail: yasun@uthsc.edu

H. Sauer et al. (eds.), *Studies on Cardiovascular Disorders*,
Oxidative Stress in Applied Basic Research and Clinical Practice,
DOI 10.1007/978-1-60761-600-9_20, 

remodeling is associated with an inflammatory reaction, which is followed by scar formation at the site of infarction and hypertrophy with interstitial fibrosis and vascular remodeling in the noninfarcted myocardium. This leads to ventricular remodeling characterized by alterations in left ventricular size, shape, and wall thickness [4–6]. Fibrous tissue that forms at the site of cardiomyocyte loss preserves the structural integrity and is integral to the heart's recovery, while the structural remodeling of viable myocardium impairs tissue behavior. Multiple factors may, in fact, contribute to left ventricular remodeling at different stages post MI. There is experimental evidence to suggest that oxidative stress mediated by ROS plays a role in the pathogenesis of myocardial repair/remodeling after MI, a hypothesis that has been earning growing recognition [7–11]. Oxidative stress results from an oxidant/antioxidant imbalance: an excess of oxidants relative to the antioxidant capacity. The heart with acute MI undergoes an increased ROS production as well as antioxidant deficit, first in the infarcted myocardium, followed by the noninfarcted myocardium. Chronic antioxidant treatment suppresses cardiac oxidative stress, attenuates ventricular remodeling, partially preserving left ventricle function and improved survival in rats or mice with experimental MI [12–14]. Experimental studies have also demonstrated that oxidative stress can induce most, if not all, of the changes that are thought to contribute to myocardial remodeling, including proinflammatory cytokine release, cardiomyocyte apoptosis [15], fibrogenesis [16], cell proliferation [17, 18], and hypertrophy [19]. In this chapter, the potential relevance of oxidative stress on apoptosis, inflammatory/fibrogenic responses, hypertrophy, and cardiac dysfunction in the infarcted heart will be discussed. The role of antioxidants in cardiac remodeling and dysfunction will be also discussed.

## 20.2 Cardiac Oxidative Stress and Antioxidant Capacity

### *20.2.1 Reactive Oxygen Species Production*

Superoxide ($O_2^-$), hydroxyl ($OH^-$), and peroxynitrite ($ONOO^-$) are simple molecules characterized by the presence of unpaired electrons. ROS can be produced intracellularly through electron leakage from mitochondria during oxidative phosphorylation and through the activation of several cellular enzymes, including NADPH oxidase, xanthine oxidase, and nitric oxide synthase [20–22]. $O_2^-$ can rapidly react with nitric oxide (NO) to form $ONOO^-$, or convert to $H_2O_2$ to form $OH^-$ [20]. ROS in low concentrations serve as signaling molecules [23]. However, these agents elicit harmful effects when produced in excess [20]. The toxicity associated with the excessive production of these compounds is prevented by antioxidant defense systems that maintain a healthy cellular environment. Living cells have both enzymatic and nonenzymatic defense mechanisms to balance the multitude of oxidative challenges presented to them. The enzymatic subgroup includes superoxide dismutase (SOD), catalase, and glutathione peroxidase (GSHPx) [24–26]. The dismutation of $O_2^-$ by SOD results in the generation of $H_2O_2$, which catalase further metabolizes into water and oxygen. The nonenzymatic group includes a variety of biologic molecules, such as vitamins E and C [27]. Oxidative stress occurs when ROS production is enhanced and/or antioxidant reserve is suppressed.

### *20.2.2 Occurrence of Cardiac Oxidative Stress Following Myocardial Infarction*

In the myocardium, as in other tissues, antioxidant enzymes protect cells by maintaining $O_2^-$ and $H_2O_2$ at low levels. Following MI, oxidative stress is developed in both infarcted and noninfarcted myocardium. Singal and colleagues have shown the evidence of progressive decrease in SOD, catalase, and GSHPx activity, as well as vitamin E levels in the rat infarcted heart, first in the infarcted myocardium, followed by the noninfarcted myocardium [7, 28]. Our study has shown reduced SOD gene and protein expression in the infarcted myocardium [9] (see Fig. 20.1). Cardiac glutathione levels are also decreased in patients with acute MI [29]. These observations raise the possibility that impaired antioxidant capacity contributes to oxidative stress in the infarcted heart. NADPH oxidase is a major source of $O_2^-$ in the heart [30]. In the infarcted heart, NADPH oxidase expression (gp22$^{phox}$ and gp91$^{phox}$ subunits) is significantly increased in the infarcted myocardium [9, 31], with neutrophils and macrophages as the primary cells expressing the enzyme (see Fig. 20.1). These findings suggest that ROS production is also enhanced in the infarcted myocardium. Malondialdehyde (MDA) is an end product in the lipid peroxidation chain reaction and is frequently used as a marker for ROS production. Our study has shown that MDA level is significantly increased in the infarcted myocardium (see Fig. 20.1). This observation further confirmed enhanced cardiac ROS production following infarction. Moreover, 3-nitrotyrosine, a marker of oxidative stress, is highly expressed in the inflammatory cells of the infarcted myocardium, supporting the occurrence of cardiac oxidative stress following MI [9] (see Fig. 20.1). Oxidative stress in noninfarcted myocardium is contributed by multiple sources; increased mitochondrial production of ROS has been suggested in noninfarcted myocardium of mice as one of them [14]. Increased ROS levels in noninfarcted myocardium also reflect increased activity of intracellular oxidase complexes, such as NADPH oxidase, xanthine oxidase, and nitric oxide synthase [32]. In addition, reduced SOD levels were observed in the failing heart with infarction [33]. These observations indicate that the imbalance between ROS production and antioxidant defense capacity contributes to oxidative stress in noninfarcted myocardium.

## 20.3 Oxidative Stress and Cardiac Remodeling and Dysfunction

### *20.3.1 Cardiomyocyte Apoptosis in the Infarcted Heart*

Loss of cardiomyocytes is an important mechanism in the development of myocardial remodeling and cardiac failure [34]. Following MI, apoptotic cardiomyocyte death occurs in the infarcted myocardium as well as the surviving portions of the

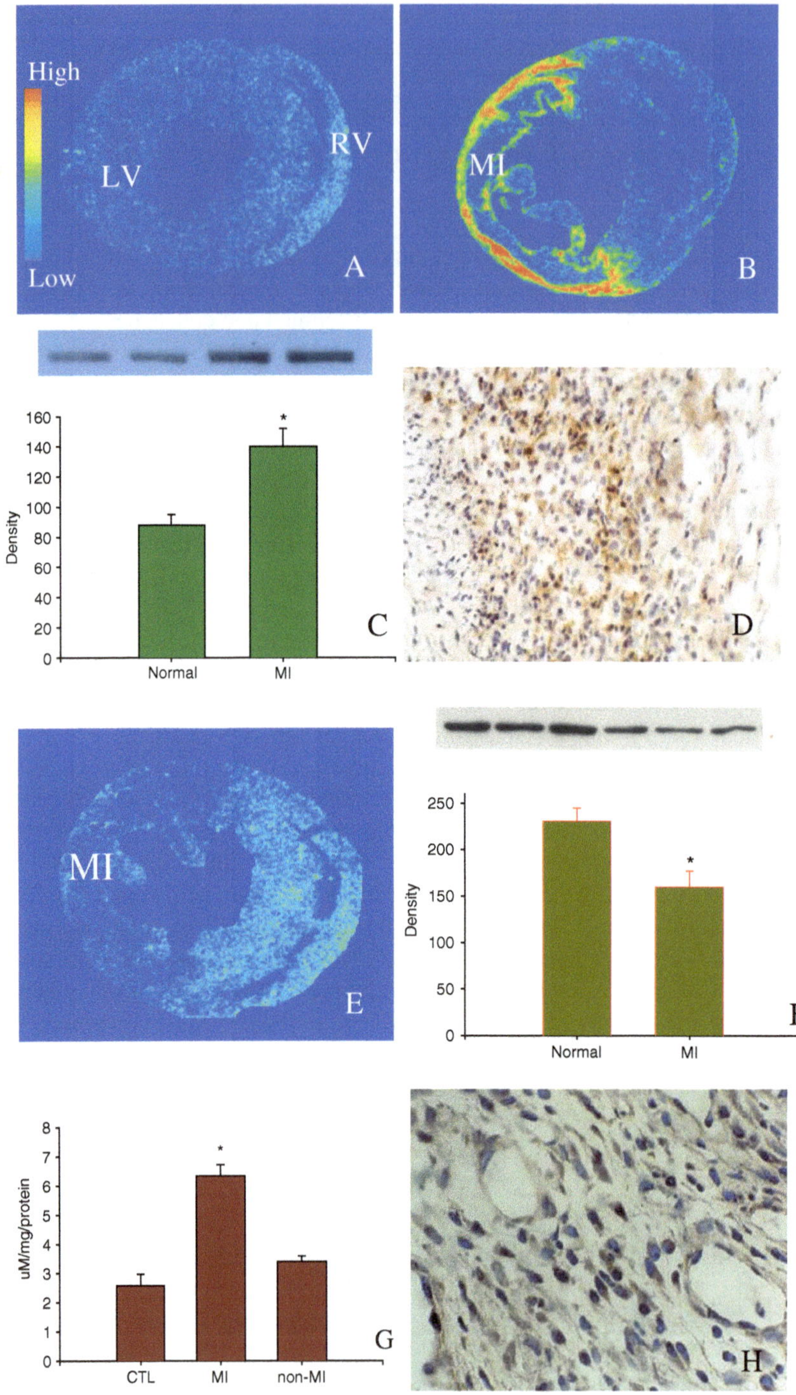


**Fig. 20.1** (continued)

wall adjacent to and remote from the infarcted myocardium [35, 36]. However, the number of apoptotic cardiomyocytes is greater in the infarcted region than in the region away from infarction. The regulation of cardiomyocyte apoptosis involves multiple mechanisms. ROS have been proven to be one of the stimulators of cardiomyocyte apoptosis [15–40]. High levels of oxidative stress have been demonstrated to cause cell necrosis, while lower levels of oxidative stress can cause cell apoptosis. In vitro studies have shown that treatment with $O_2^-$ or $H_2O_2$ in cardiomyocytes induces apoptosis [41]. In vivo studies have further demonstrated that oxidative stress triggers cardiomyocyte apoptosis in several cardiovascular diseases, including MI, ischemia/reperfusion injury, cardiomyopathy, atherosclerosis, and heart failure [42–47]. Long-term treatment with the antioxidants probucol or pyrrolidine dithiocarbamate attenuates oxidative stress and cardiomyocyte apoptosis within noninfarcted myocardium in rats [13, 37]. Another antioxidant, carvedilol, is shown to attenuate apoptosis induced by ischemia-reperfusion in the rat heart [48]. Oxidant scavengers, such as SOD and vitamin E, have been demonstrated to reduce ROS and inhibit cardiomyocyte apoptosis [49].

The mechanisms responsible for oxidative stress-mediated apoptosis in the infarcted heart are not fully understood. Multiple studies have shown that ROS can induce cardiomyocyte death by one or more mechanisms. Apoptosis is tightly controlled by a number of genes, including those primarily suppressing and those promoting apoptosis. Our previous studies identified markedly increased proapoptotic Bax expression in the infarcted heart, particularly at the site of infarction (see Fig. 20.2). Enhanced Bax expression coexists with oxidative stress and apoptosis in the infarcted heart [50]. Overexpression of antiapoptotic Bcl-2 decreases cardiomyocyte apoptosis [51]. Cesselli et al. have shown that in dog dilated cardiomyopathy, oxidative stress-induced cardiac apoptosis is related to increased $p66sh^{e}$, cytochrome c release, and activation of caspase-9 and -3 [52]. Taken together, these findings suggest that oxidative stress may trigger cardiomyocyte apoptosis via regulation of apoptotic genes.

Cytokines, such as tumor necrosis factor (TNF)-$\alpha$ and interleukin-6, are proven to stimulate cardiomyocyte apoptosis, which can be mediated by oxidative stress [53, 54]. TNF-$\alpha$-induced apoptosis is dependent on the induction of inducible nitric oxide synthase (iNOS) [55]. NO produced by iNOS appears to cause apoptosis

**Fig. 20.1** (continued) Oxidative stress in the infarcted rat heart at 1 week post MI. Detected by in situ hybridization, low levels of $gp91^{phox}$ mRNA are present in both left and right ventricles (LV, RV) of the normal heart (**panel a**). Following MI, cardiac $gp91^{phox}$ mRNA levels are largely increased, particularly at the site of MI (**panel b**). Detected by Western blot, $gp91^{phox}$ protein levels are significantly increased in the infarcted myocardium compared to normal myocardium (**panel c**). Immunohistochemistry reveals that cells expressing $gp91^{phox}$ in the infarcted myocardium are primarily inflammatory cells (**panel d**). SOD mRNA (**panel e**) and protein (**panel f**) levels are reduced in the infarcted myocardium. MDA levels are significantly increased in the infarcted myocardium (**panel g**). Immunohistochemical 3-nitrotyrosine labeling is positive in inflammatory cells of the infarcted myocardium (**panel h**)

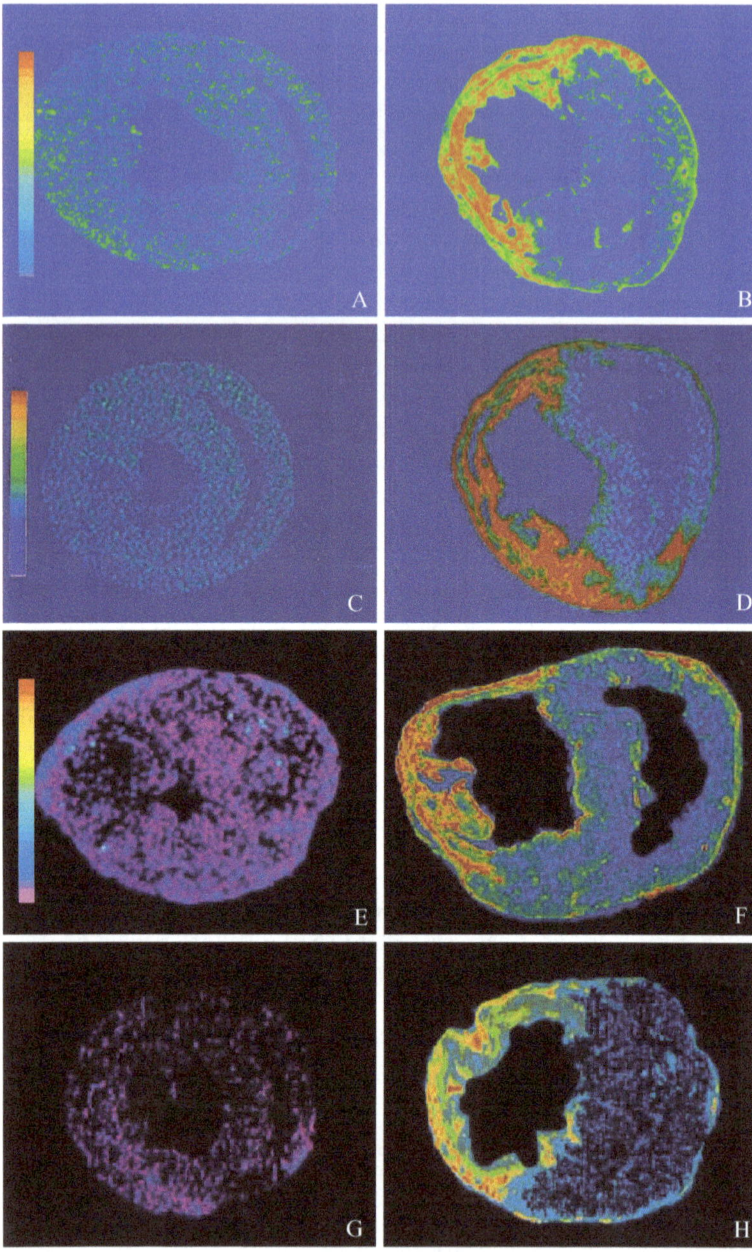


**Fig. 20.2** Bax, TGF-β, ACE, and AT1 receptor expression in the infarcted heart. By in situ hybridyzation, normal myocardium contains low levels of Bax (**panel a**) and TGF-β (**panel c**). At 7 days post MI, Bax and TGF-β gene expressions were largely increased at the site of MI and sites of damage at noninfarcted myocardium (**panels b and d**). Detected by autoradiography, binding density of ACE (**panel f**) and AT1 receptors (**panel h**) are largely increased in both infarcted and noninfarcted myocardium compared to normal heart (**panels e and g**)

through the formation of $ONOO^-$, activation of the death receptor Fas through upregulation of Fas ligand expression, inhibition of mitochrondrial ATP synthesis, and inactivation of serial antioxidant enzymes [56, 57]. In the infarcted mouse heart, iNOS expression is elevated in both infarcted and noninfarcted myocardium, whereas in iNOS knockout mice, cardiomyocyte apoptosis was reduced compared to wild-type mice with MI [58]. TNF-α-induced cardiomyocyte apoptosis can be inhibited by antioxidants such as thioredoxin and N-acetylcysteine [59]. Following acute MI, cardiac interleukin-6 expression is also significantly increased, which initiates inflammatory response, particularly at the site of infarction. Interleukin-6 is also verified to cause cardiomyocyte apoptosis, which can be suppressed by antioxidant treatment [60]. These findings suggest that oxidative stress is involved in different pathways that induce cardiomyocyte apoptosis following infarction and can be suppressed by antioxidants.

### *20.3.2 Oxidative Stress and Cardiac Inflammatory Response in the Infarcted Heart*

MI is associated with an inflammatory response, ultimately leading to healing and scar formation. Inflammatory response in the infarcted myocardium is related to the coordinated activation of a series of cytokine and adhesion molecule genes. A critical element in the regulation of these genes involves nuclear factor-kappa B (NF-κB), a redox-sensitive transcription factor. NF-κB maintains an inactive form bound to its inhibitory subunit Ikappa B under normal conditions. When tissue is injured, NF-κB can be activated by various local substances including ROS [61]. Upon activation, NF-κB stimulates inflammatory and immune responses and cellular growth by increasing the expression of specific cellular genes. NF-κB activation has been demonstrated in various models of myocardial ischemia and reperfusion [62, 63]. Activated NF-κB triggers gene expression of interstitial and vascular adhesion molecules, leading to leukocyte infiltration into the infarcted myocardium, as well as monocyte chemoattractant protein-1, and inducing recruitment of mononuclear cells. NF-κB also triggers gene expression of proinflammatory cytokines, such as TNF-α and interleukins, and initiates an inflammatory response [64]. TNF-α is not constitutively expressed in the normal heart. In rodent models of MI, TNF-α expression is significantly upregulated in the infarcted myocardium as well as in the noninfarcted myocardium [65]. TNF-α can stimulate inflammatory protein synthesis, macrophage phagocytosis, and cell growth, differentiation, and apoptosis [66]. In the infarcted myocardium, elevated NADPH oxidase is spatially coincident with activated NF-κB and enhanced TNF-α expression in the inflammatory cells [62]. In addition to the NF-κB pathway, recent studies suggest that $H_2O_2$ can directly induce cardiac TNF production via the p38 MARP pathway, and in turn mediate myocardial inflammation. Free radical scavenger treatment has been demonstrated to diminish inflammatory response and cardiac remodeling [13]. The antioxidant probucol has been shown to attenuate cardiac inflammation and improve ventricular function [67]. Thus these findings indicate that ROS serve as a proinflammatory mediator in the cardiac healing process following infarction and their role in cardiac inflammation involves several pathways.

In the infarcted heart, oxidative stress is mostly evident at the border zone, the area between the infarcted and noninfarcted myocardium [9]. ROS are known to cause oxidative damage to various cellular structures, including membranes, proteins, DNA sequences, and DNA repair enzymes and therefore have the potential to injure cardiomyocytes and vascular cells in the neighboring noninfarcted myocardium, causing additional cardiac damage/inflammation and infarct size extension. Studies have shown that antioxidant treatment reduces infarct size in the rat MI model [68, 69]. The beneficial effect of antioxidant on infarct size may be associated with the prevention of ROS-induced damage/inflammation in the border zone of the noninfarcted myocardium.

### *20.3.3 Oxidative Stress and Cardiac Fibrosis Following Infarction*

A growing bulk of evidence supports a causative role of oxidative stress in fibrogenesis in various tissues including liver, lung, arteries, nervous system, and heart [70, 71]. Extensive fibrosis is a major feature of myocardial remodeling in the infarcted heart, represented as scar at the site of infarction and interstitial fibrosis in noninfarcted myocardium [72]. Cells responsible for fibrous tissue formation at the site of infarction consist principally of phenotypically transformed fibroblast-like cells termed myofibroblasts. These cells contain α-smooth muscle actin microfilaments and are responsible for collagen synthesis and scar contraction [73]. They appear early after MI, mainly located in and around the infarcted myocardium and persist throughout healing [74]. Interstitial fibroblasts are responsible for normal collagen turnover and are considered to be a source of myofibroblasts. Myofibroblast differentiation and proliferation as well as collagen synthesis are tightly controlled by the fibrogenic cytokine transforming growth factor-beta (TGF-β). TGF-β mRNA and concentration are significantly increased in both the infarcted and noninfarcted myocardium in rats [75] (see Fig. 20.2). Oxidative stress is shown to upregulate the expression of TGF and type I collagen [76]. Following MI, enhanced expressions of TGF-β and NADPH oxidase are spatially coincident at the site of the infarcted myocardium (see Figs. 20.1 and 20.2). Treatment with the antioxidant taurine reduces oxidative stress, suppresses TGF-β gene expression and attenuates hepatic fibrosis [77]. Furthermore, in vitro studies have indicated that ROS promotes fibroblast proliferation and type I collagen gene expression in cardiac fibroblasts [78]. Chronic antioxidant treatment is shown to attenuate cardiac fibrosis [12]. However, Frantz and his coworkers have shown that targeted deletion of the NADPH oxidase subunit $gp91^{phox}$ does not affect left ventricular remodeling, including myocardial fibrosis following MI, and does not decrease the production of ROS [79]. This suggests that NADPH oxidases might not be the only important pathway for the production of ROS after MI. In addition to the heart, antioxidants d-a-tocopherol and butylated hydroxytoluene are confirmed to suppress type I collagen gene expression in the repairing liver [76]. After acute MI, progressive global left ventricular dilation occurs over months [80]. Matrix metalloproteinases (MMPs) favor this adverse remodeling. It was shown that inhibition of MMPs decreases the severity of remodeling in the infarcted heart [81]. In vitro studies have shown that ROS activates MMPs in cardiac fibroblasts [82]. Oxidative stress may, therefore, play a role in the pathogenesis of left ventricular dilation following infarction. However, in vivo studies on the regulatory role of ROS on MMPs and cardiac dilatation are lacking and further studies are required on this concept.

### *20.3.4 Oxidative Stress and Cardiac Hypertrophy Following Infarction*

Myocardial hypertrophy is often developed in noninfarcted myocardium weeks or months after MI. Extracellular stimuli such as mechanical strain, neurohormones, or cytokines have been well recognized to promote cardiomyocyte hypertrophy. Recent studies have further demonstrated that these extracellular stimuli may mediate cardiomyocyte hypertrophy via oxidative stress. ROS released acutely in large amounts have been implicated in the cell death associated with MI. The chronic release of ROS has been, however, linked to the development of left ventricular hypertrophy and heart failure progression [83, 84]. The chronic release of ROS appears to derive from the nonphagocytic NADPH oxidase and mitochondria [20]. The existence of nonphagocytic NADPH oxidase has been suggested in the cardiomyocytes. Experimental data suggest that abnormal activation of the nonphagocytic NADPH oxidase in response to neurohormones contributes to cardiomyocyte hypertrophy [85]. Acute MI is accompanied by systemic and local activation of numerous neurohumoral factors, including angiotensin (Ang)II, TNF-α, and norepinephrine. Increased AngII (AT1) receptors and angiotensin converting enzyme (ACE) are colocalized with NADPH oxidase in the infarcted heart (see Fig. 20.2) [86, 87]. Accumulating experimental evidence has demonstrated that AngII stimulates the expression of NADPH oxidase leading to the production of ROS in the repairing heart and vessels [88]. Treatment with ACE inhibitor attenuates ROS formation and prevents cardiomyocyte hypertrophy in diabetic rats to the same extent as the antioxidant N-acetylcysteine [89]. Moreover, in vitro studies further demonstrated that antioxidants prevent AngII-induced myocyte hypertrophy [90].

TNF-α is shown to provoke a hypertrophic cardiac phenotype [91]. In vitro studies have demonstrated that TNF-α-induced cardiomyocyte hypertrophy is mediated through NF-κB activation via the generation of ROS [92]. Norepinephrine-induced cardiac hypertrophy likewise also requires ROS. Treatment of cultured adult rat cardiomyocytes with norepinephrine increases NADPH oxidase and leads to cardiomyocyte hypertrophy.

The exact pathways of ROS-mediated hypertrophy remain to be determined. ROS can participate as second messengers in myocardial signaling events, activating stress-related nuclear transcription factors, leading to cellular hypertrophy [93]. These second messengers include MAP kinases, small GTP-binding proteins, the Src family of tyrosine kinases and cytokines cascades. Pretreatment with a SOD mimetic inhibits both the hypertrophy and ROS production. The overexpression of GSHPx could attenuate left ventricular hypertrophy and improve cardiac function after MI [83].

### *20.3.5 Oxidative Stress and Heart Failure*

Heart failure is initiated after cardiac damage with a resultant loss of cardiac myocytes and/or disruption of myocardium, which leads to the impairment of force generation, thereby preventing normal contractions in the heart. Heart failure may result from a variety of cardiac diseases, including MI, cardiomyopathy, and hypertensive heart disease. Accumulating evidence supports the theory that ventricular dysfunction worsens as a consequence of oxidative stress. Oxidative stress has been demonstrated in CHF animal models caused by MI, cardiomyopathy, and hypertension; antioxidant treatment in these animals is shown to improve ventricular function [12, 13, 24]. Experimental studies have demonstrated that oxidative stress may contribute to the pathogenesis of heart failure in several ways. ROS initiate myocyte apoptosis/necrosis, which may be associated with ventricular dysfunction. ROS, through lipid and protein damage, can profoundly affect the handling of calcium and other ions by myocytes and cause significant and irreversible myocardial dysfunction [94]. ROS-induced inflammation and endothelial dysfunction are also involved in the progression of heart failure. Oxidative stress–induced myocardial remodeling also contributes to the development and progression of heart failure.

The extensive body of literature in animal studies suggests that there is great potential benefit in therapies that can improve cardiac function in humans. However, the existing evidence for a role of oxidative stress in the pathogenesis of CHF in humans is not compelling. Clinical trials of antioxidant therapy for CHF are, however, few in number and so far have failed to demonstrate convincing benefits. This might be because of several potential reasons. First, ROS are derived from multi-sources in the failing heart. ROS can be produced intracellularly through electron leakage from mitochondria during oxidative phosphorylation and through the activation of several cellular enzymes, including NADPH oxidase, xanthine oxidase, and nitric oxide synthase. Treatment with a specific antioxidant, such as NADPH oxidase inhibitor, may therefore not suppress oxidative stress because of the redundant sources of ROS production. This has been confirmed by animal experiments. In gp91$^{phox}$ gene knockout mice, oxidative stress was not prevented in the infarcted myocardium. Second, the currently recognized antioxidants that were used in animals, including probucol, pyrrolidine dithiocarbamate, vitamin E, or vitamin C and N-acetyl cysteine, are either not strong enough or not specific enough with many other effects. Moreover, combinations of antioxidants, dosing and timing of treatment, and duration of the process need to be further established in patients with heart failure.

## 20.4 Summary

Experimental evidence has demonstrated that cardiac repair/remodeling is subject to oxidative stress following MI. This chapter has highlighted several mechanisms by which oxidative stress is involved in cardiac remodeling in the infarcted heart. Following acute MI, oxidative stress occurs in both the infarcted and noninfarcted myocardium; and ROS play a significant role in the initiation as well as the regulation of cardiac molecular and cellular changes, including cardiomyocyte apoptosis, inflammatory/fibrogenic responses, and hypertrophy, that contribute to myocardial repair/remodeling, leading to heart failure. These cardiac events can be significantly inhibited by antioxidants in animal models of MI. Further studies are required on the potential therapeutic interventions with antioxidants in limiting cardiac remodeling in patients with MI.

## References


1. Olivetti G, Capasso JM, Meggs LG, Sonnenblick EH, Anversa P (1991) Cellular basis of chronic ventricular remodeling after myocardial infarction in rats. Circ Res 68:856–869
2. Jugdutt BI (2003) Ventricular remodeling after infarction and the extracellular collagen matrix: when is enough enough? Circulation 108:1395–1403
3. Ertl G, Frantz S (2005) Healing after myocardial infarction. Cardiovasc Res 66:22–32
4. Anversa P, Li P, Zhang X (1993) Ischemic myocardial injury and ventricular remodeling. Cardiovasc Res 27:145–157
5. Francis GS, Mcdonald K, Chu G, Cohn JN (1995) Pathophysiologic aspects of end-stage heart failure. Am J Cardio 75:11A–16A
6. Weber KT (1997) Extracellular matrix remodeling in heart failure. Circulation 96:4065–4082
7. Hill MF, Singal PK (1996) Antioxidant and oxidative stress changes during heart failure subsequent to myocardial infarction in rats. Am J Pathol 148:291–300
8. Fukui T, Yoshiyama M, Hanatani A, Omura T, Yoshikawa J, Abe Y (2001) Expression of p22-phox an dgp91-phox, essential components of NADPH oxidase, increases after myocardial infarction. Biochem Biophys Res Commun 281:1200–1206
9. Lu L, Quinn MT, Sun Y (2004) Oxidative stress in the infarcted heart: role of de novo angiotensin II production. Biochem Biophys Res Commun 325:943–951
10. Usal A, Acarturk E, Yuregir GT, Unlukurt I, Demirci C, Kurt GI, Birand A (1996) Decreased glutathione levels in acute myocardial infarction. Jpn Heart J 37:177–182
11. Khaper N, Kaur K, Li T, Farahmand F, Singal PK (2003) Antioxidant enzyme gene expression in congestive heart failure following myocardial infarction. Mol Cell Biochem 251:9–15
12. Sia YT, Parker TG, Liu P, Tsoporis JN, Adam A, Rouleau JL (2002) Imporved post-myocardial infarction survival with probucol in rats: effects on left ventricular function, morphology, cardiac oxidative stress and cytokine expression. J Am Col Cardiol 39:148–156
13. Sia YT, Lapointe N, Parker T, Tsoporis JN, Deschepper CF, Calderone A, Pourdjabbar A, Jasmin JF, Sarrazin JF, Liu P, Adam A, Butany J, Rouleau JL (2002) Beneficial effects

of long-term use of the antioxidant probucol in heart failure in the rat. Circulation 105: 2549–2555
14. Kinugawa S, Tsutsui H, Hayashidani S, Ide T, Suematsu N, Satoh S, Utsumi H, Takeshita A (2000) Treatment with dimethylthiourea prevents left ventricular remodeling and failure after experimental myocardial infarction in mice: role of oxidative stress. Circ Res 87:392–398
15. Hare JM (2001) Oxidative stress and apoptosis in heart failure progression. Cir Res 89: 198–201
16. Poli G, Parola M (1997) Oxidative damage and fibrogenesis. Free Radical Biol Med 22: 287–305
17. Gorlach A, Kietzmann T, Hess J (2002) Redox signaling through NADPH oxidases: involvement in vascular proliferation and coagulation. Ann N Y Acad Sci 973:505–507
18. Brar SS, Kennedy TP, Sturrock AB, Huecksteadt TP, Quinn MT, Whorton AR, Hoidal JRA (2002) NAD(P)H oxidase regulates growth and transcription in melanoma cells. Am J Physiol Cell Physiol 282:C1212–C1224
19. Nakagami H, Takemoto M, Liao JKNADPH (2003) Oxidase-derived superoxide anion mediates angiotensin II-induced cardiac hypertrophy. J Mol Cell Cardiol 35:851–859
20. Sorescu D, Griendling K (2002) Reactive oxygen species, mitochondria, and NADPH oxidases in the development and progression of heart failure. Chem Herit 8:132–140
21. Droge W (2002) Free radicals in the physiological control of cell function. Physiol Rev 82: 47–52
22. Halliwell B (1991) Reactive oxygen species in living systems: source, biochemistry, and role in human disease. Am J Med 91:14S–22S
23. Finkel T (1999) Signal transduction by reactive oxygen species in non-phagocytic cells. J Leukoc Biol 65:337–340
24. Dhalla AK, Singal PK (1994) Antioxidant changes in hypertrophied and failing guinea pig heart. Am J Physiol 266:H1280–H1285
25. Young IS (2001) Woodside antioxidants in hearth and disease. J Clin Pathol 54:176–186
26. Vaziri ND, Lin C, Farmand F, Sindhu K (2003) Superoxid dismutase, catalase, glutathione eroxidase and NADPH oxidase in lead-induced hypertension. Kidney Int 63: 186–194
27. Gasparetto C, Malinverno A, Culacciati D, Gritti D, Prosperini PG, Specchia G, Ricevuti G (2005) Antioxidant vitamins reduce oxidative stress and ventricular remodeling in patients with acute myocardial infarction. Int J Immunopathol Pharmacol 18:487–496
28. Hill MF, Singal PK (1997) Right and left myocardial antioxidant responses during heart failure subsequent to myocardial infarction. Circulation 96:2414–2420
29. Usal A, Acarturk E, Yuregir GT, Unlukurt I, Demirci C, Kurt HI, Birand A (1996) Decreased glutathione levels in acute myocardial infarction. Jpn Heart J 37:177–182
30. Mohazzab-H KM, Kaminski PM, Wolin MS (1997) Lactate and PO2 modulate superoxide anion production in bovine cardiac myocytes: potential role of NADH oxidase. Circulation 15:614–620
31. Fukui T, Yoshiyama M, Hanatani A, Omura T, Yoshikawa J, Abe Y (2001) Expression of p22-phox and gp91-phox, essential components of NADPH oxidase, increases after myocardial infarction. Biochem Biophys Res Commun 281:1200–1206
32. Grieve DJ, Byrne JA, Cave AC, Shah AM (2004) Role of oxidative stress in cardiac remodeling after myocardial infarction. Heart Lung Circ 13:132–138
33. Khaper N, Kaur K, Li T, Farahmand F, Singal PK (2003) Antioxidant enzyme gene expression in congestive heart failure following myocardial infarction. Mol Cell Biochem 251: 9–15
34. Anversa P, Cheng W, Liu Y, Leri A, Redaelli G, Kajstura J (1998) Apoptosis and myocardial infarction. Basic Res Cardiol 93:8–12
35. Cheng W, Kajstura J, Nitahara JA, Li B, Reiss K, Liu Y, Clark WA, Krajewski S, Reed JC, Olivetti G, Anversa P (1996) Programmed myocyte cell death affects the viable myocardium after infarction in rats. Exp Cell Res 226:316–327

36. Kajstura J, Cheng W, Reiss K, Clark WA, Sonnenblick EH, Krajewski S, Reed JC, Olivetti G, Anversa P (1996) Apoptotic and necrotic myocyte cell deaths are independent contributing variables of infarct size in rats. Lab Invest 74:86–107
37. Li WG, Coppey L, Weiss RM, Oskarsson HJ (2001) Antioxidant therapy attenuates JNK activation and apoptosis in the remote noninfarcted myocardium after large myocardial infarction. Biochem Biophys Res Commun 280:353–357
38. Kumar D, Kirshenbaum LA, Li T, Danelisen I, Singal PK (2001) Apoptosis in adriamycin cardiomyopathy and its modulation by probucol. Antioxid Redox Signal 3:135–145
39. Pagliara P, Carla EC, Caforio S, Chionna A, Massa S, Abbro L, Dini L (2003) Kupffer cells promote lead nitrate-induced hepatocyte apoptosis via oxidative stress. Comp Hepatol 2:8–16
40. von Harsdorf R, Li PF, Dietz R (1998) Signaling pathways in reactive oxygen species-induced cardiomyocyte apoptosis. Circulation 99:2934–2941
41. Li PF, Dietz R, von Harsdorf R (1999) Superoxide induces apoptosis in cardiomyocytes, but proliferation and expression of transforming growth factor-beta1 in cardiac fibroblasts. FEBS Lett 448(2–3):206–210
42. Gottlieb RA, Burleson KO, Kloner RA, Babior BM, Engler RL (1994) Reperfusion injury induces apoptosis in rabbit cardiomyocytes. J Clin Invest 94:1621–1628
43. Itoh G, Tamura J, Suzuki M, Suzuki Y, Ikeda H, Koike M, Nomura M, Jie T, Ito KDNA (1995) Fragmentation of human infarcted myocardial cells demonstrated by the nick end labeling method and DNA agarose gel electrophoresis. Am J Pathol 146:1325–1331
44. Kajstura J, Zhang X, Liu Y, Szoke E, Cheng W, Olivetti G, Hintze TH (1995) The cellular basis of pacing-induced dilated cardiomyopathy: myocyte cell loss and myocyte cellular reactive hypertrophy. Circulation 92:2306–2317
45. Liu Y, Cigola E, Cheng W, Kajstura J, Olivetti G, Hintze TH, Anversa P (1995) Myocyte nuclear mitotic division and programmed myocyte cell death characterize the cardiac myopathy induced by rapid ventricular pacing in dogs. Lab Invest 73:771–787
46. Sharov VG, Sabbah HN, Shimoyama H, Goussev AV, Lesch M, Goldstein S (1996) Evidence of cardiocyte apoptosis in myocardium of dogs with chronic heart failure. Am J Pathol 148:141–149
47. Teiger E, Than VD, Richard L, Wisnewsky C, Tea BS, Gaboury L, Tremblay J, Schwartz K, Hamet P (1996) Apoptosis in pressure overload-induced heart hypertrophy in the rat. J Clin Invest 97:2891–2897
48. Zeng H, Liu X, Zhao H (2003) Effects of carvedilol on cardiomyocyte apoptosis and gene expression in vivo after ischemia-reperfusion in rats. J Huazhong Univ Sci Technolog Med Sci 23:127–130
49. Kumar D, Jugdutt BI (2003) Apoptosis and oxidants in the heart. J Lab Clin Med 142:288–297
50. Zhao W, Lu L, Chen SS, Sun Y (2004) Temporal and spatial characteristics of apoptosis in the infarcted rat heart. Biochem Biophys Res Commun 325:605–611
51. Hockenbery DM, Oltvai ZN, Yin XM, Milliman CL, Korsmeyer SJ (1993) Bcl-2 functions in an antioxidant pathway to prevent apoptosis. Cell 75:241–251
52. Cesselli D, Jakoniuk I, Barlucchi L, Beltrami AP, Hintze TH, Hadal-Ginard B, Kajstura J, Leri A, Anversa P (2001) Oxidative stress-mediated cardiac cell death is a major determinant of ventricular dysfunction and failure in dog dilated cardiomyopathy. Circ Res 89:279–286
53. Kubota T, Miyagishima M, Frye CS, Alber SM, Bounoutas GS, Kadokami T, Watkins SC, McTiernan CF, Feldman AM (2001) Overexpression of tumor necrosis factor-alpha activates both anti- and pro-apoptotic pathways in the myocardium. J Mol Cell Cardiol 33:1331–1344
54. Haunstetter A, Izumo S (1998) Apoptosis: basic mechanisms and implications for cardiovascular disease. Circ Res 82:1111–1129
55. Song W, Lu X, Feng Q (2000) Tumor necrosis factor-alpha induces apoptosis via inducible nitric oxide synthase in neonatal mouse cardiomyocytes. Cardiovasc Res 45:595–602
56. Almeida A, Bolanos JP (2001) A transient inhibition of mitochondrial ATP synthesis by nitric oxide synthase activation triggered apoptosis in primary cortical neurons. J Neurochem 77:676–690

57. Dobashi K, Pahan K, Chahal A, Singh I (1997) Modulation of endogenous antioxidant enzymes by nitric oxide in rat C6 glial cells. J Neurochem 68:1896–1903
58. Marfella R, Di Filippo C, Esposito K, Nappo F, Piegari E, Cuzzocrea S, Berrino L, Rossi F, Giugliano D, D'Amico M (2004) Absence of inducible nitric oxide synthase reduces myocardial damage during ischemia reperfusion in streptozotocin-induced hyperglycemic mice. Diabetes 53:454–462
59. Buttke TM, Sandstrom PA (1994) Oxidative stress as a mediator of apoptosis. Immunol Today 15:7–10
60. Wollert KC, Drexler H (2001) The role of interleukin-6 in the failing heart. Heart Fail Rev 6:95–103
61. Kabe Y, Ando K, Hirao S, Yoshida M, Handa H (2005) Redox regulation of NF-kappaB activation: distinct redox regulation between the cytoplasm and the nucleus. Antioxid Redox Signal 7:395–403
62. Lu L, Chen SS, Zhang JQ, Ramires FJ, Sun Y (2004) Activation of nuclear factor-kappaB and its proinflammatory mediator cascade in the infarcted rat heart. Biochem Biophys Res Commun 321:879–885
63. Nichols TC (2004) NF-kappaB and reperfusion injury. Drug News Perspect 17:99–104
64. Schoonbroodt S, Piette J (2000) Oxidative stress interference with the nuclear factor-kappa B activation pathways. Biochem Pharmacol 15:1075–1083
65. Jacobs M, Staufenberger S, Gergs U, Meuter K, Brandstatter K, Hafner M, Ertl G, Schorb W (1999) Tumor necrosis factor-alpha at acute myocardial infarction in rats and effects on cardiac fibroblasts. J Mol Cell Cardiol 31:1949–1959
66. Ceconi C, Curello S, Bachetti T, Corti A, Ferrari R (1998) Tumor necrosis factor in congestive heart failure: a mechanism of disease for the new millennium? Prog Cardiovasc Dis 41:25–30
67. Nakamura R, Egashira K, Machida Y, Hayashidani S, Takeya M, Utsumi H, Tsutsui H, Takeshita A (2002) Probucol attenuates left ventricular dysfunction and remodeling in tachycardia-induced heart failure: roles of oxidative stress and inflammation. Circulation 106:362–367
68. Sahna E, Acet A, Ozer MK, Olmez E (2002) Myocardial ischemia-reperfusion in rats: reduction of infarct size by either supplemental physiological or pharmacological doses of melatonin. J Pineal Res 33:234–238
69. Kim CH, Choi H, Chun YS, Kim GT, Park JW, Kim MS (2002) The protective effect of resveratrols on ischaemia-reperfusion injuries of rat hearts is correlated with antioxidant efficacy. Br J Pharmacol 135:1627–1633
70. Tsukamoto H (1993) Oxidative stress, antioxidants, and alcoholic liver fibrogenesis. Alcohol 10:465–467
71. Mastruzzo C, Crimi N, Vancheri C (2002) Role of oxidative stress in pulmonary fibrosis. Monaldi Arch Chest Dis 57:173–176
72. Sun Y, Cleutjens JP, Diaz-Arias AA, Weber KT (1994) Cardiac angiotensin converting enzyme and myocardial fibrosis in the rat. Cardiovasc Res 28:1423–1432
73. Desmouliere A, Gabbiani G (1996) The role of the myofibroblast in wound healing and fibrocontractive diseases. In: Richard AF (ed) The molecular and cellular biology of wound repair. Plenum Press, New York, NY, 391–423
74. Sun Y, Weber KT (1996) Angiotensin converting enzyme and myofibroblasts during tissue repair in the rat. J Mol Cell Cardiol 28:851–858
75. Sun Y, Zhang JQ, Zhang J, Ramires FJ (1998) Angiotensin II, transforming growth factor-beta1 and repair in the infarcted heart. J Mol Cell Cardiol 30:1559–1569
76. Lee KS, Buck M, Houglum K, Chojkier M (1995) Activation of hepatic stellate cells by TGF alpha and collagen type I is mediated by oxidative stress through c-myb expression. J Clin Invest 96:2461–2468
77. Miyazaki T, Karube M, Matsuzaki Y, Ikegami T, Doy M, Tanaka N, Bouscarel B (2005) Taurine inhibits oxidative damage and prevents fibrosis in carbon tetrachloride-induced hepatic fibrosis. J Hepatol 43:117–125

78. Murrell AC, Francis JO, Bromley L (1993) Modulation of fibroblast proliferation by oxygen free radicals. Biochem J 265:659–665
79. Frantz S, Brandes RP, Hu K, Rammelt K, Wolf J, Scheuermann H, Ertl G, Bauersachs J (2005) Left ventricular remodeling after myocardial infarction in mice with targeted deletion of the NADPH oxidase subunit gp91phox. Basic Res Cardiol 100:1–6
80. Jugdutt BI (2003) Ventricular remodeling after infarction and the extracellular collagen matrix: when is enough enough? Circulation 108:1395–1403
81. Rohde LE, Ducharme A, Arroyo LH, Aikawa M, Sukhova GH, Lopez-Anaya A, McClure KF, Mitchell PG, Libby P, Lee RT (1999) Matrix metalloproteinase inhibition attenuates early left ventricular enlargement after experimental myocardial infarction in mice. Circulation 99:3063–3070
82. Siwik DA, Pagano PJ, Colucci WS (2001) Oxidative stress regulates collagen synthesis and matrix metalloproteinase activity in cardiac fibroblasts. Am J Physiol Cell Physiol 280: C53–C60
83. Shiomi T, Tsutsui H, Matsusaka H, Murakami K, Hayashidani S, Ikeuchi M, Wen J, Kubota T, Utsumi H, Takeshita A (2004) Overexpression of glutathione peroxidase prevents left ventricular remodeling and failure after myocardial infarction in mice. Circulation 109:544–549
84. Sorescu D, Griendling KK (2002) Reactive oxygen species, mitochondria, and NAD(P)H oxidases in the development and progression of heart failure. Congest Heart Fail 8:132–140
85. Byrne JA, Grieve DJ, Cave AC, Shah AM (2003) Oxidative stress and heart failure. Arch Mal Coeur Vaiss 96:214–221
86. Fabris B, Jackson B, Kohzuki M, Perich R, Johnston CI (1990) Increased cardiac angiotensin-converting enzyme in rats with chronic heart failure. Clin Exp Pharmacol Physiol 17:309–314
87. Sun Y, Weber KT, Angiotensin II (1994) Receptor binding following myocardial infarction in the rat. Cardiovasc Res 28:1623–1628
88. Wang HD, Xu S, Johns DG, Du Y, Quinn MT, Cayatte AJ, Cohen RA (2001) Role of NADPH oxidase in the vascular hypertrophic and oxidative stress response to angiotensin II in mice. Circ Res 88:947–953
89. Fiordaliso F, Cuccovillo I, Bianchi R, Bai A, Doni M, Salio M, De Angelis N, Ghezzi P, Latini R, Masson S (2006) Cardiovascular oxidative stress is reduced by an ACE inhibitor in a rat model of streptozotocin-induced diabetes. Life Sci 79:121–129
90. Nakamura K, Fushimi K, Kouchi H, Mihara K, Miyazaki M, Ohe T, Namba M (1998) Inhibitory effects of antioxidants on neonatal rat cardiac myocyte hypertrophy induced by tumor necrosis factor-alpha and angiotensin II. Circulation 98:794–799
91. Mann DL (2002) Tumor necrosis factor-induced signal transduction and left ventricular remodeling. J Card Fail 8:S379–S386
92. Higuchi Y, Otsu K, Nishida K, Hirotani S, Nakayama H, Yamaguchi O, Matsumura Y, Ueno H, Tada M, Hori M (2002) Involvement of reactive oxygen species-mediated NF-kappa B activation in TNF-alpha-induced cardiomyocyte hypertrophy. J Mol Cell Cardiol 34:233–240
93. Nian M, Lee P, Khaper N, Liu P (2004) Inflammatory cytokines and postmyocardial infarction remodeling. Circ Res 94:1543–1553
94. Ball AM, Sole MJ (1998) Oxidative stress and the pathogenesis of heart failure. Cardiol Clin 16:665–675

# Chapter 21
# Oxidative Stress and Redox Signalling in Cardiac Remodelling

**Min Zhang, Alex Sirker, and Ajay M. Shah**

**Abstract** Cardiac remodelling describes the chronic response to stresses such as myocardial infarction (MI) or chronic hypertension, which generally becomes maladaptive over time, leading to deleterious structural and functional alterations and manifesting clinically as chronic heart failure (CHF). Although the underlying mechanisms are multifactorial, a significant body of evidence points to important roles for oxidative stress and redox signalling. Oxidative stress, occurring when excess reactive oxygen species (ROS) cannot be adequately countered by antioxidant defences, triggers cell dysfunction, energetic deficit or cell death. However, ROS may also have more subtle effects in the process of remodelling through specific modulation of redox-sensitive signalling pathways that alter gene and protein expression and function. ROS are generated from many sources including mitochondria, xanthine oxidase, uncoupled nitric oxide synthases, and NADPH oxidases; the last of these appear to be especially important in redox signalling. This chapter discusses recent advances in delineating the contribution of ROS to some of the principal alterations underlying the remodelling process (i.e., cardiomyocyte hypertrophy, apoptosis, interstitial fibrosis, contractile dysfunction, and chamber dilatation), with a particular emphasis on the role of NADPH oxidase. A better understanding of redox signalling mechanisms may enable the development of new targeted strategies for the prevention and treatment of adverse cardiac remodelling.

**Keywords** Reactive oxygen species · Cardiac remodelling · Heart failure · NADPH oxidases · Myocardial hypertrophy · Fibrosis · Apoptosis · Myocardial infarction

---

A.M. Shah (✉)
Cardiovascular Division, King's College London British Heart Foundation Centre of Research Excellence, London SE5 9NU, UK
e-mail: ajay.shah@kcl.ac.uk

H. Sauer et al. (eds.), *Studies on Cardiovascular Disorders*,
Oxidative Stress in Applied Basic Research and Clinical Practice,
DOI 10.1007/978-1-60761-600-9_21, 

## 21.1 Introduction

Cardiac remodelling describes a complex series of changes in myocardial structure and function that occur in response to stresses such as myocardial infarction (MI), chronic pressure overload (e.g., systemic hypertension), and chronic volume overload (e.g., valvular regurgitation). At the cellular level, the main components of cardiac remodelling include cardiomyocyte hypertrophy (an increase in cell size), apoptosis, abnormal excitation-contraction coupling, deficient energetics, increased interstitial fibrosis, matrix remodelling, reduced capillarisation, vascular dysfunction, and inflammation. At a molecular level, these processes are driven largely by complex changes in gene and protein expression or posttranslational modifications. These abnormalities collectively lead to substantial alterations in cardiac size, shape, coordination, wall thickness, and contractile function. Initially, some components of cardiac remodelling may be compensatory; for example, the increase in myocardial wall thickness and mass may have a beneficial adaptive function by normalizing wall stress. However, cardiac remodelling eventually leads inexorably to contractile depression, arrhythmia, ventricular dilatation, significant fibrosis, and the development of chronic heart failure (CHF), which is associated with major morbidity and mortality. Therefore, elucidation of the mechanisms underlying adverse cardiac remodelling is an important goal which may enable the development of novel therapies for patients with heart failure.

While the mechanisms underlying myocardial remodelling are undoubtedly multifactorial, a significant body of experimental and clinical data point to the important roles of increased reactive oxygen species (ROS) production in this process. For example, definitive evidence of enhanced ROS generation in the failing myocardium has been documented both in human and experimental tissue using electron spin resonance (ESR) [1]. Treatment with antioxidants attenuates chamber enlargement and the development of contractile dysfunction or left ventricular hypertrophy (LVH) in many experimental settings, supporting a role for ROS production in cardiac remodelling in vivo [2–4]. More direct molecular evidence for a causative role of ROS in cardiac remodelling and failure comes from recent studies using genetically-modified animal models in which either ROS generation or antioxidant enzyme activities have been altered—e.g., diminishment of endogenous thioredoxin activity in the heart increased oxidative stress and cardiac hypertrophy induced by aortic constriction [5]. Overexpression of glutathione peroxidase inhibited LV remodelling and failure after MI [6].

## 21.2 ROS, Oxidative Stress, and Redox Signalling

ROS include radicals such as superoxide ($O_2^{-\bullet}$) and hydroxyl ($^{\bullet}OH$) and nonradicals such as hydrogen peroxide ($H_2O_2$). The actions of $O_2^{-\bullet}$ are generally restricted to the intracellular compartment of production due to its limited diffusion capacity, but $H_2O_2$ is more stable and more cell membrane-permeable and may be more likely to act as a signalling molecule. $H_2O_2$ can also generate the highly reactive •OH via the Fenton or Haber-Weiss reactions under pathological conditions (e.g., ischaemia–reperfusion). In settings where the level of the signalling molecule nitric oxide (NO) is in the high nanomolar range, $O_2^{-\bullet}$ may react with NO to generate the potent oxidant $ONOO^{\bullet-}$, this reaction also resulting in inactivation of NO.

In health, basally generated ROS are efficiently counterbalanced by several antioxidant enzymes and nonenzymatic antioxidants, including the superoxide dismutases (SODs), catalase, glutathione peroxidase, glutathione, urate, ubiquinone, and vitamins E, C, and β-carotene. In addition, thioredoxin and thioredoxin reductase catalyze the regeneration of many antioxidant molecules and as such constitute an important antioxidant defence (as well as being involved in signalling).

The pathophysiological effects of ROS are thought to depend upon the moiety generated, its concentration and subcellular localization, and the endogenous antioxidant status [7]. Oxidative stress occurs when levels of ROS are elevated sufficiently to overwhelm the cellular antioxidant defences; in this situation, they react directly with membrane lipids, proteins, and nucleic acid, causing cellular damage, energetic deficit, dysfunction, and death (both through apoptosis and necrosis). On the other hand, the tightly regulated production of small amounts of ROS modulates the activity of diverse intracellular ion channels, pumps, protein kinases, phosphatases, and other proteins that regulate signalling cascades, thereby inducing highly specific acute and chronic changes in cell phenotype—termed "redox signalling." Such redox-regulated effects underlie the essential roles of ROS in biological processes such as normal cell proliferation and growth and may also be especially important in many of the processes that underlie adverse cardiac remodelling.

## 21.3 Cardiac Sources of ROS

All cell types within the heart, including cardiomyocytes, endothelial cells, vascular smooth muscle cells (VSMCs), fibroblasts, and infiltrating inflammatory cells can generate ROS. Potential sources of ROS in these cells include mitochondria, xanthine oxidase (XO), uncoupled NO synthases, and NADPH oxidases. The mechanisms and regulations of ROS production by these sources have been covered in recent detailed reviews [8–11]. Here, we provide a brief description of key features of these sources, while their involvement in different components of cardiac remodelling is considered in the relevant sections below.

*Mitochondrial ROS.* Cardiomyocytes have the highest volume density of mitochondria in the entire body and hence significant potential for mitochondrial ROS generation [8]. In the failing heart, this has been attributed to electron leakage mainly from complexes I [12] and II [13]. The primary mitochondrial antioxidant enzyme is manganese superoxide dismutase (MnSOD); and heart-specific MnSOD-deficient mice develop progressive heart failure with excess formation of superoxide [14]. Recently, several studies also showed important roles for other mitochondrial antioxidants, including peroxiredoxin-3 [15] and apoptosis-inducing factor (AIF)

[16]. Mitochondrial ROS may be especially important in driving apoptosis and in damage of mitochondrial DNA (mtDNA), which encodes genes such as respiratory chain complex enzymes [17, 18]. In this regard, it was reported that transgenic overexpression of the mitochondrial transcription factor A could ameliorate these defects and significantly reduce myocyte apoptosis and adverse LV remodelling post-MI [17, 18].

*Xanthine oxidase.* Xanthine oxidoreductase is a molybdoenzyme capable of catalyzing the oxidation of hypoxanthine and xanthine in the process of purine metabolism. Xanthine oxidoreductase exists as one of two interconvertible yet functionally distinct forms, namely xanthine dehydrogenase (XD) or xanthine oxidase (XO). The former reduces $NAD^+$, whereas the latter prefers molecular oxygen, leading to the production of both $O_2^{-\bullet}$ and $H_2O_2$ [10]. It has been suggested that endogenous XO synthesis in the heart may be low, but that it may be released from XO-rich organs such as liver and intestine under pathophysiological conditions and may subsequently bind to endothelial cells in situ in the heart. The ROS generated from XO could also trigger the local accumulation and activation of neutrophils (via neutrophil NADPH oxidase and via myeloperoxidase-mediated ROS generation), leading to further bursts of ROS production and ultimately cardiac dysfunction [19].

*Uncoupled NO synthases (NOS).* All three NOS isoforms (eNOS, nNOS, and iNOS) may be expressed in the failing heart; eNOS in endothelial cells and cardiomyocytes, nNOS in nerves and cardiomyocytes, and iNOS in numerous cell types. The role of NOSs is intriguing because NO has generally been regarded to be antihypertrophic [20], but the enzyme can become a ROS or peroxynitrite generator when it is deficient in its cofactor tetrahydrobiopterin ($BH_4$) or substrate L-arginine, thereby functioning as an uncoupled monomer [9]. Therefore, the role of NOS could depend upon the relative generation of ROS vs. NO.

*NADPH oxidases.* NADPH oxidases are of particular interest since they are the only enzymes discussed so far that generate ROS in a regulated manner, apparently suited for cell signalling [21]. Several homologues of the classical neutrophil NADPH oxidase have been discovered in recent years, and are recognized as major ROS sources within many tissues [22]. All NADPH oxidases contain a core Nox subunit that catalyzes electron transfer from NADPH to molecular $O_2$, thereby generating $O_2^{-\bullet}$; this core subunit exists as one of seven isoforms (Nox1–5 and Duox1–2), each of which form the basis of a distinct oxidase [23]. In the cardiovascular system, Nox2 (also known as $gp91^{phox}$ oxidase) is expressed in inflammatory cells, endothelial cells, cardiomyocytes, and fibroblasts; Nox1 is found in vascular smooth muscle cells; Nox4 is expressed in endothelial cells, cardiomyocytes, vascular smooth muscle, and fibroblasts; and Nox5 is reportedly expressed in human endothelial and vascular smooth muscle cells. Superoxide generation by Nox1 or 2 is tightly regulated; these isoforms associate with a smaller $p22^{phox}$ subunit, as well as cytosolic regulatory subunits ($p47^{phox}$, $p67^{phox}$, $p40^{phox}$, and Rac1 or homologues thereof) that are required to activate the enzyme. In contrast, Nox4 appears to be constitutively active at a low level and not to require any known cytosolic subunits for its activation.

A wide range of agonists and stimuli activate Nox1 and Nox2 oxidases, e.g., G-protein coupled receptor (GPCR) agonists such as angiotensin II, endothelin-1, α–adrenergic agonists; growth factors such as thrombin; cytokines such as tumor necrosis factor-α, (TNF-α); and mechanical forces. They are linked to diverse redox-sensitive signalling cascades, including kinases, transcription factors, matrix metalloproteinases (MMPs), and ion channels and receptors [24]. Accumulating evidence indicates important roles for NADPH oxidases in cardiac remodelling and its precursor conditions, as discussed below. NADPH oxidase-derived ROS can also trigger further ROS generation by other sources, e.g., xanthine oxidoreductase [25] or NOSs by oxidizing BH4 and therefore uncoupling the enzymes [26].

## 21.4 ROS and Cardiac Hypertrophy

Cardiomyocyte hypertrophy occurs as a response to both mechanical and neurohumoral stress [27, 28]. G-protein coupled receptor agonists such as angiotensin II, noradrenaline, and endothelin-1 are well known to induce cardiomyocyte hypertrophy. Substantial evidence indicates that ROS-regulated pathways are involved in cardiac hypertrophy in response to these stimuli.

Low levels of $H_2O_2$ can activate ERK1/2 in adult rat ventricular cardiomyocytes and produce hypertrophy [29]. Higher levels of $H_2O_2$ activate other members of the MAPK family, such as p38MAPK and JNK and may induce apoptosis [30]. ROS-dependent activation of Akt and NF-κB has also been shown to be involved in prohypertrophic effects [31, 32]. Other workers reported an involvement of ASK-1 activation upstream of NF-κB activation in GPCR agonist-induced hypertrophy [33, 34]. Cardiac hypertrophy in response to α-adrenoreceptor activation has been shown to be mediated through oxidative modification of thiol groups on Ras [35]. The redox-sensitive nature of Ras activation explains its apparent sensitivity to thioredoxin-1 (a component of the thioredoxin antioxidant system) [35].

In vitro models of mechanical stress on cultured cardiomyocytes indicate the importance of ROS in the hypertrophic response to this stimulus. Stretch induces increased protein synthesis in proportion to the elevation of ROS production, and the former is inhibited by antioxidants [36]. Stretch-induced ROS appeared to lie upstream of p38MAPK activation in other work in a similar model [37]. Other redox signalling pathways implicated in stretch-induced hypertrophy include Ras oxidation followed by activation of the Raf/MEK/ERK pathway [38].

Recently, Sadoshima's group has revealed a novel prohypertrophic mechanism involving the oxidation of Class II histone deactylases (HDAC) [39]. HDAC are key regulators of gene expression, via modification of histones, and their activity is partly determined via nuclear-cytoplasmic shuttling. It has been previously shown that thioredoxin-1 inhibits cardiac hypertrophy [5]. Thioredoxin-1 appears to act in this way by forming a multimeric protein complex with a Class II HDAC termed HDAC4, a heat shock protein termed DnaJb5 and a binding protein TBP-2. Thioredoxin-1 reduces critical cysteine residues on HDAC4, thereby retaining the latter in the nucleus. Oxidative modification of HDAC4 at cysteine residues 274/276 prevents DnaJb5-HDAC4 interaction whilst oxidation of cysteines 667/669 allows nuclear export of HDAC4 to the cytosol, thereby disinhibiting transcription factors that promote cardiac hypertrophy (i.e., leading to hypertrophy).

*Involvement of NADPH oxidases.* The expression and activity of myocardial Nox2 oxidase increases in parallel with the progression of pressure overload LVH in rodent models [40, 41] while oxidase expression and/or activity were also found to be increased in end-stage human CHF [42–45]. Interestingly, the upregulation of NADPH oxidase-mediated ROS production in human failing LV myocardium was associated with increased Rac1 activity, which is potentially inhibitable by treatment with statins [43]. In the Dahl salt-sensitive rat model, the increased NADPH oxidase expression and oxidative stress were significantly improved by chronic treatment with an aldosterone (mineralocorticoid) receptor antagonist, eplerenone, in association with improvement in LV remodelling [41].

Angiotensin II acting via $AT_1$ receptors stimulates Nox2 NADPH oxidase activity, superoxide production, and ERK1/2 activation in isolated cardiomyocytes, and hypertrophy could be inhibited by N-acetylcysteine [46]. Other studies in neonatal cardiomyocytes using specific genetic perturbations have convincingly shown an involvement of Rac1-regulated Nox2 oxidase in angiotensin II–mediated hypertrophy, through Akt activation [31].

Evidence for an involvement of Nox2 oxidase has also been found in in vivo studies. Studies in global Nox2 knockout mice subjected to subpressor angiotensin II infusion revealed that the hypertrophic response to subpressor Ang II infusion was markedly inhibited, in conjunction with reduced NADPH oxidase activation [47]. Subsequent work in mice with a cardiomyocyte-specific deletion of Rac1 showed similar results, indicating that Nox2 in cardiomyocytes was important for these effects [48]. Our group and others have also studied the involvement of Nox2 in the response to mechanical pressure overload. In this case, however, it was found that Nox2 knockout mice had a similar degree of hypertrophy to wild-type littermates [49]. Despite the lack of effect on cardiomyocyte hypertrophy in response to pressure overload, the Nox2 isoform does have effects on contractile function and fibrosis in this setting, as discussed later. Interestingly, NADPH oxidase activity was still elevated in $Nox2^{-/-}$ mice after aortic banding, attributable to increased Nox4 expression [49], leading to a possibility that this isoform could be involved in the response to pressure overload.

*Role of NOS and uncoupled NOS.* NO has been shown to exert cGMP-dependent antihypertrophic effects in cultured cardiac myocytes [50, 51]. In line with this, eNOS knockout mice were reported to exhibit enhanced left ventricular hypertrophy and dilatation in response to chronic pressure overload or MI [52–54], while transgenic mice with endothelial-specific [55] or cardiomyocyte-specific [56] overexpression of eNOS showed improved cardiac performance and reduced heart failure progression after MI. Long-term treatment with a pharmacological eNOS enhancer, AVE9488, also ameliorated LV remodelling and contractile dysfunction in rats with heart failure after infarction [57].

In contrast to these data, Takimoto et al. have reported that chronic pressure overload results in an uncoupling of eNOS and increased ROS production, which leads to enhanced hypertrophy [58]. These authors found that eNOS knockout mice had less hypertrophy and also that treatment with BH4 could ameliorate eNOS uncoupling, oxidative stress, and hypertrophy in wild-type animals. In further work, this group has shown that established cardiac hypertrophy induced by severe aortic constriction could be reversed by administration of BH4 despite ongoing aortic constriction, but that a nonspecific antioxidant, tempol, failed to exert similar effects [59].

Taken together, the above data suggest that the effects of NOS may be complex and vary according to the degree of NOS uncoupling. The precise factors that determine relative NO vs. ROS generation in specific pathological in vivo settings require further study.

*Role of xanthine oxidase.* Elevation in the expression and activity of xanthine oxidase (XO) was described in experimental animals [60] and humans [61] with heart failure. Treatment with the XO inhibitor allopurinol, which is commonly used to treat gout, reduces circulating markers of oxidative stress in CHF patients, while elevated serum levels of uric acid, a product of XO, are suggested to be a prognostic predictor in this group [62]. More direct evidence for a role of XO comes from experimental studies of cardiac remodelling. Chronic oxypurinol treatment reversed remodelling and restored contractile function in a rodent model of established hypertensive CHF [63, 64], and in dilated cardiomyopathic hamsters [65]. In chronic pressure overload in rats, it was suggested that xanthine oxidase may be more important in the advanced stages of cardiac remodelling but not during compensated hypertrophy [66]. Recently, in a study using a selective nonpurine xanthine oxidase inhibitor, febuxostat, from the day of aortic constriction, it was reported that this agent reduced markers of oxidative/nitrative stress and cardiac hypertrophy [67]. It is worth noting however that the effects of xanthine oxidase inhibition may not be purely related to ROS but may involve other known effects such as altered energy metabolism. Furthermore, a recent clinical trial of treatment with oxypurinol in patients with chronic heart failure failed to show evidence of benefit with this approach, although post hoc analyses suggested that it may be necessary to target the therapy [68].

Recent studies also suggest a link between xanthine oxidase and nNOS activity. nNOS-deficient mice were found to have persistent elevations in xanthine oxidase activity in the setting of post-MI remodelling, which was associated with enhanced hypertrophy and dilatation [69]. The authors postulated that part of the normal effect of nNOS-derived NO is to limit oxidative stress by inhibiting xanthine oxidase. It has been shown in other studies that cardiomyocyte-specific overexpression of nNOS protects against the development of maladaptive cardiac hypertrophy in the face of increased haemodynamic load, although a link to xanthine oxidase was not explored [70].

## 21.5 Extracellular Matrix Modification and Interstitial Fibrosis

The normal extracellular matrix (ECM) is crucial in maintaining the normal structure of the heart. Cardiac remodelling typically involves significant alterations of the ECM, with both a dilatation and/or shape change of the ventricle and the deposition of new and abnormally cross-linked fibrous tissue (i.e., fibrosis). Turnover of the ECM is under the control of a family of matrix metalloproteinases (MMPs), which degrade different ECM components, and specific inhibitors of these enzymes, the tissue inhibitors of matrix metalloproteinases (TIMPs) [71]. Dysregulation of the MMP/TIMP balance is well known to be involved in experimental and clinical dilated cardiomyopathy and heart failure [72–76]. Myocardial interstitial fibrosis impacts particularly on cardiac passive filling characteristics, as well as predisposing to cardiac arrhythmia, while perivascular fibrosis may impair oxygen exchange between blood vessels and parenchyma. The key cell types in fibrosis generation are resident cardiac fibroblasts and the myofibroblasts, which are transformed cells that have higher proliferative and matrix-secretory capacity. In addition, infiltrating inflammatory cells may be important in generating profibrotic stimuli through local production of cytokines [77]. Recent work has also highlighted the potential importance of the epithelial-to-mesenchymal transition (EMT) [78].

*ROS and MMPs.* ROS enhance MMP activity through several mechanisms. Firstly, redox-sensitive transcription factors (such as NF-κB, AP-1, and Ets) stimulate expression of MMP mRNA [79]. Secondly, the conversion of pro-MMPs to the active MMPs occurs via oxidative modification of specific residues in autoinhibitory regions, and studies have shown that exogenous ROS enhance this process in cultured fibroblasts [80, 81]. Thirdly, ROS may also alter the transcription of TIMPs [82].

*ROS and fibrosis.* ROS are also implicated in the processes of myofibroblast transformation and matrix collagen synthesis. It is well established that increased oxidative stress is associated with profibrotic effects, not only in the heart but also in other organs [83–85]. Furthermore, the proliferation of cultured cardiac fibroblasts is known to be ROS-dependent [86].

The relevance of different ROS sources to the development of fibrosis in vivo has been investigated in different settings. It was shown that interstitial cardiac fibrosis induced either by subpressor [47] or pressor angiotensin II infusion [87] was inhibited in Nox2 knockout mice. Similar results were also obtained in Nox2 knockout mice subjected to aortic banding or to aldosterone infusion, despite these animals having similar hypertrophy to wild-types [87, 88]. At least three mechanisms seemed to be involved in this Nox2-dependent effect: an increase in NF-κB activation, increased expression of CTGF and profibrotic genes, and increased MMP2 activation [87].

In contrast to the above studies, other in vitro work has suggested a role for Nox4 NADPH oxidase in the TGFβ–induced transformation of human cardiac fibroblasts to myofibroblasts [89]. These authors showed that Nox4 mRNA levels increased markedly following TGFβ stimulation and that siRNA-mediated knockdown of Nox4 substantially reduced TGFβ-induced α-smooth muscle actin expression. Nox4 inhibition also significantly reduced Smad 2/3 phosphorylation, suggesting that this might be the relevant target of Nox4-derived ROS. One possible reason for the differences between these findings and those from the earlier in vivo studies might be that both Nox2 and Nox4 are required for in vivo fibrosis.

Other ROS sources may also have profibrotic effects, either direct or indirect. NOS uncoupling was found to induce not only enhanced hypertrophy but also increased fibrosis after aortic constriction [58]. Similarly, studies that showed beneficial effects of xanthine oxidase inhibition in the setting of pressure overload or MI also resulted in reduced levels of fibrosis [67, 90].

*Post-MI remodelling.* Substantial alterations in MMP/TIMP balance occur in the setting of adverse LV remodelling post-MI. Clinical pathological data has shown an increase in Nox2 expression in cardiomyocytes after human MI [91], while similar findings were also reported in rodents [92]. More compelling evidence for a role of Nox2 in post-MI remodelling comes from studies in gene-modified mice. Doerries et al. studied mice deficient in $p47^{phox}$ and found that these animals had significantly reduced LV dilatation and better function and survival than wild-type littermates after MI [93]. In addition, the $p47^{phox-/-}$ animals also had significantly reduced myocyte hypertrophy, interstitial fibrosis, and apoptosis. Our group found similar results in $Nox2^{-/-}$ mice subjected to coronary ligation, with less LV dilatation, reduced fibrosis and apoptosis, better ejection fraction, greater LV contractility, and preserved diastolic relaxation [94]. Frantz et al. [95] also studied the response to MI in Nox2–deficient mice but reported no benefit of Nox2 deletion. However, it should be noted that these authors did not study littermate controls nor did they include a sham-operated group for either genotype.

The role of xanthine oxidase has also been examined in a post-MI setting. Endberding et al. reported that allopurinol reduced myocardial ROS production and attenuated ventricular dilatation and dysfunction post-MI, despite no effect on infarct size [90]. A protection against cavity dilatation and dysfunction with XO inhibition was also documented by Naumova et al. [96]. In contrast, Stull et al. found no difference in gross remodelling, but a preservation of systolic function with use of allopurinol after MI [97].

## 21.6 ROS and Apoptosis

A low but significant level of apoptotic cell death occurs both in cardiac remodelling and heart failure [98, 99], and has been convincingly shown in genetic studies to contribute to heart failure [100]. Apoptosis proceeds via one of two convergent pathways, namely receptor-based signalling leading to caspase-8 activation and mitochondrial-mediated activation of caspase-9. The convergence point of these pathways is at the level of caspase-3 activation. Counterbalancing these pathways, cells also have complex antiapoptotic or survival pathways [101]. ROS potentially impact on the apoptotic process at several levels. Redox-sensitive pathways such as the activation of ASK-1, JNK, and p38 MAPK are known to be proapoptotic [102, 103]. The direct actions of ROS on mitochondria may induce damage and release of cytochrome C into the cytosol. Redox modulation of antiapoptotic pathways may also be important, e.g., an increase in heat shock proteins [104], Akt activation, or the activation of NF-κB [105]. Therefore, ROS may potentially have "Janus"-like influences on apoptosis.

Some evidence exists for the involvement of specific sources of ROS in cardiac apoptosis. Angiotensin II increases levels of apoptosis in cultured rat embryonal and neonatal cardiomyocytes in vitro, and this effect is abrogated by pretreatment with a purported (albeit nonspecific) NADPH oxidase inhibitor, apocynin [106–108]. As discussed in the previous section, in the post-MI remodelling setting, apoptosis in the remote noninfarcted myocardium is reduced by deletion of Nox2 or $p47^{phox}$.

## 21.7 ROS, Contractile Dysfunction, and Energetics

One of the central aspects of heart failure is the occurrence of significant contractile dysfunction. This is evident not only at a global ventricular level but also at single cardiomyocyte level. Several different ROS-dependent mechanisms may affect contractile function at the level of the cardiomyocyte.

*Excitation-contraction coupling.* The redox environment influences several critical components involved in excitation-contraction coupling, largely due to the presence of critical reactive cysteine residues in ion channels and transporters that are susceptible to oxidative modification [109]. As noted previously, the physical proximity of ROS species to the channels or pumps in question are likely to determine the in vivo relevance of such interactions. The L-type calcium channel is critical in the process of calcium influx leading to calcium-induced calcium release in the myocyte. Its alpha subunit contains several cysteine residues [109] and ROS have been shown to affect Ica through the channel [110, 111]. Calcium-induced calcium release into the cytosol occurs through the ryanodine receptor. Superoxide, hydrogen peroxide, and hydroxyl have all been shown to increase the opening of these channels [112–115]. Re-uptake of calcium into the sarcoplasmic reticulum (SR) after contraction takes place through the SR calcium ATPase (SERCA). Superoxide, hydrogen peroxide, and hydroxyl all reduce calcium uptake via this pump [116, 117], whereas peroxynitrite was reported to stimulate SERCA activity through S-glutathiolation of cysteine residues [118]. The Na-Ca exchanger, which plays a role in cardiac relaxation via extrusion of calcium from the cytosol, is also redox-sensitive [119, 120]. Different potassium channels have also been shown to be modified in their action by ROS [121, 122].

ROS may modulate the function of component proteins of the contractile apparatus [123]. More recently, it has been shown that exposure of adult rat cardiomyocytes to hydrogen peroxide leads to ASK-1 activation, which causes phosphorylation of cardiac troponin T [124]. High ROS levels can also directly degrade contractile proteins; this appears important in the phenomenon of myocardial stunning after acute ischaemia, and might also be relevant in more advanced cases of chronic heart failure [125].

Despite the above data, there is relatively limited information on the actual involvement of these mechanisms in heart failure or the precise endogenous ROS sources that may be relevant. Antioxidant therapy (using edaravone) has been shown to improve disordered interdomain interaction within the ryanodine receptor in a canine model of heart failure, leading to reduced calcium leak from sarcoplasmic reticulum and improved cardiac function [126]. Another study using carvedilol in a similar model found that this drug also increased the number of free thiols of the ryanodine receptor, and this antioxidant effect was associated with preserved interdomain interaction of the receptor, reduced calcium leak, and better contractile function [127]. In studies in Nox2 knockout mice subjected to chronic pressure overload, we have shown that contractile function is better preserved than in wild-type littermates, despite the same degree of hypertrophy, but the precise subcellular mechanism involved was not investigated [88]. Nox2 has also been suggested to be involved in contractile dysfunction occurring in sepsis [128]. Casadei and colleagues have reported data implicating increased Nox2 expression and activity in the genesis of atrial fibrillation, both in humans and in a porcine model [44, 129].

*Cellular energetics.* It is known that ROS can damage various intracellular structures including mitochondria, thereby impacting on generation of energy for cellular activity. The relationship between reperfusion after ischaemia, generation of oxygen free radicals, and myocardial energy metabolism was elegantly demonstrated almost 20 years ago using NMR spectroscopy [130]. More recently, it has been shown that the oxidant stress generated by in vivo infusion of TNFα to rats results in abnormalities in both the structure and the function of the cardiac mitochondrial permeability transition pore (MPTP), and hence hyperactivation of this pore [131]. Reduced expression of the MPTP complex protein ANT (adenine nucleotide translocator) was demonstrated on Western blotting, possibly contributing to this abnormal activity.

Heart failure is known to be associated with damage to mitochondrial DNA and mitochondrial dysfunction, associated with increased generation of ROS [8]. Mitochondrial DNA is particularly susceptible to damage by ROS since, unlike nuclear DNA, it lacks a complex chromatin structure involving histones and has a more limited capacity for DNA repair [8]. This leads to alterations in transcription and protein synthesis in mitochondria. In failing hearts, there is reduced enzymatic activity of complexes I, III, and IV of the electron transport chain, causing reduced overall mitochondrial oxidative activity and energy generation for cellular needs [17]. A vicious cycle then develops, with mitochondrial ROS production leading to further mitochondrial damage [12].

ROS also modulate mitochondrial function in settings other than overt heart failure, but which are associated with cardiac remodelling. For example, in diabetic OVE26 mice, abnormal mitochondrial morphology and reduced cardiomyocyte function are found. These features can be attenuated by overexpression of catalase and also by overexpression of superoxide dismutase, implying the importance of ROS in their genesis [132, 133].

## 21.8 Therapeutic Intervention

Although the extensive animal and in vitro studies have provided consistent mechanistic support for the involvement of increased oxidative stress and redox signalling in the pathogenesis of cardiac remodelling and heart failure, the use of this information for therapeutic purposes remains elusive. Large prospective randomized clinical trials of antioxidant therapy have been largely negative [134–136]. For example, the

HOPE study failed to demonstrate any protective effect of vitamin E in reducing death or cardiovascular events in patients with risk factors for cardiovascular disease [137]. The recent OPT-CHF study of oxypurinol to treat class III to IV heart failure also found no beneficial impact on clinical outcome [68].

There may be several reasons for these disappointing findings. It should be clear from the preceding discussion that the roles of ROS in remodelling and heart failure are complex and unlikely to be amenable to individual nonspecific interventions such as antioxidant vitamins. Furthermore, assessing the degree of oxidative stress in patients with heart failure is challenging, while the dosages of antioxidants have been arbitrary. It may be necessary to design more potent and more specific agents that can better target defined ROS sources in a cell-, tissue-, and redox-sensitive pathway–specific manner, and at appropriate points during the disease stage in defined patient subsets. It is also noteworthy that many current effective therapies for heart failure (such as angiotensin-converting enzyme inhibitors, angiotensin receptor antagonists, β–blockers, and statins) all have some "antioxidant" activity, so that the opportunity to obtain further benefit with nonspecific antioxidant agents may be relatively limited.

Nevertheless, the emerging data on the complexity of redox signalling in cardiac remodelling provides an opportunity to reconsider the ways in which the therapeutic potential of "antioxidant" agents in patients with heart failure may be harnessed. Perhaps, targeting specific sources of ROS generation and/or selectively tackling detrimental aspects such as fibrosis and dilatation may be the way forward.

## 21.9 Conclusions

Oxidative stress is an important mechanism underlying the pathophysiology of adverse cardiac remodelling and CHF. Several different sources of ROS may be involved in remodelling, among which the NADPH oxidases appear to be especially important in modulating redox-sensitive signalling pathways, notably those that underlie the development of cardiomyocyte hypertrophy, interstitial fibrosis, and ventricular remodelling. Since a large body of evidence is derived from experimental settings, the type of model, the severity and chronicity of the stimulus, as well as the oxidative environment that it engenders, require consideration. The therapeutic potential of targeting ROS-mediated pathways remains to be realised. We still need to better understand how these mechanisms are regulated. Although clinical trials of antioxidant therapy have generally been negative, it seems premature to abandon this approach. A better understanding of the precise roles of oxidant stress and redox signalling mechanisms in different components of the cardiac remodelling process may enable the development of novel targeted prevention and treatment strategies for this condition.

**Acknowledgments** The authors' work is supported by the British Heart Foundation (RG/08/011/25922 and CVH/99001); the Department of Health via the National Institute for Health Research (NIHR) comprehensive Biomedical Research Centre award to Guy's &

St Thomas' NHS Foundation Trust in partnership with King's College London and King's College Hospital NHS Foundation Trust; and EU FP6 grant LSHM-CT-2005-018833, EUGeneHeart.

## References

1. Ide T, Tsutsui H, Kinugawa S et al (2000 Feb 4) Direct evidence for increased hydroxyl radicals originating from superoxide in the failing myocardium. Circ Res 86(2):152–157
2. Kinugawa S, Tsutsui H, Hayashidani S et al (2000 Sept 1) Treatment with dimethylthiourea prevents left ventricular remodeling and failure after experimental myocardial infarction in mice: role of oxidative stress. Circ Res 87(5):392–398
3. Sia YT, Lapointe N, Parker TG et al (2002 May 28) Beneficial effects of long-term use of the antioxidant probucol in heart failure in the rat. Circulation 105(21):2549–2555
4. Date M, Morita T, Yamashita N et al (2002 Mar 6) The antioxidant N-2-mercaptopropionyl glycine attenuates left ventricular hypertrophy in in vivo murine pressure-overload model. J Am Coll Cardiol 39(5):907–912
5. Yamamoto M, Yang G, Hong C et al (2003 Nov) Inhibition of endogenous thioredoxin in the heart increases oxidative stress and cardiac hypertrophy. J Clin Invest 112(9):1395–1406
6. Shiomi T, Tsutsui H, Matsusaka H et al (2004 Feb 3) Overexpression of glutathione peroxidase prevents left ventricular remodeling and failure after myocardial infarction in mice. Circulation 109(4):544–549
7. Giordano FJ (2005 Mar 1) Oxygen, oxidative stress, hypoxia, and heart failure. J Clin Invest 115(3):500–508
8. Tsutsui H, Kinugawa S, Matsushima S (2008 Oct) Mitochondrial oxidative stress and dysfunction in myocardial remodelling. Cardiovasc Res 81(3):449–456
9. Takimoto E, Kass DA (2007 Feb 1) Role of oxidative stress in cardiac hypertrophy and remodeling. Hypertension 49(2):241–248
10. Nishino T, Okamoto K, Eger BT, Pai EF, Nishino T (2008 July) Mammalian xanthine oxidoreductase – mechanism of transition from xanthine dehydrogenase to xanthine oxidase. FEBS J 275(13):3278–3289
11. Sumimoto H (2008 July) Structure, regulation and evolution of Nox-family NADPH oxidases that produce reactive oxygen species. FEBS J 275(13):3249–3277
12. Ide T, Tsutsui H, Kinugawa S et al (1999 Aug 20) Mitochondrial electron transport complex I is a potential source of oxygen free radicals in the failing myocardium. Circ Res 85(4): 357–363
13. Redout EM, Wagner MJ, Zuidwijk MJ et al (2007 Sept 1) Right-ventricular failure is associated with increased mitochondrial complex II activity and production of reactive oxygen species. Cardiovasc Res 75(4):770–781
14. Nojiri H, Shimizu T, Funakoshi M et al (2006 Nov 3) Oxidative stress causes heart failure with impaired mitochondrial respiration. J Biol Chem 281(44):33789–33801
15. Matsushima S, Ide T, Yamato M et al (2006 Apr 11) Overexpression of mitochondrial peroxiredoxin-3 prevents left ventricular remodeling and failure after myocardial infarction in mice. Circulation 113(14):1779–1786
16. van Empel VPM, Bertrand AT, van der Nagel R et al (2005 June 24) Downregulation of apoptosis-inducing factor in harlequin mutant mice sensitizes the myocardium to oxidative stress-related cell death and pressure overload-induced decompensation. Circ Res 96(12):e92–e101
17. Ide T, Tsutsui H, Hayashidani S et al (2001 Mar 16) Mitochondrial DNA damage and dysfunction associated with oxidative stress in failing hearts after myocardial infarction. Circ Res 88(5):529–535
18. Ikeuchi M, Matsusaka H, Kang D et al (2005 Aug 2) Overexpression of mitochondrial transcription factor A ameliorates mitochondrial deficiencies and cardiac failure after myocardial infarction. Circulation 112(5):683–690

19. Meneshian A, Bulkley GB (2002 July) The physiology of endothelial xanthine oxidase: from urate catabolism to reperfusion injury to inflammatory signal transduction. Microcirculation 9(3):161–175
20. Shah AM (2005 June 1) Divergent roles of endothelial nitric oxide synthase in cardiac hypertrophy and chamber dilatation? Cardiovasc Res 66(3):421–422
21. Lambeth JD (2004 Mar) Nox enzymes and the biology of reactive oxygen. Nat Rev Immunol 4(3):181–189
22. Cave AC, Brewer AC, Narayanapanicker A et al (2006) NADPH oxidases in cardiovascular health and disease. Antioxid Redox Signal 8(5–6):691–728
23. Geiszt M (2006 July 15) NADPH oxidases: New kids on the block. Cardiovasc Res 71(2):289–299
24. Anilkumar N, Sirker A, Shah AM (2009) Redox sensitive signaling pathways in cardiac remodeling, hypertrophy and failure. Front Biosci 14:3168–3187
25. McNally JS, Davis ME, Giddens DP et al (2003 Dec 1) Role of xanthine oxidoreductase and NAD(P)H oxidase in endothelial superoxide production in response to oscillatory shear stress. Am J Physiol Heart Circ Physiol 285(6):H2290–H2297
26. Landmesser U, Dikalov S, Price SR et al (2003 Apr) Oxidation of tetrahydrobiopterin leads to uncoupling of endothelial cell nitric oxide synthase in hypertension. J Clin Invest 111(8):1201–1209
27. Heineke J, Molkentin JD (2006 Aug) Regulation of cardiac hypertrophy by intracellular signalling pathways. Nat Rev Mol Cell Biol 7(8):589–600
28. Frey N, Olson EN (2003 Mar 1) Cardiac Hypertrophy: the good, the bad, and the ugly. Annu Rev Physiol 65(1):45–79
29. Xiao L, Pimentel DR, Wang J, Singh K, Colucci WS, Sawyer DB (2002 Apr 1) Role of reactive oxygen species and NAD(P)H oxidase in alpha 1-adrenoceptor signaling in adult rat cardiac myocytes. Am J Physiol Cell Physiol 282(4):C926–C934
30. Kwon SH, Pimentel DR, Remondino A, Sawyer DB, Colucci WS (2003 June 1) $H_2O_2$ regulates cardiac myocyte phenotype via concentration-dependent activation of distinct kinase pathways. J Mol Cell Cardiol 35(6):615–621
31. Hingtgen SD, Tian X, Yang J et al (2006 Sept 14) Nox2-containing NADPH oxidase and Akt activation play a key role in angiotensin II-induced cardiomyocyte hypertrophy. Physiol Genomics 26(3):180–191
32. Higuchi Y, Otsu K, Nishida K et al (2002 Feb) Involvement of reactive oxygen species-mediated NF-[kappa] B activation in TNF-[alpha]-induced cardiomyocyte hypertrophy. J Mol Cell Cardiol 34(2):233–240
33. Hirotani S, Otsu K, Nishida K et al (2002 Jan 29) Involvement of nuclear factor-{kappa}B and apoptosis signal-regulating kinase 1 in G-protein-coupled receptor agonist-induced cardiomyocyte hypertrophy. Circulation 105(4):509–515
34. Izumiya Y, Kim S, Izumi Y et al (2003 Oct 31) Apoptosis signal-regulating kinase 1 plays a pivotal role in angiotensin II-induced cardiac hypertrophy and remodeling. Circ Res 93(9):874–883
35. Kuster GM, Pimentel DR, Adachi T et al (2005 Mar 8) {alpha}-Adrenergic receptor-stimulated hypertrophy in adult rat ventricular myocytes is mediated via thioredoxin-1-sensitive oxidative modification of thiols on Ras. Circulation 111(9):1192–1198
36. Pimentel DR, Amin JK, Xiao L et al (2001 Aug 31) Reactive oxygen species mediate amplitude-dependent hypertrophic and apoptotic responses to mechanical stretch in cardiac myocytes. Circ Res 89(5):453–460
37. Aikawa R, Nagai T, Tanaka M et al (2001 Dec 14) Reactive oxygen species in mechanical stress-induced cardiac hypertrophy. Biochem Biophys Res Commun 289(4): 901–907
38. Pimentel DR, Adachi T, Ido Y et al (2006 Oct) Strain-stimulated hypertrophy in cardiac myocytes is mediated by reactive oxygen species-dependent Ras S-glutathiolation. J Mol Cell Cardiol 41(4):613–622

39. Ago T, Liu T, Zhai P et al (2008 June 13) A redox-dependent pathway for regulating class II HDACs and cardiac hypertrophy. Cell 133(6):978–993
40. Li JM, Gall NP, Grieve DJ, Chen M, Shah AM (2002 Oct 1) Activation of NADPH oxidase during progression of cardiac hypertrophy to failure. Hypertension 40(4):477–484
41. Kobayashi N, Yoshida K, Nakano S et al (2006 Apr 1) Cardioprotective mechanisms of eplerenone on cardiac performance and remodeling in failing rat hearts. Hypertension 47(4):671–679
42. Heymes C, Bendall JK, Ratajczak P et al (2003 June 18) Increased myocardial NADPH oxidase activity in human heart failure. J Am Coll Cardiol 41(12):2164–2171
43. Maack C, Kartes T, Kilter H et al (2003 Sept 30) Oxygen free radical release in human failing myocardium is associated with increased activity of Rac1-GTPase and represents a target for statin treatment. Circulation 108(13):1567–1574
44. Kim YM, Guzik TJ, Zhang YH et al (2005 Sept 30) A myocardial Nox2 containing NAD(P)H oxidase contributes to oxidative stress in human atrial fibrillation. Circ Res 97(7):629–636
45. Dudley SC Jr, Hoch NE, McCann LA et al (2005 Aug 30) Atrial fibrillation increases production of superoxide by the left atrium and left atrial appendage: Role of the NADPH and xanthine oxidases. Circulation 112(9):1266–1273
46. Nakagami H, Takemoto M, Liao JK (2003 July) NADPH oxidase-derived superoxide anion mediates angiotensin II-induced cardiac hypertrophy. J Mol Cell Cardiol 35(7):851–859
47. Bendall JK, Cave AC, Heymes C, Gall N, Shah AM (2002 Jan 22) Pivotal role of a gp91(phox)-containing NADPH oxidase in angiotensin II-induced cardiac hypertrophy in mice. Circulation 105(3):293–296
48. Satoh M, Ogita H, Takeshita K, Mukai Y, Kwiatkowski DJ, Liao JK (2006 May 9) Requirement of Rac1 in the development of cardiac hypertrophy. PNAS 103(19): 7432–7437
49. Jonathan A, Grieve DJ, Bendall JK et al (2003 Oct 31) Contrasting roles of NADPH oxidase isoforms in pressure-overload versus angiotensin II-induced cardiac hypertrophy. Circ Res 93(9):802–805
50. Wollert KC, Fiedler B, Gambaryan S et al (2002 Jan 1) Gene transfer of cGMP-dependent protein kinase I enhances the antihypertrophic effects of nitric oxide in cardiomyocytes. Hypertension 39(1):87–92
51. Calderone A, Thaik CM, Takahashi N, Chang DL, Colucci WS (1998 Feb 15) Nitric oxide, atrial natriuretic peptide, and cyclic GMP inhibit the growth-promoting effects of norepinephrine in cardiac myocytes and fibroblasts. J Clin Invest 101(4):812–818
52. Ichinose F, Bloch KD, Wu JC et al (2004 Mar 1) Pressure overload-induced LV hypertrophy and dysfunction in mice are exacerbated by congenital NOS3 deficiency. Am J Physiol Heart Circ Physiol 286(3):H1070–H1075
53. Scherrer-Crosbie M, Ullrich R, Bloch KD et al (2001 Sept 11) Endothelial nitric oxide synthase limits left ventricular remodeling after myocardial infarction in mice. Circulation 104(11):1286–1291
54. Ruetten H, Dimmeler S, Gehring D, Ihling C, Zeiher AM (2005 June 1) Concentric left ventricular remodeling in endothelial nitric oxide synthase knockout mice by chronic pressure overload. Cardiovasc Res 66(3):444–453
55. Jones SP, Greer JJM, van Haperen R, Duncker DJ, de Crom R, Lefer DJ (2003 Apr 15) Endothelial nitric oxide synthase overexpression attenuates congestive heart failure in mice. Proc Natl Acad Sci U S A 100(8):4891–4896
56. Janssens S, Pokreisz P, Schoonjans L et al (2004 May 14) Cardiomyocyte-specific overexpression of nitric oxide synthase 3 improves left ventricular performance and reduces compensatory hypertrophy after myocardial infarction. Circ Res 94(9):1256–1262
57. Fraccarollo D, Widder JD, Galuppo P et al (2008 Aug 19) Improvement in left ventricular remodeling by the endothelial nitric oxide synthase enhancer AVE9488 after experimental myocardial infarction. Circulation 118(8):818–827

58. Takimoto E, Champion HC, Li M et al (2005 May 1) Oxidant stress from nitric oxide synthase-3 uncoupling stimulates cardiac pathologic remodeling from chronic pressure load. J Clin Invest 115(5):1221–1231
59. Moens AL, Takimoto E, Tocchetti CG et al (2008 May 20) Reversal of cardiac hypertrophy and fibrosis from pressure overload by tetrahydrobiopterin: efficacy of recoupling nitric oxide synthase as a therapeutic strategy. Circulation 117(20):2626–2636
60. Ukai T, Cheng CP, Tachibana H et al (2001 Feb 6) Allopurinol enhances the contractile response to dobutamine and exercise in dogs with pacing-induced heart failure. Circulation 103(5):750–755
61. Cappola TP, Kass DA, Nelson GS et al (2001 Nov 13) Allopurinol improves myocardial efficiency in patients with idiopathic dilated cardiomyopathy. Circulation 104(20):2407–2411
62. Farquharson CAJ, Butler R, Hill A, Belch JJF, Struthers AD (2002 July 9) Allopurinol improves endothelial dysfunction in chronic heart failure. Circulation 106(2):221–226
63. Minhas KM, Saraiva RM, Schuleri KH et al (2006 Feb 3) Xanthine oxidoreductase inhibition causes reverse remodeling in rats with dilated cardiomyopathy. Circ Res 98(2):271–279
64. Yamamoto E, Kataoka K, Yamashita T et al (2007 Oct 1) Role of xanthine oxidoreductase in the reversal of diastolic heart failure by candesartan in the salt-sensitive hypertensive rat. Hypertension 50(4):657–662
65. Hayashi K, Kimata H, Obata K et al (2008 Apr) Xanthine oxidase inhibition improves left ventricular dysfunction in dilated cardiomyopathic hamsters. J Card Fail 14(3):238–244
66. de Jong JW, Schoemaker RG, de Jonge R et al (2000 Nov) Enhanced expression and activity of xanthine oxidoreductase in the failing heart. J Mol Cell Cardiol 32(11):2083–2089
67. Xu X, Hu X, Lu Z et al (2008 Nov) Xanthine oxidase inhibition with febuxostat attenuates systolic overload-induced left ventricular hypertrophy and dysfunction in mice. J Card Fail 14(9):746–753
68. Hare JM, Mangal B, Brown J et al (2008 June 17) Impact of oxypurinol in patients with symptomatic heart failure: results of the OPT-CHF study. J Am Coll Cardiol 51(24): 2301–2309
69. Saraiva RM, Minhas KM, Raju SVY et al (2005 Nov 29) Deficiency of neuronal nitric oxide synthase increases mortality and cardiac remodeling after myocardial infarction: role of nitroso-redox equilibrium. Circulation 112(22):3415–3422
70. Loyer X, Gomez AM, Milliez P et al (2008 June 24) Cardiomyocyte overexpression of neuronal nitric oxide synthase delays transition toward heart failure in response to pressure overload by preserving calcium cycling. Circulation 117(25):3187–3198
71. Spinale FG (2007 Oct 1) Myocardial matrix remodeling and the matrix metalloproteinases: influence on cardiac form and function. Physiol Rev 87(4):1285–1342
72. Creemers EEJM, Cleutjens JPM, Smits JFM, Daemen MJAP (2001 Aug 3) Matrix metalloproteinase inhibition after myocardial infarction: a new approach to prevent heart failure? Circ Res 89(3):201–210
73. Tyagi SC, Campbell SE, Reddy HK, Tjahja E, Voelker DJ (1996 Feb 9) Matrix metalloproteinase activity expression in infarcted, noninfarcted and dilated cardiomyopathic human hearts. Mol Cell Biochem 155(1):13–21
74. Tyagi SC, Kumar SG, Haas SJ et al (1996 July) Post-transcriptional regulation of extracellular matrix metalloproteinase in human heart end-stage failure secondary to ischemic cardiomyopathy. J Mol Cell Cardiol 28(7):1415–1428
75. Sivakumar P, Gupta S, Sarkar S, Sen S (2008 Jan) Upregulation of lysyl oxidase and MMPs during cardiac remodeling in human dilated cardiomyopathy. Mol Cell Biochem 307(1–2):159–167
76. Yan AT, Yan RT, Spinale FG et al (2008 Feb) Relationships between plasma levels of matrix metalloproteinases and neurohormonal profile in patients with heart failure. Eur J Heart Fail 10(2):125–128
77. Sun Y (2009 Feb 15) Myocardial repair/remodelling following infarction: roles of local factors. Cardiovasc Res 81(3):482–490

78. Zeisberg EM, Tarnavski O, Zeisberg M et al (2007 Sept) Endothelial-to-mesenchymal transition contributes to cardiac fibrosis. Nat Med 13(8):952–961
79. Sorescu D, Griendling KK (2002 May) Reactive oxygen species, mitochondria, and NAD(P)H oxidases in the development and progression of heart failure. Congest Heart Fail 8(3):132–140
80. Siwik DA, Pagano PJ, Colucci WS (2001 Jan 1) Oxidative stress regulates collagen synthesis and matrix metalloproteinase activity in cardiac fibroblasts. Am J Physiol Cell Physiol 280(1):C53–C60
81. Tyagi SC, Kumar S, Glover G (1995 July) Induction of tissue inhibitor and matrix metalloproteinase by serum in human heart-derived fibroblast and endomyocardial endothelial cells. J Cell Biochem 58(3):360–371
82. Wang Y, Rosen H, Madtes DK et al (2007 Nov 2) Myeloperoxidase inactivates TIMP-1 by oxidizing its N-terminal cysteine residue: an oxidative mechanism for regulating proteolysis during inflammation. J Biol Chem 282(44):31826–31834
83. Kinnula VL, Fattman CL, Tan RJ, Oury TD (2005 Aug 15) Oxidative stress in pulmonary fibrosis: a possible role for redox modulatory therapy. Am J Respir Crit Care Med 172(4):417–422
84. Poli G (2000 June) Pathogenesis of liver fibrosis: role of oxidative stress. Mol Aspect Med 21(3):49–98
85. Ha H, Lee HB (2003 Aug 1) Reactive oxygen species and matrix remodeling in diabetic kidney. J Am Soc Nephrol 14(90003):S246–S249
86. Chen YL, Liu JC, Loh SH et al (2008 Sept 28) Involvement of reactive oxygen species in urotensin II-induced proliferation of cardiac fibroblasts. Eur J Pharmacol 593(1–3): 24–29
87. Johar S, Cave AC, Narayanapanicker A, Grieve DJ, Shah AM (2006 July 1) Aldosterone mediates angiotensin II-induced interstitial cardiac fibrosis via a Nox2-containing NADPH oxidase. FASEB J 20(9):1546–1548
88. Grieve DJ, Byrne JA, Siva A et al (2006 Feb 21) Involvement of the nicotinamide adenosine dinucleotide phosphate oxidase isoform Nox2 in cardiac contractile dysfunction occurring in response to pressure overload. J Am Coll Cardiol 47(4):817–826
89. Cucoranu I, Clempus R, Dikalova A et al (2005 Oct 28) NAD(P)H oxidase 4 mediates transforming growth factor-{beta}1-induced differentiation of cardiac fibroblasts Into myofibroblasts. Circ Res 97(9):900–907
90. Engberding N, Spiekermann S, Schaefer A et al (2004 Oct 12) Allopurinol attenuates left ventricular remodeling and dysfunction after experimental myocardial infarction: a new action for an old drug? Circulation 110(15):2175–2179
91. Krijnen PAJ, Meischl C, Hack CE et al (2003 Mar 1) Increased Nox2 expression in human cardiomyocytes after acute myocardial infarction. J Clin Pathol 56(3):194–199
92. Fukui T, Yoshiyama M, Hanatani A, Omura T, Yoshikawa J, Abe Y (2001 Mar 16) Expression of p22-phox and gp91-phox, essential components of NADPH oxidase, increases after myocardial infarction. Biochem Biophys Res Commun 281(5):1200–1206
93. Doerries C, Grote K, Hilfiker-Kleiner D et al (2007 Mar 30) Critical role of the NAD(P)H oxidase subunit p47phox for left ventricular remodeling/dysfunction and survival after myocardial infarction. Circ Res 100(6):894–903
94. Looi YH, Grieve DJ, Siva A et al (2008 Feb 1) Involvement of Nox2 NADPH oxidase in adverse cardiac remodeling after myocardial infarction. Hypertension 51(2): 319–325
95. Frantz S, Brandes RP, Hu K et al (2006 Mar) Left ventricular remodeling after myocardial infarction in mice with targeted deletion of the NADPH oxidase subunit gp91PHOX. Basic Res Cardiol 101(2):127–132
96. Naumova AV, Chacko VP, Ouwerkerk R, Stull L, Marban E, Weiss RG (2006 Feb 1) Xanthine oxidase inhibitors improve energetics and function after infarction in failing mouse hearts. Am J Physiol Heart Circ Physiol 290(2):H837–H843

97. Stull LB, Leppo MK, Szweda L, Gao WD, Marban E (2004 Nov 12) Chronic treatment with allopurinol boosts survival and cardiac contractility in murine postischemic cardiomyopathy. Circ Res 95(10):1005–1011
98. Takemura G, Fujiwara H (2004 Oct) Role of apoptosis in remodeling after myocardial infarction. Pharmacol Ther 104(1):1–16
99. Gonzalez A, Ravassa S, Lopez B, Loperena I, Querejeta R, Diez J (2006 July) Apoptosis in hypertensive heart disease: a clinical approach. Curr Opin Cardiol 21(4):288–294
100. Wencker D, Chandra M, Nguyen K et al (2003 May) A mechanistic role for cardiac myocyte apoptosis in heart failure. J Clin Invest 111(10):1497–1504
101. Narula J, Haider N, Arbustini E, Chandrashekhar Y (2006 Dec) Mechanisms of disease: apoptosis in heart failure—seeing hope in death. Nat Clin Pract Cardiovasc Med 3(12): 681–688
102. Matsuzawa A, Ichijo H (2008 Nov) Redox control of cell fate by MAP kinase: physiological roles of ASK1-MAP kinase pathway in stress signaling. Biochim Biophys Acta Gen Subj 1780(11):1325–1336
103. Filomeni G, Ciriolo MR (2006 Nov 1) Redox control of apoptosis: an update. Antioxid Redox Signal 8(11–12):2187–2192
104. Diaz-Latoud C, Buache E, Javouhey E, Arrigo AP (2005 Mar 1) Substitution of the unique cysteine residue of murine Hsp25 interferes with the protective activity of this stress protein through inhibition of dimer formation. Antioxid Redox Signal 7(3–4):436–445
105. Shakibaei M, Schulze-Tanzil G, Takada Y, Aggarwal BB (2005 Mar 1) Redox regulation of apoptosis by members of the TNF superfamily. Antioxid Redox Signal 7(3–4): 482–496
106. Grishko V, Pastukh V, Solodushko V, Gillespie M, Azuma J, Schaffer S (2003 Dec 1) Apoptotic cascade initiated by angiotensin II in neonatal cardiomyocytes: role of DNA damage. Am J Physiol Heart Circ Physiol 285(6):H2364–H2372
107. Goldenberg I, Shainberg A, Jacobson KA, Shneyvays V, Grossman E (2003 Oct) Adenosine protects against angiotensin II-induced apoptosis in rat cardiocyte cultures. Mol Cell Biochem 252(1–2):133–139
108. Goldenberg I, Grossman E, Jacobson KA, Shneyvays V, Shainberg A (2001 Sept) Angiotensin II-induced apoptosis in rat cardiomyocyte culture: a possible role of AT1 and AT2 receptors. J Hypertens 19(9):1681–1689
109. Zima AV, Blatter LA (2006 July 15) Redox regulation of cardiac calcium channels and transporters. Cardiovasc Res 71(2):310–321
110. Goldhaber JI, Ji S, Lamp ST, Weiss JN (1989 June) Effects of exogenous free radicals on electromechanical function and metabolism in isolated rabbit and guinea pig ventricle. Implications for ischemia and reperfusion injury. J Clin Invest 83(6):1800–1809
111. Gill JS, McKenna WJ, Camm AJ (1995 Mar 16) Free radicals irreversibly decrease Ca2+ currents in isolated guinea-pig ventricular myocytes. Eur J Pharmacol 292(3–4): 337–340
112. Kawakami M, Okabe E (1998 Mar 1) Superoxide anion radical-triggered Ca2+ release from cardiac sarcoplasmic reticulum through ryanodine receptor Ca2+ channel. Mol Pharmacol 53(3):497–503
113. Zima AV, Copello JA, Blatter LA (2004 Mar 15) Effects of cytosolic NADH/NAD+ levels on sarcoplasmic reticulum Ca2+ release in permeabilized rat ventricular myocytes. J Physiol (Lond) 555(3):727–741
114. Oba T, Koshita M, Yamaguchi M (1996 Sept 1) $H_2O_2$ modulates twitch tension and increases Po of Ca2+ release channel in frog skeletal muscle. Am J Physiol Cell Physiol 271(3): C810–C818
115. Anzai K, Ogawa K, Kuniyasu A, Ozawa T, Yamamoto H, Nakayama H (1998 Aug 28) Effects of hydroxyl radical and sulfhydryl reagents on the open probability of the purified cardiac ryanodine receptor channel incorporated into planar lipid bilayers. Biochem Biophys Res Commun 249(3):938–942

116. Rowe GT, Manson NH, Caplan M, Hess ML (1983 Nov) Hydrogen peroxide and hydroxyl radical mediation of activated leukocyte depression of cardiac sarcoplasmic reticulum. Participation of the cyclooxygenase pathway. Circ Res 53(5):584–591
117. Morris TE, Sulakhe PV (1997) Sarcoplasmic reticulum Ca(2+)-pump dysfunction in rat cardiomyocytes briefly exposed to hydroxyl radicals. Free Radic Biol Med 22(1–2): 37–47
118. Adachi T, Weisbrod RM, Pimentel DR et al (2004 Nov) S-Glutathiolation by peroxynitrite activates SERCA during arterial relaxation by nitric oxide. Nat Med 10(11): 1200–1207
119. Reeves JP, Bailey CA, Hale CC (1986 Apr 15) Redox modification of sodium-calcium exchange activity in cardiac sarcolemmal vesicles. J Biol Chem 261(11):4948–4955
120. Goldhaber JI (1996 Sept 1) Free radicals enhance Na+/Ca2+ exchange in ventricular myocytes. Am J Physiol Heart Circ Physiol 271(3):H823–H833
121. Cerbai E, Ambrosio G, Porciatti F, Chiariello M, Giotti A, Mugelli A (1991 Oct) Cellular electrophysiological basis for oxygen radical-induced arrhythmias. A patch-clamp study in guinea pig ventricular myocytes. Circulation 84(4):1773–1782
122. Nakaya H, Takeda Y, Tohse N, Kanno M (1992 May) Mechanism of the membrane depolarization induced by oxidative stress in guinea-pig ventricular cells. J Mol Cell Cardiol 24(5):523–534
123. Gao WD, Liu Y, Marban E (1996 Nov 15) Selective effects of oxygen free radicals on excitation-contraction coupling in ventricular muscle: implications for the mechanism of stunned myocardium. Circulation 94(10):2597–2604
124. He X, Liu Y, Sharma V et al (2003 July 1) ASK1 associates with troponin T and induces troponin T phosphorylation and contractile dysfunction in cardiomyocytes. Am J Pathol 163(1):243–251
125. Bolli R (1998 Jun) Causative role of oxyradicals in myocardial stunning: a proven hypothesis. A brief review of the evidence demonstrating a major role of reactive oxygen species in several forms of postischemic dysfunction. Basic Res Cardiol 93(3):156–162
126. Yano M, Okuda S, Oda T et al (2005 Dec 6) Correction of defective interdomain interaction within ryanodine receptor by antioxidant is a new therapeutic strategy against heart failure. Circulation 112(23):3633–3643
127. Mochizuki M, Yano M, Oda T et al (2007 Apr 24) Scavenging free radicals by low-dose carvedilol prevents redox-dependent Ca2+ leak via stabilization of ryanodine receptor in heart failure. J Am Coll Cardiol 49(16):1722–1732
128. Peng T, Lu X, Feng Q (2005 Apr 5) Pivotal role of gp91phox-containing NADH oxidase in lipopolysaccharide-induced tumor necrosis factor-{alpha} expression and myocardial depression. Circulation 111(13):1637–1644
129. Kim YM, Kattach H, Ratnatunga C, Pillai R, Channon KM, Casadei B (2008) Association of atrial nicotinamide adenine dinucleotide phosphate oxidase activity with the development of atrial fibrillation after cardiac surgery. J Am Coll Cardiol 51(1):68–74
130. Ambrosio G, Zweier JL, Flaherty JT (1991 Dec) The relationship between oxygen radical generation and impairment of myocardial energy metabolism following post-ischemic reperfusion. J Mol Cell Cardiol 23(12):1359–1374
131. Mariappan N, Soorappan RN, Haque M, Sriramula S, Francis J (2007 Nov 1) TNF-{alpha}-induced mitochondrial oxidative stress and cardiac dysfunction: restoration by superoxide dismutase mimetic Tempol. Am J Physiol Heart Circ Physiol 293(5):H2726–H2737
132. Ye G, Metreveli NS, Donthi RV et al (2004 May 1) Catalase protects cardiomyocyte function in models of type 1 and type 2 diabetes. Diabetes 53(5):1336–1343
133. Shen X, Zheng S, Metreveli NS, Epstein PN (2006 Mar 1) Protection of cardiac mitochondria by overexpression of MnSOD reduces diabetic cardiomyopathy. Diabetes 55(3): 798–805
134. Gaziano JM (2004 Dec) Vitamin E and cardiovascular disease: observational studies. Ann N Y Acad Sci 1031:280–291

135. Sesso HD, Buring JE, Christen WG et al (2008 Nov 12) Vitamins E and C in the prevention of cardiovascular disease in men: the Physicians' Health Study II randomized controlled trial. J Am Med Assoc 300(18):2123–2133
136. Cook NR, Albert CM, Gaziano JM et al (2007 Aug 13) A randomized factorial trial of vitamins C and E and beta carotene in the secondary prevention of cardiovascular events in women: results from the Women's Antioxidant Cardiovascular Study. Arch Intern Med 167(15):1610–1618
137. The Heart Outcomes Prevention Evaluation Study Investigators. (2000 Jan 20) Vitamin E supplementation and cardiovascular events in high-risk patients. N Engl J Med 342(3): 154–160

# Chapter 22
# Oxidative Stress and Cardiovascular Fibrosis

**Subramaniam Pennathur, Louise Hecker, and Victor J. Thannickal**

**Abstract** Oxidative stress is defined as an imbalance of the generation of reactive oxygen species (ROS) in excess of the capacity of cells/tissues to detoxify or scavenge them. Such a state of oxidative stress may alter the structure/function of cellular macromolecules that eventually leads to tissue/organ dysfunction. The harmful effects of ROS have been largely attributed to its indiscriminate, stochastic effects on the oxidation of protein, lipids, or DNA. Alternatively, detrimental effects of ROS may be mediated by aberrant "redox signaling" in specific pathophysiologic contexts. Cardiovascular disease is a major cause of morbidity and mortality in industrialized nations. Acute myocardial infarction from atherosclerotic coronary artery disease often results in remodeling responses of the myocardium that may culminate in congestive heart failure. Another important cause of CHF is chronic pressure overload due to systemic hypertension. We discuss mechanisms by which oxidative stress contributes to the pathogenesis of vascular disease, endothelial dysfunction, and cardiovascular fibrosis. Identifying specific pathways for ROS generation and roles in cardiovascular fibrosis could lead to rational design of drugs that promote tissue repair/regeneration, while attenuating the progression of CHF.

**Keywords** Oxidative stress · Fibrosis · Redox signaling · Atherosclerosis · Vascular

## 22.1 Introduction

Oxidative stress is defined as an imbalance of the generation of reactive oxygen species (ROS) in excess of the capacity of cells/tissues to detoxify or scavenge them. Such a state of oxidative stress may alter the structure/function of cellular

V.J. Thannickal (✉)
Division of Pulmonary, Allergy and Critical Care Medicine, University of Alabama at Birmingham, Birmingham, Alabama, USA; Department of Internal Medicine, University of Michigan Medical Center, Ann Arbor, MI 48109, USA
e-mail: vjthan@uab.edu

H. Sauer et al. (eds.), *Studies on Cardiovascular Disorders*,
Oxidative Stress in Applied Basic Research and Clinical Practice,
DOI 10.1007/978-1-60761-600-9_22, 

macromolecules that eventually leads to tissue/organ dysfunction. Oxidative stress may be viewed as a "necessary evil" (or consequence) of the utilization of molecular oxygen ($O_2$) for biological evolution from simple aerobic eukaryotes to more complex mammalian species [1]. The harmful effects of ROS have been largely attributed to its indiscriminate, stochastic effects on oxidation of protein, lipids, or DNA. Alternatively, detrimental effects of ROS may be mediated by aberrant "redox signaling" in specific pathophysiological contexts. This latter effect may be attributed to dysregulation of the normal physiological roles of ROS in cell signaling, which may also have an evolutionary basis [1, 2].

Cardiovascular disease is a major cause of morbidity and mortality in industrialized nations. Myocardial infarction (MI) from atherosclerotic coronary artery disease (CAD) may result in remodeling responses of the myocardium that culminate in congestive heart failure (CHF) [3, 4]. Another important cause of CHF is chronic pressure overload due to systemic hypertension [3]. Here, we discuss mechanisms by which oxidative stress contributes to the pathogenesis of vascular disease, endothelial dysfunction, and cardiac fibrosis. We will also discuss the potential role of specific ROS-generating pathways in disease pathogenesis. Identifying specific pathways of reactive oxidant generation that promote cardiovascular fibrosis could ultimately lead to rational design of drugs to promote tissue repair/regeneration, thus preventing the progression of CHF.

## 22.2 Congestive Heart Failure: Epidemiology and Risk Factors

The increase in incidence and prevalence of CHF poses an urgent national and global public health priority. According to the American Heart Association, 550,000 new cases occur each year [5]. More than five million Americans are afflicted with this disorder. It remains the leading cause of hospitalizations among medicare beneficiaries. CHF has diverse etiologies and occurs commonly following acute and chronic cardiac injury. According to the Framingham investigators, the lifetime risk of developing CHF at age 40 is approximately 20% [5]. The major risk factors for CHF parallel those for CAD, and include diabetes, hypertension, hypercholesterolemia, smoking, genetic factors, and aging.

Several factors have contributed to the increasing prevalence of CHF in the United States. First, risk factors contributing to CAD are increasingly prevalent. Diabetes mellitus, in particular, is a serious worldwide health problem affecting an increasing number of individuals. A recent study by the World Health Organization supports the notion that the incidence of diabetes is increasing at epidemic proportions as the worldwide prevalence of this disease is expected to grow from 171 million in 2000 to 366 million by 2030 [6]. This is, in part, explained by the increasing incidence of obesity and metabolic syndrome, adaptation of a western diet, and a more sedentary lifestyle. Approximately 7% of the United States population is diabetic, and 54 million individuals are at risk. Moreover, type 2 diabetes is being diagnosed more frequently at both extremes of ages: young children/adolescents and geriatric populations. It is particularly prevalent in minority populations such as American Indians, African Americans, and Hispanic/Latino Americans. Developing countries are also seeing a substantial increase in the incidence of diabetes as their populations are beginning to adapt a western lifestyle. Second, there is a marked increase in the elderly population (> 65 years of age), which is expected to increase from 35 million in 2000 to 70.3 million in 2030. The risk of CHF increases markedly with advancing age. The annual rates per 1,000 population of new CHF events for white men are 15.2 for those between 65 and 74 years of age, 31.7 for those between 75 and 84 years of age, and 65.2 for those >85 years of age [5]. Even if the incidence of CHF remains the same, this demographic shift will result in doubling of the incidence of CHF. Third, improved patient outcomes following acute MI have paradoxically increased the incidence of CHF as patients who survive the acute coronary event are at greater risk for developing CHF. Finally, improved care for those patients who already have CHF increases the prevalence of the disease [7].

Endothelial dysfunction, assessed clinically as impaired endothelium-dependent vasodilation, is an early feature of atherosclerosis. Endothelial dysfunction is implicated in the progression of vascular disease and the eventual development of CHF [8, 9]. It is seen in subjects with established CAD and peripheral vascular disease. Interventions that improve outcomes in subjects with CAD, such as lipid-lowering drugs, angiotensin-converting enzyme inhibitors, and physical activity, improve endothelial function, both in coronary and peripheral arteries [8, 10]. While the pathophysiology of end-stage heart disease is complex, nearly all patients with this disorder develop cardiac fibrosis in response to either acute or chronic injury. Indeed, fibrosis in response to injury is thought to be the major mechanism of end-stage failure involving multiple organ systems, including the kidney [11], lung [12], and heart [13].

In atherosclerosis, the vascular endothelium itself may promote inflammatory responses. When healthy, the endothelium creates an antithrombotic milieu by secreting factors such as NO and prostacyclin (which mediate antiaggregatory effects on platelets); proteins C and S, and heparin (which have anticoagulant effects); and tissue plasminogen activator (which is fibrinolytic). Endothelial dysfunction is frequently accompanied by endothelial expression of vasoconstrictors and proinflammatory, proliferative, and procoagulant factors that are proatherogenic [8, 10]. Thus, it is likely that the status of an individual's endothelial function reflects that person's risk for developing atherosclerosis. Human studies have shown that traditional risk factors for atherosclerosis predispose individuals to endothelial dysfunction, and that strategies aimed at reducing cardiovascular risk factors can improve endothelial function [8, 10]. Impaired endothelium-dependent vasodilation in the coronary circulation predicts adverse CAD events and long-term outcome. A decline in NO bioavailability can result from multiple factors, including decreased eNOS expression, lack of substrates for eNOS, decreased eNOS activation, or accelerated NO degradation [14, 15].

## 22.3 Oxidative Stress in the Initiation/Progression of Vascular Disease

During the past two decades, considerable evidence has implicated oxidative stress in several distinct human disorders, including aging [16], atherosclerosis [17–19], neurodegenerative diseases [20, 21], diabetes [22–24], and end-stage renal disease [25, 26]. Accumulating evidence suggests that oxidative stress plays a central role in vascular disease, in part by promoting endothelial dysfunction [27, 28]. The endothelium forms the structural interface between the vascular wall and circulating blood. Endothelial dysfunction is a key early feature in atherogenesis and has been implicated in plaque progression/instability [8, 10]. It is characterized by a reduction in the bioavailability of vasodilators such as endothelium-derived NO and a relative/absolute abundance of vasoconstrictors. Thus, endothelium-dependent vasodilation, a hallmark of normal endothelial function, is impaired in vascular disease [8, 10].

Although oxidative stress has been associated with atherosclerosis, its origins and precise role in pathogenesis are less well understood. For example, it is not known whether oxidative stress is a primary event that occurs early in the disease, or whether it represents a secondary phenomenon that merely reflects end-stage tissue damage [29]. This distinction has important clinical relevance. If oxidative stress simply reflects tissue damage, interventions that reduce it may fail to affect the disease process. If oxidative stress promotes tissue injury, therapies that interrupt oxidative pathways early in the disease may prevent complications, and those that act later may slow disease progression.

Atherosclerosis is often characterized as a chronic inflammatory disease of the vascular wall associated with the infiltration of lipid-laden inflammatory cells, such as monocyte-derived macrophages and T-lymphocytes [9]. Although it is well known that elevated levels of low density lipoproteins (LDL) greatly increase the risk for atherosclerosis [30], in vitro studies suggest that LDL by itself is not atherogenic; oxidative modification of LDL, however, may initiate the disease process [31, 32]. This notion led to the "oxidation hypothesis," which posits that LDL must be oxidatively modified to initiate atherogenesis. Many lines of evidence support this hypothesis. Oxidized LDL is taken up by scavenger receptors of macrophages, which then become lipid-laden foam cells, the pathologic hallmark of early atherosclerotic lesions [33]. Oxidized LDL has been isolated from human and animal atherosclerotic tissue, and immunohistochemical studies have detected oxidized lipids in atherosclerotic lesions [34–36]. All major cell types involved in atherosclerosis—smooth muscle cells, endothelial cells, and macrophages—produce ROS that can oxidize LDL in vitro [37–39]. Oxidized LDL stimulates smooth muscle cells to synthesize extracellular matrix and activate a signaling cascade by interacting with the lectin-like OxLDL receptor [33, 40]. Oxidized LDL also attracts mononuclear cells that produce monocyte chemoattractant protein-1 and other inflammatory cytokines, which can lead to the conversion of fatty streaks in more complex lesions that involve the incorporation of smooth muscle-like cells that migrate from the media into the subendothelial space. Finally, several structurally unrelated lipid-soluble antioxidants that inhibit LDL oxidation in vitro inhibit atherosclerosis in hypercholesterolemic animals [41–43].

## 22.4 Oxidative Stress Pathways in Cardiovascular Disease

Proteins, lipids, and nucleic acids are important targets of oxidative stress in vivo. However, the specific oxidant-generating pathways that contribute to oxidative stress in cardiovascular diseases are not well defined. One reason for this current lack of understanding is that oxidizing intermediates are difficult to detect in vivo because they are short-lived and are generated at low levels. The evidence for a number of oxidative stress pathways activated in cardiovascular disease and related conditions are discussed in this section.

*The glycoxidation pathway*: The glycoxidation pathway is particularly relevant in high-glucose states such as diabetes mellitus, a well-established risk factor for cardiovascular disease. In its open-chain form, glucose has a carbonyl group that is a target of oxidative chemistry. Glucose auto-oxidization has been reported to mediate covalent linkage of glucose to amino groups on proteins by a hydroxyl radical–dependent mechanism [44]. Glucose also reacts nonenzymatically with proteins to form the reversible Schiff base adduct, which subsequently can rearrange itself into the stable Amadori product and advanced glycosylation end products (AGE). In vitro, free metal ions catalyze steps in a nonenzymatic glycoxidation pathway that generates AGE products. Metal-catalyzed hydroxyl radical formation can peroxidize lipids and convert phenylalanine residues of proteins into isomers of tyrosine such as *ortho*-tyrosine and *meta*-tyrosine [45–47]. Reduced, redox-active metal ions ($M^{n+}$) such as $Fe^{2+}$ and $Cu^{1+}$ generate hydroxyl radical ($HO^{\bullet}$) when they react with hydrogen peroxide ($H_2O_2$; Eq. 1).

$$M^{n+} + H_2O_2 \rightarrow M^{(n-1)+} + HO^{\bullet} + HO^{-} \tag{1}$$

AGEs can damage tissues through a number of mechanisms, including generation of oxidizing intermediates, formation of immune complexes, interaction with a cellular receptor called RAGE (Receptor for AGE), and induction of cytokine release [17, 48]. Although RAGE binds to AGE-modified proteins in vitro with high affinity, its ligands in vivo are unclear. High levels of AGEs accumulate in renal failure, even in nondiabetic patients, and this process reverses after renal transplantation, implicating the kidneys in AGE production and/or clearance [29, 49–51]. Many studies have shown that age-adjusted levels of pentosidine and Nε-carboxymethyllysine, two known AGE products, correlate with the development of diabetic complications [46, 52–55]. Thus, glycoxidation reactions represent one plausible mechanism for oxidative injury in diabetes-associated cardiovascular disease.

*The glucose-polyunsaturated fatty acid pathway*: Glucose can also generate reactive intermediates by interacting with polyunsaturated fatty acids (PUFA). When incubated with LDL or a control protein, ribonuclease (RNAse), pathophysiologically relevant concentrations of glucose induce formation of oxidatively modified amino acids in LDL even in the absence of free metal ions; in contrast, glucose exposure does not increase levels of oxidized amino acids in RNAse [56]. This study indicates that glucose is capable of generating a species resembling the hydroxyl radical by a carbonyl/PUFA pathway, a potential mechanism for localized oxidative stress in tissues vulnerable to diabetic damage [56].

*The reactive nitrogen pathway*: Reactive intermediates of nitric oxide (NO), normally produced by endothelial cells to regulate vascular tone, represents another potent source of oxidants. NO is also produced by inflammatory cells such as macrophages, which are early participants in atherosclerotic lesion formation. NO reacts with superoxide ($O_2^{\bullet-}$) to generate peroxynitrite ($ONOO^-$; Eq. 2).

$$O_2^{\bullet-} + NO^{\bullet} \rightarrow ONOO^- \tag{2}$$

Peroxynitrite is a potent oxidant that converts tyrosine residues to 3-nitrotyrosine, which may serve as a biomarker for the activation of the reactive nitrogen pathway. It has been detected in LDL and high density lipoproteins (HDL) isolated from human diabetic atherosclerotic lesions [57–59]. Additionally, plasma nitrotyrosine levels are elevated in patients with CAD [60, 61].

*NADPH oxidases*: NADPH oxidase (NOX) family enzymes are major producers of $O_2^{\bullet-}$ by cells of the myocardium and vasculature [14]. Several NOX isoforms present in the endothelium, smooth muscle cells, and fibroblasts are selectively upregulated by mediators implicated in the pathogenesis of atherosclerosis. These mediators include angiotensin II, endothelin-1, hypercholesterolemia, shear stress, nonesterified fatty acids, hyperglycemia, and growth factors. Angiotensin II may represent a pathophysiologically relevant pathway for stimulating the production of reactive intermediates by artery wall cells because inhibitors of this pathway lower the risk of cardiovascular events [62]. In human subjects, NOX activity correlates inversely with endothelial function, even after other major risk factors for atherosclerosis, including diabetes and hypercholesterolemia, are taken into account [15]. The primary enzymatic product of NOX enzymes is $O_2^{\bullet-}$ [63]. At neutral pH, $O_2^{\bullet-}$ is a reducing agent rather than an oxidant. However, $O_2^{\bullet-}$ dismutates enzymatically or nonenzymatically into $H_2O_2$, which can then oxidize thiol residues, a mechanism for cellular signaling via the inactivation of cysteine-containing phosphatases [64]. It can also function as an oxidizing substrate for heme proteins such as myeloperoxidase (MPO). $O_2^{\bullet-}$ also reacts at a diffusion-controlled rate with NO to form $ONOO^-$, a powerful reactive nitrogen species that nitrates tyrosine residues on proteins and may induce oxidative damage to other macromolecular substrates.

*Uncoupled endothelial nitric oxide synthase* (*eNOS*): eNOS synthesizes NO in endothelial cells and its uncoupling has been described in cardiovascular disorders. One proposed mechanism involves oxidation of its cofactor, tetrahydrobiopterin ($BH_4$), resulting in the transfer of electrons from eNOS to molecular oxygen, generating $O_2^{\bullet-}$ [65, 66]. An alternative mechanism for uncoupling eNOS involves overproduction of angiotensin II, which can induce dihydrofolate reductase deficiency.

Because dihydrofolate reductase maintains $BH_4$ in its reduced form, its deficiency uncouples eNOS. $BH_4$ oxidation and NOS uncoupling have been demonstrated in hypertension, diabetes, and hypercholesterolemia. Moreover, administering $BH_4$ improves endothelium-dependent vasodilation in experimental animals and humans with those conditions [15].

*The mitochondrial electron transport pathway*: Electron "leakage" across the mitochondrial electron transport chain is another source of $O_2^{\bullet-}$ and consequently of $H_2O_2$. Plasma levels of both glucose and free fatty acids promote mitochondrial generation of ROS. This has been proposed as one mechanism for cellular damage in diabetes [27]. Mitochondrial $O_2^{\bullet-}$ overproduction mediated by hyperglycemia might also increase polyol pathway activity, protein kinase C (PKC) activity, and hexosamine flux, resulting in cellular dysfunction and tissue damage [27]. Exposure of endothelial cells to exogenous oxidants leads to mitochondrial damage and augments $O_2^{\bullet-}$ production, a potential mechanism for feed-forward amplification of ROS generation [14]. Moreover, $O_2^{\bullet-}$ inhibits glyceraldehyde phosphate dehydrogenase, a key glycolytic enzyme which may lead to the accumulation of upstream metabolites. Such effects on glycolysis may promote end-organ damage by diverting metabolites into the hexosamine pathway or stimulating the polyol and diacylglycerol-PKC pathways. Benfotiamine, a lipid-soluble thiamine analog, inhibits these pathways by activating pentose phosphate pathway enzyme transketolase; furthermore, benfotiamine treatment prevents experimental diabetic retinopathy [67].

*Xanthine oxidase*: Another source of $O_2^{\bullet-}$ and $H_2O_2$ in mammalian cells is xanthine oxidase, which converts hypoxanthine and xanthine to uric acid, while reducing molecular oxygen. Hydrogen peroxide can increase levels of xanthine oxidase, further amplifying $O_2^{\bullet-}$ production. Inflammatory cytokines, such as tumor necrosis factor-α, and oxidation of cysteine residues by oxidants such as peroxynitrite can result in the conversion of xanthine dehydrogenase to xanthine oxidase. Xanthine oxidase is an important source of oxidants in a variety of pathophysiological states, including diabetes, hypertension, atherosclerosis, ischemia–reperfusion, and heart failure [15]. Endothelial levels of xanthine oxidase are elevated in humans with heart failure and subjects with CAD, and they correlate with degree of impairment in endothelium-dependent vasodilation [68].

*The myeloperoxidase pathway*: The major pathway through which macrophages and other phagocytic cells of the innate immune system generate ROS is by activation of the membrane-bound NOX homolog, NOX2, which generates $O_2^{\bullet-}$ as its primary enzymatic product (Eq. 3).

$$\mathrm{NADPH} + 2\,\mathrm{O_2} \rightarrow \mathrm{NADP^+} + \mathrm{H^+} + 2\,\mathrm{O_2^{\bullet-}} \tag{3}$$

Superoxide dismutates into hydrogen peroxide (Eq. 4).

$$2\,\mathrm{H^+} + 2\,\mathrm{O_2^{\bullet-}} \rightarrow \mathrm{H_2O_2} + \mathrm{O_2} \tag{4}$$

Hydrogen peroxide can then be used by another phagocytic enzyme, myeloperoxidase [69, 70], to convert the chloride ion to hypochlorous acid (HOCl) (Eq. 5).

$$H_2O_2 + Cl^- + H^+ \rightarrow HOCl + H_2O \qquad (5)$$

Oxidation of NO with oxygen yields nitrite ($NO_2^-$), which myeloperoxidase converts to nitrogen dioxide radical ($NO_2^\bullet$); Eq. 6), a potent nitrating intermediate [71, 72].

$$NO_2^- + H_2O_2 \rightarrow NO_2^\bullet + H_2O \qquad (6)$$

Reactive nitrogen species, including peroxynitrite and $NO_2^\bullet$, might contribute to the inflammatory process by nitrating lipoproteins and other biomolecules. Hyperglycemia can activate PKC [73, 74], which leads to phagocyte activation, secretion of myeloperoxidase, and oxidant generation. Nonesterified fatty acids (NEFAs) that commonly are overabundant in diabetes can also activate phagocytes in vitro. These changes might enhance the production of $O_2^{\bullet-}$ and $H_2O_2$, which myeloperoxidase converts into more potent cytotoxic oxidants, such as hypochlorous acid and nitrogen dioxide radical.

## 22.5 Biomarkers of Oxidative Stress Pathways

Reactive oxygen/nitrogen intermediates are difficult to detect in vivo due to their high reactivity with endogenous substrates; however, some of these oxidized substrates may serve as biomarkers for the activation of specific oxidative stress pathways. Immunohistochemistry and dihydroethidium fluorescence have been used to study oxidation-specific epitopes and oxidant production. These techniques are highly sensitive, and are useful for localization of oxidative events. However, they are nonspecific and, at best, semiquantitative. In contrast, mass spectrometry (MS) offers a powerful set of analytical tools for identifying and quantifying oxidized biomolecules.

Isotope dilution gas-chromatography/mass spectrometry (GC/MS) is a highly sensitive and specific method that has been used to quantify oxidation of specific markers. Biomolecules derived from plasma or tissue are first separated by GC, derivatized, and ionized. The mass-to-charge ratios of ions derived by fragmenting the ionized, derivatized parent compound are determined by MS. A full scan mass spectrum obtained for a target analyte can unequivocally identify a target biomolecule because each compound has a unique fragmentation pattern. The analyte is quantified by adding a stable, isotopically labeled internal standard, which is identical to the target analyte, except for the heavy isotope. With certain ionization processes, such as electron capture negative-ion chemical ionization, it is possible to detect and quantify subfemtomole levels of biomolecules. Using a combination of free radical generating systems in vitro and studying tissue and plasma samples in animal models of disease and humans, these oxidative biomarker patterns may be indicative of the activation of specific oxidative stress pathways [24, 56–61, 75–86] (Table 22.1). Similar molecular fingerprinting strategies to identify tissue-specific pathways of oxidation may provide additional insights into oxidative stress mechanisms in cardiovascular diseases.

**Table 22.1** Patterns of tyrosine oxidation products associated with the activation of specific oxidative stress pathways

| | *Ortho*-tyrosine | *Meta*-tyrosine | Dityrosine | Nitrotyrosine | Chlorotyrosine |
|---|---|---|---|---|---|
| 1. Hydroxyl radical pathway | ↑ | ↑ | ↑ | – | – |
| 2. MPO pathway | – | – | ↑ | ↑ | ↑ |
| 3. RNS by MPO | – | – | ↑ | ↑ | ↑ |
| 4. RNS by peroxynitrite | ↑ | ↑ | ↑ | ↑↑ | – |
| 5. Superoxide/hydrogen peroxide (NADPH oxidases) | ↑ | ↑ | ↑ | ↑[a] | ↑[b] |

↑, Increased; ↑↑, markedly increased; –, no change.
[a] In presence of MPO and NO.
[b] In presence of MPO [17, 18, 56–59, 71, 75–77, 79, 83–85, 87–91].

The hydroxyl group of the phenolic ring of tyrosine residues in proteins is highly reactive to electrophilic substitution reactions, which usually occur at the *ortho* position on the aromatic ring [92]. Oxidation reactions of tyrosine include chlorination (to form *chlorotyrosine*; catalyzed by myeloperoxidase), nitration (to form *nitrotyrosine*; mediated by RNS), and dimerization (to form *dityrosine*; mediated by tyrosine radicals/ROS) (Fig. 22.1). Mass spectrometry has emerged as a critical analytical tool for the detection and quantitation of posttranslational modification of tyrosine residues [93–95]. There is evidence that oxidized amino acids in plasma and urine

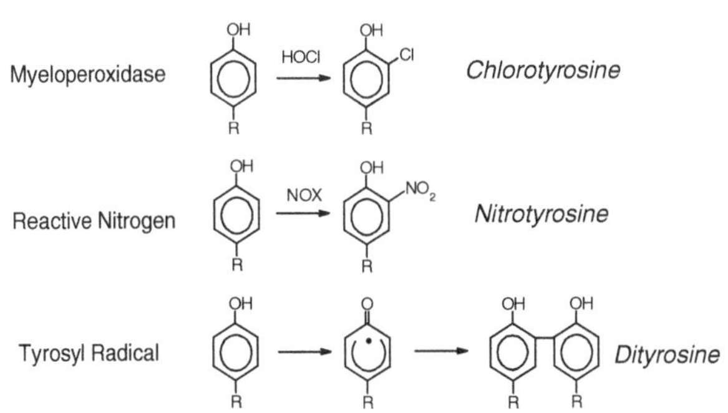


**Fig. 22.1** Oxidation products of tyrosine: specific oxidative modifications of tyrosine formed in tissues or plasma may serve as biomarkers for the activation of specific oxidative stress pathways

can serve as markers for assessing oxidative stress in vivo [59–61, 75, 76, 81, 83, 86, 96]. Steady state levels of these markers in plasma and urine are proportional to their rate of generation and can serve as indices of chronic oxidative stress in vivo. A case-control study demonstrated that systemic levels of protein-bound nitrotyrosine were significantly higher among patients with CAD when compared with healthy subjects; statin therapy lowered levels of oxidation markers in plasma, raising the possibility that statins mediate antioxidant effects in vivo [60, 61, 96]. Thus, these oxidative biomarkers may serve not only to assess degree of oxidative stress but also to monitor efficacy of therapy. Further studies are required to validate the utility of these biomarkers in response to specific therapies and interventions in human subjects.

## 22.6 NOX Enzymes and Cardiovascular Fibrosis

Tissue fibrosis in mammals is a complex multicellular process that results from a failure of tissue regeneration and that invariably progresses to end-organ failure [12, 97, 98]. Following acute MI, the development of myocardial fibrosis often leads to CHF. Oxidative stress is thought to play a key role in the initiation and/or progression of fibrosis in diverse organ systems [99–101]. Although many of the oxidative stress pathways discussed may contribute to fibrogenic responses in the cardiovascular system, we focus here on the role of NOX family enzymes in this process.

The NOX enzyme family has been linked to several cardiovascular pathologies, such as cardiac hypertrophy and fibrosis, atherosclerosis, vascular inflammation, and angiogenesis [102, 103]. Myocardial NOX activity is elevated in patients with end-stage heart failure [104, 105], as well as in experimental models of pressure overload and left ventricular hypertrophy [106]. Further, treatment with an NADPH oxidase inhibitor, apocynin, inhibits aldosterone-induced cardiac fibrosis in rats [107]. The role(s) of individual NOX family members and how they contribute to cardiovascular disease are only beginning to be elucidated, although both NOX2 and NOX4 appear to play critical roles. Angiotensin II and aldosterone have been shown to contribute to cardiac remodeling and fibrosis, and accumulating evidence suggests that NOX2 is a critical mediator of these pathways. The fibrotic effects of angiotensin II are potentiated by mineralocorticoid receptor activation which enhances cardiac NOX2-mediated oxidative stress, and receptor blockade improves NOX2-mediated cardiac oxidative stress-induced remodeling [108]. Aldosterone-induced cardiac fibrosis was shown to be inhibited in $NOX2^{-/-}$ mice [109]. $NOX2^{-/-}$ mice subjected to coronary artery ligation show enhanced cardioprotection post-MI compared to wild-type mice, including significantly decreased left ventricular dilatation/dysfunction, cardiomyocyte hypertrophy, apoptosis, and interstitial fibrosis; moreover, systolic/diastolic functions were better preserved in $NOX2^{-/-}$ mice [110]. NOX2 has also been shown to contribute to the development of cardiac contractile dysfunction and interstitial fibrosis during pressure overload [111].

Another member of the NOX gene family, NOX4, may also mediate profibrotic effects in the context of tissue injury/repair. Myofibroblasts are key effectors of fibrogenesis and tissue remodeling in multiple organ systems, including the cardiovascular system [112]. Myofibroblast differentiation has been demonstrated to be mediated by ROS generation through a NOX4-dependent mechanism [113]. Additionally, dystrophin-deficient mice show increased cardiac fibrosis, which correlates with enhanced NOX4 expression [114]. Studies of the expression of the various NOX homologs in chronic phases of cardiomyopathy in human subjects will provide important insights into the relevant NOX enzymes that may be targeted for therapeutic benefit.

## 22.7 Therapeutic Implications for Cardiovascular Fibrosis

Therapeutic strategies to inhibit oxidative stress in cardiovascular fibrosis and related conditions may involve administration of: (a) antioxidants; or (b) inhibitors of ROS-generating enzymes. Most interventional studies have, so far, focused on the former approach of augmenting tissue antioxidant capacity. The proposed role of oxidized LDL in atherogenesis suggests that a high dietary antioxidant intake might prevent premature vascular disease in humans. However, the majority of prospective, double-blind, placebo-controlled trials of one proposed lipid-soluble antioxidant, vitamin E, have failed to demonstrate any reduction of clinical events in patients with established atherosclerosis [115]. The disappointing results of such trials have led some to question the role of oxidative stress in the pathogenesis of coronary artery disease in humans. However, it is unclear if vitamin E supplementation actually mediates a significant antioxidant effect in vivo [115]. A study of healthy humans taking vitamin E dietary supplements as high as 2,000 IU/day for 8 weeks found no change in levels of three lipid oxidation products: 4-hydroxynonenal and two isoprostanes [116]. These investigators assessed products of lipid peroxidation using GC/MS, a sensitive and specific method. Such observations highlight the importance of documenting that a proposed antioxidant intervention actually inhibits oxidative reactions in vivo [18, 115, 117]. The ability to accurately quantify amino acid oxidation markers in tissue samples, plasma, and urine can provide a means of monitoring the efficacy of therapeutic interventions, in addition to providing mechanistic insights into disease pathogenesis.

Other factors that may hinder the successful translation of proof-of-concept animal studies to humans is the variability among certain mammalian species in the proposed biological pathways and the design/context of animal studies that are typically performed over relatively short time periods, in comparison to the chronic nature of the disease process in patients with cardiovascular fibrosis/CHF. Another important consideration is appropriate patient selection and genotyping/phenotyping, since there is likely to be heterogeneity in pathogenetic mechanisms that give rise to common clinical syndromes. If an antioxidant therapeutic strategy is designed, one might consider selecting a patient cohort with evidence of increased oxidative stress (for example, subjects with elevated plasma levels of oxidized amino acids). Finally, therapeutic strategies that more directly target the source of ROS generation (for example, pharmacological inhibitors of NOX2 or NOX4) may prove to be more specific and effective in comparison to antioxidant interventions for disorders such as cardiovascular fibrosis.

**Acknowledgments** Research in the authors' laboratory is supported by grants from the National Institutes of Health, R01 HL094230 (SP and VJT), R21 HL092237 (SP) and R01 HL067967 (VJT), the Michigan Metabolomics and Obesity Center, and the Biomedical Mass Spectrometry Facility, University of Michigan. SP is supported by the Doris Duke Foundation Clinical Scientist Development Award.

## References

1. Thannickal VJ (2009) Oxygen in the evolution of complex life and the price we pay. Am J Respir Cell Mol Biol 40(5):507–510; Epub 2008 Oct 31
2. Lambeth JD (2007) Nox enzymes, ROS, and chronic disease: an example of antagonistic pleiotropy. Free Radic Biol Med 43:332–347
3. Murdoch CE, Zhang M, Cave AC et al (2006) NADPH oxidase-dependent redox signalling in cardiac hypertrophy, remodelling and failure. Cardiovasc Res 71:208–215
4. Sun Y (2007) Oxidative stress and cardiac repair/remodeling following infarction. Am J Med Sci 334:197–205
5. Rosamond W, Flegal K, Furie K et al (2008) Heart disease and stroke statistics–2008 update: a report from the American Heart Association Statistics Committee and Stroke Statistics Subcommittee. Circulation 117:e25–e146
6. Wild S, Roglic G, Green A et al (2004) Global prevalence of diabetes: estimates for the year 2000 and projections for 2030. Diabetes Care 27:1047–1053
7. Schocken DD, Benjamin EJ, Fonarow GC et al (2008) Prevention of heart failure: a scientific statement from the American Heart Association Councils on Epidemiology and Prevention, Clinical Cardiology, Cardiovascular Nursing, and High Blood Pressure Research; Quality of Care and Outcomes Research Interdisciplinary Working Group; and Functional Genomics and Translational Biology Interdisciplinary Working Group. Circulation 117:2544–2565
8. Bonetti PO, Lerman LO, Lerman A (2003) Endothelial dysfunction: a marker of atherosclerotic risk. Arterioscler Thromb Vasc Biol 23:168–175
9. Ross R (1999) Atherosclerosis – an inflammatory disease. N Engl J Med 340:115–126
10. Deanfield JE, Halcox JP, Rabelink TJ (2007) Endothelial function and dysfunction: testing and clinical relevance. Circulation 115:1285–1295
11. Qian Y, Feldman E, Pennathur S et al (2008) From fibrosis to sclerosis: mechanisms of glomerulosclerosis in diabetic nephropathy. Diabetes 57:1439–1445
12. Thannickal VJ, Toews GB, White ES et al (2004) Mechanisms of pulmonary fibrosis. Annu Rev Med 55:395–417
13. Aneja A, Tang WH, Bansilal S et al (2008) Diabetic cardiomyopathy: insights into pathogenesis, diagnostic challenges, and therapeutic options. Am J Med 121:748–757
14. Cai H (2005) NAD(P)H oxidase-dependent self-propagation of hydrogen peroxide and vascular disease. Circ Res 96:818–822
15. Guzik TJ, Harrison DG, Vascular NADPH (2006) Oxidases as drug targets for novel antioxidant strategies. Drug Discov Today 11:524–533
16. Ames BN, Shigenaga MK, Hagen TM (1993) Oxidants, antioxidants, and the degenerative diseases of aging. Proc Natl Acad Sci U S A 90:7915–7922
17. Baynes JW, Thorpe SR (2000) Glycoxidation and lipoxidation in atherogenesis. Free Radic Biol Med 28:1708–1716

18. Heinecke JW (2003) Oxidative stress: new approaches to diagnosis and prognosis in atherosclerosis. Am J Cardiol 91:12A–16A
19. Pennathur S, Heinecke JW (2007) Oxidative stress and endothelial dysfunction in vascular disease. Curr Diab Rep 7:257–264
20. Lin MT, Beal MF (2006) Mitochondrial dysfunction and oxidative stress in neurodegenerative diseases. Nature 443:787–795
21. Butterfield DA (2006) Oxidative stress in neurodegenerative disorders. Antioxid Redox Signal 8:1971–1973
22. Brownlee M (1995) Advanced protein glycosylation in diabetes and aging. Annu Rev Med 46:223–234
23. Jay D, Hitomi H, Griendling KK (2006) Oxidative stress and diabetic cardiovascular complications. Free Radic Biol Med 40:183–192
24. Pennathur S, Heinecke JW (2007) Mechanisms for oxidative stress in diabetic cardiovascular disease. Antioxid Redox Signal 9:955–969
25. Cottone S, Lorito MC, Riccobene R et al (2008) Oxidative stress, inflammation and cardiovascular disease in chronic renal failure. J Nephrol 21:175–179
26. Percy C, Pat B, Poronnik P et al (2005) Role of oxidative stress in age-associated chronic kidney pathologies. Adv Chronic Kidney Dis 12:78–83
27. Brownlee M (2001) Biochemistry and molecular cell biology of diabetic complications. Nature 414:813–820
28. Houstis N, Rosen ED, Lander ES (2006) Reactive oxygen species have a causal role in multiple forms of insulin resistance. Nature 440:944–948
29. Baynes JW, Thorpe SR (1999) Role of oxidative stress in diabetic complications: a new perspective on an old paradigm. Diabetes 48:1–9
30. Brown MS, Goldstein JL (1992) Koch's postulates for cholesterol. Cell 71:187–188
31. Goldstein JL, Ho YK, Basu SK et al (1979) Binding site on macrophages that mediates uptake and degradation of acetylated low density lipoprotein, producing massive cholesterol deposition. Proc Natl Acad Sci U S A 76:333–337
32. Witztum JL, Steinberg D (1991) Role of oxidized low density lipoprotein in atherogenesis. J Clin Invest 88:1785–1792
33. Steinberg D (2002) Atherogenesis in perspective: hypercholesterolemia and inflammation as partners in crime. Nat Med 8:1211–1217
34. Haberland ME, Fong D, Cheng L (1988) Malondialdehyde-altered protein occurs in atheroma of Watanabe heritable hyperlipidemic rabbits. Science 241:215–218
35. Yla-Herttuala S, Palinski W, Butler SW et al (1994) Rabbit and human atherosclerotic lesions contain IgG that recognizes epitopes of oxidized LDL. Arterioscler Thromb 14:32–40
36. Yla-Herttuala S, Palinski W, Rosenfeld ME et al (1989) Evidence for the presence of oxidatively modified low density lipoprotein in atherosclerotic lesions of rabbit and man. J Clin Invest 84:1086–1095
37. Heinecke JW, Rosen H, Chait A (1984) Iron and copper promote modification of low density lipoprotein by human arterial smooth muscle cells in culture. J Clin Invest 74:1890–1894
38. Morel DW, DiCorleto PE, Chisolm GM (1984) Endothelial and smooth muscle cells alter low density lipoprotein in vitro by free radical oxidation. Arteriosclerosis 4:357–364
39. Steinbrecher UP, Parthasarathy S, Leake DS et al (1984) Modification of low density lipoprotein by endothelial cells involves lipid peroxidation and degradation of low density lipoprotein phospholipids. Proc Natl Acad Sci U S A 81:3883–3887
40. Glass CK, Witztum JL (2001) Atherosclerosis. The road ahead. Cell 104:503–516
41. Carew TE, Schwenke DC, Steinberg D (1987) Antiatherogenic effect of probucol unrelated to its hypocholesterolemic effect: evidence that antioxidants in vivo can selectively inhibit low density lipoprotein degradation in macrophage-rich fatty streaks and slow the progression of atherosclerosis in the Watanabe heritable hyperlipidemic rabbit. Proc Natl Acad Sci U S A 84:7725–7729

42. Kita T, Nagano Y, Yokode M et al (1987) Probucol prevents the progression of atherosclerosis in Watanabe heritable hyperlipidemic rabbit, an animal model for familial hypercholesterolemia. Proc Natl Acad Sci U S A 84:5928–5931
43. Parthasarathy S, Young SG, Witztum JL et al (1986) Probucol inhibits oxidative modification of low density lipoprotein. J Clin Invest 77:641–644
44. Hunt JV, Dean RT, Wolff SP (1988) Hydroxyl radical production and autoxidative glycosylation. Glucose autoxidation as the cause of protein damage in the experimental glycation model of diabetes mellitus and ageing. Biochem J 256:205–212
45. Huggins TG, Wells-Knecht MC, Detorie NA et al (1993) Formation of o-tyrosine and dityrosine in proteins during radiolytic and metal-catalyzed oxidation. J Biol Chem 268:12341–12347
46. Huggins TG, Staton MW, Dyer DG et al (1992) o-Tyrosine and dityrosine concentrations in oxidized proteins and lens proteins with age. Ann N Y Acad Sci 663:436–437
47. Kaur H, Halliwell B (1994) Detection of hydroxyl radicals by aromatic hydroxylation. Methods Enzymol 233:67–82
48. Stern DM, Yan SD, Yan SF et al (2002) Receptor for advanced glycation endproducts (RAGE) and the complications of diabetes. Ageing Res Rev 1:1–15
49. Miyata T, van Ypersele de Strihou C, Kurokawa K et al (1999) Alterations in nonenzymatic biochemistry in uremia: origin and significance of "carbonyl stress" in long-term uremic complications. Kidney Int 55:389–399
50. Miyata T, Ueda Y, Yoshida A et al (1997) Clearance of pentosidine, an advanced glycation end product, by different modalities of renal replacement therapy. Kidney Int 51:880–887
51. Miyata T, Fu MX, Kurokawa K et al (1998) Autoxidation products of both carbohydrates and lipids are increased in uremic plasma: is there oxidative stress in uremia? Kidney Int 54:1290–1295
52. Dyer DG, Blackledge JA, Thorpe SR et al (1991) Formation of pentosidine during nonenzymatic browning of proteins by glucose. Identification of glucose and other carbohydrates as possible precursors of pentosidine in vivo. J Biol Chem 266:11654–11660
53. McCance DR, Dyer DG, Dunn JA et al (1993) Maillard reaction products and their relation to complications in insulin-dependent diabetes mellitus. J Clin Invest 91:2470–2478
54. Monnier VM (2001) Transition metals redox: reviving an old plot for diabetic vascular disease. J Clin Invest 107:799–801
55. Sell DR, Lapolla A, Odetti P et al (1992) Pentosidine formation in skin correlates with severity of complications in individuals with long-standing IDDM. Diabetes 41:1286–1292
56. Pennathur S, Ido Y, Heller JI et al (2005) Reactive carbonyls and polyunsaturated fatty acids produce a hydroxyl radical-like species: a potential pathway for oxidative damage of retinal proteins in diabetes. J Biol Chem 280:22706–22714
57. Leeuwenburgh C, Hardy MM, Hazen SL et al (1997) Reactive nitrogen intermediates promote low density lipoprotein oxidation in human atherosclerotic intima. J Biol Chem 272:1433–1436
58. Leeuwenburgh C, Rasmussen JE, Hsu FF et al (1997) Mass spectrometric quantification of markers for protein oxidation by tyrosyl radical, copper, and hydroxyl radical in low density lipoprotein isolated from human atherosclerotic plaques. J Biol Chem 272:3520–3526
59. Pennathur S, Bergt C, Shao B et al (2004) Human atherosclerotic intima and blood of patients with established coronary artery disease contain high density lipoprotein damaged by reactive nitrogen species. J Biol Chem 279:42977–42983
60. Shishehbor MH, Aviles RJ, Brennan ML et al (2003) Association of nitrotyrosine levels with cardiovascular disease and modulation by statin therapy. J Am Med Assoc 289:1675–1680
61. Shishehbor MH, Brennan ML, Aviles RJ et al (2003) Statins promote potent systemic antioxidant effects through specific inflammatory pathways. Circulation 108:426–431
62. Heart Outcomes Prevention Evaluation Study Investigators (2000) Effects of ramipril on cardiovascular and microvascular outcomes in people with diabetes mellitus: results of the HOPE study and MICRO-HOPE substudy. Lancet 355:253–259

63. Lambeth JDNOX (2004) Enzymes and the biology of reactive oxygen. Nat Rev Immunol 4:181–189
64. Rhee SG, Bae YS, Lee SR et al (2000) Hydrogen peroxide: a key messenger that modulates protein phosphorylation through cysteine oxidation. Sci STKE 2000:PE1
65. Landmesser U, Dikalov S, Price SR et al (2003) Oxidation of tetrahydrobiopterin leads to uncoupling of endothelial cell nitric oxide synthase in hypertension. J Clin Invest 111: 1201–1209
66. Vasquez-Vivar J, Kalyanaraman B, Martasek P et al (1998) Superoxide generation by endothelial nitric oxide synthase: the influence of cofactors. Proc Natl Acad Sci U S A 95:9220–9225
67. Hammes HP, Du X, Edelstein D et al (2003) Benfotiamine blocks three major pathways of hyperglycemic damage and prevents experimental diabetic retinopathy. Nat Med 9: 294–299
68. Guzik TJ, Sadowski J, Guzik B et al (2006) Coronary artery superoxide production and nox isoform expression in human coronary artery disease. Arterioscler Thromb Vasc Biol 26:333–339
69. Clark RA, Klebanoff SJ (1977) Myeloperoxidase–$H_2O_2$–halide system: cytotoxic effect on human blood leukocytes. Blood 50:65–70
70. Klebanoff SJ (1980) Oxygen metabolism and the toxic properties of phagocytes. Ann Intern Med 93:480–489
71. Eiserich JP, Hristova M, Cross CE et al (1998) Formation of nitric oxide-derived inflammatory oxidants by myeloperoxidase in neutrophils. Nature 391:393–397
72. Gaut JP, Byun J, Tran HD et al (2002) Myeloperoxidase produces nitrating oxidants in vivo. J Clin Invest 109:1311–1319
73. Ishii H, Jirousek MR, Koya D et al (1996) Amelioration of vascular dysfunctions in diabetic rats by an oral PKC beta inhibitor. Science 272:728–731
74. King GL, Ishii H, Koya D (1997) Diabetic vascular dysfunctions: a model of excessive activation of protein kinase C. Kidney Int Suppl 60:S77–S85
75. Bergt C, Pennathur S, Fu X et al (2004) The myeloperoxidase product hypochlorous acid oxidizes HDL in the human artery wall and impairs ABCA1-dependent cholesterol transport. Proc Natl Acad Sci U S A 101:13032–13037
76. Bhattacharjee S, Pennathur S, Byun J et al (2001) NADPH oxidase of neutrophils elevates o,o'-dityrosine cross-links in proteins and urine during inflammation. Arch Biochem Biophys 395:69–77
77. Brennan ML, Hazen SL (2003) Amino acid and protein oxidation in cardiovascular disease. Amino Acids 25:365–374
78. Hazen SL, Crowley JR, Mueller DM et al (1997) Mass spectrometric quantification of 3-chlorotyrosine in human tissues with attomole sensitivity: a sensitive and specific marker for myeloperoxidase-catalyzed chlorination at sites of inflammation. Free Radic Biol Med 23:909–916
79. Lamharzi N, Renard CB, Kramer F et al (2004) Hyperlipidemia in concert with hyperglycemia stimulates the proliferation of macrophages in atherosclerotic lesions: potential role of glucose-oxidized LDL. Diabetes 53:3217–3225
80. Leeuwenburgh C, Hansen P, Shaish A et al (1998) Markers of protein oxidation by hydroxyl radical and reactive nitrogen species in tissues of aging rats. Am J Physiol 274:R453–R461
81. Leeuwenburgh C, Hansen PA, Holloszy JO et al (1999) Hydroxyl radical generation during exercise increases mitochondrial protein oxidation and levels of urinary dityrosine. Free Radic Biol Med 27:186–192
82. Leeuwenburgh C, Wagner P, Holloszy JO et al (1997) Caloric restriction attenuates dityrosine cross-linking of cardiac and skeletal muscle proteins in aging mice. Arch Biochem Biophys 346:74–80
83. Pennathur S, Heinecke JW (2004) Mechanisms of oxidative stress in diabetes: implications for the pathogenesis of vascular disease and antioxidant therapy. Front Biosci 9:565–574

84. Pennathur S, Jackson-Lewis V, Przedborski S et al (1999) Mass spectrometric quantification of 3-nitrotyrosine, ortho-tyrosine, and o,o'-dityrosine in brain tissue of 1-methyl-4-phenyl-1,2,3, 6-tetrahydropyridine-treated mice, a model of oxidative stress in Parkinson's disease. J Biol Chem 274:34621–34628
85. Pennathur S, Wagner JD, Leeuwenburgh C et al (2001) A hydroxyl radical-like species oxidizes cynomolgus monkey artery wall proteins in early diabetic vascular disease. J Clin Invest 107:853–860
86. Zheng L, Nukuna B, Brennan ML et al (2004) Apolipoprotein A-I is a selective target for myeloperoxidase-catalyzed oxidation and functional impairment in subjects with cardiovascular disease. J Clin Invest 114:529–541
87. Leeuwenburgh C, Hansen PA, Holloszy JO et al (1999) Oxidized amino acids in the urine of aging rats: potential markers for assessing oxidative stress in vivo. Am J Physiol 276: R128–R135
88. Abu-Soud HM, Hazen SL (2000) Nitric oxide is a physiological substrate for mammalian peroxidases. J Biol Chem 275:37524–37532
89. Beckman JS, Chen J, Ischiropoulos H et al (1994) Oxidative chemistry of peroxynitrite. Methods Enzymol 233:229–240
90. Brennan ML, Hazen SL (2003) Emerging role of myeloperoxidase and oxidant stress markers in cardiovascular risk assessment. Curr Opin Lipidol 14:353–359
91. Heinecke JW (2002) Oxidized amino acids: culprits in human atherosclerosis and indicators of oxidative stress. Free Radic Biol Med 32:1090–1101
92. Giulivi C, Traaseth NJ, Davies KJ (2003) Tyrosine oxidation products: analysis and biological relevance. Amino Acids 25:227–232
93. DiMarco T, Giulivi C (2007) Current analytical methods for the detection of dityrosine, a biomarker of oxidative stress, in biological samples. Mass Spectrom Rev 26:108–120
94. Witze ES, Old WM, Resing KA et al (2007) Mapping protein post-translational modifications with mass spectrometry. Nat Methods 4:798–806
95. Schoneich C, Sharov VS (2006) Mass spectrometry of protein modifications by reactive oxygen and nitrogen species. Free Radic Biol Med 41:1507–1520
96. Shishehbor MH, Hazen SL (2004) Inflammatory and oxidative markers in atherosclerosis: relationship to outcome. Curr Atheroscler Rep 6:243–250
97. Thannickal VJ, Loyd JE (2008) Idiopathic pulmonary fibrosis: a disorder of lung regeneration? Am J Respir Crit Care Med 178:663–665
98. Gurtner GC, Werner S, Barrandon Y et al (2008) Wound repair and regeneration. Nature 453:314–321
99. Urtasun R, de la Rosa LC, Nieto N (2008) Oxidative and nitrosative stress and fibrogenic response. Clin Liver Dis 12:769–790
100. Zhao W, Zhao T, Chen Y et al (2008) Oxidative stress mediates cardiac fibrosis by enhancing transforming growth factor-beta1 in hypertensive rats. Mol Cell Biochem 317:43–50
101. Kinnula VL, Fattman CL, Tan RJ et al (2005) Oxidative stress in pulmonary fibrosis: a possible role for redox modulatory therapy. Am J Respir Crit Care Med 172:417–422
102. Cave AC, Brewer AC, Narayanapanicker A et al (2006) NADPH oxidases in cardiovascular health and disease. Antioxid Redox Signal 8:691–728
103. Murdoch CE, Grieve DJ, Cave AC et al (2006) NADPH oxidase and heart failure. Curr Opin Pharmacol 6:148–153
104. Heymes C, Bendall JK, Ratajczak P et al (2003) Increased myocardial NADPH oxidase activity in human heart failure. J Am Coll Cardiol 41:2164–2171
105. Maack C, Kartes T, Kilter H et al (2003) Oxygen free radical release in human failing myocardium is associated with increased activity of rac1-GTPase and represents a target for statin treatment. Circulation 108:1567–1574
106. Li JM, Gall NP, Grieve DJ et al (2002) Activation of NADPH oxidase during progression of cardiac hypertrophy to failure. Hypertension 40:477–484

107. Park YM, Park MY, Suh YL et al (2004) NAD(P)H oxidase inhibitor prevents blood pressure elevation and cardiovascular hypertrophy in aldosterone-infused rats. Biochem Biophys Res Commun 313:812–817
108. Stas S, Whaley-Connell A, Habibi J et al (2007) Mineralocorticoid receptor blockade attenuates chronic overexpression of the renin-angiotensin-aldosterone system stimulation of reduced nicotinamide adenine dinucleotide phosphate oxidase and cardiac remodeling. Endocrinology 148:3773–3780
109. Johar S, Cave AC, Narayanapanicker A et al (2006) Aldosterone mediates angiotensin II-induced interstitial cardiac fibrosis via a Nox2-containing NADPH oxidase. FASEB J 20:1546–1548
110. Looi YH, Grieve DJ, Siva A et al (2008) Involvement of Nox2 NADPH oxidase in adverse cardiac remodeling after myocardial infarction. Hypertension 51:319–325
111. Grieve DJ, Byrne JA, Siva A et al (2006) Involvement of the nicotinamide adenosine dinucleotide phosphate oxidase isoform Nox2 in cardiac contractile dysfunction occurring in response to pressure overload. J Am Coll Cardiol 47:817–826
112. Hinz B, Phan SH, Thannickal VJ et al (2007) The myofibroblast: one function, multiple origins. Am J Pathol 170:1807–1816
113. Cucoranu I, Clempus R, Dikalova A et al (2005) NAD(P)H oxidase 4 mediates transforming growth factor-beta1-induced differentiation of cardiac fibroblasts into myofibroblasts. Circ Res 97:900–907
114. Spurney CF, Knoblach S, Pistilli EE et al (2008) Dystrophin-deficient cardiomyopathy in mouse: expression of Nox4 and Lox are associated with fibrosis and altered functional parameters in the heart. Neuromuscul Disord 18:371–381
115. Heinecke JW (2001) Is the emperor wearing clothes? Clinical trials of vitamin E and the LDL oxidation hypothesis. Arterioscler Thromb Vasc Biol 21:1261–1264
116. Meagher EA, Barry OP, Lawson JA et al (2001) Effects of vitamin E on lipid peroxidation in healthy persons. J Am Med Assoc 285:1178–1182
117. Mashima R, Witting PK, Stocker R (2001) Oxidants and antioxidants in atherosclerosis. Curr Opin Lipidol 12:411–418

# Chapter 23
# Oxidative Risk Factors for Cardiovascular Disease in Women

**Manuela Gago-Dominguez, Xuejuan Jiang, and Jose Esteban Castelao**

**Abstract** Many risk factors that promote cardiovascular disease (CVD) have been identified. These include hypertension, hypercholesterolemia, diabetes, decreased estrogen in postmenopausal women, increased homocysteine, and cigarette smoking. It has recently become clear that a mechanism common to these risk factors is oxidative stress. CVD risk factors specific to women are parity, oophorectomy, preeclampsia, and menopause. There are several proposed mechanisms to explain these women-specific associations, such as reduced lifetime exposure to estrogen and insulin resistance, but the underlying mechanism is still unclear. One fact that did not receive much attention is the role of the oxidation hypothesis in these reproductive factors–CVD associations. In fact, pregnant, oophorectomized, and postmenopausal women exhibit higher levels of lipid peroxidation than nonpregnant, nonoophorectomized, and premenopausal women, respectively. We propose that the increased levels of lipid peroxidation during these states are responsible, at least in part, for their increased risk of CVD. This review extends the concept of the oxidation hypothesis of CVD to reproductive risk factors in women. It also addresses the potential role of oxidative stress in the hyperthyroidism-CVD relationship, as hyperthyroidism is a common disorder that most frequently occurs in women. We also discuss how screening human populations for reactive oxygen species (ROS) levels could help identify groups with a high level of ROS that may be at risk of developing CVD.

**Keywords** Lipid peroxidation · Oxidative stress · Cardiovascular disease · Reproductive factors · Thyroid diseases

M. Gago-Dominguez (✉)
Department of Preventive Medicine, USC/Norris Comprehensive Cancer Center, Keck School of Medicine of the University of Southern California, Los Angeles, CA 90089-9175, USA
e-mail: mgago@usc.edu

H. Sauer et al. (eds.), *Studies on Cardiovascular Disorders*,
Oxidative Stress in Applied Basic Research and Clinical Practice,
DOI 10.1007/978-1-60761-600-9_23, © Springer Science+Business Media, LLC 2010

## 23.1 Introduction

Epidemiological, clinical, and genetic studies indicate that elevated levels of low density lipoprotein (LDL) greatly increase the risk for atherosclerosis [1]. However, LDL fails to exert potentially atherogenic effects in vitro, suggesting that it must be modified to promote vascular disease. The hypothesis that oxidized LDL (oxLDL) is necessary, if not obligatory, in the development of atherosclerotic lesions was formulated more than 20 years ago with the seminal observation that uptake of native LDL by macrophages did not result in foam cell formation. In contrast, uptake of oxLDL via scavenger receptors resulted in the unregulated accumulation of lipid [2, 3]. Since then, multiple studies have provided firm evidence of an important role of oxLDL in atherogenesis [4–7]. (1) OxLDL has been isolated from human and animal atherosclerotic lesions [8, 9], where chemical and immunohistochemical studies have detected oxidized lipids [10, 11]. (2) A number of structurally unrelated antioxidants inhibit atherogenesis in animal models of hypercholesterolemia [12–14]. (3) A wide range of epidemiological studies has suggested that increased intake of dietary antioxidants lowers the risk of atherosclerosis [15, 16]. (4) a recent clinical study has finally established a causal connection between the levels of oxLDL and the risk of coronary artery disease [17]. Tsimikas et al. [17] found higher circulating levels of oxLDL to be strongly associated with angiographically documented coronary artery disease, especially in patients 60 years old or younger, suggesting that the atherogenicity of Lp(a) lipoprotein may be mediated in part by associated proinflammatory oxidized phospholipids [17].

## 23.2 Role of Lipid Peroxidation in the Epidemiology of CVD in Women

Women-specific risk factors for CVD are parity, oophorectomy, preeclampsia, and menopause. There are several proposed mechanisms to explain these associations, such as reduced lifetime exposure to estrogen and insulin resistance, but the underlying mechanism is still unclear. One fact that did not receive much attention is the role of the oxidation hypothesis in these reproductive factors–CVD associations. In fact, pregnant, oophorectomized, and post menopausal women exhibit higher levels of lipid peroxidation than nonpregnant, nonoophorectomized, and premenopausal women, respectively. We have proposed that the increased levels of lipid peroxidation during these states may be responsible, at least in part, for their increased risk of CVD [18]. This review thus extends the concept of the oxidation hypothesis of CVD to reproductive risk factors in women.

*Parity.* The association between parity and CVD risk in women has been assessed in a number of prospective studies [19–27], and the majority [19–21, 23, 25, 26] found a positive association. These studies have emphasized two possible biological mechanisms for this association. In the first, it is proposed that each pregnancy permanently “resets” ovarian function, leading to a reduced lifetime exposure to estrogen [28]. The established inverse association between parity and breast cancer, a known estrogen-dependent disease, provides support for this hypothesis [29]. However, the role of endogenous estrogen in the cause of CVD in women is unclear [30–32]. A second biological mechanism suggests that because normal pregnancy is a state of relative insulin resistance, repeated pregnancies may result in permanent detrimental effects on lipid and glucose metabolism [33–35].

Despite these proposed mechanisms, the underlying mechanism for the parity-CVD association is still unclear. We have proposed that increased lipid peroxidation in pregnant women may be a mechanism responsible, at least in part, for their increased risk of CVD [18]. Numerous studies have found statistically significantly higher levels of serum or plasma lipid peroxides in pregnant women compared with nonpregnant subjects [36, 37]. Triglyceride levels, which have been associated with lipid peroxidation in some studies [38], are consistently found to be higher in pregnant than nonpregnant women [37]. Studies on metabolic risk factors in pregnancy lend support to this hypothesis. Several studies have shown that normal pregnancy is associated with a shift towards small, dense LDL subfractions, along with raised triacylglycerols and cholesterol [35, 39, 40]. Clinical studies have established that individuals with a small, dense LDL subfraction profile are more likely to develop coronary heart diseases [35]. These small, dense LDL subfractions are more susceptible to oxidation than larger, lighter ones and, once oxidized, promote foam cell formation, initiate endothelial dysfunction, and thereby promote atherogenesis in a variety of ways [35].

*Hypertension during pregnancy/eclampsia.* Several studies have examined the relationship between hypertension during pregnancy/eclampsia and CVD incidence/mortality [41–46]. Jonsdottir et al. [41] found that death rates from ischemic heart disease were higher in women who had hypertension in pregnancy when compared with the general population, and that this risk might be linked to increasing severity of the disease in pregnancy. These authors examined causes of death in 374 women with a history of hypertensive complications in pregnancy and noted that their death rate from complications of coronary heart disease (standardized mortality ratio 1.47; 95% confidence interval (CI) 1.05–2.02) was higher than expected from analysis of population data from public health and census reports during corresponding periods. Moreover, they noted that the relative risk of dying from ischemic heart disease (risk ratio 2.61; 95% CI 1.11–6.12) was higher among women who had had eclampsia or preeclampsia (risk ratio 1.90; 95% CI 1.02–3.52) compared to those with hypertension alone [41].

A prospective cohort study using data from the Royal College of General Practitioners' oral contraceptive study also reported that a history of preeclampsia increased the risk of cardiovascular conditions in later life. For total ischemic heart disease the relative risk was 1.7 (95% CI 1.3–2.2) [42]. A retrospective cohort study from Scotland using hospital discharge data has also recently reported an association between preeclampsia and later ischemic heart disease in the mother (risk ratio 2.0; 95% CI 1.5–2.5) [43]. Two other cohort studies have independently indicated an increased risk of myocardial ischemia and related CVD later in life in women who had hypertensive disorder [44, 45].

In a recent case-control study, death with evidence of ischemic heart disease was also more common in women who had hypertension in pregnancy (24.3%) than in normotensive control women (14.6%) (risk ratio 1.66; 95% CI 1.27–2.17). There was a trend linking increasing severity of hypertensive disease in pregnancy with death rates from ischemic heart disease [46].

We have proposed the hypothesis that increased lipid peroxidation in women suffering from hypertension during pregnancy/preeclampsia may be one mechanism responsible, at least in part, for their increased CVD risk [18]. Growing evidence suggests that placental oxidative stress is involved in the etiopathogenesis of preeclampsia. Several studies have shown that women suffering from preeclampsia exhibit elevated levels of lipid peroxidation in the maternal circulation compared to women with a normal pregnancy and in preeclamptic placenta compared with normal placenta [36]. The placenta seems to play an important role in the generation of lipid peroxides secreted into the maternal circulation [36]. The mechanism by which the placenta is an important source of lipid peroxides is unknown, but it has been proposed to be related to the process of ischemia-reperfusion injury, which is a known source of oxidative stress. Reduced perfusion as a result of abnormal placentation leads to ischemia-reperfusion injury to the placenta. Placental oxidative stress, which results from ischemia-reperfusion injury, may play a role in the etiology of preeclampsia by promoting lipid peroxidation and alterations in endothelial cells [47].

*Oophorectomy.* Several early autopsy studies showed an increase in coronary artery disease in young women who had an oophorectomy [30, 48]. In 1963, Ritterband and colleagues [49] reported that surgically intact women had less heart disease than women who had bilateral oophorectomy or a hysterectomy only, but there was no difference in CHD rates in oophorectomized vs. hysterectomized women. In a meta-analysis of observational studies [50], bilateral oophorectomy was found to increase the risk of CVD by more than two-fold (relative risk 2.62; 95% CI 2.05–3.35). The Women's Health Initiative Observational Study also observed higher prevalence and incidence of CVD in women with a hysterectomy, regardless of oophorectomy status [51]; however, no difference in CVD risk was observed between women with hysterectomy only and women with hysterectomy plus oophorectomy. The Nurses' Health Study evaluated the long-term health outcomes of ovarian conservation at the time of hysterectomy [52]. Compared with ovarian conservation, bilateral oophorectomy at the time of hysterectomy for benign diseases was associated with increased risk of incident fatal and nonfatal coronary heart disease (CHD) (multivariable hazard ratio 1.17; 95% CI 1.02–1.35) and death from CHD (1.28; 95% CI 1.00–1.64). Among those who never used estrogen therapy, oophorectomy before age 50 years was associated with an even higher risk of incident CHD (1.98; 95% CI 1.18–3.32).

We have proposed that the increased lipid peroxidation in oophorectomized women may be the mechanism responsible, at least in part, for their increased risk of CVD [18]. An increase in the level of lipid peroxides after bilateral oophorectomy has been found in serum or plasma of women [53] and female rodents [54–56]. Lipid peroxide levels in the serum of seven premenopausal women who had undergone bilateral oophorectomy were normal before the operation, but were increased 15, 30, and 60 days after the operation [36].

*Menopause.* Premenopausal women have a decreased risk of heart disease compared with men of a comparable age [57], but the protection disappears when the women go through menopause. After menopause the risk of CVD in women increases, such that by 10 years postmenopause the risk in women is similar to that in men [58]. This is true even when they have the same risk factors as men, including smoking, high blood cholesterol levels, and a family history of heart disease. This increased risk of CVD occurs, at least in part, because of changes in serum lipid and plasma fibrinogen levels following menopause [57]. Although some of these changes may be the result of aging and not solely attributable to menopause, there is general agreement that menopause has a negative impact on these CVD risk factors [57].

We have proposed that increased lipid peroxidation in postmenopausal women may be the mechanism responsible, at least in part, for their increased risk of CVD [18]. In parallel to their CVD risk, women have statistically lower levels of lipid peroxidation than men [59–61] and premenopausal women have lower levels than postmenopausal women [36]. Several studies (one with a sample size of over 900 women) comparing serum lipid peroxide levels of normal pre- and postmenopausal women consistently found that the levels in postmenopausal women were higher than those in premenopausal women [36]. Thus, it is possible that increased lipid peroxidation in postmenopausal women is the mechanism responsible, at least partially, for their increased risk of CVD.

*Hyperthyroidism* is a common disorder that most frequently occurs in women in their middle decades (about eight times more than in men). It has a prevalence of 2% in females and 0.2% in males in iodine replete areas such as the United Kingdom and the United States [62]. Hyperthyroidism has a major influence on the cardiovascular system and is known to induce many cardiovascular effects, such as sinus tachycardia, systolic hypertension, changes in ventricular systolic and diastolic function, and predisposition to dysrhythmias, especially atrial fibrillation [63].

A number of epidemiological studies have investigated the association of hyperthyroidism with cardiovascular mortality [64–68]. A follow-up study of 10,552 Swedish hyperthyroid patients treated with radioiodine [64] found an overall standardized mortality ratio of 1.47 (95% CI 1.43–1.51), and CVD was the most common cause of death (1.65; 95% CI 1.59–1.71). Similarly, a cohort study [65] of 7,209 British subjects with hyperthyroidism treated with radioiodine also identified marked excess in mortality from all causes. The excess mortality was largely accounted for by an excess of deaths because of circulatory diseases, both cardiovascular (standardized mortality ratio 1.2, 95% CI 1.2–1.3) and cerebrovascular (1.4; 95% CI 1.2–1.5) [65]. Increased risks of both all-cause mortality and CVD mortality were also observed in follow-up studies conducted in the United States [68] and Finland [67].

Subclinical hyperthyroidism has been found to be an independent risk factor for the subsequent development of atrial fibrillation [69]. In a follow-up study of 3,121 cardiac patients, subclincal hyperthyroidism was associated with a higher risk of cardiac death compared with euthyroidism [70]. Subjects with subclinical hyperthyroidism have been shown to have an increased vascular mortality when followed over a 10-year period [71]. Thyroid hormones are known to have both direct and indirect effects on the myocardium, to affect the autonomic nervous system, and to predispose to a number of arrhythmias [72–75].

We have proposed that increased lipid peroxidation in hyperthyroid patients may be the mechanism responsible, at least in part, for their increased risk of CVD [18]. We have reviewed the evidence linking lipid peroxidation and thyroid diseases in detail in a recent manuscript [76] and will summarize it next. Multiple investigations in humans and experimental animals have reported that the rate of lipid peroxidation induced by free oxygen radicals increases in hyperthyroidism [77–83] but is suppressed in hypothyroidism [77, 84–86], although evidence for the latter is more limited. In one study, for example, LDL oxidation was markedly higher in hyperthyroidism than in hypothyroidism or control subjects [83]. In that study, involving 16 patients with hyperthyroidism, 16 with hypothyroidism, and 16 age- and sex-matched healthy normolipidemic control subjects [83], hyperthyroidism was associated with significantly higher lipid peroxidation, as characterized by several lipid peroxidation markers such as a higher native LDL content in lipid peroxides, a lower lag phase, and a higher oxidation rate than in the other two groups. Several studies have also shown that thyroid hormones appear to exert a pro-oxidant activity in target cells and increase lipid peroxidation [87–90].

## 23.3 Summary

In this review, we have extended the concept of the oxidation hypothesis of CVD from established nonreproductive CVD risk factors such as hypertension, hypercholesterolemia, diabetes, increased homocysteine, smoking, decreased alcohol intake; to reproductive factors, i.e., parity, pregnancy-induced hypertension, oophorectomy, and menopause. We have also extended the oxidation hypothesis of CVD to hyperthyroidism, a condition potentially associated with increased risk of CVD. In addition, there are also studies that support the notion of ethnic differences in lipid peroxidation and oxidative stress generation. In adults with type 2 diabetes, for example, increased levels of oxidative stress were found in African-Caribbeans compared to Caucasians [91]. In another study, lipid peroxidation was found to be higher in African Americans compared to Caucasians during hyperlipidemia induced by lipid infusion [92].

Screening human populations for ROS levels could help identify groups with a high level of ROS that are at risk of developing CVD. The human population is heterogeneous regarding ROS levels and lipid peroxidation generation. People who overgenerate ROS may be at high risk for developing CVD because of the oxidative damage to cell constituents (DNA, proteins, lipids, etc.) and cell structures [93]. Intake of exogenous antioxidants (vitamins E, C, beta-carotene. and others) could protect against CVD in people with innate or acquired high levels of ROS. It is important to point out that baseline levels of ROS were not taken into account in the clinical trials that did not find a protective effect of antioxidants in CVD. Although the results of the first prospective trial with CVD as an end point were encouraging [94], subsequent trials did not find a protective effect [95–97].

**Acknowledgments** Reprinted from Castelao J.E. and Gago-Dominguez, M. "Risk factors for cardiovascular disease in women: relationship to lipid peroxidation and oxidative stress." Medical Hypotheses 71:39–44; copyright (2008) with permission from Elsevier. This study was supported in part by U.S. National Institutes of Health grant R01CA114472.

## References

1. Brown MS, Goldstein JL (1992) Koch's postulates for cholesterol. Cell 71:187–188
2. Tsimikas S, Witztum JL (2001) Measuring circulating oxidizety lipoprotein that increase its atherogenicity. Circulation 103:1930–1932
3. Steinberg D, Parthasarathy S, Carew TE, Khoo JC, Witztum JL (1989) Beyond cholesterol. Modifications of low-density lipoprotein to evaluate coronary risk. N Engl J Med 320: 915–924
4. Esterbauer H, Gebicki J, Puhl H, Jurgens G (1992) The role of lipid peroxidation and antioxidants in oxidative modification of LDL. Free Radic Biol Med 13:341–390
5. Berliner JA, Heinecke JW (1996) The role of oxidized lipoproteins in atherogenesis. Free Radic Biol Med 20:707–727
6. Steinbrecher UP, Parthasarathy S, Leake DS, Witztum JL, Steinberg D (1984) Modification of low density lipoprotein by endothelial cells involves lipid peroxidation and degradation of low density lipoprotein phospholipids. Proc Natl Acad Sci U S A 81:3883–3887
7. Heinecke JW (2002) Oxidized amino acids: culprits in human atherosclerosis and indicators of oxidative stress. Free Radic Biol Med 32:1090–1101
8. Daugherty A, Zweifel BS, Sobel BE, Schonfeld G (1988) Isolation of low density lipoprotein from atherosclerotic vascular tissue of Watanabe heritable hyperlipidemic rabbits. Arteriosclerosis 8:768–777
9. Yla-Herttuala S, Palinski W, Rosenfeld ME, Parthasarathy S, Carew TE, Butler S, Witztum JL, Steinberg D (1989) Evidence for the presence of oxidatively modified low density lipoprotein in atherosclerotic lesions of rabbit and man. J Clin Invest 84:1086–1095
10. Haberland ME, Fong D, Cheng L (1988) Malondialdehyde-altered protein occurs in atheroma of Watanabe heritable hyperlipidemic rabbits. Science 241:215–218
11. Rosenfeld ME, Palinski W, Yla-Herttuala S, Butler S, Witztum JL (1990) Distribution of oxidation specific lipid-protein adducts and apolipoprotein B in atherosclerotic lesions of varying severity from WHHL rabbits. Arteriosclerosis 10:336–349
12. Heinecke JW (1998) Oxidants and antioxidants in the pathogenesis of atherosclerosis: implications for the oxidized low density lipoprotein hypothesis. Atherosclerosis 141:1–15
13. Diaz MN, Frei B, Vita JA, Keaney JF Jr (1997) Antioxidants and atherosclerotic heart disease. N Engl J Med 337:408–416
14. Chisolm GM 3rd, Hazen SL, Fox PL, Cathcart MK (1999) The oxidation of lipoproteins by monocytes-macrophages. Biochemical and biological mechanisms. J Biol Chem 274: 25959–25962
15. Stampfer MJ, Hennekens CH, Manson JE, Colditz GA, Rosner B, Willett WC (1993) Vitamin E consumption and the risk of coronary disease in women. N Engl J Med 328: 1444–1449
16. Rimm EB, Stampfer MJ, Ascherio A, Giovannucci E, Colditz GA, Willett WC (1993) Vitamin E consumption and the risk of coronary heart disease in men. N Engl J Med 328: 1450–1456

17. Tsimikas S, Brilakis ES, Miller ER, McConnell JP, Lennon RJ, Kornman KS, Witztum JL, Berger PB (2005) Oxidized phospholipids, Lp(a) lipoprotein, and coronary artery disease. N Engl J Med 353:46–57
18. Castelao JE, Gago-Dominguez M (2008) Risk factors for cardiovascular disease in women: relationship to lipid peroxidation and oxidative stress. Med Hypotheses 71:39–44
19. Dekker JM, Schouten EG (1993) Number of pregnancies and risk of cardiovascular disease. N Engl J Med 329:1893–1894; author reply 4–5
20. Ness RB, Harris T, Cobb J, Flegal KM, Kelsey JL, Balanger A, Stunkard AJ, D'Agostino RB (1993) Number of pregnancies and the subsequent risk of cardiovascular disease. N Engl J Med 328:1528–1533
21. Green A, Beral V, Moser K (1988) Mortality in women in relation to their childbearing history. BMJ 297:391–395
22. Ness RB, Schotland HM, Flegal KM, Shofer FS (1994) Reproductive history and coronary heart disease risk in women. Epidemiol Rev 16:298–314
23. Qureshi AI, Giles WH, Croft JB, Stern BJ (1997) Number of pregnancies and risk for stroke and stroke subtypes. Arch Neurol 54:203–206
24. Colditz GA, Willett WC, Stampfer MJ, Rosner B, Speizer FE, Hennekens CH (1987) A prospective study of age at menarche, parity, age at first birth, and coronary heart disease in women. Am J Epidemiol 126:861–870
25. Haynes SG, Feinleib M (1980) Women, work and coronary heart disease: prospective findings from the Framingham Heart Study. Am J Public Health 70:133–141
26. Mant D, Villard-Mackintosh L, Vessey MP, Yeates D (1987) Myocardial infarction and angina pectoris in young women. J Epidemiol Community Health 41:215–219
27. Steenland K, Lally C, Thun M (1996) Parity and coronary heart disease among women in the American Cancer Society CPS II population. Epidemiology 7:641–643
28. Bernstein L, Pike MC, Ross RK, Judd HL, Brown JB, Henderson BE (1985) Estrogen and sex hormone-binding globulin levels in nulliparous and parous women. J Natl Cancer Inst 74:741–745
29. McPherson K, Steel CM, Dixon JM (2000) ABC of breast diseases. Breast cancer-epidemiology, risk factors, and genetics. BMJ 321:624–628
30. Barrett-Connor E (1997) Sex differences in coronary heart disease. Why are women so superior? The 1995 Ancel Keys Lecture. Circulation 95:252–264
31. Lawlor DA, Ebrahim S, Davey Smith G (2001) Sex matters: secular and geographical trends in sex differences in coronary heart disease mortality. BMJ 323:541–545
32. Lawlor DA, Ebrahim S, Davey Smith G (2002) Role of endogenous oestrogen in aetiology of coronary heart disease: analysis of age related trends in coronary heart disease and breast cancer in England and Wales and Japan. BMJ 325:311–312
33. Sattar N, Greer IA (2002) Pregnancy complications and maternal cardiovascular risk: opportunities for intervention and screening? BMJ 325:157–160
34. Greer IA (1999) Thrombosis in pregnancy: maternal and fetal issues. Lancet 353:1258–1265
35. Martin U, Davies C, Hayavi S, Hartland A, Dunne F (1999) Is normal pregnancy atherogenic? Clin Sci (Lond) 96:421–425
36. Gago-Dominguez M, Castelao J, Pike MC, Sevanian A, Haile RW (2005) Role of lipid peroxidation in the epidemiology and prevention of breast cancer. Cancer Epidemiol Biomarkers Prev 14:2829–2839
37. Sarandol E, Dirican M, Serdar Z (2004) Oxidizability of apolipoprotein B-containing lipoproteins, levels of lipid peroxidation products and antioxidants in normal pregnancy. Arch Gynecol Obstet 270:157–160
38. Rumley AG, Woodward M, Rumley A, Rumley J, Lowe GD (2004) Plasma lipid peroxides: relationships to cardiovascular risk factors and prevalent cardiovascular disease. QJM 97: 809–816
39. Sattar N, Greer IA, Louden J, Lindsay G, McConnell M, Shepherd J, Packard CJ (1997) Lipoprotein subfraction changes in normal pregnancy: threshold effect of plasma

triglyceride on appearance of small, dense low density lipoprotein. J Clin Endocrinol Metab 82:2483–2491
40. Hubel CA, Shakir Y, Gallaher MJ, McLaughlin MK, Roberts JM (1998) Low-density lipoprotein particle size decreases during normal pregnancy in association with triglyceride increases. J Soc Gynecol Investig 5:244–250
41. Jonsdottir LS, Arngrimsson R, Geirsson RT, Sigvaldason H, Sigfusson N (1995) Death rates from ischemic heart disease in women with a history of hypertension in pregnancy. Acta Obstet Gynecol Scand 74:772–776
42. Hannaford P, Ferry S, Hirsch S (1997) Cardiovascular sequelae of toxaemia of pregnancy. Heart 77:154–158
43. Smith GC, Pell JP, Walsh D (2001) Pregnancy complications and maternal risk of ischaemic heart disease: a retrospective cohort study of 129,290 births. Lancet 357:2002–2006
44. Irgens HU, Reisaeter L, Irgens LM, Lie RT (2001) Long term mortality of mothers and fathers after pre-eclampsia: population based cohort study. BMJ 323:1213–1217
45. Wilson BJ, Watson MS, Prescott GJ, Sunderland S, Campbell DM, Hannaford P, Smith WC (2003) Hypertensive diseases of pregnancy and risk of hypertension and stroke in later life: results from cohort study. BMJ 326:845–849
46. Arnadottir GA, Geirsson RT, Arngrimsson R, Jonsdottir LS, Olafsson O (2005) Cardiovascular death in women who had hypertension in pregnancy: a case-control study. BJOG 112:286–292
47. Gupta S, Agarwal A, Sharma RK (2005) The role of placental oxidative stress and lipid peroxidation in preeclampsia. Obstet Gynecol Surv 60:807–816
48. Barrett-Connor E, Bush TL (1991) Estrogen and coronary heart disease in women. J Am Med Assoc 265:1861–1867
49. Ritterband AB, Jaffe IA, Densen PM, Magagna JF, Reed E (1963) Gonadal function and the development of coronary heart disease. Circulation 27:237–251
50. Atsma F, Bartelink ML, Grobbee DE, van der Schouw YT (2006) Postmenopausal status and early menopause as independent risk factors for cardiovascular disease: a meta-analysis. Menopause 13:265–279
51. Howard BV, Kuller L, Langer R, Manson JE, Allen C, Assaf A, Cochrane BB, Larson JC, Lasser N, Rainford M, Van Horn L, Stefanick ML et al. (2005) Risk of cardiovascular disease by hysterectomy status, with and without oophorectomy: the Women's Health Initiative Observational Study. Circulation 111:1462–1470
52. Parker WH, Broder MS, Chang E, Feskanich D, Farquhar C, Liu Z, Shoupe D, Berek JS, Hankinson S, Manson JE (2009) Ovarian conservation at the time of hysterectomy and long-term health outcomes in the Nurses' Health Study. Obstet Gynecol 113:1027–1037
53. Asada Y, Komura S, Ohishi N, Kasugai M, Suganuma N, Mizutani S, Tomoda Y, Yagi K (1990) Effect of ovariectomy on serum lipid peroxide level in women. J Clin Biochem Nutr 8:247–252
54. Yagi K (1997) Female hormones act as natural antioxidants – a survey of our research. Acta Biochim Pol 44:701–709
55. Love RR, Philips J (2002) Oophorectomy for breast cancer: history revisited. J Natl Cancer Inst 94:1433–1434
56. Hernandez I, Delgado JL, Diaz J, Quesada T, Teruel MJ, Llanos MC, Carbonell LF (2000) 17beta-estradiol prevents oxidative stress and decreases blood pressure in ovariectomized rats. Am J Physiol Regul Integr Comp Physiol 279(5):R1599–R1605
57. Haddock BL, Marshak HP, Mason JJ, Blix G (2000) The effect of hormone replacement therapy and exercise on cardiovascular disease risk factors in postmenopausal women. Sports Med 29:39–49
58. Kafonek SD (1994) Postmenopausal hormone replacement therapy and cardiovascular risk reduction: a review. Drugs 47(Suppl 2):16–24
59. Dincer Y, Ozen E, Kadioglu P, Hatemi H, Akcay T (2001) Effect of sex hormones on lipid peroxidation in women with polycystic ovary syndrome, healthy women, and men. Endocr Res 27:309–316

60. Ide T, Tsutsui H, Ohashi N, Hayashidani S, Suematsu N, Tsuchihashi M, Tamai H, Takeshita A (2002) Greater oxidative stress in healthy young men compared with premenopausal women. Arterioscler Thromb Vasc Biol 22:438–442
61. Block G, Dietrich M, Norkus EP, Morrow JD, Hudes M, Caan B, Packer L (2002) Factors associated with oxidative stress in human populations. Am J Epidemiol 156:274–285
62. Tunbridge WM, Evered DC, Hall R, Appleton D, Brewis M, Clark F, Evans JG, Young E, Bird T, Smith PA (1977) The spectrum of thyroid disease in a community: the Whickham survey. Clin Endocrinol (Oxf) 7:481–493
63. Klein I, Ojamaa K (2001) Thyroid hormone and the cardiovascular system. N Engl J Med 344:501–509
64. Hall P, Lundell G, Holm LE (1993) Mortality in patients treated for hyperthyroidism with iodine-131. Acta Endocrinol (Copenh) 128:230–234
65. Franklyn JA, Maisonneuve P, Sheppard MC, Betteridge J, Boyle P (1998) Mortality after the treatment of hyperthyroidism with radioactive iodine. N Engl J Med 338:712–718
66. Franklyn JA, Sheppard MC, Maisonneuve P (2005) Thyroid function and mortality in patients treated for hyperthyroidism. J Am Med Assoc 294:71–80
67. Metso S, Auvinen A, Salmi J, Huhtala H, Jaatinen P (2008) Increased long-term cardiovascular morbidity among patients treated with radioactive iodine for hyperthyroidism. Clin Endocrinol (Oxf) 68:450–457
68. Bauer DC, Rodondi N, Stone KL, Hillier TA (2007) Thyroid hormone use, hyperthyroidism and mortality in older women. Am J Med 120:343–349
69. Sawin CT, Geller A, Wolf PA, Belanger AJ, Baker E, Bacharach P, Wilson PW, Benjamin EJ, D'Agostino RB (1994) Low serum thyrotropin concentrations as a risk factor for atrial fibrillation in older persons. N Engl J Med 331:1249–1252
70. Iervasi G, Molinaro S, Landi P, Taddei MC, Galli E, Mariani F, L'Abbate A, Pingitore A (2007) Association between increased mortality and mild thyroid dysfunction in cardiac patients. Arch Intern Med 167:1526–1532
71. Parle JV, Maisonneuve P, Sheppard MC, Boyle P, Franklyn JA (2001) Prediction of all-cause and cardiovascular mortality in elderly people from one low serum thyrotropin result: a 10-year cohort study. Lancet 358:861–865
72. Dorr M, Wolff B, Robinson DM, John U, Ludemann J, Meng W, Felix SB, Volzke H (2005) The association of thyroid function with cardiac mass and left ventricular hypertrophy. J Clin Endocrinol Metab 90:673–677
73. Biondi B, Palmieri EA, Fazio S, Cosco C, Nocera M, Sacca L, Filetti S, Lombardi G, Perticone F (2000) Endogenous subclinical hyperthyroidism affects quality of life and cardiac morphology and function in young and middle-aged patients. J Clin Endocrinol Metab 85:4701–4705
74. Kahaly GJ, Nieswandt J, Mohr-Kahaly S (1998) Cardiac risks of hyperthyroidism in the elderly. Thyroid 8:1165–1169
75. Osman F, Gammage MD, Sheppard MC, Franklyn JA (2002) Clinical review 142: cardiac dysrhythmias and thyroid dysfunction: the hidden menace? J Clin Endocrinol Metab 87: 963–967
76. Gago-Dominguez M, Castelao JE (2008) Role of lipid peroxidation and oxidative stress in the association between thyroid diseases and breast cancer. Crit Rev Oncol Hematol 68:107–114
77. Pereira B, Rosa LF, Safi DA, Bechara EJ, Curi R (1994) Control of superoxide dismutase, catalase and glutathione peroxidase activities in rat lymphoid organs by thyroid hormones. J Endocrinol 140:73–77
78. Seven A, Tasan E, Hatemi H, Burcak G (1999) The impact of propylthiouracil therapy on lipid peroxidation and antioxidant status parameters in hyperthyroid patients. Acta Medica Okayama 53:27–30
79. Dirican M, Tas S (1999) Effects of vitamin E and vitamin C supplementation on plasma lipid peroxidation and on oxidation of apolipoprotein B-containing lipoproteins in experimental hyperthyroidism. J Med Investig 46:29–33

80. Bianchi G, Solaroli E, Zaccheroni V, Grossi G, Bargossi AM, Melchionda N, Marchesini G (1999) Oxidative stress and anti-oxidant metabolites in patients with hyperthyroidism: effect of treatment. Horm Metab Res 31:620–624
81. Adali M, Inal-Erden M, Akalin A, Efe B (1999) Effects of propylthiouracil, propranolol, and vitamin E on lipid peroxidation and antioxidant status in hyperthyroid patients. Clin Biochem 32:363–367
82. Konukoglu D, Yelke HK, Hatemi H, Sabuncu T (2001) Effects of oxidative stress on the erythrocyte Na+,K+ ATPase activity in female hyperthyroid patients. J Toxicol Environ Health A 63:289–295
83. Costantini F, Pierdomenico SD, De Cesare D, De Remigis P, Bucciarelli T, Bittolo-Bon G, Cazzolato G, Nubile G, Guagnano MT, Sensi S, Cuccurullo F, Mezzetti A (1998) Effect of thyroid function on LDL oxidation. Arterioscler Thromb Vasc Biol 18:732–737
84. Chehade J, Kim J, Pinnas JL, Mooradian AD (1999) Age-related changes in the thyroid hormone effects on malondialdehyde-modified proteins in the rat heart. Proc Soc Exp Biol Med 222:59–64
85. Chehade J, Kim J, Pinnas JL, Mooradian AD (1999) Malondialdehyde binding of rat cerebral proteins is reduced in experimental hypothyroidism. Brain Res 829:201–203
86. Paller MS (1986) Hypothyroidism protects against free radical damage in ischemic acute renal failure. Kidney Int 29:1162–1166
87. Tapia G, Pepper I, Smok G, Videla LA (1997) Kupffer cell function in thyroid hormone-induced liver oxidative stress in the rat. Free Radic Res 26:267–279
88. Morini P, Casalino E, Sblano C, Landriscina C (1991) The response of rat liver lipid peroxidation, antioxidant enzyme activities and glutathione concentration to the thyroid hormone. Int J Biochem 23:1025–1030
89. Asayama K, Dobashi K, Hayashibe H, Kato K (1989) Vitamin E protects against thyroxine-induced acceleration of lipid peroxidation in cardiac and skeletal muscles in rats. J Nutr Sci Vitaminol (Tokyo) 35:407–418
90. Kose K, Terzi S, Dogan P (1997) The relationship between high plasma thyroid hormone (T4, T3) levels and oxidative damage. Ann Med Sci 6:102–106
91. Mehrotra S, Ling KL, Bekele Y, Gerbino E, Earle KA (2001) Lipid hydroperoxide and markers of renal disease susceptibility in African-Caribbean and Caucasian patients with Type 2 diabetes mellitus. Diabet Med 18:109–115
92. Lopes HF, Morrow JD, Stoijiljkovic MP, Goodfriend TL, Egan BM (2003) Acute hyperlipidemia increases oxidative stress more in African Americans than in white Americans. Am J Hypertens 16:331–336
93. Salganik RI (2001) The benefits and hazards of antioxidants: controlling apoptosis and other protective mechanisms in cancer patients and the human population. J Am Coll Nutr 20: 464S–472S
94. Stephens NG, Parsons A, Schofield PM, Kelly F, Cheeseman K, Mitchinson MJ (1996) Randomised controlled trial of vitamin E in patients with coronary disease: Cambridge Heart Antioxidant Study (CHAOS). Lancet 347:781–786
95. GISSI-Prevenzione Investigators (Gruppo Italiano per lo Studio della Sopravvivenza nell'Infarto miocardico) (1999) Dietary supplementation with n-3 polyunsaturated fatty acids and vitamin E after myocardial infarction: results of the GISSI-Prevenzione trial. Lancet 354:447–455
96. 96.The Alpha-TocopherolBCCPSG (1994) The effect of vitamin E and beta carotene on the incidence of lung cancer and other cancers in male smokers. N Engl J Med 330:1029–1035
97. Yusuf S, Dagenais G, Pogue J, Bosch J, Sleight P (2000) Vitamin E supplementation and cardiovascular events in high-risk patients. The heart outcomes prevention evaluation study investigators. N Engl J Med 342:154–160

# Chapter 24
# Protective Effects of Food on Cardiovascular Diseases

**Alfonso Giovane and Claudio Napoli**

**Abstract** Experimental and epidemiological evidence have been accumulated in the last decades demonstrating a stringent correlation between nutrition lifestyle and chronic-inflammatory diseases like cardiovascular diseases (CVD) and cancer. It is now agreed that the incidence of these diseases can be reduced by diet. The French paradox [1], or the Mediterranean diet [2], has provided a scientific explanation, namely that the antioxidants contained in red wine, vegetables, fruits, and olive oil have been found to have a beneficial effect on cellular oxidative stress. Thus food is now considered not only a resource to satisfy a caloric requirement but a way to acquire micronutrients with helpful consequences for our lifespan. Terms like "nutraceutical" and "functional food" have been coined to better describe the contribution of food to our health.

**Keyword** Healthy foods

## 24.1 Oxidative Stress

Reactive oxygen species (ROS) and other radical species are considered to play an important role in the pathogenesis of several diseases, including coronary heart disease (CHD), stroke, cancer, and shock, as well as aging [3–5]. The main sources of ROS in the cell are mitochondria, NAD(P)H oxidase, and cytochrome P450 enzymes. Cells possess an efficient molecular mechanism to operate a stringent control on the intracellular ROS level and to maintain the balance between oxidant and antioxidant molecules. Superoxide, hydrogen peroxide, and the hydroxyl radical, under normal conditions are scavenged by antioxidant enzymes, including catalase, superoxide dismutase, and glutathione peroxidase, or by antioxidant compounds like vitamin E, vitamin C, carotenoids, and polyphenols acquired with food.

---

A. Giovane (✉)
Department of Biochemistry and Biophysics, 1st School of Medicine, Second University of Naples, Naples, Italy
e-mail: alfonso.giovane@unina2

H. Sauer et al. (eds.), *Studies on Cardiovascular Disorders*,
Oxidative Stress in Applied Basic Research and Clinical Practice,
DOI 10.1007/978-1-60761-600-9_24, 

A cellular oxidative stress condition is determined by an imbalance between the generation of ROS and the antioxidant defence capacity of the cell, and can affect major cellular components, including lipids, proteins, and DNA [6]. Proteins are important targets of cellular redox, which can modulate their function through conformational changes. ROS can also influence gene expression profiles by affecting intracellular signal transduction pathways.

A fundamental role in oxidative stress is played by nitric oxide synthase (NOS), which catalyses the production of nitric oxide (NO) starting from L-arginine and NADPH. This enzyme is expressed in several isoforms, among which the endothelial isoform (eNOS) has been studied largely for its involvement in endothelial dysfunction. eNOS activity is finely regulated by its cofactor tetrahydrobiopterin (BH4), in the absence of which eNOS can generate the superoxide anion ($O_2^-$) rather than NO. Since BH4 is produced from GTP through a pathway whose rate-limiting enzyme is GTP cyclohydrolase, this enzyme can control NO or $O_2^-$ production by altering the BH4 concentration. Thus, when BH4 concentration is reduced, eNOS produces $O_2^-$ and ROS production is increased. Several enzymes like NADPH oxidase, xanthine oxidase, myeloperoxidase, cyclooxygenase, and mitochondria can reduce NO availability by increasing concentration of $O_2^-$, which converts NO into peroxynitrite. Recently [7] it has been demonstrated that the switch of eNOS to $O_2^-$ production is determined by the eNOS/BH4 stoichiometry, together with the intracellular ratio between tetrahydrobiopterin and its oxidative form dihydrobiopterin, rather than absolute concentrations of BH4, even in the absence of exogenous oxidative stress. As a consequence of these findings [8], the plasma tetrahydrobiopterin/dihydrobiopterin ratio has been proposed as a possible marker for the endothelial dysfunction.

NO is also identified [9] as a physiologic regulator of electron transfer and ATP synthesis by inhibiting cytochrome oxidase. Moreover, NO stimulates the mitochondrial production of $O_2$ active species, primarily $O_2^-$ and $H_2O_2$; and, depending on NO matrix concentration, of peroxynitrite, which is responsible for the nitrosylation and nitration of the mitochondrial components. Circulating low density lipoprotein (LDL) carrying cholesterol into the blood stream can be oxidized by ROS to oxidized LDL (LDLox), which is thought to play a pathogenic role in atherosclerosis leading to heart attack and ischemic stroke [10, 11]. More importantly, there is evidence of LDLox in early atherosclerotic lesions during human fetal development [12] and early infancy [13]. Generally, a very complex interaction occurs, regarding cholesterolemia, between maternal and fetal circulation during pregnancy (reviewed in [14] and [15]). Antioxidant nutrients are supposed to reduce the atherosclerosis progression because of their ability to inhibit the damage produced by the oxidative processes [3, 16–18]. Advanced forms of the CVD and CHD in adults can be explained by the "fetal origins" hypothesis of atherogenesis, which suggests that conditions, most likely nutritional and genetic, "program" the fetus for the development of these diseases [19–21]. The role of birth weight is difficult to understand except as a proxy for events in intrauterine life. However, lifestyle factors during adulthood make much greater contributions than birth weight with respect to risk of CVD.

## 24.2 L-arginine

Several antioxidants have shown beneficial effects against atherosclerosis and CHD in experimental models [18, 22–30]. Recently, diets containing vitamins E and C and L-arginine, the natural precursor of NO, have shown helpful effects in perturbed shear stress–induced atherosclerosis [31, 32] by enhancing the protection afforded by moderate physical exercise [33]. LDLox may induce a decreased uptake of L-arginine [34]. The decrease of the L-arginine pool may uncouple eNOS, increasing the production of superoxide radicals from oxygen, the cosubstrate of eNOS. Interestingly, glycoxidized LDL downregulates eNOS in human coronary cells [35]. Several epidemiological studies have shown that the risk of CHD can be reduced by consumption of antioxidant vitamins, such as vitamin E and ß-carotene [3, 36, 37]. Furthermore, some clinical trials also suggest a reduced risk of CHD with vitamin E [38]. However, some large-scale trials in humans have failed to confirm the protective effect of L-arginine or β-carotene and have produced questionable results with vitamin E. In the Heart Outcomes Prevention Evaluation (HOPE) Study [39, 40], and in the Heart Protection Study [41], diets supplemented with vitamin E did not result in any beneficial effects on cardiovascular events. Another randomized clinical trial on Vascular Interaction with Age in Myocardial Infarction (VINTAGE MI) demonstrated that L-arginine added to standard postinfarction therapies does not improve vascular stiffness measurements or ejection fraction, and may be associated with higher postinfarction mortality [42]. More likely, antioxidants may affect the progression of long-term lesions but not necessarily modulate the development of preexisting atherosclerotic lesions (i.e., cerebrovascular disease, peripheral arterial disease, or CHD) or reduce the clinical manifestations of plaque rupture [3]. Thus, to study the effective role of antioxidants in the inhibition of atherosclerosis, it is necessary to investigate the progression of early atherosclerotic lesions in young adults. Such intervention may also prevent proatherogenic events during fetal development [14].

## 24.3 Lycopene

Lycopene is a carotenoid naturally present in tomato and is one of the dietary antioxidants that has received much attention recently [43, 44]. An inverse relationship has been found between the intake of tomatoes and the lycopene distribution in serum and adipose tissue and the incidence of CHD [45, 46]. A number of in vitro studies show that lycopene can protect native LDL from oxidation and can suppress cholesterol synthesis [47]. However, ß-carotene but not lycopene inhibited the oxidation of LDL in endothelial cells [48]. One of the earlier studies investigating the relationship between myocardial infarctions and serum antioxidant status, including lycopene [49], reported an odds ratio of 0.75. A multicenter case-control study (EURAMIC) [45] indicated that only lycopene levels, and not β-carotene levels, were found to be protective with an odds ratio of 0.52 for the contrast of the 10th and 90th percentiles, with a P value of 0.005. A component of this EURAMIC study was further analyzed [50]. In this case-control study of adipose tissue, lycopene levels showed an odds ratio of 0.39. In another study on Atherosclerosis Risk in Communities (ARIC), 231 cases and an equal number of control subjects showed an odds ratio of 0.81 when fasting serum antioxidant levels were assessed in relationship to the intima-media thickness [51]. In an additional cross-sectional study comparing Lithuanian and Swedish populations having diverging mortality rates from CHD, lower lycopene levels in the blood were found to be associated with increased risk and mortality from CHD [52]. Furthermore, in an Austrian study on stroke prevention, lower levels of lycopene and α-tocopherol in the serum were reported in individuals from an elderly population at high risk for cerebral damage [53]. However, although the epidemiological studies conducted so far provide persuasive evidence for the role of lycopene in prevention of CHD, a causal relationship between lycopene intake and the risk of CHD is only suggestive and still far from confirmed. Such proof can be obtained only by performing controlled studies on clinical dietary intervention in which both the biomarkers of the oxidative stress status and the disease are calculated. There is a need to recognize genetic determinants or biomarkers that can predict which individual is at highest risk for chronic heart failure (CHF). In dietary intervention studies, healthy human subjects, nonsmokers not on any medication or vitamin supplements, were found to have a significant increase in serum lycopene levels and lower levels of serum lipid peroxidation, LDL cholesterol protein, and DNA oxidation when they obtained lycopene (20–150 mg/day) from traditional tomato products and nutritional supplements for one week [54]. However, recent data [55] suggest that lycopene metabolites may possess specific biological activities within several important cellular signaling pathways and molecular targets. The use of well-defined subject populations, standardized outcome measures of oxidative stress and the disease, and lycopene ingestion in long-term studies, are essential for a meaningful interpretation of the results. Adipose tissue and circulating levels of lycopene appear to be better indicators of disease prevention than dietary intake. Moreover, a better understanding of how genes and gene-environment interactions lead to the CVD is essential, together with a better focusing of ethnic/racial differences in the development and progression of CVDAs for all the micronutrients. Lycopene bioavailability in food is a critical point to be considered. Lycopene, being a hydrophobic molecule, is poorly soluble in water; and during the digestive phase remains mostly bound to the vegetable matrix. However, mild cooking and a moderate amount of fat in the diet enhance its bioavailability. Therefore, cooking tomato in olive oil (the Mediterranean diet) is the best way to increase lycopene bioavailability [56]. Lycopene may be present in all-trans or cis isomers and cis isomer forms are more bioavailable than all-trans. In fact, although about 90% of the lycopene in dietary sources is found in the all-trans conformation, human tissues contain mainly cis isomers of this carotenoid [57] and cis lycopene isomers are produced during heating and processing of tomato products. Furthermore, lycopene accumulation in some tissues is inversely related to androgen status and appears to be inversely related to energy intake.

## 24.4 Phenols and Polyphenols

Phenolic and polyphenolic compounds are characterized by an aromatic or phenolic ring in their structure and include flavonoids, phenolic acids, and lignans. In plant cells, phenolic compounds are frequently coupled to sugars to reduce their endogenous toxicity.

Phenols and polyphenols are the most abundant antioxidants present in the vegetable kingdom. The main dietary sources of phenols and polyphenols are olive oil, fruits, cereals, dry legumes, and plant-derived beverages such as fruit juices, tea, coffee, red wine, and chocolate (Table 24.1) [58]. The daily dietary intake of these antioxidants is generally 1 g/day, which is much higher than that of all other classes of phytochemicals and known dietary antioxidants. For example, the intake of vitamin C is about 10 times lower and that of vitamin E and carotenoids is about 100 times lower [59]. However, it should be considered that the nutrient composition of vegetables can be strongly affected by genetic and environmental factors like growth factors, temperature, humidity, type of soil, application of fertilizers, and stress induced by UV radiation, pesticides, or microorganisms. Further modification can also occur during transportation and storage conditions. Thus, during these stages, chemical and physical changes occurring in vegetables can lead to loss of potentially beneficial components. In addition, industrial vegetable transformation to produce juices, purees, or canned vegetables can severely reduce the antioxidant power of vegetables during heat treatments for pasteurization or concentration. Despite their wide distribution in vegetables, the healthy effects of antioxidants have been recognized by scientists only rather recently [60]. The first hypothesis about the way of action of polyphenols and other antioxidants was that through their ability in scavenging free radicals they protect cell constituents against oxidative damage. However, this idea now appears to be an oversimplified explanation of their protective activity. Some evidence has been found that cells respond to antioxidants mainly through enzymes involved in signal transduction or direct interactions with receptors, leading to modification of the cell redox status and possibly triggering a series of redox-dependent reactions [61, 62]. In a recent report from the World Health Organization [63], a reduction of disease risk by flavonoids was considered "possible" for CVD and "insufficient" for cancers. A large body of evidence on the prevention of diseases mediated by polyphenols is derived from in vitro or animal studies, which are often performed with doses higher than those to which humans are exposed through their diets [64]. Epidemiological studies seem to confirm the protective effects of polyphenol consumption against CVD [65]. It is worth noting that polyphenols possess both antioxidant and pro-oxidant properties, with contrasting effects on cell physiological processes. As antioxidants, polyphenols may improve cell survival; as pro-oxidants, they may induce apoptosis, preventing tumor growth [66]. However, the biological effects of polyphenols may extend further than the modulation of oxidative stress. There are interesting aspects regarding the action of pomegranate fruit (Punica granatum L.) [67]. The content of soluble polyphenol in pomegranate juice (PJ) varies, depending on variety, from 0.2–1.0%, and includes mainly catechins, anthocyanins, gallic and ellagic acids, and ellagic tannins [68].

**Table 24.1** Antioxidant content in major vegetable sources

| Substance | Formula | Major sources | Amount (mg/100g) |
|---|---|---|---|
| Caffeic acid | 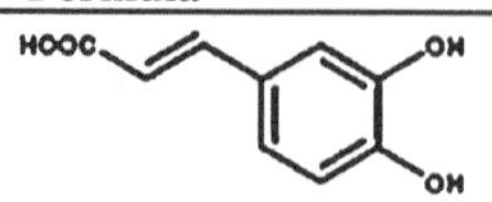 | Roasted coffee bean<br>Red wine<br>Carrot | 700-1000 [a]<br>1.0<br>0.1 |
| Gallic acid | 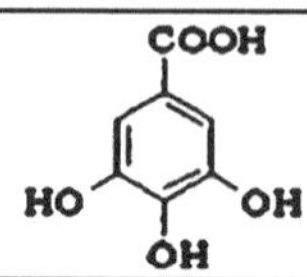 | Green tea<br>Red wine | 21.2 [b]<br>4.5 |
| Tyrosol | 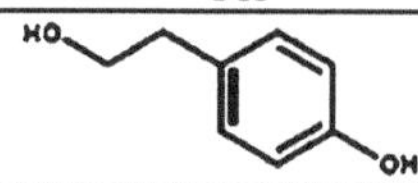 | Olive oil | 36.6 [c] |
| Hydroxytyrosol | 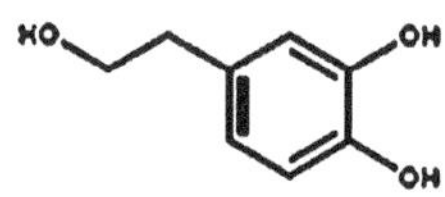 | Olive oil | 36.6 [c] |
| Oleuropein | 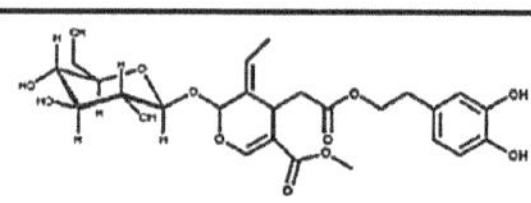 | Olive oil | 36.6 [c] |
| Epigallocatechin | 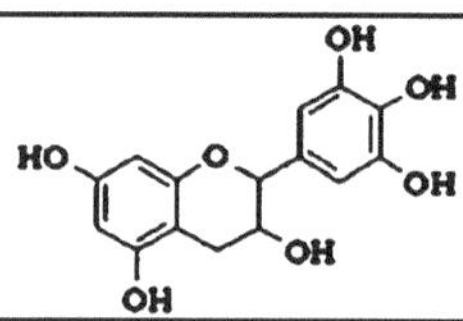 | Green tea | 351 [b] |
| Epigallocatechin gallate | 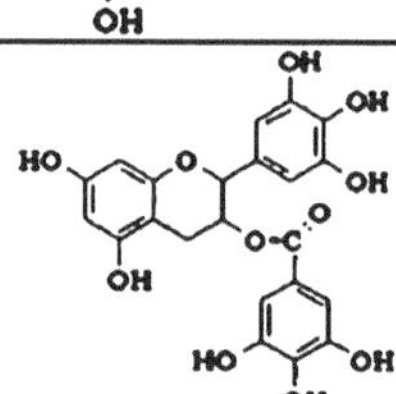 | Green tea | 540 [b] |
| Lycopene | 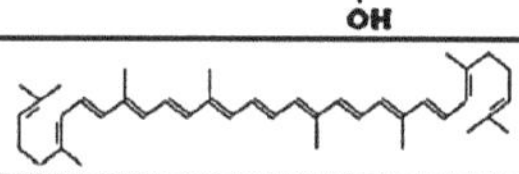 | Tomato | 3.0 |
| Campesterol | 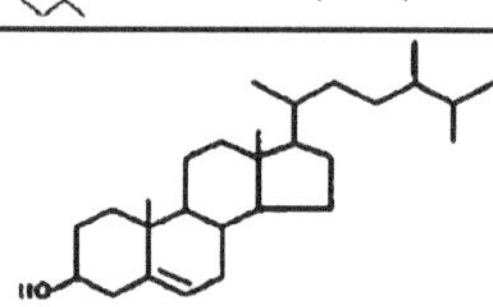 | Cauliflower<br>Brussels sprouts<br>Broccoli | 9.5<br>8.0<br>6.9 |
| Stigmasterol | 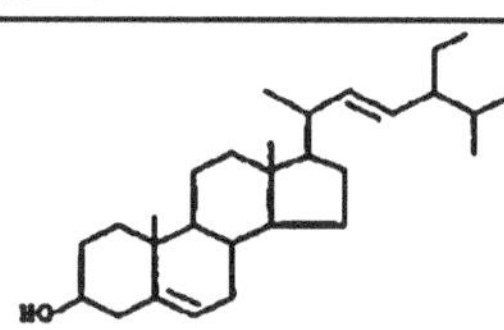 | Celery<br>Cauliflower<br>Tomato<br>Carrot<br>Broccoli | 7.0<br>3.7<br>1.7<br>2.8<br>1.1 |

[a] As chlorogenic acid
[b] In dry green tea leaf
[c] Total phenolic content

Notably, PJ possesses potent antioxidant activity that elicits antiatherogenic properties in mice [69] and can inhibit cyclooxygenases and lipoxygenases [70]. Prolonged consumption of PJ can mainly correct the perturbed shear stress-induced proatherogenic disequilibrium by decreasing the activity of transcription factors sensitive to the redox status; and by increasing eNOS activity, both in vitro in cultured human coronary artery EC, and in vivo in hypercholesterolemic mice [69]. The disturbed blood flow caused in the regions affected by atherosclerosis leads to perturbed shear stress that, in turn, causes endothelial damage [71]. The reduction in macrophage foam cell formation, oxidation-specific epitopes, and lesion area in atherosclerotic-prone lesion regions (low- and high-prone areas) in mice treated with PJ clearly confirms the correlation between the antiatherogenic properties elicited by PJ and its antioxidative effects [72]. eNOS expression and the activity modulation of transcription factors sensitive to redox status, like ELK-1 and p-JUN, are associated with antiatherogenic activity in such areas. These effects are similar to those produced by antioxidants (vitamins E and C) and L-arginine. The PJ antioxidant level was found to be higher than in other natural juices such as orange, blueberry, and cranberry, as well as in red wine [73]. Polyphenols from red wine can reduce LDL aggregation in vitro and in vivo, and PJ administered to hypertensive patients also causes a significant decline in systemic blood pressure [74]. Patients with carotid artery stenosis consuming PJ for three years showed a reduced carotid intima-media thickness, blood pressure, and LDL oxidation [75]. PJ has also been shown to revert downregulation of eNOS expression induced by LDLox in human coronary endothelial cells [76], suggesting that PJ, by enhancing the eNOS bioactivity, can produce helpful effects on the evolution of clinical vascular complications, CHD, and atherogenesis in humans. PJ, tested for its capacity to upregulate and/or activate eNOS in bovine pulmonary artery endothelial cells, elicited no effects on eNOS protein expression or catalytic activity, thus indicating that PJ possesses potent antioxidant activity that results in marked protection of NO against oxidative destruction [77]. However, it is not only PJ that contains antioxidant properties, but also pomegranate byproducts (which include the whole pomegranate fruit left after juice preparation), which have been shown to reduce atherosclerotic lesion size by up to 57% and oxidative stress in apolipoprotein E-deficient mice [78]. Daily consumption of PJ by diabetic patients resulted in antioxidative effects on serum and macrophages [79] and improved stress-induced myocardial ischemia in patients who have CHD. Thus, PJ and its byproducts may improve redox status of the arterial cells.

Tea pigment also exerted some antiatherosclerotic effects [80]. In fact, it is generally assumed that a regular consumption of tea, especially green tea, reduces the risk of cardiovascular disease [81]. Green tea contains several catechins, like epicatechin (EC), epigallocatechin (EGC), and epicatechin-3-gallate (ECG), all showing protective health effects; but the most potent for beneficial cardiovascular effects is epigallocatechin-3-gallate (EGCG) [82]. Recently, it has been demonstrated [83] that black tea possesses equivalent effects. However, it was shown that short- and long-term black tea consumption reverses endothelial dysfunction in CVD patients [84]. Similarly, the ingestion of polyphenols contained in purple grape juice had beneficial effects on endothelial function in patients with CHD [85].

These data suggest that polyphenols can protect arteries from vascular damage by means of antioxidant effects and NO restoration. However, trials using different antioxidants were not able to show any beneficial effects in terms of prevention of major CVD events. This divergence can be explained by considering that the models used in experimental studies, although very useful to investigate pathophysiological mechanisms, may not completely reflect the disease in humans. Alternatively, the dosage of antioxidants used in those studies may not have been suitable, and/or the progression of disease may have been too severe.

## 24.5 Dietary Fiber

As shown by several Epidemiological researchers, a diet rich in fiber has an inverse relationship to CVD risk, and for this reason many leading organizations recommend increased fiber in the daily diet [86]. There are essentially two different types of dietary fiber, soluble and insoluble; insoluble fiber has little to no effect on lowering cholesterol, whereas soluble fiber has a more sustained effect. Soluble fibers like pectin, beta-glucan, guar gum, and psyllium have been shown to reduce low density lipoprotein (LDL) cholesterol in the serum by 3% to 10%. However, the biologic mechanisms explaining how fiber influences lower serum LDL cholesterol levels have yet to be fully elucidated [87]. Different hypotheses have been formulated: (i) interference by water-soluble fibers with lipid and/or bile acid metabolism; (ii) inhibition of hepatic cholesterol synthesis by fermentation products; (iii) delayed absorption of carbohydrates leading to a decrease in insulin concentration.

(i) Evidence has been produced suggesting that water-soluble fibers bind bile acids in the intestinal lumen when micelles are formed [88]. On the other hand, it has been suggested that water-soluble fibers may form a water layer in the intestinal lumen that acts as a barrier which decreases the absorption of fats and cholesterol, and the reabsorbtion of bile acids. As a consequence, the production of bile acids starting from cholesterol is increased in the liver, which through its LDL cholesterol receptors increases the cholesterol uptake from serum, thus decreasing LDL cholesterol concentrations.

(ii) Water-soluble fibers are fermented in the large bowel by colonic bacteria with the production of carboxylic acids like butyrate, propionate, and acetate, usually recognized as short-chain fatty acids (SCFA). It has been suggested that SCFA production, and in particular changes in the propionate/acetate ratio, may inhibit lipid metabolism, but the mechanism by which this inhibition takes place has not yet been elucidated. There is indeed some evidence that fatty acid synthesis is inhibited by propionate in isolated rat hepatocytes. However, it seems that the effects of SCFA might depend upon the relative proportions of acetate and propionate [88]. In fact, the higher the propionate:acetate ratio, the greater the lipid-lowering effect, although this effect was found in healthy men, but not women.

(iii) Water-soluble fibers, being highly hygroscopic, may reduce the rate of glucose absorption, due to the increased retention of intestinal water together with water soluble molecules [88]. The lowered postprandial glucose concentrations results in a decrease in insulin secretion and a lesser insulin-stimulated hepatic HMG-CoA activity and hence cholesterol synthesis.

Recently an association between dietary fiber and levels of C-reactive protein (CRP) has been proposed as a clinical indicator of inflammation in research in a large national sample in the USA. Evidence of a possible metabolic effect on inflammatory markers, coupled with epidemiological indications demonstrating that high-fiber diets are beneficial, suggests that inflammation may be an important mediator in the association between dietary fiber and CVD. In a study aiming to prevent coronary heart disease in women through diet and lifestyle, subjects in the highest quintile of fiber intake (median 22.9 g/day) had an age-adjusted relative risk for major coronary events that was 47% lower than women in the lowest quintile (11.5 g/day) [89]. The relationship between the CVD risk factors and the source or type of dietary fiber intake shows that the highest total dietary fiber and insoluble dietary fiber intakes were associated with a significantly lower risk of overweight and elevated waist-to-hip ratio, blood pressure, plasma apolipoprotein (apo) B, apo B:apo A–I, cholesterol, triacylglycerols, and homocysteine [90]. Practical recommendations for CVD prevention include a food-based approach favoring increased intake of whole-grain and dietary fiber (especially soluble fiber), fruit, and vegetables, providing a mixture of different types of fibers [91].

## 24.6 Fatty Acids

The antiinflammatory benefits of regular fish consumption or fish oil ingestion have been proved to protect against CVD. The long-chain fatty acids unique to marine foods (eicosapentaenoic acid [EPA] and docosahexaenoic acid [DHA]), along with plants rich in a-linolenic acid, compete with the long-chain fatty acid, arachidonic acid, in cell membranes, interfering with the inflammatory metabolic cascades [87]. The antiinflammatory effects of daily fish oil ingestion are very similar to those demonstrated for vegetarian diets [92]. Randomized secondary prevention clinical trials with fish oils (eicosapentaenoic acid, docosahexaenoic acid) and a-linolenic acid have demonstrated reductions in risk comparable to those seen in landmark secondary prevention trials with lipid-lowering drugs. The antiinflammatory activity of fish oil may vary among different sources due to variations in EPA/DHA content [93]. PUFAs, and in particular total n-3 fatty acids, were independently associated with lower levels of proinflammatory markers (IL-6, IL-1ra, TNF-α, C-reactive protein) and higher levels of antiinflammatory markers (soluble IL-6r, IL-10, TGFα) [94]. Several mechanisms explaining the cardioprotective effect of the n-3 PUFA have been suggested, including antiarrhythmic and antithrombotic roles. A recent meta-analysis of 10 qualifying randomized controlled trials, encompassing 14,727 patients, reported that daily fish oil consumption reduced the incidence of death due to myocardial infarction by 24% and all-cause mortality by 16% [95]. n-3 PUFAs have been recently shown to directly inhibit vascular calcification via p38-MAPK and PPAR-$\gamma$ pathways, and to reduce gene expression of cyclooxygenase-2, an inflammatory gene that through the activation of some metalloproteinases is involved in plaque angiogenesis, plaque rupture, and oxidative stress reduction.

The reduced expression of proinflammatory proatherogenic genes by omega-3 fatty acids can affect the extent of leukocyte adhesion to vascular endothelium, early atherogenesis, and later stages of plaque development and plaque rupture, giving a possible explanation for the vasculoprotective effects of these nutrients [96]. The antiarrhythmic action of n-3 PUFA can be also be explained by its modulation of channel activities, like voltage-gated $Na^{(+)}$ and L-type $Ca^{(2+)}$ channels [97]. From recent randomized trials it seems clear that both fish and plant sources of n-3 PUFAs can favorably impact CV health [98]. Several international guidelines have been published for the dietary intake of n-3 PUFA, because fish is an important source of the n-3 PUFA. However, vegetable sources, including grains and oils, can substitute for fish, offering an alternative source.

## 24.7 Phytosterols

Plant sterols and stanols, known as phytosterols, are found naturally in many plant foods including vegetable oils, nuts, fruits, and vegetables [87]. The two most abundant plant sterols are campesterol and sitosterol. Phytosterols have chemical structures that are similar to cholesterol, and it seems that they affect the absorption of cholesterol by competing with the absorption of dietary and biliary cholesterol in the intestinal lumen, replaying it in the mixed micelles [99]. It has been shown that a minimum consumption of 800 mg/day is necessary to produce a significant decrease of cholesterol in the serum. Other studies showed that phytosterol esters seem to have an additive effect to statin therapy. Several food factories are now preparing functional foods containing phytosterols like yogurt drinks, orange juices, and snack bars, with a market that in recent years is increasing rapidly [100]. However, it seems that phytosterols may decrease the absorption of beta carotene, thus the consumption of large amounts of phytosterol should be balanced with an equivalent consumption of beta carotene.

## 24.8 Ethanol and Nonethanolic Components of Wine

Ethanol and nonethanolic components of wine may have a specific protective effect on the myocardium, independent of the classical risk factors implicated in vascular atherosclerosis and thrombosis in animal models [101]. Mechanisms by which the consumption of alcoholic beverages protects against cardiac injury induced by ischemia are not yet fully understood. The protective effect of alcohol has been explained essentially by its action on blood lipids (increase in high density lipoprotein levels) and platelets (decreased aggregation), resulting in a reduced rate of coronary artery obstruction [102], but other mechanisms are probably involved. In fact, the effect of alcohol on HDL cholesterol alone appears to explain 50–70% of the observed inverse association of alcohol with risk of CHD. Moderate alcohol ingestion has been shown to improve postischemic myocardial systolic and diastolic function in rats and to attenuate the postischemic reduction in coronary vascular resistance. Moderate drinking may improve the early outcomes after acute myocardial infarction and prevent sudden cardiac death [103], suggesting a direct effect of ethanol on the ischemic myocardium that has been referred to as "ethanol preconditioning." According to dietary guidelines, the daily amount (approximately 0.6 oz of alcohol) is no more than one drink per day for women and no more than two for men, preferably taken with meals [87]. So far, ethanol ingestion shows its cardioprotective effects by mediating the activity of adenosine triphosphate-dependent potassium (KATP) channels, adenosine type 1 (A[1]) receptors, alpha(1)-adrenoceptors, and the epsilon isoform of protein kinase C (PKC) [104]. Compared to other alcoholic beverages, the contribution of wine (expecially red wine) to cardioprotective effects depends both on alcohol and polyphenolic antioxidant components. Thus, in the wine a synergistic effect is realized by polyphenolic antioxidants, which provide cardioprotection by their ability to function as in vivo antioxidants; and by the alcoholic components that contribute by adapting the hearts to oxidative stress. Consumption of low doses of alcohol increases the heart oxidative stress, which induces the expression of several cardioprotective oxidative stress–inducible proteins, including heat shock protein (HSP). Other studies suggests antiinflammatory and/or antioxidant effects of moderate drinking. Increasing evidence for cardioprotective effects mediated by moderate alcohol consumption involve cellular and molecular mechanisms related to NO. Furthermore, total nitrates and nitrites were found to be increased in the blood of rats after eight weeks consumption of moderate alcohol [105]. In addition, recent experimental evidence, albeit in animals, suggests that the ethanol and/or nonethanolic fraction of wine might exert preconditioning effects in models of myocardial ischemia/reperfusion. There is no doubt that such an observation, if confirmed in human subjects, might open new perspectives in the prevention and treatment of ischemic CHD.

## References

1. Belleville J (2002) The French paradox: possible involvement of ethanol in the protective effect against cardiovascular diseases. Nutrition 18(2):173–177
2. Hardin-Fanning F (2008) The effects of a Mediterranean-style dietary pattern on cardiovascular disease risk. Nurs Clin North Am 43(1):105–115
3. Ignarro LJ, Napoli C (2004) Novel features of nitric oxide, endothelial nitric oxide synthase, and atherosclerosis. Curr Atheroscler Rep 6(4):281–287
4. Napoli C, Palinski W (2005) Neurodegenerative diseases: insights into pathogenic mechanisms from atherosclerosis. Neurobiol Aging 26(3):293–302
5. Crimi E, Sica V, Williams-Ignarro S, Zhang H, Slutsky AS, Ignarro LJ et al (2006) The role of oxidative stress in adult critical care. Free Radic Biol Med 40(3):398–406

6. Schiffrin EL (2008) Oxidative stress, nitric oxide synthase, and superoxide dismutase: a matter of imbalance underlies endothelial dysfunction in the human coronary circulation. Hypertension 51(1):31–32
7. Crabtree MJ, Tatham AL, Al Wakeel Y, Warrick N, Hale AB, Cai S et al (2009) Quantitative regulation of intracellular endothelial nitric-oxide synthase (eNOS) coupling by both tetrahydrobiopterin-eNOS stoichiometry and biopterin redox status: insights from cells with tet-regulated GTP cyclohydrolase I expression. J Biol Chem 284(2):1136–1144
8. Takeda M, Yamashita T, Shinohara M, Sasaki N, Takaya T, Nakajima K et al (2009) Plasma tetrahydrobiopterin/dihydrobiopterin ratio. Circ J 73(5):955–962
9. Carreras MC, Franco MC, Peralta JG, Poderoso JJ (2004) Nitric oxide, complex I, and the modulation of mitochondrial reactive species in biology and disease. Mol Aspects Med 25(1–2):125–139
10. Steinberg D (2005) Thematic review series: the pathogenesis of atherosclerosis. An interpretive history of the cholesterol controversy: part II: the early evidence linking hypercholesterolemia to coronary disease in humans. J Lipid Res 46(2):179–190
11. Napoli C, Lerman LO, de Nigris F, Gossl M, Balestrieri ML, Lerman A (2006) Rethinking primary prevention of atherosclerosis-related diseases. Circulation 114(23):2517–2527
12. Napoli C, D'Armiento FP, Mancini FP, Postiglione A, Witztum JL, Palumbo G et al (1997) Fatty streak formation occurs in human fetal aortas and is greatly enhanced by maternal hypercholesterolemia. Intimal accumulation of low density lipoprotein and its oxidation precede monocyte recruitment into early atherosclerotic lesions. J Clin Invest 100(11):2680–2690
13. Napoli C, Glass CK, Witztum JL, Deutsch R, D'Armiento FP, Palinski W (1999) Influence of maternal hypercholesterolaemia during pregnancy on progression of early atherosclerotic lesions in childhood: Fate of Early Lesions in Children (FELIC) Study. Lancet 354(9186):1234–1241
14. Palinski W, Napoli C (2002) The fetal origins of atherosclerosis: maternal hypercholesterolemia, and cholesterol-lowering or antioxidant treatment during pregnancy influence in utero programming and postnatal susceptibility to atherogenesis. FASEB J 16(11): 1348–1360
15. Armitage JA, Khan IY, Taylor PD, Nathanielsz PW, Poston L (2004) Developmental programming of the metabolic syndrome by maternal nutritional imbalance: how strong is the evidence from experimental models in mammals? J Physiol 561(Pt 2):355–377
16. Boullier A, Bird DA, Chang MK, Dennis EA, Friedman P, Gillotre-Taylor K et al (2001) Scavenger receptors, oxidized LDL, and atherosclerosis. Ann N Y Acad Sci 947:214–222
17. Versari D, Daghini E, Rodriguez-Porcel M, Sattler K, Galili O, Pilarczyk K et al (2006) Chronic antioxidant supplementation impairs coronary endothelial function and myocardial perfusion in normal pigs. Hypertension 47(3):475–481
18. Kalantar-Zadeh K, Anker SD, Horwich TB, Fonarow GC (2008) Nutritional and anti-inflammatory interventions in chronic heart failure. Am J Cardiol 101(11A):89E–103E
19. Liguori A, D'Armiento FP, Palagiano A, Balestrieri ML, Williams-Ignarro S, de Nigris F et al (2007) Effect of gestational hypercholesterolaemia on omental vasoreactivity, placental enzyme activity and transplacental passage of normal and oxidised fatty acids. BJOG 114(12):1547–1556
20. Palinski W, Yamashita T, Freigang S, Napoli C (2007) Developmental programming: maternal hypercholesterolemia and immunity influence susceptibility to atherosclerosis. Nutr Rev 65(12 Pt 2):S182–S187
21. Yamashita T, Freigang S, Eberle C, Pattison J, Gupta S, Napoli C et al (2006) Maternal immunization programs postnatal immune responses and reduces atherosclerosis in offspring. Circ Res 99(7):e51–e64
22. Ahn MR, Kunimasa K, Kumazawa S, Nakayama T, Kaji K, Uto Y et al (2009) Correlation between antiangiogenic activity and antioxidant activity of various components from propolis. Mol Nutr Food Res 53(5):643–651

23. Toledo F, Arancibia-Avila P, Park YS, Jung ST, Kang SG, Heo BG et al (2008) Screening of the antioxidant and nutritional properties, phenolic contents and proteins of five durian cultivars. Int J Food Sci Nutr 59(5):415–427
24. McKeever TM, Lewis SA, Smit HA, Burney P, Cassano PA, Britton J (2008) A multivariate analysis of serum nutrient levels and lung function. Respir Res 9:67
25. Haleem MA, Barton KL, Borges G, Crozier A, Anderson AS (2008) Increasing antioxidant intake from fruits and vegetables: practical strategies for the Scottish population. J Hum Nutr Diet 21(6):539–546
26. Rhone M, Basu A (2008) Phytochemicals and age-related eye diseases. Nutr Rev 66(8): 465–472
27. Terao J, Kawai Y, Murota K (2008) Vegetable flavonoids and cardiovascular disease. Asia Pac J Clin Nutr 17(Suppl 1):291–293
28. Seeram NP, Aviram M, Zhang Y, Henning SM, Feng L, Dreher M et al (2008) Comparison of antioxidant potency of commonly consumed polyphenol-rich beverages in the United States. J Agric Food Chem 56(4):1415–1422
29. Leifert WR, Abeywardena MY (2008) Cardioprotective actions of grape polyphenols. Nutr Res 28(11):729–737
30. Hooper L, Kroon PA, Rimm EB, Cohn JS, Harvey I, Le Cornu KA et al (2008) Flavonoids, flavonoid-rich foods, and cardiovascular risk: a meta-analysis of randomized controlled trials. Am J Clin Nutr 88(1):38–50
31. de Nigris F, Lerman LO, Ignarro SW, Sica G, Lerman A, Palinski W et al (2003) Beneficial effects of antioxidants and L-arginine on oxidation-sensitive gene expression and endothelial NO synthase activity at sites of disturbed shear stress. Proc Natl Acad Sci U S A 100(3):1420–1425
32. Hayashi T, Juliet PA, Matsui-Hirai H, Miyazaki A, Fukatsu A, Funami J et al (2005) l-Citrulline and l-arginine supplementation retards the progression of high-cholesterol-diet-induced atherosclerosis in rabbits. Proc Natl Acad Sci U S A 102(38):13681–13686
33. Napoli C, Williams-Ignarro S, de Nigris F, Lerman LO, Rossi L, Guarino C et al (2004) Long-term combined beneficial effects of physical training and metabolic treatment on atherosclerosis in hypercholesterolemic mice. Proc Natl Acad Sci U S A 101(23):8797–8802
34. Vergnani L, Hatrik S, Ricci F, Passaro A, Manzoli N, Zuliani G et al (2000) Effect of native and oxidized low-density lipoprotein on endothelial nitric oxide and superoxide production: key role of L-arginine availability. Circulation 101(11):1261–1266
35. Napoli C, Lerman LO (2002) de NF, Loscalzo J, Ignarro LJ. Glycoxidized low-density lipoprotein downregulates endothelial nitricoxide synthase in human coronary cells. J Am Coll Cardiol 40(8):1515–1522
36. Kohlmeier L, Kark JD, Gomez-Gracia E, Martin BC, Steck SE, Kardinaal AF et al (1997) Lycopene and myocardial infarction risk in the EURAMIC Study. Am J Epidemiol 146(8):618–626
37. Agarwal S, Rao AV (2000) Carotenoids and chronic diseases. Drug Metabol Drug Interact 17(1–4):189–210
38. Brigelius-Flohe R (2009) Vitamin E the shrew waiting to be tamed. Free Radic Biol Med 46(5):543–554
39. Yusuf S, Sleight P, Pogue J, Bosch J, Davies R, Dagenais G (2000) Effects of an angiotensin-converting-enzyme inhibitor, ramipril, on cardiovascular events in high-risk patients. The Heart Outcomes Prevention Evaluation Study Investigators. N Engl J Med 342(3):145–153
40. Lonn E, Yusuf S, Hoogwerf B, Pogue J, Yi Q, Zinman B et al (2002) Effects of vitamin E on cardiovascular and microvascular outcomes in high-risk patients with diabetes: results of the HOPE study and MICRO-HOPE substudy. Diabetes Care 25(11):1919–1927
41. MRC/BHF Heart Protection Study of antioxidant vitamin supplementation in (2002) 20,536 high-risk individuals: a randomised placebo-controlled trial. Lancet 360(9326):23–33
42. Schulman SP, Becker LC, Kass DA, Champion HC, Terrin ML, Forman S et al (2006) L-arginine therapy in acute myocardial infarction: the Vascular Interaction With Age

in Myocardial Infarction (VINTAGE MI) randomized clinical trial. J Am Med Assoc 295(1):58–64
43. Erdman JW Jr, Ford NA, Lindshield BL (2009) Are the health attributes of lycopene related to its antioxidant function? Arch Biochem Biophys 483(2):229–235
44. Hu MY, Li YL, Jiang CH, Liu ZQ, Qu SL, Huang YM (2008) Comparison of lycopene and fluvastatin effects on atherosclerosis induced by a high-fat diet in rabbits. Nutrition 24(10):1030–1038
45. Arab L, Steck S (2000) Lycopene and cardiovascular disease. Am J Clin Nutr 71(6 Suppl):1691S–1695S
46. Rao AV, Agarwal S (2000) Role of antioxidant lycopene in cancer and heart disease. J Am Coll Nutr 19(5):563–569
47. Dugas TR, Morel DW, Harrison EH (1998) Impact of LDL carotenoid and alpha-tocopherol content on LDL oxidation by endothelial cells in culture. J Lipid Res 39(5): 999–1007
48. Dugas TR, Morel DW, Harrison EH (1999) Dietary supplementation with beta-carotene, but not with lycopene, inhibits endothelial cell-mediated oxidation of low-density lipoprotein. Free Radic Biol Med 26(9–10):1238–1244
49. Street DA, Comstock GW, Salkeld RM, Schuep W, Klag MJ (1994) Serum antioxidants and myocardial infarction. Are low levels of carotenoids and alpha-tocopherol risk factors for myocardial infarction? Circulation 90(3):1154–1161
50. Kardinaal AF, Kok FJ, Ringstad J, Gomez-Aracena J, Mazaev VP, Kohlmeier L et al (1993) Antioxidants in adipose tissue and risk of myocardial infarction: the EURAMIC Study. Lancet 342(8884):1379–1384
51. Iribarren C, Folsom AR, Jacobs DR Jr, Gross MD, Belcher JD, Eckfeldt JH (1997) Association of serum vitamin levels, LDL susceptibility to oxidation, and autoantibodies against MDA-LDL with carotid atherosclerosis. A case-control study. The ARIC Study Investigators. Atherosclerosis Risk in Communities. Arterioscler Thromb Vasc Biol 17(6):1171–1177
52. Kristenson M, Zieden B, Kucinskiene Z, Elinder LS, Bergdahl B, Elwing B et al (1997) Antioxidant state and mortality from coronary heart disease in Lithuanian and Swedish men: concomitant cross sectional study of men aged 50. BMJ 314(7081):629–633
53. Schmidt R, Fazekas F, Hayn M, Schmidt H, Kapeller P, Roob G et al (1997) Risk factors for microangiopathy-related cerebral damage in the Austrian stroke prevention study. J Neurol Sci 152(1):15–21
54. Rao AV, Agarwal S (1998) Bioavailability and in vivo antioxidant properties of lycopene from tomato products and their possible role in the prevention of cancer. Nutr Cancer 31(3):199–203
55. Mein JR, Lian F, Wang XD (2008) Biological activity of lycopene metabolites: implications for cancer prevention. Nutr Rev 66(12):667–683
56. Bugianesi R, Salucci M, Leonardi C, Ferracane R, Catasta G, Azzini E et al (2004) Effect of domestic cooking on human bioavailability of naringenin, chlorogenic acid, lycopene and beta-carotene in cherry tomatoes 5. Eur J Nutr 43(6):360–366
57. Failla ML, Chitchumroonchokchai C, Ishida BK (2008) In vitro micellarization and intestinal cell uptake of cis isomers of lycopene exceed those of all-trans lycopene. J Nutr 138(3):482–486
58. Johnston C (2009) Functional foods as modifiers of cardiovascular disease. Am J Lifestyle Med 3(1 Suppl):39S–43S
59. Manach C, Scalbert A, Morand C, Remesy C, Jimenez L (2004) Polyphenols: food sources and bioavailability. Am J Clin Nutr 79(5):727–747
60. Gutteridge JM, Halliwell B (2000) Free radicals and antioxidants in the year 2000. A historical look to the future. Ann N Y Acad Sci 899:136–147
61. Scalbert A, Williamson G (2000) Dietary intake and bioavailability of polyphenols. J Nutr 130(8S Suppl):2073S–2085S

62. Scalbert A, Manach C, Morand C, Remesy C, Jimenez L (2005) Dietary polyphenols and the prevention of diseases. Crit Rev Food Sci Nutr 45(4):287–306
63. WHO/FAO (2003) Diet, nutrition, and the prevention of chronic diseases. World Health Organization, Geneva
64. Vita JA (2005) Polyphenols and cardiovascular disease: effects on endothelial and platelet function. Am J Clin Nutr 81(1 Suppl):292S–297S
65. Arts IC, Hollman PC (2005) Polyphenols and disease risk in epidemiologic studies. Am J Clin Nutr 81(1 Suppl):317S–325S
66. Lambert JD, Hong J, Yang GY, Liao J, Yang CS (2005) Inhibition of carcinogenesis by polyphenols: evidence from laboratory investigations. Am J Clin Nutr 81(1 Suppl): 284S–291S
67. Basu A, Penugonda K (2009) Pomegranate juice: a heart-healthy fruit juice. Nutr Rev 67(1):49–56
68. Ben NC, Ayed N, Metche M (1996) Quantitative determination of the polyphenolic content of pomegranate peel. Z Lebensm Unters Forsch 203(4):374–378
69. de Nigris F, Williams-Ignarro S, Lerman LO, Crimi E, Botti C, Mansueto G et al (2005) Beneficial effects of pomegranate juice on oxidation-sensitive genes and endothelial nitric oxide synthase activity at sites of perturbed shear stress. Proc Natl Acad Sci U S A 102(13):4896–4901
70. Schubert SY, Lansky EP, Neeman I (1999) Antioxidant and eicosanoid enzyme inhibition properties of pomegranate seed oil and fermented juice flavonoids. J Ethnopharmacol 66(1):11–17
71. Napoli C, Ignarro LJ (2001) Nitric oxide and atherosclerosis. Nitric Oxide 5(2):88–97
72. Kaplan M, Hayek T, Raz A, Coleman R, Dornfeld L, Vaya J et al (2001) Pomegranate juice supplementation to atherosclerotic mice reduces macrophage lipid peroxidation, cellular cholesterol accumulation and development of atherosclerosis. J Nutr 131(8):2082–2089
73. Aviram M, Dornfeld L, Kaplan M, Coleman R, Gaitini D, Nitecki S et al (2002) Pomegranate juice flavonoids inhibit low-density lipoprotein oxidation and cardiovascular diseases: studies in atherosclerotic mice and in humans. Drugs Exp Clin Res 28(2–3):49–62
74. Aviram M, Dornfeld L (2001) Pomegranate juice consumption inhibits serum angiotensin converting enzyme activity and reduces systolic blood pressure. Atherosclerosis 158(1): 195–198
75. Aviram M, Rosenblat M, Gaitini D, Nitecki S, Hoffman A, Dornfeld L et al (2004) Pomegranate juice consumption for 3 years by patients with carotid artery stenosis reduces common carotid intima-media thickness, blood pressure and LDL oxidation. Clin Nutr 23(3):423–433
76. de Nigris F, Williams-Ignarro S, Botti C, Sica V, Ignarro LJ, Napoli C (2006) Pomegranate juice reduces oxidized low-density lipoprotein downregulation of endothelial nitric oxide synthase in human coronary endothelial cells. Nitric Oxide 15(3):259–263
77. Ignarro LJ, Byrns RE, Sumi D, de Nigris F, Napoli C (2006) Pomegranate juice protects nitric oxide against oxidative destruction and enhances the biological actions of nitric oxide. Nitric Oxide 15(2):93–102
78. Rosenblat M, Volkova N, Coleman R, Aviram M (2006) Pomegranate byproduct administration to apolipoprotein e-deficient mice attenuates atherosclerosis development as a result of decreased macrophage oxidative stress and reduced cellular uptake of oxidized low-density lipoprotein. J Agric Food Chem 54(5):1928–1935
79. Rosenblat M, Hayek T, Aviram M (2006) Anti-oxidative effects of pomegranate juice (PJ) consumption by diabetic patients on serum and on macrophages. Atherosclerosis 187(2):363–371
80. Lou FQ, Zhang MF, Zhang XG, Liu JM, Yuan WL (1989) A study on tea-pigment in prevention of atherosclerosis. Chin Med J (Engl) 102(8):579–583
81. Babu PV, Liu D (2008) Green tea catechins and cardiovascular health: an update. Curr Med Chem 15(18):1840–1850

82. Brown AL, Lane J, Coverly J, Stocks J, Jackson S, Stephen A et al (2009) Effects of dietary supplementation with the green tea polyphenol epigallocatechin-3-gallate on insulin resistance and associated metabolic risk factors: randomized controlled trial. Br J Nutr 101(6):886–894
83. Lorenz M, Urban J, Engelhardt U, Baumann G, Stangl K, Stangl V (2009) Green and black tea are equally potent stimuli of NO production and vasodilation: new insights into tea ingredients involved. Basic Res Cardiol 104(1):100–110
84. Duffy SJ, Keaney JF Jr, Holbrook M, Gokce N, Swerdloff PL, Frei B et al (2001) Short- and long-term black tea consumption reverses endothelial dysfunction in patients with coronary artery disease. Circulation 104(2):151–156
85. Chou EJ, Keevil JG, Aeschlimann S, Wiebe DA, Folts JD, Stein JH (2001) Effect of ingestion of purple grape juice on endothelial function in patients with coronary heart disease. Am J Cardiol 88(5):553–555
86. King DE (2005) Dietary fiber, inflammation, and cardiovascular disease. Mol Nutr Food Res 49(6):594–600
87. Retelny VS, Neuendorf A, Roth JL (2008) Nutrition protocols for the prevention of cardiovascular disease. Nutr Clin Pract 23(5):468–476
88. Theuwissen E, Mensink RP (2008) Water-soluble dietary fibers and cardiovascular disease. Physiol Behav 94(2):285–292
89. Stampfer MJ, Hu FB, Manson JE, Rimm EB, Willett WC (2000) Primary prevention of coronary heart disease in women through diet and lifestyle. N Engl J Med 343(1): 16–22
90. Lairon D, Arnault N, Bertrais S, Planells R, Clero E, Hercberg S et al (2005) Dietary fiber intake and risk factors for cardiovascular disease in French adults. Am J Clin Nutr 82(6):1185–1194
91. Olendzki B, Speed C, Domino FJ (2006) Nutritional assessment and counseling for prevention and treatment of cardiovascular disease. Am Fam Physician 73(2):257–264
92. Ginter E (2008) Vegetarian diets, chronic diseases and longevity. Bratisl Lek Listy 109(10):463–466
93. Bhattacharya A, Sun D, Rahman M, Fernandes G (2007) Different ratios of eicosapentaenoic and docosahexaenoic omega-3 fatty acids in commercial fish oils differentially alter pro-inflammatory cytokines in peritoneal macrophages from C57BL/6 female mice. J Nutr Biochem 18(1):23–30
94. Ferrucci L, Cherubini A, Bandinelli S, Bartali B, Corsi A, Lauretani F et al (2006) Relationship of plasma polyunsaturated fatty acids to circulating inflammatory markers. J Clin Endocrinol Metab 91(2):439–446
95. Yzebe D, Lievre M (2004) Fish oils in the care of coronary heart disease patients: a meta-analysis of randomized controlled trials. Fundam Clin Pharmacol 18(5):581–592
96. De CR, Massaro M (2005) Omega-3 fatty acids and the regulation of expression of endothelial pro-atherogenic and pro-inflammatory genes. J Membr Biol 206(2):103–116
97. Xiao YF, Sigg DC, Leaf A (2005) The antiarrhythmic effect of n-3 polyunsaturated fatty acids: modulation of cardiac ion channels as a potential mechanism. J Membr Biol 206(2):141–154
98. Harper CR, Jacobson TA (2005) Usefulness of omega-3 fatty acids and the prevention of coronary heart disease. Am J Cardiol 96(11):1521–1529
99. Windler E, Zyriax BC, Kuipers F, Linseisen J, Boeing H (2009) Association of plasma phytosterol concentrations with incident coronary heart disease. Data from the CORA study, a case-control study of coronary artery disease in women. Atherosclerosis 203(1):284–290
100. Mannarino E, Pirro M, Cortese C, Lupattelli G, Siepi D, Mezzetti A et al (2009) Effects of a phytosterol-enriched dairy product on lipids, sterols and 8-isoprostane in hypercholesterolemic patients: a multicenter Italian study. Nutr Metab Cardiovasc Dis 19(2):84–90
101. Saremi A, Arora R (2008) The cardiovascular implications of alcohol and red wine. Am J Ther 15(3):265–277

102. Ruf JC (2004) Alcohol, wine and platelet function. Biol Res 37(2):209–215
103. de LM, Salen P, Martin JL, Boucher F, Paillard F, de LJ (2002) Wine drinking and risks of cardiovascular complications after recent acute myocardial infarction. Circulation 106(12):1465–1469
104. Pagel PS, Kersten JR, Warltier DC (2004) Mechanisms of myocardial protection produced by chronic ethanol consumption. Pathophysiology 10(2):121–129
105. Abou-Agag LH, Khoo NK, Binsack R, White CR, Darley-Usmar V, Grenett HE et al (2005) Evidence of cardiovascular protection by moderate alcohol: role of nitric oxide. Free Radic Biol Med 39(4):540–548

# Chapter 25
# Novel Synthetic Antioxidants and Nitrated Lipids: From Physiology to Therapeutic Implications

**Gloria V. López and Homero Rubbo**

**Abstract** Low density lipoprotein (LDL) oxidation has been proposed as an early event in the development of atherosclerosis. However, several clinical studies failed in preventing cardiovascular diseases by administration of natural antioxidants. Herein, we discuss these studies as well as the use of novel synthetic antioxidants such as hybrid compounds designed to improve the efficacy of natural antioxidants. In particular, we designed novel hybrid antioxidants (tocopherol analogs–nitric oxide donors) that share nitric oxide–releasing properties and LDL incorporation capacity, demonstrating the importance of this site-specific release of nitric oxide in the cascade of events involved in the inhibition of LDL oxidation. This offers novel approaches for the prevention of atherosclerosis and related disorders that involve reactive oxygen and nitrogen species. Secondly, we discuss the biological actions of nitrated fatty acids, which are novel antiinflammatory signaling mediators, representing another potential promising pharmacological strategy against cardiovascular diseases.

**Keywords** Antioxidants · Vitamin E · Nitric oxide donors · Lipid nitration · Atherosclerosis · LDL oxidation · ardiovascular disease

## 25.1 Introduction

Atherosclerosis and its cardiovascular complications constitute the major cause of morbidity and mortality in western countries [1]. The development and progression of the atherosclerosis basic lesion is a complex and multifactorial process which begins in the first decade of life, but could be stopped or delayed in its progression [2]. In the past decades, a series of significant studies suggested that vascular

---

G.V. López (✉)
Laboratorio de Química Orgánica, Facultad de Ciencias, Universidad de la República, 11400 Montevideo, Uruguay
e-mail: vlopez@fq.edu.uy

H. Sauer et al. (eds.), *Studies on Cardiovascular Disorders*,
Oxidative Stress in Applied Basic Research and Clinical Practice,
DOI 10.1007/978-1-60761-600-9_25, 

oxidative stress, mainly when it is associated with modification and oxidation of low density lipoprotein (LDL), represents a major pathological component in the atherogenic process. Studies have demonstrated that intact LDL per se does not present significant proatherogenic effects [3–9]. In fact, cholesterol accumulation in atherosclerotic lesions is not due to cellular uptake of native LDL through LDL receptors, but rather to the uptake of its oxidized-modified form by an alternative receptor, called a scavenger receptor, present in monocytes/macrophages and smooth muscle cells (SMC) [7]. In contrast to native LDL, oxidized LDL uptake by scavenger receptors of macrophages and SMC is not subject to negative feedback regulation; and therefore this process results in massive uptake and accumulation of cholesterol and its oxidation products, leading to the formation of foam cells (lipid-filled cells), major components of the atheroma plaque [3, 10]. In addition to promoting foam cell formation, oxidized LDL alters the endothelial production and bioavailability of nitric oxide (•NO) [11], stimulates endothelial cell apoptosis [12], has direct chemotactic effects on monocytes [13, 14], and induces vascular cells to produce growth, adhesion, and chemotactic factors, which account for the emergence of an inflammatory focus in arterial intima [6, 10, 15, 16].

## 25.2 Natural Antioxidants and Prevention of Cardiovascular Disease

In recent years, the possibility that dietary and supplemental antioxidants might prevent or delay cardiovascular disease has raised great interest in scientific and public communities. Most of the scientific studies concern the major antioxidants present in plasma: particularly, lipid-soluble α-tocopherol (vitamin E), which exerts potent antioxidant and anti-inflammatory properties [17], as well as β-carotene (a vitamin A precursor), and water-soluble vitamin C. Accordingly, the oxidative hypothesis of atherosclerosis suggests that antioxidant supplementation might prevent or retard the genesis and natural history of the atheroma [6]. Consequently, many observations and epidemiological studies have been performed to study the potential relationship between plasma levels of natural antioxidants and cardiovascular event risks (Table 25.1) [18]. Results of these studies are consistent with an inverse correlation between vitamin E plasma levels and cardiovascular mortality [19]. Although most epidemiological studies have demonstrated that vitamin E intake is inversely related with cardiovascular complications, recent studies of antioxidant supplementation in patients with or without a previous history of cardiovascular disease were unable to demonstrate its efficacy in preventing major coronary events (Table 25.2) [20–22].

### *25.2.1 Introduction to Vitamin E*

The antioxidant role of vitamin E has been attributed to its capacity to protect cells and tissues from free radical effects, by acting as a lipid-based, radical chain-breaking molecule. More recently, alternative nonantioxidant functions of

**Table 25.1** Observational studies: cardiovascular disease and antioxidant plasma levels

| Observational studies | Patients' characteristics | Year | Data analyzed | Results |
|---|---|---|---|---|
| WHO/MONICA Project [23] | More than 100,000 middle-aged men | 1991 | Cardiovascular heart disease (CHD) mortality and vitamin E plasma levels | In populations with similar values of serum cholesterol and blood pressure, an inverse correlation between ischemic heart disease (IHD) mortality and vitamin E plasma levels was observed<br>It was estimated that the threshold risk for CVD would be <25 μM. In this particular population, that threshold corresponds to <4.3 μmol vitamin E/mmol cholesterol |
| Riemersma et al. [24] | 110 Cases of angina, 394 controls | 1991 | History of angina and plasma concentrations of vitamins | Patients with a history of angina had a lower vitamin E/cholesterol ratio than controls |
| Singh et al. [25] | 595 elderly people (72 with CHD) | 1995 | Coronary artery disease and vitamins plasma levels | There was an inverse correlation between mortality by CHD and plasma levels of vitamin E and C |
| Feki et al. [26] | 62 patients with atherosclerosis and 65 controls | 2000 | Coronary artery disease and vitamin E plasma levels | There were lower plasma levels of tocopherol in patients with confirmed atherosclerosis, when compared with controls |
| Mezzetti et al. [27] | 102 healthy 80-year-old and older subjects, followed for 47.4 months | 2002 | Cardiovascular events and vitamin plasma levels | Lower vitamin E levels were associated with higher risk of future cardiovascular events. No similar association was observed for vitamin C or beta-carotene |

vitamin E have been proposed [42]. These include the role of vitamin E in cell signaling and gene activity regulation, the specific binding of proteins to α-tocopherol to redirect vitamin E to specific cell locations, and the metabolism of individual tocopherols. Natural vitamin E comprises a mixture of eight different forms: α-, β-, γ-, and δ-tocopherol; and α-, β-, γ-, and δ-tocotrienol. Tocotrienols have unsaturated side chains, whereas tocopherols contain a phytyl tail with three chiral centers (Fig. 25.1), which naturally have the 2R, 4′R, 8′R stereochemistry. Commercially available vitamin E consists of a mixture of tocopherols and tocotrienols having equal amounts of the eight possible tocopherol stereoisomers (*all-rac*), in either their

**Table 25.2** Clinical trials of vitamin E supplementation in cardiovascular disease

| Trial | Number and patient characteristics | Type of prevention[a] | Daily dose and α-tocopherol form | Follow-up (years) | Endpoints | Results |
|---|---|---|---|---|---|---|
| ATBC [28] | 1,862; male smokers, aged 50–69 years | P + S | 50 mg, all-rac[b] | 5–8 | MI[c], death due to cardiovascular events | Vitamin E showed no effect |
| CHAOS [29] | 2,002; mean age 62 years, CAD[d] confirmed | S | 400–800 IU, RRR-[e] | 1.4 | Death due to CVD[f], nonfatal MI | Vitamin E treatment reduced nonfatal MI rates |
| GISSI [30] | 11,324; with recent myocardial infarction | S | 300 mg, all-rac | 3.5 | Cardiovascular death, nonfatal MI | Vitamin E showed no effect |
| HOPE [31] | 9,541; mean age 66 years and at high risk for cardiovascular events | P + S | 400 IU, RRR- | 4–6 | Death due to CVD | Vitamin E showed no effect |
| HOPE-TOO [32] | 7,030, mean age 66 years and at high risk for cardiovascular events | P + S | 400 IU, RRR- | 7 | Death due to CVD | Vitamin E showed no effect |
| SPACE [33] | 196; hemodialysis patients with preexisting CVD | S | 800 IU, RRR- | 1.4 | Death due to CVD, MI | Vitamin E reduced CVD and MI final complications |
| HPS [34] | 20,536; aged 40–80 years, with occlusive arterial disease, or diabetes | P + S | 600 mg,[g] all-rac | 5 | Death due to CVD, MI | Vitamin E showed no effect |
| PPP [35] | 4,495, aged 55–80 years, with one or more cardiovascular risk factors | P | 300 mg, all-rac | 3.6 | Death due to CVD, MI | Results with vitamin E were inconclusive |
| ASAP [36, 37] | 520; men and menopausal women aged 45–69 years, with serum cholesterol ≥ 5.0 mmol/L | P | 272 IU,[h] RRR- | 6 | Common carotid artery intima-media thickness | Vitamins E and C retarded atherosclerosis progression |

**Table 25.2** (continued)

| Trial | Number and patient characteristics | Type of prevention[a] | Daily dose and α-tocopherol form | Follow-up (years) | Endpoints | Results |
|---|---|---|---|---|---|---|
| SU.VI.MAX [38] | 15,000 french adults | P | 30 mg | 7.5 | CVD events | |
| WHS [39] | 39,876 healthy women aged 45 years and older | P | 600 IU | 10 | MI, stroke, CVD death | Vitamin E showed no effect |
| WACS [40] | 8,171 female nurses with previous CVD | S | 400 IU | 9.4 | MI, stroke, coronary revascularization, CVD death | No overall effect of vitamin E |
| PHS II [41] | 14,641 male physicians aged 50 years and older | P | 400 IU | 10 | Major CV events | Vitamin E showed no effect |

[a]P, primary prevention; S, secondary prevention
[b]α-tocopherol acetate of synthesis without stereochemical control of C-2, C-4′, and C-8′
[c]Myocardial infarction
[d,e]2R, 4′R, 8′R-α-tocopherol acetate
[f]Cardiovascular disease
[g] + Vitamin C, 250 mg
[h] + Vitamin C, 500 mg

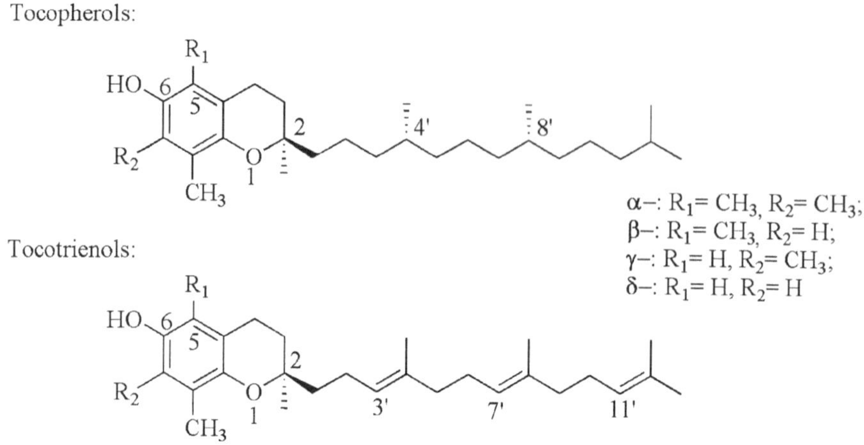


**Fig. 25.1** Vitamin E refers to one or more of the structurally related phenolic compounds named tocopherols and tocotrienols

naturally occurring or ester forms. Further, the bioavailability and bioequivalence of the different forms of vitamin E is complex: although the amount of γ-tocopherol in the diet is higher than that of α-tocopherol, plasma γ-tocopherol concentration is only 10% vs. α-tocopherol, which is the most abundant form in plasma. Dietary vitamin E is taken up by intestinal cells together with nutritional lipids and released in the lymph within chylomicrons. After passing through the lymphatic pathway the chylomicrons reach the systemic circulation, and are progressively hydrolyzed under the action of endothelial lipoprotein lipase (LPL) present in the target tissue [43]. The vitamin reaches the liver via chylomicron remnants, which are taken up mainly via the LDL receptor and the alpha-tocopherol transfer protein (α-TTP). α-TTP selectively sorts out α-tocopherol from all incoming tocopherols for incorporation into very low density lipoprotein (VLDL) [44]. A large fraction of total secreted VLDL is hydrolyzed by LPL and converted into LDL, which becomes the major carrier of vitamin E to the peripheral tissues. Excess amounts of α-tocopherol, as well as of the other tocopherols and tocotrienols, are metabolized and excreted.

### *25.2.2 Randomized and Placebo-Controlled Studies for Primary and Secondary Prevention of Atherosclerosis*

Many randomized and placebo-controlled studies were performed in patients with an increased risk for the development of atherosclerosis or manifest atherosclerosis. In several of these studies for primary or secondary prevention, several antioxidants were administered, but most were focused on vitamin E (from natural sources or synthetic administration). Results of these studies are compiled in Table 25.2. The Alpha-Tocopherol Beta-Carotene Cancer Prevention (ATBC) study was focused on supplementation of α-tocopherol (50 mg/day) and β-carotene (20 mg/day). The median follow-up was 5.3 years. The primary aim of this study was to investigate the effects of this supplementation on the incidence of lung cancer and cardiovascular diseases. In this sense, the results showed that the proportion of major coronary events in smoking men with a previous myocardial infarction was not decreased with supplements. The Cambridge Heart Antioxidant Study (CHAOS) found that among patients with angiographically proven symptomatic coronary atherosclerosis, supplementation with α-tocopherol (800 or 400 IU daily) substantially decreased the incidence of nonfatal myocardial infarction, but not the risk of cardiovascular death. On the other hand, in the GISSI-Prevenzione trial (Gruppo Italiano per lo Studio della Sopravvivenza nell'Infarto Miocardico), 11,324 patients surviving recent myocardial infarction were randomized to receive 300 mg daily of synthetic vitamin E ($n$ = 2,830): 1 g daily of n-3 PUFA ($n$ = 2,836), both ($n$ = 2,830), or none (control, $n$ = 2,828) for 3.5 years. The primary combined endpoint was death, nonfatal myocardial infarction, and stroke. It was observed that vitamin E had no effect on any of the endpoints. In contrast, the Secondary Prevention with Antioxidants of Cardiovascular Disease in Endstage Renal Disease (SPACE) trial enrolled haemodialysis patients with preexisting cardiovascular disease (n = 196), and showed that vitamin E supplementation reduced cardiovascular disease endpoints and myocardial infarction. Patients aged 40–75 years, randomized to receive 800 IU/day vitamin E or matching placebo, were followed for 519 days. Primary endpoints included myocardial infarction, ischemic stroke, peripheral vascular disease, and unstable angina. Secondary endpoints were cardiovascular mortality and total mortality. The Heart Protection Study (HPS) of antioxidant vitamin supplementation included 20,536 UK adults (aged 40–80) with coronary disease, other occlusive arterial disease, or diabetes. They were randomly allocated to receive antioxidant vitamin supplementation (600 mg vitamin E, 250 mg vitamin C, 20 mg β-carotene daily) or matching placebo. Primary outcomes were major coronary events and fatal or nonfatal vascular events. After 5 years, although this regimen increased blood vitamin concentrations substantially, it did not produce any significant benefit. The Primary Prevention Project (PPP) aimed to investigate in general practice the efficacy of vitamin E in primary prevention of cardiovascular events in people with one or more major cardiovascular risk factors. In this study, 4,495 people were randomly allocated to receive low-dose aspirin (100 mg/day) and vitamin E (300 mg/day). The primary endpoint was a composite endpoint of cardiovascular death, stroke, and myocardial infarction. After a mean follow-up of 3.6 years, the trial was prematurely stopped. Vitamin E showed no effect on any prespecified endpoint. The Antioxidant Supplementation in Atherosclerosis Prevention (ASAP) study demonstrated that supplementation with a combination of vitamin E (136 IU twice daily) and vitamin C (250 mg twice daily) slowed down atherosclerotic progression in hypercholesterolemic patients. The subjects were 520 smoking and nonsmoking men and postmenopausal women aged 45–69 years with serum cholesterol $\geq$ 5.0 mmol/L. Atherosclerosis progression was assessed ultrasonographically. The SU.VI.MAX study, a primary prevention trial, examined the role of antioxidants in reducing cardiovascular disease, evaluating for 8 years the efficacy of a balanced combination of antioxidants (including 30 mg of vitamin E). A total of 13,017 french adults (7,876 women aged 35–60 years and 5,141 men aged 45–60 years) was included. Median follow-up time was 7.5 years. This supplementation did not result in any major effect on ischemic cardiovascular disease incidence in men or women. In addition, the Women's Health Study (WHS) tested the efficacy of vitamin E supplements in preventing heart disease or cancer in apparently healthy women. They evaluated 600 IU vitamin E or placebo taken every other day for an average of 10.1 years in 39,876 women aged at least 45 years. There was no significant effect of vitamin E on cardiovascular disease prevention. However, the WHS recently reported that vitamin E supplements decreased the incidence of thromboembolism, providing a possible mechanistic basis for the decreased cardiovascular mortality due to an α-tocopherol's pharmacologic effect [45]. Simultaneously, the Women's Antioxidant Cardiovascular Study (WACS) tested the effects of vitamin E (600 IU every other day) along with other antioxidants on the combined outcome of myocardial infarction, stroke, coronary revascularization, or cardiovascular disease death among 8,171 female health professionals. They were aged 50 years or older, with history of cardiovascular disease or three or more cardiovascular disease risk factors, and were followed for a mean of 9.4 years. This study demonstrated no effect of vitamin E on cardiovascular events of women at high risk of cardiovascular disease. They also found no evidence for harm. These negative results are consistent with the majority of trials of these antioxidants in both primary and secondary prevention. Limitations of the trial include the lack of complete follow-up and compliance. At the same time, the Physicians' Health Study (PHS) II tested the effect of vitamins E and C on 14,641 healthy male physicians given daily supplements of 400 IU vitamin E and 500 mg vitamin C. Primary endpoints combined major cardiovascular events (nonfatal myocardial infarction, nonfatal stroke, and cardiovascular disease death). In this large-scale trial, after an 8-year average treatment and follow-up, neither vitamin E nor vitamin C supplementation reduced the risk of major cardiovascular events.

In summary, in contrast to the epidemiological studies, most of the placebo-controlled studies for primary and secondary prevention of atherosclerosis failed to show a protective effect of α-tocopherol supplementation, even after administration of high doses (Table 25.2). Only in subsets of patients at high risk for atherosclerosis were there suggestive beneficial effects. This lack of protective effect of vitamin E has also been confirmed for many meta-analyses [46–50]. Another noteworthy aspect of antioxidant supplementation is the profile of their side-effects, especially after administration of higher doses. Miller and coworkers [47] performed a meta-analysis of the dose-response relationship between vitamin E supplementation and all-cause mortality by using data of 19 clinical trials. Of these trials, 9 tested vitamin E alone and the remaining 10, vitamin E combined with other vitamins or minerals. The dosages of vitamin E ranged from 16.5 to 2,000 IU/day (median, 400 IU/day). Supplementation of high vitamin E doses alone or in combination with vitamin C was dose-dependently associated with increased all-cause mortality, indicating that long-time vitamin E supplementation of doses $\geq$400 IU/day for at least one year may be harmful. Benefits or risks of lower-dosage supplementation were unclear. However, it should be noted that biological activity of vitamin E compounds differs among isomer forms, and these analyses include vitamin E from different sources and administered according to distinct protocols, either alone or in combination with other substances, for observation periods ranging from 1.4 to 8.2 years. In addition, the included trials were often performed in patients of different groups (i.e., normal population, smokers, patients at cardiovascular risk, patients with chronic diseases) and also different with respect to gender and age. Recently, Bjelakovic and coworkers published the results of their meta-analysis on antioxidant supplements for primary and secondary prevention [49]. Based on 68 randomized trials, the authors concluded that treatment with β-carotene, vitamin A, and vitamin E seems to increase all-cause mortality, while the potential roles of vitamin C and selenium on mortality need further study. However, it should be noted that they included antioxidant supplements at any dose, duration, and route of administration, singly or in combination with other antioxidants bearing different mechanisms of action, biotransformation, and bioavailability. Also, the examined population varied: the effects of supplements were assessed in the general population or in patients with gastrointestinal, cardiovascular, neurological, skin, ocular, renal, endocrinological, and rheumatoid diseases. One interesting observation of this meta-analysis that could be a motive for confounding vitamin supplement effects is that only a small proportion of trials used adequate methodologies [51]. In addition, other factors could explain the failure of vitamin E supplementation to prevent atherosclerosis: (i) the inclusion of patients without biochemical evidence of increased oxidative stress; (ii) the relatively short duration and suboptimal dosages of vitamin E treatment; (iii) the poor patient compliance added to the lack of monitoring of vitamin E levels, among other factors; and (iv) differences in life style, socioeconomic status, individual risk factors, nutrition, and additional administration of other pharmaceuticals [52, 53]. Before additional large, randomized clinical trials of vitamin E are performed, specific biologic and surrogate marker effects of vitamin E in each target population must be defined carefully. Another explanation for the different results in epidemiological studies and studies performed for primary or secondary prevention is that atherosclerosis is a continuously progressive disease. It is possible that initial changes of the vascular wall are responsive to antioxidant therapy, whereas advanced lesions are not [54]. We think that the oxidative hypothesis of atherosclerosis is not necessarily disproved by the failure of antioxidant trials [55], emphasizing the importance of developing new drugs with improved antioxidant capacities in vivo.

## 25.3 Synthetic Antioxidants Represent an Exciting Novel Strategy to Prevent Cardiovascular Disease

Recently, Roberts and coworkers determined by a randomized placebo-controlled study that a daily dose of vitamin E (RRR-α-tocopherol) of more than 1,600 IU for at least 16 weeks is necessary to reduce plasma concentrations of F2-isoprostanes, a biomarker of free radical–mediated lipid peroxidation, in a population at risk for cardiovascular events [56]. Maximum suppression of plasma concentrations of

F2-isoprostanes seen in participants, who were supplemented with the largest dose of vitamin E (3,200 IU/day), was only 49%, suggesting that the antioxidant potency of vitamin E in humans does not seem to be great. Taking into account these findings, together with the above-mentioned conflicting results of antioxidant supplement trials for cardiovascular prevention, long-term treatment with high-dose vitamin E cannot be justified at present. Therefore, Roberts's results reinforce the necessity of developing more potent antioxidants than α-tocopherol. Research has focused on α-tocopherol structural modifications and their role in its antioxidant activity. Thus, many tocopherol derivatives, analogues, and related phenols have been prepared and studied with respect to their reaction with peroxyl radicals [57]. It was found that few tocopherols have higher antioxidant activity than α-tocopherol. However, even though α-tocopherol and an α-tocopherol model in which the phytyl side chain is replaced by a methyl group (Fig. 25.2) have similar antioxidant activity, the latter compound generally shows little or no vitamin E activity in vivo. The phytyl chain appears to be necessary for penetration into phospholipids. In fact, the stereochemistry of the phytyl tail is known to affect the bioactivity of α-tocopherol. Among the eight stereoisomers of α-tocopherol, the natural isomer *RRR*-α-tocopherol has been shown to be the most bioactive [58]. Indeed, the isomer with *S* stereochemistry at position 2 has about 30% of the activity of *RRR*-α-tocopherol. Membranes "recognize" *RRR*-α-tocopherol, and that explain its bioactivity. Then structural modifications of α-tocopherol were studied in order to improve its antioxidant activity (Fig. 25.2): phenols without a fused heterocyclic ring (**1**), substitution of the oxygen of the chroman ring by nitrogen (**5**), sulfur

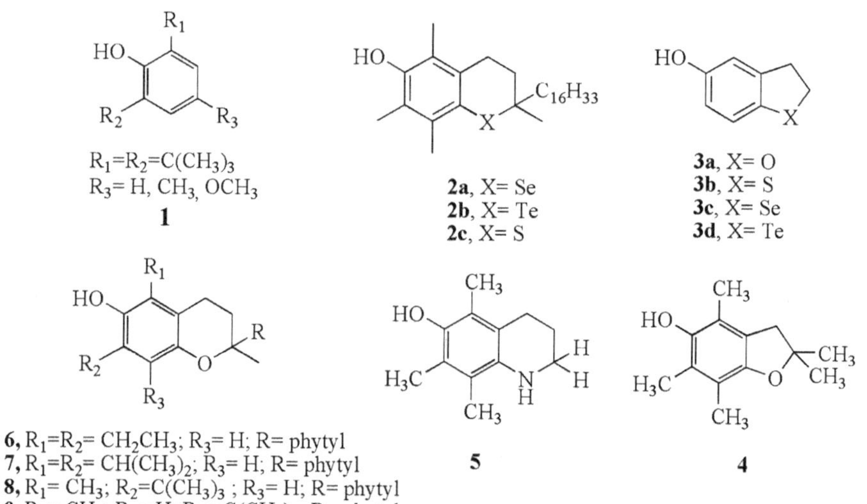


**Fig. 25.2** Structural analogs of α-tocopherol: phenols without a fused heterocyclic ring (**1**), substitution of the oxygen of the chroman ring by nitrogen (**5**), sulfur (**2c, 3b**), selenium (**2a, 3c**), tellurium (**2b, 3d**), with a 5-membered ring instead of a 6-membered ring (**4**), and different substitution patterns at ortho positions of the OH group (**6–9**)

(**2c**, **3b**), selenium (**2a**, **3c**), tellurium (**2b**, **3d**), or also compounds with a 5-membered ring instead of a 6-membered ring (**4**), and different substitution patterns at ortho positions of the OH group (**6–9**). In most cases, however, α-tocopherol continued being the most active.

In the early 1970s, probucol (Fig. 25.3), an antioxidant that significantly reduces plasma cholesterol concentrations, was discovered during a search for nontoxic hypocholesterolemic agents [59]. This compound has been shown to be able to lower plasma cholesterol concentrations in patients with hypercholesterolemia [60–62], preventing the progression of atherosclerosis in Watanabe heritable hyperlipidemic rabbits [63, 64]. The mechanism of action of probucol involves a chain-breaking lipid oxidation activity, inhibiting oxidative modifications of LDL and its uptake by macrophages [65–67]. However, since it was found that it causes adverse effects on cardiac electrophysiology, it is no longer available for clinical studies [68]. Consequently, many efforts have been made to develop compounds related to probucol that have improved efficacy without the adverse effects. Many monoester or monoether derivatives of probucol have been prepared, so they might retain the beneficial antioxidant and lipid-lowering properties of probucol through the remaining phenol group, but could have an improved safety profile based on their inability to form toxic quinone metabolites. These compounds exhibit many of the in vitro properties desirable in a molecule to treat atherosclerosis, with AGI1067 (Fig. 25.3) being the most relevant [69]. This promising finding led AtheroGenics, Inc., to perform a large clinical trial (Aggressive Reduction of Inflammation Stops Events (ARISE), a phase III clinical study) of this lead drug candidate, also known as

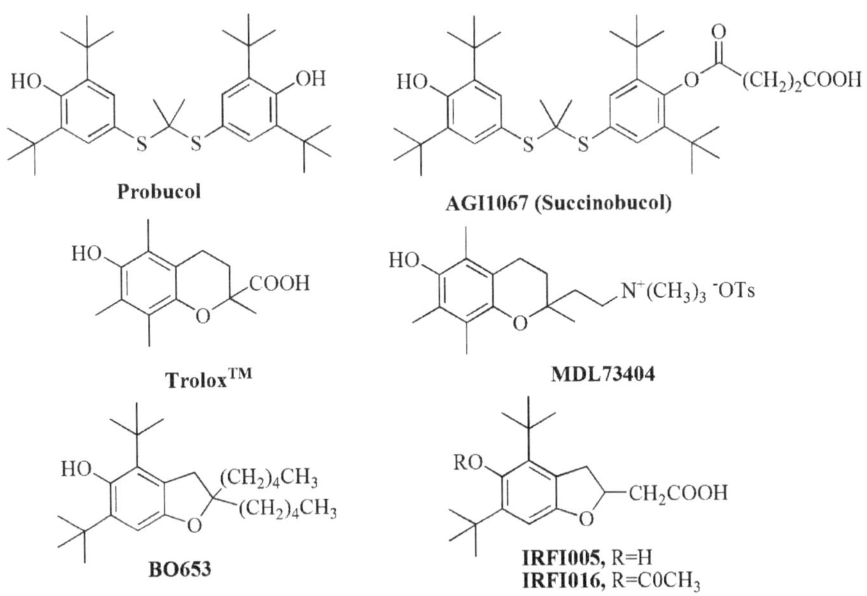


**Fig. 25.3** Structure of Probucol and its derivative succinobucol; a water-soluble α-tocopherol analogue, Trolox, and its analog MDL73404, and 2,3-dihydrobenzofuran analogs (BO653, IRFI005, IRFI016)

succinobucol. However, the results were disappointing [70]. Hence, further studies must be carried out before succinobucol is used in patients with atherosclerosis.

On the other hand, given that under conditions of oxidative stress myocardial concentrations of α-tocopherol are reduced and pretreatment with α-tocopherol reduces membrane-related alterations resulting from ischemia and reperfusion, a water-soluble α-tocopherol analogue, trolox (Fig. 25.3), has been assayed and shown to reduce infarct size in dogs subjected to myocardial ischemia followed by reperfusion [71]. Subsequently, Grisar and coworkers reported an analog (MDL73404, Fig. 25.3) which is also hydrophilic accumulating in heart tissue [72–74], reducing myocardial infarct size following reperfusion [75, 76]. Further analogs of α-tocopherol and ascorbic acid with permanent cationic substituents were synthesized, and found to scavenge lipoperoxyl and superoxide radicals in vitro and to accumulate in heart tissue (cardioselectivity) [77]. The in vivo uptake into myocytes (cardioselectivity) is common to permanently cationic compounds. However, the pharmacological value of this derivative as a cardioprotective agent remains under further investigation. Nowadays, authors are looking for analogs that sufficiently penetrate the brain, in order to develop central nervous system active compounds with application to treat neurodegenerative disorders.

In 1997, Niki and colleagues designed phenolic compounds with all structural features to improve antioxidant potency. This led to BO653 (Fig. 25.3), which showed lipid chain-breaking antioxidant activity [78, 79]. Additional studies addressed its potential application in atherosclerosis [80, 81]. The promising initial findings led to a large phase III clinical trial for safety and effectiveness in preventing restenosis in stented vessels, which has been completed, but a final report by Chugai Pharmaceuticals is not yet available (http://www.clinicaltrials.gov. Accessed 14 Dec 2008). At the same time, raxofelast (IRFI016, Fig. 25.3) and its deacetylated metabolite (IRFI005, Fig. 25.3) emerged from a series of compounds designed by Ceccarelli and coworkers, with the aim of maximizing antioxidant potency of phenols related to α-tocopherol [82]. The antioxidant and radical-scavenging activity of IRFI 005 has been demonstrated in several in vitro systems and in various models of ischemia-reperfusion injury [83–85]. In addition, raxofelast exerts multiple protective antiinflammatory effects in animal models [86–88]. Moreover, several studies in genetically diabetic mice demonstrated potential therapeutic effects [89–91]. Further studies have been performed, and currently raxofelast is in early clinical phase II (Biomedica Foscama SpA., Italy) as a potential therapeutic agent against diabetic complications and atherosclerosis. Other vitamin E derivatives were prepared in order to overcome the clinical limitations of vitamin E, such as the water-soluble tocopherol monoglucoside (TMG, Fig. 25.4) and trimetazidine (TMZ) derivatives (Fig. 25.5). The first was prepared by Murase and colleagues via biotransformation pathways, and its antioxidant activity was previously investigated [92, 93]. In addition, TMG has been shown to inhibit development of atherosclerosis in rabbit aorta at a level similar to that of probucol [94]. Moreover, TMG exerted protective antiinflammatory effects in different animal models [95, 96]. Numerous studies were also performed regarding its radioprotective capacity [97], leading to a phase I trial in patients undergoing hemibody radiation. Although

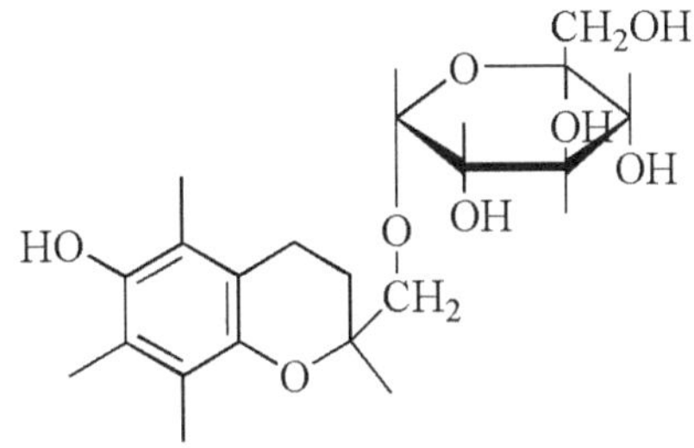

**TMG**

**Fig. 25.4** Tocopherol monoglucoside

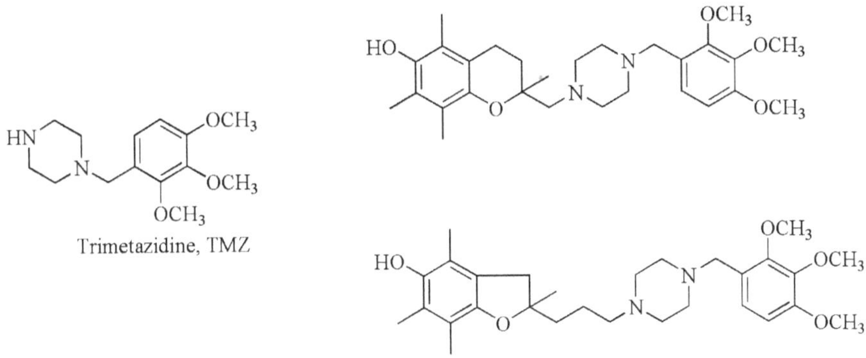

Trimetazidine, TMZ

**Fig. 25.5** Structure of trimetazidine and its derivatives

TMG has the potential to be a good cytoprotector, further clinical evaluation is needed [98].

On the other hand, since trimetazidine (TMZ, Fig. 25.5) is a well-known drug used in angina treatment, Testa and coworkers postulated that coupling TMZ structure with derivatives of tocopherol, flavonoids, coumarin, or cresol would improve the potential activity of these agents [99]. The compounds were examined for their potency to inhibit the oxidation of human serum albumin and lipid peroxidation, and results showed that TMZ derivatives having a chroman or benzofuran group (Fig. 25.5) proved to be quite active.

Since Vitamin E and carotenoids show similar and complementary properties and protect against a variety of pathological processes, and since both compounds are found in nutritional supplements to prevent several diseases, the synthetic linkage of carotenoids with vitamin E might thus increase the chemopreventive activity of the individual compounds. Therefore, Sliwka et al. prepared derivatives shown in Fig. 25.6, combining vitamin E structure, or its water-soluble analog trolox, and carotenoic acid. These antioxidant compounds revealed, in the DPPH (1,1-diphenyl-2-picrylhydrazyl) test, an additive effect, consisting of the radical quenching activity of the carotenoid and trolox [100].

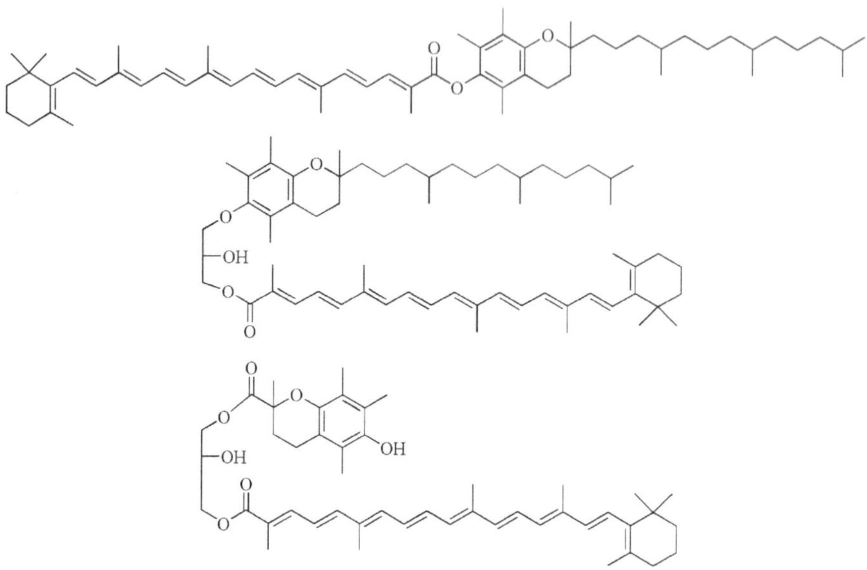

Fig. 25.6 Antioxidant agents which combine in their structure tocopherol and carotenoid moieties

On the basis of these data, which suggest that cooperative interactions may occur between carotenoids and tocopherols, Manfredini and coworkers designed FeAOX-6 (Fig. 25.7) [101], a novel antioxidant that combines into a single molecule the chroman head of tocopherols and a fragment of lycopene. The ability of FeAOX-6 to inhibit lipid peroxidation and reactive oxygen species (ROS) production induced by different free radical sources in arachidonic acid solution and in isolated thymocytes was investigated. Its antioxidant efficiency was also compared with that of α-tocopherol, lycopene, and a mixture of the two antioxidants. The results strongly suggest that FeAOX-6 may be useful as a new potential therapeutic agent in chronic disease in which free radical damage is involved [101]. Further studies were performed with the aim of fully characterizing the antioxidant potential of FeAOX-6 in biological systems [102, 103]. Moreover, considering the essential role in atherogenesis played by macrophages, Manfredini et al. evaluated the effects of the natural antioxidant α-tocotrienol and of the newly designed compound FeAOX-6 on macrophage functions involved in foam cell formation [104]. Results showed that FeAOX-6 or α-tocotrienol induce a strong dose-dependent reduction

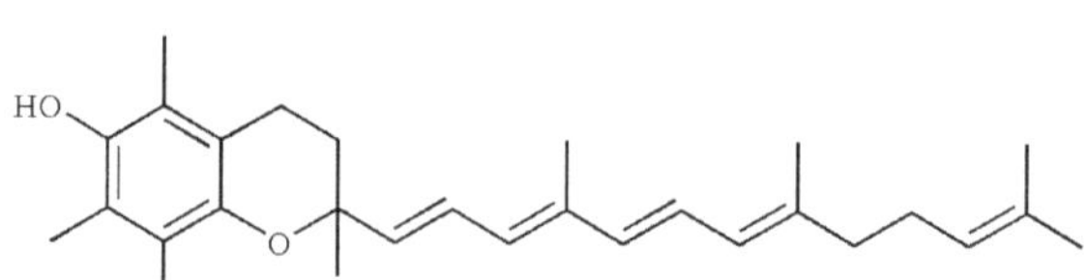

FeAOX-6

Fig. 25.7 Antioxidants which combine antioxidant structural features of both tocopherols and carotenoids into a single molecule

of cholesterol and reduce cholesterol accumulation in human macrophages. The degree of reduction with α-tocotrienol was greater than that induced by FeAOX-6 and did not correlate with their respective antioxidant capacities. In the same way, Manfredini and coworkers [127] designed hybrid analogs of vitamins C and E. In particular, molecular combinations of the pharmacophores of the two vitamins, namely the chromane and the 2,3-dihydroxy-2,3-enono-1,4-lactone rings, were studied (Fig. 25.8).

**Fig. 25.8** Hybrid analogues of vitamin E and C

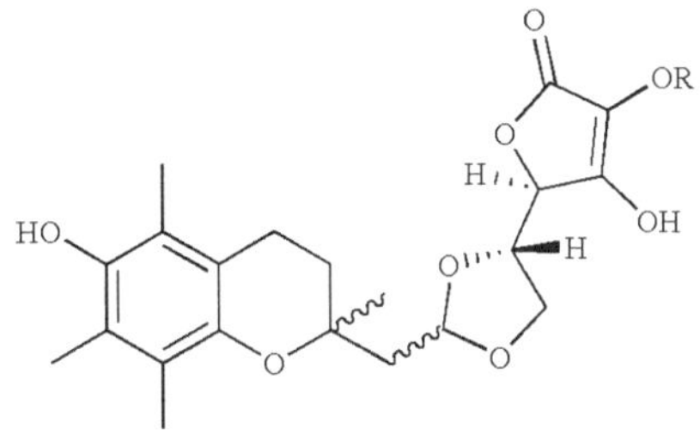


The antioxidant activity of these compounds was investigated by the evaluation of their capability to inhibit malondialdehyde (MDA) production in rat liver microsomal membranes. Moreover, they were further evaluated for ability to reduce ischemia-reperfusion damage, and proved to be effective in preventing damage to isolated rabbit heart. Altogether, these results demonstrated the potential therapeutic applications of these hybrid compounds in pathological events in which free radical damage is involved and the possible extension of this approach to other synergistic antioxidants. Koufaki and coworkers, e.g., synthesized new hybrid compounds combining the pharmacophoric redox moiety of vitamin E with another antioxidant agent such as lipoic acid (Fig. 25.9) [105], and examined their antioxidant activity and protective effects against reperfusion arrhythmias in isolated heart preparations.

All chromanol-lipoic acid hybrid compounds tested were strong inhibitors of lipid peroxidation in rat liver microsomal membranes induced by ferrous ions and ascorbate. Moreover, the new molecules reduced reperfusion arrhythmias. Gasco and coworkers described hybrid compounds of antioxidant and vasodilating agents

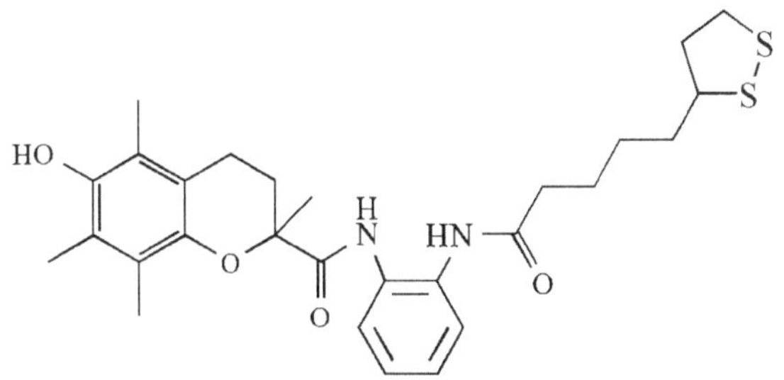


**Fig. 25.9** Relevant chromanol-lipoic hybrid compound

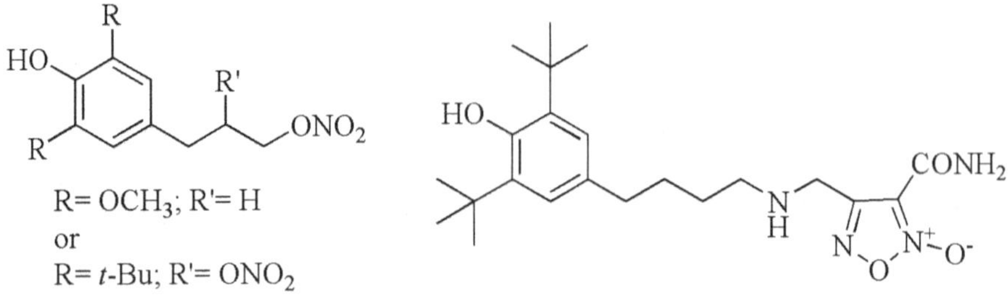


**Fig. 25.10** Hybrid compounds with balanced antioxidant and vasodilating properties

as a new class of potential antiatherogenic agents (Fig. 25.10) [106]. The products were obtained by joining phenols with either nitrooxy or furoxan moieties. In fact, some of the products behave principally as vasodilators and other as antioxidants. However, further in vivo studies should clarify whether some of these products may become preclinical candidates.

Simultaneously, our group started to work on the design, synthesis, and biological characterization of a large series of tocopherol-mimetic nitric oxide ($^{\bullet}$NO)$^{-}$ releasing molecules (Fig. 25.11) [107–109]. It is well known that α-tocopherol

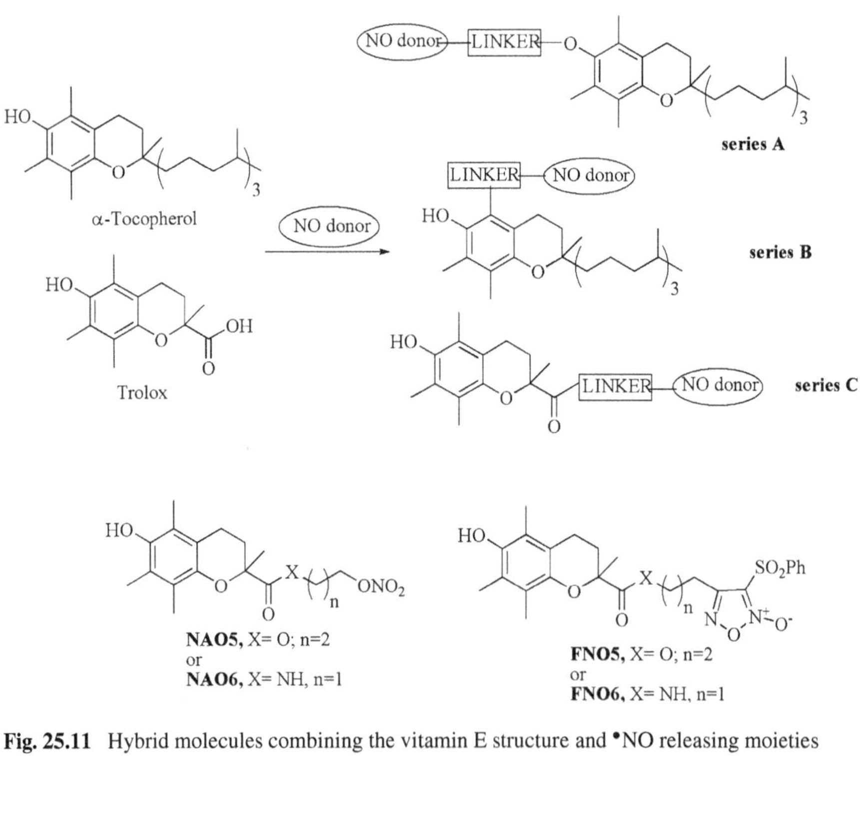


**Fig. 25.11** Hybrid molecules combining the vitamin E structure and $^{\bullet}$NO releasing moieties

is selectively targeted into LDL during its metabolism because of the action of the α-tocopherol transfer protein. Besides, •NO is a free radical species that has strong biological antioxidant actions. In fact, it has been previously shown that •NO inhibits LDL oxidation by scavenging propagating lipid radicals [110–113]. Moreover, •NO is capable of diffusing into LDL, and that makes it a potentially more effective lipid antioxidant than α-tocopherol (Fig. 25.12) [114–116]. Therefore, hybrid molecules combining the vitamin E substructure and •NO releasing moieties (furoxans, organic nitrates, nitrosothiols) to target •NO delivery in vivo, specifically into LDL, should be a possible therapeutic strategy to protect LDL from oxidative modifications.

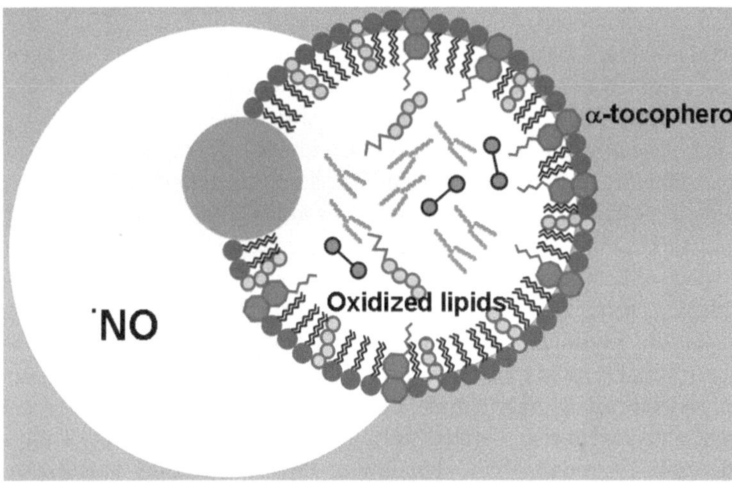


**Fig. 25.12** •NO diffusion into LDL

In recent years, pharmacologic strategies have been developed with the aim of taking advantage of the other biological actions of •NO, beyond its vasodilator role. In particular, •NO-releasing compounds have been investigated with respect to action as anti-inflammatory agents and with the ability to modulate cell proliferation and immune response. In these cases, drug hybrids were developed, coupling a carrier molecule with pharmacologic properties per se, with a •NO-releasing structure: for instance, •NO-releasing aspirin [117], •NO donor-prednisolone [118], and •NO donor-ursodeoxycholic acid [119]. None of these drugs was designed and/or studied with regard to releasing •NO in the LDL core. In contrast, the hybrid compounds developed by us are incorporated with lipoproteins, conferring local and controlled release of •NO into LDL to delay or prevent lipoprotein oxidation. These hybrid compounds release •NO because of the presence of a furoxan or a nitrooxy substructure, inhibiting platelet aggregation and exhibiting vasorelaxing properties. Moreover, they effectively protect LDL from oxidation, combining the tocopherol substructure with affinity to LDL and the antioxidant properties of the •NO-donor. These "site-specific" properties are significant, considering that the protective effects of typical •NO donors or antioxidants decrease with time and distance from biological targets, i.e., LDL. Therefore, in order to study the affinity of synthesized compounds to α-TTP, we performed theoretical and experimental binding studies (unpublished results). In this sense, FNO6 (Fig. 25.11) is the molecule with the best profile. Our observations emphasize the necessity of performing further studies to analyze the LDL-protective activity of these compounds and their atheroprotective effects.

## 25.4 Nitrolipids

Nitric oxide-derived species such as peroxynitrite ($ONOO^-$) and nitrogen dioxide ($NO_2^\bullet$) react with unsaturated fatty acids, yielding a variety of oxidized and nitrated products [112]. Nitrated fatty acids (nitroalkenes) have been detected in the plasma of healthy and hypercholesterolemic patients, as well as in human red blood cell membranes, as novel antiinflammatory mediators, having the capacity to release •NO, exert cGMP-dependent vasorelaxation, and modulate signaling pathways [120–125]. Nitrated unsaturated fatty acids are potent electrophiles that mediate reversible nitroalkylation reactions with both glutathione and Cys and His residues of proteins. This occurs both in vitro and in vivo, and is viewed as a mechanism to transduce redox- and •NO-dependent cell signaling by a covalent, thiol–reversible, post-translational modification that regulates protein structure, function, and subcellular distribution [126].

Current data reveal that nitrated fatty acids serve as mediators of physiological and pathophysiological cell signaling processes, including vascular cell and inflammatory signaling *via* PPARγ-dependent mechanisms [120, 127–130]. Thus, nitroalkenes exert receptor-dependent mediated signaling roles, thereby regulating cell differentiation and inflammatory responses. On the other hand, Cui et al. [131] reported inhibition of proinflammatory cytokines in macrophages by nitro-fatty acids via a PPARγ-independent mechanism. Modification of other transcription factors, i.e., the NF-κB subunit p65, can also explain the protective effects of nitrated lipids. In fact, NF-κB activation in activated macrophages is inhibited by nitro-fatty acids in parallel to nitroalkylation of (p65) Cys38 [131]. Recent studies have shown that heme oxygenase 1 (HO-1) plays a central role in vascular inflammatory signaling reactions and mediates a protective response. Wright et al. [130] reported that nitro-fatty acids mediate the induction of HO-1 by PPARγ-independent mechanisms, representing a novel cell signaling action of inflammation-derived nitroalkenes.

It has been well established that arachidonic acid (AA) signaling cascades and •NO pathways are intrinsically related [132]. We evaluated the ability of nitroarachidonic acid ($AANO_2$) to exert protective actions during macrophage activation by following nitric oxide synthase 2 (NOS2) induction [127]. Micromolar levels of $AANO_2$ caused a reduction of •NO generation in activated cells, while western blot analysis confirmed that NOS2 protein levels were lower in the presence of $AANO_2$

vs. controls. The observed downregulation of NOS2 by $AANO_2$ under inflammatory stimulus should contribute to the physiological shutdown of the inflammatory response in macrophages.

Should we propose nitrated lipids as a novel pharmacological strategy to prevent cardiovascular disease? Even though there are no in vivo studies evaluating the pharmacological properties of these nitrated products, preliminary data in animal models suggest a potential beneficial role of nitrated lipid administration on the development of inflammatory diseases [133–136].

## 25.5 Conclusions

Considering the above discussions in this chapter, site-directed synthetic antioxidants, as well as nitrated lipids, would represent better options than natural antioxidants to prevent cardiovascular diseases. An important difference in the biological actions exerted by synthetic antioxidants and nitrated lipids is that in the latter case, small quantities serve as mediators of potent signal transduction cascades and seem to play homeostatic roles. During oxidative stress conditions related to LDL oxidation and atherosclerosis, administration of synthetic antioxidants directed to LDL (i.e., tocopherol analogs) or nitrated lipids may serve as cyto- and tissue-protective agents, partially counteracting proinflammatory effects of oxidant exposure. This should inhibit the propagation of lipid oxidation in LDL, in part by attenuating the oxidant-dependent inflammatory response, including the downregulation of NOS2. New information from animal studies as well as clinical data is needed in order to allow identifying the "optimal" antioxidant to prevent cardiovascular disease.

**Acknowledgments** This work was supported by grants from Wellcome Trust, UK, and Programa de Desarrollo en Ciencias Básicas (PEDECIBA), Comisión Sectorial de Investigación Científica (CSIC), Fondo Clemente Estable and Programa de Desarrollo Tecnológico, Dinacyt, Uruguay.

## References

1. Schoen FJ (1994) Blood vessels. In: Robbins SL, (ed). Pathologic basis of disease. Saunders, Philadelphia
2. Ross R (1992) Atherogenesis. In: Gallin JI, Goldstein IM, Snyderman R, eds. Inflammation: Basic principles and clinical correlates. Raven Press, New York
3. Steinberg D, Parthasarathy S, Carew TE et al (1989) Beyond cholesterol. Modifications of low-density lipoprotein that increase its atherogenicity. N Engl J Med 320:915–924
4. Berliner JA, Heinecke JW (1996) The role of oxidized lipoproteins in atherogenesis. Free Radic Biol Med 20:707–727
5. Berliner JA, Navab M, Fogelman AM et al (1995) Atherosclerosis: basic mechanisms. Oxidation, inflammation, and genetics. Circulation 91:2488–2496
6. Frostegard J, Haegerstrand A, Gidlund M et al (1991) Biologically modified LDL increases the adhesive properties of endothelial cells. Atherosclerosis 90:119–126
7. Steinberg D (1997) Low density lipoprotein oxidation and its pathobiological significance. J Biol Chem 272:20963–20966

8. Steinbrecher UP, Lougheed M, Kwan WC et al (1989) Recognition of oxidized low density lipoprotein by the scavenger receptor of macrophages results from derivatization of apolipoprotein B by products of fatty acid peroxidation. J Biol Chem 264: 15216–15223
9. Yla-Herttuala S, Palinski W, Rosenfeld ME et al (1989) Evidence for the presence of oxidatively modified low density lipoprotein in atherosclerotic lesions of rabbit and man. J Clin Invest 84:1086–1095
10. Lusis AJ (2000) Atherosclerosis. Nature 407:233–241
11. Vergnani L, Hatrik S, Ricci F et al (2000) Effect of native and oxidized low-density lipoprotein on endothelial nitric oxide and superoxide production: key role of L-arginine availability. Circulation 101:1261–1266
12. Kotamraju S, Hogg N, Joseph J et al (2001) Inhibition of oxidized low-density lipoprotein-induced apoptosis in endothelial cells by nitric oxide. Peroxyl radical scavenging as an antiapoptotic mechanism. J Biol Chem 276:17316–17323
13. Quinn MT, Parthasarathy S, Fong LG et al (1987) Oxidatively modified low density lipoproteins: a potential role in recruitment and retention of monocyte/macrophages during atherogenesis. Proc Natl Acad Sci U S A 84:2995–2998
14. Quinn MT, Parthasarathy S, Steinberg D (1988) Lysophosphatidylcholine: a chemotactic factor for human monocytes and its potential role in atherogenesis. Proc Natl Acad Sci U S A 85:2805–2809
15. Parhami F, Fang ZT, Fogelman AM et al (1993) Minimally modified low density lipoprotein-induced inflammatory responses in endothelial cells are mediated by cyclic adenosine monophosphate. J Clin Invest 92:471–478
16. Witztum JL (1993) Role of oxidised low density lipoprotein in atherogenesis. Br Heart J 69:S12–S18
17. Singh U, Devaraj S, Jialal I (2005) Vitamin E, oxidative stress, and inflammation. Annu Rev Nutr 25:151–174
18. Violi F, Cangemi R, Sabatino G et al (2004) Vitamin E for the treatment of cardiovascular disease: is there a future?. Ann N Y Acad Sci 1031:292–304
19. Gey KF (1994) Optimum plasma levels of antioxidant micronutrients: ten years of antioxidant hypothesis on arteriosclerosis. Bibl Nutr Diet (51):84–99
20. Hooper L, Ness AR, Smith GD (2001) Antioxidant strategy for cardiovascular disease. Lancet 357:1704–1705
21. Eidelman RS, Hollar D, Hebert PR et al (2004) Randomized trials of vitamin E in the treatment and prevention of cardiovascular disease. Arch Intern Med 164:1552–1556
22. Shekelle PG, Morton SC, Jungvig LK et al (2004) Effect of supplemental vitamin E for the prevention and treatment of cardiovascular disease. J Gen Intern Med 19:380–389
23. Gey KF, Puska P, Jordan P et al (1991) Inverse correlation between plasma vitamin E and mortality from ischemic heart disease in cross-cultural epidemiology. Am J Clin Nutr 53:326S–334S
24. Riemersma RA, Wood DA, Macintyre CC et al (1991) Risk of angina pectoris and plasma concentrations of vitamins A, C, and E and carotene. Lancet 337:1–5
25. Singh RB, Ghosh S, Niaz MA et al (1995) Dietary intake, plasma levels of antioxidant vitamins, and oxidative stress in relation to coronary artery disease in elderly subjects. Am J Cardiol 76:1233–1238
26. Feki M, Souissi M, Mokhtar E et al (2000) Vitamin E and coronary heart disease in Tunisians. Clin Chem 46:1401–1405
27. Mezzetti A, Zuliani G, Romano F et al (2001) Vitamin E and lipid peroxide plasma levels predict the risk of cardiovascular events in a group of healthy very old people. J Am Geriatr Soc 49:533–537
28. Rapola JM, Virtamo J, Ripatti S et al (1997) Randomised trial of alpha-tocopherol and beta-carotene supplements on incidence of major coronary events in men with previous myocardial infarction. Lancet 349:1715–1720

29. Stephens NG, Parsons A, Schofield PM et al (1996) Randomised controlled trial of vitamin E in patients with coronary disease: Cambridge Heart Antioxidant Study (CHAOS. Lancet 347:781–786
30. Gruppo Italiano per lo Studio della Sopravvivenza nell'Infarto miocardico (1999) Dietary supplementation with n-3 polyunsaturated fatty acids and vitamin E after myocardial infarction: results of the GISSI-Prevenzione trial. Lancet 354:447–455
31. Yusuf S, Dagenais G, Pogue J et al (2000) Vitamin E supplementation and cardiovascular events in high-risk patients. The Heart Outcomes Prevention Evaluation Study Investigators. N Engl J Med 342:154–160
32. Lonn E, Bosch J, Yusuf S et al (2005) HOPE and HOPE-TOO Trial Investigators. Effects of long-term vitamin E supplementation on cardiovascular events and cancer. J Am Med Assoc 293(11):1338–1347
33. Boaz M, Smetana S, Weinstein T et al (2000) Secondary prevention with antioxidants of cardiovascular disease in endstage renal disease (SPACE): randomised placebo-controlled trial. Lancet 356:1213–1218
34. Heart Protection Study Collaborative Group (2002) MRC/BHF Heart Protection Study of antioxidant vitamin supplementation in 20,536 high-risk individuals: a randomised placebo-controlled trial. Lancet 360:23–33
35. de Gaetano G; Collaborative Group of the Primary Prevention Project (2001) Low-dose aspirin and vitamin E in people at cardiovascular risk: a randomised trial in general practice. Lancet 357:89–95
36. Salonen JT, Nyyssönen K, Salonen R et al (2000) Antioxidant Supplementation in Atherosclerosis Prevention (ASAP) study: a randomized trial of the effect of vitamins E and C on 3-year progression of carotid atherosclerosis. J Intern Med 248:377–386
37. Salonen RM, Nyyssonen K, Kaikkonen J et al (2003) Antioxidant Supplementation in Atherosclerosis Prevention Study Six-year effect of combined vitamin C and E supplementation on atherosclerotic progression: the Antioxidant Supplementation in Atherosclerosis Prevention (ASAP) Study. Circulation 107:947–953
38. Hercberg S, Galan P, Preziosi P et al (2004) The SU.VI.MAX Study: a randomized, placebo-controlled trial of the health effects of antioxidant vitamins and minerals. Arch Intern Med 164:2335–2342
39. I-Min Lee I-M, Cook NR, Gaziano JM et al (2005) Vitamin E in the primary prevention of cardiovascular disease and cancer. The Women's Health Study: a randomized controlled trial. J Am Med Assoc 294:56–65
40. Cook NR, Albert CM, Gaziano JM et al (2007) A randomized factorial trial of vitamins C and E and beta carotene in the secondary prevention of cardiovascular events in women. Results from the Women's Antioxidant Cardiovascular Study. Arch Intern Med 167: 1610–1618
41. Sesso HD, Buring JE, Christen WG et al (2008) Vitamins E and C in the prevention of cardiovascular disease in men: the Physicians' Health Study II randomized controlled trial. J Am Med Assoc 300:2123–2133
42. Azzi A, Ricciarelli R, Zingg JM (2002) Non-antioxidant molecular functions of alpha-tocopherol (vitamin E). FEBS Lett 519:8–10
43. Hacquebard M, Carpentier YA (2005) Vitamin E: absorption, plasma transport and cell uptake. Curr Opin Clin Nutr Metab Care 8:133–138
44. Traber MG, Arai H (1999) Molecular mechanisms of vitamin E transport. Annu Rev Nutr 19:343–355
45. Glynn RJ, Ridker PM, Goldhaber SZ et al (2007) Effects of random allocation to vitamin E supplementation on the occurrence of venous thromboembolism: report from the Women's Health Study. Circulation 116(13):1497–1503
46. Kraemer K, Koch W, Hoppe PP (2004) Is all-rac-α-tocopherol different from RRR-α-tocopherol regarding cardiovascular efficacy? A meta-analysis of clinical trials. Ann N Y Acad Sci 1031:435–438

47. Vivekananthan DP, Penn MS, Sapp SK et al (2003) Use of antioxidant vitamins for the prevention of cardiovascular disease: meta-analysis of randomised trials. Lancet 361: 2017–2023
48. Miller ER, Pastor-Barriuso R, Dalal D et al (2005) Meta-analysis: high-dosage vitamin E supplementation may increase all-cause mortality. Ann Intern Med 142:37–46
49. Bleys J, Miller ER 3rd, Pastor-Barriuso R et al (2006) Vitamin-mineral supplementation and the progression of atherosclerosis: a meta-analysis of randomized controlled trials. Am J Clin Nutr 84:880–887
50. Bjelakovic G, Nikolova D, Gluud LL et al (2007) Mortality in randomized trials of antioxidant supplements for primary and secondary prevention: systematic review and meta-analysis. J Am Med Assoc 297:842–857
51. Gluud LL (2006) Bias in clinical intervention research. Am J Epidemiol 163:493–501
52. Robinson I, de Serna DG, Gutierrez A et al (2006) Vitamin E in humans: an explanation of clinical trial failure. Endocr Pract 12:576–582
53. Siekmeier R, Steffen C, März W (2007) Role of oxidants and antioxidants in atherosclerosis: results of in vitro and in vivo investigations. J Cardiovasc Pharmacol Ther 12: 265–282
54. Engler MM, Engler MB, Malloy MJ et al (2003) Antioxidant vitamins C and E improve endothelial function in children with hyperlipidemia: Endothelial Assessment of Risk from Lipids in Youth (EARLY) Trial. Circulation 108:1059–1063
55. Steinberg D, Witztum JL (2002) Is the oxidative modification hypothesis relevant to human atherosclerosis? Do the antioxidant trials conducted to date refute the hypothesis?. Circulation 105:2107–2111
56. Roberts LJ, Oates JA, Linton MF et al (2007) The relationship between dose of vitamin E and suppression of oxidative stress in humans. Free Radic Biol Med 43:1388–1393
57. Cerecetto H, López GV (2007) Antioxidants derived from vitamin E: an overview. Mini Rev Med Chem 7:315–338
58. Weiser H, Vecchi M (1982) Stereoisomers of alpha-tocopheryl acetate II. Biopotencies of all eight stereoisomers, individually or in mixtures, as determined by rat resorption-gestation tests. Int J Vitam Nutr Res 5:351–370
59. Neuworth MB, Laufer RJ, Barnhart JW et al (1970) Synthesis and hypocholesterolemic activity of alkylidenedithio bisphenols. J Med Chem 13:722–725
60. Cortese C, Marenah CB, Miller NE et al (1982) The effects of probucol on plasma lipoproteins in polygenic and familial hypercholesterolaemia. Atherosclerosis 44:319–325
61. Dachet C, Jacotot B, Buxtorf JC (1985) The hypolipidemic action of probucol. Drug transport and lipoprotein composition in type IIa hyperlipoproteinemia . Atherosclerosis 58:261–268
62. Fellin R, Gasparotto A, Valerio G et al (1986) Effect of probucol treatment on lipoprotein cholesterol and drug levels in blood and lipoproteins in familial hypercholesterolemia. Atherosclerosis 59:47–56
63. Kita T, Nagano Y, Yokode M et al (1987) Probucol prevents the progression of atherosclerosis in Watanabe heritable hyperlipidemic rabbit, an animal model for familial hypercholesterolemia. Proc Natl Acad Sci U S A 84:5928–5931
64. Carew TE, Schwenke DC, Steinberg D (1987) Antiatherogenic effect of probucol unrelated to its hypocholesterolemic effect: evidence that antioxidants in vivo can selectively inhibit low density lipoprotein degradation in macrophage-rich fatty streaks and slow the progression of atherosclerosis in the Watanabe heritable hyperlipidemic rabbit. Proc Natl Acad Sci U S A 84:7725–7729
65. Naruszewicz M, Carew TE, Pittman RC et al (1984) A novel mechanism by which probucol lowers low density lipoprotein levels demonstrated in the LDL receptor-deficient rabbit. J Lipid Res 25:1206–1213
66. Parthasarathy S, Young SG, Witztum JL et al (1986) Probucol inhibits oxidative modification of low density lipoprotein. J Clin Invest 77:641–644

67. Mao SJT, Yates MT, Rechtin AE et al (1991) Antioxidant activity of probucol and its analogs in hypercholesterolemic Watanabe rabbits. J Med Chem 34:298–302
68. Klein L (1981) QT-interval prolongation produced by probucol. Arch Intern Med 141: 1102–1103
69. Somers PK (2000) Monoesters of probucol for the treatment of cardiovascular and inflammatory disease. US Patent 6121319
70. Tardif J-C, McMurray JJV, Klug E et al, for the Aggressive Reduction of Inflammation Stops Events (ARISE) Trial Investigators (2008) Effects of succinobucol (AGI-1067) after an acute coronary syndrome: a randomised, double-blind, placebo-controlled trial. Lancet 371:1761–1768
71. Mickle DAG, Ki R-K, Weisel RD et al (1989) Myocardial salvage with trolox and ascorbic acid for an acute evolving infarction. Ann Thorac Surg 47:553–557
72. Grisar JM, Petty MA, Bolkenius FN et al (1991) A cardioselective, hydrophilic *N,N,N*-trimethylethanaminium alpha-tocopherol analog that reduces myocardial infarct size. J Med Chem 34:257–260
73. Bolkenius FN, Grisar JM, De Jong W (1991) A water-soluble quaternary ammonium analog of alpha-tocopherol that scavenges lipoperoxyl, superoxyl and hydroxyl radicals. Free Radical Res Commun 14:363–372
74. Dow J, Petty MA, Grisar JM et al (1991) Cardioselectivity of alpha-tocopherol analogues, with free radical scavenger activity, in the rat. Drug Metab Dispos 19:1040–1045
75. Petty MA (1992) MDL 73,404: A cardioselective antioxidant that protects against myocardial reperfusion injury. Cardiovasc Drug Rev 10:413–424
76. Lukovic L, Petty MA, Bolkenius FN et al (1993) Protection of infarcted, chronically reperfused hearts by an alpha-tocopherol analogue. Eur J Pharmacol 233:63–70
77. Grisar JM, Marciniak G, Bolkenius FN et al (1995) Cardioselective ammonium, phosphonium, and sulfonium analogs of alpha-tocopherol and ascorbic acid that inhibit in vitro and ex vivo lipid peroxidation and scavenge superoxide radicals. J Med Chem 38: 2880–2886
78. Noguchi N, Iwaki Y, Takahashi M et al (1997) 2,3-Dihydro-5-hydroxy-2,2-dipentyl-4,6-di-tert-butylbenzofuran: design and evaluation as a novel radical-scavenging antioxidant against lipid peroxidation. Arch Biochem Biophys 342:236–243
79. Noguchi N, Okimoto Y, Tsuchiya J et al (1997) Inhibition of oxidation of low-density lipoprotein by a novel antioxidant, BO-653, prepared by theoretical design . Arch Biochem Biophys 347:141–147
80. Cynshi O, Kawabe Y, Suzuki T et al (1998) Antiatherogenic effects of the antioxidant BO-653 in three different animal models. Proc Nat Acad Sci USA 95:10123–10128
81. Noguchi N, Niki E (2000) Phenolic antioxidants: a rationale for design and evaluation of novel antioxidant drug for atherosclerosis. Free Rad Biol Med 28:1538–1546
82. Ceccarelli S, De Vellis P, Scuri R et al (1993) Synthesis of novel 2-substituted-5-oxycoumarans via a direct route to 2,3-dihydro-5-hydroxy-2-benzofuranacetic acids. J Heterocycl Chem 30:679–690
83. Bindoli A, Rigobello MP, Musacchio E et al (1991) Protective action of a new benzofuran derivative on lipid peroxidation and sulphydryl groups oxidation. Pharmacol Res 24: 369–375
84. Iuliano L, Pedersen JZ, Camastra C et al (1999) Protection of low density lipoprotein oxidation by the antioxidant agent IRFI005, a new synthetic hydrophilic vitamin E analogue. Free Rad Biol Med 26:858–868
85. Campo GM, Squadrito F, Campo S et al (1998) Beneficial effect of raxofelast, an hydrophilic vitamin E analogue, in the rat heart after ischemia and reperfusion injury. J Mol Cell Cardiol 30:1493–1503
86. Cuzzocre S, Costantino G, Mazzon E et al (1999) Beneficial effects of raxofelast (IRFI 016), a new hydrophilic vitamin E-like antioxidant, in carrageenan-induced pleurisy. Br J Pharmacol 126:407–414

87. Altavilla D, Famulari C, Passaniti M et al (2003) Lipid peroxidation inhibition reduces NF-kappaB activation and attenuates cerulein-induced pancreatitis. Free Radic Res 37: 425–435
88. Romeo C, Antonuccio P, Esposito M et al (2004) Raxofelast, a hydrophilic vitamin E-like antioxidant, reduces testicular ischemia-reperfusion injury. Urol Res 32:367–371
89. Galeano M, Torre V, Deodato B et al (2001) Raxofelast, a hydrophilic vitamin E-like antioxidant, stimulates wound healing in genetically diabetic mice. Surgery 129:467–477
90. Altavilla D, Saitta A, Cucinotta D et al (2001) Inhibition of lipid peroxidation restores impaired vascular endothelial growth factor expression and stimulates wound healing and angiogenesis in the genetically diabetic mouse. Diabetes 50:667–674
91. Chowienczyk PJ, Brett SE, Gopaul NK et al (2000) Oral treatment with an antioxidant (raxofelast) reduces oxidative stress and improves endothelial function in men with type II diabetes. Diabetologia 43:974–977
92. Murase H, Yamauchi R, Kato K et al (1997) Synthesis of a novel vitamin E derivative, 2-(alpha-D-glucopyranosyl)methyl-2,5,7,8-tetramethylchroman-6-ol, by alpha-glucosidase-catalyzed transglycosylation. Lipids 32:73–78
93. Murase H, Moon J-H, Yamauchi R et al (1998) Antioxidant activity of a novel vitamin E derivative, 2-(alpha-D-glucopyranosyl)methyl-2,5,7,8-tetramethylchroman-6-ol. Free Radic Biol Med 24:217–225
94. Yoshida N, Murase H, Kunieda T et al (2002) Inhibitory effect of a novel water-soluble vitamin E derivative on atherosclerosis in rabbits. Atherosclerosis 162:111–117
95. Yoshida N, Yoshikawa T, Yamaguchi T et al (1999) A novel water-soluble vitamin E derivative protects against experimental colitis in rats. Antioxid Redox Signal 1:555–562
96. Ikeda Y, Mochizuki Y, Nakamura Y et al (2000) Protective effect of a novel vitamin E derivative on experimental traumatic brain edema in rats–preliminary study. Acta Neurochir Suppl 76:343–345
97. Nair CK, Devi PU, Shimanskaya R et al (2003) Water soluble vitamin E (TMG) as a radioprotector. Indian J Exp Biol 41:1365–1371
98. Huilgol NG, Nair CK, Merhotra P et al (2005) A phase I trial of tocoferol monoglucoside in patients undergoing hemi-body radiation. J Can Res Ther 1:38–40
99. Ancerewicz J, Migliavacca E, Carrupt P-A et al (1998) Structure-property relationships of trimetazidine derivatives and model compounds as potential antioxidants. Free Radic Biol Med 25:113–120
100. Larsen E, Abendroth J, Partali V et al (1998) Combination of vitamin E with a carotenoid: α-tocopherol and trolox linked to β-Apo-8'-carotenoic acid. Chem Eur J 4:113–117
101. Palozza P, Piccioni E, Avanzi L et al (2002) Design, synthesis, and antioxidant activity of FeAOX-6, a novel agent deriving from a molecular combination of the chromanyl and polyisoprenyl moieties. Free Radic Biol Med 33:1724–1735
102. Rota C, Tomasi A, Palozza P et al (2006) Antioxidant effect of FeAOX-6 on free radical species produced during iron catalyzed breakdown of tert-butyl hydroperoxide. Free Radic Res 40:141–146
103. Palozza P, Verdecchia S, Avanzi L et al (2006) Comparative antioxidant activity of tocotrienols and the novel chromanyl-polyisoprenyl molecule FeAox-6 in isolated membranes and intact cells. Mol Cell Biochem 287:21–32
104. Napolitano M, Avanzi L, Manfredini S et al (2007) Effects of new combinative antioxidant FeAOX-6 and alpha-tocotrienol on macrophage atherogenesis-related functions. Vascul Pharmacol 46:394–405
105. Manfredini S, Vertuani S, Manfredi B et al (2000) Novel antioxidant agents deriving from molecular combinations of vitamins C and E analogues: 3,4-dihydroxy-5(R)-[2(R,S)-(6-hydroxy-2,5,7,8-tetramethyl-chroman-2(R,S)-yl-methyl)-[1,3]dioxolan-4(S)-yl]-5H-furan-2-one and 3-O-octadecyl derivatives. Bioorg Med Chem 8:2791–2801
106. Cena C, Boschi D, Tron GC et al (2004) Development of a new class of potential antiatherosclerosis agents: NO-donor antioxidants. Bioorg Med Chem Lett 14:5971–5974

107. González M, López GV, Cerecetto H et al (2004) Análogos de tocoferol dadores de óxido nítrico. UY Patent 28445.
108. López GV, Batthyány C, Blanco F et al (2005) Design, synthesis, and biological characterization of potential antiatherogenic nitric oxide releasing tocopherol analogs. Bioorg Med Chem 13:5787–5796
109. López GV, Blanco F, Hernández P et al (2007) Second generation of alpha-tocopherol analogs-nitric oxide donors: synthesis, physicochemical, and biological characterization. Bioorg Med Chem 15:6262–6272
110. Hogg N, Kalyanaraman B, Joseph J et al (1993) Inhibition of low-density lipoprotein oxidation by nitric oxide. Potential role in atherogenesis. FEBS Lett 334:170–174
111. Hogg N, Struck A, Goss SP et al (1995) Inhibition of macrophage-dependent low density lipoprotein oxidation by nitric-oxide donors. J Lipid Res 36:1756–1762
112. Rubbo H, Radi R, Trujillo M et al (1994) Nitric oxide regulation of superoxide and peroxynitrite-dependent lipid peroxidation. Formation of novel nitrogen-containing oxidized lipid derivatives. J Biol Chem 269:26066–26075
113. Rubbo H, Parthasarathy S, Barnes S et al (1995) Nitric oxide inhibition of lipoxygenase-dependent liposome and low-density lipoprotein oxidation: termination of radical chain propagation reactions and formation of nitrogen-containing oxidized lipid derivatives. Arch Biochem Biophys 324:15–25
114. Denicola A, Batthyány C, Lissi E et al (2002) Diffusion of nitric oxide into low density lipoprotein. J Biol Chem 277:932–936
115. Möller M, Botti H, Batthyány C et al (2005) Direct measurement of nitric oxide and oxygen partitioning into liposomes and low density lipoprotein. J Biol Chem 280:8850–8854
116. Rubbo H, Radi R, Anselmi D et al (2000) Nitric oxide reaction with lipid peroxyl radicals spares alpha-tocopherol during lipid peroxidation. Greater oxidant protection from the pair nitric oxide/alpha-tocopherol than alpha-tocopherol/ascorbate. J Biol Chem 275: 10812–10818
117. Gasco A, Fruttero R, Lazzarato L et al (2007) Salicylic acid derivatives. WO2007/060112.
118. Turesin F, del Soldato P, Wallace JL (2003) Enhanced anti-inflammatory potency of a nitric oxide-releasing prednisolone derivative in the rat. Br J Pharmacol 139:966–972
119. Fiorucci S, Antonelli E, Morelli O et al (2001) NCX-1000, a NO-releasing derivative of ursodeoxycholic acid, selectively delivers NO to the liver and protects against development of portal hypertension. Proc Natl Acad Sci U S A 98:8897–8902
120. Baker PR, Lin Y, Schopfer FJ et al (2005) Fatty acid transduction of nitric oxide signaling: multiple nitrated unsaturated fatty acid derivatives exist in human blood and urine and serve as endogenous peroxisome proliferator-activated receptor ligands. J Biol Chem 280: 42464–42475
121. Lima ES, Di Mascio P, Abdalla DS (2003) Cholesteryl nitrolinoleate, a nitrated lipid present in human blood plasma and lipoproteins. J Lipid Res 44:1660–1666
122. Baker PR, Schopfer FJ, Sweeney S et al (2004) Red cell membrane and plasma linoleic acid nitration products: synthesis, clinical identification, and quantitation. Proc Natl Acad Sci U S A 101:11577–11582
123. Lima ES, Bonini MG, Augusto O et al (2005) Nitrated lipids decompose to nitric oxide and lipid radicals and cause vasorelaxation. Free Radic Biol Med 39:532–539
124. Schopfer FJ, Baker PR, Giles G et al (2005) Fatty acid transduction of nitric oxide signalling. Nitrolinoleic acid is a hydrophobically stabilized nitric oxide donor. J Biol Chem 280:19289–19297
125. Villacorta L, Zhang J, García-Barrio MT et al (2007) Nitro-linoleic acid inhibits vascular smooth muscle cell proliferation via the Keap1/Nrf2 signaling pathway. Am J Physiol Heart Circ Physiol 293:H770–H776
126. Batthyány C, Schopfer FJ, Baker PR et al (2006) Reversible post-translational modification of proteins by nitrated fatty acids in vivo. J Biol Chem 281:20450–20463

127. Trostchansky A, Souza JM, Ferreira A et al (2007) Synthesis, isomer characterization, and anti-inflammatory properties of nitroarachidonate. Biochemistry 46:4645–4653
128. Schopfer FJ, Lin Y, Baker PR et al (2005) Nitrolinoleic acid: an endogenous peroxisome proliferator-activated receptor gamma ligand. Proc Natl Acad Sci U S A 102:2340–2345
129. Coles B, Bloodsworth A, Clark SR et al (2002) Nitrolinoleate inhibits superoxide generation, degranulation, and integrin expression by human neutrophils: novel anti-inflammatory properties of nitric oxide-derived reactive species in vascular cells. Circ Res 91:375–381
130. Wright MM, Schopfer FJ, Baker PR et al (2006) Fatty acid transduction of nitric oxide signaling: Nitrolinoleic acid potently activates endothelial heme oxygenase 1 expression. Proc Natl Acad Sci U S A 103:4299–4304
131. Cui T, Schopfer FJ, Zhang J et al (2006) Nitrated fatty acids: Endogenous anti-inflammatory signaling mediators. J Biol Chem 281:35686–35698
132. Salvemini D, Seibert K, Masferrer JL et al (1995) Nitric oxide activates the cyclooxygenase pathway in inflammation. Am J Ther 2:616–619
133. Rubbo H, Trostchansky A, O'Donnell VB (2009) Peroxynitrite-mediated lipid oxidation and nitration. Mechanisms and consequences. Arch Biochem Biophys 484:167–172
134. Ferreira A, Ferrari M, Trostchansky A et al (2009) Macrophage activation induces formation of the anti-inflammatory lipid cholesteryl-nitrolinoleate. Biochemical J 417:223–234
135. Trostchansky A, Moller M, Bartesagui S et al (2010) Nitric oxide redox biochemistry in lipid environments. In: Ignarro, L (ed) NITRIC OXIDE Biology and Pathobiology, 2nd edn. Academic Press, pp 27–60
136. Schopfer, FJ, Cole MP, Groeger AL et al (2010) Covalent peroxisome proliferator-activated receptor gamma adduction by nitro-fatty acids. Selective ligand activity and anti-diabetic signaling actions. J Biol Chem 285:12321–12333

# Chapter 26
# Thioredoxin in the Cardiovascular System—Towards a Thioredoxin-Based Antioxidative Therapy

**Cameron World and Bradford C. Berk**

**Abstract** Oxidative stress results from an imbalance between the production of reactive oxygen species (ROS) and the endogenous antioxidant systems, which detoxify the reactive intermediates. Diseases of the cardiovascular system, including atherosclerosis, diabetes, cardiac hypertrophy, and congestive heart disease are characterized by enhanced production of ROS. In these conditions ROS promote cardiovascular pathology in part by activating inflammatory signaling pathways. One of the principal antioxidant defense mechanisms is the small 12 kDa oxidoreductase, thioredoxin (TRX), that catalyzes the conversion of disulfide oxidized proteins to their thiol-reduced forms. TRX via this catalytic activity confers a protective effect by the ability to regulate pathological signal transduction, inhibit apoptosis, and reduce inflammation. Recent studies using transgenic mice have shown that overexpression of TRX within vascular tissue exerts a protective effect, while injection of recombinant TRX has similarly been demonstrated to possess therapeutic value for the treatment of various vascular diseases.

**Keywords** Thioredoxin · Thioredoxin-interacting protein · Cardiovascular disease · Oxidative stress · Endothelial cell · Cardiomyocyte

## 26.1 Introduction

The thioredoxin (TRX) system (TRX, TRX reductase, and NADPH) is a ubiquitous thiol oxidoreductase system that regulates cellular reduction/oxidation (redox) status (Fig. 26.1). TRX is one of the major antioxidant enzymes in cells catalyzing the reduction of oxidized proteins via its thioltransferase activity. Human TRX1 is a 104 amino acid protein with a molecular weight of 12 kDa that is predominantly

---

B.C. Berk (✉)
University of Rochester Medical Center, Rochester, NY 14642, USA
e-mail: bradford_berk@urmc.rochester.edu

H. Sauer et al. (eds.), *Studies on Cardiovascular Disorders*,
Oxidative Stress in Applied Basic Research and Clinical Practice,
DOI 10.1007/978-1-60761-600-9_26, 

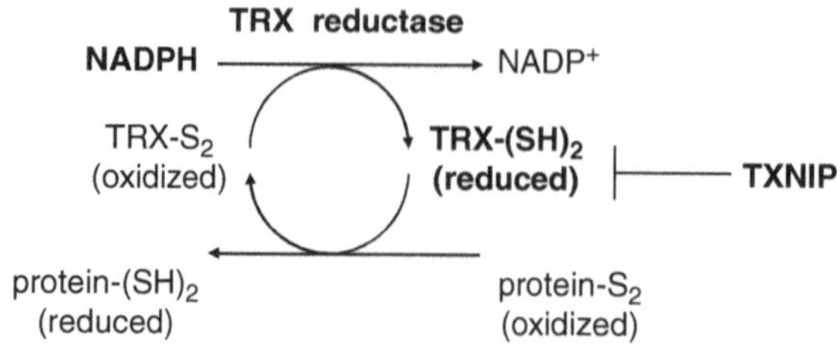


**Fig. 26.1** The redox action of TRX. TRX is an oxidoreductase, which in the reduced state catalyzes the conversion of oxidized proteins ($S_2$) to their reduced form ($SH_2$). As a consequence, TRX itself is oxidized at cysteine residues within its catalytic domain and inactivated. Activated TRX is regenerated by the action of TRX reductase and the expenditure of reducing equivalents in the form of NADPH. The endogenous TRX inhibitor TXNIP is only capable of binding to and inhibiting TRX in its reduced form

localized to the cytosol [1]. Members of the TRX family of proteins possess a conserved catalytic site (Cys32-Gly-Pro-Cys35) that undergoes reversible oxidation to the cysteine disulfide (TRX-$S_2$). As a consequence of reducing equivalents from the catalytic site, cysteine residues are transferred to the disulfide protein substrate (protein-$S_2$), which catalyzes the conversion of an oxidized substrate protein to its reduced form (Fig. 26.1). The oxidized TRX is then reduced back to the dithiol form (TRX-$(SH)_2$) by the NADPH-dependent flavoprotein TRX reductase. In addition to the two catalytic cysteine residues at positions 32 and 35, mammalian TRX1 contains three critical structural cysteine residues at positions -63, -69, and -73 that are not found in prokaryotic TRX proteins. A second TRX (TRX2, a 166 amino acid, 18 kDa protein) was identified in mitochondria [2]. TRX2 has a conserved TRX catalytic site and a consensus signal sequence for mitochondrial translocation. In the mitochondria TRX2 has its own specific TRX2 reductase, which, like that of the cytosolic TRX1 reductase, catalyzes the conversion of oxidized TRX2 to its reduced form at the expenditure of NADPH. Studies have demonstrated that TRX2, like TRX1, protects against oxidant-induced cell death, but many details of how TRX2 functions remain unknown. Thus, unless otherwise indicated, TRX refers to TRX1 in this review.

## 26.2 Actions of TRX

### *26.2.1 Antioxidant Properties*

Principally, the action of TRX is that of an oxidoreductase, through which it acts to catalyze the conversion of disulfides to thiols and as a result to limit oxidative protein damage. Most of its antioxidant properties in cells are thought to work via the ability to regulate thioredoxin peroxidase. As members of a conserved family of peroxiredoxins (PRXs), thioredoxin peroxidases in their reduced form scavenge oxidants such as $H_2O_2$ [3]. Via the ability to control intracellular $H_2O_2$ levels,

PRXs have been shown to play a role in the regulation of signal transduction [3]. Notably, PRXII binds to and regulates the phosphorylation of the platelet-derived growth factor receptor (PDGFR) [4], and PRXII expression was necessary to limit PDGF-mediated smooth muscle cell migration [4]. Furthermore, neointimal thickening in injured carotid arteries was enhanced in PRXII knockout mice compared with wild-type controls [4]. In addition, PRXs, through their action upon peroxynitrite reductase, are capable of reducing the levels of cellular peroxynitrite [5]. The significance of PRX in the pathology associated with cardiovascular disease is illustrated by the demonstration that transgenic overexpression of PRXIII inhibited left ventricular remodeling and heart failure after coronary artery ligation [6].

### *26.2.2 Signaling*

In the cytoplasm, TRX controls signal transduction during oxidative stress by its ability to bind and regulate the activity of kinases and phosphatases. Apoptosis signal-regulating kinase 1 (ASK1) is a mitogen-activated protein (MAP) kinase that activates c-Jun N-terminal kinase (JNK) and p38 MAP kinase, both of which contribute to stress-mediated apoptotic signaling [7]. In response to oxidative stress such as that generated by the inflammatory cytokine tumor necrosis factor (TNFα), TRX dissociates from ASK1 and as a result ASK1 is activated [8]. Moreover, the interaction between ASK1 and TRX was dependent on the oxidative status of TRX [8]. In endothelial cells (EC), laminar shear stress inhibits JNK activation and subsequent induction of vascular cell adhesion molecule (VCAM) expression by increasing the association between TRX and ASK1 [9]. It was demonstrated that this occurred by the ability of laminar flow to downregulate the expression of the endogenous inhibitor of TRX, TRX-interacting protein (TXNIP) [9]. Regulation of TRX by TXNIP will be discussed in further detail below. Various studies have also demonstrated that TRX acts by inhibiting distinct signaling pathways. Transactivation of the epidermal growth factor receptor (EGFR) by either TNFα or lysophosphatidic acid (LPA) is inhibited by overexpression of TRX [10, 11]. The mechanism by which this occurs has not been elucidated, though it is likely to involve the antioxidant activity of TRX because the nonspecific antioxidant N-acetyl cysteine (NAC) exerts a similar inhibitory effect on EGFR transactivation [11]. Furthermore, TRX has been demonstrated to bind to and inhibit the activity of many isoforms of protein kinase C (PKC) [12]. Increased PKC activity (especially of PKC-α, -β, and -δ) has been reported in diabetes [13]. Inhibition of PKC activity prevents extracellular matrix accumulation, increases in vascular permeability, and abnormal angiogenesis associated with the pathology of diabetes [13]. PKCς exhibited elevated activity in EC at regions of the vasculture exposed to disturbed flow [14]. These regions, such as those found at vessel bifurcations or curvatures, preferentially exhibit focal development of atherosclerosis [15]. At these sites it is likely that PKCς contributes to the pathology associated with atherosclerosis, including the ability of PKCς to mediate JNK and caspase-3 activation in response to TNFα stimulation [16]. Finally, TRX regulates the redox status of phosphatases, including PTEN [17], cdc25 [18], calcineurin [19], and PTP1B [20]. The ability of ROS generated in response to growth factors has long been recognized to involve the oxidation of critical catalytic thiol groups of phosphatases [21]. The recovery of phosphatase activity involves the reduction of the disulfide bond formed between two cysteine residues, a reaction catalyzed by TRX. Thus TRX acts to limit the extent and duration of signaling associated with oxidative stress.

### *26.2.3 Transcription*

Though TRX is largely found within the cytoplasm, it can translocate to the nucleus in response to ROS-producing agents such as UV light, phorbol esters, TNFα [22, 23], and low doses of $H_2O_2$ [24]. TRX contains no recognizable nuclear localization signal, though translocation of TRX into the nucleus has been demonstrated to require the activity of TRX reductase, which maintains TRX in the reduced dithiol state [25]. A more recent study has demonstrated a role for lysine 81 and glutamate 82 of TRX, which mediate an interaction with the import receptor karopherin α [24]. In addition, nitric oxide (NO) has been shown to induce TRX translocation to the nucleus. The NO donor S-nitroso-N-acetylpenicillamine (SNAP)-mediated increase of TRX in nuclear extracts was shown to require the redox-sensitive cysteine 118 of the small GTPase Ras [26], a site that is nitrosylated in response to NO and promotes Ras activation [27]. The study of Schroeder et al. [23] showed that Trx itself was nitrosylated at cysteine 69 in response to NO, which was necessary for TRX nuclear localization. The role of nitrosylation to regulate the posttranscriptional activity of TRX will be discussed below. Once in the nucleus, TRX selectively regulates the DNA binding of several transcription factors. The ability of TRX to regulate the activity of NFκB appears to be dependent on its subcellular localization [23]. Overexpression of wild-type TRX suppressed NFκB-dependent gene expression in response to UV irradiation. In contrast, overexpression of nuclear targeted TRX enhanced luciferase activity [23]. In the cytoplasm, TRX was found to bind and block the proteasomal degradation of the NFκB inhibitor IκB [23]. However, in the nucleus, TRX promotes NFκB DNA binding by maintaining the p50 subunit of NFκB in the reduced state [28]. These results suggest that drugs inhibit nuclear localization of TRX and inhibit NFκB-mediated transcription would be a logical approach to treat inflammatory disease, including atherosclerosis, because of the role of NFκB to regulate the expression of various inflammatory genes, including adhesion molecules such as vascular cell adhesion molecule (VCAM), which controls the interaction of monocytes to the endothelium.

TRX has also been demonstrated to bind redox factor 1 (Ref1), an apurinic/apyrimidinic endonuclease, which in turn transfers reducing thiols to AP1 to stimulate its DNA binding [22]. AP1 DNA binding and activity have been shown to require the activity of TRX reductase [25]. The glucocorticoid receptor [29] and estrogen receptor [30] are nuclear receptor transcription factors whose activity is also regulated by the activity of TRX. In the case of the glucocorticoid receptor it was shown that it possessed critical redox-sensitive cysteine residues that, when oxidized, inhibited ligand [31] and DNA binding [32], and were reduced by the action of TRX. The role of TRX in regulating the transcriptional activity of the estrogen receptor may underlie the mechanism by which hormone therapy in the form of estradiol treatment inhibits inflammation, particularly in the endothelium, where it regulates the production of vasodilators such as nitric oxide and prostacyclin [33].

The catalytic activity of TRX was also shown to be necessary to inhibit the activity and binding of the peroxisome proliferator–acativated receptor (PPAR)-α to the PPARα DNA response element [34]. PPARα influences cell proliferation, inflammation, differentiation, and apoptosis, and its expression is enhanced within the myocardium of hypertrophic hearts induced by isoproterenol [35]. In addition, the PPARα signaling pathway is activated in rodent models of streptozotocin-induced diabetes and in the obese db/db mouse [36]. In line with these findings, an inhibitor of PPARα, fenofibrate, inhibited cardiac hypertrophy in response to endothelin [37]. Under hypoxic conditions, TRX, via the action of Ref1, reduces the cysteine thiols of HIF1α, which in turn leads to heterodimerization with HIF1β and binds to the DNA response elements that regulate the expression of VEGF, erythropoietin, and iNOS [38]. TRX by this mechanism regulates angiogenesis, and as such may represent a strategy for the treatment of hypoxic-mediated cardiomyocyte apoptosis within the hypertrophic heart.

### 26.2.4 Survival

Oxidative stress is a key mediator of cell death, and TRX, by its ability to regulate the levels of intracellular ROS, possesses potent antiapoptotic activity. TRX is induced in various human cancers such as neoplastic cervical squamous epithelial cells [39] and hepatocellular carcinoma [40], and downregulated in the dexamethasone-induced apoptosis of a murine thyoma-derived cell line [41]. Overexpression of TRX in WEHI7.2 cells protected cells from drug-induced apoptosis; and when injected into immunodeficient SCID mice, formed tumors that exhibited greater growth than tumors formed by wild-type cells [42]. In human EC, TRX expression was enhanced following exposure to either laminar shear stress or a low dose (10 μM) of $H_2O_2$ concomitant with inhibition of serum-depletion–induced apoptosis [43]. In contrast, specific knockdown of TRX expression with siRNA was found to induce apoptosis in unstimulated cells and reversed the cell survival promoted by $H_2O_2$ treatment [43]. Interestingly, under conditions of severe oxidative stress, TRX translocates to lysosomes where it is degraded by the protease cathepsin D, resulting in enhanced cell death [44]. In contrast, upon exposure to a low dose of $H_2O_2$, TRX translocates to the nucleus, where it increases transcription factor binding to antioxidant response elements and upregulates the expression of the cell survival gene glutathione S-transferase P1 [24]. The ability of TRX to exert an antiapoptotic effect likely occurs through several mechanisms. In addition to acting as an antioxidant, TRX acts as an antiapoptotic protein via the interaction and inhibition of the proinflammatory kinase ASK1 as discussed above. TRX also acts as an antiapoptotic factor by its ability to catalyze the nitrosylation of caspase-3, which inhibits caspase-3 activity [45, 46]. TRX binds procaspase-3 directly and transfers a NO moiety to it, thereby preventing its proteolytic cleavage and activation. Independent studies have further demonstrated a role for TRX in regulating the activation of ERK1/2 and promoting cell survival following exposure to nitrosative stress [47].

## 26.3 Regulation of TRX Activity

TRX contains five redox-reactive, thiol-containing cysteine residues, all of which appear important for function. The thioltransferase catalytic domain involves Cys32 and 35 and is highly conserved throughout evolution. In addition Cys62, 69, and 73, which are not present in prokaryotic TRX or TRX2, are also functionally important [48, 49]. The role of these cysteine residues is evident from the demonstration that they exhibit posttranslational modification, including oxidation, nitrosylation, nitration, and glutathionylation. As a result, an alteration to the extent and duration of posttranslational modification regulates the catalytic activity of TRX.

### *26.3.1 Oxidation*

The catalytic cysteine residues Cys32 and Cys35 are necessary for the thioltransferase activity of TRX. These residues are oxidized and a disulfide formed when TRX catalyzes the reduction of target proteins through the transfer of reducing equivalents from the active site residues to the disulfide substrate. TRX is inactivated by oxidation and requires the flavoprotein TRX reductase to regenerate reduced TRX via the conversion of NADPH to $NADP^+$. In addition, a second disulfide is formed between cysteines 62 and 69 [48], residues not found in either the bacterial TRX or the mitochondrial TRX2. The formation of a nonactive site disulfide between cyteines 62 and 69 is predicted to disrupt the site at which TRX binds TRX reductase. As a result, it was shown that the nonactive site disulfide decreased the rate at which the active site of TRX and thus its activity was regenerated by TRX reductase [48].

### *26.3.2 Nitrosylation*

S-nitrosylation is the reversible covalent binding of nitric oxide (NO) to the thiol group of a reactive cysteine, which may regulate protein function. Several studies have demonstrated that TRX is nitrosylated on cysteine residues 69 and 73

[46, 50–52]. Nitrosylation of TRX on Cys69 maintains TRX activity and promotes the antiapoptotic activity of TRX in EC [50]. This was later confirmed when it was shown that statins increased TRX nitosylation, which enhanced TRX activity [51]. The study of Haendeler et al. [50] also demonstrated that overexpression of TRX increased total cellular nitrocysteine content, while overexpression of a mutant TRX (in which cysteine at position 69 was mutated to a serine) decreased total cellular nitrocysteine content independent of the catalytic cysteines 32 and 35. The authors speculated that nitrosylated TRX on cysteine 69 acted as a transnitrosylation molecule, which inhibited activators of apoptosis by delivering NO to their active sites. It had previously been shown that NO and S-nitrosylation inhibited several members of the caspase family [53], and the ability of TRX to facilitate this reaction was confirmed [45 46]. A role for TRX to inhibit caspase activity via transnitrosylation was shown to require cysteine 73 and involve a direct interaction between TRX and procaspase-3 [46]. Furthermore, inhibition of caspase activity and apoptosis by nitrosylated TRX was shown to be more efficient than with either GSNO or GSH alone [46].

### *26.3.3 Glutathionylation*

Protein glutathionylation, the formation of a disulfide between a protein cysteine residue and the cysteine in the tripeptide glutathione (GSH), is a modification induced in cells by oxidative stress. Glutathionylation can occur by direct oxidation of a protein and GSH, by a thiol-disulfide exchange between a protein cysteine residue and oxidized glutathione (GSSG), and also by S-nitrosoglutathione (GSNO) [54, 55]. Protein glutathionylation is reversible with deglutathionylation catalyzed by glutaredoxin (GRX) [56]. Glutathionylation of TRX occurs at Cys72 in response to both oxidized glutathione (GSSG) and GSNO [57]. Furthermore, it was demonstrated that glutathionylation inhibited TRX activity in a cell-free, in vitro, insulin-reducing assay [57]. To date, a role for the regulation of TRX activity by glutathionylation in vivo remains to be investigated. However, reports have demonstrated that there is an increase in protein glutathionylation in vascular cells in response to inflammatory mediators such as angiotensin II and oxidized LDL [58–60]. In particular, glutathionylation of the small GTPase Ras at cysteine 118 resulted in its activation, promoting the phosphorylation of downstream signaling components such as Akt and p38, and resulting in an induction of protein synthesis in vascular smooth muscle cells stimulated with angiotensin II [58]. Glutathionylation of TRX has been shown to be transient; the mechanism proposed is that reduced TRX is capable of deglutathionylating itself by virtue of its ability to reduce mixed disulfides [57]. Therefore it is likely, though yet to be established, that TRX may exert an additional therapeutic benefit in disease characterized by elevated oxidative stress, such as atherosclerosis and diabetes, via its ability to deglutathionylate kinases and regulate inflammatory signaling.

### *26.3.4 Nitration*

Nitration, the process of an adding of a nitro (-NO2) group to tyrosine residues, has long been used as a marker of reactive nitrogen species in vivo, particularly of peroxynitrite ($ONOO^-$). To this end, it has been shown that there is an increase in the amount of nitrated proteins associated with oxidative stress and apoptosis in the diabetic heart [61]. A recent study demonstrated that TRX was nitrated at tyrosine 49, which inhibited TRX activity and resulted in dissociation of TRX from the inflammatory kinase ASK1 [62]. As a result, nitrated TRX promoted the activation of ASK1 and increased cell apoptosis.

## 26.4 Perturbation of TRX in Vascular Disease

TRX expression is regulated in response to many stimuli, including hypoxia [63], phorbol ester stimulation [64], and UV irradiation [65]. Plasma levels of secreted TRX have been used as markers of inflammatory disease, and are elevated in patients infected with human immunodeficiency virus (HIV) [66], and in the synovial fluid of patients with rheumatoid arthritis [67]. In a similar manner, TRX expression and plasma serum levels are disrupted as a consequence of various cardiovascular diseases.

### *26.4.1 Plasma Levels*

Serum TRX levels have been shown to be significantly elevated in patients with dilated cardiomyopathy and acute coronary syndromes [68]. Wahlgren and Pekkari [69] showed that TRX plasma levels were significantly elevated in patients following angioplasty. Exhaustion of tissue TRX as a result of secretion following chronic oxidative stress is likely to represent a mechanism independent of expression, by which to reduce the presence of antioxidant enzymes such as TRX within vascular tissues. In addition, there has been demonstrable elevation of plasma TRX levels as a clinical marker of oxidative stress associated with cardiovascular disease. Plasma TRX levels and the formation of small platelet aggregates were shown to be higher in patients with acute myocardial infarction compared to those with stable exertional angina [70]. Furthermore, patients with myocardial infarction and high plasma TRX levels demonstrated an increased likelihood of having impaired cardiac function [70]. In addition, plasma TRX levels were found to be elevated in patients with unstable angina compared to those with stable angina [71]. Within this group of patients, those with high TRX plasma levels exhibited a significant difference in the incidence of recurrent angina attacks at rest after treatment [71]. Thus, elevated TRX plasma levels have been considered as a predictive risk factor for various cardiovascular diseases characterized by high oxidative stress.

### 26.4.2 Expression

TRX expression is dynamically regulated in the aorta of the spontaneously hypertensive rat (SHR); it was decreased with increased protein and DNA oxidation [72]. In human atherosclerotic plaques, TRX expression was dramatically enhanced in CD68-positive macrophages and the overlying endothelial cell layer [73]. Furthermore, in balloon-injured rat carotid arteries TRX immunoreactivity begins to increase starting 2 weeks after injury and remains elevated at 6 weeks [73]. It is possible that elevated TRX facilitates endothelial regrowth, which has been shown in an in vitro culture system [74]. As a result, regrowth of the endothelium by enhanced TRX expression is able to limit neointima formation by limiting the continued growth of the underlying smooth muscle cell layers [75]. In cultured endothelial cells, an increase in TRX expression has been observed in response to NO donors such as sodium nitroprusside and nitrosoglutathione [73], low doses of hydrogen peroxide [43], and laminar shear stress [43]. Enhanced TRX expression in response to 10 μM hydrogen peroxide was demonstrated to mediate a prosurvival effect against serum deprivation [43]. In the heart exposed to ischemia-reperfusion, a condition characterized by acutely enhanced ROS generation, TRX expression is reduced [76]. In contrast, when hearts are subjected to short episodes of ischemia, each followed by a small duration of reperfusion, a process known as ischemia preconditioning (which prevents myocardial damage in response to subsequent ischemia), TRX expression is dramatically enhanced [76]. Very recently it has been demonstrated that TRX expression is enhanced in the shoulder of stable human carotid plaques [77], and coronary vessels of patients with unstable angina pectoris [78]. In both of these studies enhanced TRX expression colocalized with increased oxidative stress and was predominately associated with CD68-positive macrophages.

## 26.5 Expression and Actions of a TRX Inhibitor

### 26.5.1 TRX-Interacting Protein

Early studies identified a 50 kDa protein that interacts with TRX in a yeast two hybrid screen [79, 80]. This protein, termed TRX binding protein 2 (TBP2), was found to be identical to vitamin D upregulated protein (VDUP1), which was originally reported as an upregulated gene in HL-60 cells treated with 1α,25-dihydroxy vitamin $D_3$ [81]. Furthermore, TBP2/VDUP1 bound to reduced TRX but not to oxidized TRX nor to mutant TRX, in which the two redox active cysteine residues are substituted by serine [79, 80]. In line with this finding was the demonstration that TBP2/VDUP1 inhibited TRX activity in an insulin-reducing assay [79]. Therefore, TBP2 (now known as TRX-interacting protein [TXNIP]) was considered an inhibitor of TRX activity. It was shown that TXNIP expression was induced by various stress stimuli that mediate an increase in oxidative stress, such as $H_2O_2$

[82]. Overexpression of TXNIP inhibited TRX activity and in parallel increased the activity of the inflammatory kinase ASK1 and its substrate JNK [82]. Furthermore, inhibition of TRX activity by TXNIP mediated an increase in apoptotic cell death in the presence of oxidative stress [82]. Recently, TXNIP has been found to be a biomechanical-sensitive protein regulated by stretch in cardiomyocytes [83] and by laminar shear stress in endothelial cells [9]. TXNIP downregulation in the heart increased cardiomyocyte survival [83], and in the endothelium inhibited inflammation [9]. In support of these conclusions, it was found that TXNIP expression was significantly increased in rat hearts following acute myocardial ischemia [84]. Inhibition of myocardial TXNIP mRNA expression reduced cardiomyocyte apoptosis and activation of the inflammatory kinase ASK1 in response to myocardial ischemia [84]. In addition, inhibition of TXNIP expression in ischemic hearts reduced the cardiac expression of procollagen type I $\alpha 2$ and myocardial scar formation, and significantly improved cardiac function [84]. Studies with a tissue-specific TXNIP-knockout mouse showed TXNIP knockout hearts exhibited attenuated cardiac hypertrophy and preserved left ventricular contractility through four weeks of pressure overload [85]. Some of the beneficial effects of TXNIP inhibition may not be due to increased TRX activity, but rather to altered myocardial energy homeostasis.

Recent data regarding TXNIP regulation of glucose and lipid metabolism suggest multiple mechanisms by which TXNIP may regulate cell homeostasis. Specifically, in mesangial cells TXNIP expression has been found to be one of the most highly inducible genes in response to high glucose [86, 87]. In smooth muscle cells it has been shown that hyperglycemia promotes oxidative stress by its ability to inhibit TRX activity via an induction of TXNIP [88]. In an animal model of streptozotocin-induced diabetes mellitus there was a similar demonstrable decrease in TRX activity that paralleled an induction in TXNIP expression and an increase in oxidative stress within the vessel wall [88]. Furthermore, it was shown that incubation of pulmonary artery smooth muscle cells with the NO donor GSNO suppressed TXNIP expression and as a result increased TRX activity [89]. This observation was found to be independent of TRX nitrosylation (as discussed above) and occurred by an inhibition of TXNIP transcription [89]. Thus, protective mediators such as NO donors may exert some of their effects by decreasing TXNIP expression.

## 26.6 Genetic Manipultion of TRX Expression

### *26.6.1 Transgenic Mice*

An early transgenic study demonstrated that cardiac-specific overexpression of a dominant-negative human TRX (DN-hTRX), in which the disulfide oxidoreductase activity of TRX is selectively inhibited, exhibited significant increases in oxidative stress and cardiac hypertrophy in response to pressure overload [90]. Biochemically, it was demonstrated in DN-hTRX hearts that there were increases in the ERK1/2 activity and Ras thiolation [90], which have been demonstrated to play a role in

ERK activation and subsequent cardiomyocyte growth [91]. It was later confirmed that α-adrenergic receptor-stimulated ventricular myocyte hypertrophy involved the oxidative modification of free thiols on Ras, which was abolished by adenoviral overexpression of TRX [92]. In addition, TRX overexpression was found to inhibit ERK1/2 activation and prevent cellular hypertrophy associated with α-adrenergic receptor stimulation [92]; this occurred independently of the ability of TRX to act as an antioxidant and scavenge ROS. These experiments demonstrated a role for endogenous TRX as a regulator of oxidative stress, signaling, and the pathology associated with cardiac hypertrophy. Transgenic mice in which TRX was overexpressed demonstrated a significant decrease in infarct size, which correlated with improved cardiac function [93]. In addition, TRX overexpressing transgenic mice were protected against adriamycin-induced cardiotoxcity [94]. Adriamycin (ADR), an antitumor agent, has been associated with cardiac failure due to generation of ROS in cardiomyocytes [95]. In TRX transgenic mice, ADR-induced protein oxidation in heart lysates was attenuated, and animal death following chronic ADR exposure was inhibited [94]. Furthermore, electron microscopy analysis of the left ventricular wall revealed that ADR-mediated morphological changes, such as swelling of mitochondria, cristae loss, and disorganized myofilaments, were diminished in TRX transgenic mice [94]. In addition to these studies, overexpression of TRX in transgenic mice also protected against development of both type 1 [96] and type 2 diabetes [97]. Overexpression of TRX in the pancreatic beta cells of nonobese diabetic mice significantly reduced the incidence of diabetes [96]. The same study showed that expression of TRX protected against STZ-induced pancreatic beta cell cytotoxicity [96]. Overxpression of TRX in a mouse model of type 2 diabetes, the db/db mouse, significantly inhibited weight loss and blood glucose levels and increased insulin immunoreactivity within the pancreas [97]. These studies suggest that enhancing TRX function is a therapeutic strategy for the treatment of cardiac and cardiovascular disease.

## 26.7 Therapeutic Use of TRX

Early experiments performed with recombinant TRX (rTRX) demonstrated therapeutic intervention was possible for the treatment of ischemia-reperfusion lung injury. These studies showed that rTRX increased arterial oxygen tension following reperfusion [102–105]; inhibited intraalveolar exudation, interstitial thickening, and cellular infiltration; [105] and inhibited the production of lipid peroxides generated by ROS [104]. Additional studies with rTRX showed benefits in renal ischemia-reperfusion injury [98], arthritis [99], retinal photooxidative damage [100], and hepatic fibrosis [101]. These investigations support the concept that rTRX exerts a therapeutic benefit via its antioxidant properties. Many cardiovascular diseases now are characterized by an increase in the production of ROS (reviewed in [106]). Therefore an increase in the activity of antioxidants such as TRX may represent a mechanism by which to counteract increased levels of oxidative stress, and as a consequence ameliorate the pathology associated with vascular disease.

In a model of reperfusion-induced arrhythmias in isolated rat hearts, rTRX more effectively reduced the incidence of ventricular fibrillation than the administration of recombinant superoxide dismutase (SOD) [107]. In cardiac ischemia-reperfusion, the intraperitoneal injection of rTRX significantly reduced myocyte apoptosis and myocardial infarct size [108]. However, administration of rTRX that had been inactivated by nitration at tyrosine 49 (described above) failed to exert a cardioprotective effect [62]. In the ischemic/reperfused heart, rTRX markedly upregulated the expression of manganese SOD (mnSOD) and reduced tissue superoxide content. This work provides an in vivo demonstration of TRX-mediated induction of mnSOD expression, which in previous studies had been shown with isolated cells [109] and shown to require TRX translocation to the nucleus and activation of CREB [110]. Recently this work was confirmed in a rat model of ischemic injury in which rTRX significantly inhibited myocardial infarct size, systemic oxidative stress levels, and infiltration of inflammatory cells [111]. In addition, rTRX reduced myocyte apoptosis as evidenced by a reduced release of mitochondrial cytochrome c, inhibition of caspase-9 activity, and induction of the prosurvival protein BcL2 [111]. A similar finding was demonstrated in TRX transgenic mice that showed enhanced protection against hypoxia-induced lung injury in which upregulation of BcL2 was considered to be one of the major antiapoptotic effects of TRX [112].

As discussed above, TRX activity is enhanced when nitrosylated on Cys69, which in turn increases the ability of TRX to inhibit the apoptosis of endothelial cells [50]. In line with this, studies showed rTRX inhibited myocardial ischemia, promoted myocyte survival, and inhibited activation of the stress-regulated kinase p38 when rTRX was S-nitrosated by preincubation with S-nitrosoglutathione [113]. Thus the use of nitrosylated TRX as a therapeutic strategy, rather than drugs that stimulate the nitrosylation of endogenous TRX, may represent a novel manner in which to treat diseases characterized by oxidative stress and cell death.

## 26.8 Summary and Conclusions

The small thiolreductase TRX was originally described for its ability to regulate the synthesis of deoxyribonucleotides during S phase by ribonucleotide reductase in E. coli [114]. It has since been shown to be the principal defense mechanism against elevated oxidative stress in eukaryotic cells by its ability to catalyze the conversion of oxidized cysteine residues to reduced thiols. Many of the features of inflammatory cardiovascular disease, including heart failure, atherosclerosis, and diabetes, are characterized by elevated oxidative stress. Overexpression of TRX in transgenic mice inhibited cardiac hypertrophy, reduced infarct size, and improved cardiac function. Furthermore, the administration of recombinant TRX exerted a cardioprotective effect by promoting cardiomyocyte survival, reducing oxidative stress, attenuating inflammatory cell infiltration, and inhibiting infarct size. Recent data show that the actions of TRX are not due simply to its antioxidant functions. Rather, changes in signal transduction regulate cell survival, gene expression, and inflammation; and as a result limit the pathology of various cardiovascular diseases such as cardiac hypertrophy and atherosclerosis, making TRX ideal as a therapeutic strategy that warrants further intensive investigation.

## References


1. Powis G, Mustacich D, Coon A (2000) The role of the redox protein thioredoxin in cell growth and cancer. Free Radic Biol Med 29:312–322
2. Miranda-Vizuete A, Damdimopoulos AE, Spyrou G (2000) The mitochondrial thioredoxin system. Antioxid Redox Signal 2:801–810
3. Fourquet S, Huang ME, D'Autreaux B, Toledano MB (2008) The dual functions of thiol-based peroxidases in $H_2O_2$ scavenging and signaling. Antioxid Redox Signal 10:1565–1576
4. Choi MH, Lee IK, Kim GW et al (2005) Regulation of PDGF signalling and vascular remodelling by peroxiredoxin II. Nature 435:347–353
5. Uwayama J, Hirayama A, Yanagawa T et al (2006) Tissue Prx I in the protection against Fe-NTA and the reduction of nitroxyl radicals. Biochem Biophys Res Commun 339:226–231
6. Matsushima S, Ide T, Yamato M et al (2006) Overexpression of mitochondrial peroxiredoxin-3 prevents left ventricular remodeling and failure after myocardial infarction in mice. Circulation 113:1779–1786
7. Ichijo H, Nishida E, Irie K et al (1997) Induction of apoptosis by ASK1, a mammalian MAPKKK that activates SAPK/JNK and p38 signaling pathways. Science 275:90–94
8. Saitoh M, Nishitoh H, Fujii M et al (1998) Mammalian thioredoxin is a direct inhibitor of apoptosis signal-regulating kinase (ASK) 1. EMBO J 17:2596–2606
9. Yamawaki H, Pan S, Lee RT, Berk BC (2005) Fluid shear stress inhibits vascular inflammation by decreasing thioredoxin-interacting protein in endothelial cells. J Clin Invest 115:733–738
10. Hirota K, Murata M, Itoh T, Yodoi J, Fukuda K (2001) Redox-sensitive transactivation of epidermal growth factor receptor by tumor necrosis factor confers the NF-kappa B activation. J Biol Chem 276:25953–25958
11. Hirota K, Murata M, Itoh T, Yodoi J, Fukuda K (2001) An endogenous redox molecule, thioredoxin, regulates transactivation of epidermal growth factor receptor and activation of NF-kappaB by lysophosphatidic acid. FEBS Lett 489:134–138
12. Watson JA, Rumsby MG, Wolowacz RG (1999) Phage display identifies thioredoxin and superoxide dismutase as novel protein kinase C-interacting proteins: thioredoxin inhibits protein kinase C-mediated phosphorylation of histone. Biochem J 343(Pt 2):301–305
13. Das Evcimen N, King GL (2007) The role of protein kinase C activation and the vascular complications of diabetes. Pharmacol Res 55:498–510
14. Magid R, Davies PF (2005) Endothelial protein kinase C isoform identity and differential activity of PKCzeta in an athero-susceptible region of porcine aorta. Circ Res 97:443–449
15. Gimbrone MA Jr, Topper JN, Nagel T, Anderson KR, Garcia-Cardena G (2000) Endothelial dysfunction, hemodynamic forces, and atherogenesis. Ann N Y Acad Sci 902:230–239; discussion 239–240
16. Garin G, Abe J, Mohan A et al (2007) Flow antagonizes TNF-alpha signaling in endothelial cells by inhibiting caspase-dependent PKC zeta processing. Circ Res 101:97–105
17. Lee SR, Yang KS, Kwon J et al (2002) Reversible inactivation of the tumor suppressor PTEN by $H_2O_2$. J Biol Chem 277:20336–20342
18. Sohn J, Rudolph J (2003) Catalytic and chemical competence of regulation of cdc25 phosphatase by oxidation/reduction. Biochemistry 42:10060–10070
19. Bogumil R, Namgaladze D, Schaarschmidt D et al (2000) Inactivation of calcineurin by hydrogen peroxide and phenylarsine oxide. Evidence for a dithiol-disulfide equilibrium and implications for redox regulation. Eur J Biochem 267:1407–1415

20. Lee SR, Kwon KS, Kim SR, Rhee SG (1998) Reversible inactivation of protein-tyrosine phosphatase 1B in A431 cells stimulated with epidermal growth factor. J Biol Chem 273:15366–15372
21. Cho SH, Lee CH, Ahn Y et al (2004) Redox regulation of PTEN and protein tyrosine phosphatases in H(2)O(2) mediated cell signaling. FEBS Lett 560:7–13
22. Hirota K, Matsui M, Iwata S et al (1997) AP-1 transcriptional activity is regulated by a direct association between thioredoxin and ref-1. Proc Natl Acad Sci U S A 94:3633–3638
23. Hirota K, Murata M, Sachi Y et al (1999) Distinct roles of thioredoxin in the cytoplasm and in the nucleus. A two-step mechanism of redox regulation of transcription factor NF-kappaB. J Biol Chem 274:27891–27897
24. Schroeder P, Popp R, Wiegand B, Altschmied J, Haendeler J (2007) Nuclear redox-signaling is essential for apoptosis inhibition in endothelial cells—important role for nuclear thioredoxin-1. Arterioscler Thromb Vasc Biol 27:2325–2331
25. Karimpour S, Lou J, Lin LL et al (2002) Thioredoxin reductase regulates AP-1 activity as well as thioredoxin nuclear localization via active cysteines in response to ionizing radiation. Oncogene 21:6317–6327
26. Arai RJ, Masutani H, Yodoi J et al (2006) Nitric oxide induces thioredoxin-1 nuclear translocation: possible association with the p21Ras survival pathway. Biochem Biophys Res Commun 348:1254–1260
27. Lander HM, Hajjar DP, Hempstead BL et al (1997) A molecular redox switch on p21(ras). Structural basis for the nitric oxide-p21(ras) interaction. J Biol Chem 272:4323–4326
28. Matthews JR, Wakasugi N, Virelizier JL, Yodoi J, Hay RT (1992) Thioredoxin regulates the DNA binding activity of NF-kappa B by reduction of a disulphide bond involving cysteine 62. Nucleic Acids Res 20:3821–3830
29. Makino Y, Yoshikawa N, Okamoto K et al (1999) Direct association with thioredoxin allows redox regulation of glucocorticoid receptor function. J Biol Chem 274:3182–3188
30. Hayashi S, Hajiro-Nakanishi K, Makino Y et al (1997) Functional modulation of estrogen receptor by redox state with reference to thioredoxin as a mediator. Nucleic Acids Res 25:4035–4040
31. Chakraborti PK, Garabedian MJ, Yamamoto KR, Simons SS Jr (1992) Role of cysteines 640, 656, and 661 in steroid binding to rat glucocorticoid receptors. J Biol Chem 267: 11366–11373
32. Hutchison KA, Matic G, Czar MJ, Pratt WB (1992) DNA-binding and non-DNA-binding forms of the transformed glucocorticoid receptor. J Steroid Biochem Mol Biol 41: 715–718
33. Mikkola T, Ranta V, Orpana A, Ylikorkala O, Viinikka L (1996) Effect of physiological concentrations of estradiol on PGI2 and NO in endothelial cells. Maturitas 25:141–147
34. Liu GH, Qu J, Shen X (2006) Thioredoxin-mediated negative autoregulation of peroxisome proliferator-activated receptor alpha transcriptional activity. Mol Biol Cell 17:1822–1833
35. Yuan J, Wu J, Hang ZG et al (2008) Role of peroxisome proliferator-activated receptor alpha activation in acute myocardial damage induced by isoproterenol in rats. Chin Med J (Engl) 121:1569–1573
36. Finck BN, Lehman JJ, Leone TC et al (2002) The cardiac phenotype induced by PPARalpha overexpression mimics that caused by diabetes mellitus. J Clin Invest 109:121–130
37. Liang F, Wang F, Zhang S, Gardner DG (2003) Peroxisome proliferator activated receptor (PPAR)alpha agonists inhibit hypertrophy of neonatal rat cardiac myocytes. Endocrinology 144:4187–4194
38. Ema M, Hirota K, Mimura J et al (1999) Molecular mechanisms of transcription activation by HLF and HIF1alpha in response to hypoxia: their stabilization and redox signal-induced interaction with CBP/p300. EMBO J 18:1905–1914
39. Fujii S, Nanbu Y, Nonogaki H et al (1991) Coexpression of adult T-cell leukemia-derived factor, a human thioredoxin homologue, and human papillomavirus DNA in neoplastic cervical squamous epithelium. Cancer 68:1583–1591

40. Kawahara N, Tanaka T, Yokomizo A et al (1996) Enhanced coexpression of thioredoxin and high mobility group protein 1 genes in human hepatocellular carcinoma and the possible association with decreased sensitivity to cisplatin. Cancer Res 56:5330–5333
41. Briehl MM, Cotgreave IA, Powis G (1995) Downregulation of the antioxidant defence during glucocorticoid-mediated apoptosis. Cell Death Differ 2:41–46
42. Baker A, Payne CM, Briehl MM, Powis G (1997) Thioredoxin, a gene found overexpressed in human cancer, inhibits apoptosis in vitro and in vivo. Cancer Res 57:5162–5167
43. Haendeler J, Tischler V, Hoffmann J, Zeiher AM, Dimmeler S (2004) Low doses of reactive oxygen species protect endothelial cells from apoptosis by increasing thioredoxin-1 expression. FEBS Lett 577:427–433
44. Haendeler J, Popp R, Goy C et al (2005) Cathepsin D and $H_2O_2$ stimulate degradation of thioredoxin-1: implication for endothelial cell apoptosis. J Biol Chem 280:42945–42951
45. Mitchell DA, Marletta MA (2005) Thioredoxin catalyzes the S-nitrosation of the caspase-3 active site cysteine. Nat Chem Biol 1:154–158
46. Mitchell DA, Morton SU, Fernhoff NB, Marletta MA (2007) Thioredoxin is required for S-nitrosation of procaspase-3 and the inhibition of apoptosis in Jurkat cells. Proc Natl Acad Sci U S A 104:11609–11614
47. Arai RJ, Ogata FT, Batista WL et al (2008) Thioredoxin-1 promotes survival in cells exposed to S-nitrosoglutathione: Correlation with reduction of intracellular levels of nitrosothiols and up-regulation of the ERK1/2 MAP Kinases. Toxicol Appl Pharmacol 233(2):227–237
48. Watson WH, Pohl J, Montfort WR et al (2003) Redox potential of human thioredoxin 1 and identification of a second dithiol/disulfide motif. J Biol Chem 278:33408–33415
49. Weichsel A, Gasdaska JR, Powis G, Montfort WR (1996) Crystal structures of reduced, oxidized, and mutated human thioredoxins: evidence for a regulatory homodimer. Structure 4:735–751
50. Haendeler J, Hoffmann J, Tischler V et al (2002) Redox regulatory and anti-apoptotic functions of thioredoxin depend on S-nitrosylation at cysteine 69. Nat Cell Biol 4:743–749
51. Haendeler J, Hoffmann J, Zeiher AM, Dimmeler S (2004) Antioxidant effects of statins via S-nitrosylation and activation of thioredoxin in Endothelial Cells: A novel vasculoprotective function of statins. Circulation 110:856–861
52. Hashemy SI, Holmgren A (2008) Regulation of the catalytic activity and structure of human thioredoxin 1 via oxidation and S-nitrosylation of cysteine residues. J Biol Chem 283:21890–21898
53. Li J, Billiar TR, Talanian RV, Kim YM (1997) Nitric oxide reversibly inhibits seven members of the caspase family via S-nitrosylation. Biochem Biophys Res Commun 240: 419–424
54. Li J, Huang FL, Huang KP (2001) Glutathiolation of proteins by glutathione disulfide S-oxide derived from S-nitrosoglutathione. Modifications of rat brain neurogranin/RC3 and neuromodulin/GAP-43. J Biol Chem 276:3098–3105
55. Klatt P, Pineda Molina E, Perez-Sala D, Lamas S (2000) Novel application of S-nitrosoglutathione-Sepharose to identify proteins that are potential targets for S-nitrosoglutathione-induced mixed-disulphide formation. Biochem J 349:567–578
56. Holmgren A, Johansson C, Berndt C et al (2005) Thiol redox control via thioredoxin and glutaredoxin systems. Biochem Soc Trans 33:1375–1377
57. Casagrande S, Bonetto V, Fratelli M et al (2002) Glutathionylation of human thioredoxin: a possible crosstalk between the glutathione and thioredoxin systems. Proc Natl Acad Sci U S A 99:9745–9749
58. Adachi T, Pimentel DR, Heibeck T et al (2004) S-glutathiolation of Ras mediates redox-sensitive signaling by angiotensin II in vascular smooth muscle cells. J Biol Chem 279:29857–29862
59. Clavreul N, Adachi T, Pimental DR et al (2006) S-glutathiolation by peroxynitrite of p21ras at cysteine-118 mediates its direct activation and downstream signaling in endothelial cells. FASEB J 20:518–520

60. Clavreul N, Bachschmid MM, Hou X et al (2006) S-glutathiolation of p21ras by peroxynitrite mediates endothelial insulin resistance caused by oxidized low-density lipoprotein. Arterioscler Thromb Vasc Biol 26:2454–2461
61. Turko IV, Li L, Aulak KS et al (2003) Protein tyrosine nitration in the mitochondria from diabetic mouse heart. Implications to dysfunctional mitochondria in diabetes. J Biol Chem 278:33972–33977
62. Tao L, Jiao X, Gao E et al (2006) Nitrative inactivation of thioredoxin-1 and its role in postischemic myocardial apoptosis. Circulation 114:1395–1402
63. Berggren M, Gallegos A, Gasdaska JR et al (1996) Thioredoxin and thioredoxin reductase gene expression in human tumors and cell lines, and the effects of serum stimulation and hypoxia. Anticancer Res 16:3459–3466
64. Nakamura H, Matsuda M, Furuke K et al (1994) Adult T cell leukemia-derived factor/human thioredoxin protects endothelial F-2 cell injury caused by activated neutrophils or hydrogen peroxide. Immunol Lett 42:75–80
65. Sachi Y, Hirota K, Masutani H et al (1995) Induction of ADF/TRX by oxidative stress in keratinocytes and lymphoid cells. Immunol Lett 44:189–193
66. Nakamura H, De Rosa S, Roederer M et al (1996) Elevation of plasma thioredoxin levels in HIV-infected individuals. Int Immunol 8:603–611
67. Yoshida S, Katoh T, Tetsuka T et al (1999) Involvement of thioredoxin in rheumatoid arthritis: its costimulatory roles in the TNF-alpha-induced production of IL-6 and IL-8 from cultured synovial fibroblasts. J Immunol 163:351–358
68. Kishimoto C, Shioji K, Nakamura H et al (2001) Serum thioredoxin (TRX) levels in patients with heart failure. Jpn Circ J 65:491–494
69. Wahlgren CM, Pekkari K (2005) Elevated thioredoxin after angioplasty in peripheral arterial disease. Eur J Vasc Endovasc Surg 29:281–286
70. Miyamoto M, Kishimoto C, Shioji K et al (2003) Cutaneous arteriolar thioredoxin expression in patients with heart failure. Circ J 67:116–118
71. Hokamaki J, Kawano H, Soejima H et al (2005) Plasma thioredoxin levels in patients with unstable angina. Int J Cardiol 99:225–231
72. Tanito M, Nakamura H, Kwon YW et al (2004) Enhanced oxidative stress and impaired thioredoxin expression in spontaneously hypertensive rats. Antioxid Redox Signal 6:89–97
73. Takagi Y, Gon Y, Todaka T et al (1998) Expression of thioredoxin is enhanced in atherosclerotic plaques and during neointima formation in rat arteries. Lab Invest 78:957–966
74. Farina AR, Tacconelli A, Cappabianca L et al (2003) Thioredoxin inhibits microvascular endothelial capillary tubule formation. Exp Cell Res 291:474–483
75. Haudenschild CC, Schwartz SM (1979) Endothelial regeneration. II. Restitution of endothelial continuity. Lab Invest 41:407–418
76. Das DK, Maulik N (2004) Conversion of death signal into survival signal by redox signaling. Biochemistry (Mosc) 69:10–17
77. Okami N, Kawamata T, Yamamoto G et al (2008) Laser microdissection-based analysis of hypoxia- and thioredoxin-related genes in human stable carotid plaques. Cardiovasc Pathol 18(5):294–300
78. Nishihira K, Yamashita A, Imamura T et al (2008) Thioredoxin in coronary culprit lesions: Possible relationship to oxidative stress and intraplaque hemorrhage. Atherosclerosis 201(2):360–367
79. Nishiyama A, Matsui M, Iwata S et al (1999) Identification of thioredoxin-binding protein-2/vitamin D(3) up-regulated protein 1 as a negative regulator of thioredoxin function and expression. J Biol Chem 274:21645–21650
80. Yamanaka H, Maehira F, Oshiro M et al (2000) A possible interaction of thioredoxin with VDUP1 in HeLa cells detected in a yeast two-hybrid system. Biochem Biophys Res Commun 271:796–800
81. Chen KS, DeLuca HF (1994) Isolation and characterization of a novel cDNA from HL-60 cells treated with 1,25-dihydroxyvitamin D-3. Biochim Biophys Acta 1219:26–32

82. Junn E, Han SH, Im JY et al (2000) Vitamin D3 up-regulated protein 1 mediates oxidative stress via suppressing the thioredoxin function. J Immunol 164:6287–6295
83. Wang Y, De Keulenaer GW, Lee RT (2002) Vitamin D(3)-up-regulated protein-1 is a stress-responsive gene that regulates cardiomyocyte viability through interaction with thioredoxin. J Biol Chem 277:26496–26500
84. Xiang G, Seki T, Schuster MD et al (2005) Catalytic degradation of vitamin D up-regulated protein 1 mRNA enhances cardiomyocyte survival and prevents left ventricular remodeling after myocardial ischemia. J Biol Chem 280:39394–39402
85. Yoshioka J, Imahashi K, Gabel SA et al (2007) Targeted deletion of thioredoxin-interacting protein regulates cardiac dysfunction in response to pressure overload. Circ Res 101:1328–1338
86. Cheng DW, Jiang Y, Shalev A et al (2006) An analysis of high glucose and glucosamine-induced gene expression and oxidative stress in renal mesangial cells. Arch Physiol Biochem 112:189–218
87. Qi W, Chen X, Gilbert RE et al (2007) High glucose-induced thioredoxin-interacting protein in renal proximal tubule cells is independent of transforming growth factor-beta1. Am J Pathol 171:744–754
88. Schulze PC, Yoshioka J, Takahashi T et al (2004) Hyperglycemia promotes oxidative stress through inhibition of thioredoxin function by thioredoxin-interacting protein. J Biol Chem 279:30369–30374
89. Schulze PC, Liu H, Choe E et al (2006) Nitric oxide-dependent suppression of thioredoxin-interacting protein expression enhances thioredoxin activity. Arterioscler Thromb Vasc Biol 26:2666–2672
90. Yamamoto M, Yang G, Hong C et al (2003) Inhibition of endogenous thioredoxin in the heart increases oxidative stress and cardiac hypertrophy. J Clin Invest 112:1395–1406
91. Pimentel DR, Adachi T, Ido Y et al (2006) Strain-stimulated hypertrophy in cardiac myocytes is mediated by reactive oxygen species-dependent Ras S-glutathiolation. J Mol Cell Cardiol 41:613–622
92. Kuster GM, Pimentel DR, Adachi T et al (2005) Alpha-adrenergic receptor-stimulated hypertrophy in adult rat ventricular myocytes is mediated via thioredoxin-1-sensitive oxidative modification of thiols on Ras. Circulation 111:1192–1198
93. Turoczi T, Chang VW, Engelman RM et al (2003) Thioredoxin redox signaling in the ischemic heart: an insight with transgenic mice overexpressing Trx1. J Mol Cell Cardiol 35:695–704
94. Shioji K, Kishimoto C, Nakamura H et al (2002) Overexpression of thioredoxin-1 in transgenic mice attenuates adriamycin-induced cardiotoxicity. Circulation 106:1403–1409
95. Doroshow JH (1983) Effect of anthracycline antibiotics on oxygen radical formation in rat heart. Cancer Res 43:460–472
96. Hotta M, Tashiro F, Ikegami H et al (1998) Pancreatic beta cell-specific expression of thioredoxin, an antioxidative and antiapoptotic protein, prevents autoimmune and streptozotocin-induced diabetes. J Exp Med 188:1445–1451
97. Yamamoto M, Yamato E, Toyoda S et al (2008) Transgenic expression of antioxidant protein thioredoxin in pancreatic beta cells prevents progression of type 2 diabetes mellitus. Antioxid Redox Signal 10:43–49
98. Kasuno K, Nakamura H, Ono T, Muso E, Yodoi J (2003) Protective roles of thioredoxin, a redox-regulating protein, in renal ischemia/reperfusion injury. Kidney Int 64:1273–1282
99. Tsuji G, Koshiba M, Nakamura H et al (2006) Thioredoxin protects against joint destruction in a murine arthritis model. Free Radic Biol Med 40:1721–1731
100. Tanito M, Masutani H, Nakamura H et al (2002) Attenuation of retinal photooxidative damage in thioredoxin transgenic mice. Neurosci Lett 326:142–146
101. Okuyama H, Nakamura H, Shimahara Y et al (2003) Overexpression of thioredoxin prevents acute hepatitis caused by thioacetamide or lipopolysaccharide in mice. Hepatology 37:1015–1025

102. Yokomise H, Fukuse T, Hirata T et al (1994) Effect of recombinant human adult T cell leukemia-derived factor on rat lung reperfusion injury. Respiration 61:99–104
103. Yagi K, Liu C, Bando T et al (1994) Inhibition of reperfusion injury by human thioredoxin (adult T-cell leukemia-derived factor) in canine lung transplantation. J Thorac Cardiovasc Surg 108:913–921
104. Fukuse T, Hirata T, Yokomise H et al (1995) Attenuation of ischaemia reperfusion injury by human thioredoxin. Thorax 50:387–391
105. Okubo K, Kosaka S, Isowa N et al (1997) Amelioration of ischemia-reperfusion injury by human thioredoxin in rabbit lung. J Thorac Cardiovasc Surg 113:1–9
106. Lefer DJ, Granger DN (2000) Oxidative stress and cardiac disease. Am J Med 109:315–323
107. Aota M, Matsuda K, Isowa N et al (1996) Protection against reperfusion-induced arrhythmias by human thioredoxin. J Cardiovasc Pharmacol 27:727–732
108. Tao L, Gao E, Hu A et al (2006) Thioredoxin reduces post-ischemic myocardial apoptosis by reducing oxidative/nitrative stress. Br J Pharmacol 149:311–318
109. Das KC, Lewis-Molock Y, White CW (1997) Elevation of manganese superoxide dismutase gene expression by thioredoxin. Am J Respir Cell Mol Biol 17:713–726
110. Chiueh CC, Andoh T, Chock PB (2005) Induction of thioredoxin and mitochondrial survival proteins mediates preconditioning-induced cardioprotection and neuroprotection. Ann N Y Acad Sci 1042:403–418
111. Wu XW, Teng ZY, Jiang LH et al (2008) Human thioredoxin exerts cardioprotective effect and attenuates reperfusion injury in rats partially via inhibiting apoptosis. Chin Med J (Engl) 121:819–826
112. Yamada T, Iwasaki Y, Nagata K et al (2007) Thioredoxin-1 protects against hyperoxia-induced apoptosis in cells of the alveolar walls. Pulm Pharmacol Ther 20:650–659
113. Tao L, Gao E, Bryan NS et al (2004) Cardioprotective effects of thioredoxin in myocardial ischemia and reperfusion: role of S-nitrosation [corrected]. Proc Natl Acad Sci U S A 101:11471–11476
114. Laurent TC, Moore EC, Reichard P (1964) Enzymatic synthesis of deoxyribonucleotides. IV. Isolation and characterization of thioredoxin, the hydrogen donor from Escherichia coli B. J Biol Chem 239:3436–3444

# Chapter 27
# The Protective Effect of Melatonin on the Heart

**Amanda Lochner**

**Abstract** Melatonin (N-acetyl-5-methoxytryptamine) is synthesized mainly by the pineal gland and regulates mammalian and circadian rhythms. It is a highly conserved molecule found in organisms from unicells to vertebrates. Melatonin is a potent free radical scavenger and an antioxidant, and highly effective in protecting against oxidative stress. The hormone is involved in many physiological systems, including the cardiovascular system. This review summarizes the current knowledge regarding the effects of melatonin on the ischaemic-reperfused heart, focusing on a number of newer aspects of melatonin research, namely, its effects on mitochondria, the role of the melatonin receptor, its antiadrenergic actions in cardioprotection, and its effects on intracellular calcium handling. The potential of melatonin as a therapeutic agent for humans is strengthened by the finding that it ameliorates tissue damage in ischaemia/reperfusion in a number of organs. In addition, melatonin is a cheap drug, which can be obtained without prescription. It is absorbed rapidly, which makes it an ideal drug for use in conditions characterized by the production of copious amounts of free radicals.

**Keywords** Melatonin receptors · Free radical · Scavenger · Anti-oxidant · Cardioprotection · Ischaemia/reperfusion injury · Mitochondrial · Permeability transition pore · Antiadrenergic actions

## 27.1 Melatonin and the Heart

Melatonin (N-acetyl-5-methoxytryptamine) is synthesized mainly by the pineal gland and regulates mammalian and circadian rhythms. It is a highly conserved molecule found in organisms from unicells to vertebrates [1]. Its metabolism, synthetic rate-limiting enzymes, sites of synthesis, regulatory mechanisms, mechanism

A. Lochner (✉)
Division of Medical Physiology, Department of Biomedical Sciences, Faculty of Health Sciences, University of Stellenbosch, Tygerberg 7505, Republic of South Africa
e-mail: alo@sun.ac.za

H. Sauer et al. (eds.), *Studies on Cardiovascular Disorders*, Oxidative Stress in Applied Basic Research and Clinical Practice,
DOI 10.1007/978-1-60761-600-9_27, 

of breakdown, and the function of its metabolites have recently been reviewed by Tan and coworkers [2]. The hormone is localized ubiquitously in the cytosol, membrane, and nuclear compartments, with the highest concentration in the mitochondria [3]. Melatonin can easily reach all cellular compartments because of its small size and amphiphilic nature.

The finding made by Tan and coworkers [4] that melatonin is a potent free radical scavenger and an antioxidant has since been confirmed by several workers, and several reviews have appeared on this particular topic (see for example [2, 5–8]). The major metabolites of melatonin, N1-acetyl-N2-formyl-5-methoxykynuramine (AFMK) and N-acetyl-5-methoxykynuramine (AMK), also exhibit free radical scavenging ability. It has been calculated that, considering the cascade of reactions that includes AFMK, one melatonin molecule can scavenge up to 10 reactive oxygen (ROS) or nitrogen (RNS) species [2]. This cascade of reactions therefore makes melatonin highly effective, even at low concentrations, in protecting against oxidative stress.

Melatonin possesses an electron-rich aromatic indole ring and functions as an electron donor, thereby reducing and repairing electrophilic radicals [9]. Thus it has been shown to directly scavenge oxygen-based free radicals such as $^{\bullet}OH$ [4, 10, 11], superoxide [6], $H_2O_2$ [8], hypochlorous acid (HOCl) [12], singlet oxygen [13], nitric oxide, and peroxynitrite anion [6, 14–16]. It also reduces electron leakage and free radical formation by mitochondria [3, 17]. The potent free radical scavenging and antioxidant actions of melatonin are also demonstrated by the significant reduction in oxidative stress in streptozotocin-induced diabetic hearts when treated with melatonin [18].

Interestingly, melatonin is 5 and 14 times more effective at scavenging the highly toxic hydroxyl radical than glutathione and mannitol, respectively [19], and twice as efficient as vitamin E in detoxifying the peroxyl radical [20].

In addition to melatonin's activities as a free radical scavenger, it upregulates antioxidant enzymes (e.g., SOD, catalase, glutathione peroxidase) and downregulates pro-oxidant enzymes (e.g., nitric oxide synthase, 5- and 12-lipoxygenases) [21–24]. Melatonin has also been reported to bind quinone reductase 2, which is considered a melatonin receptor and important in the detoxification of quinones [25, 26]. The direct free radical scavenging effects of melatonin are receptor-independent [7], but its indirect antioxidative functions may be mediated by receptors, located either in the cell membrane or nucleus [10, 27]. However, the mechanism whereby melatonin stimulates antioxidant enzyme activities is still unclear. It has been proposed that melatonin acts via its interaction with calmodulin [28, 29], by antagonizing the physiological effects of calmodulin [30]. This has been suggested to lead to inactivation of the RORα receptor, a nuclear transcription factor, and to a subsequent increment in antioxidant enzyme activities [22]. This, however, still needs to be verified.

Apart from the above, melatonin is involved in many other physiological systems, including the cardiovascular system [31, 32]. A link between the pineal gland, melatonin, and the heart was indicated by a number of observations made in pinealectomized animals: (i) Pinealectomy increased serum cholesterol and myocardial malondialdehyde (MDA) levels as well as heart weight. These changes could be reversed by melatonin administration. The myocardial fibrosis and myxomatous degeneration of cardiac valves which were seen in all rats could, however, not be reversed by the short-term application of melatonin [33]. (ii) Pinealectomy renders the heart more susceptible to ischaemic/reperfusion injury, as shown by an increase in infarct size in hearts from pinealectomized animals subjected to ischaemia/reperfusion [34]. (iii) Similarly, reducing melatonin levels by continuous light exposure caused increases in body and heart weights and myocardial collagen I/III ratio [35].

Acute myocardial infarction appears to be associated with reduced nocturnal melatonin levels. Significantly lower serum melatonin levels [36] and reduced melatonin synthesis [37] were reported in patients with coronary heart disease. A later study on patients diagnosed with acute myocardial infarction revealed a reduced nocturnal melatonin elevation, associated with increased oxidative stress [38]. Similar reduced nocturnal melatonin release in coronary artery disease patients [39] and in patients with cardiac syndrome X [40] were also reported.

It is well established that myocardial ischaemia/reperfusion causes free radical generation, which may lead to intracellular damage and cell death (the so-called reperfusion injury). Thus there is continued interest in discovering and characterizing new free radical scavengers and antioxidants of high potency and low toxicity. Melatonin fulfills most of these criteria, and the virtual absence of toxicity renders it a good candidate for long-term use [11]. The possible beneficial effects of melatonin on the ischaemic reperfused heart have received much attention during the past few years, as reflected by a number of recent reviews [35, 41–43].

In view of the above, this review will summarize our current knowledge regarding the effects of melatonin on the ischaemic-reperfused heart, focusing on a number of newer aspects of melatonin research, namely, its effects on mitochondria, the role of melatonin receptor-mediated events, and its antiadrenergic actions in cardioprotection. The effect of melatonin on the regulation of blood pressure has recently been reviewed by Paulis and Simko [35] and will not be discussed in this review.

## 27.2 Melatonin and Ischaemia/Reperfusion Injury

Most of the studies thus far have used rats as experimental animals, applying pharmacological concentrations of the hormone. Using the isolated, perfused rat heart as a model, melatonin was shown to reduce premature ventricular contractions and ventricular fibrillation [44–46], improve functional recovery during reperfusion [46–49], and reduce infarct size [47–50]. It was also shown that melatonin-induced cardioprotection was associated with attenuated myeloperoxidase activity and MDA levels, suggesting a reduction in lipid peroxidation and cell injury [51]. In view of the generation of large numbers of free radicals at the onset of reperfusion [52], it is not surprising that the time of administration of melatonin is crucial: using the perfused rat heart, it has to be administered either before and after ischaemia or only during reperfusion after ischaemia to induce cardioprotection [48]. However, it has also been shown that melatonin administered in vivo before the ischaemic insult was effective. It was assumed that, under these circumstances, melatonin was still present in sufficient quantities at the onset of reperfusion [50, 53].

Pharmacological concentrations were used in the above studies, showing a dose-dependent effect of melatonin on cardioprotection [45, 48]. However, two recent reports showed that reduction of endogenous circulating melatonin by pinealectomy increased myocardial injury and reduced the survival rate of animals. These findings suggest that relatively low endogenous melatonin levels are protective of the heart during episodes of hypoxia and reoxygenation [34]. A later study also showed that infarct size was larger in the ischaemia/reperfused heart when rats were melatonin-deficient [50], also confirming the effectiveness of physiological concentrations of melatonin. The findings are difficult to reconcile with the melatonin dose-dependency observed in perfused hearts [48], where concentrations of $< 50\ \mu M$ melatonin were found to be ineffective cardioprotectants. Clearly this has to be investigated further.

In most of the above studies on the cardioprotective actions of melatonin, only the acute effects of melatonin were studied. In view of the question whether melatonin has a future as a cardioprotective agent [54], it became necessary to evaluate the effects of long-term treatment of the hormone on ischaemic/reperfusion injury. Long-term treatment of melatonin per se has been shown to be without toxic effects [48, 55]. For example, the long-term effects with melatonin treatment were evaluated by determining infarct size and functional recovery after global ischaemia, 24 h after intraperitoneal administration of melatonin (2.5 or 5.0 mg/kg) or after oral administration in the drinking water for 7 days (40 $\mu$g/ml). Both interventions caused a significant reduction in the infarct size of hearts subjected to 35 min of regional ischaemia. Moreover, the cardioprotection persisted for 2–4 days after discontinuation of treatment [48], suggesting the potential of long-term treatment with melatonin.

## 27.3 Role of the Melatonin Receptors in Cardioprotection

Despite the overwhelming evidence for the involvement of free radical scavenging and antioxidant activity of melatonin in cardioprotection, convincing findings have also been reported suggesting a role for its receptors in this regard.

Melatonin receptors have been found to be widely distributed not only in the brain but in many other organs as well [56]. Two high–affinity, G-protein-coupled melatonin receptors have been identified in mammals: namely, the MT1 (formerly Mel 1a or mt1) and MT2 (Mel 1b) receptors [57, 58]. The MT1 receptor was identified in chicken [59] and human [60] coronary arteries, as well as in chicken [61] and rat [62] hearts, coronary arteries, and aortas [60]. This receptor was shown to be associated with various second messengers: for example, $G_i$ protein-coupled decrease in cAMP levels [63–65], Gq-coupled phospholipase C activation [66], or G-coupled activation of the Kir 3 K-channels [67]. The MT2 receptor also couples with the Gq-protein. The third receptor type, MT3, which has a lower affinity for melatonin, is probably not coupled with a G protein [68] and belongs to the family of quinine reductases [25, 69]. However, the precise localization of these receptors is not yet known. In view of the fact that melatonin readily passes through membranes, it is possible that these receptors are localized intracellularly. Nuclear binding sites/receptors for melatonin have been found in several tissues, and belong to the RZR/ROR orphan receptor family, which includes the products of three genes: splicing variants of RORα, RZRβ, and RORγ [70, 71]. According to a recent review by Reiter et al. [72], this wide-spread distribution of cellular membrane receptors and the existence of nuclear receptors could indicate that melatonin functions in every cell it comes in contact with, particularly since there are no morphophysiological barriers to the hormone.

Despite the current knowledge regarding the putative role of the melatonin receptor in the heart, very little is known about its intracellular signaling pathways in this tissue. It is generally accepted that melatonin activation of the MT1/2 receptor, via the $G_i$ protein, stimulates the phospholipase C pathway and causes activation of the cAMP response element binding protein (CREB), which, in turn, modulates early gene transcription [42]. However, most of the current knowledge regarding the intracellular signaling of melatonin comes from tissues other than the heart. It also appears that some of its effects are dose- and tissue-dependent. For example, its antiproliferative action on HUVECs occur via inactivation of the ERK/Akt/PKC pathways [73]. Similarly, its oncostatic effects on glioma cells are mediated, at least in part, by the inhibition of PKC activity which, in turn, results in Akt and NFkappaB inhibition [74]. In other tissues, melatonin acts by activation of PKC, for example, in NIE-115 cells [75], MDCK cells [76], and PC3 cells [77]. Other second messengers that have been demonstrated are inositol trisphosphate in splenocytes [78] and INS1 insulinoma cells [79], as well as IRS-1/PI3-kinase in the hypothalamus [80]. Melatonin's antiapoptotic effects in U937 monocytic cells also occur via its receptor, phospholipase C activation, and calcium influx [81]. However, the presence and significance of the latter pathways in the heart need to be established. Iuvone and Gan [82, 83], using chick retinal cultures, showed that melatonin receptors are coupled to inhibitory G proteins, reduced cAMP accumulation, and the dopamine receptor-regulated adenylyl cyclase. It is now known that melatonin has definite antiadrenergic actions in the heart (see antiadrenergic actions of melatonin).

Recent studies suggest a role for the melatonin receptors in the cardioprotective actions of the hormone [48, 84]. Sallinen and coworkers [84] showed that continuous long-term postinfarction exogenous melatonin administration increased the number of MT2 receptors in the left ventricle of rats (after myocardial infarction), suggesting the importance of functional MT2 receptors in these conditions. Interestingly, an age-related reduction in mRNA of MT1 and MT2 expression levels, as well as MT1 protein expression, were observed in several tissues, including the heart [85]. These observations, together with the reduction in melatonin levels in aging tissues, suggest a role for melatonin in the onset of aging.

More direct evidence for the melatonin receptor in its cardioprotective actions was the observation that administration of the high affinity melatonin receptor antagonist, luzindole, together with melatonin, before and during reperfusion after ischaemia, completely abolished the reduction in infarct size observed with melatonin administration alone [48]. It was also recently found that luzindole administration could abolish the long-term cardioprotective effects of melatonin: luzindole coadministered intraperitoneally with melatonin for one week before experimentation abolished the long-term protection exerted by melatonin alone, using infarct size as endpoint (Lochner and Genade, unpublished data). These observations are surprising in view of the convincing evidence for the free radical scavenging and antioxidant actions of melatonin. The significance of the melatonin receptor in its actions on the heart was also suggested by the observation that the protective effect of melatonin against cyclosporine A toxicity could be abolished by luzindole [86].

A recent study by Genade and coworkers [87] was the first to show that melatonin-induced cardioprotection in the isolated perfused rat heart was associated with activation of the survival kinase PKB/Akt and a reduction of the proapoptotic p38MAPKinase during early reperfusion. Whether these changes were the cause or the result of the protection remains to be established.

## 27.4 Antiadrenergic Actions of Melatonin

It is well established that the activation of the adrenergic nervous system and generation of the second messenger, cAMP, play a crucial role in the genesis of dysrhythmias associated with myocardial infarction [88]. It has also been known for many years that melatonin may exert antiarrhythmic activity by decreasing cardiac sympathetic activity [89, 90]. Indeed, several workers have recently reported that melatonin reduces the incidence of ventricular arrhythmias and fibrillation during ischaemia/reperfusion of isolated rat hearts [44, 46, 47]. It is thus possible that the antidysrhythmic effects of melatonin could be attributed to its antiadrenergic actions, as was demonstrated by its effects on the contractility of isolated papillary muscles [62, 91]. It has also been shown that the $EC_{50}$ for isoproterenol in papillary muscle was increased approximately tenfold by melatonin [62]. The antiadrenergic actions of melatonin were recently further investigated by Genade and coworkers [87], using tissue cAMP levels as indicator. Their results showed that melatonin, at a concentration of 50 $\mu$M, significantly counteracted the powerful beta-adrenergic stimulation by both isoproterenol and forskolin under normoxic conditions. These responses were receptor-dependent, since the specific melatonin receptor blocker, luzindole, abolished this effect.

Nitric oxide, guanylyl cyclase, and PKC were shown to be involved in the antiadrenergic actions of melatonin, since the inhibitors L-NAME, ODQ, and bisindolylmaleimide respectively, significantly counteracted the antiadrenergic effects of melatonin [87]. However, the exact mechanism whereby NO mediates these actions of melatonin is unclear, since this mediation can occur if melatonin causes release of NO, which is unlikely since melatonin has been shown to scavenge NO [14–16] and to prevent induction of iNOS [92, 93]. It is possible that melatonin potentiates the effects of NOS stimulation during beta-adrenergic stimulation. Evidence exists for control of beta-adrenergic mechanisms in generation of NO in cardiomyocytes; for example, isoproterenol upregulates NOS expression [94] and activates eNOS via $G_i$-alpha, causing an increase in cGMP [95]. However, further work is necessary to elucidate the role of NO in the antiadrenergic effects of melatonin.

The question subsequently arose whether these antiadrenergic actions of melatonin are involved in its cardioprotective effects. It is well established that activation of the beta-adrenergic pathway during ischaemia is harmful and that attenuation of these effects by beta-blockade is cardioprotective [88]. It was shown that simultaneous administration of melatonin and L-NAME or ODQ, which could abolish the potent effects of isoproterenol on tissue cAMP, also abolished cardioprotection. Both these observations suggest that melatonin-induced cardioprotection could be mediated via its antiadrenergic actions, with NO playing an important role, probably by attenuating beta-adrenergic stimulation during ischaemia.

Melatonin has also been shown to protect the heart against myocardial injury induced by high doses of isoproterenol [96]. Isoproterenol-induced release of cardiac troponin T and troponin I was significantly reduced by melatonin pretreatment, while these hearts showed fewer histological changes. While the melatonin-induced changes could be attributed to scavenging of free radicals or by reduction of intracellular calcium accumulation, they may also be due to the potent antiadrenergic effects of the hormone.

As discussed above, melatonin activates PKC in a number of cell types [75, 97]. Should this also take place in heart muscle, it could be responsible for counteracting isoproterenol-induced cAMP generation via inhibitory cross-talk [98, 99].

## 27.5 Melatonin and Mitochondria

Mitochondria play an important role in cell death and apoptosis in ischaemia/reperfusion, as well as in cardioprotection. It is well established that mitochondria and the electron transport chain sustain progressive damage during myocardial ischaemia [100–103]. Mitochondria have been shown to be a primary source for ROS production during ischaemia [104, 105], with complex III as the major site of production [106, 107]. It is also well established that ROS production occurs even more abundantly during early reperfusion [108–110]. Thus mitochondria, being a major source of ROS production in ischaemia/reperfusion, could also be a major target for a free radical attack.

The discovery that the mitochondrion is a target for melatonin action opened up new avenues for research on the effects of this hormone [111]. For example, melatonin has protective effects in a number of conditions, such as Parkinson's disease, Alzheimer's disease, sepsis, etc., in all of which mitochondrial dysfunction plays an important role [112]. As apoptosis is characteristic of these diseases, it was expected that melatonin should have antiapoptotic effects [113]. The effects of melatonin on apoptosis have been summarized in a recent review by Leon and coworkers [111]. For example, the hormone inhibits apoptosis in models of Parkinsonism [114], Alzheimer's disease [115], and ischaemia-reperfusion of neuronal tissue [116, 117]. Indeed, melatonin has been reported to have antiapoptotic effects also in the ischaemic-reperfused heart [49].

It has been suggested that the reduction in the activity of the respiratory chain complexes observed in mitochondria from ischaemic-reperfused hearts [118–120] is due to ROS-induced cardiolipin oxidation [118, 121], which, in turn, may increase the electron leak from the electron transport chain, perpetuating a cycle of oxygen-radical–induced damage. The possibility that these mitochondrial changes could be reversed by melatonin, in view of its free radical scavenger properties and lipophilic nature, was investigated by Petrosillo and coworkers [122]. Administration of melatonin (50 μM before and after 30 min global ischaemia), attenuated the decline in mitochondrial complex I and III activity, decreased $H_2O_2$ production, and prevented ischaemia/reperfusion–induced cardiolipin loss and the increase in its level of peroxidation. Melatonin treatment also protected mitochondrial integrity and functional capacity, as indicated by the state 3 and 4 respiration rates and respiratory control ratios. This study therefore provided strong evidence for the possibility that the target for the free radical scavenging actions of melatonin may indeed be the mitochondrion.

Further evidence for the significance of the mitochondrial targeting of melatonin was provided by the use of senescence-accelerated mice [123]. It has been proposed that free radicals produced during aerobic respiration may induce oxidative damage which accumulates over time [124–126]. Heart mitochondria from senescent-accelerated mice showed significant age-induced increases in lipid peroxidation, a reduction in glutathione peroxidase (GPx) and glutathione reductase (GRd) activity, as well as complex II, III, and IV activities and mitochondrial ATP content. All these changes could be counteracted by administration of melatonin in drinking water. A follow-up study by the same group [124] confirmed that use of melatonin as a single antioxidant therapy prevented the age-dependent production of free radicals and oxidative damage to both cardiac and diaphragm mitochondria. Their data supported the efficacy of long-term melatonin treatment in preventing age-dependent mitochondrial oxidative stress.

Thus it is generally accepted that the antioxidant and free radical scavenging capacity of melatonin protects proteins of the electron transport chain and mitochondrial DNA from ROS/RNS-induced damage, while it also interacts with lipid bilayers, reducing lipid peroxidation and stabilizing mitochondrial inner membranes (for a review, see ref 111). However, it has recently been suggested that the antiapoptotic properties of melatonin may be due to its direct interaction with the mitochondrial permeability transition pore (MPTP). The MPTP is a complex located between the inner and outer mitochondrial membranes, consisting of several mitochondrial proteins, including the voltage-dependent anion channels, adenine nucleotide translocase, and cyclophilin D [127–129]. Opening of this pore by an increase in intracellular [$Ca^{2+}$] and ROS allows passage of molecules >1,500 Da across the inner mitochondrial membrane, rapid passage of protons and depolarization of the mitochondria, uncoupling of oxidative phosphorylation, and loss of ATP, which could lead to mitochondrial swelling and release of cytochrome

C [130]. It has recently been demonstrated that melatonin directly inhibits MPTP currents in hepatic mitoplasts, using a patch-clamping technique [128]: it strongly inhibits MPTP currents in a dose-dependent manner with an $IC_{50}$ of 0.8 μM. These workers showed in a follow-up experiment that melatonin significantly decreased cytochrome release and depressed activation of caspase 3 in ischaemia/reperfusion of the brain, also suggesting an effect of melatonin on the MPTP. The antiapoptotic effects of melatonin on the infarcted heart suggest that it may well keep the pore in a closed confirmation.

## 27.6 Melatonin and Intracellular $Ca^{2+}$ Handling

Melatonin is thought to modulate intracellular $Ca^{2+}$ via activation of its G-protein–coupled membrane receptors or through a direct interaction with calmodulin. Fluorescence and nuclear magnetic resonance (NMR) studies showed that melatonin itself weakly interacts with calmodulin, but this does not necessarily indicate that this does not occur in biological media [131]. Early studies indicated that melatonin-induced vasorelaxation [132] and inhibition of contraction [133] were due to inhibition of the sarcolemmal $Ca^{2+}$ channels. It was also found that melatonin stimulates the cardiac sarcolemmal $Ca^{2+}$ pump ($Ca^{2+}$-$Mg^{2+}$-dependent ATPase) [134], which could lead to a reduction in intracellular calcium. Using confocal microscopy and the fluorophore fluo 3, it has been shown that melatonin is indeed capable of reducing intracellular calcium in cardiomyocytes exposed to chemical hypoxia [135]. This was associated with a reversal of hypoxia-induced morphological changes. Whether this was due to a direct effect on myocardial $Ca^{2+}$ movements or merely a reflection of melatonin-induced cardioprotection was not determined in this study.

A similar reduction in intracellular calcium by melatonin was observed in isolated ventricular myocytes subjected to metabolic inhibition and anoxia [136]. These workers showed that melatonin treatment mitigates sarcoplasmic (SR) – $Ca^{2+}$ handling and $Ca^{2+}$ homeostasis in cardiomyocytes in chronic hypoxic conditions [136]. Amongst others, melatonin preserved the sarcoplasmic calcium content. The protein expression of SERCA, but not the ryanodine receptor and the sodium-calcium exchanger, was significantly attenuated in the chronic hypoxic rat heart, suggesting that the lowered SR $Ca^{2+}$ uptake causes a reduction in the SR-$Ca^{2+}$ content and impaired $Ca^{2+}$ homeostasis. This downregulation of SERCA expression was not seen in the melatonin-treated hearts, showing that the mechanistic effect on the ameliorated SR-Ca re-uptake and $Ca^{2+}$ could be mediated by a transcriptional regulation of SERCA expression. Whether the antioxidant effects of the hormone or its interaction with its receptors are involved in SERCA regulation remains to be determined.

An inhibitory effect of melatonin on the opening of the L-type calcium channels and calcium influx was observed in striatal neurons [137].

Interestingly, melatonin, when administered continuously for two weeks postinfarction, caused a significant reduction in the expression of the LV dihydropyridine receptor and ryanodine receptor, as well as SERCA mRNA, compared with vehicle-treated rats [84]. The amounts of these proteins were not changed, contrary to the reduced expression of the mRNAs, and it was suggested that melatonin might have a protective effect on the dihydropyridine and ryanodine receptors as well as SERCA proteins after myocardial infarction, possibly by reducing the oxidation of these proteins through its antioxidant actions. These changes were also associated with increases in the number of MT2 receptors in melatonin-treated animals, supporting the role of the melatonin receptor in its cardioprotective actions. Unfortunately, these observations were not correlated with SERCA function and intracellular $Ca^{2+}$ homeostasis.

The above study also showed for the first time a positive relationship between melatonin and LV atrial natriuretic peptide (ANP) levels [84]: a fivefold increase in ANP was observed when melatonin was administered for two weeks postinfarction. This finding adds one more possible way in which melatonin could protect the heart. However, the release of ANP is a $Ca^{2+}$ dependent process [138], while melatonin has been shown to reduce intracellular $Ca^{2+}$-accumulation. Clearly, the role of ANP in melatonin-induced cardioprotection needs to be investigated.

Finally, melatonin appears to affect almost all role players involved in myocardial calcium fluxes: namely, the sarcolemma, the sarcoplasmic reticulum, and the mitochondria.

## 27.7 Reversal of Harmful Effects of Clinically Used Drugs

A number of clinically useful drugs have side effects mediated by free radicals and related reactants. For example, cyclosporine A, an immunosuppresant agent, has significant harmful effects on the kidney, liver, and heart [86]. Cyclosporine A causes a significant increase in thiobarbituric acid–reactive substances (TBARS) and a depletion of cardiac antioxidative enzymes. Simultaneous administration of cyclosporine and melatonin caused a reduction in TBARS and normalized the altered myocardial morphology. Interestingly, this protective effect of melatonin was lost during coadministration with luzindole, suggesting a role for the melatonin receptor in this scenario.

Anthracycline antibiotics, such as doxorubicin, are potent, broad-spectrum chemotherapeutic agents which are highly effective [139–141], but their clinical use is seriously limited by their cardiotoxicity [142–145]. Doxorubicin causes extensive morphological changes, e.g., destruction and disorganization of myofibrils, mitochondrial degeneration, and lipid accumulation. A combination of doxorubicin and melatonin treatment reduced these morphological changes [143–145]. Melatonin treatment reduced malondialdehyde (MDA) levels and increased reduced glutathione (GSH) levels, suggesting increased protection by antioxidative enzymes [146]. It has also been reported that both physiological and pharmacological concentrations of melatonin protect the heart from doxorubicin-induced damage [144].

Epirubicin is a new anthracycline analogue which appears to be less cardiotoxic than the parent compound doxorubicin [140, 147]. Nevertheless, it caused ultrastructural changes in the heart, associated with increased nitrite/nitrate production, while having no significant effect on MDA and GSH levels. Melatonin partially reduced the ultrastructural changes and nitrite/nitrate levels in epirubicin-treated rats.

## 27.8 Melatonin as Cardioprotective Agent in Humans?

Several years ago Duncker and Verdouw [54] posed the question whether melatonin has a future as a cardioprotective agent. Since then, several convincing experimental studies have appeared, elaborating on the beneficial effects of melatonin on the ischaemic/reperfused heart. Indeed, our knowledge regarding its mode of action has also increased remarkably. The potential of melatonin as a therapeutic agent was further strengthened by findings that melatonin ameliorates tissue damage and the severity of compromised function after ischaemia/reperfusion of other organs, e.g., the brain, liver, kidney, gastrointestinal tract, lung, and placenta (for a review, see ref. 41). In addition, melatonin can protect against cardiotoxicity induced by chemotherapeutic drugs, for example, doxorubicin [141, 143]. A recent study showed that melatonin intake is associated with amelioration of metabolic and morphological abnormalities associated with obesity in an animal model [148]; although the underlying mechanisms are unclear, they may include antioxidant and receptor-mediated effects.

During and after surgical procedures, there is a well-defined stress response, modification of which has been shown to reduce morbidity and mortality in patients (for a review, see [149]). The effect of melatonin in relation to surgery in humans has been tested in only one trial on newborns [150], where it was shown to reduce stress markers. The drug should also be tested in the adult surgical population.

With regard to the myocardium, most of the animal studies done so far have focused on the acute effects of melatonin administration on ischaemia/reperfusion injury in experimental models. A recent study by Lochner and coworkers [48] showed that melatonin administered in the drinking water of rats had long-term protective effects which continued for at least two days after cessation of treatment. In this regard, Sallinen et al. [84] showed that two weeks continuous postinfarction subcutaneous melatonin supply contributed to the postinfarction cardioprotective actions. Melatonin also regulates collagen accumulation in the scars of infarcted hearts, which, in turn, could improve the tensile strength and retard the development of complications [151]. While recognizing the shortcomings of the animal models used thus far, it is surprising that only one clinical trial is currently in progress. The MARIA trial is a unicentre, randomized, double-blind, placebo-controlled study of melatonin as an adjunct in patients with acute myocardial infarction undergoing primary angioplasty [152]. Clearly, the outcome of such a study may play a pivotal role in the ultimate decision whether or not to use melatonin in patients with ischaemic heart disease or, perhaps more importantly, to use the drug prophylactically. Another possible application would be to administer melatonin at the onset of reperfusion after open-heart surgery. As far as we know, this has not been done before. Apart from the beneficial effects listed above, chronic administration of melatonin has been found to be not toxic [55]. It is cheap, can be obtained without prescription, and is rapidly absorbed when given orally, which makes it an ideal drug for use in conditions characterized by the production of copious amounts of free radicals.

## References

1. Hardeland RFB (1996) Ubiquitous melatonin-presence and effects in unicells, plants and animals. Trends Comp Biochem Physiol 2:25–45
2. Tan DX, Manchester LC, Terron MP et al (2007) One molecule, many derivatives; a never-ending interaction of melatonin with reactive oxygen and nitrogen species. J Pineal Res 42:28–42
3. Acuna-Castroviejo D, Martin M, Macias M et al (2001) Melatonin, mitochondria and cellular bioenergetics. J Pineal Res 30:64–74
4. Tan DX, Reiter RJ, Manchester LC et al (2002) Chemical and physical properties and potential mechanisms: melatonin as a broad spectrum anti-oxidant and free radical scavenger. Curr Top Med Chem 2:181–197
5. Reiter RJ (1998) Melatonin, active oxygen species and neurological damage. Drug News Perspect 11:291–296
6. Reiter RJ, Tan DX, Mayo JC et al (2003) Melatonin as an anti-oxidant: biochemical mechanisms and pathophysiological implications in humans. Acta Biochim Pol 50:1129–1146
7. Reiter RJ, Tan DX, Terron MP et al (2007) Melatonin and its metabolites: new findings regarding their production and their radical scavenging actions. Acta Biochimica 54:1–9
8. Tan DX, Manchester LC, Reiter RJ et al (2000) Melatonin directly scavenges hydrogen peroxide; a potentially new metabolic pathway of melatonin biotransformation. Free Rad Biol Med 29:1177–1185
9. Martinez GR, Almeida EA, Klitzke CF et al (2005) Measurement of melatonin and its metabolites: importance for the evaluation of their biological roles. Endocrine 27:111–118
10. Karbownik M, Reiter RJ (2000) Antioxidative effects of melatonin in protection against cellular damage caused by ionizing radiation. Proc Soc Exp Biol Med 225:9–22
11. Reiter RJ, Tan DX, Burkhardt S (2002) Reactive oxygen and nitrogen species and cellular and organismal decline: amelioration with melatonin. Mech Ageing Dev 123:1007–1018
12. Zavodnik IB, Lapshina EA, Zavodnik LB et al (2004) Hypochlorous acid-induced oxidative stress in Chinese hamster B14 cells: viability, DNA and protein damage and protective action of melatonin. Mutat Res 559:31–45
13. Matuszak Z, Bilska MA, Reszka RJ et al (2003) Interaction of singlet oxygen with melatonin and related indoles. Photochem Photobiol 78:449–458
14. Noda Y, Mori A, Liburdy R, Packer L (2000) Melatonin and its precursors scavenge nitric oxide and peroxynitrite. J Pineal Res 29:184–192
15. Turjanski A, Chaia ZD, Doctorovich F et al (2000) N-nitrosomelatonin. Acta Cyrstallogr C 56:682–683
16. Turjanski A, Leonik F, Rosenstein RE et al (2000) Scavenging of NO by melatonin. J Am Chem Soc 122:10468–10469

17. Reiter RJ, Tan DX, Manchester LC, El-Sawi MR (2002) Melatonin reduces oxidant damage and promotes mitochondrial respiration: implications for ageing. Ann N Y Acad Sci 979:238–250
18. Aksoy N, Vural H, Sabuncu T, Aksoy S (2003) Effects of melatonin on oxidative-antioxidative status of tissues in streptozotocin-induced diabetic rats. Cell Biochem Funct 21:121–125
19. Hardeland R, Reiter RJ, Poeggeler B, Tan DX (1993) The significance of the metabolism of the neurohormone melatonin: anti-oxidative protection and formation of bioactive substances. Neurosci Biobehav Rev 17:347–357
20. Pieri C, Marra M, Moroni F et al (1994) Melatonin; a peroxyl radical scavenger more effective than vitamin E. Life Sci 55:PL271–PL276
21. Hardeland R (2005) Antioxidative protection by melatonin: multiplicity of mechanisms from radical detoxification to radical avoidance. Endocrine 27:119–130
22. Tomas-Zapico C, Coto-Montes AA (2005) Proposed mechanism to explain the stimulatory effect of melatonin on antioxidative enzymes. J Pineal Res 39:99–104
23. Hardeland R, Pandi-Perumal SR, Cardinali DP (2006) Melatonin. Int J Biochem Cell Biol 38:313–316
24. Suzen S, Bozkaya P, Coban T, Nebiogu D (2006) Investigation of the in vitro antioxidant behaviour of some 2-phenylindole derivatives: discussion on possible antioxidant mechanisms and comparison with melatonin. J Enzyme Inhib Med Chem 21:405–411
25. Nosjean O, Ferro M, Coge F et al (2000) Investigation of the melatonin binding site MT3 as the quinone reductase 2. J Biol Chem 275:31311–31317
26. Witt-Enderby PA, Bennet J, Jarzynaka MM et al (2003) Melatonin receptors and their regulation: biochemical and structural mechanisms. Life Sci 72:2183–2198
27. Reiter RJ, Tan DX, Gitto E et al (2004) Pharmacological utility of melatonin in reducing oxidative cellular and molecular damage. Pol J Pharmacol 56:159–170
28. Benitez-King G, Huerto-Delgadillo L, Anton-Tay F (1991) Melatonin modifies calmodulin cell levels in MDCK and NIE-115 cell lines and inhibits phosphodiesterase activity in vitro. Brain Res 557:289–292
29. Benitez-King G, Huerto-Delgadillo L, Anton-Tay F (1993) Binding of 3H melatonin to calmodulin. Life Sci 53:201–207
30. Romero MP, Garcia-Perganeda A, Guerrero JM et al (1998) Membrane-bound calmodulin in Xenopus laevis oocytes as a novel binding site for melatonin. FASEB J 12: 1401–1408
31. Vazan R, Styk J, Beder I, Pancza D (2003) Effect of melatonin on the isolated heart in the standard perfusion conditions and in the condition of the calcium paradox. Gen Physiol Biophys 22:41–50
32. Vazan R, Beder I, Styk J (2004) Melatonin and the heart. Gesk Fysiol 53:29–33
33. Mizrat B, Parlakpinar H, Acet A, Turkoz Y (2004) Effects of pinealectomy and exogenous melatonin on rat hearts. Acta Histochem 106:29–36
34. Sahna E, Olmez E, Acet A (2002) Effects of physiological and pharmacological concentrations of melatonin on ischemia-reperfusion arrhythmias in rats: can the incidence of sudden cardiac death be reduced? J Pineal Res 32:194–198
35. Paulis L, Simko F (2007) Blood pressure modulation and cardiovascular protection by melatonin: potential mechanisms behind. Physiol Res 56:671–684
36. Brugger P, Markti W, Herld M (1995) Impaired nocturnal secretion of melatonin in coronary heart disease. Lancet 345:1408
37. Sakotnik A, Liebmann PM, Stoschitzky K et al (1999) Decreased melatonin synthesis in patients with coronary artery disease. Eur Heart J 20:1314–1317
38. Dominguez-Rodriguez A, Abreu-Gonzalez P, Garcia MJ et al (2002) Decreased nocturnal melatonin levels during acute myocardial infarction. J Pineal Res 33:248–252
39. Yaprak M, Altun A, Vardar A et al (2003) Decreased nocturnal synthesis of melatonin in patients with coronary artery disease. Int J Cardiol 89:103–107

40. Altun A, Yaprak M, Altoz M et al (2002) Impaired nocturnal synthesis of melatonin in patients with cardiac syndrome X. Neurosci Lett 327:143–145
41. Reiter RJ, Tan D-X (2003) Melatonin: a novel protective agent against oxidative injury of the ischemic/reperfused heart. Cardiovasc Res 58:10–19
42. Tengattini S, Reiter RJ, Tan D-X et al (2008) Cardiovascular diseases: protective effect of melatonin. J Pineal Res 44:16–25
43. Giacomo CG, Antonio M (2007) Melatonin in cardiac ischemia/reperfusion-induced mitochondrial adaptive changes. Cardiovasc Haematol Disord Drug Targets 7:163–169
44. Tan D-X, Manchester LC, Reiter RJ et al (1998) Ischemia/reperfusion-induced arrhythmias in the isolated rat heart: Prevention by melatonin. J Pineal Res 25:184–191
45. Szarszoi O, Asemu G, Vanecek J et al (2001) Effects of melatonin on ischemia and reperfusion injury of the rat heart. Cardiovasc Drugs Therap 5:251–257
46. Kaneko S, Okumura K, Numaguchi Y et al (2000) Melatonin scavenges hydroxyl radical and protects isolated rat hearts from ischemic reperfusion injury. Life Sci 67:101–112
47. Lagneux C, Joyeux M, Demenge P et al (2000) Protective effect of melatonin against ischemia-reperfusion injury in the isolated rat heart. Life Sci 66:503–509
48. Lochner A, Genade S, Davids A et al (2006) Short- and long-term effects of melatonin on myocardial post-ischemic recovery. J Pineal Res 40:56–63
49. Dobsak P, Siegelova J, Eicher JC et al (2003) Melatonin protects against ischemia-reperfusion injury and inhibits apoptosis in isolated working rat heart. Pathophysiology 9:179–187
50. Sahna E, Parlakpinar H, Turkoz Y, Acet A (2005) Protective effects of melatonin on myocardial ischemia/reperfusion induced infarct size and oxidative changes. Physiol Res 54:491–495
51. Kacmaz A, User EY, Sehirli AO et al (2005) Protective effects of melatonin against ischemia/reperfusion-induced oxidative remote organ injury in the rat. Surg Today 35: 744–750
52. Opie LH (2004) Heart physiology; from cell to circulation, 4th edn. Williams & Wilkens, Lippincott, pp 575–582
53. Lee YM, Chen HR, Hsiao G et al (2002) Protective effects of melatonin on ischemia/reperfusion injury in vivo. J Pineal Res 33:72–80
54. Dunker DJ, Verdouw PD (2001) Has melatonin a future as a cardioprotective agent? Cardiovasc Drugs Therap 15:205–207
55. Prevo M, Johnson P, Chester A (2000) Twenty-eight day toxicity study of melatonin infused subcutaneously in rats via an osmotic pump. Int J Toxicol 19:107–118
56. Witt-Enderby PA, Radio NM, Doctor JS, David VL (2006) Therapeutic treatments potentially mediated by melatonin receptors: potential clinical uses in the prevention of osteoporosis, cancer and as an adjuvant therapy. J Pineal Res 41:297–305
57. Dubocovich ML (1995) Melatonin receptors: are there multiple subtypes? Trends Pharmacol Sci 16:50–56
58. Dubocovich ML, Cardinali DP, Guardiola-Lemaitre B et al (1998) Melatonin receptors. The IUPHAR Compendium of Receptor characterization and classification. IUPHAR Media, London, pp 187–193
59. Pang CS, Xi SC, Brown GM et al (2002) 2[$^{125}$I]Iodomelatonin binding and interaction with beta-adrenergic signalling in chick heart/coronary artery physiology. J Pineal Res 32: 243–252
60. Ekmekcioglu C, Thalhammer T, Humpeler S et al (2003) The melatonin receptor subtype MT2 is present in the human cardiovascular system. J Pineal Res 35:40–44
61. Pang CS, Brown GM, Tang PL et al (1993) 2-[$^{125}$I]Iodomelatonin binding sites in the lung and heart: a link between the photo-periodic signal, melatonin and the cardiopulmonary system. Biol Signals 2:228–236
62. Abete P, Bianco S, Calabrese C et al (1997) Effects of melatonin in isolated rat papillary muscle. FEBS Lett 412:79–85

63. Capsoni S, Viswanathan M, De Oliveira AM, Saavedra JM (1994) Characterization of melatonin receptors and signal transduction system in rat arteries forming the circle of Willis. Endocrinology 135:373–378
64. Reppert SM, Weaver DR, Godson C (1996) Melatonin receptors step into the light: cloning and classification of subtypes. Trends Pharmacol Sci 17:100–102
65. Witt-Enderby PA, Dubocovich ML (1996) Characterization and regulation of the human ML1A melatonin receptor stably expressed in Chinese hamster ovary cells. Mol Pharmacol 50:166–174
66. Brydon L, Roka F, Petit L et al (1999) Dual signaling of human Mel 1a melatonin receptors via $G_{i2}$, $G_{i3}$ and $G_{q11}$ proteins. Mol Endocrin 13:2025–2038
67. Nelson CS, Marino JL, Allen CN (1996) Melatonin receptors activate heteromeric G-protein coupled Kir 3 channels. NeuroReport 7:717–720
68. Mor M, Plazzi PV, Spadoni G, Tarzia G (1999) Melatonin. Curr Med Chem 6:501–518
69. Nosjean O, Nicoilas JP, Klupsch F et al (2001) Comparative pharmacological studies of melatonin receptor: MT1, MT2 and MT3/QR2. Tissue distribution of MT3/QR2. Biochem Pharmacol 61:1369–1379
70. Carrillo-Vico A, Garcia-Pergoneda A, Naji L et al (2003) Expression of membrane and nuclear melatonin receptor mRNA and protein in mouse immune system. Cell Mol Life Sci 60:2272–2278
71. Naji L, Carrillo-Vico A, Guerrero JM, Calvo JR (2004) Expression of membrane and nuclear melatonin receptors in mouse peripheral organs. Life Sci 74:2227–2236
72. Reiter RJ, Tan DK, Tou MJ et al (2008) Biogenic amines in the reduction of oxidative stress: melatonin and its metabolites. Neuro Endocrinol Lett 29:391–398
73. Cui P, Yu M, Luo Z et al (2008) Intracellular signalling pathways involved in cell growth inhibition of human umbilical vein endothelial cells by melatonin. J Pineal Res 44:107–114
74. Martin V, Herrera F, Garcia-Santos G et al (2007) Involvement of protein kinase C in melatonin's oncostatic effect in C6 glioma cells. J Pineal Res 43:239–244
75. Bellon A, Ortiz-Lopez L, Ramirez-Rodriguez G (2007) Melatonin induces neuritogenesis at early stages in NIE-115 cells through actin rearrangements via activation of protein kinase C and Rho-associated kinase. J Pineal Res 42:214–221
76. Ramirez-Rodriguez G, Ortiz-Lopez L, Benitez-King G (2007) Melatonin increases stress fibers and focal adhesions in MDCK cells via PKC. J Pineal Res 42:180–190
77. Sampson SR, Lupowitz Z, Braiman L, Zisapel N (2006) Role of protein kinase Calpha in melatonin signal transduction. Mol Cell Endocrinol 252:82–87
78. Markowska M, Mrozkowiak A, Pawlak J et al (2004) Intracellular second messengers involved in melatonin signal transduction in chicken splenocytes in vitro. J Pineal Res 37:207–212
79. Bach AG, Wolgast S, Muhlbauer E, Peschke E (2005) Melatonin stimulates inositol-1,4, 5-trisphosphate and $Ca^{2+}$ release from INS1 insulinoma cells. J Pineal Res 39:316–323
80. Ahne GF, Caperuto LC, Pereira-Da-Silva M et al (2004) In vivo activation of insulin receptor tyrosine kinase by melatonin in the rat hypothalamus. J Neurochem 90:559–566
81. Radogna F, Paternoster L, Albertini MC et al (2007) Melatonin antagonizes apoptosis via receptor interaction in U937 monocytic cells. J Pineal Res 43:154–162
82. Iuvone PM, Gan J (1994) Melatonin receptor-mediated inhibition of cyclic AMP accumulation in chick retinal cell cultures. J Neurochem 63:118–124
83. Iuvone PM, Gan J (1995) Functional interaction of melatonin receptors and D1 dopamine receptors in cultured chick retinal neurons. J Neurosci 15:2179–2185
84. Sallinen P, Mänttäri S, Leskinen H, Vakkuri O et al (2008) Long-term postinfarction melatonin administration alters the expression of DHPR, $RyR_2$, SERCA2 and $MT_2$ and elevates the ANP level in the rat left ventricle. J Pineal Res 45:61–69
85. Sanchez-Hidalgo M, Guerrero Montavez JM, Carrascosa-Salmoral MD et al (2009) Decreased MT1 and MT2 melatonin receptor expression in extrapineal tissues of the rat during physiological ageing. J Pineal Res 46(1):29–35

86. Rezzani R, Rodella LF, Bonomini F, Tengattini S et al (2006) Beneficial effects of melatonin in protecting against cyclosporine A-induced cardiotoxicity are receptor mediated. J Pineal Res 41:288–295
87. Genade S, Genis A, Ytrehus K, Huisamen B, Lochner A (2008) Melatonin receptor-mediated protection against myocardial ischaemia/reperfusion injury: role of its anti-adrenergic actions. J Pineal Res 45:449–458
88. Lubbe WF, Podzuweit T, Opie LH (1992) Potential arrhythmogenic role of cyclic adenosine monophosphate (cAMP) and cytosolic calcium overload: implications for prophylactic effects of beta-blockers in myocardial infarction and proarrhythmic effects of phosphodiesterase inhibitors. J Am Coll Cardiol 19:1622–1633
89. Blatt CM, Rabinowitz SH, Lown B (1979) Central serotonergic agents raise the repetitive extrasystole threshold of the vulnerable period of the canine ventricular myocardium. Circ Res 44:723–730
90. Verrier RL (1986) Neurochemical approaches to the prevention of ventricular fibrillation. Fed Proc 45:2191–2196
91. Arvola L, Bertelsen E, Lochner A, Ytrehus K (2006) Sustained anti-β-adrenergic effect of melatonin in guinea pig heart papillary muscle. Scand Cardiovasc J 40:37–42
92. Mahal HS, Sharma HS, Mukherjee T (1999) Anti-oxidant properties of melatonin: a pulse radiolysis study. Free Radic Biol Med 26:557–565
93. Blanchard B, Pompom D, Ducroq C (2000) Nitrosation of melatonin by nitric oxide and peroxynitrite. J Pineal Res 29:184–192
94. Slezak J, Buchwalon IB, Schulze W et al (2004) Cellular control of nitric oxide synthase expression and activity in rat cardiomyocytes. Antioxid Redox Signal 6:345–352
95. Balligand J-L (1999) Regulation of cardiac β-adrenergic response by nitric oxide. Cardiovasc Res 43:607–620
96. Acikel M, Buyukokuroglu ME, Aksoy H et al (2003) Protective effects of melatonin against myocardial injury induced by isoproterenol in rats. J Pineal Res 35:75–79
97. Ramirez-Rodriguez G, Ortis-Lopez L, Benitez-King G (2007) Melatonin increases stress fibres and focal adhesions in MDCK cells. J Pineal Res 42:180–190
98. Perlini S, Khoury EP, Norton GR et al (1998) Adenosine mediates sustained adrenergic desensitization in the rat heart via activation of protein kinase C. Circ Res 83:761–771
99. Arvola L, Hassef D, Melnikov AL et al (2002) Adenosine induces prolonged anti-beta-adrenergic effects in guinea-pig papillary muscle. Acta Physiol Scand 175:11–17
100. Duan J, Karmazyn M (1989) Relationship between oxidative phosphorylation and adenine nucleotide translocase activity of two populations of cardiac mitochondria and mechanical recovery of ischemic hearts following reperfusion. Can J Physiol Pharmacol 67:704–709
101. Lesnefsky EJ, Gudz TI, Migita CT et al (2001) Ischemic injury to mitochondrial electron transport in the ageing heart: damage to the iron-sulfur protein subunit of electron transport complex III. Arch Biochem Biophys 385:117–128
102. Flameng W, Andres J, Ferdinandy P et al (1991) Mitochondrial function in myocardial stunning. J Mol Cell Cardiol 23:1–11
103. Lesnefsky EJ, Tandler B, Ye J et al (1997) Myocardial ischemia decreases oxidative phosphorylation through cytochrome oxidase in subsarcolemmal mitochondria. Am J Physiol Heart Circ Physiol 273:H1544–H1554
104. Becker LB, van den Hoek TL, Shao ZH et al (1999) Generation of superoxide in cardiomyocytes during ischaemia before reperfusion. Am J Physiol Circ Physiol 277:H2240–H2246
105. Davies MJ (1989) Direct detection of radical production in the ischaemic and reperfused myocardium: current status. Free Rad Res Commun 7:275–284
106. Chen Q, Vasquez EJ, Moghaddas S et al (2003) Production of reactive oxygen species by mitochondria: central role of complex III. J Biol Chem 278:36027–36031
107. Han D, Antunes F, Canali R et al (2003) Voltage-dependent anion channels control the release of superoxide anion from mitochondria to cytosol. J Biol Chem 278:5557–5563
108. Das DK (1994) Cellular, biochemical and molecular aspects of reperfusion injury. Ann N Y Acad Sci 723:116–127

109. Zweier JL, Flaherty JT, Weisfeldt ML (1987) Direct measurement of free radical generation following reperfusion of ischemic myocardium. Proc Natl Acad Sci U S A 84:1404–1407
110. Ambrosio G, Zweier JL, Flaherty JT (1991) The relationship between oxygen radical generation and impairment of myocardial energy metabolism following post-ischemic reperfusion. J Mol Cell Cardiol 23:1359–1374
111. Leon J, Acuna-Castroviejo D, Escames G et al (2005) Melatonin mitigates mitochondrial malfunction. J Pineal Res 38:1–9
112. Acuna-Castroviejo D, Martin M, Macias M et al (2001) Melatonin, mitochondria and cellular bioenergetics. J Pineal Res 30:65–74
113. Acuna-Castroviejo D, Escames G, Carazo A et al (2002) Melatonin, mitochondrial homeostasis and mitochondrial-related diseases. Curr Top Med Chem 2:133–151
114. Mayo JC, Sainz RM, Uria H et al (1998) Melatonin prevents apoptosis induced by 6-hydroxydopamine in neuronal cells: implications for Parkinson's disease. J Pineal Res 24:179–192
115. Pappolla MA, Sos M, Omar RA et al (1997) Melatonin prevents death of neuroblastoma cells exposed to the Alzheimer amyloid peptide. J Neurosci 17:1683–1690
116. Cheung RT (2003) The utility of melatonin in reducing cerebral damage resulting from ischemia and reperfusion. J Pineal Res 34:153–160
117. Pei Z, Cheung RT (2003) Melatonin protects SHSY5Y neuronal cells but not cultured astrocytes from ischemia due to oxygen and glucose deprivation. J Pineal Res 34:194–201
118. Paradies G, Petrosillo G, Pistolese M et al (1999) Lipid peroxidation and alterations to oxidative metabolism in mitochondria isolated from rat heart subjected to ischemia and reperfusion. Free Radic Biol Med 27:42–50
119. Lesnefsky EJ, Chen Q, Moghaddas S et al (2004) Blockade of electron transport during ischemia protects cardiac mitochondria. J Biol Chem 279:47961–47967
120. Petrosillo G, Ruggiero FM, Di Venosa N, Paradies G (2003) Decreased complex III activity in mitochondria isolated from rat heart subjected to ischemia and reperfusion: role of reactive oxygen species and cardiolipin. FASEB J 17:714–716
121. Paradies G, Petrosillo G, Pistolese M et al (2004) Decrease in mitochondrial Complex I activity in ischemic/reperfused heart. Involvement of reactive oxygen species and cardiolipin. Circ Res 94:53–59
122. Petrosillo G, Di Venosa N, Pistolese M et al (2006) Protective effect of melatonin against mitochondrial dysfunction associated with cardiac ischemia-reperfusion: role of cardiolipin. FASEB J 20:269–276
123. Rodriguez MI, Carretero M, Escames G et al (2007) Chronic melatonin treatment prevents age-dependent cardiac mitochondrial dysfunction in senescence mice. Free Rad Res 41: 15–24
124. Rodriguez MI, Escames G, Lopez LC et al (2007) Melatonin administration prevents cardiac and diaphragmatic mitochondrial oxidative damage in senescence-accelerated mice. J Endocrinol 194:637–643
125. Beckman KB, Ames BN (1998) The free radical theory of aging matures. Physiol Rev 78:547–581
126. Shoal RS, Mockett AJ, Orr WC (2002) Mechanisms of ageing: an appraisal of the oxidative stress hypothesis. Free Rad Biol Med 33:575–586
127. Crompton M (1999) The mitochondrial permeability transition pore and its role in cell death. Biochem J 341:233–249
128. Andrabi SA, Sayeed I, Siemen D et al (2004) Direct inhibition of the mitochondrial permeability transition pore: a possible mechanism responsible for antiapoptotic effects of melatonin. FASEB J 18:869–871
129. Costa AD, Garlid K (2008) Intramitochondrial signaling: interactions among mitoKATP, PKC epsilon, ROS and MPT. Am J Physiol Heart Circ Physiol 295:H953–H961
130. Murphy E, Steenbergen C (2008) Mechanisms underlying acute protection from cardiac ischemia-reperfusion injury. Physiol Rev 88:581–609

131. Turjanski AG, Estrin DA, Rosenstein RE et al (2004) NMR and molecular dynamic studies of the interaction of melatonin with calmodulin. Protein Sci 13:2925–2938
132. Shibata S, Satake N, Takagi T, Usui H (1989) Vasorelaxing action of melatonin in rabbit basilar artery. Gen Pharmacol 20:677–680
133. Satake N, Shibata S, Takagi T (1986) The inhibitory action of melatonin on the contractile response to 5-hydroxytryptamine in various isolated vascular smooth muscles. Gen Pharmacol 17:553–558
134. Chen LD, Tan D-X, Reiter RJ et al (1993) In vivo and in vitro effects of the pineal gland and melatonin on ($Ca^{2+}Mg^{2+}$)-dependent ATPase in cardiac sarcolemma. J Pineal Res 14: 178–183
135. Salie R, Harper I, Cillie C et al (2001) Melatonin protects against ischemic-reperfusion myocardial damage. J Mol Cell Cardiol 33:343–357
136. Yeung HM, Hung MW, Fung ML (2008) Melatonin ameliorates calcium homeostasis in myocardial and ischemia-reperfusion injury in chronically hypoxic rats. J Pineal Res 45:373–382
137. Escames G, Macias M, Leon J et al (2001) Calcium-dependent effects of melatonin inhibition of glutamatergic response in rat striatum. J Neuroendocrinol 13:459–466
138. Laine M, Id L, Vuolteenaho O et al (1996) Role of calcium in stretch-induced release and mRNA synthesis of natriuretic peptides in isolated rat atrium. Pfugers Arch 432:953–960
139. Booser DJ, Hortobagyi GN (1994) Anthracyclin antibiotics in cancer therapy. Focus on drug resistance. Drugs 47:223–258
140. Omrod D, Holm K, Goa K, Spencer C (1999) Epirubicin: a review of its efficacy as adjuvant therapy and in the treatment of metastatic disease in breast cancer. Drugs Aging 15:389–416
141. Ahmed HH, Mannaa F, Elmegeed GA, Doss SH (2005) Cardioprotective activity of melatonin and its novel synthesized derivatives on doxorubicin-induced cardiotoxicity. Bioorg Med Chem 13:1847–1857
142. Kitakaze M (2001) Doxorubicin, to treat cancers or to cause cardiac injury, that is the question. Cardiovasc Drugs Therap 15:199–200
143. Balli E, Mete UO, Tuli A et al (2004) Effect of melatonin on the cardiotoxicity of doxorubicin. Histol Histopathol 19:1101–1108
144. Sahna E, Parlakpinar H, Ozer MK et al (2003) Melatonin protects against myocardial doxorubicin toxicity in rats: role of physiological concentrations. J Pineal Res 35:257–261
145. Oz E, Erbas D, Surucu HS, Duzgun E (2006) Prevention of doxorubicin-induced cardiotoxicity by melatonin. Mol Cell Biochem 282:31–37
146. Dziegiel P, Muruwska-Cialowicz E, Jethon Z et al (2003) Melatonin stimulates the activity of protective antioxidative enzymes in myocardial cells of rats in the course of doxorubicin intoxication. J Pineal Res 35:183–187
147. Andersson BS, Eksborg S, Vidal RF et al (1999) Anthraquinone-induced cell injury: acute toxicity of carminimycin, epirubicin, idorubicin and mitoxantrone in isolated cardiomyocytes. Toxicology 135:11–20
148. Hussein MR, Ahmed OG, Hassan AF, Ahmed MA (2007) Intake of melatonin is associated with amelioration of physiological changes, both metabolic and morphological pathologies associated with obesity: an animal model. Int J Exp Pathol 88:19–29
149. Kücükakin B, Gögenur I, Reiter RJ, Rosenberg J (2009) Oxidative stress in relation to surgery: is there a role for the antioxidant melatonin. J Surg Res 152(2):338–347
150. Gitto E, Romeo C, Reiter RJ et al (2004) Melatonin reduces oxidative stress in surgical neonates. J Pediatr Surg 39:184–189
151. Drobnik J, Karbownik-Lewinska M, Szczepanowska A et al (2008) Regulatory influence of melatonin on collagen accumulation in the infarcted heart scar. J Pineal Res 45(3):285–290
152. Dominguez-Rodriguez A, Abreu-Gonzalez P, Garcia-Gonzalez MJ et al (2007) A unicenter, randomized double-blind, parallel-group, placebo-controlled study of melatonin as an adjunct in patients with acute myocardial infarction undergoing primary angioplasty. The melatonin adjunct in the acute myocardial infarction treated with angioplasty (MARIA) trial: study design and rationale. Contemp Clin Trials 28:532–539

# Chapter 28
# Exercise-Induced Cardioprotection: Overview with an Emphasis on the Role of Antioxidants

**Karyn L. Hamilton and John C. Quindry**

**Abstract** A major goal in the treatment of cardiovascular disease (CVD) is developing effective interventions for preserving cardiac function and cardiac tissue viability following a myocardial ischemia-reperfusion (IR) event. Over the last few decades, it has become widely accepted that reactive oxygen species (ROS) play a critical role in mediating cellular damage and dysfunction in CVD. In the search for viable therapies against IR damage due to ROS, exercise has received growing attention. A wealth of recent studies demonstrates that the physiological stimulus of endurance exercise is overwhelmingly cardioprotective. By virtue of its sustainable nature, exercise as an experimental model in the study of cardioprotective mediators may prove to be invaluable in the discovery of translatable interventions against IR injury. Much of this optimism is founded on the fact that the exercised myocardium is better adapted to preserve the redox environment during an ischemic event. In this review we present a summary of the evidence that supports exercise as an effective means of promoting a cardioprotected phenotype. The focus of the discussion will be on the role of endogenous antioxidants in mediating protection and secondarily on the protective mechanisms peripheral to redox control.

**Keywords** Exercise · Ischemia · Reperfusion · Oxidants · Antioxidants · Cardiovascular

## 28.1 Introduction

Cardiovascular disease (CVD) is the leading cause of death in industrialized countries. While CVD can have many faces, ischemic heart disease is among the most prevalent cardiac manifestation [1]. Ischemic heart disease can also be directly

K.L. Hamilton (✉)
Human Performance Clinical Research Laboratory, Applied Human Sciences, Colorado State University, Fort Collins, CO 80523-1582, USA
e-mail: karynh@cahs.colostate.edu

H. Sauer et al. (eds.), *Studies on Cardiovascular Disorders*,
Oxidative Stress in Applied Basic Research and Clinical Practice,
DOI 10.1007/978-1-60761-600-9_28, 

linked to many other cardiac manifestations with high morbidity and mortality, including ventricular arrhythmias and congestive heart failure. It is clear from a large pool of published research that cardiomyocyte injury results not only from ischemia, or blood flow inadequate for meeting the demand of the tissue, but also from reperfusion, or reinstitution of adequate blood flow to previously ischemic tissue. The damage associated with these events is called ischemia-reperfusion (IR) injury and will be the primary focus for this review [2]. While no shortage of investigations into IR injury exists, with over 23,000 citations resulting from a typical data base query for research related to myocardial ischemia-reperfusion, the difficulty has been in discovering cardioprotective mediators that ultimately translate to effective clinical interventions. This review provides an overview of myocardial IR injury and summarizes the evidence demonstrating that endurance exercise is a stimulus capable of affording a cardioprotected phenotype. The primary focus will be on the role of oxidants in mediating cardiac injury, and the potential role of exercise-induced endogenous antioxidants in mediating protection against IR injury.

## 28.2 Principles of Myocardial IR Injury

As stated previously, myocardial injury sustained as a result of ischemic heart disease is a significant clinical problem. Ischemia, commonly defined as a mismatch in supply of blood to the myocardium and demand for blood by this chronically active muscle, creates an imbalance in oxygen and energy provision and inadequate removal of metabolic waste leading to cellular dyshomeostasis [3]. While reestablishing adequate blood flow is critical for salvaging myocardial tissue, doing so results in additional cellular injury via complex and interrelated mechanisms [4, 5]. The extent of the damage to the cardiac tissue is at least in part dictated by the duration of the ischemic period [6]. The classic notion of a "heart attack" implies heart muscle death. This phenomenon occurs when the ischemic duration extends beyond about 20 min. It has become clear that the loss of cardiomyocytes is due to both necrotic and apoptotic cell death [7–9]. However, periods of myocardial ischemia lasting just 1–5 min can produce life-threatening ventricular arrhythmias without deficits in ventricular pump function. This fact is highlighted by the multitude of lives saved over the last decade by placement of automated external defibrillators in public places. The approximate 5–20 min time window between tissue death and the onset of arrhythmias is marked by a phenomenon commonly referred to as myocardial stunning. Stunning is characterized by disturbances in cardiac contractility, but without evidence of significant cardiomyocyte death [5, 6]. Among the mediators of the injury and dysfunction resulting from ischemia and reperfusion are disruptions in intracellular calcium, mitochondrial dysfunction, activation of proteases, alterations of membrane function and permeability, inflammation, and accelerated oxidant production or "oxidative stress." While the focus of this review is oxidative stress, it is noteworthy to highlight the fact that altered redox balance is closely related to the other mediators of IR injury [5].

## 28.3 Oxidative Stress in Myocardial IR Injury

To understand oxidative stress during IR injury, one must have a fundamental knowledge of free radicals and where they are produced. Free radicals are highly reactive molecules that possess unpaired electrons in their outer orbitals and are capable of acting as oxidizing agents in biological systems. Nonradical derivatives of oxygen can also act as oxidizing agents; this has led to widespread use of the term "reactive oxygen species" (ROS) to encompass both radical and nonradical derivatives of oxygen that can impact redox balance toward oxidative stress. Classically, oxidative stress has served as a generic term used to describe either ROS-mediated oxidative damage to cellular components or an imbalance in the production of oxidants and availability of antioxidants. Recent discoveries, however, indicate that oxidative stress is really the outcome of chronic disruption of cellular redox signaling and control systems to the detriment of cell health [10]. This new understanding of oxidative stress helps explain why small concentrations of ROS play critical roles in cellular homeostasis and signaling, while myocardial IR results in markedly accelerated oxygen radical production contributing to all levels of myocardial dysfunction, damage, and cell death. The primary sources of ROS production, at least in contractile tissues, are the mitochondria, cellular oxidases, and production of nitric oxide by nitric oxide synthase. Notably, nitric oxide can proceed to form other highly reactive nitrogen species. Other sources for ROS production include autoxidation of catecholamines, ROS generation by phagocytic white cells, and formation of reactive species due to the disruption of metal containing proteins, particularly iron and copper.

Defining theoretical sources of ROS and the potential impact on cellular systems is easy. However, measuring ROS production and quantifying cellular oxidant stress are not at all trivial endeavors. Direct measurement and characterization of free radicals produced during IR has been achieved via electron paramagnetic resonance (EPR) spectroscopy and spin trapping [11]. Using these sensitive direct measurements, it has been demonstrated in vivo that oxygen radicals contribute to myocardial damage during IR [12]. Evidence suggests that oxidative damage occurs during both ischemia and reperfusion, though the majority of IR induced oxidative stress is imposed by reperfusion. As mentioned previously, the mitochondria are a chief site of ROS production, and this is certainly the case in cardiac muscle. During an ischemic insult, mitochondrial respiration is diminished. However, the rapid stop-start of mitochondrial electron transport complexes during ischemia and subsequent reperfusion results in a free radical "burst," as respiratory electron carriers are reoxidized [13]. During energy deficiency, a source of superoxide generation is $Ca^{2+}$-mediated activation of xanthine oxidase and NADPH/NADH oxidases [14]. Cellular damage invokes an inflammatory response to the ischemic region, thereby exacerbating the oxidant stress via infiltrating ROS-producing immune cells. Inflammation also accelerates superoxide production via activation of cyclooxygenase-2 and the inducible isoform of nitric oxide synthase leading to increased nitric oxide and subsequent formation of other reactive species, most notably peroxynitrite.

Oxidant stress associated with IR has been shown to have widespread effects, including alterations in cardiomyocyte metabolism [15], signal transduction [16], and gene expression [17]. Cell damage mediated by ROS can impact nearly any cellular component and generally involves modification of lipids, protein, and nucleic acids. For example, when ROS react with polyunsaturated fatty acids in cell and subcellular membranes, altered membrane fluidity and permeability, and ultimately impaired cell function can ensue. Protein oxidation tends to occur at sites of "susceptible" amino acids (i.e., those with accessible thiol groups) and can result in altered function of enzymes, receptors, and proteins responsible for transport, structural integrity, and contraction [14, 18, 19]. Finally, oxidative damage to DNA can occur particularly at susceptible guanine bases, resulting in altered genetic code both directly and indirectly via subsequent DNA repair and replication mechanisms [13, 14]. Overall, the accelerated production of ROS during IR can rapidly overwhelm endogenous antioxidant defenses, resulting in oxidative damage which leads to cell dysfunction and death [5, 13, 14]. Fig. 28.1 summarizes the primary oxidant stresses encountered by cardiac myocytes undergoing an ischemic stress.

### *28.3.1 Calcium Regulation, Proteolysis, Membrane Integrity, and Inflammation*

While the focus of this review is on alterations in redox balance during myocardial IR, oxidative stress is closely related to the other mediators known to have a hand in IR injury. For example, while calcium homeostasis is normally tightly regulated, during ischemia calcium dyshomeostasis occurs quickly, resulting in uncharacteristic increases in free intracellular calcium [20, 21]. Accelerated production of ROS during IR contributes to altered membrane permeability, thereby impeding removal of sodium through various ion transport systems. In turn, increasing cytosolic sodium concentrations give rise to subsequent calcium influx which may, in a vicious cycle, lead to increased production of ROS by triggering calcium-activated oxidases [5, 21]. Dysregulation of cytosolic calcium can also contribute to activation of proteases called calpains [22]. Calpains can contribute to proteolytic cleavage of cell proteins, including those activated by the proteolytic processing, and subsequently trigger cellular apoptosis [23]. Finally, much of the oxidative stress experienced during IR comes from inflammatory sources. This fact is particularly true during reperfusion and scenarios of prolonged intermittent ischemia. Much of the oxidative stress introduced by the immune system is delivered by neutrophils chemotactically attracted to the vascular beds of damaged heart tissue. Activated neutrophils migrate into the tissue, creating free radical bursts via NADPH oxidase activity, by release of proteolytic enzymes into the affected area, and by signaling other arms of the immune system [24].

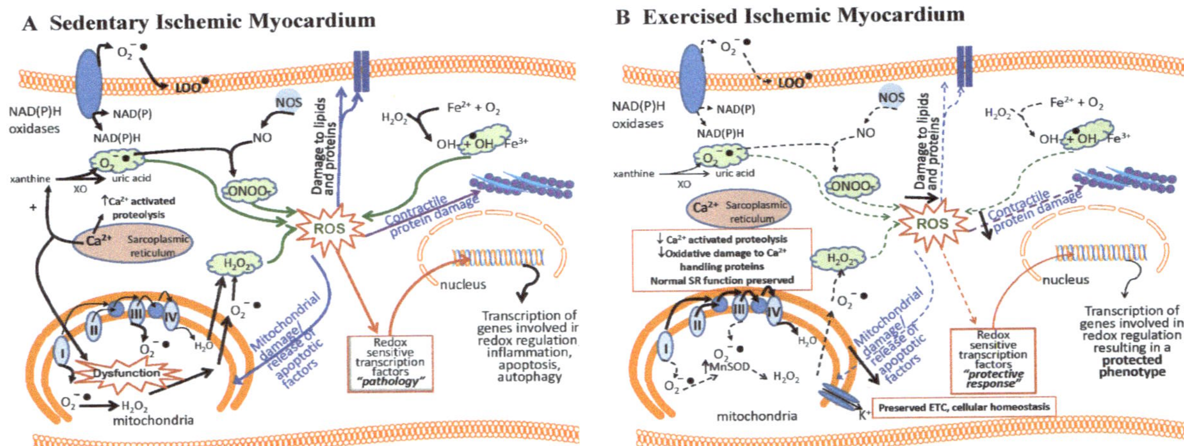


**Fig. 28.1** The bioenergetic supply:demand mismatch during ischemia rapidly creates hypoxic conditions within the contracting cardiac muscle (**panel a** – Sedentary ischemic myocardium). Hypoxia leads to diminished electron transport activity as limited cellular $O_2$ stores are depleted. Consequently, superoxide is produced at greater rates in complexes I and III. Superoxide becomes $H_2O_2$ primarily through enzymatic conversion and diffuses to other cellular locations, resulting in damage throughout the cell. Cell damage and low ATP concentrations lead to calcium stress within the cytosol and mitochondrial matrix. Additional superoxide is generated from xanthine and NAD(P)H oxidases, contributing to the collective cellular oxidative stress. Apoptotic processes, inflammation, and transcription of stress genes are among the resulting pathological outcomes. In the exercised ischemic myocardium (**panel b**), the same ischemic stressors are met with a variety of known and unknown defenses. Somewhat preserved mitochondrial function limits superoxide formation, resulting in better control of mitochondrial and cellular homeostasis. These beneficial outcomes in the exercised heart effectively break the chain reaction of calcium overload and superoxide generation peripheral to the mitochondria. The cellular environment of exercised heart muscle is preserved, yielding improved clinical outcomes en masse

## 28.4 Exercise-Induced Protection Against Myocardial IR Injury: Overview of Putative Mechanisms

It has long been recognized that endurance exercise is associated with a protective effect on the heart. For example, in 1978 McElroy and colleagues performed a well-controlled experiment in which chronically swim trained rats were shown to have greater protection against myocardial infarction compared to sedentary counterparts [25]. Since that time, endurance exercise has been shown to afford protection against myocardial stunning [26–36], myocardial infarction [25, 28, 37–40], protease activation [41, 42], and apoptosis [42, 43]. The protection that endurance exercise mediates occurs independently of advancing age [27, 43–45]. Perhaps most intriguing is the observation that both chronic exercise training (i.e., ten weeks) and acute endurance exercise (i.e., 1–10 days) result in protection during myocardial ischemia and reperfusion. The resultant protection is evident for approximately nine days following cessation of exercise training [29]. Importantly, as few as three consecutive days of either moderate intensity (i.e., ~50% $VO_2max$) or high intensity (i.e., ~75% $VO_2max$) endurance exercise provide similar cardioprotection [33]. Table 28.1 summarizes the existing literature demonstrating the role of exercise in protection against cardiac injury.

While the mechanisms responsible for exercise-induced cardioprotection remain incompletely understood, several have been investigated. These include adaptations in the coronary vasculature, induction of myocardial heat shock proteins, regulation of calcium homeostasis, altered ATP-sensitive potassium channels, regulation of calcium homeostasis and calcium activated proteolytic activity, and upregulation of endogenous antioxidants [70, 71].

### *28.4.1 Coronary Circulation*

In some species, development of collateral coronary circulation may occur in response to prolonged (i.e., months to years) endurance exercise training [72–74]. However, the cardioprotective benefits of short-term exercise described by many of the investigations referenced here are not attributed to alterations in collateral circulation. Therefore, the protection observed following acute endurance exercise is instead due to changes in the expression or activity of cardioprotective proteins and other constituents of the exercised heart. Readers are referred to other excellent reviews for discussions on coronary collateral circulation and cardioprotection [73].

### *28.4.2 Myocardial Heat Shock Proteins*

In response to a wide variety of stressors, cells induce increased expression of cytoprotective chaperone proteins known as heat shock proteins (HSP). In addition

**Table 28.1** Summary of published literature demonstrating exercise-induced cardioprotection

| Exercise Mode | Experimental Model (Species-IR Method) | Cardiac Performance Outcome in Exercised Hearts | Proposed Mechanism | Reference |
|---|---|---|---|---|
| T-20 m/min, 5 days/week, 11–16 weeks | R-IP | S | NE-calcium control | [46] |
| T-30 m/min, 10 weeks | R-IV | N | NE | [32] |
| T-30 m/min, 1 day | R-IV | N | MnSOD | [39] |
| S-up to 40 min/day, 6 weeks | R-IP | S | NE-hypothesized norepinephrine | [47] |
| T-30 m/min, 3 days | R-IV | B,S | NE-hypothesized antioxidant defenses | [35] |
| T-30 m/min, 3–5 days | R-IV | B,N | NE-short term exercise protective | [27] |
| T-single max exercise | D-IV | A | NE-hypothesized NO | [48] |
| T-28 m/min, 6 weeks | R,C-hypoxia | Metabolic stress | NE-hypothesized sarcoKATP channel | [49] |
| T-30 m/min, 1 day | R-IV | N | NE | [50] |
| T-14 and 20 m/min, 6% grade | R-IP | S | NE | [44] |
| T-30 m/min, 3 days | R-IV | B,N | NE-hypothesized antioxidant activity | [40] |
| T-18 m/min, 12 weeks | R-IP | S | NE | [51] |
| T-30 m/min, 2 days | R-IV | N | HSP-72 not essential | [28] |
| T-30 m/min, 3 days | R-IV | A | MnSOD | [36] |
| T-25 m/min, 4 days/week, 10 weeks | R-IP | | NE-bioenergetic enzymes | [52] |
| T-30 m/min, 3 days | R-IP | S | NE | [29] |
| T-30 m/min, 3 days | R-IP | S | MnSOD | [34] |
| T-single max exercise | D-IV | A | COX-2 | [53] |
| W-daily, 6–8 weeks | R-IV | A | NE-myocardial ion control | [54] |
| T-up to 30 m/min, 5 days | R-IP | N | NE-hypothesized sex-dependent mechanisms | [38] |
| T-25 m/min | R-IP | S | NE-hypothesized HSP-72 | [55] |
| T-single max exercise | D-IV | A | iNOS | [56] |
| T-3 days, 30 m/min young, 20 m/min old | R-IP | P-young and old hearts | NE | [57] |
| T-30 m/min, 3 days | R-IP | S | NE-preservation of SERCA-2A channels | [41] |
| T-up to 30 m/min, 5 days | R-IP | N | Sarco KATP – gender-dependent effect | [58] |
| T-25 m/min, 5 day/week, 16 weeks | R-IP, hypoperfusion | S | NE-cardiac remodeling, HSP-72 | [51] |

**Table 28.1** (continued)

| Exercise Mode | Experimental Model (Species-IR Method) | Cardiac Performance Outcome in Exercised Hearts | Proposed Mechanism | Reference |
|---|---|---|---|---|
| T-5 days/week, 10 weeks, up to 35 m/min | R-IP, chronic nandrolone | S,N | NE-hypothesized antioxidant defenses | [59] |
| T-3 days/week, 55% VO2max, 14 weeks | R-IV | N | NE-hypothesized products of gene expression | [60] |
| T-30 m/min | R-IP | N | Sarco KATP channel | [61] |
| T-65% VO2max, 7 days | M-IV | B | eNOS, iNOS (not conclusive) | [62] |
| S-3 h/day, 5 days/week | R-IV | A | Akt, PI3K | [63] |
| T-30 m/min, 3 days | R-IV | N, P | HSP-70 not essential for protection against IR-mediated cell death | [64] |
| T-25 m/min | R-IP | N | Opioid receptors | [65] |
| T-25 min, 6 km/h | D-IV | A,N | NADPH oxidase | [66] |
| T-30 m/min | R,C-$H_2O_2$ challenge oisolated mitochonria | P | NE – hypothesized mitochondrial adaptations | [67] |
| T-30 m/min | R-IV | B,N,P | MnSOD essential mediator | [42] |
| T-25 m/min, 5% grade | R-IP, doxorubicin induced cardiotoxicity | B,S | NE-hypothesized protection due to antioxidant defenses | [68] |
| T-30 m/min, 2% grade | R-IP | N | PKC (α,δ,ε)-mediated activation-mediated | [69] |

Keys: T, treadmill; S, swim; W, wheel running; C, cells; D, dog; M, mouse; R, rat; IP, isolated perfused; IV, in vivo IR; Protection against: A, arrhythmia; B, stress/damage biomarkers; S, stunning; N, necrotic death; P, apoptotic cell death; NE, not examined directly.

to the classic stress, hyperthermia, HSPs are induced in response to ROS exposure and acute damage to native cellular proteins; all of these stresses occur in response to a typical bout of acute exercise [75, 76]. While various HSPs have been reported to provide protection against cellular challenges, evidence indicates that elevated myocardial HSP72 is a valuable defense against IR injury [77], and has been shown to increase as much as fivefold in the heart following a few bouts of endurance exercise [28, 30, 35, 40, 78]. However, most recent evidence confirms that induction of HSP72, while undoubtedly a protective event, is not essential for exercise-induced cardioprotection, as protection is observed in the absence of an increase in myocardial HSP72 [35, 40, 78]. To add support to the finding that HSP72 is not essential for exercise-induced cardioprotection, exercise provides protection regardless of advancing age, while aging attenuates HSP72 induction by both exercise and thermal stress [28, 44, 57]. It should be noted, however, that there is not complete agreement on the lack of essentiality of HSP72 for exercise-related protection against IR injury [31, 79], though clearly other important mediators are playing a role in providing protection in the absence of an increase in myocardial HSP72.

### *28.4.3 Regulation of Calcium*

As mentioned previously, dysregulation of cytosolic calcium concentrations is a hallmark of IR injury. Data demonstrating that exercise improves myocardial calcium handling during IR are sparse but noteworthy. Impaired calcium handling associated with myocardial stunning has been shown to be attenuated by previous exercise training [46]. This finding relates well to the observation that endurance exercise attenuates the IR-induced activity of the calcium-activated protease calpain [41]. Whether exercise-induced alterations in calpain activity are directly related to improved calcium handling or to another component of calpain regulation remains undetermined. Nonetheless, sarcoplasmic/endoplasmic calcium ATPase (SERCA-2A), which has been shown to be decreased following IR, remains unchanged in exercised myocardium exposed to IR [41]. As will be discussed in more detail later, this preservation of calcium handling proteins appears to relate to maintenance of improvement of antioxidant defenses resulting from endurance exercise. Further investigation is warranted to elucidate the contribution of exercise-mediated alterations to calcium handling as a component of cardioprotection.

### *28.4.4 ATP-Sensitive Potassium Channels and Protein Kinase C*

Intriguing evidence implicates involvement of ATP-sensitive potassium channels in the mediation of cardioprotection during IR. Sarcolemmal ATP-sensitive potassium channels of cardiac myocytes are inhibited by ATP, which acts from the cytosolic side of the sarcolemma to inhibit potassium channel opening [80]. Using pharmacologic agents, it has been demonstrated that opening these channels allows potassium to flow out of the myocyte, providing cardioprotection during myocardial IR [81, 82]. This protection is thought to be related to shortening of the cardiac action potential, thus inhibiting calcium entry into the cell via L type calcium channels, protecting the cardiac myocyte during IR by abrogating the cytosolic calcium overload associated with IR [81, 82]. The role that sarcolemmal ATP-sensitive potassium channels play in exercise-induced cardioprotection remains relatively unknown, but existing evidence does suggest that they may contribute to protection against myocardial infarction [58, 61]. Mitochondrial ATP-sensitive potassium channels, located on the inner mitochondrial membrane [83], open in response to transient ischemia, resulting in activation and translocation of specific PKC isoforms, along with increases in adenosine [81, 82, 84]. Pharmacological evidence demonstrates that opening these mitochondrial channels also relates to protection against IR injury [81, 82, 83, 85]. The mechanism of this protection remains incompletely understood but may relate to attenuated mitochondrial calcium accumulation during ischemia, decreased ROS production during reperfusion, and enhanced mitochondrial energy production following the ischemic episode [81, 82, 84, 86]. The role that mitochondrial ATP-sensitive potassium channels may play in exercise-induced cardioprotection remains unclear. One investigation concluded that opening these mitochondrial channels is not essential for exercise-induced protection against myocardial infarction [58].

As inferred above, activation of protein kinase C (PKC) can occur subsequent to opening of ATP-sensitive potassium channels as well as in response to many other growth, proliferation, and survival stimuli [82, 84]. Indeed, PKC activation in heart muscle has been shown to promote a cardioprotected phenotype [87]. Evidence suggests that acute exercise results in activation of PKC$\alpha$, $\delta$, and $\varepsilon$ [88, 69], while repeated exercise bouts increase PKC$\varepsilon$ [88]. Nonspecific PKC inhibition has recently been shown to suppress acute exercise-induced protection against myocardial infarction [69]. Therefore, the essentiality and isoform specificity of PKC activation for exercise-induced protection against IR warrants further investigation.

## 28.5 Role of Antioxidants in Exercise-Induced Cardioprotection

### *28.5.1 An Introduction to Intrinsic Defenses*

With an appreciation that oxygen can act as a damaging oxidizing species, one must also realize that aerobic organisms rely on oxygen for survival and thrive because they have well evolved antioxidant defenses. Endogenous antioxidant defenses against ROS include enzymatic and nonenzymatic antioxidants. Antioxidant defenses include: (1) compounds that catalytically remove reactive species; these consist of several isoforms of superoxide dismutase (SOD), catalase

(CAT), and glutathione peroxidase (GPx); (2) proteins that minimize the participation of oxidants such as iron, copper, and heme; these include transferrins, haptoglobins, metallothionein, and proteins that oxidize ferrous iron; (3) proteins that protect biomolecules against oxidative damage; this group could include molecular chaperone proteins mentioned in the previous section of this review; and (4) compounds that scavenge ROS and reactive nitrogen species such as peroxynitrite; this category could include glutathione, uric acid, bilirubin, and compounds from the diet such as α-tocopherol and ascorbic acid. The reader is referred to one of many excellent texts and reviews for a more thorough review of the biology of oxidant-antioxidant chemistry [14].

As mentioned earlier, aerobic species have developed complex intrinsic protective mechanisms to counteract the damaging effects of oxidants and electrophiles. Among the endogenous protective mechanisms developed by aerobes are a group of enzymes collectively referred to as "phase 2 enzymes" [89]. Included in this group are glutathione transferases, glutathione peroxidase, catalase, thioredoxin, heme oxygenase-1, UDP-glucuronosyltransferases, NAD(P)H:quinone oxidoreductase 1, and paroxonases. A growing body of evidence demonstrates that phase 2 genes can be transcriptionally induced. Phase 2 gene transcriptional regulation may be an effective strategy for achieving protection against the stresses associated with many chronic pathological conditions, including cardiovascular diseases. Nuclear factor-erythroid-2-related factor 2 (Nrf2) plays a crucial role in the coordinated induction of the genes encoding phase 2 enzymes and does so by regulating transcription of genes with an antioxidant response element (ARE) in their promoter regions [90]. Nrf2 is maintained in an inactive state in the cytosol by association with Keap1. When Keap1 is modified at one or more of its cysteine thiol groups, Nrf2 is activated and translocates to the nucleus. Phosphorylation of specific serine or threonine residues in Nrf2 may also help with nuclear localization and transcriptional regulation by Nrf2. Multiple upstream kinases, such as mitogen-activated protein kinases, phosphatidylinositol-3-kinase/Akt, and protein kinase C may play a role in Nrf2 phosphorylation and transcriptional regulation of phase 2 enzymes. Following nuclear translocation of activated Nrf2, transcription of phase 2 genes is initiated, impacting the cellular capacity for intrinsic redox regulation [90]. Indeed, evidence suggests that Nrf2 signaling is important for regulation of intrinsic antioxidant defenses and oxidative stress in the cardiac tissue [91, 92].

### 28.5.2 *Exercise, Enzymatic Antioxidants, and Cardioprotection*

Endurance exercise demands greater electron flux through the mitochondrial electron transport chain to meet the greater energy demands of muscular work. Primarily as a result of increased oxidative phosphorylation and incomplete reduction of oxygen, generation of reactive oxygen species increases during exercise [93, 94]. Hence, exercise itself can be considered a contributor to oxidative stress. This oxidative stress is obviously a nonlethal eustress widely believed to be the stimulus that results in an upregulation of cellular antioxidant defenses, including the enzymatic antioxidants [95]. In this discussion of intrinsic antioxidant defenses in exercised hearts, additional commentary details the current speculative rationale about yet-to-be-identified mechanisms of exercise-induced protection.

To date, little is known about exercise and Nrf2-mediated induction of phase 2 enzymes via ARE binding. What is known, however, represents efforts to better understand links between exercise-induced mitochondrial biogenesis, and Nrf2 activation, along with Nrf1 and some transcriptional coactivators involved in this process [96]. Because acute exercise accelerates ROS production in muscle, it is logical to hypothesize that Nrf2 activation stems from oxidative modification of exposed cysteine residues. Activation occurs following dissociation of Nrf2 from its regulatory protein Keap1 [90]. Similarly, because exercise leads to activation of upstream kinases known to phosphorylate Nrf2, endurance exercise is a logical stimulus for Nrf2 nuclear translocation. While it is known that Nrf2 activation and binding at ARE regions occurs in cardiac cells [92], it is not known whether this is the mechanism by which exercise increases intrinsic antioxidant enzyme activity. The remaining focus of this section will detail what is known about exercise and myocardial antioxidant enzyme activity and its contribution to cardioprotection.

Over the last decade of exercise-induced cardioprotection research, catalytic removal of oxidizing species during IR has received the most attention. In mammals, the enzymatic antioxidant SOD exists in cytosolic, mitochondrial, and extracellular isoforms. In form, the primary difference between these SOD isoforms involves the particular metal cofactors, though in function all isoforms are effective in the decomposition of the free radical superoxide to hydrogen peroxide. The cytosolic isoform requires Cu and Zn as cofactors, whereas Mn is the cofactor in the mitochondrial isoform [14, 18, 97]. Both CAT and GPx catalyze reactions that convert hydrogen peroxide to oxygen and water. Glutathione peroxidase utilizes the abundant nonprotein thiol glutathione (GSH) as a reducing agent, producing oxidized glutathione (GSSG). Glutathione reductase then recycles GSSG to the reduced form, GSH, using other reducing compounds such as $\alpha$-tocopherol or ascorbic acid.

The effect of endurance exercise on myocardial antioxidant enzyme activities has been widely investigated. While some reports of exercise-related increases in GPx activity exist [60, 98, 99], most findings support the notion that myocardial GPx activity is not altered by exercise [27, 35, 40, 79, 100, 55]. Investigating the optimal means by which exercise could increase cardiac GPx could be well-spent efforts, because GPx overexpression reportedly results in significant protection against IR-induced cardiac dysfunction [101]. With respect to the effect of exercise on myocardial CAT, results are equivocal and seemingly dependent on the length of exercise training. Some studies demonstrate increased myocardial CAT activity in the exercised heart [29, 33, 34, 44, 79, 98], whereas others report no increase [27, 35, 36, 40]. However, it is generally held that even short-term endurance exercise results in a rapid increase in myocardial MnSOD activity [27, 29, 32, 34–36, 39, 40, 60].

The strongest evidence directly linking exercise-induced increases in myocardial antioxidants that catalytically remove ROS, and protection against IR, implicates a contributory role for exercise-induced increases in SOD. Exercise has been shown to increase expression and activity of SOD [27, 29, 32, 34–36, 39, 40, 60, 67, 102], with the most consistent increases observed in the mitochondrial isoform, MnSOD. To test the essentiality of increases in MnSOD for exercise-induced protection, several studies employed antisense oligonucleotide technology as a gene silencing technique to prevent the exercise-induced increase in myocardial MnSOD. The first report to silence the exercise-mediated increase in MnSOD also resulted in abrogated exercise-induced protection against myocardial infarction [39]. A subsequent investigation confirmed these findings and demonstrated that exercise-induced MnSOD plays an important protective role against IR-induced ventricular arrhythmias [36]. Moreover, recent investigations demonstrate that an exercise-induced increase in myocardial MnSOD activity is at least partially responsible for the protective effect of exercise against myocardial infarction and apoptotic cell death [42]. Interestingly, the protection afforded by exercise-induced alterations in MnSOD activity appears to be directly related to preservation of proteins involved in calcium handling and prevention of calcium-activated proteolysis by calpains [42]. That is, exercise provides protection against damage to calcium handling proteins during IR, and attenuates IR-related activation of calpains. When MnSOD is prevented from increasing in response to endurance exercise, this preservation of calcium handling apparatus and prevention of calpain activation is lost [42]. In contrast to these investigations, MnSOD was not shown to be essential for the exercise-induced protection against in vitro myocardial stunning–induced deficits in ventricular contractility [34].

Exercise has been shown to increase other phase 2 enzymes, but without well-controlled studies to investigate the essentiality of these exercise-induced adaptations for protection against ischemia or other tissue injury. For example, glutathione S-transferase has been shown to be greater in heart and other tissues following exercise training [103–105]. Glutathione S-transferases catalyze the conjugation of reduced glutathione via the sulfhydryl group, to electrophilic centers on a wide variety of substrates. Some glutathione S-transferases exhibit GPx-like activity, catalytically removing organic peroxides [14]. Therefore, increasing the activity of glutathione S-transferases could be useful in the detoxification of endogenous compounds such as peroxidized lipids and other organic peroxides. Similar increases have been observed in cardiac heme oxygenase-1, another potentially important and highly inducible component of redox systems in the heart [60]. Heme oxygenase, the rate-limiting enzyme in heme breakdown, and the metabolites resulting from heme catabolism, may play important roles in regulating protection against oxidative stress. The mechanisms by which heme oxygenases exert antioxidant effects are thought to be via degradation of the pro-oxidative heme, the release of biliverdin and subsequent conversion to bilirubin, both of which have significant antioxidant properties under physiologic scenarios [106, 107].

In summary, compounds that catalytically remove ROS are of clear importance in minimizing injury due to oxidative events including myocardial IR; and evidence suggests that endurance exercise is a stress capable of increasing such compounds. Some evidence suggests that increases in enzymatic antioxidants may represent an "essential" component of exercise-induced protection against myocardial IR. However, much remains to be understood about the power of exercise to regulate phase 2 enzymes and, therefore, cellular antioxidant networks.

### *28.5.3 Exercise and Nonenzymatic Compounds That Scavenge ROS*

In addition to compounds that catalytically remove ROS and other oxidizing species, the intrinsic antioxidant defense system includes compounds that can directly quench ROS via nonenzymatic means. Perhaps one of the most prevalent of these compounds is the most abundant biological nonprotein thiol, glutathione [108]. While evidence shows that acute or short-term exercise consistently increases GSH [27], extended exercise training (e.g., weeks to months) fairly consistently increases myocardial reduced and total glutathione content [95, 109]. In addition to its role in directly quenching ROS, the reduced form of glutathione (GSH) serves as an electron donor for enzymatic antioxidants and also, importantly, restores other nonenzymatic antioxidant compounds such as vitamins C and E back to their reduced forms [14, 97]. Accordingly, pharmacological depletion of GSH impairs cardiac function and survival during IR [13, 110, 111]. Because acute exercise imparts cardioprotection without increases in myocardial GSH concentrations, GSH does not likely represent an essential component of exercise preconditioning.

In addition to exercise-induced increases in GSH, other direct nonenzymatic antioxidants that could contribute to exercise-induced cardioprotection exist. For example, exercise may increase catabolism of heme compounds to other direct antioxidants, bilirubin, and biliverdin via increases in the rate-limiting enzyme heme oxygenase-1 [60, 106, 107, 112]. Another compound with strong reducing capacity is uric acid. In humans and higher primates, uric acid is the final oxidation product of purine metabolism and is excreted in urine. Uric acid, like ascorbic acid, is a strong reducing agent and, therefore, a potent antioxidant. In humans, it is believed that a significant portion of the antioxidant capacity of blood plasma comes from uric acid. It is a strong scavenger of both oxygen and nitrogen radicals and may protect plasma proteins against oxidation [14]. It has been demonstrated that accelerated purine metabolism during acute exercise and exercise training results in greater uric acid concentrations in plasma and saliva [113–116]. It is also reported that uric acid can move into muscle compartments such that it could, theoretically, exert antioxidant effects in contractile tissues [117–119]. However, no investigation has sought to elucidate whether exercise-induced changes in uric acid can contribute to protection of cardiac muscle during IR. Further, it is somewhat paradoxical that higher serum uric acid concentration has been reported as a strong and independent predictor of metabolic syndrome [120]. This paradox demonstrates current limitations in understanding of whether high uric acid is part of a defensive strategy to improve antioxidant capacity or merely reflects chronic oxidative stress.

Vitamins E and C are widely investigated compounds that can directly scavenge ROS in lipid and aqueous compartments of cells [121]. As humans are not capable of biosynthesizing these compounds, they must be derived from the diet or supplements. Exercise training alone will, therefore, obviously not alter intake of vitamins E and C. It is tempting, however, to speculate that exercise-induced increases in intrinsic antioxidants and/or decreases in oxidative stress could allow for more vitamin E and vitamin C to be maintained in the reduced form of these antioxidants and, therefore, enhance ROS-quenching capacity. Unfortunately, the scientific support for this notion is lacking. Though it is beyond the scope of this review, it is worth noting that a critical evaluation of investigations of vitamin E and C supplementation for preventing cardiac damage due to ischemic heart disease leads to the conclusion that the benefit is minimal at best [121]. Evaluation of clinical trials undertaken in Europe and North America do not conclude that supplementation with vitamins E and C decrease cardiovascular and ischemic heart disease mortality and nonfatal myocardial infarction [122, 123]. Collectively, data to support the use of these exogenous vitamins to favorably impact the risk of cardiovascular events, or to suggest that exercise training enhances antioxidant need for supplemental vitamins E and C are unconvincing.

In summary, the mechanisms by which exercise affords consistent protection against myocardial IR injury remain incompletely understood. The exercised heart appears to be resistant to injury because of fortifications against the primary means of IR mediated cellular damage. Current findings indicate that exercise improves myocardial antioxidant defenses and that this improvement relates to better regulation of calcium homeostasis and prevention of calcium-activated proteolysis.

## 28.6 Oxidative Stress Prevention in the Exercised Heart: A Unique Form of Cardioprotection

In this summary of the literature, a case has been presented whereby exercised hearts are resistant to IR injury by endogenous antioxidant enzyme fortification as well as by activation of mechanisms that blunt the free radical load experienced during IR (summarized in Table 28.1 and Fig. 28.1). That all the known mechanisms of exercise-mediated cardioprotection are directly or indirectly linked to redox mechanisms is a logical conclusion, considering the role oxidative stress plays in IR injury. Within the paradigm of cardiac preconditioning, a term that means heart protection which persists after therapeutic condition has dissipated, the exercise stimulus appears to be different from other means of eliciting a protected phenotype. For example, the classic paper on cardiac preconditioning by Murry et al. first demonstrated preconditioning in response to short intervals of experimentally-induced ischemia and reperfusion, termed "ischemic preconditioning." In this original study, anesthetized dogs received four repetitions of short-duration ischemia separated by intervals of perfusion. Subsequent to this, an extended 40 min ischemic challenge was performed, and it was observed that the ischemic preconditioned hearts were protected against myocardial tissue death to a greater extent than hearts that were not preconditioned [124]. In the years since, thousands of studies have been undertaken to more fully understand the mechanisms of injury as well as protection. Thus, with knowledge of disease-causing and protection-giving mechanisms in hand, it would theoretically be possible to elicit a preconditioned phenotype using pharmacologic interventions. Unfortunately, however, findings from ischemic preconditioning research have not resulted in a viable therapy against IR damage [125]. There are several important reasons to explain why IR resistance cannot currently be delivered pharmaceutically with any reliability. First and foremost, it is virtually impossible to predict when a myocardial infarction is going to occur, even in individuals with a long history of ischemic heart disease. Second, the mechanisms of protection elicited by ischemic preconditioning rely primarily on signaling pathways common to inflammation. Thus, while these mechanisms of ischemic preconditioning may prove beneficial in the short term (days), they are not biologically intended as a long-term solution for tissue viability [126].

Thus, we arrive once again at exercise as a mode of preconditioning the heart, and the question that begs to be asked is whether the protection afforded by exercise can be elicited pharmaceutically. Currently, the answer to that question is not known, though several research groups are currently seeking the answer. Perhaps more important to the present, it should be restated that exercise appears to be a unique stimulus for mediating protection against IR damage. In contrast to the aforementioned ischemic preconditioning, years of clinical evidence support the notion that exercise is clearly sustainable as a viable intervention against IR injury. Again, exercised hearts seem to be protected by adaptations resulting in enhanced redox regulation. These adaptations occur via improved bioenergetics, attenuated production of ROS, and intrinsic antioxidant fortification. Whether exercise as an experimental model can be used in the translation of preconditioning research into a therapeutic intervention against IR damage is a matter for future research.

## 28.7 Conclusions and Summary

A wealth of research using animal models demonstrates the efficacy of exercise as an intervention against all forms of myocardial IR injury. This body of evidence supporting the beneficial role of exercise for heart health grows, paradoxically, as the population continues to become more sedentary. Fortunately, even short-term exercise exposure appears sufficient in preventing much of the damage incurred during IR. Importantly, cardioprotection resulting from exercise appears to be long-lasting and is certainly more cost-effective than most pharmacological approaches could hope to be. Perhaps most importantly, exercise also imparts a litany of other positive effects both physical and psychological. The challenge, however, is the difficulty in promoting adherence to exercise, particularly in a formerly inactive population. Finally, as mechanisms of exercise-induced cardioprotection remain incompletely understood, it is evident and imperative that more efforts toward identifying the stimuli and cellular alterations imposed by exercise, key for acquisition of the cardioprotected phenotype, are warranted.

## References


1. American Heart Association (2004) Heart disease and stroke statistics – 2004 update. American Heart Association, Dallas
2. Buja LM (2005) Myocardial ischemia and reperfusion injury. Cardiovasc Pathol 14(4): 170–175
3. Kloner RA, Bolli R, Marban E et al (1998) Medical and cellular implications of stunning, hibernation, and preconditioning: an NHLBI workshop. Circulation 97(18):1848–1867
4. Vetterlein F, Muhlfeld C, Cetegen C et al (2006) Redistribution of connexin43 in regional acute ischemic myocardium: influence of ischemic preconditioning. Am J Physiol Heart Circ Physiol 291(2):H813–H819
5. Bolli R, Marban E (1999) Molecular and cellular mechanisms of myocardial stunning. Physiol Rev 79(2):609–634
6. Downey JM (1990) Free radicals and their involvement during long-term myocardial ischemia and reperfusion. Annu Rev Physiol 52:487–504
7. Baines CP, Molkentin JD (2005) STRESS signaling pathways that modulate cardiac myocyte apoptosis. J Mol Cell Cardiol 38(1):47–62
8. Kubasiak LA, Hernandez OM, Bishopric NH et al (2002) Hypoxia and acidosis activate cardiac myocyte death through the Bcl-2 family protein BNIP3. Proc Natl Acad Sci USA 99(20):12825–12830
9. Garg S, Narula J, Chandrashekhar Y (2005) Apoptosis and heart failure: clinical relevance and therapeutic target. J Mol Cell Cardiol 38(1):73–79
10. Jones DP (2008) Radical-free biology of oxidative stress. Am J Physiol Cell Physiol 295(4):C849–C868
11. Zweier JL, Flaherty JT, Weisfeldt ML (1987) Direct measurement of free radical generation following reperfusion of ischemic myocardium. Proc Natl Acad Sci U S A 84(5):1404–1407
12. Grill HP, Zweier JL, Kuppusamy P et al (1992) Direct measurement of myocardial free radical generation in an in vivo model: effects of postischemic reperfusion and treatment with human recombinant superoxide dismutase. J Am Coll Cardiol 20(7):1604–1611
13. Werns SW, Fantone JC, Ventura A et al (1992) Myocardial glutathione depletion impairs recovery of isolated blood-perfused hearts after global ischaemia. J Mol Cell Cardiol 24(11):1215–1220
14. Halliwell B, Gutteridge J (2007) Free radicals in biology and medicine, 4th edn. Oxford University Press, New York
15. Marsin AS, Bertrand L, Rider MH et al (2000) Phosphorylation and activation of heart PFK-2 by AMPK has a role in the stimulation of glycolysis during ischaemia. Curr Biol 10(20):1247–1255
16. Hausenloy DJ, Yellon DM (2006) Survival kinases in ischemic preconditioning and postconditioning. Cardiovasc Res 70(2):240–253
17. Nordlie MA, Wold LE, Simkhovich BZ et al (2006) Molecular aspects of ischemic heart disease: ischemia/reperfusion-induced genetic changes and potential applications of gene and RNA interference therapy. J Cardiovasc Pharmacol Ther 11(1):17–30
18. Yu BP (1994) Cellular defenses against damage from reactive oxygen species. Physiol Rev 74(1):139–162
19. Sun J, Huang SH, Zhu YC et al (2005) Anti-oxidative stress effects of Herba leonuri on ischemic rat hearts. Life Sci 76(26):3043–3056

20. Carrozza JP Jr, Bentivegna LA, Williams CP et al (1992) Decreased myofilament responsiveness in myocardial stunning follows transient calcium overload during ischemia and reperfusion. Circ Res 71(6):1334–1340
21. Dhalla N, Temsah R, Netticadan T et al (2001) Calcium overload in ischemia/reperfusion injury. In: Sperelakis, N (ed) Heart physiology and pathophysiology, 4th edn. Academic Press, San Diego
22. Perrin BJ, Huttenlocher A (2002) Calpain. Int J Biochem Cell Biol 34(7):722–725
23. Kroemer G, Martin SJ (2005) Caspase-independent cell death. Nat Med 11(7):725–730
24. Ambrosio G, Tritto I (1999) Reperfusion injury: experimental evidence and clinical implications. Am Heart J 138(2 Pt 2):S69–S75
25. McElroy CL, Gissen SA, Fishbein MC (1978) Exercise-induced reduction in myocardial infarct size after coronary artery occlusion in the rat. Circulation 57(5):958–962
26. Bowles DK, Farrar RP, Starnes JW (1992) Exercise training improves cardiac function after ischemia in the isolated, working rat heart. Am J Physiol 263(3 Pt 2):H804–H809
27. Demirel HA, Powers SK, Zergeroglu MA et al (2001) Short-term exercise improves myocardial tolerance to in vivo ischemia-reperfusion in the rat. J Appl Physiol 91(5):2205–2212
28. Demirel HA, Hamilton KL, Shanely RA et al (2003) Age and attenuation of exercise-induced myocardial HSP72 accumulation. Am J Physiol Heart Circ Physiol 285(4): H1609–H1615
29. Lennon SL, Quindry J, Hamilton KL et al (2004) Loss of exercise-induced cardioprotection after cessation of exercise. J Appl Physiol 96(4):1299–1305
30. Locke M, Tanguay RM, Klabunde RE et al (1995) Enhanced postischemic myocardial recovery following exercise induction of HSP 72. Am J Physiol 269(1 Pt 2):H320–H325
31. Paroo Z, Haist JV, Karmazyn M et al (2002) Exercise improves postischemic cardiac function in males but not females: consequences of a novel sex-specific heat shock protein 70 response. Circ Res 90(8):911–917
32. Powers SK, Demirel HA, Vincent HK et al (1998) Exercise training improves myocardial tolerance to in vivo ischemia-reperfusion in the rat. Am J Physiol 275(5 Pt 2):R1468–R1477
33. Lennon SL, Quindry JC, French JP et al (2004) Exercise and myocardial tolerance to ischaemia-reperfusion. Acta Physiol Scand 182(2):161–169
34. Lennon SL, Quindry JC, Hamilton KL et al (2004) Elevated MnSOD is not required for exercise-induced cardioprotection against myocardial stunning. Am J Physiol Heart Circ Physiol 287(2):H975–H980
35. Hamilton KL, Powers SK, Sugiura T et al (2001) Short-term exercise training can improve myocardial tolerance to I/R without elevation in heat shock proteins. Am J Physiol Heart Circ Physiol 281(3):H1346–H1352
36. Hamilton KL, Quindry JC, French JP et al (2004) MnSOD antisense treatment and exercise-induced protection against arrhythmias. Free Radic Biol Med 37(9):1360–1368
37. Brown DA, Jew KN, Sparagna GC et al (2003) Exercise training preserves coronary flow and reduces infarct size after ischemia-reperfusion in rat heart. J Appl Physiol 95(6):2510–2518
38. Brown DA, Lynch JM, Armstrong CJ et al (2005) Susceptibility of the heart to ischaemia-reperfusion injury and exercise-induced cardioprotection are sex-dependent in the rat. J Physiol 564(Pt 2):619–630
39. Yamashita N, Hoshida S, Otsu K et al (1999) Exercise provides direct biphasic cardioprotection via manganese superoxide dismutase activation. J Exp Med 189(11):1699–1706
40. Hamilton KL, Staib JL, Phillips T et al (2003) Exercise, antioxidants, and HSP72: protection against myocardial ischemia/reperfusion. Free Radic Biol Med 34(7):800–809
41. French JP, Quindry JC, Falk DJ et al (2005) Ischemia-reperfusion induced calpain activation and SERCA2a degradation are attenuated by exercise training and calpain inhibition. Am J Physiol Heart Circ Physiol 290(1):H128–H136
42. French JP, Hamilton KL, Quindry JC et al (2008) Exercise-induced protection against myocardial apoptosis and necrosis: MnSOD, calcium-handling proteins, and calpain. FASEB J 22(8):2862–2871

43. Quindry J, French J, Hamilton K et al (2005) Exercise training provides cardioprotection against ischemia-reperfusion induced apoptosis in young and old animals. Exp Gerontol 40(5):416–425
44. Starnes JW, Taylor RP, Park Y (2003) Exercise improves postischemic function in aging hearts. Am J Physiol Heart Circ Physiol 285(1):H347–H351
45. Le Page C, Noirez P, Courty J et al (2008) Exercise training improves functional post-ischemic recovery in senescent heart. Exp Gerontol 44(3):177–182
46. Bowles DK, Starnes JW (1994) Exercise training improves metabolic response after ischemia in isolated working rat heart. J Appl Physiol 76(4):1608–1614
47. Abete P, Calabrese C, Ferrara N et al (2000) Exercise training restores ischemic preconditioning in the aging heart. J Am Coll Cardiol 36(2):643–650
48. Babai L, Szigeti Z, Parratt JR et al (2002) Delayed cardioprotective effects of exercise in dogs are aminoguanidine sensitive: possible involvement of nitric oxide. Clin Sci (Lond) 102(4):435–445
49. Jew KN, Moore RL (2002) Exercise training alters an anoxia-induced, glibenclamide-sensitive current in rat ventricular cardiocytes. J Appl Physiol 92(4):1473–1479
50. Hoshida S, Yamashita N, Otsu K et al (2002) Repeated physiologic stresses provide persistent cardioprotection against ischemia-reperfusion injury in rats. J Am Coll Cardiol 40(4):826–831
51. Reger PO, Barbe MF, Amin M et al (2006) Myocardial hypoperfusion/reperfusion tolerance with exercise training in hypertension. J Appl Physiol 100(2):541–547
52. Burelle Y, Wambolt RB, Grist M et al (2004) Regular exercise is associated with a protective metabolic phenotype in the rat heart. Am J Physiol Heart Circ Physiol 287(3):H1055–H1063
53. Nagy O, Hajnal A, Parratt JR et al (2004) Delayed exercise-induced protection against arrhythmias in dogs–effect of celecoxib. Eur J Pharmacol 499(1–2):197–199
54. Collins HL, Loka AM, Dicarlo SE (2005) Daily exercise-induced cardioprotection is associated with changes in calcium regulatory proteins in hypertensive rats. Am J Physiol Heart Circ Physiol 288(2):H532–H540
55. Moran M, Blazquez I, Saborido A et al (2005) Antioxidant defences and ecto-5′-nucleotidase are not involved in the exercise-training induced protection against ischemia/reperfusion injury in the rat heart. Exp Physiol 9(4):507–517
56. Hajnal A, Nagy O, Litvai A et al (2005) Nitric oxide involvement in the delayed antiarrhythmic effect of treadmill exercise in dogs. Life Sci 77(16):1960–1971
57. Quindry J, French J, Hamilton K et al (2005) Exercise training provides cardioprotection against ischemia-reperfusion induced apoptosis in young and old animals. Exp Gerontol 40:416–425
58. Brown DA, Chicco AJ, Jew KN et al (2005) Cardioprotection afforded by chronic exercise is mediated by the sarcolemmal, and not the mitochondrial, isoform of the KATP channel in the rat. J Physiol 569(Pt 3):913–924
59. Chaves EA, Pereira-Junior PP, Fortunato RS et al (2006) Nandrolone decanoate impairs exercise-induced cardioprotection: role of antioxidant enzymes. J Steroid Biochem Mol Biol 99(4–5):223–230
60. Marini M, Lapalombella R, Margonato V et al (2007) Mild exercise training, cardioprotection and stress genes profile. Eur J Appl Physiol 99(5):503–510
61. Chicco AJ, Johnson MS, Armstrong CJ et al (2007) Sex-specific and exercise-acquired cardioprotection is abolished by sarcolemmal KATP channel blockade in the rat heart. Am J Physiol Heart Circ Physiol 292(5):H2432–H2437
62. Akita Y, Otani H, Matsuhisa S et al (2007) Exercise-induced activation of cardiac sympathetic nerve triggers cardioprotection via redox-sensitive activation of eNOS and upregulation of iNOS. Am J Physiol Heart Circ Physiol 292(5):H2051–H2059
63. Zhang KR, Liu HT, Zhang HF et al (2007) Long-term aerobic exercise protects the heart against ischemia/reperfusion injury via PI3 kinase-dependent and Akt-mediated mechanism. Apoptosis 12(9):1579–1588

64. Quindry JC, Hamilton KL, French JP et al (2007) Exercise-induced HSP-72 elevation and cardioprotection against infarct and apoptosis. J Appl Physiol 103(3):1056–1062
65. Dickson EW, Hogrefe CP, Ludwig PS et al (2008) Exercise enhances myocardial ischemic tolerance via an opioid receptor-dependent mechanism. Am J Physiol Heart Circ Physiol 294(1):H402–H408
66. Sanchez G, Escobar M, Pedrozo Z et al (2008) Exercise and tachycardia increase NADPH oxidase and ryanodine receptor-2 activity: possible role in cardioprotection. Cardiovasc Res 77(2):380–386
67. Kavazis AN, McClung JM, Hood DA et al (2008) Exercise induces a cardiac mitochondrial phenotype that resists apoptotic stimuli. Am J Physiol Heart Circ Physiol 294(2): H928–H935
68. Wonders KY, Hydock DS, Schneider CM et al (2008) Acute exercise protects against doxorubicin cardiotoxicity. Integr Cancer Ther 7(3):147–154
69. Melling CW, Thorp DB, Milne KJ et al (2009) Myocardial Hsp70 phosphorylation and PKC-mediated cardioprotection following exercise. Cell Stress Chaperones 14(2):141–150
70. Powers SK, Lennon SL, Quindry J et al (2002) Exercise and cardioprotection. Curr Opin Cardiol 17(5):495–502
71. Quindry JC, Hamilton KL (2007) Ischemic preconditioning and exercise induced protection during myocardial ischemia-reperfusion: A critical comparison. Curr Cardiol Rev 3(40):225–263
72. Laughlin MH, McAllister RM (1992) Exercise training-induced coronary vascular adaptation. J Appl Physiol 73(6):2209–2225
73. Firoozan S, Forfar JC (1996) Exercise training and the coronary collateral circulation: is its value underestimated in man? Eur Heart J 17(12):1791–1795
74. Lu X, Wu T, Huang P et al (2008) Effect and mechanism of intermittent myocardial ischemia induced by exercise on coronary collateral formation. Am J Phys Med Rehabil 87(10): 803–814
75. Knowlton AA (1995) The role of heat shock proteins in the heart. J Mol Cell Cardiol 27(1):121–131
76. Morimoto RI, Santoro MG (1998) Stress-inducible responses and heat shock proteins: new pharmacologic targets for cytoprotection. Nat Biotechnol 16(9):833–838
77. Plumier JC, Currie RW (1996) Heat shock-induced myocardial protection against ischemic injury: a role for Hsp70? Cell Stress Chaperones 1(1):13–17
78. Taylor RP, Harris MB, Starnes JW (1999) Acute exercise can improve cardioprotection without increasing heat shock protein content. Am J Physiol 276(3 Pt 2):H1098–H1102
79. Harris MB, Starnes JW (2001) Effects of body temperature during exercise training on myocardial adaptations. Am J Physiol Heart Circ Physiol 280(5):H2271–H2280
80. Noma A (1983) ATP-regulated K+ channels in cardiac muscle. Nature 305(5930):147–148
81. Gross ER, Peart JN, Hsu AK et al (2003) K(ATP) opener-induced delayed cardioprotection: involvement of sarcolemmal and mitochondrial K(ATP) channels, free radicals and MEK1/2. J Mol Cell Cardiol 35(8):985–992
82. Gross GJ, Peart JN (2003) KATP channels and myocardial preconditioning: an update. Am J Physiol Heart Circ Physiol 285(3):H921–H930
83. Inoue I, Nagase H, Kishi K et al (1991) ATP-sensitive K+ channel in the mitochondrial inner membrane. Nature 352(6332):244–247
84. O'Rourke B (2004) Evidence for mitochondrial K+ channels and their role in cardioprotection. Circ Res 94(4):420–432
85. Kong X, Tweddell JS, Gross GJ et al (2001) Sarcolemmal and mitochondrial K(atp)channels mediate cardioprotection in chronically hypoxic hearts. J Mol Cell Cardiol 33(5):1041–1045
86. Costa AD, Quinlan CL, Andrukhiv A et al (2006) The direct physiological effects of mitoK(ATP) opening on heart mitochondria. Am J Physiol Heart Circ Physiol 290(1): H406–H415

87. Hausenloy DJ, Yellon DM (2007) Reperfusion injury salvage kinase signalling: taking a RISK for cardioprotection. Heart Fail Rev 12(3–4):217–234
88. Carson LD, Korzick DH (2003) Dose-dependent effects of acute exercise on PKC levels in rat heart: is PKC the heart's prophylactic? Acta Physiol Scand 178(2):97–106
89. Dinkova-Kostova AT, Talalay P (2008) Direct and indirect antioxidant properties of inducers of cytoprotective proteins. Mol Nutr Food Res 52(Suppl 1):S128–S138
90. Surh YJ, Kundu JK, Na HK (2008) Nrf2 as a master redox switch in turning on the cellular signaling involved in the induction of cytoprotective genes by some chemopreventive phytochemicals. Planta Med 74(13):1526–1539
91. Zhu H, Jia Z, Misra BR et al (2008) Nuclear factor E2-related factor 2-dependent myocardiac cytoprotection against oxidative and electrophilic stress. Cardiovasc Toxicol 8(2):71–85
92. Piantadosi CA, Carraway MS, Babiker A et al (2008) Heme oxygenase-1 regulates cardiac mitochondrial biogenesis via Nrf2-mediated transcriptional control of nuclear respiratory factor-1. Circ Res 103(11):1232–1240
93. Finaud J, Lac G, Filaire E (2006) Oxidative stress: relationship with exercise and training. Sports Med 36(4):327–358
94. Reid MB, Haack KE, Franchek KM et al (1992) Reactive oxygen in skeletal muscle. I. Intracellular oxidant kinetics and fatigue in vitro. J Appl Physiol 73(5):1797–1804
95. Powers SK, Hamilton K (1999) Antioxidants and exercise. Clin Sports Med 18(3):525–536
96. Wright DC, Han DH, Garcia-Roves PM et al (2007) Exercise-induced mitochondrial biogenesis begins before the increase in muscle PGC-1alpha expression. J Biol Chem 282(1):194–199
97. Powers SK, Ji LL, Leeuwenburgh C (1999) Exercise training-induced alterations in skeletal muscle antioxidant capacity: a brief review. Med Sci Sports Exerc 31(7):987–997
98. Husain K, Somani SM (2005) Interaction of exercise and adenosine receptor agonist and antagonist on rat heart antioxidant defense system. Mol Cell Biochem 270(1–2):209–214
99. Gunduz F, Senturk UK, Kuru O et al (2004) The effect of one year's swimming exercise on oxidant stress and antioxidant capacity in aged rats. Physiol Res 53(2):171–176
100. Criswell D, Powers S, Dodd S et al (1993) High intensity training-induced changes in skeletal muscle antioxidant enzyme activity. Med Sci Sports Exerc 25(10):1135–1140
101. Dabkowski ER, Williamson CL, Hollander JM (2008) Mitochondria-specific transgenic overexpression of phospholipid hydroperoxide glutathione peroxidase (GPx4) attenuates ischemia/reperfusion-associated cardiac dysfunction. Free Radic Biol Med 45(6):855–865
102. Powers SK, Quindry JC, Kavazis AN (2008) Exercise-induced cardioprotection against myocardial ischemia-reperfusion injury. Free Radic Biol Med 44(2):193–201
103. Melikoglu MA, Kaldirimci M, Katkat D et al (2008) The effect of regular long term training on antioxidant enzymatic activities. J Sports Med Phys Fitness 48(3):388–390
104. Souza-Rabbo MP, Silva LF, Auzani JA et al (2008) Effects of a chronic exercise training protocol on oxidative stress and right ventricular hypertrophy in monocrotaline-treated rats. Clin Exp Pharmacol Physiol 35(8):944–948
105. Husain K (2003) Interaction of exercise training and chronic NOS inhibition on blood pressure, heart rate, NO and antioxidants in plasma of rats. Pathophysiology 10(1):47–56
106. Ryter SW, Otterbein LE, Morse D et al (2002) Heme oxygenase/carbon monoxide signaling pathways: regulation and functional significance. Mol Cell Biochem 234–235(1–2):249–263
107. Stocker R, Yamamoto Y, McDonagh AF et al (1987) Bilirubin is an antioxidant of possible physiological importance. Science 235(4792):1043–1046
108. Meister A, Anderson ME (1983) Glutathione. Annu Rev Biochem 52:711–760
109. Reddy Avula CP, Fernandes G (1999) Modulation of antioxidant enzymes and lipid peroxidation in salivary gland and other tissues in mice by moderate treadmill exercise. Aging (Milano) 11(4):246–252
110. Leichtweis S, Ji LL (2001) Glutathione deficiency intensifies ischaemia-reperfusion induced cardiac dysfunction and oxidative stress. Acta Physiol Scand 172(1):1–10

111. Paroo Z, Meredith MJ, Locke M et al (2002) Redox signaling of cardiac HSF1 DNA binding. Am J Physiol Cell Physiol 283(2):C404–C411
112. Abraham NG, Kappas A (2008) Pharmacological and clinical aspects of heme oxygenase. Pharmacol Rev 60(1):79–127
113. Seifi-Skishahr F, Siahkohian M, Nakhostin-Roohi B (2008) Influence of aerobic exercise at high and moderate intensities on lipid peroxidation in untrained men. J Sports Med Phys Fitness 48(4):515–521
114. Mastaloudis A, Leonard SW, Traber MG (2001) Oxidative stress in athletes during extreme endurance exercise. Free Radic Biol Med 31(7):911–922
115. Gonzalez D, Marquina R, Rondon N et al (2008) Effects of aerobic exercise on uric acid, total antioxidant activity, oxidative stress, and nitric oxide in human saliva. Res Sports Med 16(2):128–137
116. Green HJ, Fraser IG (1988) Differential effects of exercise intensity on serum uric acid concentration. Med Sci Sports Exerc 20(1):55–59
117. Hellsten Y, Sjodin B, Richter EA et al (1998) Urate uptake and lowered ATP levels in human muscle after high-intensity intermittent exercise. Am J Physiol 274(4 Pt 1):E600–E606
118. Hellsten Y, Svensson M, Sjodin B et al (2001) Allantoin formation and urate and glutathione exchange in human muscle during submaximal exercise. Free Radic Biol Med 31(11):1313–1322
119. Hellsten Y, Tullson PC, Richter EA et al (1997) Oxidation of urate in human skeletal muscle during exercise. Free Radic Biol Med 22(1–2):169–174
120. Sui X, Church TS, Meriwether RA et al (2008) Uric acid and the development of metabolic syndrome in women and men. Metabolism 57(6):845–852
121. Hamilton KL (2007) Antioxidants and cardioprotection. Med Sci Sports Exerc 39(9):1544–1553
122. Robinson I, de Serna DG, Gutierrez A et al (2006) Vitamin E in humans: an explanation of clinical trial failure. Endocr Pract 12(5):576–582
123. Dagenais GR, Marchioli R, Yusuf S et al (2000) Beta-carotene, vitamin C, and vitamin E and cardiovascular diseases. Curr Cardiol Rep 2(4):293–299
124. Murry CE, Jennings RB, Reimer KA (1986) Preconditioning with ischemia: a delay of lethal cell injury in ischemic myocardium. Circulation 74(5):1124–1136
125. Bolli R, Becker L, Gross G et al (2004) Myocardial protection at a crossroads: the need for translation into clinical therapy. Circ Res 95(2):125–134
126. Wilson EM, Diwan A, Spinale FG et al (2004) Duality of innate stress responses in cardiac injury, repair, and remodeling. J Mol Cell Cardiol 37(4):801–811

# Chapter 29
# Antioxidative Properties of Statins in the Heart

**Oliver Adam and Ulrich Laufs**

**Abstract** The overall benefits observed with the lipid-lowering HMG-CoA reductase inhibitors (statins) appear to be greater than might be expected from changes in lipid levels alone; and the positive effects have only partially been reproduced with other lipid lowering drugs, suggesting effects in addition to cholesterol lowering. In experimental models, many of the cholesterol-independent effects of statins are mediated by inhibition of isoprenoids, which serve as lipid attachments for intracellular signaling molecules such as small Rho GTP-binding proteins, whose membrane localization and function are dependent on isoprenylation. This chapter summarizes one of the hallmarks of the cardiac effects of statins, namely their antioxidative properties in the heart.

**Keywords** Statin · Prevention · Oxidative stress · Small G proteins

## 29.1 Introduction

Lowering of serum cholesterol levels via inhibition of hepatic cholesterol synthesis and subsequent upregulation of LDL receptors in the liver is the primary mechanism of HMG-CoA reductase inhibitors (statins) [1]. However, recent evidence suggests that statins have beneficial effects beyond cholesterol lowering, and in extrahepatic tissues. Experimental and clinical evidence revealed that several important regulators of the cardiovascular system can be regulated by statins. Prominent candidates are the endothelial NO synthase (eNOS), endothelin, free oxygen radicals, MHC-II, the protein kinase Akt, and the metalloproteinases. This article reviews the antioxidative effects of statins in the heart.

O. Adam (✉)
Klinik für Innere Medizin III, Kardiologie, Angiologie und Internistische Intensivmedizin, Universitätsklinikum des Saarlandes, D-66421 Homburg/Saar, Germany
e-mail: o.adam@freenet.de

H. Sauer et al. (eds.), *Studies on Cardiovascular Disorders*, Oxidative Stress in Applied Basic Research and Clinical Practice,
DOI 10.1007/978-1-60761-600-9_29, 

## 29.2 Properties of HMG-CoA Reductase Inhibitors (Statins)

### *29.2.1 Mechanism Mediating Cholesterol-Dependent Effects of Statins*

HMG-CoA reductase is the rate-limiting enzyme for cholesterol biosynthesis in the liver [1]. HMG-CoA reductase catalyzes the four-electron reductive deacylation of HMG-CoA to CoA and mevalonate, which is the committed and rate-limiting step in cholesterol biosynthesis [Fig. 29.1] [2].The liver is the primary site of action of all HMG-CoA reductase (statins). All statins share an HMG-like moiety and inhibit the reductase by the same mechanism. Statins block the access of the substrate HMG-CoA to the reductase by occupying the HMG-binding pocket and part of the binding surface for CoA [3]. The tight binding of statins is due to the large number of van der Waals interactions between the inhibitor and HMG-CoA reductase. The extrahepatic plasma concentration and permeability, e.g., into vascular cells, differs between statins and depends mainly on their lipophilicity [4, 5, 6, 7]. Inhibition of cholesterol synthesis in hepatocytes leads to upregulation of hepatic LDL receptor expression. As a consequence, LDL and its precursors are cleared from the circulation [1]. Furthermore, plasma concentration of antiatherogenic HDL and apo A-I are increased by statin treatment [8]. Additionally, inhibition of the HMG-CoA reductase reduces the synthesis of intermediates of the mevalonate pathway [1].

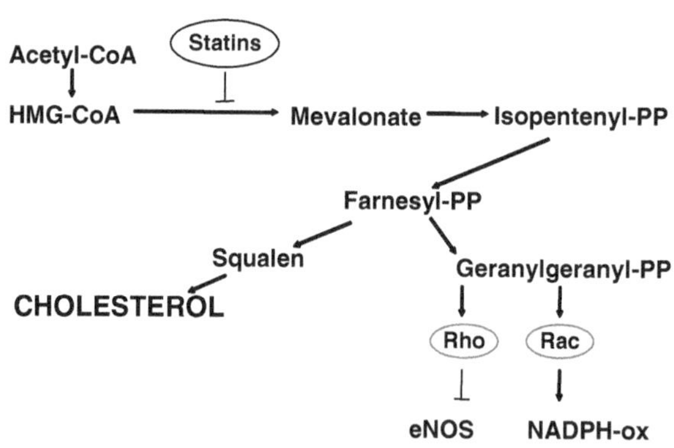


**Fig. 29.1** Effects of statins on the HMG-CoA reductase pathway. Inhibition of mevalonate synthesis not only blocks the synthesis of cholesterol but also the isoprenoid intermediates of the cholesterol pathway. The isoprenoid geranylgeranylpyrophosphate plays an important role in the posttranslational modification of proteins. The membrane translocation and activity of the small GTP-binding proteins Rho and Rac depend on their geranylgeranylation. Statins inhibit small G protein isoprenylation and function

### *29.2.2 Mechanism Mediating Cholesterol-Independent Effects of Statins*

An important mechanism underlying cholesterol-independent effects of statins is the inhibition of isoprenoid intermediates of the cholesterol synthesis pathway [Fig. 29.1] [3, 9]. Statins competitively inhibit the synthesis of L-mevalonic acid, the immediate product of HMG-CoA reductase. At the same time, statins prevent the synthesis of other important isoprenoid intermediates of the cholesterol biosynthetic pathway, such as farnesylpyrophosphate (FPP) and geranylgeranylpyrophosphate (GGPP) [1] [Fig. 29.1], which are important lipid attachments for the posttranslational modification of a variety of proteins, such as the small guanosine triphosphate (GTP)-binding proteins Ras, Rho, Rab, Rac, Ral, or Rap [10]. Thus, protein isoprenylation permits the covalent attachment, subcellular localization, and intracellular trafficking of membrane-associated proteins. Members of the Rho GTPase family are major substrates for posttranslational modification by prenylation [11, 10]. These small GTP-binding proteins cycle between the inactive GDP-bound state and active GTP-bound state. Rho translocation from the cytoplasm to the plasma membrane is dependent on geranylgeranylation, whereas Ras translocation is dependent on farnesylation, in endothelial cells [12, 13]. The inhibition of Rac, Ras, and Rho isoprenylation statins leads to the accumulation of inactive Rac, Ras, and Rho in the cytoplasm. Because Rho is a major target of geranylgeranylation, inhibition of Rho and its downstream target, Rho-kinase, is a likely mechanism mediating some of the cholesterol-independent effects of statins on the vascular wall [14]. The members of the Rho GTPase family (RhoA, Rac, and Cdc42) serve specific functions in terms of cell shape, motility, secretion, and proliferation, although overlapping functions between the members could be observed in overexpressed systems. The distinct but complementary functions of Rho family members also extend to their effects on cell signaling. It is therefore not surprising to find that Rho-induced changes in the actin cytoskeleton and gene expression are related [15].

## 29.3 Pathophysiology of Oxidative Stress

Reactive oxygen species (ROS), including superoxide ($O_2^-$), hydroxyl (OH), and hydrogen peroxide ($H_2O_2$), can induce oxidation and damage to DNA, membranes, proteins, and other macromolecules, if they are present in excess. Therefore, diverse specific and nonspecific antioxidant defence systems exist to scavenge and degrade ROS to nontoxic molecules [16, 17]. The imbalance between ROS production and their removal by antioxidant systems in favour of excess ROS is termed *oxidative stress*. $O_2^-$ is normally produced in small amounts as a byproduct of the use of molecular oxygen during mitochondrial oxidative phosphorylation. A family of superoxide dismutase enzymes rapidly converts $O_2^-$ to $H_2O_2$, which is itself broken down by glutathione peroxidase and catalase to water. The pathophysiological effects of ROS depend on the type, concentration, and specific site of production.

High concentrations of ROS tend to react with numerous protein centres, DNA, cell membranes, and other molecules, causing considerable cellular damage, as well as generating other more reactive radicals. Local targeted production of ROS at lower levels serves as a second-messenger system that transmits biological information through the highly specific modulation of intracellular signaling molecules, enzymes, and proteins—which is termed redox signaling. These processes are involved in the activation of many signal transduction protein kinases and transcription factors, the stimulation of DNA synthesis, the expression of growth-related genes [18, 16], and the regulation of myocardial excitation–contraction coupling [19]. The reaction of $O_2^-$ with the signaling molecule nitric oxide, which in health has a central role in vascular homeostasis as well as in modulating cardiac function, is another ROS-related pathophysiological mechanism [20, 16, 17, 21]. Oxidative stress is a hallmark of chronic heart failure [22, 23]. In patients with chronic heart failure, increased oxidative stress is associated with reduced left ventricular (LV) function and correlates with the severity of the disease [22, 23]. Furthermore, cell culture and animal studies suggest that reactive oxygen species (ROS) may be important mediators of cardiac hypertrophy and the development of contractile dysfunction [24, 25, 26, 27]. The progression of cardiac hypertrophy to LV dysfunction can be prevented by application of antioxidants in animal models [24, 25, 28]. In human failing myocardium, increased NADPH oxidase-related ROS production is associated with enhanced membrane expression and activity of the small G-protein Rac1 [29].

## 29.4 Antioxidative Effects of Statins in the Myocardium

### *29.4.1 Effects of Statins on Ventricular Myocardium and Cardiac Function*

Animal and human studies suggest that statins may have antioxidative effects on the ventricular myocardium. Cholesterol lowering itself contributes to the antioxidative effects of HMG-CoA reductase inhibitors. Importantly, LDL cholesterol and its oxidized forms are reduced in the presence of statins [30, 5].

Animal studies suggest that a phagocyte-type NADPH oxidase may be a relevant source of ROS in the ventricular myocardium [31, 32, 33, 34]. In cardiomyocytes, two of the five components, $p22^{phox}$ (phox, for phagocyte oxidase) and $gp91^{phox}$, are bound to the membranes. The other three components, $p40^{phox}$, $p47^{phox}$, and $p67^{phox}$, exist in the cytosol, forming a complex. The entire cytosolic complex migrates to the membrane after phosphorylation of the cytosolic components by various stimuli. Importantly, not only the core subunits but also two low-molecular-weight guanine nucleotide binding proteins, Rac1 and Rap, are required for activation. During activation, Rac1 binds GTP and migrates to the membrane with the core cytosolic complex. Therefore, Rac1 is critically involved in the activation of cardiovascular NADPH oxidase. ROS production by NADPH oxidase is involved in cardiac hypertrophy in response to pressure overload [35, 34], stretch [36], angiotensin II infusion [32], and α-adrenergic stimulation [37].

Rac1 is required for NADPH oxidase activity, and cardiac hypertrophy is mediated, in part, by myocardial oxidative stress [32, 38, 24, 25, 35, 34, 39, 26]. Indeed, experimental models show that statins inhibit cardiac hypertrophy through an antioxidant mechanism involving inhibition of Rac1 geranylgeranylation [40, 13, 41, 9,29,42] [Fig. 29.2]. Statins inhibit angiotensin II-induced oxidative stress and cardiac hypertrophy in rodents [28]. Clinical studies confirm these observations, in that statins inhibit cardiac hypertrophy in humans with hypercholesterolemia [43]. In left ventricular myocardium from patients with heart failure due to ischemic cardiomyopathy (ICM) or dilated cardiomyopathy (DCM), NADPH oxidase activity is increased compared with nonfailing controls. In failing myocardium, increased oxidative stress is associated with increased translocation of Rac1 from the cytosol to the membrane and increased Rac1-GTPase activity. In samples of the right atrial myocardium from patients undergoing cardiac surgery who were prospectively treated with atorvastatin or pravastatin (40 mg/day, 4 weeks), Rac1-GTPase activity was significantly decreased to 67.9 and 65.6% compared to patients without statin. Both atorvastatin and pravastatin significantly reduced angiotensin II-stimulated NADPH oxidase activity. These data suggest that extrahepatic effects of statins can be observed in humans and may be beneficial for patients with chronic heart failure [29].

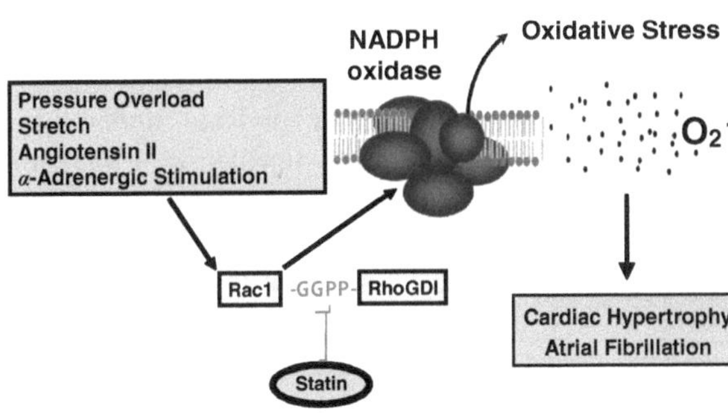


**Fig. 29.2** Inhibition of Rac1 activity by statins exerts antioxidative effect via inhibition of NADPH oxidase. Angiotensin II, stretch, alpha-adrenergic activation, or pressure overload induce superoxide production by the NADPH oxidase via activation of Rac1 GTPase. Statins are able to inhibit NADPH oxidase through inhibition of the association of RhoGDIα with Rac1, which depends on its geranylgeranylation

### *29.4.2 Effects of Statins on Atrial Myocardium and Atrial Fibrillation*

In isolated atrial myocytes from human right atrial appendages, the NADPH oxidase is a main source of atrial superoxide production [44]. NADPH oxidase–derived ROS may have several pathologic effects in the atrial myocardium, including oxidative degradation of endocardial NO, local activation of coagulation cascade components, prothrombotic molecules such as PAI-1 and tissue factor, induction of fibrosis, inflammatory responses, and alteration of ion channel function [45, 46, 47, 48, 44, 49, 50, 51, 52]. Several animal and human studies suggest that increased atrial oxidative stress may play an important role in inducing and maintaining AF [46, 50, 53]. AF induced by rapid atrial pacing in pigs is characterized by increased NAD(P)H oxidase activity and superoxide production in the left atrium (LA) [47]. Furthermore, treatment of AF induced by rapid atrial pacing in chronically instrumented dogs, with ascorbate, an antioxidant and peroxynitrite decomposition catalyst, attenuated atrial pacing-induced peroxynitrite formation and electrical remodeling [54].

Right human atrial appendages of patients with AF undergoing the maze procedure exhibit higher levels of the oxidative markers 3-nitrotyrosine and protein carbonyls compared to patients with sinus rhythm undergoing cardiac surgery [50]. Myofibrillar creatine kinase (MM-CK), a controller of myocyte contractility sensitive to oxidative injury, was reduced in AF patients, potentially contributing to altered myofibrillar energetics and contractile dysfunction [50]. NO synthase also contributed to atrial superoxide production in fibrillating atria, suggesting that increased oxidative stress may lead to NOS "uncoupling." These findings indicate that NADPH oxidase significantly contributes to superoxide production in AF [44]. Indeed, left atrial tissue of patients with atrial fibrillation is characterized by upregulation of total protein expression of Rac1, the membrane content of Rac, Rac1-GTPase activity, and the superoxide-producing NADPH oxidase, compared to patients with sinus rhythm [55].

Mice with cardiac specific overexpression of Rac1 under the control of the αMHC promoter (RacET), developed atrial fibrillation with aging, which was associated with an increased NADPH oxidase activity. Notably, treatment with HMG-CoA reductase inhibitors inhibits Rac1 activation by inhibiting its geranylgeranylation and membrane translocation [55, 56]. Indeed, oral treatment of the Rac1-overexpressing mice with statins inhibited Rac1, lowered NADPH oxidase activity, and markedly reduced the incidence of atrial fibrillation [55]. Interestingly, prospective short-term oral statin treatment of patients with ischemic heart disease is able to downregulate Rac1 activation and NADPH oxidase activity in the right atrium, suggesting that relevant antioxidative atrial effects of statins may also occur in humans [29]. Another important piece of evidence is provided by the recent ARMYDA-3 (Atorvastatin for Reduction of Myocardial Dysrhythmia After Cardiac Surgery) study, a randomized, prospective trial that investigated the effect of statin therapy on the prevalence of postoperative AF in patients without a history of AF undergoing cardiac surgery [57]. Treatment with atorvastatin started 1 week prior to surgery was associated with a 60% reduction in risk of postoperative AF.

## 29.5 Potential Effects of Statin Withdrawal

In experimental models, statins improve endothelial function through increased NO production and decreased oxidative stress, in addition to cholesterol lowering.

Withdrawal of statin treatment confers overshoot activation of the small G-proteins Rho and Rac, causing production of reactive oxygen species and suppression of NO bioavailability [58, 59].

The mechanism underlying the overshoot activation of Rac after statin withdrawal is that geranylgeranyl pyrophosphate and Rac are expressed excessively under statin treatment. After statin withdrawal, geranylgeranyl pyrophosphate becomes available and the active Rac is anchored in the membrane and activates NADPH oxidase [60]. Via this mechanism, abrupt discontinuation of statin medication may exert negative effects in patients at acute vascular risk, for example, during acute coronary syndromes or ischemic stroke; while in stable vascular patients discontinuation appears to be safe [58].

## 29.6 Summary and Conclusions

In summary, evidence from experimental and clinical studies shows that statins have additional properties in the heart beyond cholesterol lowering, including beneficial effects on oxidative stress. Recent evidence suggests that some of these effects of statins on the myocardium are mediated via inhibition of isoprenoid synthesis and subsequent inhibition of small G proteins such as Rac1 GTPase in myocardial cells. However, further clinical trials are warranted to better characterize the clinical importance of the antioxidative effects of statins in the myocardium.

## References

1. Goldstein JL, Brown MS (1990) Regulation of the mevalonate pathway. Nature 343:425–430
2. Rodwell VW, Nordstrom JL, Mitschelen JJ (1976) Regulation of HMG-CoA reductase. Adv Lipid Res 14:1–74
3. Istvan ES, Deisenhofer J (2001) Structural mechanism for statin inhibition of HMG-CoA reductase. Science 292:1160–1164
4. Blum CB (1994) Comparison of properties of four inhibitors of 3-hydroxy-3-methylglutaryl-coenzyme A reductase. Am J Cardiol 73:3D–11D
5. Corsini A, Maggi FM, Catapano AL (1995) Pharmacology of competitive inhibitors of HMG-CoA reductase. Pharmacol Res 31:9–27
6. Lea AP, McTavish D (1997) Atorvastatin. A review of its pharmacology and therapeutic potential in the management of hyperlipidaemias. Drugs 53:828–847
7. McTaggart F, Buckett L, Davidson R et al (2001) Preclinical and clinical pharmacology of Rosuvastatin, a new 3-hydroxy-3-methylglutaryl coenzyme A reductase inhibitor. Am J Cardiol 87:28B–32B
8. Vega GL, Grundy SM (1998) Effect of statins on metabolism of apo-B-containing lipoproteins in hypertriglyceridemic men. Am J Cardiol 81:36B–42B
9. Liao JK, Laufs U (2005) Pleiotropic effects of statins. Annu Rev Pharmacol Toxicol 45: 89–118
10. Van AL, D'Souza-Schorey C (1997) Rho GTPases and signaling networks. Genes Dev 11:2295–2322
11. Hall A (1998) Rho GTPases and the actin cytoskeleton. Science 279:509–514
12. Laufs U, La Fata V, Plutzky J et al (1998) Upregulation of endothelial nitric oxide synthase by HMG CoA reductase inhibitors. Circulation 97:1129–1135

13. Laufs U, Liao JK (1998) Post-transcriptional regulation of endothelial nitric oxide synthase mRNA stability by Rho GTPase. J Biol Chem 273:24266–24271
14. Laufs U, Endres M, Stagliano N et al (2000) Neuroprotection mediated by changes in the endothelial actin cytoskeleton. J Clin Invest 106:15–24
15. Tapon N, Hall A (1997) Rho, Rac and Cdc42 GTPases regulate the organization of the actin cytoskeleton. Curr Opin Cell Biol 9:86–92
16. Li JM, Shah AM (2004) Endothelial cell superoxide generation: regulation and relevance for cardiovascular pathophysiology. Am J Physiol Regul Integr Comp Physiol 287:R1014–R1030
17. Seddon M, Looi YH, Shah AM (2007) Oxidative stress and redox signalling in cardiac hypertrophy and heart failure. Heart 93:903–907
18. Finkel T (1999) Signal transduction by reactive oxygen species in non-phagocytic cells. J Leukoc Biol 65:337–340
19. Gao WD, Liu Y, Marban E (1996) Selective effects of oxygen free radicals on excitation-contraction coupling in ventricular muscle. Implications for the mechanism of stunned myocardium. Circulation 94:2597–2604
20. Dixon LJ, Morgan DR, Hughes SM et al (2003) Functional consequences of endothelial nitric oxide synthase uncoupling in congestive cardiac failure. Circulation 107:1725–1728
21. Verhaar MC, Westerweel PE, van Zonneveld AJ et al (2004) Free radical production by dysfunctional eNOS. Heart 90:494–495
22. Belch JJ, Bridges AB, Scott N et al (1991) Oxygen free radicals and congestive heart failure. Br Heart J 65:245–248
23. Mallat Z, Philip I, Lebret M et al (1998) Elevated levels of 8-iso-prostaglandin F2alpha in pericardial fluid of patients with heart failure: a potential role for in vivo oxidant stress in ventricular dilatation and progression to heart failure. Circulation 97:1536–1539
24. Dhalla NS, Temsah RM, Netticadan T (2000) Role of oxidative stress in cardiovascular diseases. J Hypertens 18:655–673
25. Kinugawa S, Tsutsui H, Hayashidani S et al (2000) Treatment with dimethylthiourea prevents left ventricular remodeling and failure after experimental myocardial infarction in mice: role of oxidative stress. Circ Res 87:392–398
26. Siwik DA, Tzortzis JD, Pimental DR et al (1999) Inhibition of copper-zinc superoxide dismutase induces cell growth, hypertrophic phenotype, and apoptosis in neonatal rat cardiac myocytes in vitro. Circ Res 85:147–153
27. Wassmann S, Hilgers S, Laufs U et al (2002) Angiotensin II type 1 receptor antagonism improves hypercholesterolemia-associated endothelial dysfunction. Arterioscler Thromb Vasc Biol 22:1208–1212
28. Takemoto M, Node K, Nakagami H et al (2001) Statins as antioxidant therapy for preventing cardiac myocyte hypertrophy. J Clin Invest 108:1429–1437
29. Maack C, Kartes T, Kilter H et al (2003) Oxygen free radical release in human failing myocardium is associated with increased activity of Rac1-GTPase and represents a target for statin treatment. Circulation 108:1567–1574
30. Anderson TJ, Meredith IT, Yeung AC et al (1995) The effect of cholesterol-lowering and antioxidant therapy on endothelium-dependent coronary vasomotion. N Engl J Med 332: 488–493
31. Aikawa R, Nawano M, Gu Y et al (2000) Insulin prevents cardiomyocytes from oxidative stress-induced apoptosis through activation of PI3 kinase/Akt. Circulation 102:2873–2879
32. Bendall JK, Cave AC, Heymes C et al (2002) Pivotal role of a gp91(phox)-containing NADPH oxidase in angiotensin II-induced cardiac hypertrophy in mice. Circulation 105: 293–296
33. Cave A, Grieve D, Johar S et al (2005) NADPH oxidase-derived reactive oxygen species in cardiac pathophysiology. Philos Trans R Soc Lond B Biol Sci 360:2327–2334
34. MacCarthy PA, Grieve DJ, Li JM et al (2001) Impaired endothelial regulation of ventricular relaxation in cardiac hypertrophy: role of reactive oxygen species and NADPH oxidase. Circulation 104:2967–2974

35. Li JM, Gall NP, Grieve DJ et al (2002) Activation of NADPH oxidase during progression of cardiac hypertrophy to failure. Hypertension 40:477–484
36. Aikawa R, Komuro I, Yamazaki T et al (1999) Rho family small G proteins play critical roles in mechanical stress-induced hypertrophic responses in cardiac myocytes. Circ Res 84: 458–466
37. Xiao L, Pimentel DR, Wang J et al (2002) Role of reactive oxygen species and NAD(P)H oxidase in alpha(1)-adrenoceptor signaling in adult rat cardiac myocytes. Am J Physiol Cell Physiol 282:C926–C934
38. Bokoch GM, Diebold BA (2002) Current molecular models for NADPH oxidase regulation by Rac GTPase. Blood 100:2692–2696
39. Pracyk JB, Tanaka K, Hegland DD et al (1998) A requirement for the rac1 GTPase in the signal transduction pathway leading to cardiac myocyte hypertrophy. J Clin Invest 102: 929–937
40. Custodis F, Eberl M, Kilter H et al (2006) Association of RhoGDI alpha with Rac1 GTPase mediates free radical production during myocardial hypertrophy. Cardiovasc Res 71: 342–351
41. Laufs U, Liao JK (2003) Isoprenoid metabolism and the pleiotropic effects of statins. Curr Atheroscler Rep 5:372–378
42. Liao JK (2005) Effects of statins on 3-hydroxy-3-methylglutaryl coenzyme A reductase inhibition beyond low-density lipoprotein cholesterol. Am J Cardiol 96:24–33
43. Lee TM, Chou TF, Tsai CH (2002) Association of pravastatin and left ventricular mass in hypercholesterolemic patients: role of 8-iso-prostaglandin f2alpha formation. J Cardiovasc Pharmacol 40:868–874
44. Kim YM, Guzik TJ, Zhang YH et al (2005) A myocardial Nox2 containing NAD(P)H oxidase contributes to oxidative stress in human atrial fibrillation. Circ Res 97:629–636
45. Allessie M, Ausma J, Schotten U (2002) Electrical, contractile and structural remodeling during atrial fibrillation. Cardiovasc Res 54:230–246
46. Cai H, Li Z, Goette A et al (2002) Downregulation of endocardial nitric oxide synthase expression and nitric oxide production in atrial fibrillation: potential mechanisms for atrial thrombosis and stroke. Circulation 106:2764–2766
47. Dudley SC Jr, Hoch NE, McCann LA et al (2005) Atrial fibrillation increases production of superoxide by the left atrium and left atrial appendage: role of the NADPH and xanthine oxidases. Circulation 112:1266–1273
48. Gassanov N, Brandt MC, Michels G et al (2006) Angiotensin II-induced changes of calcium sparks and ionic currents in human atrial myocytes: potential role for early remodeling in atrial fibrillation. Cell Calcium 39:175–186
49. Kumagai K, Nakashima H, Saku K (2004) The HMG-CoA reductase inhibitor atorvastatin prevents atrial fibrillation by inhibiting inflammation in a canine sterile pericarditis model. Cardiovasc Res 62:105–111
50. Mihm MJ, Yu F, Carnes CA et al (2001) Impaired myofibrillar energetics and oxidative injury during human atrial fibrillation. Circulation 104:174–180
51. Nattel S (2002) New ideas about atrial fibrillation 50 years on. Nature 415:219–226
52. Neuberger HR, Schotten U, Blaauw Y et al (2006) Chronic atrial dilation, electrical remodeling, and atrial fibrillation in the goat. J Am Coll Cardiol 47(3):644–653
53. Nattel S, Khairy P, Roy D et al (2002) New approaches to atrial fibrillation management: a critical review of a rapidly evolving field. Drugs 62:2377–2397
54. Carnes CA, Chung MK, Nakayama T et al (2001) Ascorbate attenuates atrial pacing-induced peroxynitrite formation and electrical remodeling and decreases the incidence of postoperative atrial fibrillation. Circ Res 89:E32–E38
55. Adam O, Frost G, Custodis F et al (2007) Role of Rac1 GTPase activation in atrial fibrillation. J Am Coll Cardiol 50:359–367
56. Laufs U, Kilter H, Konkol C et al (2002) Impact of HMG CoA reductase inhibition on small GTPases in the heart. Cardiovasc Res 53:911–920

57. Patti G, Chello M, Candura D et al (2006) Randomized trial of atorvastatin for reduction of postoperative atrial fibrillation in patients undergoing cardiac surgery: results of the ARMYDA-3 (Atorvastatin for Reduction of MYocardial Dysrhythmia After cardiac surgery) study. Circulation 114:1455–1461
58. Endres M, Laufs U (2006) Discontinuation of statin treatment in stroke patients. Stroke 37:2640–2643
59. Laufs U, Endres M, Custodis F et al (2000) Suppression of endothelial nitric oxide production after withdrawal of statin treatment is mediated by negative feedback regulation of rho GTPase gene transcription. Circulation 19(102):3104–3110
60. Vecchione C, Brandes RP (2002) Withdrawal of 3-hydroxy-3-methylglutaryl coenzyme A reductase inhibitors elicits oxidative stress and induces endothelial dysfunction in mice. Circ Res 91:173–179

# Index




H. Sauer et al. (eds.), *Studies on Cardiovascular Disorders*,
Oxidative Stress in Applied Basic Research and Clinical Practice,
DOI 10.1007/978-1-60761-600-9, © Springer Science+Business Media, LLC 2010

**I**

**O**

**Q**



**R**

FSC
www.fsc.org
MIX
Papier | Fördert
gute Waldnutzung
FSC® C083411

Kolibri 360 GmbH
Ferdinand-Jühlke-Straße 7
99095 Erfurt, Deutschland
produktsicherheit@kolibri360.de